ORGANIC SOLAR CELLS

ORGANIC SOLAR CELLS

Fundamentals, Devices, and Upscaling

edited by
Barry P. Rand
Henning Richter

PAN STANFORD PUBLISHING

Published by

Pan Stanford Publishing Pte. Ltd.
Penthouse Level, Suntec Tower 3
8 Temasek Boulevard
Singapore 038988

Email: editorial@panstanford.com
Web: www.panstanford.com

British Library Cataloguing-in-Publication Data
A catalogue record for this book is available from the British Library.

Organic Solar Cells: Fundamentals, Devices, and Upscaling

ISBN 978-981-4463-65-2 (Hardcover)
ISBN 978-981-4463-66-9 (eBook)

Printed in the USA

Contents

Preface

When considering how best to present a volume on organic solar cells, we quickly realized the inherent complexity and depth of the topic. On one hand, it is a technology on the verge of commercialization, while on the other hand, there still remain a large range of fundamental and technological questions to be answered. These include a better understanding of the physical processes occurring within and at the interfaces of active and interlayers, device architectures optimized simultaneously for performance and lifetime, and remaining challenges related to large-scale cost-efficient manufacturing. In order to enable industrial implementation of organic photovoltaics (OPV), parallel progress in these and many other pertinent areas is required. To achieve this goal, a close collaboration of scientists and engineers with different backgrounds such as organic and physical chemistry, physics, materials science, and electrical engineering is necessary. Furthermore, it is well accepted that experimental, theoretical, and computational approaches all need to contribute.

With this need for complementary knowledge in different areas in mind, the 15 chapters of this book cover all aspects pertinent for the continued successful development of OPV, including materials design and synthesis, device architecture and characterization, quantum chemical and optical modeling, as well as topics related to roll-to-roll processing, transparent conductors, substrates, module design and fabrication, and finally lifetime and stability. Different, and sometimes even competing approaches, are addressed, such as the use of small molecules compared to polymers as electron donor materials or liquid processing and vapor deposition. We are grateful to the authors from around the world, all leading experts in the areas addressed in the different chapters. While a comprehensive overview is provided, extensive reference sections give the interested scientist the ability to consult the original literature if desired.

The book is intended to be used by advanced undergraduate or graduate students and scientists interested or already working in OPV. Particularly, we aim to allow for easy access to information complementary to the specific area of the reader but necessary for further progress. For example, organic chemists working on electron-donor or -acceptor materials can gain additional understanding of characteristics needed from a device perspective while physicists developing new device architectures can more easily check the compatibility of new architectures with large-scale manufacturing. As such, with this collected knowledge, we hope that advances may be accelerated such that companies with different focuses and skill sets, ranging from materials suppliers to device manufacturers, can fully develop the value chain of a new emerging industry.

Barry P. Rand
Henning Richter
April 2014

Part I

Materials and Device Architectures

Chapter 1

Solution-Processed Donors

Beate Burkhart and Barry C. Thompson
Department of Chemistry and Loker Hydrocarbon Institute,
University of Southern California, Los Angeles, CA 90089-1661, USA
barrycth@usc.edu

1.1 Introduction

With the world population growing rapidly, the energy demand is predicted to increase up to 40% by 2035 and one of the biggest tasks in recent history is to find technologies that can meet this challenge. Even though fossil fuels currently dominate the energy market, it is projected that renewable energy sources will play a much bigger role by 2035 due to increasing prices and limited resources of nonrenewables as well as environmental concerns [1].

Enough solar energy reaches the earth within 1 hour to meet the energy demand of the human population for an entire year [2], but electricity generation from nonhydro-based renewable energy sources was only about 3% in 2009, with solar being a small fraction of that. The primary reason for this is that the photovoltaic market is currently dominated by polycrystalline silicon solar cells (>80%)

Organic Solar Cells: Fundamentals, Devices, and Upscaling
Edited by Barry P. Rand and Henning Richter

ISBN 978-981-4463-65-2 (Hardcover), 978-981-4463-66-9 (eBook)
www.panstanford.com

that have a high cost, are slow to manufacture, and take up a lot of space making them unattractive for the consumer [3–6].

On the other hand, the vision of organic photovoltaics is that of low-cost, flexible, and lightweight solar cells that are easily integrated into existing infrastructure and visually pleasing [7–9]. In the last decade, this motivation has spurred enormous research efforts on conjugated semiconducting materials and power conversion efficiencies have improved from 2.5% in 2001 to over 9% in 2012 [10–23].

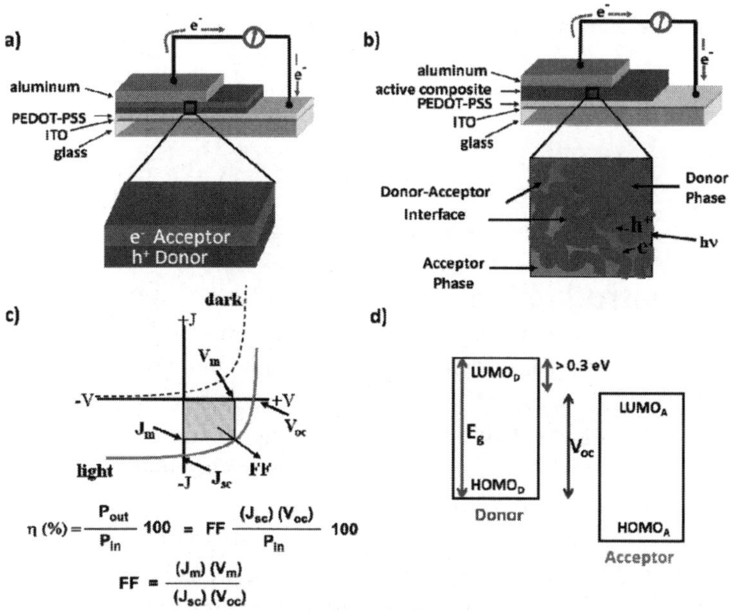

Figure 1.1 Schematic illustration of (a) a donor–acceptor bilayer solar cell; (b) a BHJ solar cell; (c) typical current–voltage curves for dark and light currents in solar cells with the output parameters: J_{sc} (short-circuit current density), V_{oc} (open-circuit voltage), FF (fill factor), and J_m and V_m (current and voltage at maximum power point)—the power conversion efficiency (η) and FF are defined below the J–V curve; and (d) simplified energy-level diagram illustrating the HOMO and LUMO energies of a donor and acceptor pair, the bandgap of the donor (E_g), the V_{oc} which is empirically related to the $HOMO_D$–$LUMO_A$ offset and the necessary driving force for exciton dissociation ($LUMO_D$–$LUMO_A > 0.3$ eV).

Compared to inorganic semiconductors, organic materials such as polymers and small molecules have several advantages. The very high absorption coefficients (likely to exceed 10^5 cm^{-1}) are ~1000 times higher than for polycrystalline silicon, which makes it possible to use very thin films of active material (between 50 and 200 nm) and still absorb photons efficiently [24,25]. Additionally, organic materials have much lower densities than inorganic semiconductors (poly(3-hexylthiophene) = 1.1 vs. silicon 2.33 g/cm^2) [26,27]. The combination of these two factors ensures that the amount of material used in organic solar cells is considerably smaller than for inorganic solar cells, thus reducing cost and weight.

In inorganic semiconductors, photon absorption produces free electrons and holes (due to the high dielectric constant of the material), which are then collected at the respective electrodes. On the other hand, organic semiconductors have a much lower dielectric constant (generally between 2 and 4) and upon absorption of a photon with appropriate wavelength a so-called exciton (a tightly bound electron–hole pair) is formed [28]. In 1986, Tang introduced a bilayer solar cell (the general architecture of a current bilayer solar cell is shown in Fig. 1.1a featuring an interface between an electron donor (copper phthalocyanine) and electron acceptor (a perylene tetracarboxylic derivative that has a higher electron affinity) providing the necessary driving force for exciton separation [29]. Due to short singlet exciton diffusion lengths (generally believed to be ~10 nm), the efficiency of bilayer solar cells is inherently limited, as the excitons either recombine before reaching the donor–acceptor interface or the film thickness is too thin to absorb a relevant portion of the sunlight [28]. This problem was solved by the introduction of the bulk heterojunction (BHJ) solar cell (Fig. 1.1b) by Yu and Heeger and Halls et al. in 1995, where the donor and acceptor materials are intimately mixed forming a bicontinuous network with small-scale phase separation [30,31]. In the ideal case, this small-scale phase separation allows all excitons to reach a donor–acceptor interface and dissociate into electrons and holes even with active layer thicknesses of 100–200 nm, which are thick enough to absorb the majority of incident photons [32]. The BHJ is currently by far the dominant solar cell geometry for organic solar cells offering the advantage of a single-solution processing step to form the active layer, which is easily adaptable for large-scale fabrication at low temperature and ambient pressure using minimal energy [33,34]. In this chapter, we will consequently focus on binary blend BHJ solar cells with a single, solution-processable active layer, whereas other solution-processed device architectures, such as tandem solar cells, are discussed in Chapter 6.

Since Tang's seminal discovery [29], fullerenes have emerged as the most important class of electron acceptors and are currently used almost exclusively. For solution-processed solar cells, functionalized fullerenes such as [6,6]-phenyl-C_{61}-butyric acid methyl ester ($PC_{61}BM$) are predominantly used, with other acceptors (small molecules, polymers, and inorganic materials, c.f. Chapter 3) showing much lower efficiencies in BHJ solar cells [35–49]. The considerable improvement of organic solar cell efficiencies over the last decade is thus predominantly due to huge efforts focused on the development of new donor materials. In this chapter, we will give an overview of the donor material characteristics required for high efficiencies and the progress toward embodying these characteristics as well as the design principles employed for the synthesis of novel donor materials. In addition to the synthetic approach, further optimization of efficiencies is frequently achieved through modification of the active layer morphology (e.g., using solvent additives) as well as solar cell architecture (introduction of interlayers, improved electrodes, inverted device structure), and these strategies are thoroughly covered in Chapters 4, 5, 6, 7 and 10 [50–55].

To understand the guiding principles of donor design, their characteristics can be correlated to the solar cell output parameters which are briefly introduced in the following paragraph. The power conversion efficiency (η or PCE) of an organic solar cell is determined by several parameters, namely the short-circuit current density (J_{sc}), the open-circuit voltage (V_{oc}), as well as the fill factor (FF) and can be expressed as η (%) = ($J_{sc} \times V_{oc} \times$ FF)/(input power) when standard AM1.5G conditions with light intensity of 100 mW/cm^2 are used (equations and generalized J–V curve in Fig. 1.1c) [56]. When designing new semiconducting materials the goal is to simultaneously optimize all parameters and consequently maximize η.

Necessary characteristics of the donor component in solution-processable BHJ solar cells can be divided into electronic requirements (such as a low bandgap for high J_{sc} and correct positioning of frontier energy levels for high V_{oc} and sufficient driving force for exciton dissociation) and morphological requirements (such as optimal mixing with fullerenes, a certain degree of crystallinity, and high charge carrier mobility). The control over the electronic properties of donor materials has been the driving force behind the synthesis of a staggering number of new materials, whereas principles governing the morphology of the active layer are less understood and have been less frequently targeted. Below, we will briefly discuss the influence the bandgap as well as frontier energy levels (highest occupied molecular orbital (HOMO) and lowest unoccupied molecular orbital (LUMO)) have on the output parameters of a solar cell, such as J_{sc} and V_{oc}, and in the following subchapter we will introduce the design principles available for the synthetic chemist to tune the electronic properties of a material and thus optimize those parameters.

To achieve high J_{sc} the active layer needs to absorb sunlight broadly and intensely across the solar spectrum (from the UV/vis into the near infrared (NIR), photon flux shown in Fig. 1.2) because J_{sc} is proportional to the product of spectral absorption breadth and absorption intensity of the active layer [11]. As illustrated by the red curve in Fig. 1.2, a donor material with an absorption onset of 650 nm can only reach a limited maximum theoretical J_{sc} of 16.77 mA/cm^2, whereas a donor with a low bandgap and an absorption onset of 950 nm can theoretically achieve a J_{sc} of 35.72 mA/cm^2 (numbers are calculated by integrating the area underneath the solar photon flux in Fig. 1.2 assuming an external quantum efficiency (EQE) of 100% across the absorption breadth). Considering that most fullerenes primarily absorb in the UV, and often to a lesser degree in the visible region of the solar spectrum, an important design principle for novel donor materials is to lower the bandgap, allowing absorption of longer wavelength light, and thus high J_{sc}. It has to be mentioned though that $PC_{71}BM$ in particular, and in contrast to $PC_{61}BM$, can contribute considerably to the photocurrent of solar cells because it exhibits significant visible light absorption with subsequent hole transfer to the donor material [57,58]. Often polymers with a bandgap lower than poly(3-hexylthiophene) (P3HT), which is still considered the benchmark for conjugated polymers and has a bandgap of 1.9 eV (see Section 1.3.1), are called low bandgap.

Originally though, the term was used for polymers with bandgaps lower than 1.5 eV, which are able to absorb NIR photons [59]. Mixing of donor materials with the fullerene acceptor at favorable ratios is also beneficial for high J_{sc} because, in combination with the fact that fullerenes do not absorb in the NIR, the absorption coefficients of donor materials in the visible spectra are generally considerably higher than for fullerenes. Within the constraints of a defined thickness of the active layer, dictated by the relatively low charge carrier mobilities of organic materials, a large amount of fullerene would thus be unfavorable because it dilutes the absorption of the polymer potentially leading to lower J_{sc}.

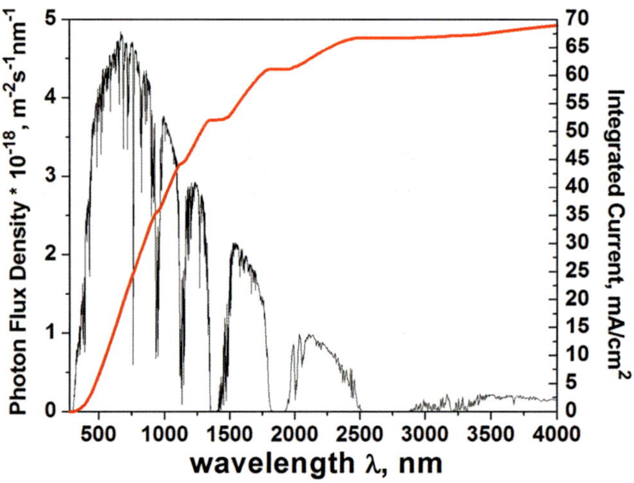

Figure 1.2 Photon flux density from the sun (AM1.5G) as a function of wavelength. The red curve represents the integrated short-circuit current density which is on the y-axis at the right.

It also has to be taken into account that the frontier energy levels of the donor material (HOMO and LUMO) have to be positioned correctly relative to the fullerene acceptor in order to obtain maximum V_{oc} and to provide a sufficient driving force for exciton dissociation. The V_{oc} is empirically related to the magnitude of the $HOMO_D$–$LUMO_A$ (Fig. 1.1d) offset, although this is a very simplified picture and multiple reports point to the importance of the charge transfer state in determining the V_{oc} [60–63]. Based on this premise, newly designed polymers tend to have lower HOMO levels than P3HT (which is most commonly reported as 5.2 eV) in order to increase V_{oc} but also retain a $LUMO_D$–$LUMO_A$ offset >0.3 eV (Fig. 1.1d) which is thought necessary as a downhill driving force for exciton dissociation (and thus efficient charge generation) [14]. An alternative approach to increase the V_{oc} is the use of an acceptor which has a higher lying LUMO than commonly used PCBM. This has recently been shown successfully with ICBA (a *bis*-indene-fullerene adduct) which in combination with P3HT shows a V_{oc} of 0.84 V instead of 0.63 V for P3HT:PCBM [40,41,64–66]. So far though, efficient solar cells combining ICBA with other polymers than P3HT remain relatively rare [67]. As mentioned above, the V_{oc} is related to the $HOMO_D$–$LUMO_A$ offset and the HOMO and LUMO energies of donor materials are generally determined using cyclic voltammetry (CV). Considerable inconsistencies in the literature regarding the use of the standard redox couple ferrocenium/ferrocene (Fc^+/Fc) in CV measurements make it very difficult to compare values reported for HOMO and LUMO levels [68]. Consequently, during our discussion of state-of-the-art donor materials (Section 1.3), we will in most cases not mention exact numbers but rather comment on general trends and the more reliable value of V_{oc} itself.

Finally, the FF is affected by the hole mobility of the active material and is strongly tied to the morphology of the donor:fullerene blend (degree of phase separation between donor and acceptor, aggregation, crystallinity), with a high fill factor (FF > 0.6) indicating balanced, trap-free charge transport through the bulk and favorable morphology. Additionally, the quality of the interfaces between the active layer and the electrodes has a significant effect on the FF, where decreased shunt resistance and FF are indicative of increased charge recombination, leakage current and space charge build-up due to low quality interfaces (see Chapter 4 for further details) [69]. The FF can be improved by extensive optimization of device fabrication through methods such as solvent annealing, thermal annealing, or the use of additives and optimization of interfaces (see Chapters 7 and 10 for a detailed discussion on the BHJ morphology) [50,52,53].

This brief overview illustrates that for the optimization of solar cell efficiencies a careful fine-tuning of the energy levels of donor materials, also with respect to fullerene energy levels, is necessary in order to achieve an optimal balance between J_{sc} and V_{oc} because maximization of either one would result in a decrease of the other parameter. For example, in order to drastically lower the bandgap of a donor material (which would maximize the theoretically attainable J_{sc}) and still obtain a working solar cell, the HOMO energy level has to eventually shift upward to retain a large enough driving force for exciton dissociation ($LUMO_D$–$LUMO_A$ offset>0.3 eV), resulting in a loss of V_{oc} and likely no overall improvement of PCE. From theoretical calculations it was estimated that optimal donor materials will have a bandgap of ~1.5 eV and HOMO level of ~5.4 eV when properly balanced and can reach efficiencies over 10–12% in binary blend BHJ solar cells with a single active layer [14,70,71]. Our focus here is on the design principles for donor materials that aim to achieve this balance.

1.2 Design Principles for Bandgap and Energy-Level Control of Donor Materials

In the following we will introduce the design principles which are available for the synthetic chemist to control the electronics of donor materials, specifically energy levels and bandgap, through structural modifications and thus tune solar cell output parameters.

Conjugated polymers possess an inherent bandgap because of the bond length alternation (BLA) between single and double bonds in the polymer backbone. In the (hypothetical) case of a polymer with completely delocalized electrons the bandgap would be zero, but such systems are unstable (Peierls instability) leading to the localization of electrons and consequently BLA [72–74].

Several design principles have been established in order to decrease the BLA (and thus the bandgap) in conjugated polymers, specifically the stabilization of the polymer quinoidal form and the donor/acceptor (D/A) approach. In aromatic polymers two nondegenerate ground states exist, with the quinoidal form having a higher energy but lower bandgap than the aromatic form (Fig. 1.3a). In order to stabilize the quinoidal form (and lower its energy below the aromatic form) fused ring systems such as isothianaphthene (ITN), thienopyrazine (TP), and thienothiophene (TT) have been developed. In the first two cases, the aromatic stabilization energy is lower for thiophene than for the fused ring system (benzene or pyrazine), leading to dearomatization of thiophene (and consequently aromatization of benzene or pyrazine) and increased delocalization of electrons along the backbone (Fig. 1.3a) shows poly(isothianaphthene) as an example) [75,76]. For poly(thienothiophene), however, it was predicted that the aromatic and quinoidal forms have approximately the same energy and are thus both contributing to the polymer structure [77,78]. This contribution of the quinoidal form not only decreases BLA but also planarizes the polymer backbone, with both effects resulting in a lowered bandgap.

As mentioned, another design principle that can be used is the D/A approach illustrated in Fig. 1.3b, also called push–pull approach, where electron-rich monomers and electron-poor monomers (common examples are shown in Fig. 1.4) are incorporated in an alternating fashion into the polymer backbone [72–74,80,81]. This alternating structure can stabilize the quinoidal form because it increases the double bond character between the monomers by favoring the charge-separated resonance structure (positive charge on the electron-rich monomer and negative charge on the electron-poor monomer). The D/A approach also offers the advantage that the HOMO and LUMO level of polymers can, to a certain degree, be individually tuned because they are located on different parts of the polymer, with the LUMO almost exclusively on the acceptor unit whereas the HOMO is more delocalized but often predominantly on the donor unit [17,82]. The D/A approach is currently by far the most dominant approach to make efficient low bandgap polymers and small molecules and Fig. 1.3b shows one such example (D/A copolymer PBTTPD with an efficiency of 7.3%) [79].

As mentioned above, the backbone of conjugated polymers consists of alternating single and double bonds with the p_z orbitals of the carbon atoms overlapping. If the backbone is twisted out of plane around single bonds the p_z orbitals no longer overlap and the delocalization of electrons is even further limited, resulting in an increased bandgap [72–74]. In addition to stabilizing the quinoidal form and applying the D/A approach several other design principles have been developed to ensure planarity of the polymer backbone and the most common approaches are discussed below.

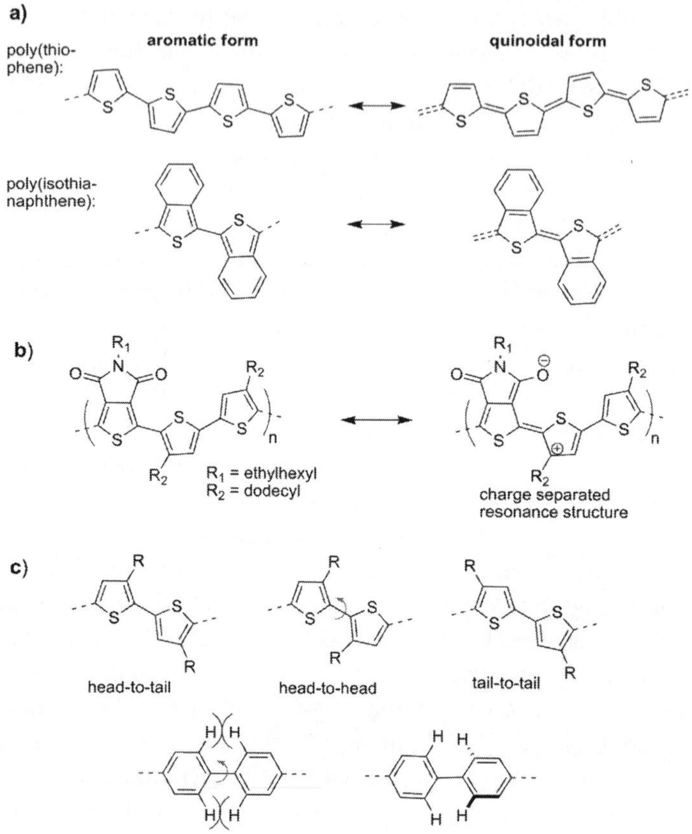

Figure 1.3 (a) Nondegenerate ground states of conjugated polymers: the aromatic and quinoidal forms of poly(thiophene) and poly(isothianaphthene) are shown as examples, (b) representative charge-separated structure of a donor/acceptor (D/A) copolymer (PBTTPD [79]), where thienopyrroledione (TPD) is the acceptor and dodecylthiophene the donor unit, (c) head-to-tail, head-to-head, and tail-to-tail connection of substituted thiophenes and steric hindrance between o-hydrogen atoms of neighboring phenyl units (torsional twists around single bonds are indicated with a curved arrow).

Although most monomers are designed to be symmetric, automatically resulting in regiosymmetric polymers, in some important cases monomers are unsymmetric. Unsymmetrical monomers substituted with an alkyl chain (e.g., 3-hexylthiophene) are generally copolymerized in a regioregular (rr) fashion (with the side chains predominantly oriented head-to-tail) as opposed to randomly (where the backbone contains head-to-tail, head-to-head, and tail-to-tail linkages) in order to reduce unfavorable steric interactions between side chains (Fig. 1.3c). The best example for this is P3HT, where regiorandom (ra) P3HT not only has a much higher bandgap (due to torsional twists) than rr-P3HT but other properties such as the hole mobility are also negatively impacted [83,84]. On the other hand, the nature of the side chains is important for both symmetric and unsymmetric monomers, with linear side chains introducing less steric hindrance than bulky side chains. Often though, a trade-off is required between planarity of the backbone and solubility of the polymer in common organic solvents to ensure processability [17]. One approach to planarize the backbone, even when branched, bulky side chains are used, is the introduction of spacers, such as a nonsubstituted thiophene, between the monomers [85]. This is also important when neighboring monomers both contain phenyl groups in the backbone in order to avoid torsional twists between o-hydrogen atoms (e.g., carbazole donor copolymerized with benzothiadiazole (BTD) acceptor) (Fig. 1.3c) [51,86]. Additionally, fused ring systems such as cyclopentadithiophene (CPDT), carbazole, and benzodithiophene (BDT), which rigidify the backbone and suppress rotational disorder around single bonds (ladder polymers take this concept to the extreme), are often used in D/A materials to planarize the polymer backbone [19,51,87].

In general, the most direct influence a synthetic chemist has on the energy levels (HOMO and LUMO) of conjugated polymers and small molecules is through the use of substituents. When an electron-donating group, such as an alkoxy group, is used an electron density is pushed into the conjugated backbone and the HOMO level of the material is raised, often accompanied by a reduction in bandgap because the LUMO is less affected. On the other hand, when electron-withdrawing groups are used, such as a cyano group or fluorine atom, the electron density in the backbone is reduced lowering the HOMO level and increasing the electron affinity [20,88]. Beyond the relatively straightforward influence alkyl side chains have on the solubility as well as planarity of polymers they can also have a pronounced impact on material energy levels, and thus V_{oc}, although the effect is less understood [89,90].

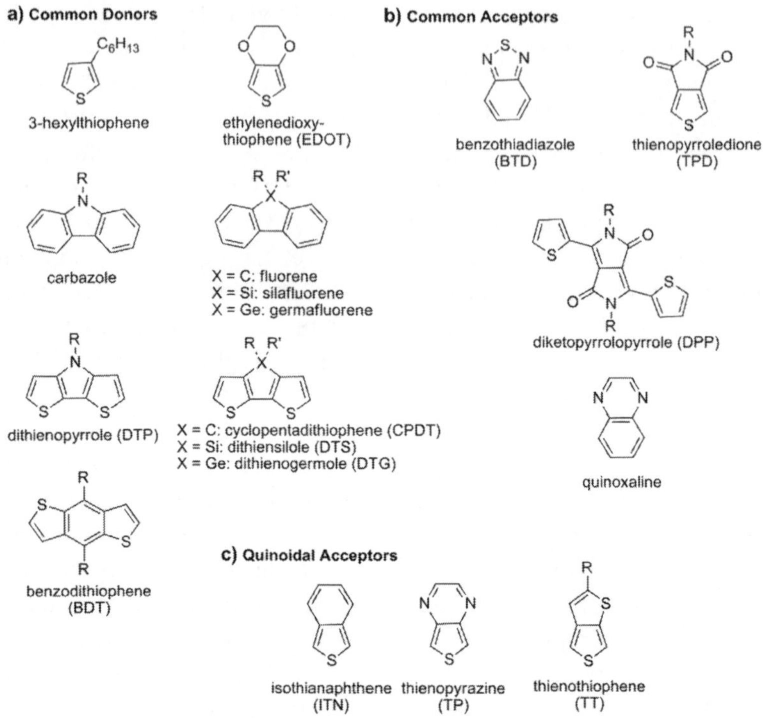

Figure 1.4 (a) Common donor monomers, (b) common acceptor monomers, and (c) quinoidal acceptors.

Finally, if considering a polymer in the solid state, favorable intermolecular interactions can also have a considerable influence on the bandgap by further planarizing the backbone. This is often apparent through a considerable red shift of the UV/vis absorption spectrum when going from solution to thin film. Strategies outlined above to planarize the polymer backbone generally also improve intermolecular packing leading to higher degrees of crystallinity. Another important factor when considering intermolecular packing is the molecular weight and polydispersity index (PDI) of polymers, which has been shown to have considerable impact on the morphology of the active layer and thus solar cell efficiency and can be influenced by choosing an appropriate polymerization and purification procedure [51].

The application of the above design principles in the pursuit of optimal donor materials will be discussed in the following sections.

1.3 State-of-the-Art Polymer and Small Molecule Donors for Organic Solar Cells

In the last paragraphs, we discussed the electronic characteristics necessary for donor materials in order to achieve higher solar cell efficiencies (low bandgap, correct positioning of frontier energy levels) and introduced the important design principles available to tune these characteristics and optimize solar cell output parameters (especially J_{sc} and V_{oc}). In the following sections, we will now focus on demonstrating the progress toward embodying those characteristics by summarizing the state-of-the-art polymer and small molecule donors. Polymer donors are divided into homopolymers (containing either only one monomer or only donor monomers), perfectly alternating D/A copolymers (where a donor monomer is copolymerized with an acceptor monomer resulting in a perfectly alternating monomer sequence in the polymer backbone), and finally random and semirandom D/A copolymers (which also contain donor and acceptor monomers but are less ordered than perfectly alternating D/A copolymers).

1.3.1 Homopolymers

Poly(3-alkylthiophenes), which are the most common homopolymers, are generally made either via Grignard metathesis (GRIM) or Rieke polymerization methods [84,91,92], both of which give polymers with high regioregularities (percentage of head-to-tail connections in the polymer backbone) and molecular weights. GRIM, which is predominantly used on the laboratory scale, is a nickel-based living chain-growth polymerization and affords low-PDI polymers. Other methods such as Stille and Suzuki [93–96] coupling have also been used for poly(alkylthiophenes) but have the disadvantage that monomer synthesis is more complicated. Table 1.1 summarizes the solar cell performance of some of the most effective P3ATs reported to date.

P3HT, which has a HOMO level of 5.2 eV and E_g of 1.9 eV is by far the most studied polymer for organic solar cells and, with PCBM as the acceptor, efficiencies of up to 5% (Table 1.1) have been achieved but most reported PCE values are between 3% and 4% [65,100,101].

Higher efficiencies have been published with indene-C_{60}-bisadduct ($IC_{60}BA$) (6.5%) and indene-C_{70}-bisadduct ($IC_{70}BA$) (7.4%) due to the increased V_{oc} because of the higher acceptor LUMO level [66,97]. This dominance of P3HT is due to a combination of favorable properties such as semicrystallinity, high hole mobility, as well as the formation of a small-scale phase-separated, stable BHJ morphology which makes it an almost ideal candidate for organic solar cells. Out of all investigated poly(3-alkylthiophenes) the hexyl side chain of P3HT seems to offer the best compromise between solubility (processability), morphology (crystallinity and polymer:fullerene mixing), and chromophore density of the polymer with only poly(3-pentylthiophene) achieving similar solar cell efficiencies [102–106]. As mentioned above, ra-P3HT is amorphous and has much lower hole mobility than rr-P3HT due to the unfavorable steric interactions of the side chains [83]. Even though it has been claimed that the regioregularity of P3HT has to be very high (~95%) in order to achieve high efficiencies, no significant effect on the performance can be detected for regioregularity between 90% and 98% [100,107]. On the contrary, it has been found that lower rr-P3HT (86%) still has sufficient electronic properties for high efficiency in combination with a higher thermal stability of the polymer:fullerene BHJ [108,109]. Another factor that influences the performance of P3HT solar cells is the molecular weight. It has been reported that a M_n (number average molecular weight) of over 10,000 g/mol is necessary in order to achieve high efficiencies predominantly because of low hole mobility for low M_n P3HT [110–112]. On the other hand, when M_n gets too high processability decreases due to gellation of the polymer solution and it has been suggested that a mixture of high- and low-M_n polymer performs best [65].

rr-P3HT has several shortcomings which prevent it from achieving even higher efficiencies such as the moderate V_{oc} (~0.6 V with a HOMO energy of 5.2 eV) as well as the narrow absorption breadth (absorption onset ~650 nm) resulting in a limited J_{sc}. Several attempts have been made to improve the efficiency of poly(3-alkylthiophenes) by improving the V_{oc} and some approaches are discussed below (attempts to broaden the absorption fall under the category of D/A copolymers). Hou et al. reported that by decreasing the alkyl side chain density on a thiophene backbone using a long-branched alkyl chain on every third ring (P3HDTTT) they were able to increase V_{oc} considerably compared to P3HT. However, no improvement of efficiency was achieved compared to P3HT:PCBM solar cells (Table 1.1) [98]. Ko et al. used a similar strategy by only putting alkyl chains on the 3- and 4-position of every third thiophene ring, which resulted in a lowered HOMO energy and increased V_{oc} (0.75 V). In this case the obtained efficiencies of up to 4.23% with $PC_{71}BM$ (PDOTT, Table 1.1) were found to exceed those observed for the reference solar cell with P3HT [99]. A follow-up study focused on the degree of twisting along the backbone in a family of poly(hexylthiophenes) where polymers only differ by the number and placement of the alkyl chains. They found that a certain degree of twisting is beneficial due to an increase in the V_{oc} together with almost unchanged J_{sc}, FF, and hole mobility and thus increased PCE. On the other hand, larger degrees of twisting led to increased bandgaps because of the reduced conjugation length and thus reduced J_{sc} [113]. Instead of reducing the side chain content Burkhart et al. took a different approach by introducing increasing amounts of branched ethylhexyl side chains into P3HT. Polymers with 50% or less ethylhexyl side chains had the same bandgap and similar absorption properties as P3HT but a considerably increased V_{oc} because of the lowered HOMO levels. The combination of higher V_{oc}, slightly increased J_{sc}, and high FF ultimately led to an efficiency of 3.85% for $P3HT_{75}$-co-EHT_{25} (Table 1.1) exceeding that measured for P3HT (3.48%).

1.3.2 Perfectly Alternating Donor/Acceptor Copolymers

Palladium-catalyzed cross-coupling reactions are used almost exclusively for the synthesis of well-defined D/A copolymers and current high-performing D/A polymers are either synthesized by Stille (stanyllated monomers) or Suzuki-Miyaura (boronic monomers) (Fig. 1.5) coupling which allow controlled C_{sp2}–C_{sp2} bond formation [93–96]. The resulting polymers generally have high molecular weights and are of high quality, whereas certain drawbacks such as unreacted end-groups (such as trimethyltin) as well as residual catalyst can be solved by end-capping and careful purification [114,115]. Nevertheless, the monomer synthesis often involves toxic reagents and direct arylation (Fig. 1.5) has recently been introduced as an economically and ecologically attractive alternative where unfunctionalized heteroaromatic monomers can be directly coupled through C–H bond cleavage [116]. Although it has been reported that the properties of a polymer made by direct arylation can be considerably different from the properties of the same polymer made by conventional cross-coupling polymerization it has recently been shown for the first time that a polymer made by direct arylation can be successfully (6.1%) used in solar cells, which is promising for future use of this approach [117,118].

Table 1.1 Structure and solar cell performance of select homopolymers

Polymer	J_{sc} (mA/cm^2)	V_{oc} (V)	FF	η_{max} (η_{avg}) %	Ref.
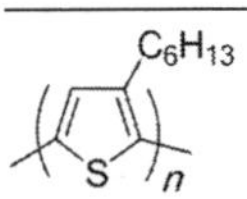 P3HT:PC$_{61}$BM (1:0.8)	9.5	0.63	0.68	5.08 (AM1.5G, 80 mW/cm^2)	[65]
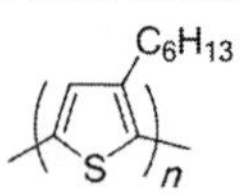 P3HT:IC$_{70}$BA (1:1)	11.35	0.87	0.75	7.4 (7.2)	[97]
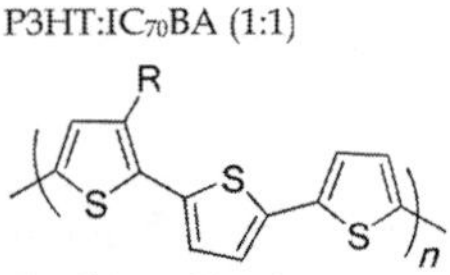 R = 2-hexyldecyl P3HDTTT:PC$_{61}$BM (1:1)	6.33	0.82	0.66	3.40	[98]
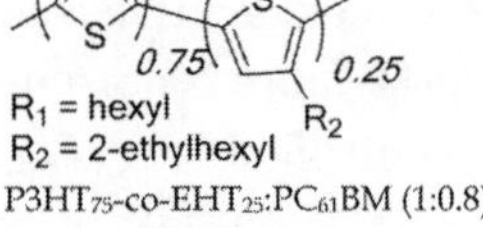 PDOTT:PC$_{71}$BM (1:1)	7.90	0.75	0.71	4.23 (4.03)	[99]
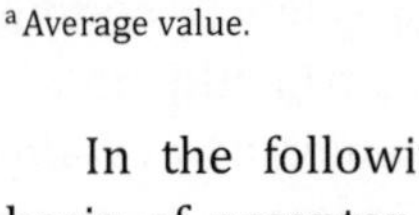 R$_1$ = hexyl R$_2$ = 2-ethylhexyl P3HT$_{75}$-co-EHT$_{25}$:PC$_{61}$BM (1:0.8)	9.85[a]	0.69[a]	0.57[a]	3.85[a]	[90]

[a] Average value.

In the following pages, D/A polymers are classified on the basis of acceptor monomer, and we further distinguish between quinoidal (predominantly TT) and nonquinoidal acceptors (BTD, TPD, diketopyrrolopyrrole (DPP), isoindigo, and quinoxaline).

Acceptors can be qualified by their electron-accepting ability which can be empirically determined based on the E_g of homologous D/A polymers (common donor unit): the smaller the E_g, the stronger the electron-accepting ability [122]. Here, we will focus predominantly on strong- and medium-strength acceptors (rather than weak acceptors) as the corresponding copolymers generally give better photovoltaic performance.

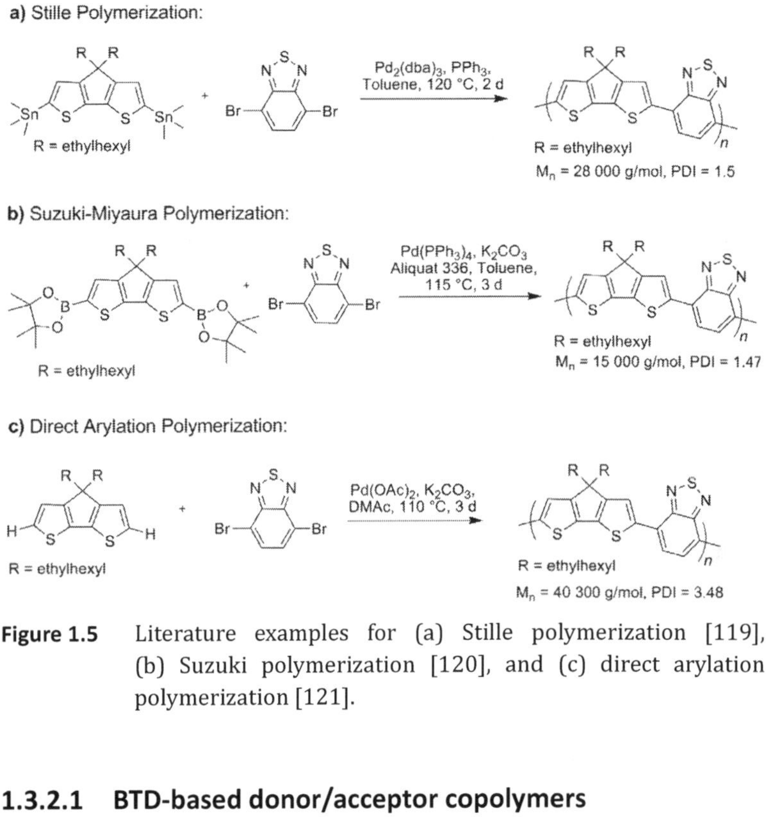

Figure 1.5 Literature examples for (a) Stille polymerization [119], (b) Suzuki polymerization [120], and (c) direct arylation polymerization [121].

1.3.2.1 BTD-based donor/acceptor copolymers

BTD (Fig. 1.4b) is a strong acceptor monomer due to the presence of the electron-withdrawing imine nitrogens. It is likely the most used acceptor monomer for conjugated D/A copolymers and can, depending on the donor comonomer, lead to polymers with very low bandgaps (and high J_{sc}) or very low HOMO levels (and high V_{oc}). BTD is symmetric, compact, and can be functionalized on the benzene unit in order to tune energy levels or increase solubility.

PCPDTBT shown in Table 1.2 contains CPDT as the planar, strong donor monomer, and BTD as the acceptor monomer and was first introduced by Mühlbacher et al. [129] showing promising properties such as a low bandgap of 1.40 eV due to the strong D/A effect, an EQE onset beyond 900 nm and PCE of ~3.2%. The solar cell performance was further optimized by Peet et al. [123] using 1,8-octanedithiol as an additive to improve the morphology of the BHJ achieving an average PCE of 5.5% and J_{sc} of 16.2 mA/cm^2 (Table 1.2), thus proving that low bandgap polymers can be very efficient in harvesting the solar spectrum and are able to achieve high efficiencies. In order to increase the V_{oc} of PCPDTBT the fluorinated analog of BTD was used (PCPDTFBT), which resulted in a downshift of the HOMO energy because of the electron-withdrawing nature of fluorine. Additionally, the presence of the fluorine atom improved the morphology of the BHJ considerably, allowing for efficient (5.51%) solar cell performance without the use of additives. The PCE was further increased to 5.81% (Table 1.2) by introducing a fullerene-based electron-collecting layer [124].

The silicon analog of PCPDTBT, where the bridge C-atom in the donor monomer CPDT has been replaced by an Si-atom, has been studied in detail by several groups (and named respectively PSBTBT [126,130] or Si-PCPDTBT [125]). While the absorption profile stays approximately constant the morphology of the material changes considerably. PCPDTBT is an amorphous material whereas PSBTBT is semicrystalline with short π–π stacking distances and hole mobilities twice as high. The better packing of PSBTBT is attributed to the significantly longer C–Si bond which reduces the steric hindrance between the bulky side chains and the adjacent thiophene rings. Solar cell efficiencies of 5.1% (PSBTBT) and 5.24% (Si-PCPDTBT) were achieved (Table 1.2) compared to 5.5% for PCPDTBT [125,126,130]. On the other hand, Coffin et al. not only exchanged the bridge atom of the donor from C to Si but also studied the impact of the alkyl side chains by changing them from branched (2-ethylhexyl) to linear (dodecyl) which lowered the bandgap slightly and resulted in an average PCE of 5.9% for P2 (Table 1.2) due to extremely high J_{sc} of 17.3 mA/cm^2. They also showed the considerable impact the molecular weight has on solar cell efficiency with an increase from 1.2% to 5.9% when the M_n was increased from 7000 to 34,000 g/mol [127].

Table 1.2 Structure and solar cell performance of select PCPDTBT analogs

Polymer	J_{sc} (mA/cm^2)	V_{oc} (V)	FF	η_{max} (η_{avg}) %	Ref.
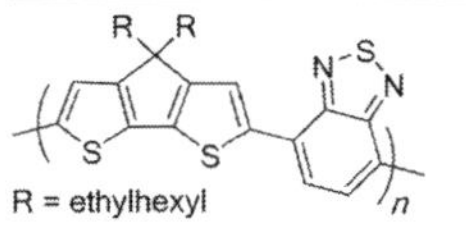 PCPDTBT:$PC_{71}BM$ (1:2 to 1:3)	16.2[a]	0.62[a]	0.55[a]	5.8 (5.5)	[123]
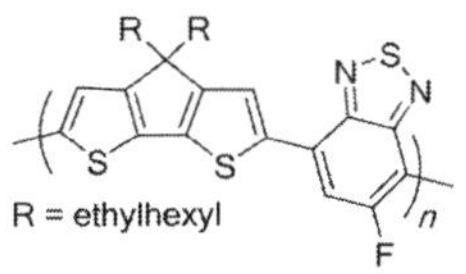 PCPDTFBT:$PC_{71}BM$ (1:2)	15.0	0.76	0.51	5.81	[124]
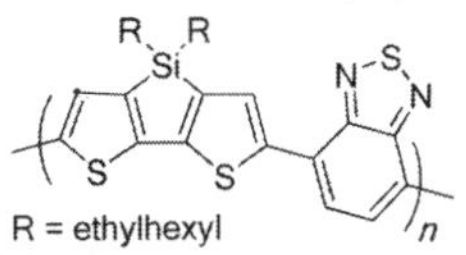 PSBTBT:$PC_{71}BM$ (1:1)	14.9 12.7[a]	0.57 0.68[a]	0.61 0.55[a]	5.24 5.1 (4.7)	[125] [126]
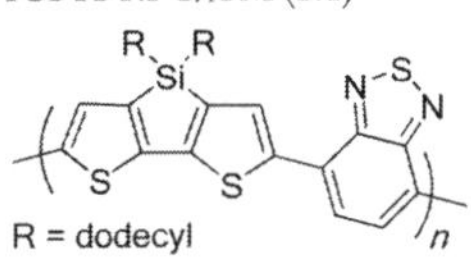 P2:$PC_{71}BM$ (1:1)	17.3[a]	0.57[a]	0.61[a]	6.0 (5.9)	[127]
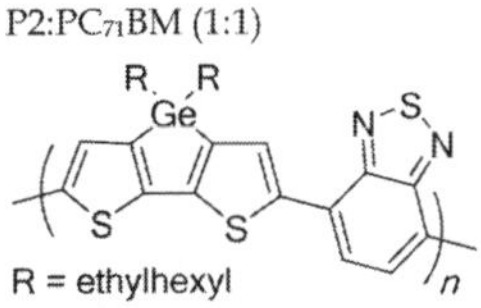 PGeBTBT:$PC_{71}BM$ (1:1)	18.4[a]	0.56[a]	0.42[a]	4.32[a]	[128]

[a] Average value.

The germanium analog of PCPDTBT, PGeBTBT, has recently been synthesized and studied in polymer:$PC_{71}BM$ solar cells. The C–Ge bond is less polarized than the C–Si bond due to the smaller electronegativity difference between C and Ge making arylgermanes more stable against bases and nucleophiles than aryl silanes and allowing the use of environmentally friendlier Suzuki polymerization. The C–Ge bond is slightly longer than the C–Si bond also resulting in semicrystalline packing in the solid state with a short π–π stacking distance (3.51 Å). The combination of high hole mobility (0.08 cm^2/(Vs)) and low bandgap (1.47 eV) is reflected in the solar cell performance with unprecedented J_{sc} of 18.4 mA/cm^2 and η = 4.5% (Table 1.2) [128].

Dithienopyrrole (DTP), where the C-bridge atom of CPDT is replaced by electron-rich nitrogen, has also been used in PCPDTBT analogs but because of its stronger electron-donating ability HOMO levels are higher leading to decreased V_{oc} and PCE even though the bandgaps are reduced [131]. Gibson et al. took a different approach by systematically replacing the sulfur atom on BTD in PCPDTBT with selenium and tellurium. They showed that the bandgap of polymers in solution systematically decreases from 1.59 to 1.06 eV when replacing sulfur with the heavier chalcogens, but no solar cell data were reported [132].

PCDTBT shown in Table 1.3 contains thiophene-flanked carbazole as the donor unit and BTD as the acceptor and was introduced by Blouin et al. [133] showing an efficiency of 3.6% and high V_{oc} of 0.89 V. Since its initial report the solar cell performance of PCDTBT has been optimized several times, mainly by improving the device architecture, resulting in a maximum PCE of 7.2% (Table 1.3) [54,134,135]. Carbazole is a dibenzene unit bridged by a nitrogen atom and is a weaker donor than DTP or CPDT due to the stronger aromaticity of the benzene units, resulting in polymers with lower HOMOs and higher V_{oc}. Carbazole is usually flanked by two unsubstituted thiophene units in order to avoid unfavorable steric interactions between monomers which could cause twists in the polymer backbone. Analogs of PCDTBT containing fluorene, silafluorene or germafluorene as the donor monomer instead of carbazole have also been synthesized (BisDMO-PFDTBT, PSiF-DBT, and PGFDTBT). Even though considerable V_{oc} of up to 1 V was reported, they show limited efficiencies, except for PSiF-DBT (5.4%, Table 1.3), likely due to very low hole mobilities impeding charge transport [136–138]. In order to improve the packing in the solid state, Du et al. replaced fluorene with 9-alkylidene-9H-fluorene where the bridge sp^3-C is replaced with a sp^2-hybridized C atom planarizing the monomer. In combination with two octyloxy chains on BTD this ensured that PAFDTBT is both soluble in commonly used organic solvents and also has a high hole mobility resulting in a PCE of 6.2% (Table 1.3) [139].

Chen et al. copolymerized BTD with a donor monomer thiophene–phenylene–thiophene (TPT), which is a hybrid between fluorene and CPDT resulting in *a*-PTPTBT. The V_{oc} is with 0.85 V intermediate between BisDMO-PFDTBT (0.97 V) and PCPDTBT (0.62 V) but the

FF is very high (0.67) because of the improved morphology. Solvent-annealed *a*-PTPTBT:$PC_{61}BM$ BHJ solar cells gave a maximum efficiency of 6.4% (Table 1.3) exceeding both BisDMO-PFDTBT (Table 1.3) and PCPDTBT (Table 1.2) [140].

Table 1.3 Structure and solar cell performance of selected PCDTBT analogs

Polymer	J_{sc} (mA/cm^2)	V_{oc} (V)	FF	η_{max} (η_{avg}) %	Ref.
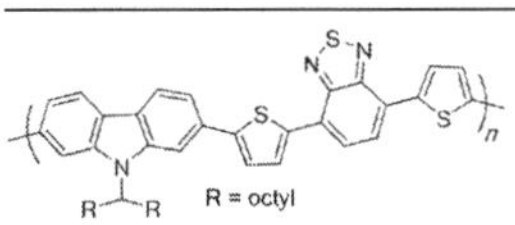 PCDTBT:$PC_{71}BM$ (1:4)	11.95	0.91	0.66	7.2 (6.5)	[135]
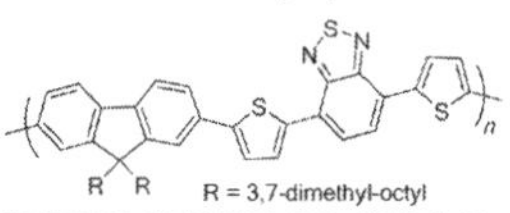 BisDMO-PFDTBT: $PC_{71}BM$ (1:3)	9.1	0.97	0.51	4.5	[136]
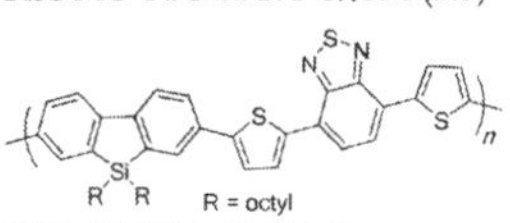 PSiF-DBT:$PC_{61}BM$ (1:2)	9.5	0.90	0.51	5.4 (AM1.5G, 80 mW/cm^2)	[138]
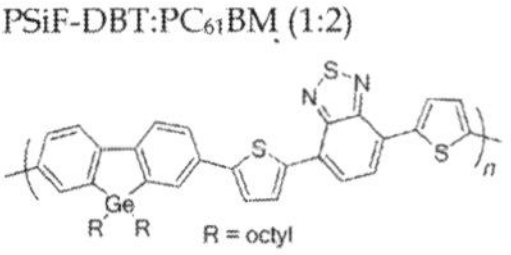 PGFDTBT: $PC_{71}BM$	6.9	0.79	0.51	2.8	[137]
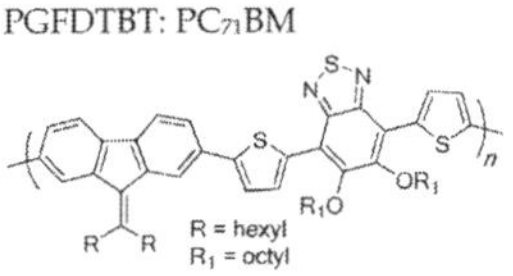 PAFDTBT: $PC_{71}BM$ (1:3)	9.9	0.89	0.70	6.2	[139]
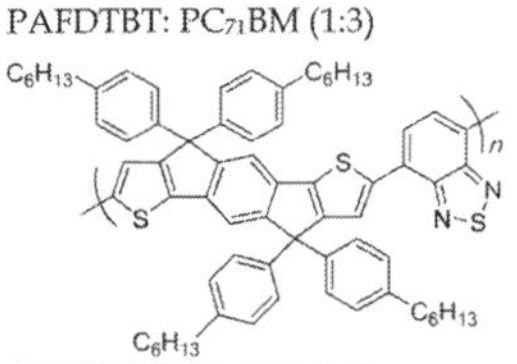 *a*-PTPTBT: $PC_{71}BM$ (1:1)	11.2	0.85	0.67	6.41 (6.1)	[140]

BDT is a relatively weak donor monomer which has been used in multiple copolymers with BTD and which allows for low HOMO levels in the resulting polymers as well as enhanced delocalization of electrons due to backbone planarization [141–143]. Zhou et al.

copolymerized BDT with difluorinated dithienyl-BTD resulting in PBnDT-DTffBT which showed an efficiency of 7.2% in the polymer: $PC_{61}BM$ BHJ solar cell (Table 1.4). Compared to nonfluorinated polymer PBnDT-DTBT, not only the V_{oc} was increased (0.91 vs. 0.87 V) due to the lowered HOMO level but also the J_{sc} (12.91 vs. 10.03 mA/cm^2) due to increased EQE, even though the bandgap remained practically identical [143]. Both Huo et al. [141] and Peng et al. [142] attached thiophene groups to the central phenyl ring of BDT in order to increase the conjugation of the polymer backbone. Huo et al. [141] synthesized PBDTTBT and achieved an efficiency of 5.66% with $PC_{71}BM$ as the acceptor (Table 1.4). On the other hand, Peng et al. [142] noticed that the introduction of a fluorine atom on BTD lowers the HOMO without increasing the bandgap, thus resulting in a slightly higher efficiency for the fluorinated polymer PBDT-FBT (6.21%, Table 1.4) compared to the alkoxy-substituted polymer PBDT-OBT (5.64%).

Table 1.4 Structure and solar cell performance of BDT/BTD copolymers

Polymer	J_{sc} (mA/cm^2)	V_{oc} (V)	FF	η_{max} (η_{avg}) %	Ref.
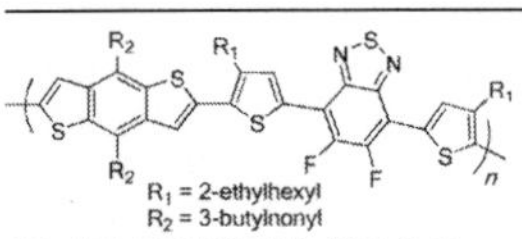 PBnDT-DTffBT:$PC_{61}BM$ (1:1)	12.91	0.91	0.61	7.2 (6.9)	[143]
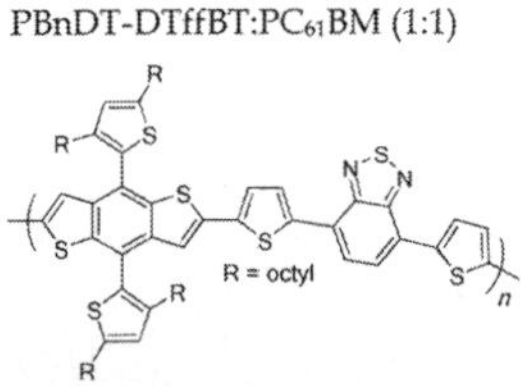 PBDTTBT:$PC_{71}BM$ (1:2)	10.7	0.92	0.57	5.66	[141]
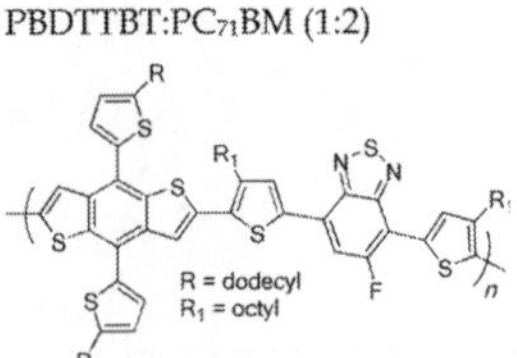 PBDT-FBT:$PC_{71}BM$ (1:2)	12.05	0.86	0.59	6.21 (6.08)	[142]
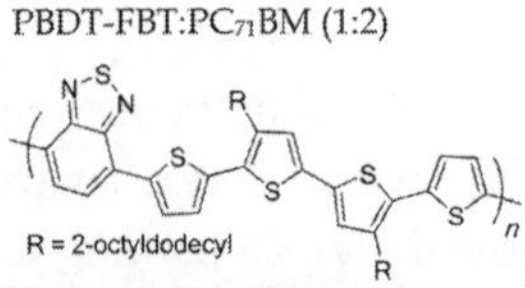 POD2T-DTBT:$PC_{71}BM$ (1:1)	12.3	0.72	0.70	6.26	[144]

Finally, Ong et al. [144] synthesized a copolymer based on dialkylquarterthiophene and BTD which not only retains the high hole mobility of poly(thiophenes) but also has a reduced bandgap because of the D/A effect between BTD and electron-rich alkylthiophene resulting in a solar cell efficiency of 6.26% (Table 1.4).

Multiple BTD analogs (Fig. 1.6), which can be either stronger or weaker acceptors than BTD, have been successfully used in D/A copolymers (Table 1.5). Zhou et al. used pyridalthiadiazole (PT), which is a stronger acceptor than BTD because pyridine is more electron deficient than benzene, and copolymerized it with several weak donor monomers. The resulting copolymers had low bandgaps (1.51–1.56 eV) and the highest efficiency (6.32%, Table 1.5) was achieved with PBnDT-DTPyT containing BDT as the donor monomer [145]. PT is an unsymmetric acceptor and simple copolymerization with donor monomers results in regiorandom copolymers. Ying et al. [146] showed that the hole mobility of regiorandom PT copolymers is several orders of magnitude lower than that of regioregular copolymers (0.005 vs. 0.6 $cm^2/(Vs)$), emphasizing the importance of the regioregularity when using unsymmetric monomers.

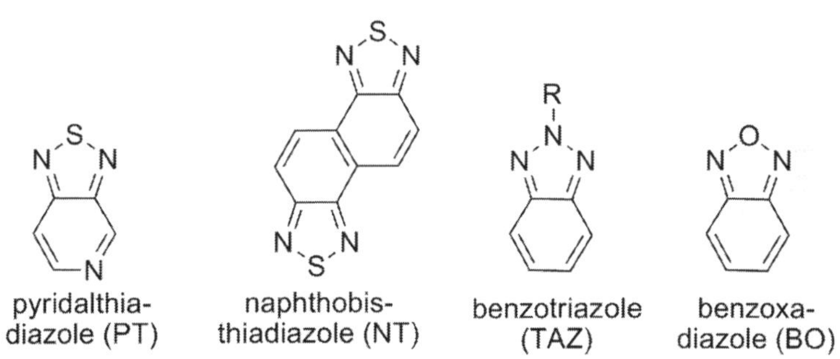

Figure 1.6 BTD analogs used in efficient polymer:fullerene BHJ solar cells.

Naphthobisthiadiazole (NT) is a slightly stronger acceptor than BTD and additionally has an enlarged planar aromatic structure which may facilitate interchain packing by rigidifying the polymer backbone. Wang et al. [147] compared analogous NT- and BTD-based copolymers and found that the bandgap of PBDT-DTNT was reduced by 0.15 eV compared to PBDT-DTBT due to the larger acceptor strength of NT. After considerable device optimization and introduction of a conjugated polyelectrolyte interlayer, efficiencies of up to 8.6% were achieved with PBDT-DTNT (Table 1.5) [148]. On the other hand, benzotriazole (TAZ) is a weaker acceptor, because the lone electron pair on the nitrogen atom is more basic (stronger positive mesomeric effect) than the lone pair on sulphur in BTD thus increasing the electron density in the ring [149,150]. This causes TAZ-based copolymers to have a wider bandgap than BTD-based copolymers. Price et al. [150] successfully used fluorinated TAZ to make PBnDT-FTAZ (with BDT as the donor monomer) which, even though it has a larger bandgap than P3HT (2vs. 1.9 eV), shows efficiencies of up to 7.1% with $PC_{61}BM$ because of a considerable increase in V_{oc} and very high FF (Table 1.5). Nie et al. [151] successfully used benzoxadiazole (BO) in a D/A copolymer with BDT which also resulted in a polymer (P1) with high V_{oc} and a maximum efficiency of 6.05% (Table 1.5).

Table 1.5 Structure and solar cell performance of polymers with BTD analogs as the acceptor

Polymer	J_{sc} (mA/cm^2)	V_{oc} (V)	FF	η_{max} (η_{avg}) %	Ref.
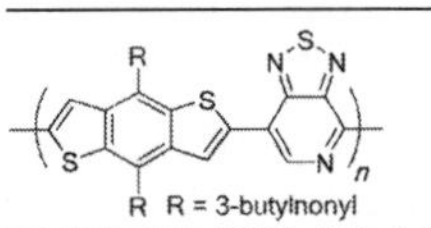 PBnDT-DTPyT:$PC_{61}BM$ (1:1)	12.78	0.85	0.58	6.32 (6.11)	[145]
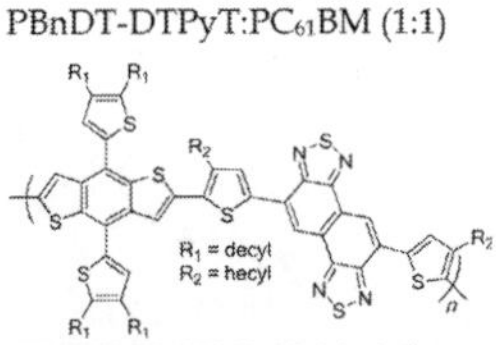 PBDT-DTNT:$PC_{71}BM$ (1:1.5)	17.4[a]	0.75[a]	0.61[a]	8.6 (8.4) (AM1.5G, 95 mW/cm^2)	[148]
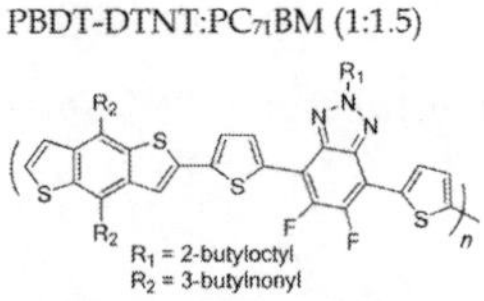 PBnDT-FTAZ:$PC_{61}BM$ (1:2)	12.45	0.79	0.72	7.1 (6.81)	[150]
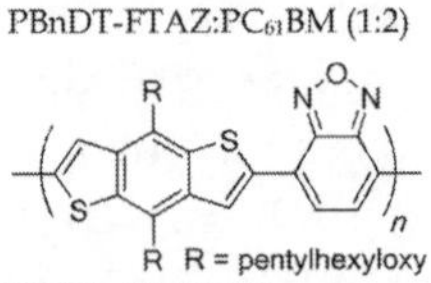 P1:$PC_{71}BM$ (1:1.5)	13.6	0.88	0.51	6.05 (5.65)	[151]

[a] Average value.

1.3.2.2 TPD-based donor/acceptor copolymers

TPD is an acceptor of medium strength due to the electron-withdrawing carbonyl groups (Fig. 1.4b). TPD-based copolymers tend to have very low HOMO levels (and consequently high V_{oc}) but due to the medium acceptor strength bandgaps are rarely very low and with most donor monomers they tend to be ~1.7 eV. TPD can be substituted with alkyl chains on the nitrogen atom in order to increase the solubility of polymers.

In 2010, Zou et al. [152] and Piliego et al. [153] almost simultaneously reported PBDTTPD (simply called P3 by Piliego et al.) which is a copolymer of BDT and TPD shown in Table 1.6. Both groups reported the same high V_{oc} (0.85 V) but Piliego et al. achieved a higher PCE (6.8% vs. 5.5%) possibly due to a combination of much smaller device area (0.03 vs. 1 cm^2) and more thorough device optimization. Piliego et al. also studied the influence that the alkyl side chain attached to TPD has on the device performance, concluding that a long, linear side chain, such as octyl, gives better properties than branched side chains because of increased order in the thin film. They also found that these TPD-based polymers have a face on orientation on the substrate (with the backbones parallel to the substrate) which is beneficial for charge transport in the device [153]. Recently Aïch et al. [154] succeeded in increasing the efficiency of PBDTTPD to 7.1% (Table 1.6) for solar cells with an active area of 1 cm^2 and $PC_{61}BM$ as the acceptor by using solvent additives to fine-tune the morphology. Li et al. replaced BDT with the stronger donor CPDT in order to reduce the bandgap and achieve a higher J_{sc}. But, as mentioned in the introduction, even though J_{sc} was increased (14.1 vs. 10.89 mA/cm^2 for PBDTTPD) V_{oc} simultaneously decreased (0.75 vs. 0.93 V for PBDTTPD) because of the higher HOMO level and the efficiency of PCPDTTPD-Oc:$PC_{71}BM$ (6.41%, Table 1.6) was not improved compared to PBDTTPD [155]. Replacing CPDT with dithienosilole (DTS), analogous to PSBTBT described above, gave PDTSTPD which has a slightly larger bandgap (1.73 eV) but lower HOMO level compared to PCPDTTPD-Oc. By blending PDTSTPD with $PC_{71}BM$ a solar cell efficiency of 7.3% was achieved [156]. A careful study on the influence of the molecular weight of PDTSTPD and the length of the alkyl side chain on TPD showed that the octyl side chain is optimal and that a high molecular weight is necessary in order to obtain high efficiencies. Whereas a M_n of 10,000 g/mol gave a low PCE of 3.1%, M_n increase to 31,000 g/mol allowed for an efficiency of 7.7% (Table 1.6) predominantly due to a simultaneous increase in hole mobility by two orders of magnitude and improved morphology [157]. Amb et al. argued that replacing the Si bridge atom with the larger Ge would result in further enhancement of ordering because the C–Ge bond is even longer and steric hindrance even more reduced. The Ge-based polymer (PDTG-TPD) has a somewhat higher HOMO level compared to the Si-based polymer together with a slightly lowered bandgap [158]. In an inverted device geometry with $PC_{71}BM$ as the acceptor PDTG-TPD initially achieved a solar cell efficiency of 7.3% which was further increased to 8.5% (Table 1.6) by the addition of an electron-transporting interlayer [23,158].

Yuan et al. successfully copolymerized TPD and the symmetric donor unit bi(dodecyl)thiophene and obtained PBTTPD which is semicrystalline and has a low HOMO level. Even though the polymer had a low M_n of 9700 g/mol the solar cell efficiency obtained with $PC_{61}BM$ was up to 4.7% [159]. They further optimized this efficiency to 7.3% (Table 1.6) by switching to $PC_{71}BM$ and optimizing the bulk morphology by using solvent additives [79]. Jo et al. synthesized a very similar polymer (P2, Table 1.6) using a relatively novel approach, direct arylation polymerization, which is a low-cost, green alternative to the standard methods such as Stille and Suzuki polymerization. They were not only able to achieve high M_n (41,000 g/mol), but also demonstrated for the first time that polymers made by this approach can be successfully used in very efficient solar cells (up to 6.1%) [117]. Finally, Najari et al. showed that the TPD unit can be used in combination with carbazole as the donor monomer in order to make polymers with very low HOMO and high V_{oc}. They obtained PCDTTPD (Table 1.6) which has an impressive V_{oc} of 1.07 V but only moderate PCE in combination with $PC_{61}BM$ [160].

1.3.2.3 Diketopyrrolopyrrole- and isoindigo-based donor/acceptor copolymers

Diketopyrrolopyrrole (DPP) has been widely used as a pigment in inks, paints and plastics and is a very strong electron acceptor which is most commonly flanked by two thiophene units [161,162].

DPP containing polymers generally have very low bandgaps (1.5 eV and lower), are semicrystalline and have high charge carrier mobilities (both hole and electron) which make them ideal candidates for OPV and field effect transistor (FET) devices. DPP can be substituted with alkyl side chains on both nitrogen atoms in order to tune the solubility of the polymers.

One of the first efficient DPP containing D/A copolymers was published by Wienk et al. in 2008. Electron-rich quarterthiophene (substituted with alkyl side chains) was combined with electron poor DPP to give pBBTDPP2 (Table 1.7) which has a bandgap of 1.4 eV. Despite this low bandgap the J_{sc} was only 11.3 mA/cm^2 which, in combination with moderate FF and V_{oc}, restricted the PCE to 4% [163]. Removing the alkyl substituents on the quaterthiophene unit induces additional planarity in the polymer backbone which can enhance packing and charge carrier mobility. In order to retain solubility DPP was substituted with much longer, branched alkyl side chains and the resulting polymer, pDPP (Table 1.7), achieved a high J_{sc} of almost 15 mA/cm^2 and PCE of 5.62% [164].

Bijleveld et al. used a terthiophene unit instead of quaterthiophene and PDPP3T (Table 1.7) exhibited high hole and electron mobilities (0.04 and 0.01 cm^2/(Vs)) but only a relatively moderate PCE of 4.7% [165]. Replacement of one of the thiophene units with a phenyl ring gave PDPPTPT (Table 1.7) which has a reduced HOMO level and increased bandgap compared to PDPP3T. This resulted in a considerably increased V_{oc} (0.8 vs. 0.65 V) and an improved solar cell efficiency of 5.5% [166]. Jung et al. replaced a thiophene unit with dithienothiophene which, due to its highly planar heteroarene structure, favors strong π–π interactions and thus high charge carrier mobilities. PDTTDPP (Table 1.7) has a very low bandgap with an absorption onset at 1055 nm and high hole mobility of up to 0.68 cm^2/(Vs). In combination with $PC_{71}BM$ as the acceptor a maximum efficiency of 6.05% and J_{sc} of 13.7 mA/cm^2 was obtained [167]. Very recently Gevaerts et al. have synthesized PDPP5T (Table 1.7) which, compared to PDPP3T, contains two additional alkyl-substituted thiophene units. In a single junction solar cell with $PC_{61}BM$ an efficiency of 5.3% was obtained and in a tandem solar cell (see Chapter 6) in combination with high bandgap polymer PCDTBT (Table 1.3) an efficiency of 7.5% was reached [168].

Table 1.6 Structure and solar cell performance of TPD-based copolymers

Polymer	J_{sc} (mA/cm^2)	V_{oc} (V)	FF	η_{max} (η_{avg}) %	Ref.
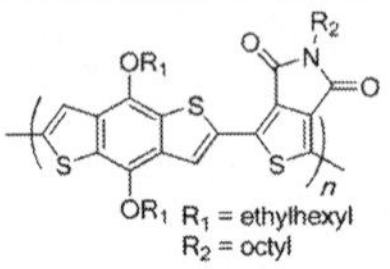 PBDTTPD:$PC_{61}BM$ (1:1.5)	10.89	0.93	0.70	7.1	[154]
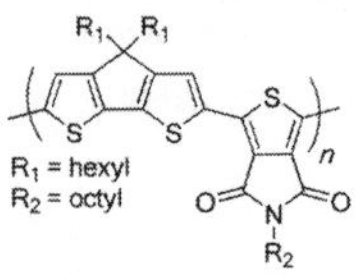 PCPDTTPD-Oc:$PC_{71}BM$ (1:2)	14.1	0.75	0.61	6.41 (6.31)	[155]
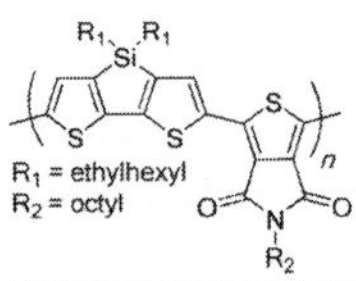 PDTSTPD:$PC_{71}BM$ (1:2)	12.13	0.91	0.70	7.7	[157]
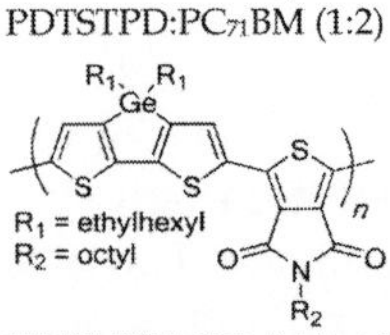 PDTG-TPD:$PC_{71}M$ (1:1.5)	14.4	0.86	0.69	8.5 (8.1)	[23]
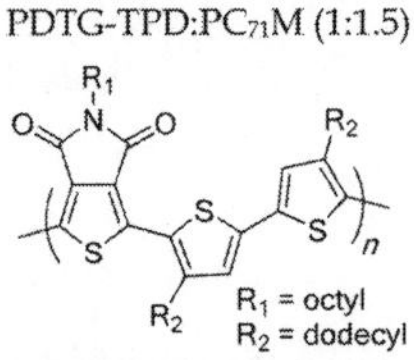 PBTTPD:$PC_{71}BM$ (1:1.5)	13.1	0.92	0.61	7.3 (7.1)	[79]
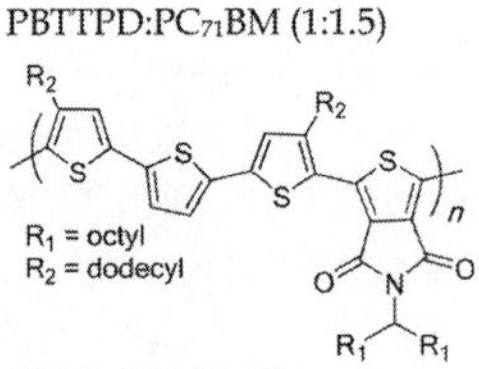 P2:$PC_{71}BM$ (1:1.2)	10.51	0.92	0.63	6.1	[117]
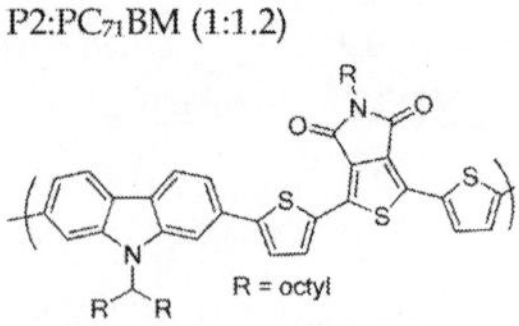 PCDTTPD:$PC_{61}BM$ (1:1)	4.72	1.07	0.36	1.82	[160]

Table 1.7 Structure and solar cell performance of diketopyrrolopyrrole-based copolymers

Polymer	J_{sc} (mA/cm^2)	V_{oc} (V)	FF	η_{max} (η_{avg}) %	Ref.
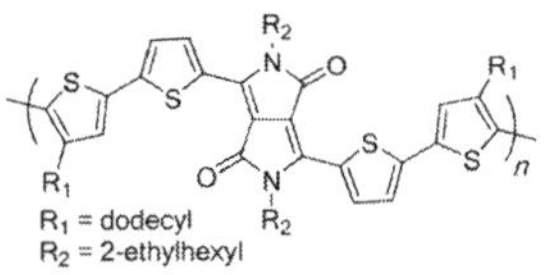 pBBTDPP2:PC$_{71}$BM (1:2)	11.5	0.61	0.58	4.0	[163]
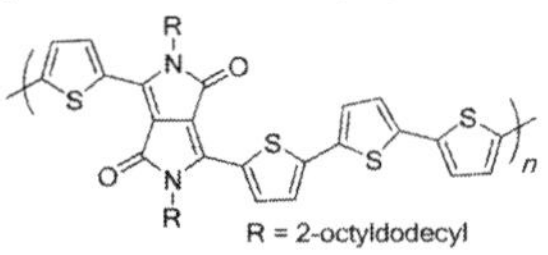 pDDP: PC$_{71}$BM (1:1)	14.84	0.63	0.60	5.62	[164]
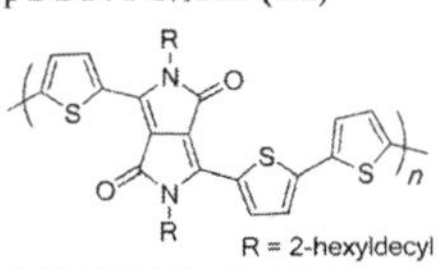 PDPP3T:PC$_{71}$BM (1:2)	11.8	0.65	0.60	4.7	[165]
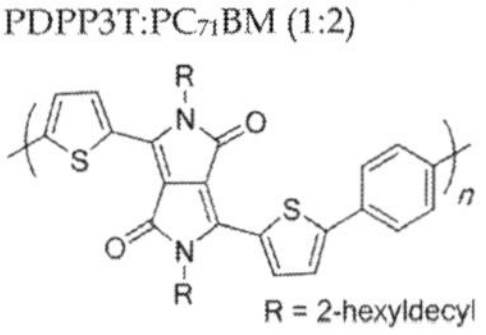 PDPPTPT:PC$_{71}$BM (1:2)	10.8[a]	0.80[a]	0.65[a]	5.6 (5.5)	[166]
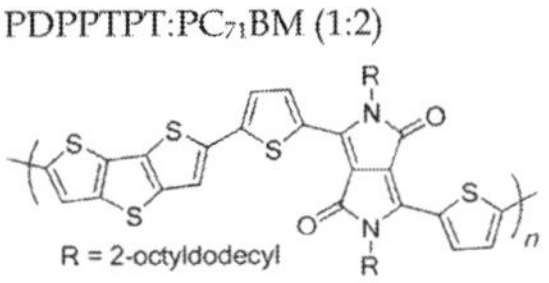 PDTTDPP:PC$_{71}$BM (1:1.5)	13.7[a]	0.66[a]	0.66[a]	6.05[a]	[167]
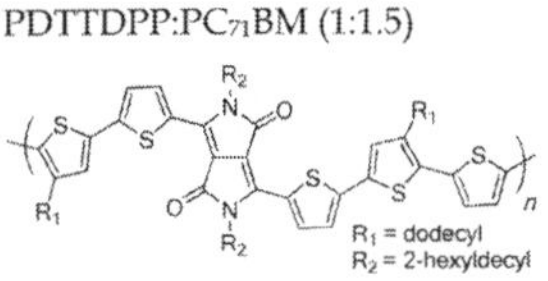 PDPP5T:PC$_{61}$BM (1:2)	14.0	0.58	0.65	5.3	[168]
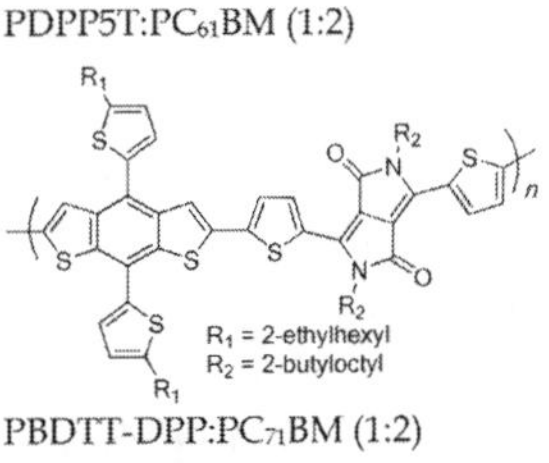 PBDTT-DPP:PC$_{71}$BM (1:2)	14.0	0.73	0.65	6.6 (6.5)	[169]

(*Continued*)

Table 1.7 (*Continued*)

Polymer	J_{sc} (mA/cm^2)	V_{oc} (V)	FF	η_{max} (η_{avg}) %	Ref.
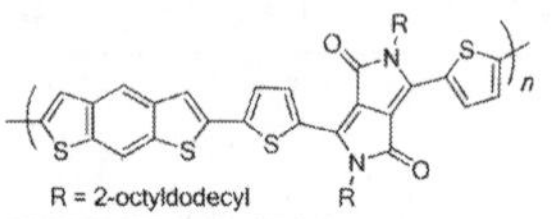 PBDTDPP:$PC_{71}BM$ (1:1)	10.49	0.82	0.60	5.16	[170]
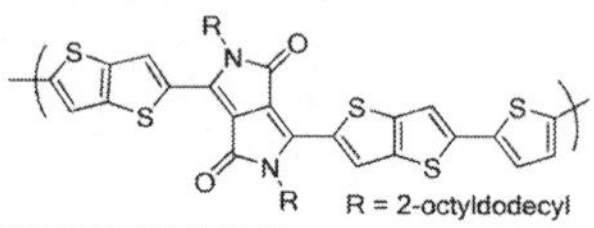 P1:$PC_{71}BM$ (1:2)	15.0	0.58	0.61	5.4	[171]
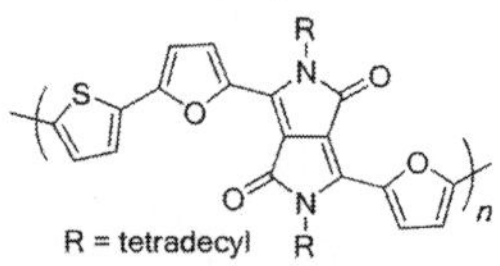 PDPP2FT-C_{14}:$PC_{71}BM$ (1:3)	14.8[a]	0.65[a]	0.64[a]	6.5 (6.2)	[172]
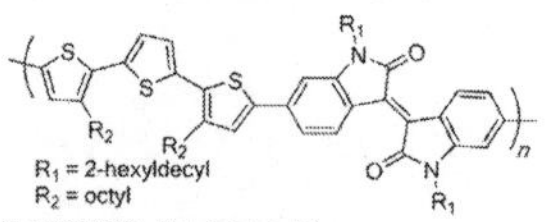 P3TI:$PC_{71}BM$ (1:1.5)	13.1	0.70	0.69	6.3	[173]

[a] Average value.

Dou et al. synthesized a family of four D/A copolymers, using a combination of DPP and two different BDT donor monomers (substituted on the central phenyl ring with either alkylated phenyl or thienyl units), in order to find an optimal low bandgap polymer for tandem solar cells. BDT is not as strongly electron donating as thiophene resulting in polymers with lower HOMO levels and consequently higher V_{oc} than those of DPP-based D/A polymers discussed above. It is interesting to note that the nature of the alkyl side chains on both DPP and BDT was critical for the solar cell performance and only the specific combination of 2-ethylhexyl chains on BDT and 2-butyloctyl on DPP gave high efficiencies. For the single layer BHJ solar cells with $PC_{71}BM$ a maximum efficiency of 6.6% was achieved for PBDTT-DPP (Table 1.7) with three of the synthesized polymers showing efficiencies of 5.8% or higher. Inverted tandem solar cells, with P3HT as the high bandgap polymer showed maximum efficiency of 8.8% [169]. Jung et al. completely removed the substituents on BDT resulting in PBDT-DPP (Table 1.7) which has an even lower HOMO level and high V_{oc} of 0.82 V but due to the considerably reduced J_{sc} the PCE was limited to 5.16% [170].

As mentioned above, the electron deficient DPP core is almost always flanked by two thiophene units and very few studies focus on the effect of modifying these units. Bronstein et al. recently replaced the thiophenes with larger TT units which can enhance the coplanarity of the polymer and improve the charge carrier mobility. As a result P1 (Table 1.7) shows a FET hole mobility of almost 2 $cm^2/(Vs)$ and a PCE of 5.4% with $PC_{71}BM$ as the acceptor (primarily limited by the moderate V_{oc}) [171]. Yiu et al. replaced the DPP flanking thiophenes with furan units which significantly increased the solubility of the polymers, allowing them to use linear instead of branched alkyl side chains on the DPP unit. Polymers with linear side chains were shown to have a shorter π–π stacking distance than those with branched side chains and outperformed them considerably in BHJ solar cells. A maximum PCE of 6.5% was obtained for PDPP2FT-C_{14} (Table 1.7) which is almost 2% higher than for the thiophene-based analog PDPP3T (Table 1.7) substituted with hexyldecyl side chains for sufficient solubility [165,172].

Isoindigo is another strong acceptor monomer that has been used in the dye industry and can be easily synthesized from widely available precursors. Isoindigo was used for the first time in D/A copolymers by Stalder et al. and they showed that, depending on the electron-donating strength of the comonomers the polymers have promising optical and electrochemical properties [174]. Despite this promising first report, solar cell efficiencies of isoindigo-based D/A polymers have mostly stayed well below those reported for DPP-based copolymers [175–178]. One notable exception is P3TI (Table 1.7) containing a planar terthiophene unit as the donor which is beneficial for π–π stacking of the polymer backbones. P3TI has a low bandgap of 1.5 eV in combination with a low HOMO energy and with $PC_{71}BM$ as the acceptor an efficiency of 6.3% was achieved [173].

1.3.2.4 Quinoxaline-based donor/acceptor copolymers

Quinoxaline is a strong acceptor due to the electron-withdrawing imine groups and was first used in a semiconducting polymer in 2007 by Gadisa et al. where it was copolymerized with fluorene to give

APFO-15 (Table 1.8). Due to the very large V_{oc} a respectable efficiency of 3.7% was achieved [179]. Interestingly, no further efficiency increase in quinoxaline-based polymers was reported until 2010 when Wang et al. published TQ1 (Table 1.8) [180–182]. TQ1 is a very simple structure synthesized by copolymerizing quinoxaline (substituted with two octyloxyphenyl groups) with thiophene and has a medium bandgap (1.70 eV) combined with a very low HOMO level (5.7 eV). A maximum efficiency of 6.0% was achieved with $PC_{71}BM$ as the acceptor [182]. Zhang et al. argued that if the phenyl substituents on quinoxaline could be connected with a single bond between the ortho positions the resulting monomer would not only be a stronger acceptor but also facilitate intermolecular packing. D/A copolymer PIDT-phanQ (using indacenodithiophene as the donor monomer) had a lower bandgap and slightly higher hole mobility than the analogous polymer PIDT-diphQ where the phenyl substituents on quinoxaline are not connected and consequently showed an improved PCE of 6.24% compared to 5.69% for PIDT-diphQ [183]. This efficiency was subsequently improved multiple times up to 6.63% (Table 1.8) by extensive device optimization [184,185]. In a similar fashion Zhang et al. used quinoxaline with a BDT unit fused on top and copolymerized it with indacenodithiophene. The resulting polymer P1 had an efficiency of 6.06% (Table 1.8) [186]. He et al. copolymerized thiophene flanked quinoxaline with 4,5-ethylene-2,7-carbazole resulting in PECz-DTQx which, after extensive device optimization and incorporation of a polyfluorene interlayer at the cathode, gave a PCE of 6.07% (Table 1.8) [187].

1.3.2.5 Quinoidal acceptors: TT, TP, and ITN

TT, ITN and TP are all monomers which stabilize the quinoidal form of conjugated polymers and are thus able to lower the bandgap by decreasing the BLA. On the other hand, only TP is an acceptor monomer in the actual sense of the word because it contains electron-withdrawing imine groups. Neither TT nor ITN are inherently electron poor but TT is frequently substituted with electron-withdrawing groups such as ester, ketone, fluorine and sulfonyl (or a combination of those) to increase its acceptor strength and lower the HOMO level of resulting polymers. TP based conjugated D/A copolymers, and to a lesser degree ITN based copolymers, have been studied extensively for the application in organic solar cells but only low to moderate efficiencies have been achieved so far [59,188–191]. On the other hand, TT-based polymers are currently some of the highest performing polymers and are discussed in detail below [22,192–194].

Table 1.8 Structure and solar cell performance of quinoxaline-based copolymers

Polymer	J_{sc} (mA/cm^2)	V_{oc} (V)	FF	η_{max} (η_{avg}) %	Ref.
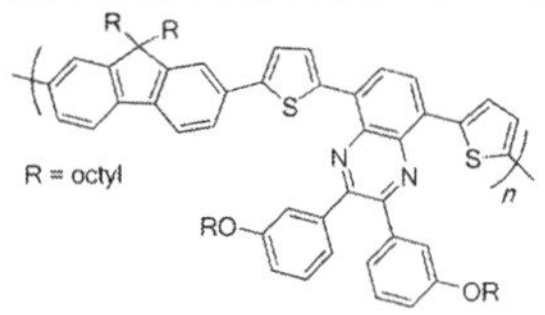 APFO-15:PC$_{61}$BM (1:3)	6.0	1.0	0.63	3.7	[179]
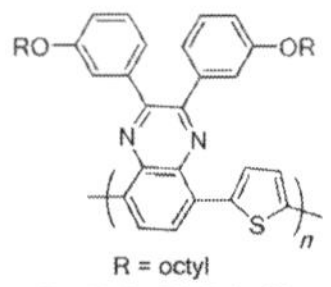 TQ1:PC$_{71}$BM (1:3)	10.5	0.89	0.64	6.0	[182]
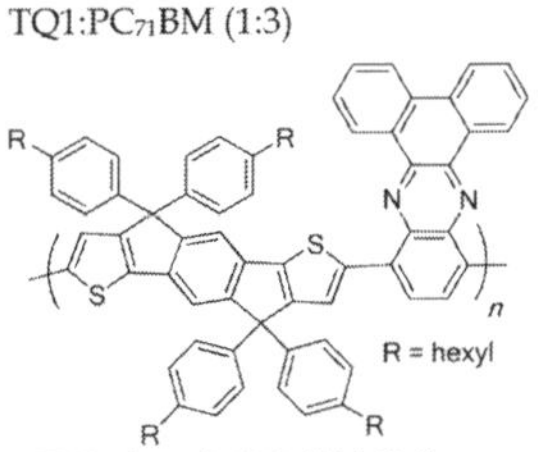 PIDT-phanQ: PC$_{71}$BM (1:3)	11.41	0.88	0.66	6.63	[184]
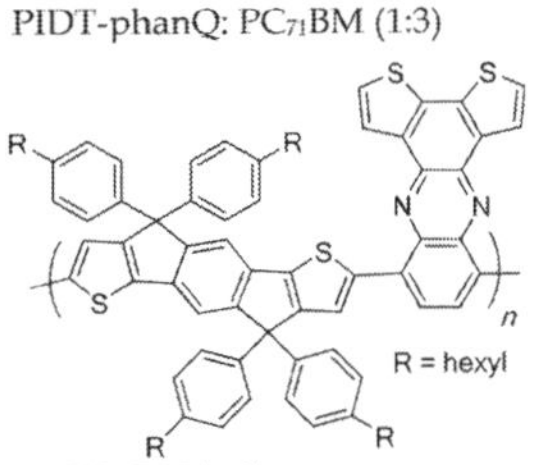 P1:PC$_{71}$BM (1:3)	11.6	0.83	0.63	6.06	[186]
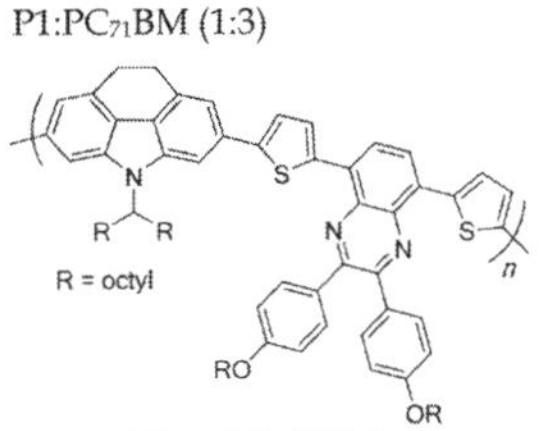 PECz-DTQx:PC$_{71}$BM (1:4)	11.4	0.81	0.66	6.07 (6.00)	[187]

Yu et al. developed a large family of copolymers PTBX (Fig. 1.7 and Table 1.9) based on alkoxy-substituted BDT and ester-substituted TT. Whereas for PTB1 linear alkyl side chains were used both on the BDT and TT units the solubility of PTB2-PTB7 was considerably enhanced by switching to branched side chains such as 2-ethylhexyl. A detailed study also showed that the thin film morphology and the π–π stacking distance are strongly influenced by the type of side chains used in the polymers. A positive correlation was found between strong π–π interactions and high FF in the BHJ solar cells and it was concluded that branched side chains have no negative impact on the π–π stacking distance if they are attached to the TT unit [195]. In general, strong π–π interactions between polymer chains are expected to be beneficial for solar cell performance because charge carrier transport occurs via a hopping process in the π–π direction as well as along chains [167,196,197].

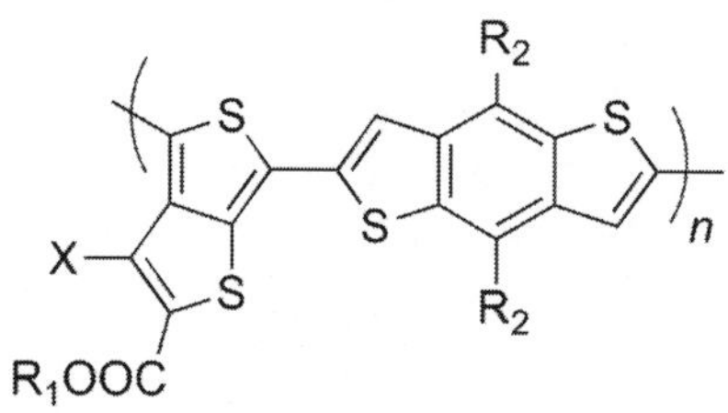

PTB1: X = H, R_1 = dodecyl, R_2 = octyloxy
PTB3: X = H, R_1 = 2-ethylhexyl, R_2 = octyl
PTB4: X = F, R_1 = octyl, R_2 = 2-ethylhexyloxy
PTB7: X = F, R_1 = 2-ethylhexyl, R_2 = 2-ethylhexyloxy
PTB9: X = H, R_1 = 2-ethylhexyl, R_2 = 2-ethylhexyloxy

Figure 1.7 Structures for PTB1, PTB3, PTB4, PTB7 and PTB9 [194,198,199].

Table 1.9 Solar cell performance of select PTB copolymers

Polymer	J_{sc} (mA/cm^2)	V_{oc} (V)	FF	η_{max} (η_{avg}) %	Ref.
PTB1:$PC_{61}BM$ (1:1.2)	15.6	0.56	0.63	5.6	[198]
PTB3:$PC_{61}BM$ (1:1)	13.9	0.72	0.58	5.85	[199]
PTB4:$PC_{61}BM$ (1:1)	13.0	0.74	0.61	5.90	[199]
PTB7:$PC_{71}BM$ (1:1.5)	14.5	0.74	0.69	7.40	[194]
	15.75	0.75	0.70	8.37	[192]
	17.5	0.75	0.70	9.2	[21]
PTB9:PC71BM (1:1.5)	14.3	0.60	0.66	5.66 (5.54)	[201]

Table 1.10 Structure and solar cell performance of TT-based copolymers

Polymer	J_{sc} (mA/cm^2)	V_{oc} (V)	FF	η_{max} (η_{avg}) %	Ref.
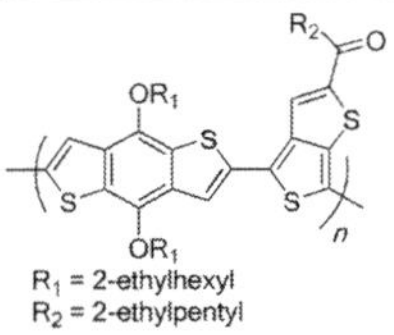 PBDTTT-C:PC$_{71}$BM (1:1.5)	14.7	0.70	0.64	6.58 (6.30)	[202]
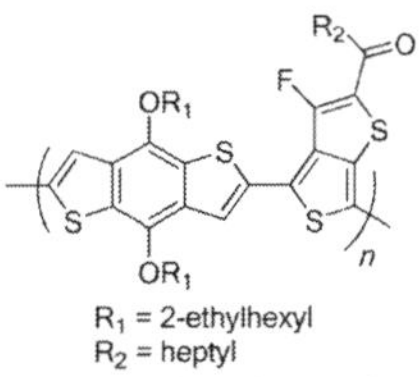 PBDTTT-CF:PC$_{71}$BM (1:1.5)	15.2	0.76	0.67	7.73 (7.40)	[203]
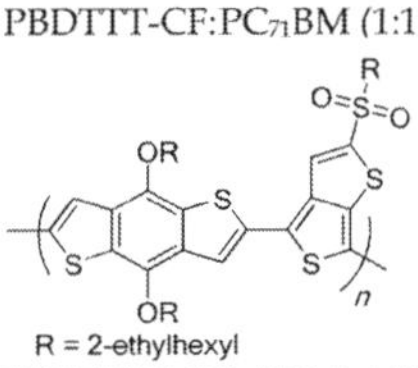 PBDTTT-S:PC$_{71}$BM (1:1.5)	14.1	0.76	0.58	6.22	[204]
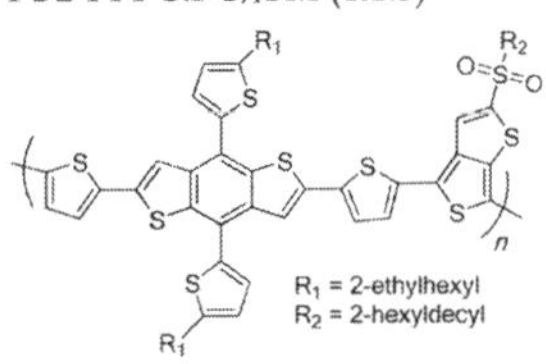 PBDTDTTT-S-T:PC$_{71}$BM (1:1)	16.35[a]	0.69[a]	0.66[a]	7.81 (7.48)	[193]
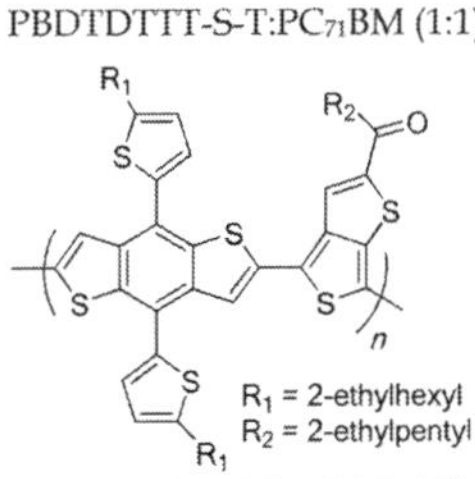 PBDTTT-C-T:PC$_{71}$BM (1:1.5)	17.48 18.39[a]	0.74 0.76[a]	0.59 0.63[a]	7.59 8.79[a]	[205] [22]
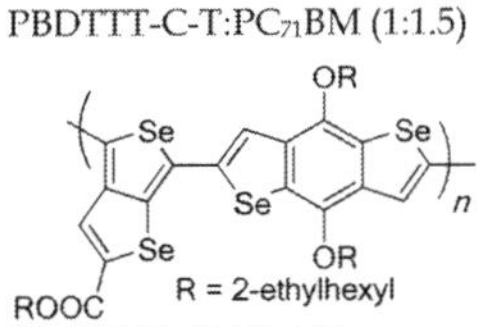 PSeB2:PC$_{71}$BM (1:1.2)	16.8	0.64	0.64	6.87 (6.46)	[201]

[a] Average value.

Multiple approaches were used in order to lower the high HOMO levels of PTBX polymers and consequently increase the V_{oc}. In PTB3 the alkoxy side chains on BDT were replaced by less electron-donating alkyl side chains, whereas in PTB4 and PTB7 the TT unit was substituted with a fluorine atom in order to decrease the HOMO level without interrupting the packing of the polymer [194,198,199]. Detailed studies showed that monofluorination of TT results in the best device performances whereas perfluorination of the polymer backbone greatly reduce PCE, probably due to poor compatibility with $PC_{71}BM$ molecules [200]. This careful optimization process resulted in a PCE of 7.4% for PTB7 (Table 1.9), which was further improved to 8.4% by introducing a cathode interlayer in the solar cell and very recently to 9.2% by using an inverted device structure and a polymer interlayer [21,192,194].

Replacing the ester group on TT with a ketone group is an alternative to substitution with fluorine which, even though it is effective in lowering the HOMO level, is also costly and synthetically more challenging. PBDTTT-C (Table 1.10) displayed an increased V_{oc} of 0.7 V (compared to 0.6 eV for the ester-substituted analog) and a PCE of 6.58% with $PC_{71}BM$ [202]. Additional substitution of TT with a fluorine atom as well as a linear alkyl chain instead of branched 2-ethylpentyl resulted in PBDTTT-CF (Table 1.10) which has an even further decreased HOMO level as well as higher hole mobility (a higher hole mobility than for most other PTBX copolymers was also observed for fluorine-substituted PTB4 and PTB7 likely due to improved packing) [194,199,203]. Consequently, an increased efficiency of 7.73% was observed for PBDTTT-CF [203].

Huang et al. used the strongly electron-withdrawing sulfonyl group in order to replace the combination of ester (ketone)/fluorine on TT and obtained PBDTTT-S (Table 1.10) which showed a high V_{oc} (0.76 V) and a PCE of 6.22% [204]. PBDTTT-S is an amorphous solid because of the disorder introduced by the bulky sulfonyl side chain and in order to improve packing alkylthienyl substituents were introduced on the central phenyl ring of BDT as well as thiophene units between BDT and TT. The thiophene spacers between monomers decrease steric hindrance between side chains whereas the alkylthienyl substituents on BDT can increase the two-dimensional conjugation and intermolecular π–π overlap. PBDTDTTT-S-T (Table 1.10) is, opposite to PBDTT-S, a semicrystalline solid with a hole mobility more than three times as high which led to an increased PCE of 7.81% [193]. Using a similar strategy Huo et al. replaced alkoxy-substituted BDT in PBDTTT-C with alkylthienyl-substituted BDT resulting in PBDTTT-C-T (Table 1.10), which not only has a lower bandgap as a consequence of increased conjugation but also an improved thermal and oxidative stability. Additionally, the hole mobility of PBDTTT-C-T was considerably increased compared to PBDTTT-C resulting in an efficiency of 7.59% [205]. Extensive device optimization by Li et al. through addition of gold nanoparticles and the use of a silver nanograting electrode led to further efficiency increase and a very high PCE of 8.79% [22].

Saadeh et al. described a new series of D/A copolymers based on selenophenoselenophene (the Se-based analog of TT) as well benzodiselenophene (the Se-based analog of BDT) and compared them with the analogous sulfur based polymer PTB9 (Fig. 1.7 and Table 1.9). PSeB2 (Table 1.10), where all sulfur atoms are replaced by selenium atoms, has a considerably lowered bandgap compared to PTB9 as well as much higher hole mobility probably due to the higher polarizability of Se compared to S and thus increased Se/Se interchain interactions. For PSeB2 a PCE of 6.87% was achieved which is a 21% increase over the analogous S containing polymer PTB9 [201].

1.3.3 Random and Semirandom Donor/Acceptor Copolymers

As opposed to perfectly alternating D/A copolymers, random and semirandom copolymers have a randomized sequence distribution of donor and acceptor monomers. Even though the perfectly alternating D/A copolymers described above have lowered bandgaps and often have good overlap with the maximum solar photon flux (at ~700 nm) leading to improved efficiencies compared to P3HT, the absorption is generally not truly broadened and short-wavelength photons are sacrificed, which in some cases limits the achievable J_{SC} or necessitates high loadings of fullerene. On the other hand, it has been shown that random structures are an effective way of broadening the absorption profiles of conjugated polymers due to the presence of multiple chromophores in the polymer backbones [119,206–209]. In 2008, Chen et al. randomly copolymerized TPT with BTD as the acceptor and thiophene as the third monomer (necessary to balance the number of functional groups) (Fig. 1.8). They obtained polymers which had lower bandgaps than P3HT (1.7 vs. 1.9 eV) without diminishing the absorption of short-wavelength photons. Even though they used TPT which is a planar monomer and should enhance π–π–conjugation as well as intermolecular stacking the measured hole mobilities were an order of magnitude below values measured for P3HT. The maximum PCE achieved for PTPTBT with $PC_{71}BM$ was 4.4% (Table 1.11) [208]. In a follow-up study, the same group looked at the use of other acceptor monomers and achieved efficiencies of over 4% (Fig. 1.8 and Table 1.11) using analogous quinoxaline and DPP-based random copolymers. PTPTDPP showed an especially broad and intense absorption profile but interestingly this did not translate into uniform photoresponse with the majority of current due to short-wavelength photons [210].

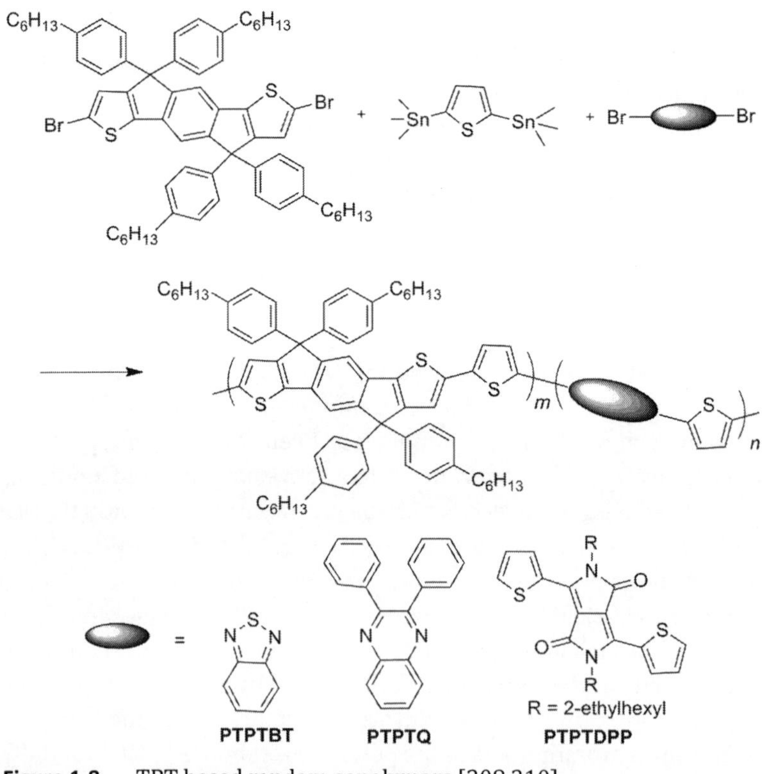

Figure 1.8 TPT-based random copolymers [208,210].

Table 1.11 Solar cell performance of TPT-based random copolymers

Polymer	J_{sc} (mA/cm^2)	V_{oc} (V)	FF	η_{avg} (η_{max}) %	Ref.
PTPTBT:$PC_{71}BM$ (1:3)	10.1	0.80	0.53	4.3 (4.4)	[208]
PTPTQ:$PC_{71}BM$ (1:3)	9.0	0.81	0.57	4.2	[210]
PTPTDPP:$PC_{71}BM$ (1:3)	10.3	0.75	0.54	4.2	[210]

More recently, the TPT core was extended even further by attaching another phenyl and thiophene unit (TPTPT) in hopes of lowering the bandgap and increasing hole mobility by planarizing the polymer backbone (Fig. 1.9). TPTPT was copolymerized with either DPP or BTD and thiophene as the third comonomers (analogous to Fig. 1.8) and the ratio between acceptor and donor monomers was adjusted in order to find the copolymer with optimal performance. PTPTPTDPP12 (which has a DPP to TPTPT ratio of 1:2) has a low bandgap (1.37 eV), broad absorption profile and high efficiency of 5.1% with $PC_{71}BM$ (Table 1.12). On the other hand, the SCLC hole mobility is low (2.6×10^{-5} cm^2/(Vs)) and, opposite to what the UV/vis absorption profile suggests, the EQE shows only very low photoresponse at longer wavelengths limiting the achievable efficiency [211].

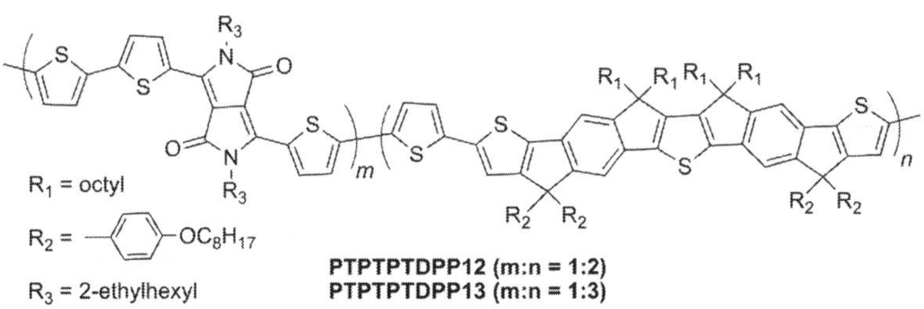

Figure 1.9 Ladder-type TPTPT-based random copolymers [211].

Table 1.12 Solar cell performance of TPTPT-based copolymers

Polymer	J_{sc} (mA/cm^2)	V_{oc} (V)	FF	η_{avg} (%)	Ref.
PTPTPTDPP12:$PC_{71}BM$ (1:4)	11.71	0.70	0.62	5.1	[211]
PTPTPTDPP13:$PC_{71}BM$ (1:3)	10.97	0.68	0.61	4.6	[211]

Li et al. copolymerized 3,5-dialkyldithienothiophene with DPP and thiophene and the resulting copolymer PDPP-T-DTT (Table 1.13)

has a low bandgap (1.37 eV) as well as a very uniform absorption profile from 370 to 900 nm. Interestingly the hole mobility of PDPP-T-DTT was an order of magnitude lower than for an analogous perfectly alternating copolymer, which the authors attributed to the disorder introduced by the random monomer linkages. With $PC_{71}BM$ an efficiency of 5.02% was achieved for PDPP-T-DTT which was predominantly limited by the moderate V_{oc} (0.58 V) [212].

Table 1.13 Structure and solar cell performance of PDPP-T-DPP

Polymer	J_{sc} (mA/cm^2)	V_{oc} (V)	FF	η (%)	Ref.
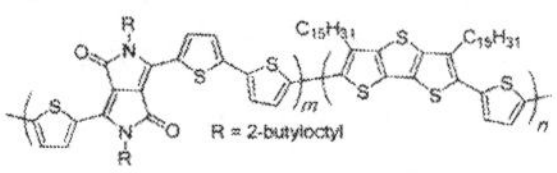 PDPP-T-DTT:$PC_{71}BM$ (1:2)	12.76	0.58	0.67	5.02	[212]

The above-mentioned examples of random copolymerizations have in common that of the three monomers used in a polymer only one is an acceptor. Nielsen et al. chose to take a different approach by introducing a second acceptor monomer into the polymer backbone. Using benzotrithiophene as the donor monomer and DPP as the acceptor and then adding either BTD or TPD (or alternatively increasing the benzotrithiophene content) allowed them to fine tune optical and electronic properties of the random copolymers (Fig. 1.10). All four novel random polymers not only show a higher efficiency than the perfectly alternating reference polymer BTT-DPP but also an increased EQE in both the short- and long-wavelength region. Copolymer P1 containing benzotrithiophene, DPP, and BTD achieved an efficiency of 5.14% in a BHJ solar cell with $PC_{71}BM$, whereas BTT-DPP only has a PCE of 2.68% (Table 1.14) [213].

Recently, Thompson et al. reported a new class of D/A copolymers called semirandom polymers mainly consisting of rr-P3HT with small amounts (between 5% and 17.5%) of acceptor monomer randomly distributed in the backbone [214–217]. Similar to completely random copolymers the randomized sequence distribution of monomers in semirandom copolymers generates broadly absorbing, multichromophoric polymers. Importantly though, the semirandom polymers retain a larger degree of structural order than random copolymers due to regiospecific placement of the reactive functional groups resulting in a restricted linkage pattern of the monomers in the polymer backbone which inherently avoids sterically unfavorable linkages (Fig. 1.11).

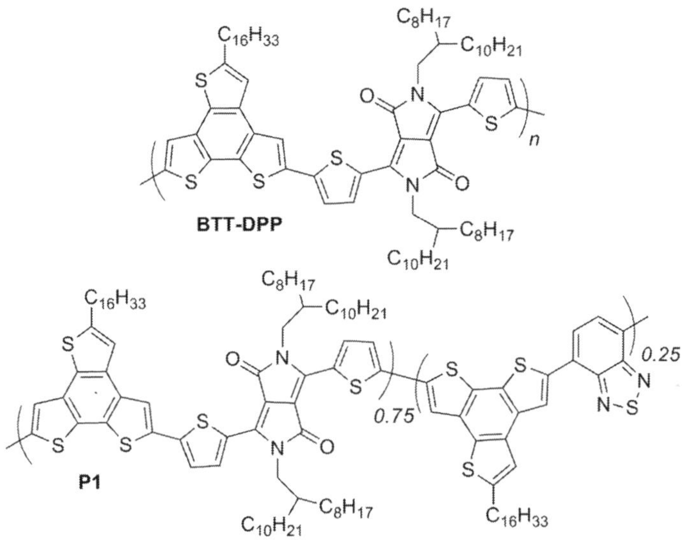

Figure 1.10 Benzotrithiophene-based D/A copolymers: BTT-DPP (perfectly alternating) and P1 (random) [213].

Table 1.14 Solar cell performance of benzotrithiophene-based polymers

Polymer	J_{sc} (mA/cm^2)	V_{oc} (V)	FF	H (%)	Ref.
BTT-DPP:PC$_{71}$BM (1:2)	6.30	0.71	0.60	2.68	[213]
P1:PC$_{71}$BM (1:2)	10.95	0.68	0.69	5.14	[213]

Not only do the resulting polymers in almost all cases show very broad and intense absorption profiles, but due to the small amounts of acceptor moieties they also retain many attractive properties of rr-P3HT such as semicrystallinity, high-hole mobility on the order of P3HT and optimal blending with PC$_{61}$BM at favorable ratios. This is in contrast to many of the above-mentioned random D/A copolymers, which despite broadened absorption, often suffer from low-hole mobilities due to a disordered polymer structure as well as the necessity of unfavorable polymer:fullerene ratios with high fullerene loadings. After the initial proof of concept (using BTD or TP as the acceptors) which resulted in polymers with very promising optical and electronic properties but only moderate solar cell efficiencies, DPP was used for a second generation of efficient semirandom copolymers [214,215]. The influence of the acceptor content was studied and it was shown that P3HTT-DPP-10% with 10% DPP strikes the best balance between low bandgap (1.51 eV) and a broad and uniform absorption profile whereas for P3HTT-DPP-15% the absorption at shorter wavelengths is already considerably reduced. All three polymers are semicrystalline and have high-hole mobilities (~2×10^{-4} $cm^2/(Vs)$) with P3HTT-DPP-10% reaching a maximum efficiency of 5.73% in a BHJ solar cell with $PC_{61}BM$ (Table 1.15) [217].

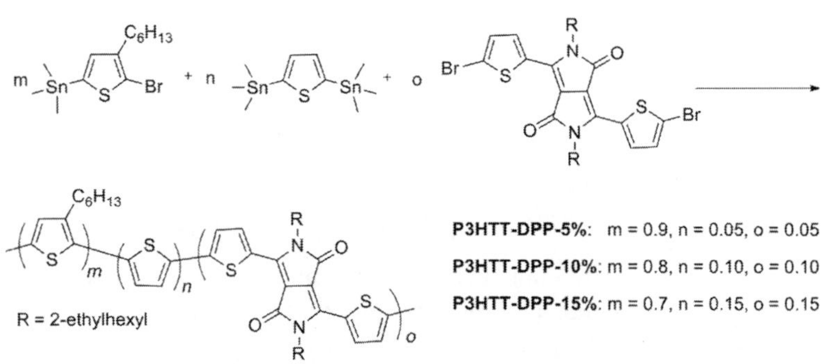

Figure 1.11 Synthesis and structures of DPP-based semirandom copolymers [214].

Table 1.15 Solar cell performance of semirandom copolymers

Polymer	J_{sc} (mA/cm^2)	V_{oc} (V)	FF	η_{avg} (η_{max}) %	Ref.
P3HTT-DPP-5%: $PC_{61}BM$ (1:1)	9.57	0.66	0.58	3.60	[214]
P3HTT-DPP-10%: $PC_{61}BM$ (1:1.3)	14.62	0.59	0.64	5.53 (5.73)	[217]
P3HTT-DPP-15%: $PC_{61}BM$ (1:2.6)	14.28	0.51	0.65	4.66 (4.72)	[217]
P3HTT-TPD-DPP (1:1): $PC_{61}BM$ (1:1.7)	15.26	0.51	0.64	4.93 (5.03)	[217]
P3HTT-TPD-DPP (2:1): $PC_{61}BM$ (1:1.5)	11.67	0.55	0.62	3.94 (4.11)	[217]
P3HTT-TPD-DPP (1:2): $PC_{61}BM$ (1:2.0)	16.37	0.50	0.61	4.92 (4.97)	[217]

Of special interest among the semirandom copolymers are those containing multiple distinct acceptor monomers because they have especially intense and broad UV/vis absorption profiles. Burkhart et al. synthesized a family of two-acceptor copolymers containing TPD and DPP where the acceptor content was kept constant at 15% but the ratio of acceptors was varied (Fig. 1.12). Importantly, those semirandom two-acceptor copolymers showed that the broad absorption profiles and high absorption coefficients can translate into a strong and broad photocurrent response from 350 nm into the NIR with EQE values of up to 40% at 800 nm for P3HTT-TPD-DPP (1:2). Additionally, it was shown that changing the acceptor ratio, even at such a small overall acceptor content, allowed to fine tune optical and electronic properties of the polymers resulting in optimized solar cell performance. Efficiencies of ~5% and currents of over 16 mA/cm^2 (Table 1.15) were achieved which are among the highest currents ever reported with $PC_{61}BM$ as the acceptor [217].

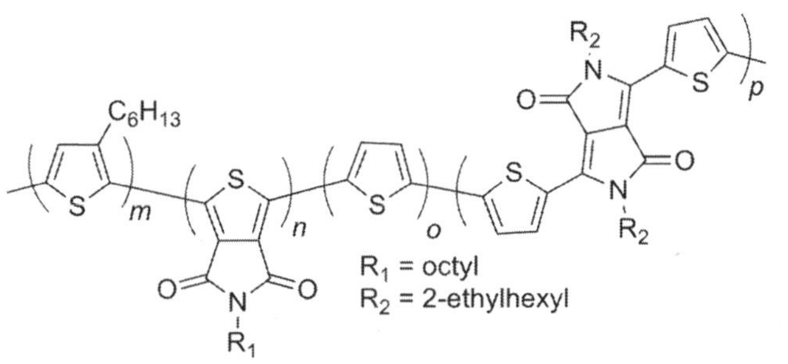

P3HTT-TPD-DPP (1:1)
m = 0.7
n = 0.075
o = 0.15
p = 0.075

P3HTT-TPD-DPP (2:1)
m = 0.7
n = 0.1
o = 0.15
p = 0.05

P3HTT-TPD-DPP (1:2)
m = 0.7
n = 0.05
o = 0.15
p = 0.1

Figure 1.12 Structure of semirandom two-acceptor copolymers containing TPD and DPP as the acceptors.

1.3.4 Small Molecule Donors

The field of solution processable organic BHJ solar cells has long been mainly dominated by conjugated polymers, not only by the volume of work reported but also by the achieved champion efficiencies. Recently though, small molecule BHJ solar cells have been catching up with a rapid pace partly motivated by the fact that the ubiquitous acceptor PCBM is a small molecule. The synthesis of small molecule donors uses many of the same design principles that apply to polymer solar cells in order to control properties such as bandgap and energy levels. More specifically, alkylthiophene-based small molecules are often designed to be regioregular in order to avoid unfavorable steric interactions and the D/A approach discussed in Section 1.2 is also successfully employed for efficient small molecule donors [218–222]. However, small molecules offer several potential advantages over polymers such as relatively simple, reproducible synthesis and purification and the monodispersity of the resulting material. On the other hand, even though molecular donors can be processed from solution their film forming properties are often worse than for polymers resulting in solar cells with lower FF and J_{sc}. Instead of discussing in detail representative classes of small molecules (such as oligothiophenes, acenes or push–pull chromophores), which can be found in several very comprehensive reviews, we will only focus on some recent champion results of solution processable small molecule BHJ solar cells [223–226].

One of the first efficient small molecules (Fig. 1.13) was synthesized by Walker et al. in 2009 and is based on a DPP core flanked by two benzofuran moieties. Even though, through careful optimization of the thin film morphology by adjusting the DPP(TBFu)$_2$:PC$_{71}$BM ratio and annealing temperatures they were able to achieve a PCE of 4.4% (Table 1.16), this value is limited by a low FF of 0.48 [227]. Zhang et al. argued that using the same design principles as for polymers (planar structure for high mobility and π–π interaction, D/A approach for low bandgaps and good film forming ability due to extended conjugated backbone) would also lead to highly efficient small molecule donors. Based on these principles they synthesized small molecule DCAO3TSi (Fig. 1.13) where the central unit is dithienosilole which is flanked by several thiophene units and endcapped with cyanoacetate and achieved an efficiency of 5.84% (Table 1.16) with a very high FF of 0.64 [228].

Liu et al. designed a family of small molecules based on an oligothiophene unit end-capped with electron-withdrawing alkyl cyanoacetate groups (Fig. 1.14). With PC$_{61}$BM as the acceptor, all molecules showed PCE above 4.5% and DCAO7T (substituted with an octyl chain on the cyanoacetate group) achieved an efficiency of 5.08% (Table 1.17) with a high FF of 0.55 [220]. The same group subsequently replaced the central thiophene unit in DCAO7T with planar BDT resulting in DCAO3T(BDT)3T (Fig. 1.14). The solar cell performance of this novel molecule was increased to 5.44% (Table 1.17) predominantly due to increased V_{oc} and higher FF (0.60) compared to DCAO7T [219]. Another variation of DCAO7T was published recently where instead of replacing the central thiophene unit the electron-withdrawing cyanoacetate groups were replaced by the electron-withdrawing dye 3-ethylrhodamine (Fig. 1.14). This molecule showed stronger and slightly broader light absorption due to the contribution of the dye which resulted in a considerably increased J_{sc} (13.98 vs. 10.74 mA/cm^2 for DCAO7T). Even with the FF reduced to 0.47 a very high efficiency of 6.1% (Table 1.17) was achieved for DERHD7T [218].

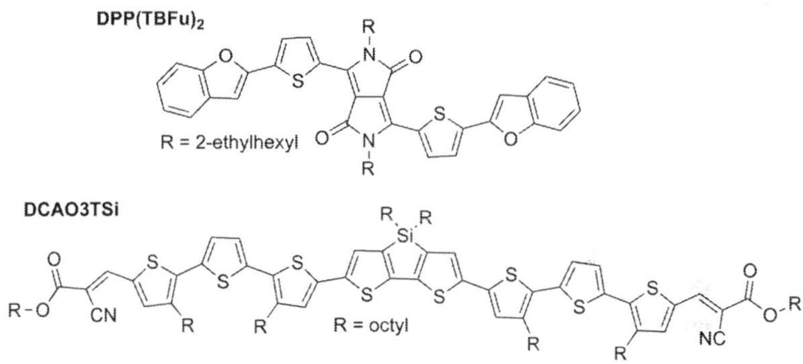

Figure 1.13 Structures of DPP(TBFu)$_2$ and DCAO3TSi [227,228].

Table 1.16 Solar cell performance of DPP(TBFu)$_2$ and DCAO3TSi

Polymer	J_{sc} (mA/cm^2)	V_{oc} (V)	FF	η_{avg} (η_{max}) %	Ref.
DPP(TBFu)$_2$:PC$_{71}$BM (60:40)	10.0	0.92	0.48	4.4	[227]
DCAO3TSi:PC$_{61}$BM (1:0.8)	11.51[a]	0.80[a]	0.64[a]	5.84[a]	[228]

[a] Maximum value.

Sun et al. also used dithienosilole as the central unit but flanked it with the strong electron acceptor PT (Fig. 1.6) which was then endcapped with electron-donating alkylated bithiophene units in order to extend the conjugation and improve film forming properties (Fig. 1.15). Due to the strong D/A effect p-DTS(PTTh$_2$)$_2$ has a low bandgap (1.5 eV) and achieved a high J_{sc} of 14.4 mA/cm^2 in a BHJ solar cell with PC$_{71}$BM. The overall efficiency was drastically improved by the use of a solvent additive during solar cell fabrication and a maximum PCE of 6.70% (Table 1.18) was achieved [222]. Henson et al. extensively studied structure–property relationships in PT-based small molecules concluding that, similar to polymers, small structural changes can lead to considerable differences in optical and electronic properties of the molecules. Fine-tuning of the molecule structures should thus in the future allow for even further increasing efficiencies [229]. Replacing PT with fluorinated BTD, which is also strongly electron withdrawing and asymmetric gave p-DTS(FBTTh$_2$)$_2$ (Fig. 1.15). After extensive optimization of the morphology using solvent additive diiodooctane a very high FF of 0.68 suggesting very efficient charge extraction together with a PCE of 7.0% (Table 1.18) was achieved [221].

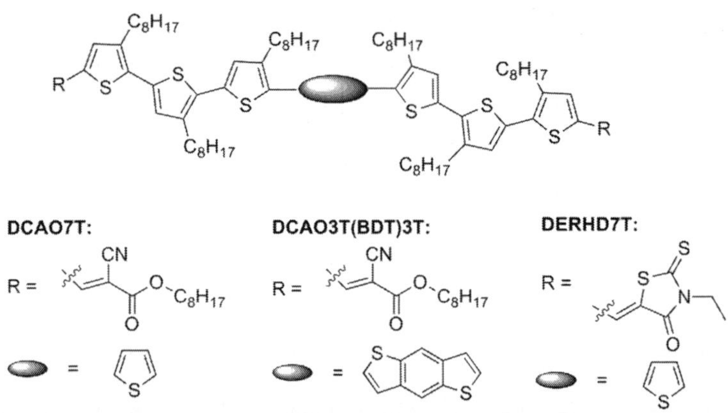

Figure 1.14 Thiophene-based push–pull molecules endcapped with either cyanoacetate or rhodamine [218–220].

Table 1.17 Solar cell performance of thiophene-based push–pull molecules endcapped with cyanoacetate or rhodamine

Polymer	J_{sc} (mA/cm^2)	V_{oc} (V)	FF	η_{max} (%)	Ref.
DCAO7T:PC$_{61}$BM (1:0.5)	10.74	0.86	0.55	5.08	[220]
DCAO3T(BDT)3T:PC$_{61}$BM (1:0.5)	9.77	0.93	0.60	5.44	[219]
DERHD7T:PC$_{61}$BM (1:0.5)	13.98	0.92	0.47	6.10	[218]

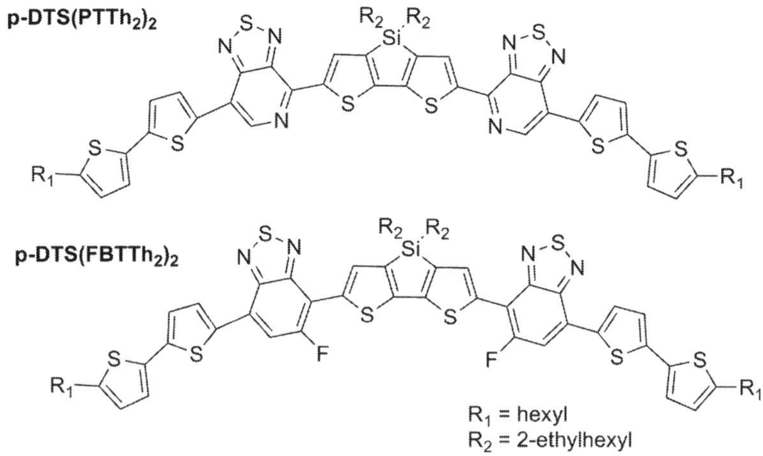

Figure 1.15 Structures of p-DTS($PTTh_2$)$_2$ and p-DTS($FBTTh_2$)$_2$.

Table 1.18 Solar cell performance of p-DTS($PTTh_2$)$_2$ and p-DTS($FBTTh_2$)$_2$

Polymer	J_{sc} (mA/cm^2)	V_{oc} (V)	FF	η_{max} (η_{avg}) %	Ref.
p-DTS($PTTh_2$)$_2$: $PC_{71}BM$ (70:30)	14.4	0.78	0.59	6.70 (6.20)	[222]
p-DTS($FBTTh_2$)$_2$: $PC_{71}BM$ (60:40)	12.8	0.81	0.68	7.0	[221]

1.4 Conclusion and Outlook

Over the last decade, tremendous progress has been made in the field of BHJ solar cells, to the point that currently efficiencies of over 6% are published frequently whereas even 4 years ago they were still relatively rare. Through careful fine-tuning of electronic and morphological properties, V_{oc} of over 1 V, J_{sc} of over 18 mA/cm^2, and FF of up to 0.75 have been achieved in BHJ cells with fullerene acceptors, although not with the same donor material [22,97,128,160,179]. The current champion efficiency of 9.2% (J_{sc} = 17.5 mA/cm^2, V_{oc} = 0.75 V, and FF = 0.70) for BHJ solar cells was achieved with PTB7 (which is a perfectly alternating D/A copolymer composed of TT and BDT) after very careful device optimization [21]. Even though reported efficiencies of over 9% are quickly approaching theoretically predicted limits of 10–12%

and commercialization of organic solar cells is currently under way, considerable challenges remain for the synthetic chemist [21,70,71]. So far it seems that the ideal donor material, which unites favorable optoelectronic properties such as a low bandgap, broad absorption, and low HOMO level with optimal morphological properties, has not been found. Often very high efficiencies are primarily achieved through extensive device optimization such as solvent additives and interlayers and these more complicated solar cell architectures are potentially difficult to realize on an industrial scale and also remove some of the important advantages of solution-processable BHJ solar cells such as easy and fast fabrication as well as low cost [21–23,192]. Not only is the ideal donor material still elusive but it has also been predicted that by increasing the dielectric constant of organic materials efficiencies of over 20% can be achieved and so far few advances have been made in this direction [230]. Additionally, areas such as efficient water-soluble donor materials for truly environmentally friendly organic solar cells as well as approaches to increase the long-term stability of polymer:fullerene composites through cleavable or cross-linkable polymers still have much room for improvement [231,232]. Furthermore, not only does the large-scale fabrication of organic solar cells remain an important research area (see Chapter 13 for a detailed discussion), but it also requires the availability of donor materials in sufficiently large quantities and thus possibly challenging scale-up of current reaction protocols [233]. Finally, the application of donor materials in ternary blend BHJ solar cells as well as tandem solar cells has the potential to lead to considerably increased efficiencies in the future [70,234–236].

References

1. International Energy Agency, and Organisation for Economic Cooperation and Development (2011) *World energy outlook 2011*. IEA, International Energy Agency, OECD, Paris.
2. Jäger-Waldau A (2011) Photovoltaics: Status and perspectives until 2020, *Green*, **1**, 277–290.
3. García-Valverde R, Cherni JA, Urbina A (2010) Life cycle analysis of organic photovoltaic technologies, *Prog Photovolt Res Appl*, **18**, 535–558.

4. Wolden CA, Kurtin J, Baxter JB, et al. (2011) Photovoltaic manufacturing: Present status, future prospects, and research needs, *J Vac Sci Technol A*, **29**, 030801.
5. Avrutin V, Izyumskaya N, Morkoç H (2011) Semiconductor solar cells: Recent progress in terrestrial applications, *Superlattices Microstruct*, **49**, 337–364.
6. Nayak PK, Garcia-Belmonte G, Kahn A, Bisquert J, Cahen D (2012) Photovoltaic efficiency limits and material disorder, *Energy Environ Sci*, **5**, 6022–6039.
7. Espinosa N, Hösel M, Angmo D, Krebs FC (2012) Solar cells with one-day energy payback for the factories of the future, *Energy Environ Sci*, **5**, 5117–5132.
8. Lunt RR, Osedach TP, Brown PR, Rowehl JA, Bulović V (2011) Practical roadmap and limits to nanostructured photovoltaics, *Adv Mater*, **23**, 5712–5727.
9. Lizin S, Van Passel S, De Schepper E, Vranken L (2012) The future of organic photovoltaic solar cells as a direct power source for consumer electronics, *Sol Energy Mater Sol Cells*, **103**, 1–10.
10. Shaheen SE, Brabec CJ, Sariciftci NS, Padinger F, Fromherz T, Hummelen JC (2001) 2.5% efficient organic plastic solar cells, *Appl Phys Lett*, **78**, 841–843.
11. Bundgaard E, Krebs F (2007) Low bandgap polymers for organic photovoltaics, *Sol Energy Mater Sol Cells*, **91**, 954–985.
12. Thompson BC, Fréchet JMJ (2008) Polymer–fullerene composite solar cells, *Angew Chem Int Ed*, **47**, 58–77.
13. Gendron D, Leclerc M (2011) New conjugated polymers for plastic solar cells, *Energy Environ Sci*, **4**, 1225–1237.
14. Thompson BC, Khlyabich PP, Burkhart B, et al. (2011) Polymer-based solar cells: State-of-the-art principles for the design of active layer components, *Green*, **1**, 29–54.
15. Chochos CL, Choulis SA (2011) How the structural deviations on the backbone of conjugated polymers influence their optoelectronic properties and photovoltaic performance, *Prog Polym Sci*, **36**, 1326–1414.
16. Beaujuge PM, Fréchet JMJ (2011) Molecular design and ordering effects in π-functional materials for transistor and solar cell applications, *J Am Chem Soc*, **133**, 20009–20029.
17. Zhou H, Yang L, You W (2012) Rational design of high performance conjugated polymers for organic solar cells, *Macromolecules*, **45**, 607–632.

18. Li G, Zhu R, Yang Y (2012) Polymer solar cells, *Nat Photon*, **6**, 153–161.
19. Bian L, Zhu E, Tang J, Tang W, Zhang F (2012) Recent progress in the design of narrow bandgap conjugated polymers for high-efficiency organic solar cells, *Prog Polym Sci*, **37**, 1292–1331.
20. Li Y (2012) Molecular design of photovoltaic materials for polymer solar cells: Toward suitable electronic energy levels and broad absorption, *Acc Chem Res*, **45**, 723–733.
21. He Z, Zhong C, Su S, Xu M, Wu H, Cao Y (2012) Enhanced power-conversion efficiency in polymer solar cells using an inverted device structure, *Nat Photon*, **6**, 593–597.
22. Li X, Choy WCH, Huo L, et al. (2012) Dual plasmonic nanostructures for high performance inverted organic solar cells, *Adv Mater*, **24**, 3046–3052.
23. Small CE, Chen S, Subbiah J, Amb CM, et al. (2011) High-efficiency inverted dithienogermole–thienopyrrolodione-based polymer solar cells, *Nat Photon*, **6**, 115–120.
24. Günes S, Neugebauer H, Sariciftci NS (2007) Conjugated polymer-based organic solar cells, *Chem Rev*, **107**, 1324–1338.
25. Walker B, Tamayo AB, Dang X-D, et al. (2009) Nanoscale phase separation and high photovoltaic efficiency in solution-processed, small-molecule bulk heterojunction solar cells, *Adv Funct Mater*, **19**, 3063–3069.
26. Yang X, Loos J, Veenstra SC, et al. (2005) Nanoscale morphology of high-performance polymer solar cells, *Nano Lett*, **5**, 579–583.
27. Van Bavel S, Sourty E, De With G, Frolic K, Loos J (2009) Relation between photoactive layer thickness, 3D morphology, and device performance in P3HT/PCBM bulk-heterojunction solar cells, *Macromolecules*, **42**, 7396–7403.
28. Blom PWM, Mihailetchi VD, Koster LJA, Markov DE (2007) Device physics of polymer:fullerene bulk heterojunction solar cells, *Adv Mater*, **19**, 1551–1566.
29. Tang CW (1986) Two-layer organic photovoltaic cell, *Appl Phys Lett*, **48**, 183–185.
30. Halls JJM, Walsh CA, Greenham NC, et al. (1995) Efficient photodiodes from interpenetrating polymer networks, *Nature*, **376**, 498–500.
31. Yu G, Heeger AJ (1995) Charge separation and photovoltaic conversion in polymer composites with internal donor/acceptor heterojunctions, *J Appl Phys*, **78**, 4510–4515.

32. Coakley KM, McGehee MD (2004) Conjugated polymer photovoltaic cells, *Chem Mater,* **16**, 4533–4542.

33. Krebs FC (2009) Fabrication and processing of polymer solar cells: A review of printing and coating techniques, *Sol Energy Mater Sol Cells,* **93**, 394–412.

34. Søndergaard R, Hösel M, Angmo D, Larsen-Olsen TT, Krebs FC (2012) Roll-to-roll fabrication of polymer solar cells, *Mater Today,* **15**, 36–49.

35. Hummelen JC, Knight BW, LePeq F, Fred W (1995) Preparation and characterization of fulleroid and methanofullerene derivatives, *J Org Chem,* **60**, 532–538.

36. Kraabel B, Hummelen JC, Vacar D, et al. (1996) Subpicosecond photoinduced electron transfer from conjugated polymers to functionalized fullerenes, *J Chem Phys,* **104**, 4267.

37. Brabec CJ, Zerza G, Cerullo G, et al. (2001) Tracing photoinduced electron transfer process in conjugated polymer/fullerene bulk heterojunctions in real time, *Chem Phys Lett,* **340**, 232–236.

38. Wienk MM, Kroon JM, Verhees WJH, et al. (2003) Efficient methano[70] fullerene/MDMO-PPV bulk heterojunction photovoltaic cells, *Angew Chem Int Ed,* **42**, 3371–3375.

39. Ai X, Beard MC, Knutsen KP, Shaheen SE, Rumbles G, Ellingson RJ (2006) Photoinduced charge carrier generation in a poly(3-hexylthiophene) and methanofullerene bulk heterojunction investigated by time-resolved terahertz spectroscopy, *J Phys Chem B,* **110**, 25462–25471.

40. He Y, Chen HY, Hou J, Li Y (2010) Indene–C_{60} bisadduct: A new acceptor for high-performance polymer solar cells, *J Am Chem Soc,* **132**, 1377–1382.

41. He Y, Zhao G, Peng B, Li Y (2010) High-yield synthesis and electrochemical and photovoltaic properties of indene-C_{70} bisadduct, *Adv Funct Mater,* **20**, 3383–3389.

42. He Y, Li Y (2011) Fullerene derivative acceptors for high performance polymer solar cells, *Phys Chem Chem Phys,* **13**, 1970–1983.

43. Li C-Z, Chueh CC, Yip HL, Zou J, Chen W-C, Jen AKY (2012) Evaluation of structure–property relationships of solution-processible fullerene acceptors and their n-channel field-effect transistor performance, *J Mater Chem,* **22**, 14976–14981.

44. Holcombe TW, Woo CH, Kavulak DFJ, Thompson BC, Fréchet JMJ (2009) All-polymer photovoltaic devices of poly(3-(4-n-octyl)-

phenylthiophene) from Grignard Metathesis (GRIM) polymerization, *J Am Chem Soc*, **131**, 14160–14161.

45. McNeill CR, Greenham NC (2009) Conjugated-polymer blends for optoelectronics, *Adv Mater*, **21**, 3840–3850.
46. Zhou E, Cong J, Wei Q, Tajima K, Yang C, Hashimoto K (2011) All-polymer solar cells from perylene diimide based copolymers: Material design and phase separation control, *Angew Chem Int Ed*, **50**, 2799–2803.
47. Mori D, Benten H, Ohkita H, Ito S, Miyake K (2012) Polymer/polymer blend solar cells improved by using high-molecular-weight fluorene-based copolymer as electron acceptor, *ACS Appl Mater Interfaces*, **4**, 3325–3329.
48. Bloking JT, Han X, Higgs AT, et al. (2011) Solution-processed organic solar cells with power conversion efficiencies of 2.5% using benzothiadiazole/imide-based acceptors, *Chem Mater*, **23**, 5484–5490.
49. Sonar P, Fong Lim JP, Chan KL (2011) Organic non-fullerene acceptors for organic photovoltaics, *Energy Environ Sci*, **4**, 1558–1574.
50. Peet J, Senatore ML, Heeger AJ, Bazan GC (2009) The role of processing in the fabrication and optimization of plastic solar cells, *Adv Mater*, **21**, 1521–1527.
51. Brabec CJ, Heeney M, McCulloch I, Nelson J (2011) Influence of blend microstructure on bulk heterojunction organic photovoltaic performance, *Chem Soc Rev*, **40**, 1185–1199.
52. Lou SJ, Szarko JM, Xu T, Yu L, Marks TJ, Chen LX (2011) Effects of additives on the morphology of solution phase aggregates formed by active layer components of high-efficiency organic solar cells, *J Am Chem Soc*, **133**, 20661–20663.
53. Liu F, Gu Y, Jung JW, Jo WH, Russell TP (2012) On the morphology of polymer-based photovoltaics, *J Polym Sci B Polym Phys*, **50**, 1018–1040.
54. Park SH, Roy A, Beaupré S, et al. (2009) Bulk heterojunction solar cells with internal quantum efficiency approaching 100%, *Nat Photon*, **3**, 297–302.
55. Murray IP, Lou SJ, Cote LJ, et al. (2011) Graphene oxide interlayers for robust, high-efficiency organic photovoltaics, *J Phys Chem Lett*, **2**, 3006–3012.
56. Kumar P, Chand S (2012) Recent progress and future aspects of organic solar cells, *Prog Photovolt Res Appl*, **20**, 377–415.
57. Burkhard GF, Hoke ET, Scully SR, McGehee MD (2009) Incomplete exciton harvesting from fullerenes in bulk heterojunction solar cells, *Nano Lett*, **9**, 4037–4041.

58. Dimitrov SD, Nielsen CB, Shoaee S, et al. (2012) Efficient charge photogeneration by the dissociation of $PC_{70}BM$ excitons in polymer/fullerene solar cells, *J Phys Chem Lett*, **3**, 140–144.

59. Rasmussen SC, Schwiderski RL, Mulholland ME (2011) Thieno[3,4-b]pyrazines and their applications to low bandgap organic materials, *Chem Commun*, **47**, 11394–11410.

60. Vandewal K, Gadisa A, Oosterbaan WD, et al. (2008) The relation between open-circuit voltage and the onset of photocurrent generation by charge-transfer absorption in polymer:fullerene bulk heterojunction solar cells, *Adv Funct Mater*, **18**, 2064–2070.

61. Veldman D, Meskers SCJ, Janssen RAJ (2009) The energy of charge-transfer states in electron donor–acceptor blends: Insight into the energy losses in organic solar cells, *Adv Funct Mater*, **19**, 1939–1948.

62. Clarke TM, Durrant JR (2010) Charge photogeneration in organic solar cells, *Chem Rev*, **110**, 6736–6767.

63. Piliego C, Loi MA (2012) Charge transfer state in highly efficient polymer–fullerene bulk heterojunction solar cells, *J Mater Chem*, **22**, 4141.

64. Laird DW, Stegamat R, Richter H, et al. (July 10, 2012) Organic photovoltaic devices comprising fullerenes and derivatives thereof, United States Patent No.: US 8,217,260 B2.

65. Ma W, Yang C, Gong X, Lee K, Heeger AJ (2005) Thermally stable, efficient polymer solar cells with nanoscale control of the interpenetrating network morphology, *Adv Funct Mater*, **15**, 1617–1622.

66. Zhao G, He Y, Li Y (2010) 6.5% efficiency of polymer solar cells based on poly(3-hexylthiophene) and indene-C_{60} bisadduct by device optimization, *Adv Mater*, **22**, 4355–4358.

67. Xin H, Subramaniyan S, Kwon T-W, Shoaee S, Durrant JR, Jenekhe SA (2012) Enhanced open circuit voltage and efficiency of donor–acceptor copolymer solar cells by using indene-C_{60} bisadduct, *Chem Mater*, **24**, 1995–2001.

68. Cardona CM, Li W, Kaifer AE, Stockdale D, Bazan GC (2011) Electrochemical considerations for determining absolute frontier orbital energy levels of conjugated polymers for solar cell applications, *Adv Mater*, **23**, 2367–2371.

69. Kim MS, Kim B-G, Kim J (2009) Effective variables to control the fill factor of organic photovoltaic cells, *ACS Appl Mater Interfaces*, **1**, 1264–1269.

70. Kotlarski JD, Blom PWM (2011) Ultimate performance of polymer:fullerene bulk heterojunction tandem solar cells, *Appl Phys Lett*, **98**, 053301.

71. Scharber MC, Mühlbacher D, Koppe M, et al. (2006) Design rules for donors in bulk-heterojunction solar cells—towards 10% energy-conversion efficiency, *Adv Mater,* **18**, 789–794.
72. Roncali J (1997) Synthetic principles for bandgap control in linear pi-conjugated systems, *Chem Rev,* **97**, 173–205.
73. Roncali J (2007) Molecular engineering of the bandgap of π-conjugated systems: Facing technological applications, *Macromol Rapid Commun,* **28**, 1761–1775.
74. Kroon R, Lenes M, Hummelen J, Blom P, De Boer B (2008) Small bandgap polymers for organic solar cells (polymer material development in the last 5 years), *Polymer Rev,* **48**, 531–582.
75. Wudl F, Kobayashi M, Heeger AJ (1984) Poly(isothianaphthene), *J Org Chem,* **49**, 3382–3384.
76. Nayak K, Marynick DS (1990) The interplay between gemetrie and electronic structures in polyisothianaphthene, polyisonaphthothiophene, polythieno(3,4-b)thienopyrazine, and polythieno(3,4-b)quinoxaline, *Macromolecules,* **23**, 2237–2245.
77. Hong SY, Marynick DS (1992) Understanding the conformational stability and electronic structures of modified polymers based on polythiophene, *Macromolecules,* **25**, 4652–4657.
78. Pomerantz M, Gu X (1997) Poly(2-decylthieno[3,4-b]thiophene). A soluble low-bandgap conducting polymer, *Synth Met,* **84**, 243–244.
79. Su M-S, Kuo C-Y, Yuan M-C, Jeng U-S, Su C-J, Wei K-H (2011) Improving device efficiency of polymer/fullerene bulk heterojunction solar cells through enhanced crystallinity and reduced grain boundaries induced by solvent additives, *Adv Mater,* **23**, 3315–3319.
80. Havinga EE, Ten Hoeve W, Wynberg H (1992) A new class of small bandgap organic polymer conductors, *Polym Bull,* **29**, 119–126.
81. Havinga EE, Ten Hoeve W, Wynberg H (1993) Alternate donor–acceptor small-band-gap semiconducting polymers; polysquaraines and polycroconaines, *Synth Met,* **55**, 299–306.
82. Mondal R, Ko S, Norton JE, et al. (2009) Molecular design for improved photovoltaic efficiency: Bandgap and absorption coefficient engineering, *J Mater Chem,* **19**, 7195–7197.
83. Mauer R, Kastler M, Laquai F (2010) The impact of polymer regioregularity on charge transport and efficiency of P3HT:PCBM photovoltaic devices, *Adv Funct Mater,* **20**, 2085–2092.

84. Osaka I, McCullough RD (2008) Advances in molecular design and synthesis of regioregular polythiophenes, *Acc Chem Res,* **41**, 1202–1214.

85. Ku S-Y, Liman CD, Burke DJ, et al. (2011) A facile synthesis of low-band-gap donor–acceptor copolymers based on dithieno[3,2-b:2′,3′-d] thiophene, *Macromolecules,* **44**, 9533–9538.

86. Boudreault P-LT, Najari A, Leclerc M (2011) Processable low-bandgap polymers for photovoltaic applications, *Chem Mater,* **23**, 456–469.

87. Scherf U (1999) Ladder-type materials, *J Mater Chem,* **9**, 1853–1864.

88. Uy RL, Price SC, You W (2012) Structure-property optimizations in donor polymers via electronics, substituents, and side chains toward high efficiency solar cells, *Macromol Rapid Commun,* **33**, 1162–1177.

89. Yang L, Zhou H, You W (2010) Quantitatively analyzing the influence of side chains on photovoltaic properties of polymer–fullerene solar cells, *J Phys Chem C,* **114**, 16793–16800.

90. Burkhart B, Khlyabich PP, Thompson BC (2012) Influence of the ethylhexyl side-chain content on the open-circuit voltage in rr-poly(3-hexylthiophene-co-3-(2-ethylhexyl)thiophene) copolymers, *Macromolecules,* **45**, 3740–3748.

91. Okamoto K, Luscombe CK (2011) Controlled polymerizations for the synthesis of semiconducting conjugated polymers, *Polym Chem,* **2**, 2424.

92. Stefan MC, Bhatt MP, Sista P, Magurudeniya HD (2012) Grignard Metathesis (GRIM) polymerization for the synthesis of conjugated block copolymers containing regioregular poly(3-hexylthiophene), *Polym Chem,* **3**, 1693–1701.

93. Babudri F, Farinola GM, Naso F (2004) Synthesis of conjugated oligomers and polymers: The organometallic way, *J Mater Chem,* **14**, 11–34.

94. Carsten B, He F, Son HJ, Xu T, Yu L (2011) Stille polycondensation for synthesis of functional materials, *Chem Rev,* **111**, 1493–1528.

95. Cheng Y-J, Luh T-Y (2004) Synthesizing optoelectronic heteroaromatic conjugated polymers by cross-coupling reactions, *J Organomet Chem,* **689**, 4137–4148.

96. Hoyos ML, Turner M, Navarro O (2011) Recent advances in polythiophene synthesis by palladium-catalyzed cross-coupling reactions, *Curr Org Chem,* **15**, 3263–3290.

97. Guo X, Cui C, Zhang M, et al. (2012) High efficiency polymer solar cells based on poly(3-hexylthiophene)/indene-C_{70} bisadduct with solvent additive, *Energy Environ Sci,* **5**, 7943–7949.

98. Hou J, Chen TL, Zhang S, Huo L, Sista S, Yang Y (2009) An easy and effective method to modulate molecular energy level of poly(3-alkylthiophene) for high-V_{OC} polymer solar cells, *Macromolecules,* **42**, 9217–9219.

99. Ko S, Verploegen E, Hong S, et al. (2011) 3,4-Disubstituted polyalkylthiophenes for high-performance thin-film transistors and photovoltaics, *J Am Chem Soc,* **133**, 16722–16725.

100. Dang MT, Hirsch L, Wantz G (2011) P3HT:PCBM, best seller in polymer photovoltaic research, *Adv Mater,* **23**, 3597–3602.

101. Marrocchi A, Lanari D, Facchetti A, Vaccaro L (2012) Poly(3-hexylthiophene): Synthetic methodologies and properties in bulk heterojunction solar cells, *Energy Environ Sci,* **5**, 8457–8474.

102. Gadisa A, Oosterbaan WD, Vandewal K, et al. (2009) Effect of alkyl side-chain length on photovoltaic properties of poly(3-alkylthiophene)/PCBM bulk heterojunctions, *Adv Funct Mater,* **19**, 3300–3306.

103. Nguyen LH, Hoppe H, Erb T, Günes S, Gobsch G, Sariciftci NS (2007) Effects of annealing on the nanomorphology and performance of poly(alkylthiophene):fullerene bulk-heterojunction solar cells, *Adv Funct Mater,* **17**, 1071–1078.

104. Troshin P, Susarova D, Khakina E, et al. (2012) Material solubility and molecular compatibility effects in the design of the fullerene/polymer composites for organic bulk heterojunction solar cells, *J Mater Chem,* **22**, 18433–18441.

105. Sun Y, Cui C, Wang H, Li Y (2012) High-efficiency polymer solar cells based on poly(3-pentylthiophene) with indene-C_{70} bisadduct as an acceptor, *Adv Energy Mater,* **2**, 966–969.

106. Wu P-T, Xin H, Kim FS, Ren G, Jenekhe SA (2009) Regioregular poly(3-pentylthiophene): Synthesis, self-assembly of nanowires, high-mobility field-effect transistors, and efficient photovoltaic cells, *Macromolecules,* **42**, 8817–8826.

107. Kim Y, Cook S, Tuladhar SM, et al. (2006) A strong regioregularity effect in self-organizing conjugated polymer films and high-efficiency polythiophene:fullerene solar cells, *Nat Mater,* **5**, 197–203.

108. Sivula K, Luscombe CK, Thompson BC, Fréchet JMJ (2006) Enhancing the thermal stability of polythiophene:fullerene solar cells by decreasing effective polymer regioregularity, *J Am Chem Soc,* **128**, 13988–13989.

109. Woo CH, Thompson BC, Kim BJ, Toney MF, Fréchet JMJ (2008) The influence of poly(3-hexylthiophene) regioregularity on fullerene-composite solar cell performance, *J Am Chem Soc,* **130**, 16324–16329.

110. Schilinsky P, Asawapirom U, Scherf U, Biele M, Brabec CJ (2005) Influence of the molecular weight of poly(3-hexylthiophene) on the performance of bulk heterojunction solar cells, *Chem Mater,* **17**, 2175–2180.

111. Kline RJ, McGehee MD, Kadnikova EN, Liu J, Fréchet JMJ, Toney MF (2005) Dependence of regioregular poly(3-hexylthiophene) film morphology and field-effect mobility on molecular weight, *Macromolecules,* **38**, 3312–3319.

112. Chang J-F, Clark J, Zhao N, et al. (2006) Molecular-weight dependence of interchain polaron delocalization and exciton bandwidth in high-mobility conjugated polymers, *Phys Rev B,* **74**, 115318.

113. Ko S, Hoke ET, Pandey L, et al. (2012) Controlled conjugated backbone twisting for an increased open-circuit voltage while having a high short-circuit current in poly(hexylthiophene) derivatives, *J Am Chem Soc,* **134**, 5222–5232.

114. Park JK, Jo J, Seo JH, et al. (2011) End-capping effect of a narrow bandgap conjugated polymer on bulk heterojunction solar cells, *Adv Mater,* **23**, 2430–2435.

115. Camaioni N, Tinti F, Franco L, et al. (2012) Effect of residual catalyst on solar cells made of a fluorene-thiophene-benzothiadiazole copolymer as electron-donor: A combined electrical and photophysical study, *Org Electron,* **13**, 550–559.

116. Facchetti A, Vaccaro L, Marrocchi A (2012) Semiconducting polymers prepared by direct arylation polycondensation, *Angew Chem Int Ed,* **51**, 3520–3523.

117. Jo J, Pron A, Berrouard P, et al. (2012) A new terthiophene-thienopyrrolodione copolymer-based bulk heterojunction solar cell with high open-circuit voltage, *Adv Energy Mater,* **2**, 1397–1403.

118. Rudenko AE, Wiley CA, Stone SM, Tannaci JF, Thompson BC (2012) Semi-random P3HT analogs via direct arylation polymerization, *J Polym Sci A Polym Chem,* **50**, 3691–3697.

119. Zhu Z, Waller D, Gaudiana R, et al. (2007) Panchromatic conjugated polymers containing alternating donor/acceptor units for photovoltaic applications, *Macromolecules,* **40**, 1981–1986.

120. Bijleveld JC, Shahid M, Gilot J, Wienk MM, Janssen RAJ (2009) Copolymers of cyclopentadithiophene and electron-deficient

aromatic units designed for photovoltaic applications, *Adv Funct Mater,* **19**, 3262–3270.

121. Kowalski S, Allard S, Scherf U (2012) Synthesis of poly(4,4-dialkyl-cyclopenta[2,1-b:3,4-b′]dithiophene-alt-2,1,3-benzothiadiazole) (PCPDTBT) in a direct arylation scheme, *ACS Macro Lett,* **1**, 465–468.

122. Zhang Z, Wang J (2012) Structures and properties of conjugated donor–acceptor copolymers for solar cell applications, *J Mater Chem,* **22**, 4178–4187.

123. Peet J, Kim JY, Coates NE, et al. (2007) Efficiency enhancement in low-bandgap polymer solar cells by processing with alkane dithiols, *Nat Mater,* **6**, 497–500.

124. Zhang Y, Zou J, Cheuh C-C, Yip H-L, Jen AK-Y (2012) Significant improved performance of photovoltaic cells made from a partially fluorinated cyclopentadithiophene/benzothiadiazole conjugated polymer, *Macromolecules,* **45**, 5427–5435.

125. Scharber MC, Koppe M, Gao J, et al. (2010) Influence of the bridging atom on the performance of a low-bandgap bulk heterojunction solar cell, *Adv Mater,* **22**, 367–370.

126. Hou J, Chen H-Y, Zhang S, Li G, Yang Y (2008) Synthesis, characterization, photovoltaic properties of a low bandgap polymer based on silole-containing polythiophenes and 2,1,3-benzothiadiazole, *J Am Chem Soc,* **130**, 16144–16145.

127. Coffin RC, Peet J, Rogers J, Bazan GC (2009) Streamlined microwave-assisted preparation of narrow-bandgap conjugated polymers for high-performance bulk heterojunction solar cells, *Nat Chem,* **1**, 657–661.

128. Fei Z, Kim JS, Smith J, et al. (2011) A low bandgap co-polymer of dithienogermole and 2,1,3-benzothiadiazole by Suzuki polycondensation and its application in transistor and photovoltaic cells, *J Mater Chem,* **21**, 16257–16263.

129. Mühlbacher D, Scharber M, Morana M, et al. (2006) High photovoltaic performance of a low-bandgap polymer, *Adv Mater,* **18**, 2884–2889.

130. Chen H-Y, Hou J, Hayden AE, Yang H, Houk KN, Yang Y (2010) Silicon atom substitution enhances interchain packing in a thiophene-based polymer system, *Adv Mater,* **22**, 371–375.

131. Yue W, Zhao Y, Shao S, et al. (2009) Novel NIR-absorbing conjugated polymers for efficient polymer solar cells: Effect of alkyl chain length on device performance, *J Mater Chem,* **19**, 2199–2206.

132. Gibson GL, McCormick TM, Seferos DS (2012) Atomistic bandgap engineering in donor–acceptor polymers, *J Am Chem Soc,* **134**, 539–547.

133. Blouin N, Michaud A, Leclerc M (2007) A low-bandgap poly(2,7-carbazole) derivative for use in high-performance solar cells, *Adv Mater,* **19**, 2295–2300.

134. Liu J, Shao S, Fang G, Meng B, Xie Z, Wang L (2012) High-efficiency inverted polymer solar cells with transparent and work-function tunable MoO_3-Al composite film as cathode buffer layer, *Adv Mater,* **24**, 2774–2779.

135. Sun Y, Takacs CJ, Cowan SR, et al. (2011) Efficient, air-stable bulk heterojunction polymer solar cells using MoOx as the anode interfacial layer, *Adv Mater,* **23**, 2226–2230.

136. Chen M-H, Hou J, Hong Z, et al. (2009) Efficient polymer solar cells with thin active layers based on alternating polyfluorene copolymer/fullerene bulk heterojunctions, *Adv Mater,* **21**, 4238–4242.

137. Allard N, Aïch RB, Gendron D, et al. (2010) Germafluorenes: New heterocycles for plastic electronics, *Macromolecules,* **43**, 2328–2333.

138. Wang E, Wang L, Lan L, et al. (2008) High-performance polymer heterojunction solar cells of a polysilafluorene derivative, *Appl Phys Lett,* **92**, 033307.

139. Du C, Li C, Li W, et al. (2011) 9-Alkylidene-9H-fluorene-containing polymer for high-efficiency polymer solar cells, *Macromolecules,* **44**, 7617–7624.

140. Chen Y-C, Yu C-Y, Fan YL, Hung LI, Chen C-P, Ting C (2010) Low-bandgap conjugated polymer for high efficient photovoltaic applications, *Chem Commun,* **46**, 6503–6505.

141. Huo L, Hou J, Zhang S, Chen HY, Yang Y (2010) A polybenzo[1,2-b:4,5-b′]dithiophene derivative with deep homo level and its application in high-performance polymer solar cells, *Angew Chem Int Ed,* **49**, 1500–1503.

142. Peng Q, Liu X, Su D, Fu G, Xu J, Dai L (2011) Novel benzo[1,2-b:4,5-b′]dithiophene-benzothiadiazole derivatives with variable side chains for high-performance solar cells, *Adv Mater,* **23**, 4554–4558.

143. Zhou H, Yang L, Stuart AC, Price SC, Liu S, You W (2011) Development of fluorinated benzothiadiazole as a structural unit for a polymer solar cell of 7% efficiency, *Angew Chem Int Ed,* **50**, 2995–2998.

144. Ong K-H, Lim S-L, Tan H-S, et al. (2011) A versatile low bandgap polymer for air-stable, high-mobility field-effect transistors and efficient polymer solar cells, *Adv Mater,* **23**, 1409–1413.

145. Zhou H, Yang L, Price SC, Knight KJ, You W (2010) Enhanced photovoltaic performance of low-bandgap polymers with deep LUMO levels, *Angew Chem Int Ed,* **49**, 7992–7995.

146. Ying L, Hsu BBY, Zhan H, et al. (2011) Regioregular pyridal[2,1,3] thiadiazole π-conjugated copolymers, *J Am Chem Soc,* **133**, 18538–18541.

147. Wang M, Hu X, Liu P, et al. (2011) Donor–acceptor conjugated polymer based on naphtho[1,2-c:5,6-c]bis[1,2,5]thiadiazole for high-performance polymer solar cells, *J Am Chem Soc,* **133**, 9638–9641.

148. Yang T, Wang M, Duan C, et al. (2012) Inverted polymer solar cells with 8.4% efficiency by conjugated polyelectrolyte, *Energy Environ Sci,* **5**, 8208–8214.

149. Patel DG (Dan), Feng F, Ohnishi Y, et al. (2012) It takes more than an imine: The role of the central atom on the electron-accepting ability of benzotriazole and benzothiadiazole oligomers, *J Am Chem Soc,* **134**, 2599–2612.

150. Price SC, Stuart AC, Yang L, Zhou H, You W (2011) Fluorine substituted conjugated polymer of medium bandgap yields 7% efficiency in polymer–fullerene solar cells, *J Am Chem Soc,* **133**, 4625–4631.

151. Nie W, MacNeill CM, Li Y, Noftle RE, Carroll DL, Coffin RC (2011) A soluble high molecular weight copolymer of benzo[1,2-b:4,5-b′] dithiophene and benzoxadiazole for efficient organic photovoltaics, *Macromol Rapid Commun,* **32**, 1163–1168.

152. Zou Y, Najari A, Berrouard P, et al. (2010) A thieno[3,4-c]pyrrole-4,6-dione-based copolymer for efficient solar cells, *J Am Chem Soc,* **132**, 5330–5331.

153. Piliego C, Holcombe TW, Douglas JD, Woo CH, Beaujuge PM, Fréchet JMJ (2010) Synthetic control of structural order in n-alkylthieno[3,4-c]pyrrole-4,6-dione-based polymers for efficient solar cells, *J Am Chem Soc,* **132**, 7595–7597.

154. Aïch BR, Lu J, Beaupré S, Leclerc M, Tao Y (2012) Control of the active layer nanomorphology by using co-additives towards high-performance bulk heterojunction solar cells, *Org Electron,* **13**, 1736–1741.

155. Li Z, Tsang S-W, Du X, et al. (2011) Alternating copolymers of cyclopenta[2,1-b;3,4-b′]dithiophene and thieno[3,4-c]pyrrole-4,6-dione for high-performance polymer solar cells, *Adv Funct Mater,* **21**, 3331–3336.

156. Chu T-Y, Tsang S-W, Zhou J, et al. (2012) High-efficiency inverted solar cells based on a low bandgap polymer with excellent air stability, *Sol Energy Mater Sol Cells*, **96**, 155–159.

157. Chu T-Y, Lu J, Beaupré S, et al. (2012) Effects of the molecular weight and the side-chain length on the photovoltaic performance of dithienosilole/thienopyrrolodione copolymers, *Adv Funct Mater*, **22**, 2345–2351.

158. Amb CM, Chen S, Graham KR, et al. (2011) Dithienogermole as a fused electron donor in bulk heterojunction solar cells, *J Am Chem Soc*, **133**, 10062–10065.

159. Yuan M-C, Chiu M-Y, Liu S-P, Chen C-M, Wei K-H (2010) A thieno[3,4-c]pyrrole-4,6-dione-based donor–acceptor polymer exhibiting high crystallinity for photovoltaic applications, *Macromolecules*, **43**, 6936–6938.

160. Najari A, Berrouard P, Ottone C, et al. (2012) High open-circuit voltage solar cells based on new thieno[3,4-c]pyrrole-4,6-dione and 2,7-carbazole copolymers, *Macromolecules*, **45**, 1833–1838.

161. Hao Z, Iqbal A (1997) Some aspects of organic pigments, *Chem Soc Rev*, **26**, 203–213.

162. Qu S, Tian H (2012) Diketopyrrolopyrrole (DPP)-based materials for organic photovoltaics, *Chem Commun*, **48**, 3039–3051.

163. Wienk MM, Turbiez M, Gilot J, Janssen RAJ (2008) Narrow-bandgap diketo-pyrrolo-pyrrole polymer solar cells: The effect of processing on the performance, *Adv Mater*, **20**, 2556–2560.

164. Liu F, Gu Y, Wang C, et al. (2012) Efficient polymer solar cells based on a low bandgap semi-crystalline DPP polymer-PCBM blends, *Adv Mater*, **24**, 3947–3951.

165. Bijleveld JC, Zoombelt AP, Mathijssen SGJ, et al. (2009) Poly(diketopyrrolopyrrole–terthiophene) for ambipolar logic and photovoltaics, *J Am Chem Soc*, **131**, 16616–16617.

166. Bijleveld JC, Gevaerts VS, Di Nuzzo D, et al. (2010) Efficient solar cells based on an easily accessible diketopyrrolopyrrole polymer, *Adv Mater*, **22**, E242–E246.

167. Jung JW, Liu F, Russell TP, Jo WH (2012) A high mobility conjugated polymer based on dithienothiophene and diketopyrrolopyrrole for organic photovoltaics, *Energy Environ Sci*, **5**, 6857–6861.

168. Gevaerts VS, Furlan A, Wienk MM, Turbiez M, Janssen RAJ (2012) Solution processed polymer tandem solar cell using efficient small and wide bandgap polymer:fullerene blends, *Adv Mater*, **24**, 2130–2134.

169. Dou L, Gao J, Richard E, et al. (2012) Systematic investigation of benzodithiophene- and diketopyrrolopyrrole-based low-bandgap polymers designed for single junction and tandem polymer solar cells, *J Am Chem Soc,* **134**, 10071–10079.

170. Woong Jung J, Woong Jo J, Liu F, Russell TP, Ho Jo W (2012) A low band-gap polymer based on unsubstituted benzo[1,2-b:4,5-b′] dithiophene for high performance organic photovoltaics, *Chem Commun,* **48**, 6933–6935.

171. Bronstein H, Chen Z, Ashraf RS, et al. (2011) Thieno[3,2-b] thiophene–diketopyrrolopyrrole-containing polymers for high-performance organic field-effect transistors and organic photovoltaic devices, *J Am Chem Soc,* **133**, 3272–3275.

172. Yiu AT, Beaujuge PM, Lee OP, Woo CH, Toney MF, Fréchet JMJ (2012) Side-chain tunability of furan-containing low-band-gap polymers provides control of structural order in efficient solar cells, *J Am Chem Soc,* **134**, 2180–2185.

173. Wang E, Ma Z, Zhang Z, et al. (2011) An easily accessible isoindigo-based polymer for high-performance polymer solar cells, *J Am Chem Soc,* **133**, 14244–14247.

174. Stalder R, Mei J, Reynolds JR (2010) Isoindigo-based donor–acceptor conjugated polymers, *Macromolecules,* **43**, 8348–8352.

175. Stalder R, Grand C, Subbiah J, So F, Reynolds JR (2012) An isoindigo and dithieno[3,2-b:2′,3′-d]silole copolymer for polymer solar cells, *Polym Chem,* **3**, 89–92.

176. Ma Z, Wang E, Jarvid ME, et al. (2012) Synthesis and characterization of benzodithiophene–isoindigo polymers for solar cells, *J Mater Chem,* **22**, 2306–2314.

177. Zhang G, Fu Y, Xie Z, Zhang Q (2011) Synthesis and photovoltaic properties of new low bandgap isoindigo-based conjugated polymers, *Macromolecules,* **44**, 1414–1420.

178. Liu B, Zou Y, Peng B, et al. (2011) Low bandgap isoindigo-based copolymers: Design, synthesis and photovoltaic applications, *Polym Chem,* **2**, 1156–1162.

179. Gadisa A, Mammo W, Andersson LM, et al. (2007) A new donor–acceptor–donor polyfluorene copolymer with balanced electron and hole mobility, *Adv Funct Mater,* **17**, 3836–3842.

180. Zhang F, Bijleveld J, Perzon E, et al. (2008) High photovoltage achieved in low bandgap polymer solar cells by adjusting energy levels of a polymer with the LUMOs of fullerene derivatives, *J Mater Chem,* **18**, 5468–5474.

181. Lindgren LJ, Zhang F, Andersson M, et al. (2009) Synthesis, characterization, and devices of a series of alternating copolymers for solar cells, *Chem Mater,* **21**, 3491–3502.

182. Wang E, Hou L, Wang Z, et al. (2010) An easily synthesized blue polymer for high-performance polymer solar cells, *Adv Mater,* **22**, 5240–5244.

183. Zhang Y, Zou J, Yip HL, et al. (2011) Indacenodithiophene and quinoxaline-based conjugated polymers for highly efficient polymer solar cells, *Chem Mater,* **23**, 2289–2291.

184. Li CZ, Chueh CC, Yip HL, O'Malley KM, Chen WC, Jen AKY (2012) Effective interfacial layer to enhance efficiency of polymer solar cells via solution-processed fullerene-surfactants, *J Mater Chem,* **22**, 8574–8578.

185. Zou J, Yip HL, Zhang Y, et al. (2012) High-performance inverted polymer solar cells: Device characterization, optical modeling, hole-transporting modifications, *Adv Funct Mater,* **22**, 2804–2811.

186. Zhang Y, Zou J, Yip HL, et al. (2011) Synthesis, characterization, charge transport, and photovoltaic properties of dithienobenzoquinoxaline- and dithienobenzopyridopyrazine-based conjugated polymers, *Macromolecules,* **44**, 4752–4758.

187. He Z, Zhang C, Xu X, et al. (2011) Largely enhanced efficiency with a PFN/Al bilayer cathode in high efficiency bulk heterojunction photovoltaic cells with a low bandgap polycarbazole donor, *Adv Mater,* **23**, 3086–3089.

188. Shaheen SE, Vangeneugden D, Kiebooms R, et al. (2001) Low band-gap polymeric photovoltaic devices, *Synth Met,* **121**, 1583–1584.

189. Vangeneugden DL, Vanderzande DJM, Salbeck J, et al. (2001) Synthesis and characterization of a poly(1,3-dithienylisothianaphthene) derivative for bulk heterojunction photovoltaic cells, *J Phys Chem B,* **105**, 11106–11113.

190. Qin Y, Kim JY, Frisbie CD, Hillmyer MA (2008) Distannylated isothianaphthene: A versatile building block for low bandgap conjugated polymers, *Macromolecules,* **41**, 5563–5570.

191. Douglas JD, Griffini G, Holcombe TW, et al. (2012) Functionalized isothianaphthene monomers that promote quinoidal character in donor–acceptor copolymers for organic photovoltaics, *Macromolecules,* **45**, 4069–4074.

192. He Z, Zhong C, Huang X, et al. (2011) Simultaneous enhancement of open-circuit voltage, short-circuit current density, and fill factor in polymer solar cells, *Adv Mater,* **23**, 4636–4643.

193. Huang Y, Guo X, Liu F, et al. (2012) Improving the ordering and photovoltaic properties by extending π-conjugated area of electron-donating units in polymers with D-A structure, *Adv Mater,* **24**, 3383–3389.

194. Liang Y, Xu Z, Xia J, et al. (2010) For the bright future-bulk heterojunction polymer solar cells with power conversion efficiency of 7.4%, *Adv Mater,* **22**, E135–E138.

195. Szarko JM, Guo J, Liang Y, et al. (2010) When function follows form: Effects of donor copolymer side chains on film morphology and BHJ solar cell performance, *Adv Mater,* **22**, 5468–5472.

196. Coropceanu V, Cornil J, Da Silva Filho DA, et al. (2007) Charge transport in organic semiconductors, *Chem Rev,* **107**, 926–952.

197. He M, Li J, Sorensen ML, et al. (2009) Alkylsubstituted thienothiophene semiconducting materials: Structure–property relationships, *J Am Chem Soc,* **131**, 11930–11938.

198. Liang Y, Wu Y, Feng D, et al. (2009) Development of new semiconducting polymers for high performance solar cells, *J Am Chem Soc,* **131**, 56–57.

199. Liang Y, Feng D, Wu Y, et al. (2009) Highly efficient solar cell polymers developed via fine-tuning of structural and electronic properties, *J Am Chem Soc,* **131**, 7792–7799.

200. Son HJ, Wang W, Xu T, et al. (2011) Synthesis of fluorinated polythienothiophene-co-benzodithiophenes and effect of fluorination on the photovoltaic properties, *J Am Chem Soc,* **133**, 1885–1894.

201. Saadeh HA, Lu L, He F, et al. (2012) Polyselenopheno[3,4-b]selenophene for highly efficient bulk heterojunction solar cells, *ACS Macro Lett,* **1**, 361–365.

202. Hou J, Chen HY, Zhang S, et al. (2009) Synthesis of a low bandgap polymer and its application in highly efficient polymer solar cells, *J Am Chem Soc,* **131**, 15586–15587.

203. Chen HY, Hou J, Zhang S, et al. (2009) Polymer solar cells with enhanced open-circuit voltage and efficiency, *Nat Photon,* **3**, 649–653.

204. Huang Y, Huo L, Zhang S, et al. (2011) Sulfonyl: A new application of electron-withdrawing substituent in highly efficient photovoltaic polymer, *Chem Commun,* **47**, 8904–8906.

205. Huo L, Zhang S, Guo X, et al. (2011) Replacing alkoxy groups with alkylthienyl groups: A feasible approach to improve the properties of photovoltaic polymers, *Angew Chem Int Ed,* **50**, 9697–9702.

206. Liang Y, Xiao S, Feng D, Yu L (2008) Control in energy levels of conjugated polymers for photovoltaic application, *J Phys Chem C*, **112**, 7866–7871.

207. Beaujuge PM, Ellinger S, Reynolds JR (2008) The donor–acceptor approach allows a black-to-transmissive switching polymeric electrochrome, *Nat Mater*, **7**, 795–799.

208. Chen CP, Chan SH, Chao TC, Ting C, Ko BT (2008) Low-bandgap poly(thiophene-phenylene-thiophene) derivatives with broaden absorption spectra for use in high-performance bulk-heterojunction polymer solar cells, *J Am Chem Soc*, **130**, 12828–12833.

209. He Y, Wang X, Zhang J, Li Y (2009) Low bandgap polymers by copolymerization of thiophene with benzothiadiazole, *Macromol Rapid Commun*, **30**, 45–51.

210. Yu CY, Chen CP, Chan SH, Hwang GW, Ting C (2009) Thiophene/phenylene/thiophene-based low-bandgap conjugated polymers for efficient near-infrared photovoltaic applications, *Chem Mater*, **21**, 3262–3269.

211. Chen CH, Cheng YJ, Chang CY, Hsu CS (2011) Donor–acceptor random copolymers based on a ladder-type nonacyclic unit: Synthesis, characterization, and photovoltaic applications, *Macromolecules*, **44**, 8415–8424.

212. Li J, Ong KH, Lim SL, Ng GM, Tan HS, Chen ZK (2011) A random copolymer based on dithienothiophene and diketopyrrolopyrrole units for high performance organic solar cells, *Chem Commun*, **47**, 9480–9482.

213. Nielsen CB, Ashraf RS, Schroeder BC, et al. (2012) Random benzotrithiophene-based donor–acceptor copolymers for efficient organic photovoltaic devices, *Chem Commun*, **48**, 5832–5834.

214. Khlyabich PP, Burkhart B, Ng CF, Thompson BC (2011) Efficient solar cells from semi-random P3HT analogues incorporating diketopyrrolopyrrole, *Macromolecules*, **44**, 5079–5084.

215. Burkhart B, Khlyabich PP, Cakir Canak T, LaJoie TW, Thompson BC (2011) "Semi-random" multichromophoric rr-P3HT analogues for solar photon harvesting, *Macromolecules*, **44**, 1242–1246.

216. Burkhart B, Khlyabich PP, Thompson BC (2012) Solar cells based on semi-random P3HT analogues containing dithienopyrrole: Influence of incorporating a strong donor, *J Photon Energy*, **2**, 021002.

217. Burkhart B, Khlyabich PP, Thompson BC (2012) Influence of the acceptor composition on physical properties and solar cell performance in semi-random two-acceptor copolymers, *ACS Macro Lett,* **1**, 660–666.

218. Li Z, He G, Wan X, et al. (2012) Solution processable rhodanine-based small molecule organic photovoltaic cells with a power conversion efficiency of 6.1%, *Adv Energy Mater,* **2**, 74–77.

219. Liu Y, Wan X, Wang F, et al. (2011) High-performance solar cells using a solution-processed small molecule containing benzodithiophene unit, *Adv Mater,* **23**, 5387–5391.

220. Liu Y, Wan X, Wang F, et al. (2011) Spin-coated small molecules for high performance solar cells, *Adv Energy Mater,* **1**, 771–775.

221. Van der Poll TS, Love JA, Nguyen TQ, Bazan GC (2012) Non-basic high-performance molecules for solution-processed organic solar cells, *Adv Mater,* **24**, 3646–3649.

222. Sun Y, Welch GC, Leong WL, Takacs CJ, Bazan GC, Heeger AJ (2011) Solution-processed small-molecule solar cells with 6.7% efficiency, *Nat Mater,* **11**, 44–48.

223. Lloyd M, Anthony J, Malliaras G (2007) Photovoltaics from soluble small molecules, *Mater Today,* **10**, 34–41.

224. Walker B, Kim C, Nguyen TQ (2011) Small molecule solution-processed bulk heterojunction solar cells, *Chem Mater,* **23**, 470–482.

225. Roncali J (2009) Molecular bulk heterojunctions: An emerging approach to organic solar cells, *Acc Chem Res,* **42**, 1719–1730.

226. Mishra A, Bäuerle P (2012) Small molecule organic semiconductors on the move: Promises for future solar energy technology, *Angew Chem Int Ed,* **51**, 2020–2067.

227. Walker B, Tamayo AB, Dang XD, et al. (2009) Nanoscale phase separation and high photovoltaic efficiency in solution-processed, small-molecule bulk heterojunction solar cells, *Adv Funct Mater,* **19**, 3063–3069.

228. Zhou J, Wan X, Liu Y, et al. (2011) A planar small molecule with dithienosilole core for high efficiency solution-processed organic photovoltaic cells, *Chem Mater,* **23**, 4666–4668.

229. Henson ZB, Welch GC, Van der Poll T, Bazan GC (2012) Pyridalthiadiazole-based narrow bandgap chromophores, *J Am Chem Soc,* **134**, 3766–3779.

230. Koster LJA, Shaheen SE, Hummelen JC (2012) Pathways to a new efficiency regime for organic solar cells, *Adv Energy Mater,* **2**, 1246–1253.

231. Helgesen M, Søndergaard R, Krebs FC (2010) Advanced materials and processes for polymer solar cell devices, *J Mater Chem,* **20**, 36–60.

232. Søndergaard R, Helgesen M, Jørgensen M, Krebs FC (2011) Fabrication of polymer solar cells using aqueous processing for all layers including the metal back electrode, *Adv Energy Mater,* **1**, 68–71.

233. Seyler H, Jones DJ, Holmes AB, Wong WWH (2012) Continuous flow synthesis of conjugated polymers, *Chem Commun,* **48**, 1598–1600.

234. Ameri T, Min J, Li N, et al. (2012) Performance enhancement of the P3HT/PCBM solar cells through NIR sensitization using a small-bandgap polymer, *Adv Energy Mater,* **2**, 1198–1202.

235. Khlyabich PP, Burkhart B, Thompson BC (2012) Compositional dependence of the open-circuit voltage in ternary blend bulk heterojunction solar cells based on two donor polymers, *J Am Chem Soc,* **134**, 9074–9077.

236. Yang L, Zhou H, Price SC, You W (2012) Parallel-like bulk heterojunction polymer solar cells, *J Am Chem Soc,* **134**, 5432–5435.

Chapter 2

Small-Molecule and Vapor-Deposited Organic Photovoltaics

Richard R. Lunt[a] **and Russell J. Holmes**[b]

[a]*Department of Chemical Engineering & Materials Science, Michigan State University, East Lansing, MI 48864, USA*

[b]*Department of Chemical Engineering and Materials Science, University of Minnesota, Minneapolis, MN 55455, USA*

rlunt@msu.edu, rholmes@umn.edu

In this chapter, we review key principles crucial in the operation and architectural construction of small-molecule organic photovoltaics (OPV). Many of the characteristics discussed are broadly applicable to other organic and small-molecular systems, but nuances and distinctions pertinent in vapor-deposited single junctions will be highlighted.

2.1 Introduction

There are four general categories of solids distinguished by the levels of bonding: ionic, covalent, metallic, and molecular (i.e., van der Waals). Organic molecules are materials that consist of carbon and hydrogen structural elements and can contain other

Organic Solar Cells: Fundamentals, Devices, and Upscaling
Edited by Barry P. Rand and Henning Richter

ISBN 978-981-4463-65-2 (Hardcover), 978-981-4463-66-9 (eBook)
www.panstanford.com

elements such as S, N, O, or metal atoms. This classification of organic materials includes polymers, hydrocarbons, alkanes, etc. as outlined in Fig. 2.1 and specifically excludes materials such as C_{60} and carbon nanotubes (both allotropes of carbon), which belong to the broader category of small-molecular solids which nonetheless exhibit many similar properties of organic molecules and are also of interest in this chapter. Organic and molecular compounds exhibit aspects of both covalent-(intramolecular) and molecular-(intermolecular) bonding. An even more important subsection of molecular compounds, and the focus of this chapter, are conjugated small-molecule compounds that are characterized by highly conjugated π-electron systems. These materials are of particular interest due to the delocalization of the π–π electron system that results in electronic excitation energies in the range of a few electronvolts. These π-electrons are delocalized above and below the plane of the atoms, creating a "localized cloud of electrons" that are essentially confined by the boundary of the molecule and imparting many of the physical (optical and electronic) semiconductor characteristics. The key distinguishing properties for small-molecule semiconductors include (1) high degree of excitation localization and low dielectric permittivity that leads to the presence of excitons, (2) lack of complete band structure where conduction and valence bands are replaced, in name, by molecular orbitals and charge conduction typically proceeds via hopping mechanisms, and (3) notably high absorption coefficients.

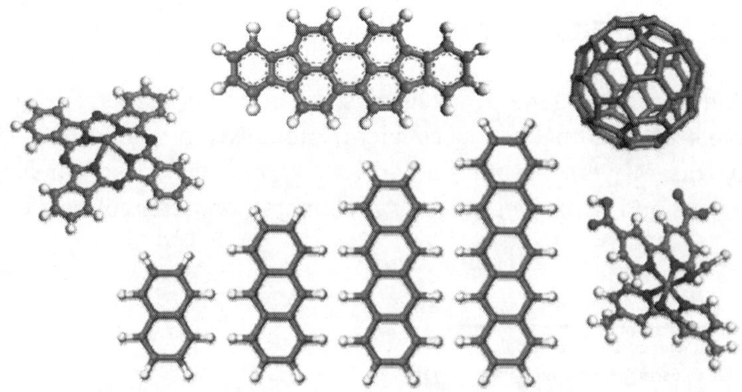

Figure 2.1 Structural diagram of several canonical conjugated small-molecular semiconductors.

In general, conventional solar cells convert various wavelengths of absorbed light to electrical power. The most basic organic solar cell structure is a metal–semiconductor Schottky diode, where a rectifying junction is formed at a semiconductor–metal interface. A more efficient organic solar cell was first demonstrated by Tang incorporating a bilayer planar heterojunction composed of thermally evaporated electron-donor and electron-acceptor layers.[1] Optical absorption in this device may occur in either layer, producing a tightly bound exciton. The potential difference across the structure that results from the asymmetry in the contact work functions provides a weak driving force for separating this exciton. By introducing a heterojunction with significant energy level offsets, the driving force for exciton dissociation at the interface is greatly enhanced. Consequently, excitons must diffuse to this interface in order to be dissociated. Once dissociated, the exciton forms a polaron pair, which itself may dissociate into mobile charge carriers that may migrate to the electrodes via a chemical potential driving force to produce current. Efficient carrier collection can then take place provided both the donor and acceptor are well-suited for carrier transport, i.e., they exhibit good charge-carrier mobility. The energetic anisotropy of the electrodes then confines the region of low- and high-potential, leading to a solar cell voltage that allows for power production. The key processes for current generation are summarized in Eq. (2.1), where the external quantum efficiency is separated into the component mechanisms.

$$\eta_{\mathrm{EQE}} = \eta_{\mathrm{A}} \eta_{\mathrm{EDf}} \eta_{\mathrm{CT}} \eta_{\mathrm{DS}} \eta_{\mathrm{CC}} \tag{2.1}$$

η_{A} is the absorption efficiency, η_{EDf} is the exciton diffusion efficiency, η_{CT} is the efficiency of generating CT states at the interface (i.e., electron or hole-transfer probability), η_{DS} is the CT dissociation efficiency (i.e., dissociation prior to geminate recombination), and η_{CC} is the charge collection efficiency (i.e., collection prior to bimolecular recombination). The distinction between η_{DS} and η_{CC} is primarily in the mechanism for recombination, geminate versus non-geminate; nonetheless, these terms are often lumped together as η_{CC}. Although organic semiconductors typically exhibit relatively large absorption coefficients, or small optical penetration lengths ($1/\alpha \sim 100$ nm), the corresponding exciton diffusion lengths are typically much smaller ($L_D \sim 10$ nm), setting up a trade-off between η_A and η_{EDf} in terms of optimum active layer thickness. Bulk-heterojunction (BHJ) solar cells, first demonstrated by Yu et al.,[2] aim to decouple these two limiting constituent efficiencies through the use of interpenetrating networks of donor and acceptor (shown schematically in Fig. 2.2), but often at the cost of potentially lower η_{CC}. Alternatively, by understanding the factors that determine the efficiency of exciton diffusion, it is possible to overcome this solar cell bottleneck via molecular and architecture tuning, thereby simultaneously increasing η_A, η_{EDf}, and η_{CC}.

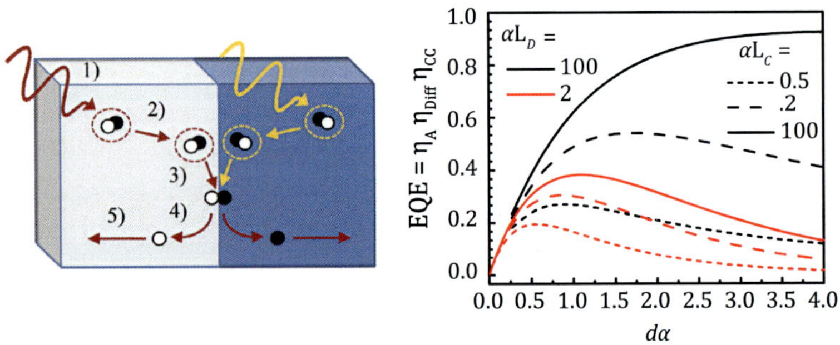

Figure 2.2 (left) Schematic of the charge generation process in OPVs: (1) photon absorption, (2) exciton diffusion, (3) charge transfer, (4) charge dissociation, (5) charge collection. (right) Calculation of the EQE in the absence of optical interference as a function of absorption coefficient (α), thickness (d), exciton diffusion length (L_D), and charge collection length (L_C).

Despite limitations in exciton diffusion efficiency and low charge-carrier mobilities, organic solar cells are steadily making gains in power conversion efficiency. A brief history of solar cell efficiencies is shown in Fig. 2.3a for both small-molecule and polymer cells. For comparison, the highest certified inorganic solar cells as a function of bandgap are included in Fig. 2.3b. Such progress leads to the question "what is the realistic upper limit for these materials?" which we will touch on in Section 2.2. While power conversion efficiencies continue to improve and have even recently surpassed 10%, the ability of OPVs to penetrate the energy markets will depend critically on additional parameters such as operational lifetime and cost discussed in greater detail in the last section of this book, Chapters 12–15. The key figure of merit will be the levelized energy cost (LEC) in $/kW-hr, which accounts for many factors including operational lifetime of the solar cell. In the remainder of this chapter we discuss the critical aspects of OPV performance and review canonical small-molecule systems.

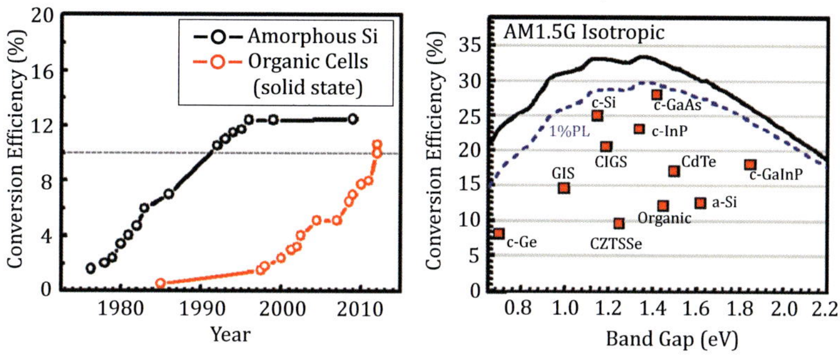

Figure 2.3 Highest reported power conversion efficiencies for (a) organic cell as a function of year and (b) inorganic cell as a function of band gap. Note in (b) the black solid line is the Shockley–Queisser thermodynamic single junction limit and the dashed blue line is the thermodynamic limit accounting for a low photoluminescence quantum yield of 1%.

2.2 Photovoltaic Characteristics

The power conversion efficiency of a photovoltaic is measured as

$$\eta_p = \frac{J_{SC} V_{OC} \mathrm{FF}}{P_0}, \tag{2.2}$$

where P_0 is the incident power density, J_{SC} is the short-circuit current density, V_{OC} is the open-circuit voltage, FF is the fill factor described by $\mathrm{FF} = (J_M V_M)/(J_{SC} V_{OC})$, J_M is the maximum power point current, and V_M is the maximum power point voltage. Example current–voltage characteristics of these OPV structures are shown in Fig. 2.4. The ideal diode equation describing the current-density (J)–voltage (V) characteristics in P–N junctions is

$$J = \frac{R_P}{R_S + R_P}\left\{ J_S\left[\exp\left(\frac{q(v - JR_S)}{nkT}\right) - 1\right] + \frac{V}{R_P}\right\} - J_{SC}, \tag{2.3}$$

where R_P is the shunt (or parallel) resistance, R_S is the series resistance, V is the voltage, J_{SC} is the photocurrent, n is the ideality factor, k is the Boltzmann constant, T is the temperature, and J_s is the reverse saturation dark current. Equation (2.3) has been used to describe P–N junctions from Si to thin film $CuIn_xGa_{(1-x)}Se_2$, GaAs, amorphous silicon, molecular and organic diodes. Indeed Eq. (2.3) can be utilized to fit the J–V curves of many organic solar cells, where R_P, R_S, n, and J_S are fitting parameters. Shown in Fig. 2.4 are a set of J–V curves for CuPc-C_{60} and CuPc-PTCBI junctions at various light intensities fit with Eq. (2.3) showing good agreement. The ideality factor, which takes the value of 1 and 2 for diffusion limited and bimolecular recombination, respectively, often is found to have values >2 for OPVs as well as other thin film solar cells (e.g., CdTe[3]). Accordingly, not all OPV systems (nor many other thin-film photovoltaics) are well-described by Eq. (2.3). Examples include semiconductor PVs with high trap densities (and high parallel photoconductivities), CdTe,[3] OPVs where buffer or blocking layer thicknesses are large enough to create reverse junctions in series,[4]

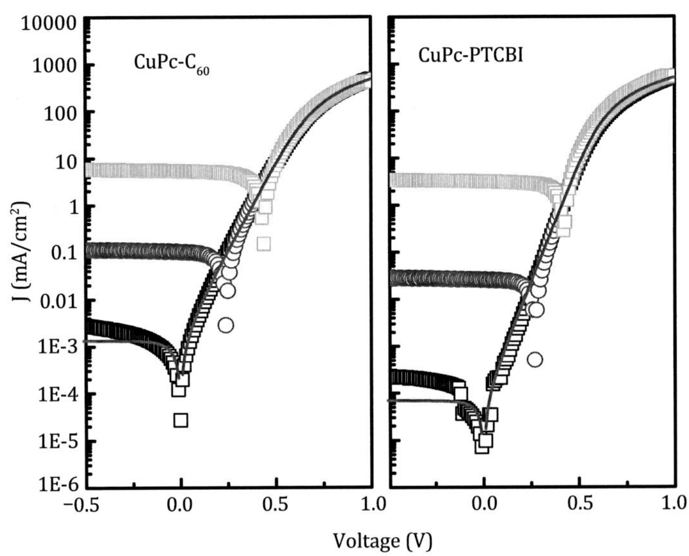

Figure 2.4 Current density–voltage (J–V) characteristics of planar vapor-deposited CuPc-C_{60} (left) and CuPc-PTCBI (right) as a function of illumination. The dark J–V is well-fit using Eq. (2.3).

and photoconductive cells. Indeed vapor-deposited CdTe solar cells, for example, are often modeled with a two-diode model where one diode stems from the contact energy barrier and the Shockley parameters are often given illumination dependencies.[3] Recently, equations analogous to Eq. (2.3) have been derived in the context of organic semiconductors based on polaron-pair recombination from a donor–acceptor heterojunction.[5] While these equations help to model greater non-idealities in OPV junctions, particularly at low temperature and low voltage, they do so with a greater number of fitting variables. It is difficult therefore to assess the model validity conclusively, particularly compared to other simple parallel or series diode models,[3] but the model of Ref. [5] could prove to be informative, particularly if it can be shown to reduce to Eq. (2.3) in limiting cases.

2.2.1 Thermodynamics and Open-Circuit Voltage

Rearranging Eq. (2.3) for open-circuit conditions in the limit of ohmic contacts, and assuming that R_P is very large (but not necessarily larger than R_S), the V_{OC} can be expressed as

$$V_{OC} = \frac{nkT}{q}\ln\left[\left(\frac{R_S + R_P}{R_P}\right)\left(\frac{J_{SC}}{J_S} + 1\right)\right] \tag{2.4}$$

so that the open-circuit voltage is primarily a function of the dark current, J_S, which in turn depends on the dominant recombination current mechanism. It is also worth noting that this expression assumes that the photocurrent is independent of forward applied voltage, a potentially gross approximation in OPVs. In the Shockley–Queisser (SQ) limit, the dark current is determined by radiative recombination as

$$J_S = qg\int_{E_G}^{\infty} \frac{E^2/\Phi_L}{\exp(E/nkT) - 1}\,dE, \tag{2.5}$$

where $g = 2\pi(n^2 + 1)/(c^2h^3)$, c is the speed of light, Φ_L is the luminescence quantum efficiency, h is Planck's constant, and E_G is the smallest active layer bandgap. Equation (2.5) implies that decreasing the bandgap will increase the dark current and therefore monotonically decrease the photovoltage. This equation also implies the use of highly luminescent materials are necessary to reach the highest voltages and efficiencies; that is, a highly luminescent semiconductor signals the absence of other non-radiative losses. Finally, the short-circuit current is related to the external quantum efficiency (EQE) via:

$$J_{SC} = q\int S(\lambda)EQE(\lambda)d\lambda, \tag{2.6}$$

where $S(\lambda)$ is the solar flux in typical units of photon/m^3 s. Now Eqs. (2.1)–(2.6) give a limit for the photovoltage in the case of radiative recombination for a single bandgap. Extending this equation to an organic donor–acceptor junction requires replacing $J_S(E_G)$ with $J_S(I_G)$, where I_G is the "interface gap" between the HOMO$_D$ and the LUMO$_A$, and V_{OC} necessarily must be less than I_G:

$$J_S = qg\int_{I_G}^{\infty} \frac{E^2/\Phi_L}{\exp(E/nkT)-1}dE \tag{2.7}$$

This is shown schematically in Fig. 2.5a where the recombination is ideally limited by interface recombination from injected electrons on the acceptor, and injected holes from the donor. In real devices, the open-circuit voltage has been shown to depend experimentally on both the electrode properties and the interface gap between the donor and acceptor.[6] When contacts are not Ohmic, a reverse diode (e.g., Schottky barrier) can form, reducing the open-circuit voltage of the equivalent circuit as charge accumulates at that barrier and reduces the quasi-Fermi level splitting.[4,7–11] This effect can be seen with OPVs with varied electrode materials[6,12–14] (e.g., Al, Ag, Ca, ITO) and with exciton blocking layers that become too thick to allow efficient carrier tunneling.[4] Both doped and buffer layers have therefore become an important aspect of OPV optimization used to tune electrode charge collection. Buffer layers will be further discussed in Chapter 4. In some cases, series resistance can also contribute to the enhancement of the open-circuit voltage as seen in Eq. (2.4), albeit at the expensive of FF and typically η_P.

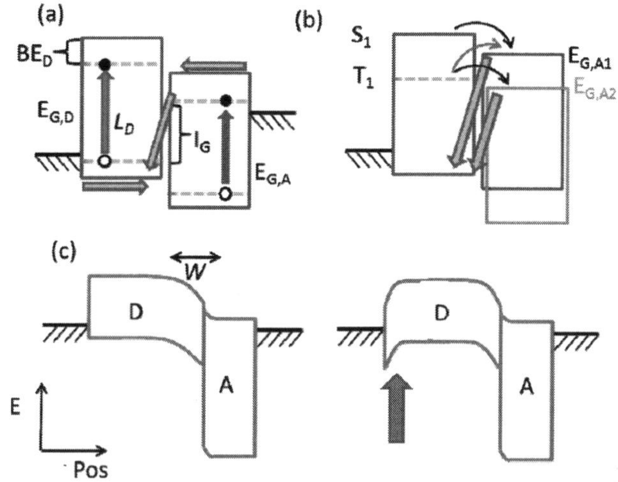

Figure 2.5 Flat band energy diagram in the absence of dipole shifts for bilayer OPVs. (a) Schematic of thermodynamically limiting recombination pathways through the interface energy gap, highlighted with the green arrows. (b) Schematic energy level positioning needed to dissociate either triplet or singlet excitons while maximizing voltage (c) Schematic of flat band energy levels with band bending at the interface due to the finite depletion width, without (left) and with (right) Schottky barriers that can create opposing series junctions and hinder charge collection.

Recently, a number of groups have looked to define limits on the ultimate efficiency for organic-based solar cells. One of the key steps in determining this limit is defining the upper limit to the achievable open-circuit voltage for donor–acceptor heterojunctions capable of separating excitons. A number of groups have convincingly linked the achievable voltage to the interface gap (LUMO of the acceptor minus the HOMO of the donor) for both small-molecule and polymer[6,15] solar cells. For example, Rand et al.[16] demonstrated this connection in vapor deposited small-molecule systems as shown in Fig. 2.6 that had been previously described in other polymer systems.[6,15,17] In that work, it was necessary to test under high-illumination and low-temperature in order to determine the maximum achievable voltage for a particular donor–acceptor pairing. Over the past decade a variety of empirical relationships have been extrapolated for the upper voltage limit including (1) V_{OC}(max) = LUMO$_A$–HOMO$_D$ – C where C is an empirical constant found to be C = 0.3 eV in Ref. [15] and C = 0.47 eV in Ref. [18], (2) V_{OC}(max) = $AE_g - C$ where E_g is the smaller bandgap of the donor and acceptor and A = 0.7, C = 0 in Ref. [19], and A = 1, C = 0.6 in Ref. [18] Alternatively, Lunt et al.[20] surveyed a range of heterojunction technologies from the past two decades including polymers, small molecules, colloidal quantum dots, and dye-sensitized cells and found that the best empirical form for the limiting open-circuit voltage under 1 Sun followed $V_{OC\_SQ} \times 0.8$ (see Fig. 2.7) where V_{OC_SQ} is the SQ limiting voltage that is a function of the smaller of the two bandgaps (see Eq. (2.7)). The factor of 0.8 in this case accounts for the energetic offset needed to drive dissociation. From this analysis, the maximum practical efficiency limits were determined to be 17% for single-junction cells and 24% for two-cell tandem cells for optically limited EQE of 80% and FF of 75%. Giebink et al.[21] estimated the upper theoretical limits based on a minimum free-energy cost for exciton dissociation of 0.3 eV, resulting in a theoretical upper limit efficiency of 24%. This value is close to the value determined by Lunt et al.[20] when accounting for optically limited EQE of 0.8 (i.e., IQE = 1.0).

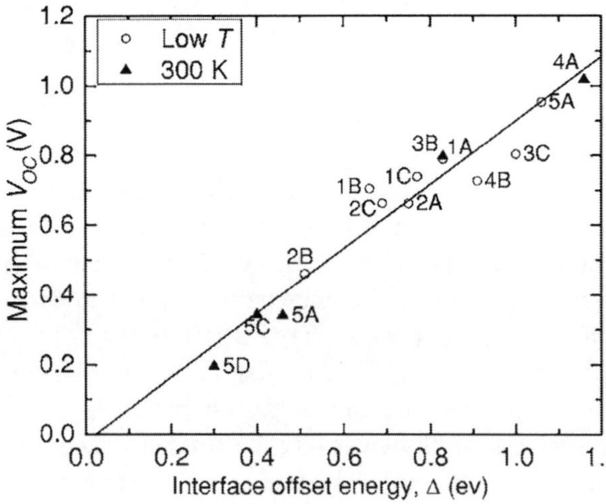

Figure 2.6 Maximum open-circuit voltage for different donor–acceptor vapor-deposited small-molecule pairs: 1 is CuPc, 2 is pentacene, 3 is α-NPD, 4 is SubPc, 5 is Ru(acac)$_3$, A is C_{60}, B is C_{70}, C is PTCBI, and D is PTCDA. Reproduced with permission from Ref. [16].

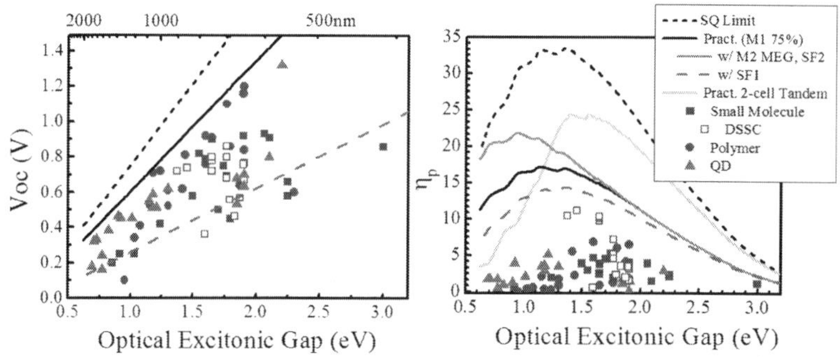

Figure 2.7 Survey of donor–acceptor heterojunctions for a range of structures including small molecules, polymers, dye-sensitized, and colloidal quantum dot solar cells across a range of bandgaps for the open-circuit voltage (left) and power conversion efficiency (right). Note that the bandgap (optical excitonic gap) is taken as the smaller of the donor–acceptor gaps. Calculations of the theoretical limits are included as solid and dashed lines. Reproduced with permission from Ref. [20].

2.3 Excitons

Excitons are a primary energy carrier in OPVs and therefore play a paramount role in their operation. We discuss the photophysical properties of excitons and review aspects of their migration in small-molecule systems. The radius of an exciton is defined as the average separation of the electron from its bound hole (as described by Eq. (2.8)). The Wannier–Mott exciton binding energy of excitons, related to this radius, is a function of the static dielectric constant according to[22]:

$$E_{\mathrm{BE}} = \frac{I_{\mathrm{H}}}{n^2 \varepsilon^2}\left(\frac{\mu_{\mathrm{r}}}{\mu_{\mathrm{e}}}\right), \tag{2.8}$$

where I_{H} is the ionization potential of hydrogen (−13.3 eV), ε is the dielectric constant of the molecular material, n is an integer state number, μ_{r} is the reduced mass of the exciton, and μ_{e} is the mass of an electron. In general, excitons are classified by their binding energy or radius. Wannier–Mott excitons, associated with a large dielectric permittivity, exhibit a separation distance much larger than the lattice constant and are common in inorganic semiconductors. Their binding energies are on the order of magnitude of kT (i.e., 0.0256 eV at 25 °C) so that they are easily dissociated at room temperature. At the other end of the spectrum, Frenkel excitons are highly localized with a radius <0.5 nm, and are typically isolated on 1–2 molecules. Accordingly, Frenkel exciton binding energies are on the order of 0.4–1.2 eV, and cannot be thermally dissociated at room temperature (Fig. 2.8). Between these two extremes are charge-transfer (CT) excitons whose radius is within one or two nearest-neighbor sites. Frenkel excitons describe the majority of excitons found in organic and molecular semiconductors, although CT states are also important. While the oscillator strength for CT transitions is generally weak, they play a role as intermediates in carrier recombination, affect exciton migration, and can lead to large conductivities in CT molecular salts. The migration properties of Frenkel and CT excitons will be discussed in Section 2.4.

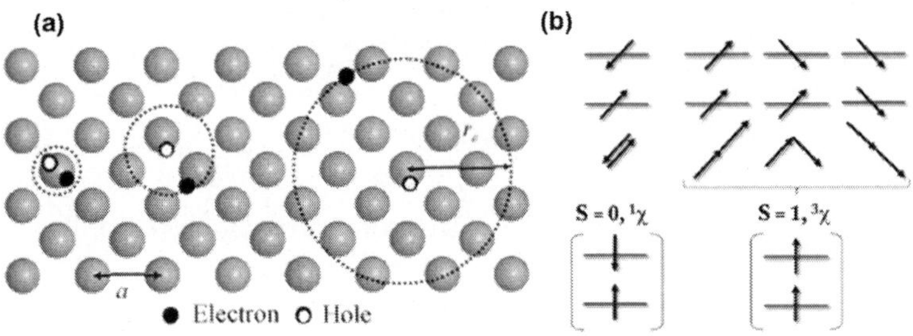

Figure 2.8 (a) Schematic of the exciton radii in a crystal lattice for Frenkel (left), charge-transfer (CT) (middle), and Wannier–Mott (right) excitons. (b) Schematic of the electron spin for antisymmetric (singlet) and symmetric (triplet) excited states. The shorthand drawing is provided in parentheses.

2.3.1 Spin and Intersystem Crossing

Exciton spin plays an important role in OPV operation. It can affect (1) exciton lifetime, (2) exciton diffusion, (3) charge transfer, (4) radiative and non-radiative recombination rates, (5) degradation kinetics, (6) oxygen sensitivity, and (7) maximum achievable open-circuit voltage for a given donor–acceptor pairing. In the zeroth order approximation, excitations from $S_0 \rightarrow T_1$ are forbidden and most organic solar cells therefore start with populations of singlet excitons under solar illumination. However, first-order transitions between singlets and triplets (intersystem crossing—ISC) can become allowed (indeed, quite efficient) if an interaction for mixing spin states is present, including symmetry and spin–orbit coupling. Accordingly, it is critical to identify the spin character of both donor/acceptor materials to have a complete understanding of the device operation and design higher functionality architectures. For example, the most common molecular acceptor utilized in high-efficiency OPVs is fullerene C_{60} and its derivatives. Molecular C_{60} is well known to have singlet-to-triplet ISC conversion efficiency close to unity (>96%) stemming from the symmetry forbidden S_1–S_0 transition and low singlet–triplet splitting.[23] While this conversion efficiency in solid films has not been directly measured, the presence of triplet excitons in bulk C_{60} has been confirmed by electron spin resonance (ESR) spectroscopy where the triplet relaxation times in solution and solid were similar (1–2 ms).[24] Indeed, many of the solid-state properties measured for fullerene compounds result directly from isolated molecular characteristics. Nonetheless, this aspect of fullerene spin character has been neglected in many experimental and theoretical treatments. Likewise, many of the donor materials utilized in OPV have at least some triplet character, even if unidentified. For example, the low fluorescence quantum efficiency of many acenes, metal porphyrins and phthalocyanines (e.g., CuPc, ZnPc, etc.) stem from combinations of both phonon emission and ISC to the triplet state.[25]

In general, triplet excited states are lower in energy than singlets due to electron repulsion interactions from electron exchange. This can be derived as

$$\Delta E_{ST} = S_1 - T_1 = 2J, \quad (2.9)$$

where S_1 is the singlet energy, T_1 is the triplet energy, and the matrix element (J) describing Pauli-principle electron exchange is >0 by convention for interaction of two particles of the same charge and therefore so is ΔE_{ST}. Di-radical and CT compounds are notable exceptions, where triplet bandgaps can be greater than the singlet bandgap. The effect of spin repulsion also plays an important role in the exciton radius and binding energy. That is, triplet excitons (parallel spins) actually have lower repulsive energy because of the avoidance of electrons to fill states with like spin. Consequently the binding energy and exciton radius of singlet/triplet excitons can be markedly different.

2.3.2 Singlet Fission

If the energy difference ΔE_{ST} becomes large enough that the triplet level is less than half the energy of the singlet, the singlet can spontaneously (thermally) decay into two triplets as $S_1 \rightarrow T_1 + T_1$. This process, known as singlet fission and analogous to the phenomenon of multiple-exciton generation (MEG) that has been reported in inorganic semiconductors,[26,27] has sparked interest as a means to surpass the SQ single-junction efficiency limit as suggested by Nozik.[28] Historically, this process has been observed for a range of polyacene crystals including pentacene ($E_S \sim 1.9$ eV, $E_T \sim 0.9$ eV), tetracene ($E_S \sim 2.3$ eV, $E_T \sim 1.3$ eV), and rubrene ($E_S \sim 2.3$ eV, $E_T \sim 1.2$ eV).[29] This phenomenon was first invoked to explain the photophysics of anthracene[30] in 1965 and the low quantum yield in tetracene.[31] Pentacene, having $T_1 < 1/2\ S_1$, has been shown to have a singlet fission efficiency near unity.[32] In contrast, tetracene has a triplet level just greater than $1/2\ S_1$ so that the process is thermally activated and "frozen out" below 150 K.[32] Although there have been a handful of studies of fission in polymers and covalently linked dimers, the fission efficiencies of both have been notoriously low. Highly disordered diphenyl tetracene films were recently shown to undergo fission, but with a singlet to triplet generation fraction of 22% that stemmed from singlet migration to fission dimer sites.[33] Nonetheless, the highest efficiency singlet fission is generally found in highly crystalline configurations since the mechanism for fission directly stems from highly correlated intra-molecular coupling.[29,32,34] Another question that arises is the possibility of observing efficient singlet fission if the S_1 and T_1 are closely matched so that the fission process proceeds through hot excitation route. Should this be shown to be efficient, it would result in theoretical limiting efficiencies better than the SQ single-junction limit and identical to the limit for solar cells with activated MEG.[20] This requires that the rate of fission be faster than vibrational relaxation from the hot excited state to the bottom of the band, and the triplet level position close to the singlet level. To date, this type of direct optically excited singlet fission has been shown to be orders of magnitude less efficient than thermally activated fission.[29] Typical exciton relaxation rates are around hundreds of femtoseconds, making it unlikely (but not impossible) that such a route to fission will be competitively efficient.

While singlet fission materials have been incorporated in a variety of vapor-deposited OPVs (e.g., pentacene[35–37], tetracene,[38,39] and rubrene[40]) for a number of years, it has only been recently recognized as impacting photocurrent generation[41–43] and at the same time was shown to be possible to enhance the internal quantum efficiencies beyond 100% in photodetector architectures incorporating layers of pentacene and C_{60}.[42] Unless both donor and acceptor exhibit high-efficiency singlet fission, care must be taken to design optical absorption profiles with high-energy photons preferentially absorbed in the singlet fission material; this could limit the overall impact of such phenomena in bulk heterojunction blends where selective optical control is a greater challenge.

2.3.3 Implication of Oxygen Ground State

Closed-shell molecules generally obey the behavior of exhibiting a singlet ground state. Open-shell molecules, on the other hand (e.g., O_2), have unpaired electrons in their HOMO and are not spin restricted. Therefore, they can have singlet or triplet ground states, and due to the lower energy of the triplet state generally have a triplet ground state. This triplet character of O_2 makes it a good triplet quencher when exposed to films containing triplet excitons and has important consequences in air-stable (and O_2 sensing) organic electronics. A triplet exciton can be quenched by triplet O_2 to form highly reactive singlet O_2 capable of hydrogen abstraction, cycloaddition, and other oxidative decomposition. This sensitivity actually suggests avenues for creating air-stable OPVs through prevention of triplet formation and/or triplet scavenging.

2.4 Energy Transfer and Exciton Migration

The optimization of organic thin-film photonic devices requires understanding of, and control over, energy transport. Numerous techniques have been employed to characterize exciton motion and migration. However, many of these techniques have drawbacks that are difficult to overcome, resulting in inaccurate and overestimated exciton diffusion lengths. We will review these techniques and place them in the larger context of energy transfer. Finally, we will discuss the impact of aggregation crystal orientation, crystal size, and resonant donor–acceptor energy transfer on the diffusion length for organic thin films.

Starting from Fermi's golden rule describing optical perturbations, Förster first derived the rate of dipole–dipole energy transfer in the weak coupling limit from a donor to acceptor. The golden rule for the transfer rate from one state to another, $K_{1\rightarrow 2}$, can be expressed as

$$K_{1\rightarrow 2} = \frac{2\pi}{\hbar}\left\langle \Psi_{a_2 b_1} \middle| H'_{12} \middle| \Psi_{a_2 b_1} \right\rangle \rho_S, \tag{2.10}$$

where H'_{12} is the perturbation matrix element, with $\Psi a_2 b_1$ the wavefunction for molecule a in the excited state (state 2) and molecule b in the unexcited state (state 1), $\hbar$ is Planck's constant, and ρ_S is the density of overlapping states. H'_{12} is the product of the vibrational overlap in molecule a between the two states (known as the Condon factor) and the electronic interaction matrix element that can be expanded into a multipole series. Substituting Einstein's expressions for absorption and emission with the first multipole term, the well-known Förster expression is

$$K^{\mathrm{F}}_{a\rightarrow b} = \frac{9\kappa^2\Phi}{8\pi\tau R^6_{ab}} \int \frac{c^4}{\omega^4 n^4} F_a(\omega)\sigma_b(\omega)d\omega, \tag{2.11}$$

where $F_a(\omega)$ is the normalized fluorescence spectrum of the donor (molecule a) on a frequency scale, $\sigma_b(\omega)$ is the normalized acceptor absorption cross section, n is the index of refraction at the peak overlap between $F_a\sigma_b$, τ is the lifetime of the donor (molecule a), and Φ is the quantum yield of the donor. $F_a(\omega)$ is normalized as $F_a = PL(\omega)/\int PL(\omega)d\omega$ and $\sigma_b = \alpha MW/\rho N$ where α is the absorption coefficient, MW is the molecular weight of molecule b (acceptor), ρ is the density, and N_L is Avogadro's number. Equation (2.11) describes transfer between both like and unlike molecules provided the weak coupling limit is met, i.e., that the vibrational bandwidth (ΔE) is larger than the interaction energy between states, $\Delta E > |H_{12}|$.

Equation (2.11) is often expressed with a characteristic Förster radius, with (R_0) as

$$K^{\mathrm{F}}_{a\to b} = \frac{1}{\tau}\left(\frac{R_0}{R}\right)^6. \tag{2.12}$$

In the strong coupling (coherence) limit, $\Delta E << |H_{12}|$ and the transition rate is $K_{a\to b} = 4|H_{12}|/\hbar$ which is proportional to $1/R^3_{ab}$. However, fast vibrational relaxations in the excited state manifold result in dephasing of the excited states. Therefore, coherence is not typically observed at room temperature and is, in fact, rarely observed except at cryogenic temperatures.[44]

This analysis was extended to spin-forbidden energy transitions and approximating the wavefunctions in the perturbation element (H_{12}) as hydrogenic wave functions by Dexter as

$$K^{\mathrm{D}}_{a\to b} = \frac{2\pi K}{\hbar R^2_{ab}} e^{\left(-\frac{2R^2_{ab}}{\xi}\right)} \int F_a(E) F_b(E)\, dE, \tag{2.13}$$

where $F_a(E)$ is the normalized phosphorescence spectrum of donor (molecule a), $F_b(E)$ is the normalized absorption spectrum of the acceptor (molecule b). ξ is the average orbital Bohr radius of the final and initial electronic states, and K is a constant that lumps hydrogen wavefunction normalization scaling constants as well as the spin exchange component so that $K<<1$ and is not easily calculated.[12] This normalization proceeds as[44] $F_a = PL_a(E)/\left[E^3 \int PL_a(E)/E^3 dE\right]$ and $F_b = \alpha_b(E)/\left[E \int \alpha_b(E)/E dE\right]$. Davidovich and Knox[45] demonstrated that this approach for the estimation of the density of states is valid if the spectra arise from transitions that have singlet character due to spin–orbit coupling and not vibronic induction. While Dexter transfer can contribute to singlet transfer, the transfer rate is typically negligible when compared to Förster rates for these dipole-allowed transition rates.

For higher order transfer, such as quadrapole perturbation interaction,[46] the transfer rate is

$$K^{M}_{a\to b} = \frac{135 c^6 \kappa^2 \Phi}{4\pi\tau n^4 R^8_{ab}} \int \frac{F_a(\omega)\sigma_b(\omega)}{\omega^6} d\omega, \tag{2.14}$$

where κ is a dipole–quadrapole orientation factor. These higher order perturbation interactions decay very rapidly, and are therefore negligible when considered for allowed transitions, but could be competitive with electron exchange mechanisms. This higher multipole transfer mechanism is also allowed for triplets.

The diffusion analysis is highly simplified by considering a cubic lattice of lattice constant, a, where diffusion is assumed to be isotropic and limited to the six nearest-neighbor molecule interactions. Since the coherence time for exciton motion is generally much smaller than the exciton hopping time at room temperature, the diffusion coefficient can be defined as

$$D \equiv \frac{L_{\mathrm{D}}^2}{Z\tau} = \frac{a^2}{6\tau_{\mathrm{H}}}, \tag{2.15}$$

where we take Z = 1 (for comparison to literature values), τ is the exciton lifetime, and τ_{H} is the hopping time. The diffusion process for singlet excitons can occur for any of the mechanisms described above, but primarily occurs via electrostatic dipole–dipole coupling between like molecules since it usually has the fastest rate. When Förster self-transfer is the dominant mechanism for transfer, $1/\tau_{\mathrm{H}}$ is estimated by the Förster rate, k_F, given in Eq. (2.11). From Eqs. (2.12) and (2.15) the diffusion length can be written as

$$L_{\mathrm{D}} = \sqrt{\frac{\kappa^2 \Phi_{\mathrm{F}} \sigma}{8\pi n^2 a^4}} = \frac{1}{\sqrt{6}} \frac{R_0^3}{a^2}. \tag{2.16}$$

Although there has been speculation about the accuracy of the Förster expression at the scale of molecular separations in crystals (i.e., weak coupling limit), we can nonetheless use this as a starting point for theoretical comparisons.

Equation (2.16) highlights several key aspects of singlet diffusion by Förster transfer. For example, it is possible to estimate the diffusion length based solely on the estimation of R_0 and the characteristic molecular spacing. This indicates that for crystalline films, the directions of closest spacing will have the largest diffusion length, which is what has been observed in crystalline systems such as anthracene, tetracene, and DIP. This also suggests that it is preferable to grow crystalline films with molecules lying flat to the substrate (in OPV structures) to increase Förster transfer and therefore the L_D. Finally, the diffusion length depends on the quantum yield which may be counter-intuitive. In a solar cell, for example, one would expect luminescence to act as a loss mechanism, but this would be true only if the exciton is not within a distance L_D of an interface and not actually reabsorbed. Indeed, the thermodynamic SQ limit with only radiative recombination can only be approached for quantum yields approaching 1 (see Fig. 2.3).

For nearest-neighbor diffusion by concerted electron exchange hopping, the transfer rate is described by Dexter theory as described in Eq. (2.13). From Eqs. (2.13) and (2.15), the Dexter-mediated diffusion length is simply

$$L_D = \sqrt{\frac{\tau K \sigma_D}{3\hbar}} e^{-a/\xi}, \tag{2.17}$$

where ξ is typically of the order of hydrogen Bohr radius (i.e., 0.053 nm), and has been estimated for PtOEP to be 0.065 ± 0.02 nm.[47]

Inclusion of hopping to non-nearest neighbors for cubic lattices introduces a sum over longer jumps

$$L_D = \sum_{n=1}^{\infty} \frac{1}{\sqrt{6}} \frac{R_0^3}{(na)^2} = \frac{\pi^2}{6\sqrt{6}} \frac{R_0^3}{a^2}, \tag{2.18}$$

where n is an integer number of lattice intervals in non-diagonal larger jumps and Eq. (2.18) simply adds a prefactor to Eq. (2.16). A similar prefactor can be derived for Eq. (2.16) to be $1/1 - e^{(-a/\xi)}$. Further modifications to these simplified models of exciton diffusion can be made, including (1) corrections for non-cubic crystal lattices (e.g., full diffusion tensor) and (2) replacement of the Förster transfer expression with generalized Förster theory,[48–50] which describes intermediate coupling regimes where the dipole–dipole approximation breaks down for singlet energy migrations. However, these corrections add considerable complexity, and as will be shown below are not necessarily needed. In the case of Dexter-mediated diffusion, no such correction is necessary for triplet migration, but depends very sensitively on the orbital length (ξ) which is difficult to predict a priori or from measureable photophysical quantities.

Finally, we consider the trivial energy transfer process involving the emission and reabsorption of a photon between a donor and an acceptor molecule. Unlike the previous mechanisms, the molecules do not interact directly: the acceptor simply intercepts the emitted photon. Therefore, the transfer rate is a function of the luminescence quantum yield (Φ) of the donor and the absorption coefficient (α) of the acceptor. The mean displacement, $\langle R_d \rangle$, of this process for single hops at a fixed frequency in one dimensions is

$$\langle R_d \rangle = \frac{\int_0^\infty (R_d) \cdot e^{-\alpha R_d} dR_d}{\int_0^\infty e^{-\alpha R_d} dR_d} = \frac{1}{\alpha} \tag{2.19}$$

Treating this radiative migration process as a diffusion process with the non-radiative rate, k_{NR}, that defines time scale before diffusion stops and the radiative rate, k_R, defining the hopping rate ($k_{NR} = k_R(1/\Phi - 1)$), the diffusivity is

$$D_R = \frac{1}{6} \langle R_d \rangle^2 k_R \tag{2.20}$$

and the diffusion length becomes

$$L_D \equiv \sqrt{D_R \tau_{NR}} = \sqrt{\frac{1}{6\alpha^2} \frac{\Phi}{(1-\Phi)}} \tag{2.21}$$

This equation inherently accounts for multiple remission and reabsorption events where there are k_R/k_{NR} number of hops. In the limit of zero luminescence quantum yield, excitons cannot move by a trivial energy-transfer mechanism and so $L_D = 0$. At a quantum yield approaching unity, excitons can move indefinitely as they will be continuously emitted and reabsorbed without loss. Equation (2.21) can be further integrated over all wavelengths so that L_D will again become a function of the overlap integral, similar to the Förster expression of Eq. (2.13).

Table 2.1 Measured exciton diffusion lengths for a range of vapor-deposited and small-molecule semiconductors

Material	Exciton	Crystallinity (Orient.)	L_D (nm)	Method[b)]/ Reference
Alq_3	S	Amorphous	20	
Anthracene	S*	Cryst.	200	SR-PC[53]
	S*	Cryst.	130//b	SPC[54]
	S*	Cryst.	100//a	SPC[54]
	S*	Cryst.	44//ab	SPC[55]
	S*	Cryst.	88//ab	SPC[55]
CBP	S	Amorphous	17.0 (±0.8) [a)]	SR-PLQ[51]
ClAlPc	–	–	8 (±4)	EQE[56]
CoPc	–	–	1.6	EQE[57]
	–	–	15.4	EQE[57]
DIP	S	P. Cryst—30 nm (flat)	22.0 (±0.6)	SR-PLQ[51]
FePc	–	–	1.0	EQE[57]
NiPc	–	–	9.0	EQE[57]
α-NPD	S	Amorphous	4.9	OLED[58]
	S	Amorphous	6.5 (±0.3)	EQE[59]
	S*	–	13.4	EQE[60]
	S	–	7.7 [a)]	Spacer T-PLQ[61]
PTCDA	S	Amorphous	8.0 (±0.7)	SR-PLQ[51]
	S	P. Cryst—100 nm	10.5 (±1.0) //(102)	SR-PLQ[62]
	S	P. Cryst—400 nm	21.5 (±0.5) //(102)	SR-PLQ[62]
PTCBI	S	–	3.0 (±0.3)	T-PLQ[63]
PTCBI-*cis*	S	–	2.8 (±0.2)	T-PLQ[64]
PTCBI-*trans*	S	–	4.3 (±0.3)	T-PLQ[64]
Tetracene	S*	Cryst.	29	SPC[54]
TnBuPP	S*	Cryst.	13 (±1)	EQE[65]
ZnPc	–	–	15.0	EQE[57]
ZnPc	S	(upright)	26 (±2)	SR-PLQ[66]
ZnPc	S	(flat)	15 (±2)	SR-PLQ[66]
C_{60}	T*	–	40 (±5)	EQE[63]
α-NPD	T	–	87 (±3)	S-PLE[67]
PtTPBP	T	Amorphous	6.0 (±0.8)	SR-PLQ[51]
pq-Pt(acac)	T	P. Cryst—>150 nm (upright)	21.5 (±0.5)	SR-PLQ[47]
PtOEP	T	Amorphous	6.5 (±1.5)	SR-PLQ[47]
PtOEP	T—Mon.	P. Cryst—(upright)	18.0 (±0.6)	SR-PLQ[51]
PtOEP	T—Aggre.	P. Cryst—(upright)	13.0 (±0.5)	SR-PLQ[51]

[a)]Corrected for long-range energy transfer to quenching layer for cases with significant spectral overlap.

[b)]SR-PLQ—spectrally resolved photoluminescence quenching, SPC—spectral photoconductivity, T-PLQ—thickness photoluminescence quenching, S-PLE—Sensitized Photoluminescence Emission, EQE—external quantum efficiency.

*Tentatively assigned spin character. Conductive measurements often do not make clear spin distinctions and can have contribution from both singlet and triplets.

A range of exciton diffusion lengths for vapor-deposited small molecules and their associated measurement techniques are summarized in Table 2.1. Values measured by Schottky diode-based photocurrent measurements without consideration of optical interference are not reported since this technique notoriously overestimates the diffusion length.[51,52] Determination of L_D from EQE can be confounded with the charge collection efficiency and multiple fitting parameters but provide a lower limit estimate.

2.4.1 Exciton Diffusion and Crystalline Order

While charge-carrier mobilities have been convincingly linked to the extent of crystalline order, very few studies have explored similar effects with regard to exciton diffusion, another key energy carrier. Routes to increase crystalline ordering during thin-film growth will be discussed in Section 1.5. Recently Lunt et al.[62] demonstrated that both singlet and triplet exciton diffusion can be impacted by crystalline domain size. For the case of singlet diffusion in 3,4,9,10-perylenetetracarboxylic dianhydride (PTCDA), the exciton diffusion was found to be a monotonic function of the extent of crystalline order up to a critical grain size, where grain boundaries were found to act as quenching sites (see Fig. 2.9); however, in contrast to the case of the charge transport, these changes are not necessarily due to changes in diffusivities for singlet excitons. This grain-size dependence was seen in both the exciton diffusion length and quantum yield for fluorescence where larger grain films resulted in a reduction in the non-radiative rate. The diffusion length was found to saturate at a grain size 10–20× the diffusion length in the amorphous limit. This is due to the generation of a substantial population of excitons near quenching grain boundaries up to this size.

2.4.2 Exciton Diffusion and Photophysical Relaxation

Diffusion by energy transfer ultimately competes with all other relaxation processes available to the exciton. In Section 2.4.1, the rate of exciton quenching at grain boundaries was found to compete with the rate of energy transfer. In materials where exciton diffusion occurs by Förster transfer, a reduction in the non-radiative decay rate (and increase in the photoluminescence efficiency) leads to an enhanced rate of energy transfer and an increase in L_D. Menke et al. have recently demonstrated this correlation by examining the dependence of L_D in SubPc as a function of dilution in the wide energy-gap material *p*-bis(triphenylsilyl)benzene (UGH2).[68] Dilution serves to increase the spacing between molecules of SubPc, and affect many of the parameters responsible for energy transfer, most notably, the photoluminescence efficiency. Interestingly, a ~50% increase in L_D is observed for a 25 wt.% film of SubPc as compared to a neat film, despite a concomitant increase in the intermolecular spacing. While highly efficient OPVs have been demonstrated using dilute donor layers, a more practical prospect would be to realize the same effect in neat film through thoughtful molecular design with consideration between the balance in the photoluminescence efficiency (or non-radiative decay) and the hopping distance (transfer rate) as a function of molecular spacing. To this end, Mullenbach et al. have examined similar effects in neat film where chemical modifications are used to spatially separate molecules of rubrene.[69] Modification of rubrene at the para-phenyl position with $-CH_3$ and $-CF_3$ leads to an increase in L_D from 6.4 nm for rubrene to 10.2 nm for a substituted rubrene. The increase in L_D again comes from a concomitant increase in the photoluminescence efficiency. This work highlights the potential to a priori design materials for enhanced exciton diffusion.

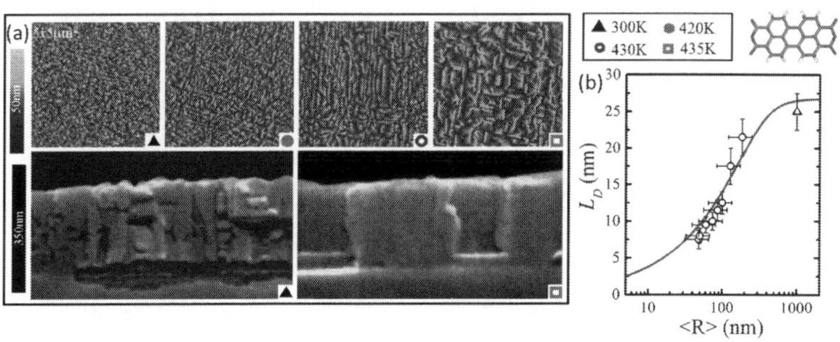

Figure 2.9 (a) Polycrystalline morphology (AFM, top—SEM cross section, bottom) of PTCDA grown at various substrate temperatures. (b) Measured diffusion length (SR-PLQ) as a function of mean grain size measured in AFM from (a). From Ref. [62] with permission.

2.4.3 Exciton Diffusion and Long-Range Energy Transfer

Resonant energy transfer (RET), i.e., the (spatial) relocation of the acquired electronic energy, immediately following photo-excitation, can be useful (1) in describing exciton migration between like molecules, (2) in deconvoluting confounding processes for intrinsic diffusion length measurements with dissimilar donor–acceptor systems, or (3) as a mechanism to effectively enhance diffusion between dissimilar materials.

For situations where there is overlap in the absorption spectrum of the quenching layer with the emission spectrum of the layer under study, non-diffusive Förster energy transfer to the quenching layer must be considered in order to measure the intrinsic L_D. In this case, Eq. (2.22) is modified for a sample with a thin quenching layer as

$$L_D^2 \frac{\partial^2 n(x)}{\partial x^2} - n(x) + G(x) - n(x)\left[\rho_A \int \left(\frac{R_Q^6}{R(x)^6}\right) dA\right] = 0, \qquad (2.22)$$

where the Förster transfer rate is integrated over the entire quenching interface of area, A, R_Q is the Förster radius, $R(x)$ is the distance from a point in the film to any molecule within the quenching layer, and ρ_A is the number of quenching molecules per unit area. Equation (2.22) is then

$$L_D^2 \frac{\partial^2 n(x)}{\partial x^2} - n(x) + G(x) - \rho_A \frac{2\pi}{5}\left(\frac{R_Q^6}{x^4}\right) n(x) = 0, \qquad (2.23)$$

which can be solved numerically to extract the intrinsic diffusion length.

Several groups have exploited long-range energy transfer to overcome limited exciton diffusion lengths by transferring from low- to high-diffusion materials. For example, Scully et al. have utilized combinations of DOW Red and poly(*N*-dodecyl-2,5,-bis(2′-thienyl)-pyrrole-2,1,3-benzothiadiazole) (PTPTB) to effectively enhance the exciton diffusion length of DOW Red from 3 to 27 nm.[70] Coffey et al.[71] combined P3HT and CuPc donor layers in heterojunction devices to demonstrate P3HT energy transfer to the CuPc that allowed for multiple donor photocurrent generation and improved NIR response over the P3HT-only device. Moreover, Luhman et al.[61] and Lunt et al.[51] have demonstrated that this mechanism is already active in many of the donor–acceptor system pairings commonly employed in OPVs such as SubPc-C_{60} and α-NPD-C_{60}, where the donor diffusion lengths can be effectively enhanced by transfer to the longer diffusion length fullerene. The work by Coffey et al. demonstrated that this effect is similarly present in P3HT-C_{60} junctions, while CuPc-C_{60} junctions demonstrated very little long-range transfer, possibly due to the low energy or triplet character of CuPc.[71] Ultimately, exploiting long-range energy transfer can enhance photocurrent production; however, the downward energy progression required to promote such transfer has to be balanced with any potential reductions in the open-circuit voltage resulting from placing a smaller bandgap material at the dissociating interface.

2.5 Vapor Deposition Methods

Film morphology, crystalline ordering, phase, orientation, roughness, and purity can all strongly impact the properties of organic semiconductors including absorption, carrier mobility, exciton diffusion, energy levels (HOMO/LUMO), fission rate, and luminescence properties. It is therefore important to understand the mechanisms for growth of these materials to better engineer and optimize structures for useful applications. In this section, we review the techniques commonly employed in vapor deposition of small molecules and discuss the role that ordering can play in OPV performance.

2.5.1 Vacuum Thermal Evaporation

Vacuum thermal evaporation (VTE) is a commonly employed technique for growth of organics, inorganics, and metals. A schematic of VTE is provided in Fig. 2.10. Material is loaded into source boats or crucibles in the deposition chamber and then typically pumped down to a pressure of about 10^{-6}–10^{-8} torr. One of the advantages of VTE lies in the fact that the high vacuum conditions aid in material purification, particularly as a function of source cell heating cycles.[72] After loading, the source cells' material is heated/evaporated resulting in ballistic deposition onto the substrate placed directly above the source. The thickness profile (h) resulting from point-source growth with source-dependent exponent, m, and a peak film thickness, d, can be described as

$$h = d\cos(\theta)^m. \tag{2.24}$$

To achieve uniform films θ must be kept small. This reduces the overall material yield and increases operating costs. Indeed, the material utilization efficiency in these systems is typically around 0.1%–5% depending on the material, chamber configuration, size, etc.[73] Material utilization can be enhanced substantially through the use of linear, nozzle sources. Wall coating can lead to flaking and contamination of source cells. Additionally, evaporation of material at the walls of the source cell could result in collapse or shifting of the powder source, causing fluctuations in the evaporation area and, subsequently, the deposition rate. The deposition rate in VTE is typically controllable by one operating parameter, the source temperature:

$$r_{\text{evap}} = r_{\text{dep}} \cdot TF = \frac{\overline{v} \cdot p_{\text{v}}^{\text{sat}}(T)A(t)}{RT}, \tag{2.25}$$

where v is the mean thermal velocity, A is the exposed area of source material which varies in time, and TF is the tooling factor. Since the vapor pressure is ~$\exp(-\Delta H_{\text{vap}}/kT)$ precise control over the deposition rate is difficult, particularly for varying organic exposure areas (A). Furthermore, the ability to scale VTE to large substrate areas can be difficult as this typically requires large chambers to obtain uniformity. As already noted, one solution to this problem is the use of linear source boats, which allow for smaller chamber configurations.[74]

Organic molecular beam deposition (OMBD), a technique similar to VTE, utilizes a base pressure of 10^{-8}–10^{-12} torr with a thermally evaporated organic flux collimated into a beam aimed at the substrate. The difference in background pressure can have influence on the incorporation of impurities and therefore has been used to produce high-purity growth. The ultra-high vacuum (UHV) environment permits in situ experiments such as reflection high-energy electron diffraction (RHEED), ultraviolet photoemission spectroscopy (UPS), x-ray photoemission spectroscopy (XPS), scanning-tunneling microscopy (STM), etc. and has made OMBD a common growth method as well best suited for laboratory-scale development.

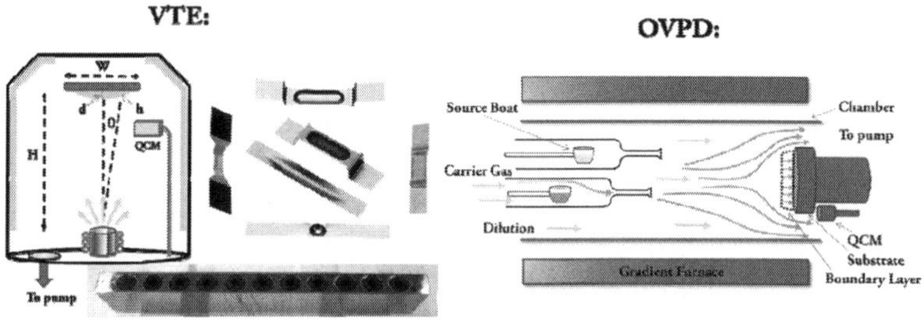

Figure 2.10 (left) Schematic of vacuum thermal evaporation (VTE) and (right) organic vapor-phase deposition (OVPD). Reproduced with permission from Ref. [47]. Linear VTE source boats are also shown.

Despite its drawbacks, vacuum deposition methods remain one of the most popular research and commercially applied techniques. Indeed, the industry standard for many ubiquitous products including LCD televisions, low-emissivity window coatings, metalized food packaging, CdTe PVs, and compact discs has layered components deposited by vacuum deposition technologies either by thermal or sputter deposition. An example large-area in-line window coating facility (vacuum sputtering[75]) is shown in Fig. 2.11. While it is not clear whether vacuum or solution techniques will ultimately be the lowest cost commercial fabrication method, there are certainly advantages to both. Solution deposition can potentially reduce capital equipment costs compared to vacuum systems, but are also likely to increase processing costs since large quantities of solvent will be required (i.e., think purified, evaporated, and condensed).

2.5.2 Organic Vapor-Phase Deposition

The technique of low-vacuum organic vapor-phase deposition (OVPD) has been gaining interest as a technique of producing uniform thin films for a variety device structures. In OVPD, unlike VTE, the process of evaporation and transport is essentially decoupled.

Organic material is thermally evaporated into a heated carrier gas and then transported through a hot wall reactor to a cooled substrate where growth is promoted by a temperature gradient. A schematic of the OVPD system is provided in Fig. 2.10. In many ways the transition from UHV OMBD deposition to OVPD is analogous to the transition from UHV MBE to CVD with inorganic semiconductor growth. There are four characteristics of OVPD growth: (1) low material waste, (2) high-precision control of the deposition process, (3) improved doping capabilities and (4) (relatively) high-pressure (low-vacuum) growth. Growth by OVPD promotes high material use efficiency because thin film formation occurs only at the cooled substrate. This is accomplished through heating the entire deposition chamber, thus preventing (wasteful) condensation of material on walls. Deposition yields of up to 50% for growth in OVPD systems are compared to the typical <10% in VTE systems.[73] In Ref. [73] Lunt et al. demonstrate that this level of material utilization is indeed possible, and further discuss the scaling of such systems.

Figure 2.11 Photograph of high-speed vacuum sputtering system used to deposit multilayer coatings on full-size glass windows. Reproduced from Ref. [75] with permission.

Two control variables, the carrier gas flow rate (Q) and the source cell temperature (T_{cell}), provide a fine and rough tuning for controlling the organic concentration in the carrier gas stream and thus the deposition rate. Although the source temperature impacts the flow through the source barrel and vice versa, these effects are small enough that the evaporation and transport processes are essentially decoupled. While different source boats can be maintained at different temperatures, fluctuations in chamber temperature are experienced by both boats simultaneously. This results in reduced fluctuations of the doping concentration that is particularly important for blended architectures.

More recently, roll-to-roll, in-line,[76] and large-area[73] OVPD systems have been developed. Furthermore, the high material utilization efficiency, excellent morphology control, and the demonstration of in-line processing capabilities with OVPD indicate that this method is a promising candidate for low-cost, large-area, and high-performance OPV fabrication.

2.5.3 Crystalline Morphology of Vapor Deposition

Covalent and ionic bonding characterize inorganic structures while organic molecules interact primarily via polarization forces, collectively referred to as van der Waals forces. The van der Waals forces of individual atoms are relatively weak ($\sim 10^{-3}$–10^{-1} eV/bond), but summed over many atoms in a molecule add up to a significant total energy.

Inorganic commensurate growth, or epitaxy, is achieved for lattice-matched materials and near-equilibrium growth conditions. Because of the former requirement, there is a significant limitation to possible material combinations. For organic growth, lattice matching is not strictly required and the modes of epitaxy are uniquely distinct. Organic epitaxy can occur for both chemisorbed materials (conventional epitaxy) and weakly binding physisorbing materials (van der Waals epitaxy (vdWE)). With the latter, unmistakable azimuthal relationships have been observed for many organic molecules grown on inorganic substrates despite a lack of true commensurate "epitaxial" relationships.[77–81] As a result, the term "quasi-epitaxy" is utilized for describing these systems that clearly lack short-range commensurability, and yet exhibit a preferred azimuthal configuration. Depending on the strength of the substrate–adsorbate interactions, chemisorption vs. physisorption, three growth modes are commonly observed: layer-by-layer (Frank–van der Merwe), layer-plus-islanding (Stranski–Krastinov), and islanding (Vollmer–Weber). In general, Stranski–Krastinov is the dominant growth mode for many organic systems except in rare cases. Both Stranski–Krastinov and Vollmer–Weber growth modes can arise from an accumulation of lattice strain (inorganic) or thermodynamic surface energies (organic). Organic systems exhibiting covalent epitaxy (e.g., on metal surfaces[82,83]) typically build up strain very quickly, resulting in amorphous, polycrystalline, and three-dimensional powder growth after only a few monolayers. Alternatively, quasi-epitaxial (QE) growth of organic semiconductors on weakly interacting and vicinal substrates can support ordered growth well beyond several monolayers.[77,84]

The key differences between organic and inorganic vapor growth are summarized below:

1) Size: Organic molecules are large (~0.2–3 nm) and have additional vibrational and rotational degrees of freedom. For example, molecules can rotate around axes of molecular symmetry. This has implications on epitaxial relationships and can act to promote or inhibit ordering. The large size and shape of molecules also leads to highly anisotropic optical and electrical properties, for example, charge-carrier mobility can vary by an order of magnitude in different crystallographic directions.
2) Crystal Structure: Molecular crystal unit cells of organic molecules are often much larger than inorganic substrates. This smears out the potential energy landscape more so than with atomic adsorbates. Additionally, the majority of molecular solids crystallize in low symmetry (monoclinic, triclinic) crystal structures. This often results in a high degree of polymorphism and low degrees of QE symmetry.
3) Bonding: The interactions between organic molecules and molecule–substrates typically involve weak van der Waals forces, although the molecule–substrate interactions can sometimes involve covalent bonding. This implies that the surface energies are generally smaller than for inorganic growth. Hence kinetic processes such as diffusion and

evaporation occur at much lower temperatures than for inorganic compounds.

Bulk single crystals of organic materials have been straightforwardly grown, historically, via zone refining[85] or gradient sublimation,[85] where the crystal shapes, sizes, and thickness are dependently linked thermodynamically, i.e., the thickness cannot be independently controlled from the lateral dimensions of such crystals. While there have been a handful of demonstrations of long-range and single-crystalline *thin-film* organic growth from the vapor and even a few demonstrations in small-area solar cell devices,[86] such control remains limited. Typical vapor deposition results in organic and molecular films ranging from completely amorphous to nanocrystalline and polycrystalline across a variety of substrates. Representative crystalline morphologies of several archetypal molecules are shown in Figs. 2.9 and 2.21.

2.6 Advances in Device Architecture

The excitonic character of organic semiconductors has strongly dictated both the understanding and evolution of device architecture for OPVs. These activities can be broadly grouped into (1) modifications of active materials for enhanced exciton diffusion in a simple bilayer planar heterojunction and (2) modifications in the area of the dissociating interface using materials that are diffusion transport limited. In this section, we discuss recent progress along both lines of investigation.

2.6.1 Engineering Planar Heterojunction OPVs

The various processes responsible for photoconversion in a planar or bilayer heterojunction OPV were described in Section 2.1. These devices are typically limited by the unfavorable trade-off between the exciton diffusion length and the optical absorption length. Consequently, planar heterojunction OPVs are much thinner than would be optimum from an optical absorption perspective in order to make both exciton diffusion and charge collection more efficient. Planar heterojunction OPVs often show power conversion efficiencies that lag counterparts constructed from mixed or bulk heterojunctions. However, planar heterojunction OPVs continue to rightly receive attention for their potentially large charge collection efficiencies facilitated by a reduced rate of recombination for photogenerated carriers. In addition, the neat planar interface present in these devices may be attractive from the perspective of realizing long-term stability, reproducibility, yield, and device lifetimes. Consequently, there continues to be interest in the development of new materials and device architectures for realizing high efficiency in a planar heterojunction.

Among the techniques aimed at overcoming the short L_D that limits the efficiency of planar OPVs, the use of phosphorescent materials offers potential opportunities. While phosphorescent materials have been previously used in planar OPVs as a means to harvest the long-lived triplet exciton state for their potentially longer L_D, these approaches have not resulted in notably higher efficiency.[87,88] Such demonstrations are usually limited by the low absorption efficiency of the phosphor, and many of the demonstrations have utilized triplet materials with L_Ds of 10–20 nm still comparable to the best singlet materials (see Table 2.1). Work by Xu et al.,[89] Luhman et al.[59] and Rand et al.[90,91] demonstrated an alternative approach to the integration of phosphors in OPVs, utilizing a triplet-sensitized, or cascaded, architecture. In a phosphor-sensitized OPV device, absorption (and singlet formation) occurs on a fluorescent host donor. This initial absorption event is followed by energy transfer to a phosphorescent guest present in the donor layer at low concentration. Excitons on the phosphorescent guest then transfer to the triplet level of the host, permitting the population of the long-lived triplet state of a fluorescent material (Fig. 2.12). Using the combination of 4,4-bis[*N*-(1-naphthyl)-*N*-phenyl-amino]biphenyl (α-NPD):iridium, tris(2-phenylpyidine) [$Ir(ppy)_3$] as the donor layer and C_{60} as the acceptor, Luhman et al.[59] were able to show a 50% increase in the exciton diffusion length of the donor layer by transporting along the triplet state versus the singlet state (inferred from device data). This increase in L_D was also found to increase both the external quantum efficiency response from α-NPD. Rand et al.[90] and Xu et al.[89] have found similar behavior using polymer donor materials. In the former, the L_D of phenyl-substituted poly(*p*-phenylenevinylene) (PPV, Super Yellow) increased from (4 ± 1) nm to (9 ± 1) nm when the phosphor platinum octaethylporphyrin (PtOEP) was included in the film. In subsequent work, the viability of phosphor sensitization in polymer BHJs was demonstrated as well, relaxing the optimum domain size of the donor and acceptor phases.[91] Singlet fission, which has been shown to be efficient in a range of acene molecules,[29] could also provide another pathway to triplet sensitization that would minimize some of the inherent thermal losses in these approaches (see Section 2.3.2).

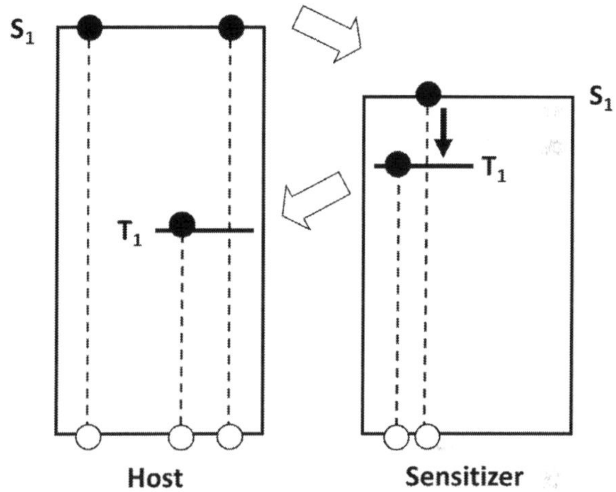

Figure 2.12 Host–guest energy level alignment for a phosphor-sensitized OPV. In this structure, optical excitation on the more concentrated host leads to the formation of singlet excitons. Host singlets are ultimately converted into host triplet excitons via energy transfer through a sensitizer species. Diffusion to the dissociating donor–acceptor interface ultimately occurs as a host triplet.

While successful in demonstrating the ability to engineer the exciton diffusion length in a planar heterojunction OPV, additional insight into energy-level design is needed for such techniques to realize the efficiencies observed in optimized bulk heterojunction OPVs. In addition, care must be taken to minimize any thermal losses that could limit optimal acceptor pairing for maximum open-circuit voltage.

2.6.2 Improving Active Layer Properties with Structural Templating

Recently, a number of groups have demonstrated and exploited the ability to template and orient polycrystalline vapor-deposited films through QE ordering (organic–organic and organic–inorganic) to improve active layer performance in OPVs. Organic semiconductors can be highly anisotropic. For example, many organic crystals exhibit order of magnitude higher charge mobilities, larger exciton diffusion lengths, and larger absorption coefficients in the cofacially stacked direction (direction of highest π–π overlap). Karl[92] has shown that the ratio of the mobilities in the b and c crystallographic directions (μ_b/μ_c), for example, are ~6 for holes in naphthalene, 1.5 for holes in anthracene, and 7 for electrons in perylene. Multiple groups have accordingly looked to exploit this anisotropy through structural templating during growth. Several examples for controlled donor orientation include copper phthalocyanine (CuPc),[93] chloroaluminumphthalocyanine (ClAlPc),[94] zinc phthalocyanine (ZnPc),[66] and lead phthalocyanine (PbPc)[95] where layers such as PTCDA,[96] diindenoperylene (DIP), CuI,[66,95] and highly ordered pyrolitc graphite (HOPG)[97] have been utilized to orient the donor crystal structures with the molecules cofacially aligned to the substrate (see Fig. 2.13). PTCDA was also utilized to similarly template the acceptor, F_{16}CuPc.[64,98] The co-facial orientation (closest packing direction) is often advantageous for OPV configurations since this orientation typically exhibits the highest charge carrier mobility,[92] exciton diffusion length,[51,64] and absorption coefficient.[93,96] Such control has led to enhancements in PV performance ranging from 10% to 100%. The first QE control of monolayers of phthalocyanines on ordered layers of PTCDA were demonstrated with OMBD by Forrest et al.[84] and extended to thicker layers in OVPD by Lunt et al.[78] where the growths were monitored with in-situ diffraction techniques. Sullivan et al.[96] subsequently showed that this oriented control of CuPc could be advantageously applied in CuPc-C_{60} heterojunctions. In their work, the CuPc contribution to the EQE was improved by roughly 60% for the flat-lying orientation versus the upright orientation, although the overall power conversion efficiency enhancement was limited to 10%, in part, due to the deep energy levels of PTCDA. This method was improved by Lassiter et al.[93] with a multilayer template that was initiated with PTCDA, followed by DIP which could support the QE ordering while simultaneously preventing the formation of a reverse junction between the PTCDA/CuPc. Lassiter et al. also showed an improvement for hybrid planar-mixed devices where the donor–acceptor were blended (co-evaporated) after an initial templated planar layer and the template was shown to impact (at least to some degree) the orientation of the blended region. This observation likely stems from partial phase segregation during the mixed growth that exposes additional donor molecules to the already oriented planar layer. Subsequently, this templating has been effectively shown for other planar-mixed systems including ClAlPc.[99] Simultaneous control over crystal orientation, domain size, and decoupling of the QE templating from the ability to form Ohmic contacts (without reverse junctions) remains an important challenge that can offer a simple and effective route to reliable performance enhancement in both planar and BHJ structures.

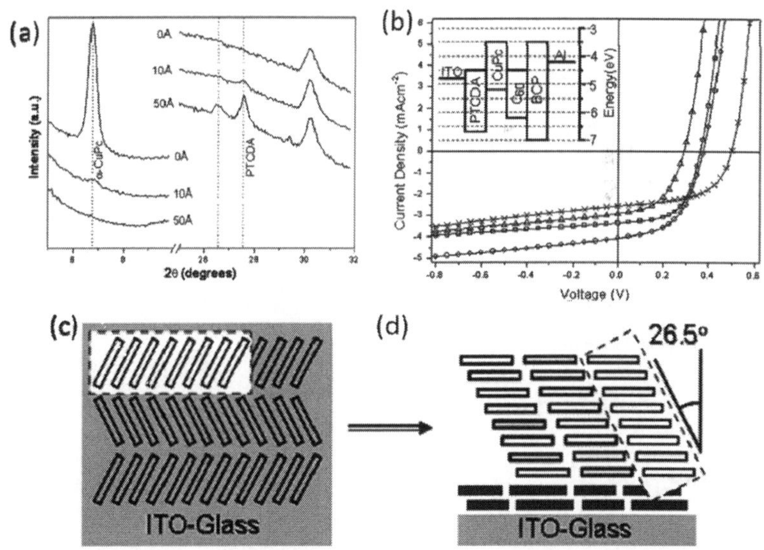

Figure 2.13 Structural templating of a CuPc-C_{60} OPV. (a) XRD, (b) current–voltage curves, (c) schematic diagram of CuPc orientation on ITO, and (d) templated with PTCDA. Adapted, with permission, from Ref. [96].

2.6.3 Engineered Nanostructure for Enhanced Dissociation—Bulk Heterojunction OPVs

The positive attributes of planar structures make the push to resolve and overcome the short exciton diffusion length a meaningful and potentially worthwhile line of investigation. In this section, we turn our focus to a widely practiced solution to the problem of poor exciton diffusion, namely, the use of a BHJ.

The core concept behind such structures is to maximize the interface area between the electron donating and accepting phases while attempting to preserve pathways for the collection of dissociated charge carriers. A BHJ can consist of a simple, uniform mixture of the donor and acceptor phases or, could be considerably more complex, involving phase separation of crystalline, amorphous, neat, and miscible domains. Well-optimized BHJs can lead to a near unity efficiency of exciton diffusion and dissociation. In all such systems, the key challenge is the need to simultaneously optimize the exciton diffusion and charge collection efficiencies.

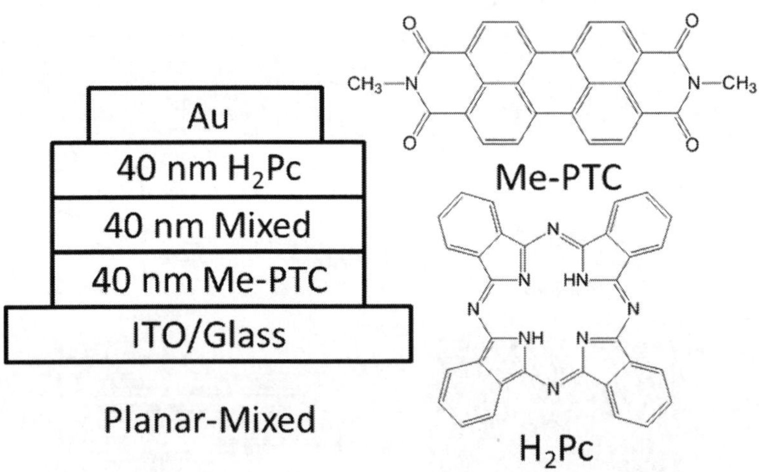

Figure 2.14 Planar-mixed architecture developed by Hiramoto et al. Here, the mixed layer is used to dissociate excitons created both in the mixture and also in the adjacent planar layers.

For small molecules, early work on uniformly mixed donor–acceptor heterojunctions was carried out by Hiramoto et al.[100] These devices consisted (see Fig. 2.14) of a uniformly mixed layer of electron-donating metal-free phthalocyanine (H_2Pc) and an electron-accepting perylene derivative sandwiched between neat layers of each material. This type of hybrid planar-mixed architecture is designed to offer a balance between the attractive attributes of the mixture (high exciton diffusion efficiency) with those of a planar architecture (low leakage current, low recombination). In this device, all three layers are optically active with exciton dissociation occurring in the uniformly mixed layer. Dissociated charge carriers diffuse out of the mixture and are transported along the neat donor and acceptor layers to the electrodes.

A key challenge in mixed devices is the observation of low fill factors. The fill factor is limited in a mixture by a combination of increased bimolecular (non-geminate) electron–hole recombination and a reduction of the charge-carrier mobilities upon dilution. As such, substantial efforts have gone into improving both of these parameters through the use of optimized film crystallinity and morphology, and the use of high conductivity transport layers to increase carrier mobilities. Here, we highlight a variety of these architectural enhancements that offer performance that exceeds that of planar heterojunction OPVs discussed above.

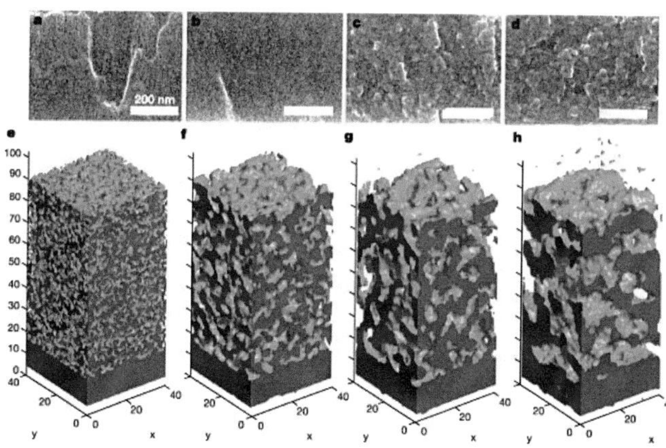

Figure 2.15 (a) Scanning electron micrograph (SEM) for a sample with no annealing. (b–d) SEMs for samples annealed for (b), 15 min at 450 K, (c) 500 K, and (d) 550 K. (e–h) Simulated effects of annealing on the interface morphology of a CuPc-PTCBI mixture. The donor–acceptor interface between is shown in green, CuPc is shown in red, and PTCBI is left transparent. The vertical axis represents film thickness. Adapted with permission from Ref. [101].

An early approach by Peumans et al.[101] focused on optimizing the degree of donor–acceptor phase separation to maximize both the exciton diffusion and charge collection efficiencies. In this work, mixed donor–acceptor films of CuPc-PTCBI were thermally annealed as a means to induce phase separation and realize a bi-continuous, BHJ morphology. With annealing, the CuPc donor was found to crystallize, leading to an increase in the FF and power conversion efficiency of the OPV. Figure 2.15a–d illustrates the change in morphology observed in this work through the use of cross-section scanning electron microscopy. Figure 2.15e–h shows simulations of the film morphology, illustrating the phase separation observed upon thermal annealing.

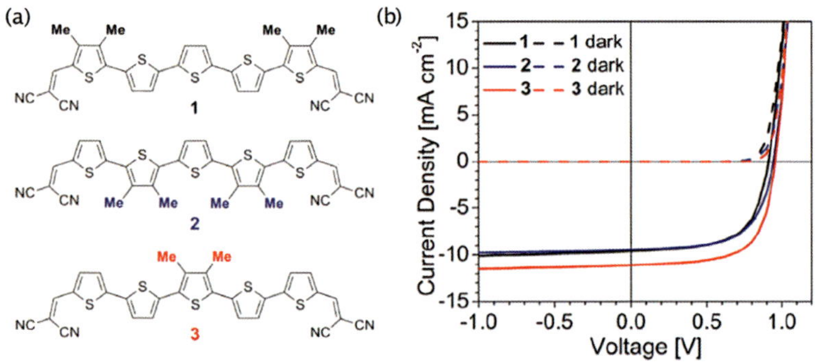

Figure 2.16 (a) Molecular structures for DCV5T-Me donor materials. (b) Current-density voltage characteristics for OPVs constructed using the molecules in (a) with a C_{60} acceptor. Adapted with permission from Ref. [102].

Work to further enhance the charge collection efficiency and FF in mixed and planar mixed-type OPVs has focused on enhancing film crystallinity as a means to improve charge transport. Recent work from Fitzner and colleagues has combined this approach with the use of charge-transporting layers that are doped for enhanced conductivity.[102,103] As an example of this approach, Fig. 2.16a shows a series of methyl-substituted dicyanovinyl (DCV)-capped A–D–A quinquethiophenes (DCV5T-Me). In Fig. 2.16b, current-density–voltage characteristics for devices constructed using DCV5T-Me donors under simulated solar illumination are shown. The structures for these devices are ITO/C_{60} (15 nm)/donor:C_{60} (30 nm, 2:1 by volume, co-evaporated at 90 °C substrate temperature)/BPAPF (5 nm)/BPAPF:NDP9 (50 nm, 90:10 by volume)/NDP9 (1 nm)/Au (50 nm). Here, the triarylamine BPAPF is p-doped with 10 wt.% NDP9 to form a hole transport layer. Devices based on compounds 1 and 2 are very similar in terms of V_{OC} (>0.9 V), J_{SC}, and FF (>0.62), leading to a power conversion efficiency of 4.8%. Devices based on compound 3 show enhanced J_{SC} and FF (0.66), leading to a power conversion efficiency of 6.1%. Further enhancements in the design of hole- and electron-transporting layers have permitted compound 3 to realize peak efficiencies of 6.9% under 1 sun illumination.

Interestingly, a complementary examination (by grazing incidence x-ray diffraction, GIXD) of crystallinity in blends of the compounds in Fig. 2.16a has shown some correlation between donor crystallite size in the blend and cell performance. In particular, compound 3 is characterized by the largest crystallites (>40 nm), suggesting that the associated D–A film morphology is more amenable to the formation of percolation pathways for charge carrier collection. This important result highlights the potential for targeted changes in molecular structure to affect changes in film morphology and ultimately, OPV performance.

In addition to work described above that combines the use of planar and mixed organic active layers, substantial investigation has also gone into devices that rely on stand-alone mixtures for photoconversion. While such structures have been explored for a number of years, recent variations on this theme have brought very interesting insights to the field of small molecule OPVs. Of important note is recent work from multiple groups indicating that mixed OPVs show peak efficiency when the composition is very acceptor-rich, particularly when utilizing fullerene acceptors. At the extreme end of the spectrum is work by Zhang et al.[104] where mixed heterojunction OPVs with a power efficiency exceeding 5% are realized using the donor–acceptor pairing of di-[4-(N,N-ditolyl-amino)-phenyl] cyclohexane (TAPC)-C_{70} (see Fig. 2.17). In this work, the authors show that essentially only the C_{70} contributes the photocurrent, and that high performance relies on the formation of a pseudo Schottky junction between the ITO/MoO_x anode and the bulk heterojunction. In this way, the authors demonstrate that many donor materials paired with C_{70} could offer similar performance provided the donor concentration is kept low.

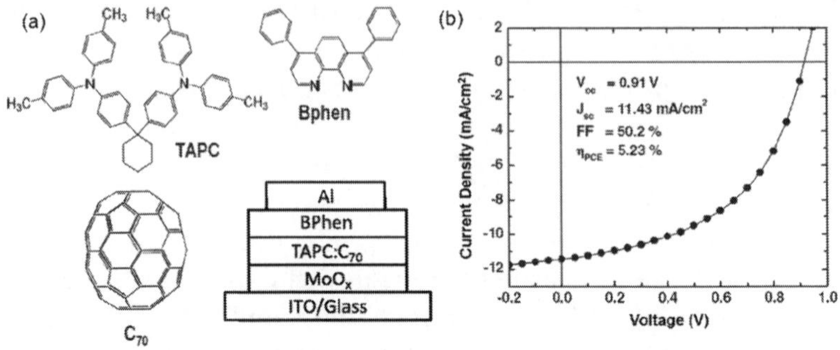

Figure 2.17 (a) Active materials for the study of acceptor-rich mixed heterojunction OPVs where C_{70} is the primary contributor to the photoresponse. (b) Current density–voltage characteristics for the device in (a). Adapted with permissionfrom Ref. [104].

This work followed previous efforts from the same group examining the performance of MoO_x/C_{60} Schottky junctions as a means to realize efficient photoconversion in a simple planar device architecture.[105] In this work, Zhang et al. found that an efficient device could in fact be constructed but that its performance was very sensitive to the processing of the MoO_x, and specifically to the oxidation state of Mo. These works suggest not only that it is possible to realize efficient photoconversion in a Schottky cell, but also that the MoO_x interlayer could play a much more active role in photocurrent generation than is often assumed.

Similar works by Pandey et al. have also examined the dependence of OPV performance in a mixed heterojunction on composition. The behavior of the SubPc-C_{60}[106] and SubPc-C_{70}[107] systems is similar to those discussed above for the TAPC-C_{70} system in that both are optimized for high acceptor compositions of about 80 wt.% fullerene (Fig. 2.18). Interestingly, for the SubPc-C_{60} system, the hole mobility is noted to increase with the addition of C_{60}, reflecting an underlying change in microstructure and the crystallization of SubPc upon C_{60} blending. This change in charge transport leads to enhancement of the fill factor, short-circuit current, and power conversion efficiency at a composition of 80 wt.% C_{60}. Separate investigations using a combination of transmission electron microscopy (TEM) and selected-area electron diffraction (SAED) have shown that SubPc crystallizes in films containing 80 wt.% C_{60}, resulting in improved hole transport and device efficiency (Fig. 2.18c).[106] A similar optimum has also been found for the SubPc-C_{70} system with co-deposited bulk heterojunction OPVs showing power conversion efficiencies >4.5%.[107] However, in contrast to the work reported by Zhang et al. for TAPC-C_{70}, devices based on SubPc-C_{70} show very clear and efficient photoresponse from both SubPc and C_{70}.

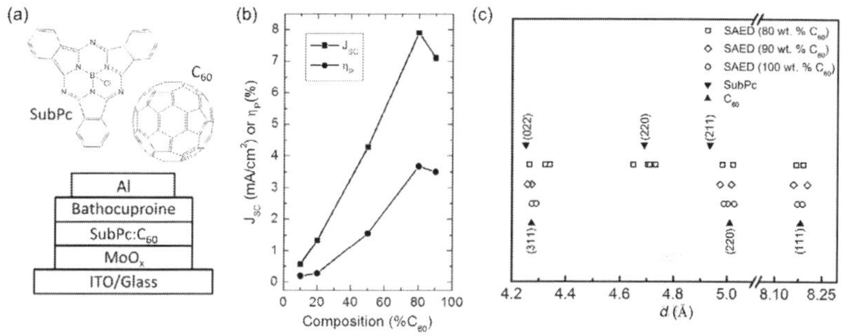

Figure 2.18 (a) Device architecture for an OPV based on uniform mixtures of SubPc and C_{60}. (b) Dependence of J_{SC} and the power conversion efficiency (η_P) on film composition. (c) Selected area electron diffraction (SAED) maxima observed for SubPc-C_{60} films of varying composition. Scattering from the (220) plane of orthorhombic SubPc is observed in films containing 80 wt.% C_{60}, suggesting some degree of SubPc crystallization. Adapted with permission from Ref. [106].

The results described above for the SubPc-C_{60} system reflect a significant change in the nanoscale crystallinity for SubPc with changes in composition. Just as is the case in polymer bulk heterojunctions where active materials are often annealed (thermal or solvent) as a means to induce phase separation, similar attempts at phase separation in small-molecule mixtures are often attempted through variations in composition. When optimized, such "nanocrystalline" OPVs can show high power conversion efficiencies.

While polymer bulk heterojunction OPVs can often exhibit crystalline domains that help in efficient charge collection, such structures are often also characterized by substantial vertical phase separation upon annealing. This vertical phase separation can be beneficial as it limits the magnitude of the dark current (by leakage) sustaining the V_{OC} of the cell. As such, the vertical composition of many polymer BHJ OPVs could typically be thought of more as a gradient than a uniform mixture. Given the ability to closely control the composition of mixtures of small-molecule materials during film deposition, there have been several attempts to realize a similar composition profile during the growth process. Early work on graded composition OPVs focused on the use of "step-gradient" composition profiles in which the gradient was constructed from a series of thin mixtures of varying composition.[108–110]

Recently, work by Pandey et al. has focused on the use of designed, continuous graded composition heterojunctions (GHJs) as a means to optimize the spatial composition of the OPV for high efficiency (Fig. 2.19a).[107,111,112] In particular, work combining SubPc with an acceptor layer of either C_{60} or C_{70} has been promising, with power conversion efficiencies of 4.5% and 5.4% for each acceptor, respectively (Fig. 2.19b).[107,112] The external quantum efficiencies of devices based on C_{70} are ~80% across most of the visible portion of the electromagnetic spectrum (Fig. 2.19c). In optimized systems, the GHJ has been shown to offer both enhanced charge collection[113] and absorption efficiencies[114] relative to uniformly mixed counterparts. More recent work by Lassiter et al.[115] has implemented the GHJ approach in tandem OPVs, combining a vapor-deposited SubPc-C_{70} GHJ with a solution-processed squaraine OPV to realize a power efficiency >6%.

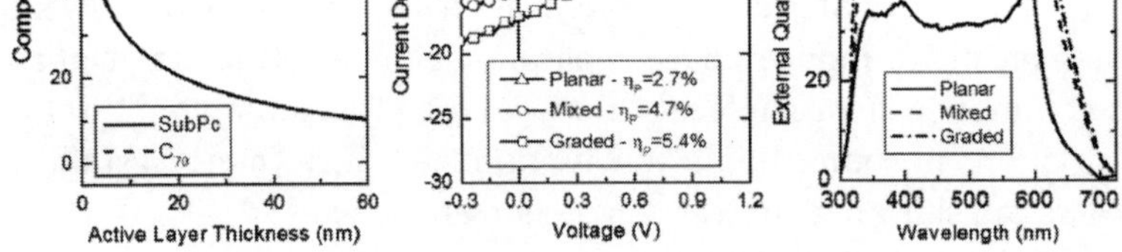

Figure 2.19 (a) Optimum composition profile for a graded heterojunction OPV based on the donor–acceptor pairing of SubPc-C_{70}. (b) Current density–voltage characteristics under AM1.5G simulated solar illumination for optimized planar (bilayer), uniformly mixed and graded heterojunction OPVs. (c) The external quantum efficiency for the devices in (b). Adapted with permission from Ref. [107].

Finally, the planar-mixed architecture first popularized by Hiramoto has been combined with the development of new active materials to realize high power conversion efficiency. One promising series of donor molecules are donor–acceptor–acceptor (D–A–A) compounds developed by the group of K.-T. Wong and described recently by Chiu et al.[116] Figure 2.20 shows an example of a novel donor featuring an electron-donating ditolylaminothienyl moiety connected to electron-accepting pyrimidine and dicyanovinylene blocks, here named DTDCTP.[116–118] This type of structure allows for significant tuning of the optical absorption spectrum and also of the position of the HOMO and LUMO of the molecule. In this way, D–A–A systems like DTDCTP offer the optimization handles needed to simultaneously maximize J_{SC} and V_{OC}.

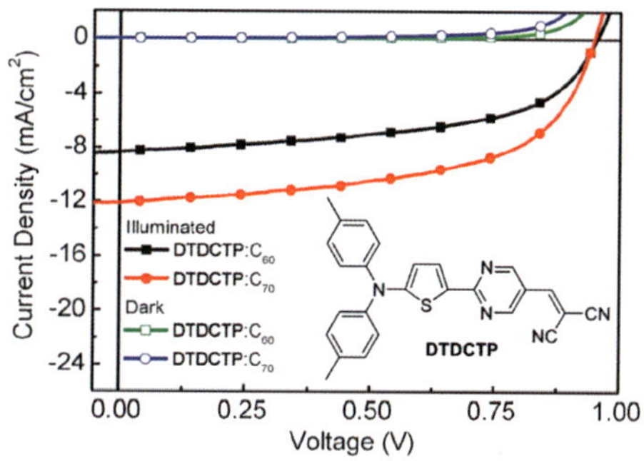

Figure 2.20 Current density–voltage characteristics for planar-mixed OPVs based on a donor–acceptor–acceptor type electron donor (DTDCTP) developed by Chiu et al. and an acceptor layer of either C_{60} or C_{70}. Figure adapted from Refs. [116,117] with permission.

The data shown in Fig. 2.20 were collected for planar-mixed devices having the following structures, ITO/MoO_x (20 nm)/DTDCTP (7 nm)/DTDCTP:C_{60} (1:1, by volume, 40 nm)/C_{60} (20 nm)/BCP (10 nm)/Ag and ITO/MoO_x (20 nm)/DTDCTP (7 nm)/DTDCTP:C_{70} (1:1, by volume, 40 nm)/C_{70} (6 nm)/BCP(10 nm)/Ag. Devices using a C_{60} acceptor show a V_{OC} of 0.95 V, J_{SC} = 8.3 mA/cm^2, FF = 0.54, leading to a power conversion efficiency of 4.3% under AM 1.5G, 1 sun illumination. Devices based on C_{70} show markedly higher efficiencies due to enhanced optical absorption on C_{70} relative to C_{60}. Specifically, DTDCTP:C_{70} showed a power conversion efficiency of 6.4% with a V_{OC} of 0.95 V, J_{SC} = 12.1 mA/cm^2, and FF = 0.56.

This efficiency is among the highest reported to date for a simple planar-mixed device constructed using thermally evaporated small molecule materials. Indeed, in addition to new OPV architectures and deposition routes, novel material development will likely continue to drive significant advances in the field.

2.6.4 Vapor Deposition Routes to Induced Bulk-Heterojunction Formation

A number of vapor deposition techniques have been employed to increase the active interface area and nanostructured morphology directly during vapor deposition beyond direct coevaporation blending discussed above.

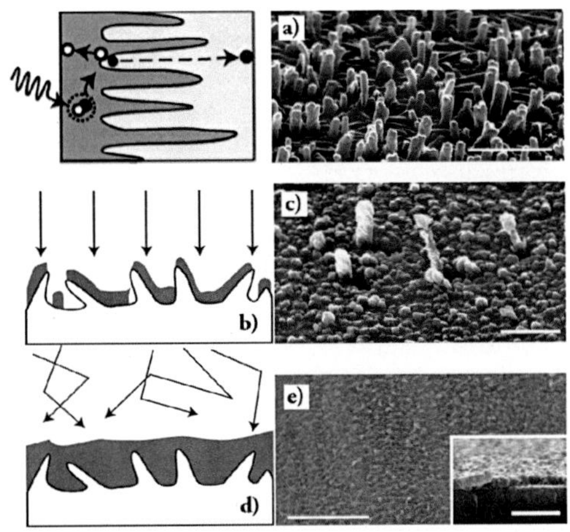

Figure 2.21 Demonstrated morphology control of high interface area CuPc-PTCBI heterojunctions grown via OPVD. From Ref. [121] with permission.

One successful approach has been the use of OVPD described in Section 2.5.2. Among the early demonstrations of OPVs constructed using OVPD, Yang et al.[119–121] demonstrated high-efficiency crystalline OPVs consisting of heterojunctions of both CuPc/PTCBI and CuPc/C_{60} grown by OVPD. BHJ structures of CuPc/PTCBI were formed by initially controlling the CuPc morphology to grow crystalline nanowires (see Fig. 2.21). PTCBI was then deposited above the CuPc, where the diffusive nature of OVPD allowed for complete filling of the CuPc nanowire mesh layer. A three-fold increase in power conversion was observed for the crystalline BHJ as compared to planar VTE control cells. Further demonstration of nanocrystalline BHJs were grown by OVPD for CuPc/C_{60}. By switching between each source material during growth, nanocrystalline formation was promoted at the substrate. An optimum device consisted of six pairs of alternating CuPc/C_{60}, and exhibited a power efficiency of ~5%.[119] These results highlighted the morphology control afforded by OVPD.

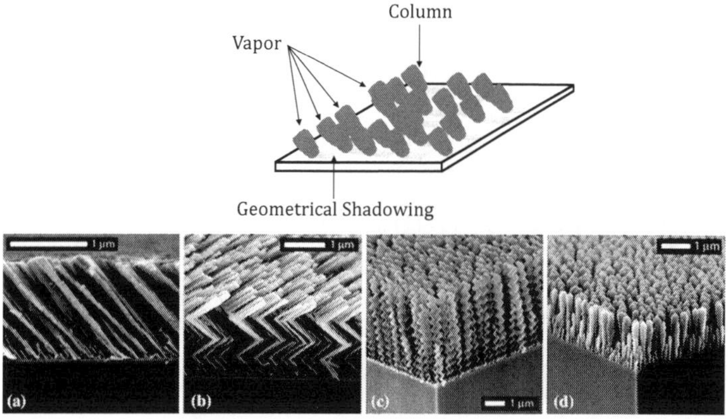

Figure 2.22 (Top) Schematic of glancing angle deposition. (bottom, a–d) Example morphologies accessible to glancing angle deposition from titled arrays (a), zig-zags (b), helical arrays (c), and straight nanowires (d). Each morphology is generated during vapor deposition by fixing, or altering, the flux orientation during growth. Adapted with permission from Ref. [75].

Another approach that has recently been employed is the use of oblique angle vapor deposition or glancing angle deposition (GLAD).[75,122,123] In this method, the incident vapor flux is directed toward the substrate at non-normal incidence, inducing molecular-scale self-shadowing that results in titled columnar structures that can result in a range of controllable nanostructures from tilted arrays to helical "wires" as shown in Fig. 2.22 Li et al. utilized this deposition scheme for ClAlPc-C_{60} heterojunctions, where ClAlPc was grown at various incidence angles (0–60°) and subsequently filled with C_{60} at normal incidence.[124] By structuring the donor as titled columns with a width on the length scale of the diffusion length, the photocurrent was shown to increase 30%–50% with a corresponding increase in the efficiency from

2.0 ± 0.1% to 2.8 ± 0.1%. The resulting change in the spectral response is shown in Fig. 2.23. The authors demonstrated a similar effect with CuPc as the donor. Zheng et al. also employed this technique with CuPc to show similar enhancements with GLAD-grown CuPc nanorods that were subsequently filled with solution-deposited PCBM.[125]

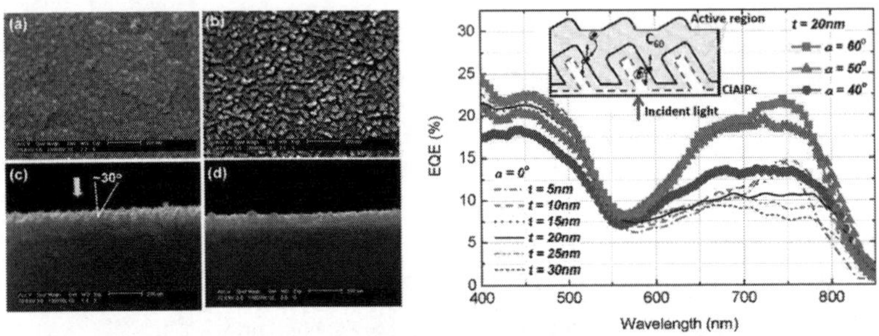

Figure 2.23 Example ClAlPc morphologies (left, a–d) grown via glancing angle deposition and the subsequent impact on the spectral external quantum efficiency (right) of these layers incorporating into heterojunction devices. Reproduced with permission from Ref. [124].

2.7 Conclusions

Despite limitations, small-molecule organic solar cells are steadily making gains in increasing efficiency beyond 10%. In this chapter, we reviewed the methods of small-molecule photovoltaic growth, and operation, with particular focus on methods for overcoming the exciton diffusion bottleneck through (1) novel architectures, (2) exploitation of growth phenonema, and (3) materials design. Because of the inherent anisotropy of organic molecules, the determination of the anisotropic properties and control over orientation, morphology and photophysical performance is paramount in the ultimate optimization of these devices. While power conversion efficiencies are likely to approach 15–18% within the next decade, the ability of vapor deposited small-molecule OPV to penetrate solar energy markets will ultimately depend critically on parameters beyond just efficiency such as operational lifetime and cost.

Acknowledgement

RRL acknowledges funding support from the National Science Foundation (CBET-1254662) and RJH acknowledges support from the National Science Foundation (DMR-1006566, CBET-1067681, DMR-1307066).

References

1. Tang CW (1986) Two-layer organic photovoltaic cell. *Appl Phys Lett,* **48**, 183.
2. Yu G, Gao J, Hummelen JC, Wudl F, Heeger AJ (1995) Polymer photovoltaic cells—enhanced efficiencies via a network of internal donor–acceptor heterojunctions. *Science,* **270**, 1789.
3. Burgelman MP, Arkipov JV (2007) (eds), *Thin Film Solar Cells Fabrication, Characterization and Applications,* John Wiley and Sons, Ltd, Hoboken, NJ.
4. Wang JC, Ren XC, Shi SQ, Leung CW, Chan PKL (2011) Charge accumulation induced S-shape *J–V* curves in bilayer heterojunction organic solar cells. *Org Electron,* **12**, 880.
5. Giebink NC, Wiederrecht GP, Wasielewski MR, Forrest SR (2010) Ideal diode equation for organic heterojunctions. I. Derivation and application. *Phys Rev B,* **82**, 155305.
6. Brabec CJ, Cravino A, Meissner D, et al. (2001) Origin of the open circuit voltage of plastic solar cells. *Adv Funct Mater,* **11**, 374.
7. Rauh D, Wagenpfahl A, Deibel C, Dyakonov V (2011) Relation of open circuit voltage to charge carrier density in organic bulk-heterojunction solar cells. *Appl Phys Lett,* **98**, 133301.
8. Brown PR, Lunt RR, Zhao N, et al. (2011) Improved current extraction from ZnO/PbS quantum dot heterojunction photovoltaics using MoO_3 interfacial layer. *Nano Lett,* **11**, 2955.
9. Wagner J, Gruber M, Wilke A, et al. (2012) Identification of different origins for s-shaped current voltage characteristics in planar heterojunction organic solar cells. *J Appl Phys,* **111**, 054509.
10. Zhang ML, Wang H, Tang CW (2011) Hole transport limited S-shape *I–V* curves in organic photovoltaic cells. *Appl Phys Lett,* **99**, 213506.
11. Lo MF, Ng TW, Liu TZ, et al. (2010) Limits of open circuit voltage in organic photovoltaic devices. *Appl Phys Lett,* **96**, 113303.

12. Frohne H, Shaheen SE, Brabec CJ, et al. (2002) Influence of the anodic work function on the performance of organic solar cells. *ChemPhysChem,* **3**, 795.
13. Tong XR, Lassiter BE, Forrest SR (2010) Inverted organic photovoltaic cells with high open-circuit voltage. *Org Electron,* **11**, 705.
14. Kang H, Hong S, Lee J, Lee K (2012) Electrostatically self-assembled nonconjugated polyelectrolytes as an ideal interfacial layer for inverted polymer solar cells. *Adv Mater,* **24**, 3005.
15. Scharber MC, Wuhlbacher D, Koppe M, et al. (2006) Design rules for donors in bulk-heterojunction solar cells—towards 10% energy-conversion efficiency. *Adv Mater,* **18**, 789.
16. Rand BP, Burk DP, Forrest SR (2007) Offset energies at organic semiconductor heterojunctions and their influence on the open-circuit voltage of thin-film solar cells. *Phys Rev B,* **75**, 115327.
17. Gadisa A, Svensson M, Andersson MR, Inganas O (2004) Correlation between oxidation potential and open-circuit voltage of composite solar cells based on blends of polythiophenes/fullerene derivative. *Appl Phys Lett,* **84**, 1609.
18. Veldman D, Meskers SCJ, Janssen RAJ (2009) The energy of charge-transfer states in electron donor–acceptor blends: Insight into the energy losses in organic solar cells. *Adv Funct Mater,* **19**, 1939.
19. Minnaert B, Burgelman M (2007) Efficiency potential of organic bulk heterojunction solar cells. *Prog Photovolt Res Appl,* **15**, 741.
20. Lunt RR, Osedach TP, Brown PR, Rowehl JA, Bulović V (2011) Practical roadmap and limits to nanostructured photovoltaics. *Adv Mater,* **23**, 5712–5727.
21. Giebink NC, Wiederrecht GP, Wasielewski MR, Forrest SR (2011) Thermodynamic efficiency limit of excitonic solar cells. *Phys Rev B,* **83**, 195326.
22. Kalinowski J (2005) *Organic Light-Emitting Diodes: Principles, Characteristics, and Processes,* Marcel Dekker, New York.
23. Arbogast JW, Darmanyan AP, Foote CS, et al. (1991) Photophysical Properties of C_{60}. *J Phys Chem,* **95**, 11.
24. Colligiani A, Taliani C (1994) ESR study on the doublet and triplet species present in pristine C_{60} fullerene powder. *Chem Mater,* **6**, 1633.
25. Turro NJ (1991) *Modern Molecular Photochemistry,* University Science Books, Sausalito, CA.

26. Beard MC (2011) Multiple exciton generation in semiconductor quantum dots. *J Phys Chem Lett,* **2**, 1282.

27. Nozik AJ (2008) Multiple exciton generation in semiconductor quantum dots. *Chem Phys Lett,* **457**, 3.

28. Hanna MC, Nozik AJ (2006) Solar conversion efficiency of photovoltaic and photoelectrolysis cells with carrier multiplication absorbers. *J Appl Phys,* **100**, 074510.

29. Smith MB, Michl J (2010) Singlet fission. *Chem Rev,* **110**, 6891.

30. Singh S, Jones WJ, Siebrand W, Stoicheff BP, Schneider WG (1965) Laser generation of excitons and fluorescence in anthracene crystals. *J Chem Phys,* **42**, 330.

31. Swenberg CE, Stacy WT (1968) Biomolecular radiationless transitions in crystalline tetracene. *Chem Phys Lett,* **2**, 327.

32. Thorsmolle VK, Averitt RD, Demsar J, et al. (2009) Morphology effectively controls singlet–triplet exciton relaxation and charge transport in organic semiconductors. *Phys Rev Lett,* **102**, 017401.

33. Roberts ST, McAnally RE, Mastron JN, et al. (2012) Efficient singlet fission found in a disordered acene film. *J Am Chem Soc,* **134**, 6388.

34. Zimmerman PM, Zhang ZY, Musgrave CB (2010) Singlet fission in pentacene through multi-exciton quantum states. *Nat Chem,* **2**, 648.

35. Potscavage WJ, Yoo S, Domercq B, Kippelen B (2007) Encapsulation of pentacene/C_{60} organic solar cells with Al_2O_3 deposited by atomic layer deposition. *Appl Phys Lett,* **90**, 253511.

36. Signerski R, Jarosz G, Godlewski J (2004) Photoelectric properties of tetracene-pentacene heterojunction. *Macromol Symp,* **212**, 369.

37. Signerski R, Jarosz G, Godlewski J (1998) Photoelectric properties of heterojunctions formed from di-(pyridyl)-perylenetetracarboxylic diimide and copper phthalocyanine or pentacene. *Synth Met,* **94**, 135.

38. Chu CW, Shao Y, Shrotriya V, Yang Y (2005) Efficient photovoltaic energy conversion in tetracene-C_{60} based heterojunctions. *Appl Phys Lett,* **86,** 243506.

39. Signerski R, Kalinowski J, Koropecky I, Nespurek S (1984) Injection photoconductivity and photovoltage in metal tetracene metal sandwich system. *Thin Solid Films,* **121**, 175.

40. Taima T, Sakai J, Yamanari T, Saito K (2006) Realization of large open-circuit photovoltage in organic thin-film solar cells by controlling measurement environment. *Jpn J Appl Phys,* **45**, L995.

41. Jadhav P, Mohanty A, Sussman J, Lee J, Baldo MA (2011) Singlet exciton fission in nanostructured organic solar cells. *Nano Lett,* **11**, 1495.
42. Lee J, Jadhav P, Baldo MA (2009) High efficiency organic multilayer photodetectors based on singlet exciton fission. *Appl Phys Lett,* **95**, 033301.
43. Ehrler B, Wilson MWB, Rao A, Friend RH, Greenham NC (2012) Singlet exciton fission-sensitized infrared quantum dot solar cells. *Nano Lett,* **12**, 1053.
44. Pope M, Swenberg CE (1982) *Electronic Processes in Organic Crystals and Polymers,* Oxford University Press, New York.
45. Davidovich MA, Knox RS (1979) On the rate of triplet excitation transfer in the diffusive limit. *Chem Phys Lett,* **68**, 391.
46. Dexter DL (1953) A theory of sensitized luminescence in solids. *J Chem Phys,* **21**, 836.
47. Lunt RR (2009) Ph.D. Thesis, Princeton University, Princeton, NJ.
48. Scholes GD, Jordanides XJ, Fleming GR (2001) Adapting the Förster theory of energy transfer for modeling dynamics in aggregated molecular assemblies. *J Phys Chem B,* **105**, 1640.
49. Wiesenhofer H, Beljonne D, Scholes GD, Hennebicq E, Bredas JL, Zojer E (2005) Limitations of the Förster description of singlet exciton migration: The illustrative example of transfer to ketonic defects in ladder-type poly(para-phenylenes). *Adv Funct Mater,* **15**, 155.
50. Beljonne D, Curutchet C, Scholes GD, Silbey RJ (2009) Beyond Förster resonance energy transfer in biological and nanoscale systems. *J Phys Chem B,* **113**, 6583.
51. Lunt RR, Giebink NC, Belak AA, Benziger JB, Forrest SR (2009) Exciton diffusion lengths of organic semiconductor thin films measured by spectrally resolved photoluminescence quenching. *J Appl Phys,* **105**, 053711.
52. Rim SB, Peumans P (2008) The effects of optical interference on exciton diffusion length measurements using photocurrent spectroscopy. *J Appl Phys,* **103**, 124515.
53. Eremenko VV, Medvedev VS (1961) *Sov Phys—Solid State,* **2**, 1426.
54. Steketee JW, Dejonge J (1962) *Philips Res Rep,* **17**, 363.
55. Mulder BJ (1966) *Philips Res Rep,* **21**, 283.
56. Lunt RR, Bulovic V (2011) Transparent, near-infrared organic photovoltaic solar cells for window and energy-scavenging applications. *Appl Phys Lett,* **98**, 113305.

57. Terao Y, Sasabe H, Adachi C (2007) Correlation of hole mobility, exciton diffusion length, and solar cell characteristics in phthalocyanine/ fullerene organic solar cells. *Appl Phys Lett,* **90**, 103515.

58. Choukri H, Fischer A, Forget S, et al. (2006) White organic light-emitting diodes with fine chromaticity tuning via ultrathin layer position shifting. *Appl Phys Lett,* **89**, 183513.

59. Luhman WA, Holmes RJ (2009) Enhanced exciton diffusion in an organic photovoltaic cell by energy transfer using a phosphorescent sensitizer. *Appl Phys Lett,* **94**, 153304.

60. Gommans H, Verreet B, Rand BP, et al. (2008) On the role of bathocuproine in organic photovoltaic cells. *Adv Funct Mater,* **18**, 3686.

61. Luhman WA, Holmes RJ (2011) Investigation of energy transfer in organic photovoltaic cells and impact on exciton diffusion length measurements. *Adv Funct Mater,* **21**, 764.

62. Lunt RR, Benziger JB, Forrest SR (2010) Relationship between crystalline order and exciton diffusion length in molecular organic semiconductors. *Adv Mater,* **22**, 1233.

63. Peumans P, Yakimov A, Forrest SR (2003) Small molecular weight organic thin-film photodetectors and solar cells. *J Appl Phys,* **93**, 3693.

64. Rim SB, Fink RF, Schoneboom JC, Erk P, Peumans P (2007) Effect of molecular packing on the exciton diffusion length in organic solar cells. *Appl Phys Lett,* **91**, 173504.

65. Huijser A, Savenije TJ, Kotlewski A, Picken SJ, Siebbeles LDA (2006) Efficient light-harvesting layers of homeotropically aligned porphyrin derivatives. *Adv Mater,* **18**, 2234.

66. Rand BP, Cheyns D, Vasseur K, et al. (2012) The impact of molecular orientation on the photovoltaic properties of a phthalocyanine/ fullerene heterojunction. *Adv Funct Mater,* **22**, 2987.

67. Zhou YC, Wu Y, Ma LL, et al. (2006) Exciton migration in organic thin films. *J Appl Phys,* **100**, 023712.

68. Menke SM, Luhman WA, Holmes RJ (2013) Tailored exciton diffusion in organic photovoltaic cells for enhanced power conversion efficiency. *Nat Mater,* **12**, 152.

69. Mullenbach TK, McGarry KA, Luhman WA, Douglas CJ, Holmes RJ (2013) Connecting molecular structure and exciton diffusion length in rubrene derivatives. *Adv Mater,* **25**, 3689.

70. Scully SR, Armstrong PB, Edder C, Frechet JMJ, McGehee MD (2007) Long-range resonant energy transfer for enhanced exciton harvesting for organic solar cells. *Adv Mater,* **19**, 2961.

71. Coffey DC, Ferguson AJ, Kopidakis N, Rumbles G (2010) Photovoltaic charge generation in organic semiconductors based on long-range energy transfer. *ACS Nano,* **4**, 5437.
72. Salzman RF, Xue JG, Rand BP, Alexander A, Thompson ME, Forrest SR (2005) The effects of copper phthalocyanine purity on organic solar cell performance. *Org Electron,* **6**, 242.
73. Lunt RR, Lassiter BE, Benziger JB, Forrest SR (2009) Organic vapor phase deposition for the growth of large area organic electronic devices. *Appl Phys Lett,* **95**, 233305.
74. Shtein M (2004) Ph.D. Thesis, Princeton University, Princeton, NJ.
75. Granqvist CG (2012) Preparation of thin films and nanostructured coatings for clean tech applications: A primer. *Sol Energy Mater Sol Cells,* **99**, 166.
76. Rolin C, Vasseur K, Genoe J, Heremans P (2010) Growth of pentacene thin films by in-line organic vapor phase deposition. *Org Electron,* **11**, 100.
77. Forrest SR (1997) Ultrathin organic films grown by organic molecular beam deposition and related techniques. *Chem Rev,* **97**, 1793.
78. Lunt RR, Benziger JB, Forrest SR (2007) Growth of an ordered crystalline organic heterojunction. *Adv Mater,* **19**, 4229.
79. Lunt RR, Sun K, Kroger M, Benziger JB, Forrest SR (2011) Ordered organic–organic multilayer growth. *Phys Rev B,* **83**, 064114.
80. Hillier AC, Ward MD (1996) Epitaxial interactions between molecular overlayers and ordered substrates. *Phys Rev B,* **54**, 14037.
81. Hooks DE, Fritz T, Ward MD (2001) Epitaxy and molecular organization on solid substrates. *Adv Mater,* **13**, 227.
82. Grand JY, Kunstmann T, Hoffmann D, et al. (1996) Epitaxial growth of copper phthalocyanine monolayers on Ag(111). *Surf Sci,* **366**, 403.
83. Lukas S, Sohnchen S, Witte G, Woll C (2004) Epitaxial growth of pentacene films on metal surfaces. *ChemPhysChem,* **5**, 266.
84. Forrest SR, Burrows PE, Haskal EI, So FF (1994) Ultrahigh-vacuum quasiepitaxial growth of model van der Waals thin films. II. Experiment. *Phys Rev B,* **49**, 11309.
85. Karl N (1989) Organic semiconductors: Purification and crystal growth. *Mol Cryst Liq Cryst,* **171**, 157.
86. Tseng RJ, Chan R, Tung VC, Yang Y (2008) Anisotropy in organic single-crystal photovoltaic characteristics. *Adv Mater,* **20**, 435.

87. Shao Y, Yang Y (2005) Efficient organic heterojunction photovoltaic cells based on triplet materials. *Adv Mater,* **17**, 2841.

88. Peumans P, Forrest SR (2001) Very-high-efficiency double-heterostructure copper phthalocyanine/C_{60} photovoltaic cells. *Appl Phys Lett,* **79**, 126.

89. Xu Z, Hu B (2008) Photovoltaic processes of singlet and triplet excited states in organic solar cells. *Adv Funct Mater,* **18**, 2611.

90. Rand BP, Schols S, Cheyns D, et al. (2009) Organic solar cells with sensitized phosphorescent absorbing layers. *Org Electron,* **10**, 1015.

91. Rand BP, Girotto C, Mityashin A, et al. (2009) Photocurrent enhancement in polymer:fullerene bulk heterojunction solar cells doped with a phosphorescent molecule. *Appl Phys Lett,* **95**, 173304.

92. Karl N (1985–2000) *Organic Semiconductors,* Landolt Börnstein/New Series Group III, **17**, 106–218 (Springer, Berlin).

93. Lassiter BE, Lunt RR, Renshaw CK, Forrest SR (2010) Structural templating of multiple polycrystalline layers in organic photovoltaic cells. *Opt Express,* **18**, A444.

94. Chauhan KV, Sullivan P, Yang JL, Jones TS (2010) Efficient photovoltaic device performance through structural modification in phthalocyanine heterojunctions. *J Phys Chem C,* **114**, 3304.

95. Kim HJ, Shim HS, Kim JW, Lee HH, Kim JJ (2012) CuI interlayers in lead phthalocyanine thin films enhance near-infrared light absorption. *Appl Phys Lett,* **100**, 263303.

96. Sullivan P, Jones TS, Ferguson AJ, Heutz S (2007) Structural templating as a route to improved photovoltaic performance in copper phthalocyanine/fullerene (C60) heterojunctions. *Appl Phys Lett,* **91**, 233114.

97. Roy SS, Bindl DJ, Arnold MS (2012) Templating highly crystalline organic semiconductors using atomic membranes of graphene at the anode/organic interface. *J Phys Chem Lett,* **3**, 873.

98. Yang JL, Schumann S, Jones TS (2011) Nanowire-array films of copper hexadecafluorophthalocyanine (F_{16}CuPc) fabricated by templated growth. *J Mater Chem,* **21**, 5812.

99. Verreet B, Muller R, Rand BP, Vasseur K, Heremans P (2011) Structural templating of chloro-aluminum phthalocyanine layers for planar and bulk heterojunction organic solar cells. *Org Electron,* **12**, 2131.

100. Hiramoto M, Fujiwara H, Yokoyama M (1991) Three-layered organic solar cell with a photoactive interlayer of codeposited pigments. *Appl Phys Lett,* **58**, 1062.

101. Peumans P, Uchida S, Forrest SR (2003) Efficient bulk heterojunction photovoltaic cells using small-molecular-weight organic thin films. *Nature,* **425**, 158.

102. Fitzner R, Mena-Osteritz, E, Mishra, A, et al. (2012) Correlation of π-conjugated oligomer structure with film morphology and organic solar cell performance. *J Am Chem Soc,* **134**, 11064.

103. Fitzner R, Reinold E, Mishra A, et al. (2011) Dicyanovinyl-substituted oligothiophenes: Structure–property relationships and application in vacuum-processed small molecule organic solar cells. *Adv Funct Mater,* **21**, 897.

104. Zhang M, Wang H, Tian H, Geng Y, Tang CW (2011) Bulk heterojunction photovoltaic cells with low donor concentration. *Adv Mater,* **23**, 4960.

105. Zhang M, Irfan, Ding H, Gao Y, Tang CW (2010) Organic Schottky barrier photovoltaic cells based on MoO_x/C_{60}. *Appl Phys Lett,* **96**, 183301.

106. Pandey R, Gunawan AA, Mkhoyan KA, Holmes RJ (2012) Efficient organic photovoltaic cells based on nanocrystalline mixtures of boron subphthalocyanine chloride and C_{60}. *Adv Funct Mater,* **22**, 617.

107. Pandey R, Zou Y, Holmes RJ (2012) Efficient, bulk heterojunction organic photovoltaic cells based on boron subphthalocyanine chloride-C_{70}. *Appl Phys Lett,* **101**, 033308.

108. Pradhan B, Pal AJ (2005) Organic photovoltaic devices: Concentration gradient of donor and acceptor materials in the molecular scale. *Synth Met,* **155**, 555.

109. Sullivan P, Heutz S, Schultes SM, Jones TS (2004) Influence of codeposition on the performance of CuPc-C_{60} heterojunction photovoltaic devices. *Appl Phys Lett,* **84**, 1210.

110. Chen L, Tang Y, Fan X, et al. (2009) Improvement of the efficiency of CuPc/C_{60}-based photovoltaic cells using a multistepped structure. *Org Electron,* **10**, 724.

111. Pandey R, Holmes RJ (2010) Organic photovoltaic cells based on continuously graded donor-acceptor heterojunctions—invited. *IEEE J Sel Top Quant Electron,* **16**, 1537.

112. Pandey R, Holmes RJ (2010) Graded donor-acceptor heterojunctions for efficient organic photovoltaic cells. *Adv Mater,* **22**, 5301.

113. Pandey R, Holmes RJ (2012) Characterizing the charge collection efficiency in bulk heterojunction organic photovoltaic cells. *Appl Phys Lett,* **100**, 083303.

114. Tress W, Leo K, Riede M (2011) Effect of concentration gradients in ZnPc: C_{60} bulk heterojunction organic solar cells. *Sol Energy Mater,* **95**, 2981.

115. Lassiter BE, Zimmerman JD, Panda A, Xiao X, Forrest SR (2012) Tandem organic photovoltaics using both solution and vacuum deposited small molecules. *Appl Phys Lett,* **101**, 063303.

116. Chiu SW, Lin LY, Lin HW, et al. (2012) A donor–acceptor–acceptor molecule for vacuum-processed organic solar cells with a power conversion efficiency of 6.4%. *Chem Commun,* **48**, 1857.

117. Chen YH, Lin LY, Lu CW, et al. (2012) Vacuum-deposited small-molecule organic solar cells with high power conversion efficiencies by judicious molecular design and device optimization. *J Am Chem Soc,* **134**, 13616.

118. Lin LY, Chen YH, Huang ZY, et al. (2011) A low-energy-gap organic dye for high-performance small-molecule organic solar cells. *J Am Chem Soc,* **133**, 15822.

119. Yang F, Sun K, Forrest SR (2007) Efficient solar cells using all-organic nanocrystalline networks. *Adv Mater,* **19**, 4166.

120. Yang F, Shtein M, Forrest SR (2005) Morphology control and material mixing by high-temperature organic vapor-phase deposition and its application to thin-film solar cells. *J Appl Phys,* **98**, 014906.

121. Yang F, Shtein M, Forrest SR (2005) Controlled growth of a molecular bulk heterojunction photovoltaic cell. *Nat Mater,* **4**, 37.

122. Hawkeye MM, Brett MJ (2007) Glancing angle deposition: Fabrication, properties, and applications of micro-and nanostructured thin films. *J Vac Sci Technol A,* **25**, 1317.

123. Steele JJ, Brett MJ (2007) Nanoribbon. *J Mater Sci-Mater Electron,* **18**, 367.

124. Li N, Forrest SR (2009) Tilted bulk heterojunction organic photovoltaic cells grown by oblique angle deposition. *Appl Phys Lett,* **95**, 123309.

125. Zheng Y, Bekele R, Ouyang JM, Xue JG (2009) Organic photovoltaic cells with vertically aligned crystalline molecular nanorods. *Org Electron,* **10**, 1621.

Chapter 3

Acceptor Materials for Solution-Processed Solar Cells

Youjun He
Institute of Chemistry, Chinese Academy of Sciences, Zhongguancun, Beijing 100190, China
Heyoujun214@gmail.com

In the past decades, polymer solar cells (PSCs) composed of a conjugated polymer or small molecule as donor and a fullerene derivative as acceptor have attracted great attention from chemical, physical, and material scientists. The power conversion efficiency (PCE) of single heterojunction PSCs has reached levels over 10%. Although this value is still lower than that of inorganic-based solar cells, and also likely not high enough for commercial applications, performance continues to increase at a steady and rapid pace. PSCs have witnessed a rapid development, which was mainly achieved by designing and synthesizing novel polymer donor materials and developing novel device architectures and processing techniques. However, less attention has been devoted to the development of fullerene acceptor materials, an aspect which is very important for further improving the PCE of single- and multiple-junction PSCs. Therefore, we will give a summary about the history, status, and future prospect of fullerene acceptor materials in this chapter. In addition,

Organic Solar Cells: Fundamentals, Devices, and Upscaling
Edited by Barry P. Rand and Henning Richter

ISBN 978-981-4463-65-2 (Hardcover), 978-981-4463-66-9 (eBook)
www.panstanford.com

inorganic nanoparticle acceptors for application in hybrid organic/inorganic bulk heterojunction solar cells will be summarized.

3.1 Introduction

Since their discovery in 1985, fullerene materials have been the subject of significant attention. Fullerenes are large carbon clusters C_n with $n = 30–190$ [1]. These kinds of molecules were named fullerene after the architect Buckminster Fuller, who designed geodesic domes with a similar shape. Among the fullerene family, C_{60} and C_{70} are the easiest members to be synthesized and thus have been utilized most prominently. The structures of C_{60} and C_{70} are shown in Fig. 3.1. C_{60} is composed of 12 pentagons and 20 hexagons, and has I_h symmetry. C_{70} is made up of 12 pentagons and 25 hexagons, and has D_{5h} symmetry. The high number fullerenes C_M contain $2(10 + M)$ carbon atoms corresponding to exactly 12 pentagons and M hexagons. For reducing surface tension and stabilizing the molecule, all pentagons in fullerenes are surrounded by hexagons. The bonds at the junction of two hexagons are shorter than the bonds at the junction of a hexagon and a pentagon.

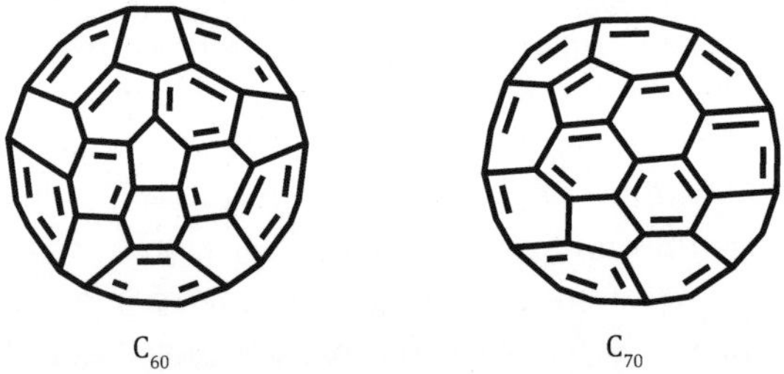

Figure 3.1 Structure of C_{60} and C_{70}.

The absorption peaks of C_{60} and C_{70} in hexane solution are located at 190 nm and 410 nm, respectively, complemented by some additional forbidden transitions in the visible part of the spectrum. The absorptions between 190 and 410 nm are due to allowed $^1T_{1u}$–^1A_g transitions, and those between 410 nm and 620 nm are due to orbital forbidden singlet–singlet transitions. The latter absorption features in the visible are responsible for the purple color of C_{60} and red color of C_{70} [2].

Fullerenes are fairly electronegative molecules as shown by their electrochemical characterization. At room temperature, they show three stepwise reduction peaks in their electrochemical spectra tested by cyclic voltammetry. At low temperatures, there are six reduction potentials [3]. As shown in the electrochemical spectra of C_{60} and C_{70} in Fig. 3.2. The reduction potential of C_{60} and C_{70} are given in Table 3.1.

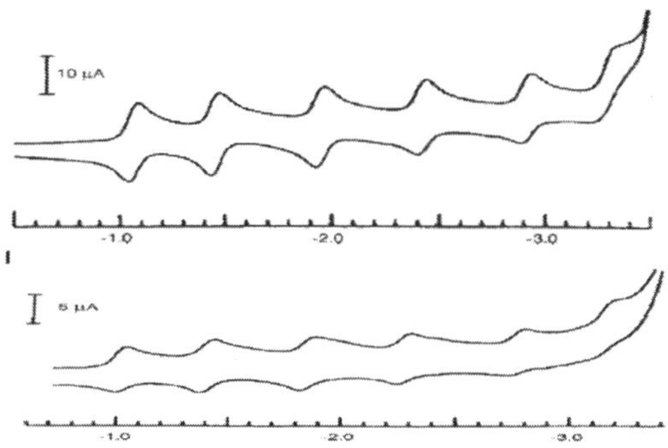

Figure 3.2 Cyclic voltammogram of C_{60} (top) and C_{70} (bottom) in acetonitrile–toluene with $TBAPF_6$ as supporting electrolyte at −10 °C using CV at a 100 mV/s scan rate (mV vs Fc/Fc^+).

Table 3.1 Reduction potentials of C_{60} and C_{70} in acetonitrile-toluene with $TBAPF_6$ as supporting electrolyte at −10 °C

Fullerene	$C_{60}{}^{0/1-}$	$C_{60}{}^{1-/2-}$	$C_{60}{}^{2-/3-}$	$C_{60}{}^{3-/4-}$	$C_{60}{}^{4-/5-}$	$C_{60}{}^{5-/6-}$
C_{60}	−0.98	−1.37	−1.87	−2.35	−2.85	−3.26
C_{70}	−0.97	−1.34	−1.78	−2.21	−2.70	−3.07

In 1992, Professor Heeger's group at UC Santa Barbara demonstrated the presence of photo-induced electron transfer from donor-type conjugated polymers to fullerenes [4]. In other words, excitons on a donor-type conjugated polymer could separate at the interface between the polymer and fullerene, resulting in an electron on the lowest unoccupied molecular orbital (LUMO) level of fullerene and a hole on the polymer highest occupied molecular orbital (HOMO) energy level. Furthermore, the forward electron transfer rate from polymer to fullerene is much faster than the back electron transfer from fullerene to polymer, implying a long-lived charge separated state with minimal recombination.

Their results therefore demonstrated that fullerenes could act as an electron-accepting material within a photovoltaic structure. Then, in 1993, Sariciftci et al. [5] fabricated bilayer organic solar cells using a soluble derivative of poly(phenylene—vinylene) (PPV) as a donor and C_{60} as an acceptor material. Improving upon that demonstration, in 1995, Yu et al. [6] fabricated bulk heterojunction (BHJ) organic solar cells based on PPV and a C_{60} derivative [6,6]-phenyl-C_{61}-butyric acid methyl ester ($PC_{60}BM$), the PCE of the device reached 2.9% when illuminated with 20 mW/cm^2 at 430 nm. Nearly at the same time, Professor Friend's group also reported an organic photovoltaic (OPV) cell based on a BHJ structure but using an electron-donating PPV polymer (MEH-PPV) as donor and an electron-withdrawing PPV polymer (CN-PPV) as an acceptor [7]. While representing a significant accomplishment, achieving sufficiently high PCE over the whole solar spectrum and not only at one specific wavelength has remained a major challenge. Therefore, many other novel soluble fullerene derivatives have been developed since then as acceptor materials for the application in PSCs, in addition to new and better absorbing electron-donor materials. Next to fullerene derivatives, there are three other kinds of acceptor materials for OPV: n-type conjugated polymers, organic small molecules, and inorganic semiconductors. Devices based on two polymers as donor and acceptor are called all-PSCs using up to now mostly n-type polymers based on 3,4,9,10-perylenetetracarboxylic diimide units or n-type PPV derivatives as electron acceptor materials [8,9]. However, particularly a rather lower electron mobility in the mixed films resulting in a poor PCE has made the development of all-PSCs challenging. Nevertheless, 6.4% performance based on all-PSCs was reported by Polyera in April 2013 using a polymer electron acceptor of unidentified composition [10]. This result certainly indicates the potential of all-PSCs. The performance of solar cells using small-molecule acceptors is lower than 4.5%, significantly less than that of fullerene acceptor-based OPVs [11–17]. While structure–property relationships are potentially easier to be obtained, particularly in comparison to polymeric acceptor materials, all yet reported PCEs have been below 3%. Significant additional work, also involving the development of more sophisticated design rules, is likely to be required before competing with other, for instance fullerenic, acceptor materials.

Inorganic semiconductors also have been used as acceptor materials in OPVs, due to their electron transporting properties. The OPV cells with organic polymer or small molecule as donor materials and inorganic semiconductors as acceptors were called organic/inorganic hybrid solar cells. While the cost of inorganic semiconductors can be lower than that of fullerene materials, the photovoltaic performance of hybrid solar cells has been found to be inferior to that of the polymer–fullerene-based devices. Details of such hybrid solar cells will be discussed below.

Until now, phenyl-C_{61}-butyric-acid-methyl ester ([60]PCBM or $PC_{60}BM$) and its C_{70} equivalent, $PC_{70}BM$, are the most common acceptor materials due to their good solubility in solvents commonly used for device fabrication, their good electron mobility, as well as their miscibility with many polymer donor materials. However, taking into account that the commercial availability of electron acceptor materials other than PCBM has been limited, new molecules, more effective particularly with the new generation of currently developed p-phase materials, are expected to play a significant role in the further improvement of the performance and emerging commercialization of OPVs.

Therefore, in the following sections, we will give a detailed overview of fullerenic and inorganic semiconductor acceptor materials including their history, current situation, and future prospects.

3.2 Fullerene Acceptor Materials

Fullerenes typically possess high electron mobility and electron affinity. Forward electron transfer rates at subpicosecond time scales from electron-donating semiconductors to fullerenes have been observed [4]. Due to this, and owing to the fact that they can be easily solution-processed and tend to self-aggregate, fullerene derivatives have been to date the most commonly used acceptor materials in OPVs. In the beginning, bilayer structures were largely utilized. As shown in Fig. 3.3, such devices consist of one metal cathode, two active layers (one is donor layer and another is acceptor layer), and one anode (typically indium tin oxide, ITO). However, an exciton diffusion length in the donor phase of only ca. 15 nm limits greatly its thickness, light absorption and therefore the short-circuit current (J_{sc}) of the device. In addition, exciton separation is limited by the interface between donor and acceptor phases, also reducing J_{sc}. Usually, the acceptor layer in a bilayer structure device is C_{60} itself which has lower LUMO levels than its derivatives and thus leads to a smaller open-circuit voltage (V_{oc}) of the device when using a given p-phase material. In bilayer devices, C_{60} layer has usually been vapor-deposited on the top of donor layer. The use of a derivative instead of unfunctionalized C_{60}, allows increasing V_{oc} but, unfortunately, limited thermal stability makes vapor deposition difficult. Liquid processing, using, e.g., spin-coating, including the selection of a suitable solvent which is not destroying the bottom donor layer can be challenging.

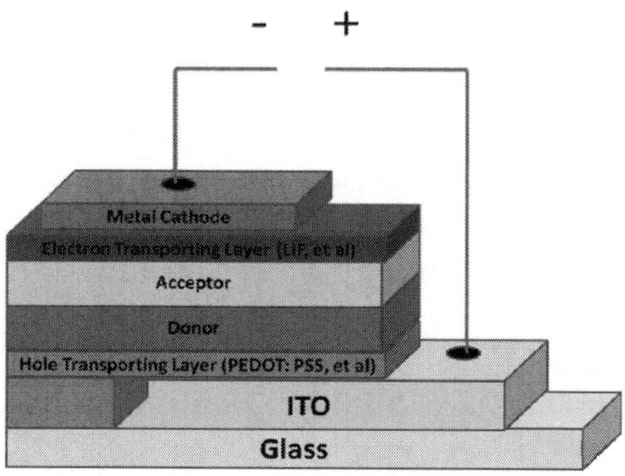

Figure 3.3 Device structure of bilayer OPV.

The above problems have been largely overcome by the use of an alternative device structure, the BHJ, shown schematically in Fig. 3.4. In a BHJ structure device, the active layer, consisting of an interpenetrating network formed by a liquid-processed blend of electron donor and acceptor materials, is sandwiched between a metal cathode and an ITO anode. In addition, hole-transporting and electron-transporting layers are usually introduced between electrodes and the active layer to improve charge collection (for more details, see Chapter 4). Being a significant advantage of the BHJ structure, the short diffusion length of excitons can be overcome by a proper nanoscale phase separation in the active layer. Therefore, the solubilities of the donor and the acceptor in the processing solvent, their crystal properties, the miscibility between them, the solvent evaporation speed, and other factors which affect the film morphology and therefore the phase separation in the active layer have significant influence on J_{sc} and performance of the device (for more detail, see Chapter 7).

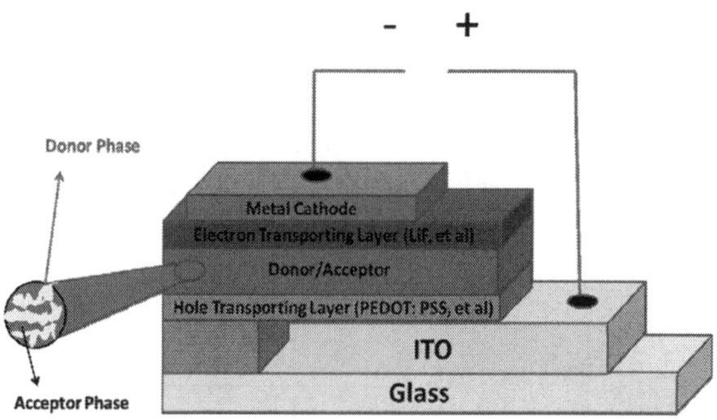

Figure 3.4 Device structure of bulk heterojunction solar cells.

In the next section, the history of fullerene-acceptor materials, factors affecting their electronic, physical and chemical characteristics, and our prospect of the future development of fullerene-acceptor materials will be discussed.

3.2.1 The History of Fullerene Acceptor Materials

While C_{60} possesses a high electron mobility (1 $cm^2/V \cdot s$) and, compared to many photoactive polymers, low energy levels (–4.0 eV LUMO and –6.0 eV HOMO), the solubility of C_{60} in organic solvents commonly used for OPV device fabrication (chloroform, dichlorobenzene, etc.) is not very high, limiting their application in BHJ solar cells. The solubility of C_{60} in various organic solvents is shown in Table 3.2 [18,19]. While similar systematic studies investigating the solubility of fullerene derivatives are not available, suitable chemical functionalization can indeed enable enhanced solubility in targeted solvents. Therefore, most reported fullerene-acceptor materials in liquid-processed BHJ organic solar cells are fullerene derivatives and not C_{60} itself.

However, taking into account that, compared to donor materials, fullerene-acceptor materials show only limited absorption of the solar light, recent improvements of the device J_{sc} were mainly caused by innovative donor materials as well as better active layer morphology. For instance, decreasing the bandgap of the donor material results in improved light absorption.

Table 3.2 Solubility of C_{60} in common organic solvents at 25 °C

Solvent	Solubility (μg/mL)	Reference
Isooctane	26	[18]
Pentane	4	[18]
Hexane	40	[18]
Octane	25	[18]
Decane	70	[18]
Dodecane	91	[18]
Tetradecane	126	[18]
Decalins	4600	[19]
Cyclohexane	51	[18]
Carbon tetrachloride	447	[18]
Mesitylene	997	[18]
Toluene	2150	[18]
Benzene	1440	[18]
Tetralin	16 000	[19]
Benzonitrile	410	[19]
Anisole	5600	[19]
Methylene chloride	254	[18]
Tetrahydrofuran	0	[19]
Chloroform	160	[19]
Dioxane	41	[18]
Chlorobenzene	7000	[19]
1,2-dichlorobenzene	27 000	[19]
1-methylnaphthalene	33 000	[19]
1-chloronaphthalene	51 000	[19]
Acetone	1	[19]
Carbon disulfide	7900	[19]
Methanol	0	[19]

Research work on fullerene acceptor materials conducted to date has mainly focused on the up-shifting the LUMO by means of chemical functionalization. The V_{oc}, being determined—at least as a first approximation—by the difference between the LUMO level of the fullerene acceptor material and the HOMO level of the donor, less negative LUMO levels result in its increase. There are several methods to up-shift LUMO levels of the fullerene derivatives:

(a) Disrupting the π-electron system of the conjugated double-bonds of the fullerene cage by attaching functional groups. An electron-donating character of such functional groups enhances the up-shift of the LUMO level, whereas electron-withdrawing groups reduce it. Multiple functionalizations, for instance the synthesis of bis-adducts, lead to a further increase of the LUMO level.

b) Synthesizing endohedral fullerene derivatives.
c) The use of fullerene–metal complexes.

Until now, most of the work reported on improved fullerenic electron-acceptor materials for OPV applications has focused on a range of functionalization schemes. While very limited availability, high prices, and challenging syntheses have been prohibitive for more exhaustive work on endohedral fullerenes, the thermal stability of fullerene–metal complexes has been found to be a concern.

In the next paragraphs, we will describe the following groups of fullerene-based electron acceptors, developed in the past decades: (1) PCBM and PCBM-like derivatives; (2) PCBM multi-adducts and PCBM-like multi-adducts; (3) Non-PCBM-like mono-adducts; (4) Non-PCBM-like bis-adducts and multi-adducts.

3.2.1.1 PCBM and PCBM-like derivatives

As part of an effort to develop soluble fullerene derivatives for applications to physics and biology [20], phenyl-C_{61}-butyric acid methyl ester ($PC_{60}BM$) has been first synthesized in 1995 by the group of Prof. Wudl at the University of California in Santa Barbara [21]. It was $PC_{60}BM$ which—due to its sufficient solubility in solvents used for donor-type polymers—allowed for the first successful demonstration of a BHJ device [6], followed by steadily improved performance of $PC_{60}BM$-based devices.

The solubility of $PC_{60}BM$ in toluene and chlorobenzene is 35 and 50 mg/ml, respectively. As compared to C_{60}, the absorption spectrum of C_{70} is significantly extended in the visible part of the spectrum, its derivatives have the potential of increased contribution to photocurrent and therefore better performance. Therefore, $PC_{70}BM$ was synthesized [22]. The chemical structures of $PC_{60}BM$ and $PC_{70}BM$ are shown in Fig. 3.5. Unlike $PC_{60}BM$, $PC_{70}BM$ was found to be a mixture of three different isomers (α, β, and γ, as shown in Fig. 3.6). At 80 mg/ml, in chlorobenzene, its solubility was found to be better than that of $PC_{60}BM$. Since the discovery of $PC_{60}BM$ and $PC_{70}BM$, they have been the most utilized acceptor materials and used with a large range of donors, resulting in encouraging device characteristics in many cases. Although many new acceptors have been synthesized and evaluated in OPV devices, no general replacement for PCBM allowing for higher performance has been identified to date. The commercial availability of PCBM and the resulting huge body of work optimizing devices and materials (including the p-phase) in combination with PCBM has made the quick development of alternatives challenging.

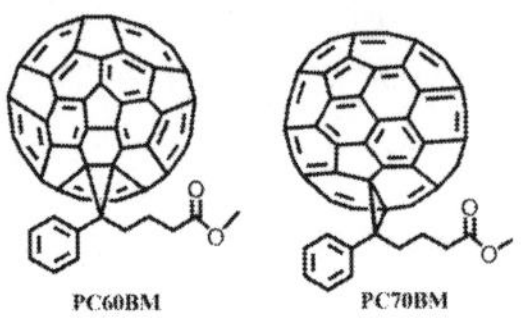

Figure 3.5 Chemical structures of $PC_{60}BM$ and $PC_{70}BM$.

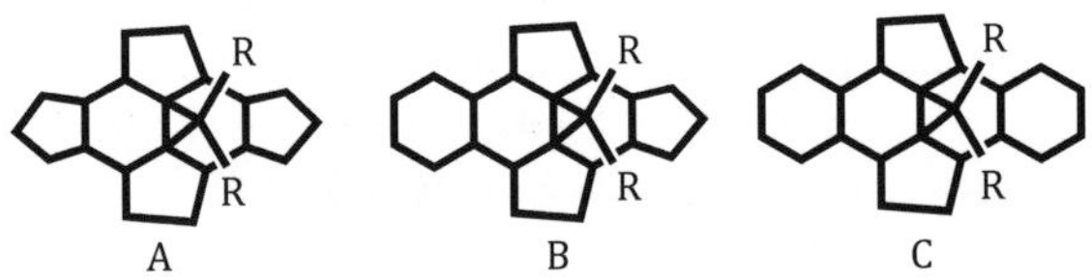

Figure 3.6 Three different isomers of C_{70} monoadduct. A (α), B (β) and C (γ).

PCBM is synthesized by [3 + 2] cycloaddition reaction between fullerene and diazomethane [21,22]. The synthetic route of $PC_{60}BM$ is shown in Fig. 3.7. While a [5,6] fulleroid, i.e., with a C–C bond cleaved in the fullerene cage, is formed in the initial addition, isomerization, accelerated, for example, by reflux in o-dichlorobenzene, leads to the final, thermodynamically stable [6, 6]-methanofullerene.

$PC_{70}BM$ was synthesized by a similar method with the exception of using C_{70} to replace C_{60} [22]. As mentioned above, due to the asymmetric shape of C_{70}, three different [6, 6] addition isomers for its monoadduct are formed. The percent of α isomer was found to be 85%, while β and γ, combined, is 15% of the compound. Due to its improved absorption characteristics [22], the acceptor material in many high-performance low-bandgap PSCs has been $PC_{70}BM$.

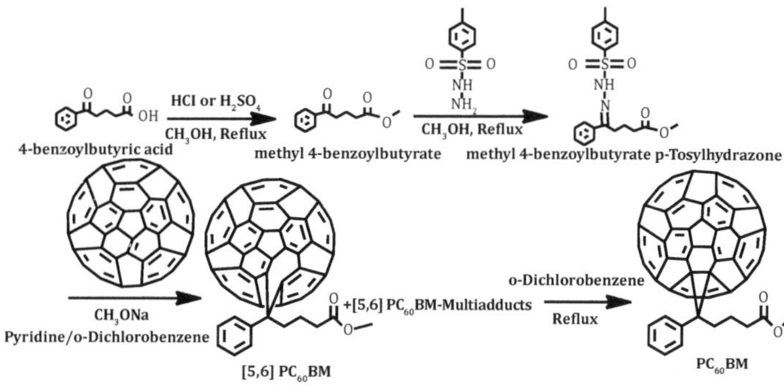

Figure 3.7 Synthetic route of $PC_{60}BM$.

A lot of modifications to $PC_{60}BM$ and $PC_{70}BM$ have been done and many novel fullerene acceptor materials with PCBM-like structures have been designed and synthesized, as shown in Figs. 3.8 and 3.9. Modifications include the change of the ester group, substitutions on the phenyl ring attached to the cyclopropane carbon (the methano-bridge) or the replacement of the phenyl, e.g., by thiophene. Also, PCBM-based endohedral fullerenes or the replacement of the methano-bridge by nitrogen have been investigated. Electrochemical reduction potentials and LUMO levels of some fullerene derivatives are summarized in Table 3.3, while the photovoltaic performances based on these materials are given in Table 3.4. In the following part, we will talk about the personalities and properties of these materials in detail. Zheng et al. [23] synthesized fullerene derivatives F1–F4. As shown in Table 3.3, these substituted PCBM derivatives F1–F4 have almost identical reduction potentials, which indicate that the variation of the alkyl chain length of the ester group does not affect the reduction potential of PCBM. Therefore, their LUMO levels are nearly same. In a subsequent set of experiments, MEH-PPV was used as donor and these materials were used as acceptors to fabricate PSC devices. The devices using acceptor material with longer alkyl group showed lower J_{sc}, indicating a reduction of the charge-carrier mobility. Seri et al. [24] reported another PCBM derivative F5, bearing a t-butyl-ester group. Having the same LUMO and HOMO levels as PCBM, the purpose of designing F5 was to achieve efficient electron transport, good solubility in common organic solvents, and enhanced molecular hindrance able to influence the intermolecular interactions in the active layer. However, the PCE of P3HT: F5 was only 3.03% which is lower than the 3.48% observed for P3HT:PCBM under the same conditions (AM 1.5, 100 mW/cm^2), which was mainly caused by the lower J_{sc} in the former system [24]. Mobility results show PCBM and F5 possessing the same electron mobility (7.0×10^{-4} cm^2/V·s) in their mixed films with P3HT whereas the mobilities of pristine F5 and PCBM films were found to be 2.1×10^{-3} cm^2/V·s and 2.5×10^{-3} cm^2/V·s, respectively. The solubility of F5 in chlorobenzene and toluene is not very good resulting in a low-quality film when spin-coated from these solutions and therefore showing a decreased device J_{sc}. Yang et al. [25] designed and synthesized F6 and F7. These two materials are readily soluble in common organic solvents (toluene, dichlorobenzene, etc.). They both have more negative reduction potentials than that of PCBM (i.e., less negative LUMO levels) which was attributed to the electron-donating (+I) effect of the alkoxy-group. Compared to PCBM, higher LUMO levels of F6 and F7 are consistent with the increased V_{oc} shown in Table 3.4. P3HT:PCBM films appear to have a better morphology than the corresponding active layers consisting of F6 or F7 mixed with P3HT, resulting in a higher J_{sc} and overall performance of P3HT:PCBM devices. In agreement with this finding, OFET mobilities of PCBM, F6 and F7 are 2.85×10^{-2} cm^2/V·s, 1.02×10^{-2} cm^2/V·s and 1.59×10^{-3} cm^2/V·s, respectively.

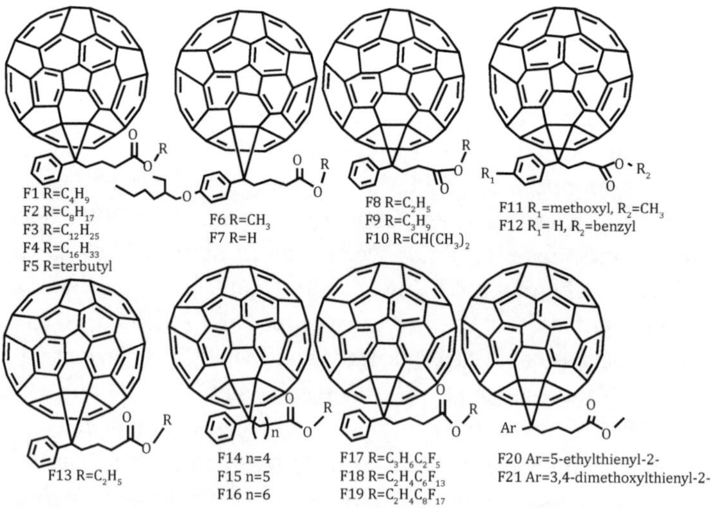

Figure 3.8 Chemical structures of PCBM-like monoadducts.

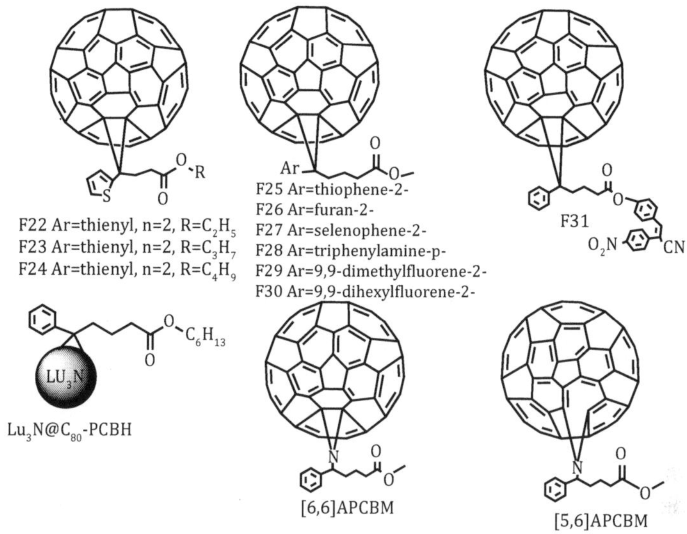

Figure 3.9 Chemical structures of some PCBM-like monoadducts.

Table 3.3 Reduction potentials of some fullerene derivatives

	E^1(red) (v)	E^2(red) (v)	E^3(red) (v)	Reference
PCBM	−0.64	−1.05	−1.56	[23]
F1	−0.63	−1.05	−1.55	[23]
F2	−0.67	−1.16	−1.69	[23]
F3	−0.64	−1.09	−1.57	[23]
F4	−0.65	−1.09	−1.62	[23]
F5	−1.101	−1.409	−1.996	[24]
PCBM	−1.076	–	–	[24]
F6	−1.171	−1.544	−2.048	[25]
F7	−1.181	−1.565	−2.087	[25]
PCBM	−1.163	−1.538	−2.040	[25]
F14	−0.89	−1.28	−1.79	[28]
F15	−0.88	−1.29	−1.84	[28]
F16	−0.90	−1.30	−1.81	[28]
PCBM	−0.88	−1.27	−1.78	[28]
F21	−0.58	−1.07	−1.51	[31]
PCBM	−0.57	−0.98	−1.51	[31]

Renz et al. [26] have investigated the fullerene solubility–current density relationship in PSCs. They synthesized fullerene derivatives F8, F9, F10 and F12, with solubilities (in chlorobenzene) of 5 g/L, 30 g/L, 22 g/L and 106 g/L, respectively. Films of these fullerene derivatives mixed with P3HT were investigated.

Films with low-solubility fullerenes show micrometer-sized fullerene aggregates, and a significant increase in size or number of those aggregates was observed after annealing. With rising fullerene solubility, these microscopic aggregates become smaller and vanish. Solar cells made with lower solubility fullerene acceptors were found to have low J_{sc}, but which increased with increasing solubility until reaching a solubility of approximately 25 mg/ml where the current density saturates. Troshin et al. [27] systematically investigated the fullerene solubility-device performance relationship in PSCs. They synthesized F11, F13, F22, F23, F24, F25 and F26, tested their solubility and photovoltaic performance mixed with P3HT. Solubilities of these compounds in chlorobenzene have been reported to be 5 mg/ml, 19 mg/ml, 23 mg/ml, 45 mg/ml, 70 mg/ml, 36 mg/ml and 58 mg/ml, respectively. The J_{sc} of devices with these acceptors mixed with P3HT increased with rising fullerene solubility. They found the solubility of the fullerene acceptor in chlorobenzene should be in the range from 30 mg/ml to 80 mg/ml for achieving high J_{sc} and FF. Zhao et al. [28] synthesized fullerene derivative F14, F15 and F16. They have nearly the same reduction potentials and LUMO levels as PCBM. The photovoltaic performance of F14 and F16 was found to be inferior to that of PCBM devices, while the PCE when using F15 is similar to that of PCBM when using P3HT as donor material. Moriwaki et al. [30] synthesized fullerene derivative F20, EThCBM, in which the phenyl of PCBM is replaced by 5-ethylthiophene. The first reduction potentials for F20 and PCBM are −1.147 V and −1.153 V, respectively, corresponding to nearly identical LUMO levels and also resulting in very similar power conversion efficiencies in conjunction with P3HT. F21, bearing a 3,4-dimethoxythiophene group instead of phenyl, in a structure otherwise identical to PCBM, was synthesized by Choi et al. [31] and a first reduction potential again nearly identical to that of PCBM was found. However, mixed with P3HT introducing two methoxyl groups, the device performance was greatly reduced. Zhang et al. [33] synthesized two fullerene derivatives F28 and F29. Unlike PCBM, which is crystalline, F28 and F29 are amorphous materials. In order to suppress crystallization, the phenyl group in PCBM was replaced by bulky triphenyl amine (TPA) or 9, 9-dimethylfluorene (MF) resulting in $TPAC_{60}$ (F28) and MFC_{60} (F29). OFET mobilities of PCBM, F28 and F29 are 1.6×10^{-2} $cm^2/V \cdot s$, 1.1×10^{-2} $cm^2/V \cdot s$ and 5.4×10^{-3} $cm^2/V \cdot s$, respectively. The thermal stability of photovoltaic devices based on F28 and F29 was found to be better than that of PCBM but PCE was slightly lower when used with P3HT. The thermal stability of active layers consisting of amorphous fullerene derivatives and P3HT has been confirmed by Kim et al. [34] using a dihexyl-fluorene-modified PCBM (F30).

Mikroyannidis et al. [35] designed and synthesized a novel PCBM derivative F31 in which the methyl group of the ester was replaced by a larger dye-like unit (Fig. 3.9). F31 has better solubility than PCBM in common organic solvents. Its solubility in THF is ca. 40 mg/ml, compared to the 25 mg/ml for PCBM. The absorption of F31 in the range from 300 to 900 nm is significantly stronger than that of PCBM, due to the contribution of the dye part. The film absorption coefficient ranged from 5.3×10^4 to 2.5×10^4 M^{-1} cm^{-1} for F31 and from 2.4×10^4 to 1.2×10^4 M^{-1} cm^{-1} for PCBM. F31 has 0.2 eV higher LUMO level than PCBM and, correspondingly, the device V_{oc} increased from 0.68 to 0.81 V using P3HT as donor material in both cases. Park et al. [36] synthesized two PCBM-like derivatives in which the bridging carbon of the cyclopropane group is replaced by nitrogen. The electron mobilities of these two iminofullerene isomers [5,6], APCBM (an open azafulleroid) and [6,6] APCBM (a closed aziridinofullerene), were found to be 4.1×10^{-2} and 2.3×10^{-2} $cm^2/V \cdot s$, respectively. Consistent with this observation, PCEs of devices having in addition P3HT in the active layer were reported to be 2.8% for the [5,6] and 2.3% for [6,6] isomer.

Ross et al. [37] designed an endohedral fullerene derivative $Lu_3N@C_{80}$-PCBH (phenyl-$Lu_3N@C_{81}$-butyric-acid-hexyl ester) bearing the same functional group as PCBM (but with a hexyl-ester). Compared with conventional, hollow, fullerenes, endohedral fullerenes have higher LUMO levels due to the interaction between the metals and the fullerene cage. The hexyl ester has been chosen because of its similarities in solubility and miscibility to PCBM (a methyl ester) when processing P3HT-based OPV devices. The first reduction potentials of $Lu_3N@C_{80}$-PCBH and PCBM are reported by Ross et al. [37] to be –1.500 and –1.220 V, respectively, indicating that the former material has a significantly higher LUMO level than the latter one. Their mobilities were found to be 4.0×10^{-4} and 1.4×10^{-3} $cm^2/V \cdot s$, respectively.

Table 3.4 Photovoltaic performance of fullerene acceptors

	V_{oc} (V)	J_{sc} (mA/cm^2)	FF (%)	PCE (%)	Reference
F1	0.85	5.22	43.2	2.45	[23]
F2	0.80	3.48	35.8	1.27	[23]
F3	0.80	2.86	35.4	1.04	[23]
F4	0.50	0.48	34.0	0.11	[23]
PCBM	0.85	4.61	39.9	2.00	[23]
F5	0.603	7.64	66.0	3.03	[24]
PCBM	0.588	8.85	67.0	3.48	[24]
F6	0.65	4.43	35.0	1.01	[25]
F7	0.64	5.83	46.0	1.73	[25]
PCBM	0.60	6.98	43.0	1.81	[25]
F8	0.36	0.70	36.0	0.10	[26]
F9	0.62	8.10	43.0	2.20	[26]
F10	0.64	8.40	52.0	2.80	[26]
F11	0.35	0.40	35.0	0.05	[27]
F12	0.62	8.40	48.0	2.50	[26]
F13	0.64	7.90	53.0	2.70	[27]
F14	0.535	8.10	53.2	2.30	[28]
F15	0.596	9.90	61.5	3.60	[28]
F16	0.540	9.30	56.4	2.80	[28]
PCBM	0.571	9.60	64.6	3.50	[28]
F17	0.52	7.82	56.0	2.31	[29]
F18	0.50	7.23	54.0	2.00	[29]
F19	0.51	7.59	49.0	1.88	[29]
F20	0.628	6.10	58.7	2.25	[30]
PCBM	0.619	6.12	58.5	2.22	[30]
F21	0.53	2.13	29.0	0.33	[31]
PCBM	0.62	10.9	62.0	4.18	[31]
F22	0.60	8.40	50.0	2.50	[27]
F23	0.60	10.16	54.0	3.40	[27]
F24	0.60	9.00	53.0	2.90	[27]
F25	0.60	10.6	58.0	3.70	[27]
F26	0.60	7.90	33.0	1.20	[27]
PCBM	0.64	10.60	55.0	3.70	[27]
F27	0.59	9.00	66.0	3.54	[32]
F28	0.65	9.90	62.0	4.00	[33]
F29	0.65	9.80	59.0	3.80	[33]
PCBM	0.63	10.4	64.1	4.20	[33]
F30	0.68	8.90	52.0	3.16	[34]
F31	0.81	10.30	63.0	5.25	[35]
PCBM	0.68	8.0	54.0	2.93	[35]
[5,6]APCBM	0.58	8.10	60.0	2.80	[36]
[6,6]APCBM	0.58	7.14	56.0	2.30	[36]
$Lu_3N@C_{80}$-PCBH	0.81	8.64	61.0	4.20	[37]
PCBM	0.63	8.90	61.0	3.40	[37]

In general, in the above-described investigations, LUMO energy levels of acceptor materials were tuned by changing the electron donating and withdrawing ability of the units attached to the fullerene core while all other parameters, particularly the donor material, remained the same. The V_{oc} of the devices then changed correspondingly. Due to reasons such as limited solubility, electron mobility or miscibility with donor materials, J_{sc} and FF of most photovoltaic devices based on such new electron-acceptor materials were found to be lower than those of PCBM devices. Among the molecules discussed to this point, as summarized in Tables 3.3 and 3.4, F31 and $Lu_3N@C_{80}$-PCBH appear to be the most interesting. F31, described above, is a PCBM-like molecule in which the methyl- of the ester group has been replaced by an organic dye, increasing the absorption of the material and therefore J_{sc}. $Lu_3N@C_{80}$-PCBH is an endohedral fullerene derivative, and its LUMO energy level is significantly higher than that of PCBM, leading to improved performance, driven by a higher V_{oc} when combined with P3HT. However, limited scalability of the synthesis of endohedral fullerenes, currently carried out by electric arc, and therefore rather high prices, are expected to be major hurdles for the commercial use of endohedral fullerenes and their derivatives.

3.2.1.2 PCBM multi-adducts and PCBM-like multi-adducts

By adding more substituents onto the fullerene core, more double bonds in the fullerene cages will be destroyed leading to up-shifted LUMO levels of the corresponding fullerenes. Therefore, fullerene multi-adducts have higher LUMO levels than the corresponding mono-adducts, which can be helpful for producing solar cells with higher V_{oc}. The research work on PCBM- and PCBM-like multi-adducts are summarized in Table 3.5. The chemical structures of some PCBM-like multi-adducts are shown in Fig. 3.10. In addition to fullerene adducts bearing multiple but identical functional groups, also "hybrids" with different functional groups such as IC_{60}PCBM (indene-C_{60}PCBM) [41] and 1,2-dihydromethano-[60] PCBM [42] have been investigated. Consistent with the higher LUMO levels, multi-adducts lead usually to devices with higher V_{oc} compared to those made from mono-adducts. However, multiple side chains are likely to inhibit the close packing of fullerenes which may slow the intermolecular electron transfer, induce energetic disorder and electron trapping and thus decrease electron mobility [38,40]. Low electron mobility of multi-adducts can then decrease the performance of resulting devices. For tri-adducts, the electron mobility was found to be even smaller than for bis-adducts, further reducing charge collection and transport. As multi-adducts are usually mixtures of many isomers which is likely to introduce additional disorder into both the molecular packing and energy levels, devices based on them have generally a lower J_{sc} than devices based on mono-adducts [38,40]. In order to determine the degree of disorder of multi-adducts, Lenes et al. have investigated the temperature dependence of the zero-field mobility as determined from various electron-only devices. Based on their results, the extent of the disorder of bis-adducts is significantly larger than that of mono-adducts [40]. As the electron quasi Fermi level increased above the LUMO of the lower lying isomers, fewer pathways are available to inject electrons into the electrode. As a result, the contribution of these molecules to the current decreases, yielding devices with higher V_{oc} compared with mono-adduct devices but with lower fill factors [38,40]. In summary, it remains a challenge to identify fullerene multi-adducts which not only allow for higher V_{oc} but also increase (or at least preserve) J_{sc} and FF.

Table 3.5 Photovoltaic performance of fullerene acceptors

	V_{oc} (V)	J_{sc} (mA/cm^2)	FF (%)	PCE (%)	Reference
PCBM-bis-adduct	0.724	9.14	68.0	4.50	[39]
PCBM	0.58	10.4	63.0	3.8	[39]
PCBM-bis-adduct	0.73	7.30	63.0	2.40	[40]
$PC_{70}BM$-bis-adduct	0.75	7.03	62.0	2.30	[40]
PCBM-tris-adduct	0.91	0.99	37.0	0.21	[40]
ThCBM–bis-adduct	0.72	7.31	66.0	2.50	[40]
$ThC_{70}BM$-bis-adduct	0.72	7.54	59.0	2.30	[40]
PCBM	0.61	8.94	60.0	2.40	[40]
ThCBM-bis-adduct	0.72	5.91	41.0	1.72	[31]
ThCBM-tris-adduct	0.64	1.88	28.0	0.34	[31]
ThCBM-tetra-adduct	0.57	0.48	32.0	0.09	[31]
PCBM	0.62	10.9	62.0	4.18	[31]
$IPC_{60}BM$	0.72	9.49	64.0	4.39	[41]
PCBM	0.58	9.41	64.0	3.49	[41]
1,2-dihydromethano-[60]PCBM	0.69	8.03	70.0	3.81	[42]
PCBM	0.58	7.42	70.0	3.02	[42]

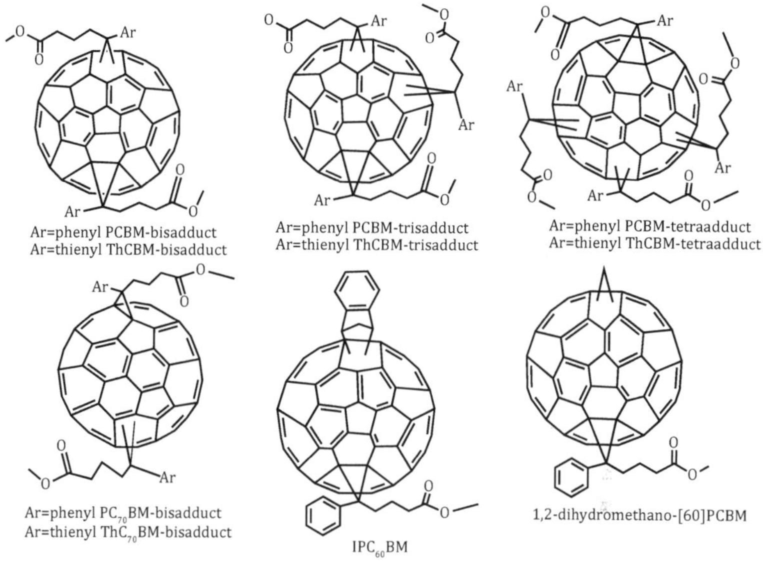

Figure 3.10 Chemical structures of some PCBM-like multi-adducts.

3.2.1.3 Non-PCBM-like mono-adducts

In addition to PCBM-like mono-adducts, many other fullerene mono-adduct acceptors have been investigated. Some of their chemical structures are shown in Figs. 3.11 and 3.12, whereas reduction potentials of some of them are given in Table 3.6. Resulting photovoltaic performances are provided in Table 3.7. As it can be seen from the data, most of these materials yield device performances inferior to that of PCBM, often due to poor solubility, low electron mobility and/or poor morphology. Troshin et al. [27] investigated the fullerene solubility-photovoltaic performance relationship. They synthesized 27 different C_{60} and C_{70} derivatives possessing various aryl or heteroaryl and/or alkyl groups that are appended to the fullerene cage via a cyclopropane moiety and, together with P3HT as donor material, and their use in BHJ OPVs is reported. F32, F33 and F34, given in Fig. 3.11 with device data provided in Table 3.7 are examples. Their solubilities (in chlorobenzene) are 31 mg/ml, 23 mg/ml and 25 mg/ml, respectively. It is shown that even slight variations of the molecular structure can cause a significant change in physical properties, particularly solubility in organic solvents. It is found that the solubility of a fullerene derivative strongly affects the morphology of its composite with P3HT. Power conversion efficiencies of solar cells comprising the fullerene derivatives investigated by Troshin et al. [27] range from negligibly low (0.02%) to considerably high (4.1%). It was concluded that the best material combinations are those where donor and acceptor components are of similar and sufficiently high solubility in the solvent used for deposition of the active layer [27].

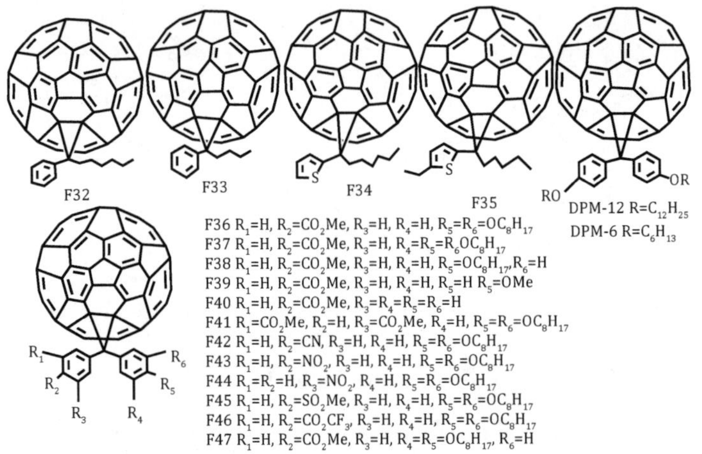

Figure 3.11 Chemical structures of non-PCBM-like mono-adducts.

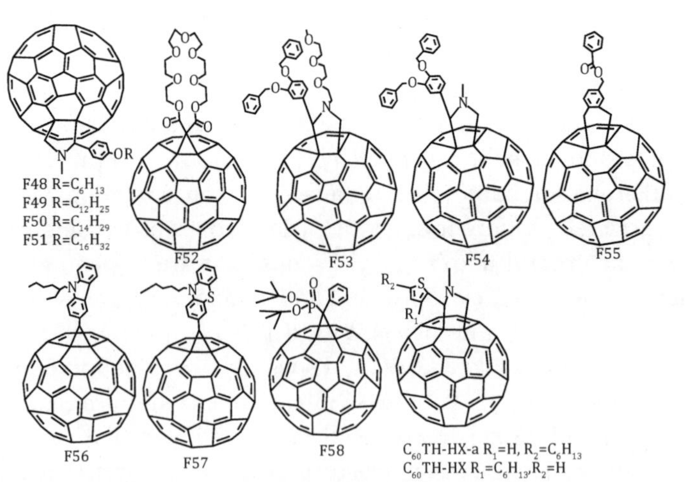

Figure 3.12 Chemical structures of some non-PCBM-like mono-adducts.

Table 3.6 Reduction potentials of some non-PCBM like mono-adducts

	E^1_{red} (V)	E^1_{red} (V)	E^1_{red} (V)	Reference
F35	−1.162	—	—	[30]
PCBM	−1.153	—	—	[30]
DPM-6	−1.084	−1.481	−1.996	[44]
PCBM	−1.077	−1.469	−1.980	[44]
F36	−0.900	−1.205	−1.420	[45]
F37	−0.945	−1.376	−1.885	[45]
F38	−0.905	−1.285	−1.505	[45]
F39	−0.835	−1.210	−1.715	[45]
F41	−0.915	−1.290	−1.530	[45]
F42	−0.905	−1.455	−1.836	[45]
F43	−0.890	−1.270	−1.770	[45]
F45	−0.883	−1.253	−1.780	[45]
F46	−0.897	−1.279	−1.737	[45]
F47	−0.915	−1.385	−1.565	[45]
PCBM	−1.049	−1.290	−1.535	[45]

Moriwaki et al. [30] synthesized fullerene material F35. The first reduction potentials of F35 (n-pentyl-substituted 1-(5-ethylthiophene-2-yl)-[6,6]-methanofullerene [60]), F25 (thionyl-analog of PCBM), and PCBM in dichloromethane were found to be −1.162, −1.143, and −1.153 V, respectively. This observation, originating from a reduced electron-withdrawing effect, indicates that alkyl chains without ester-bearing end groups result in higher (less negative) LUMO levels, higher V_{oc}, and also better device performance after blending with P3HT [30].

Riedel et al. [43] and Bolink et al. [44] utilized the di-4-alkoxyphenylmethano-fullerene derivatives DPM-12 and DPM-6, bearing one $C_{12}H_{25}$-O- or C_6H_{13}-O-group on each phenyl ring, respectively. These two materials have the same LUMO levels as that of PCBM. The field-effect electron mobility in pure DPM-12 has been found to be $\mu_e = 2 \times 10^{-4}$ cm^{2}V^{-1}s^{-1}, that is, forty times lower than the one measured in PCBM ($\mu_e = 8 \times 10^{-3}$ cm^{2}V^{-1}s^{-1}).

Interestingly, although DPM-12 and PCBM have the same LUMO levels, the V_{oc} of a P3HT: DPM-12 device is 0.1 V higher than that of similar P3HT:PCBM devices. In this case, the increased V_{oc} is thought to be mainly caused by the different morphology in P3HT: DPM12-compared to P3HT:PCBM-based active layers. However, although DPM-12-based devices have a higher V_{oc} than PCBM-based ones, the J_{sc} of the former is lower than that of the latter. In order to increase the electron mobility of DPM-12 and therefore J_{sc} of the corresponding devices, the chain length of the alkoxy-group attached to the phenyl-units was reduced from C_{12} to C_6. The three reduction potentials of DPM-6 were −1.084 V, −1.481 V and −1.996 V, very similar to PCBM for which −1.077 V, −1.469 V and −1.980 V are reported and confirming similar LUMO levels [44]. The electron mobility was determined by means of an electron single carrier device made by sandwiching a pure DPM-6 layer in between a PEDOT: PSS anode and a Ba/Ag cathode. The resulting mobility of 6×10^{-4} $cm^2 V^{-1}s^{-1}$ is significantly higher than the values reported for DPM-12 [43] but still three to four times lower than that obtained for PCBM using a similar method [44]. Compared to PCBM, the reduced performance is explained by the lower tendency of DPM-6 to crystallize and phase separate. However, Bolink et al. [44] hypothesize that under these circumstances it is not unlikely that further optimization of the processing conditions, such as the use of a higher annealing temperature, could result in improved morphology and hence efficiency. Furthermore, the more amorphous nature of the active layer is expected to result in enhanced thermal stability. Sukeguchi et al. [45] synthesized a series of diphenylmethanofullerenes, F36 to F47, all substituted at the 3-, 4- and 5-positions of both phenyl-groups and evaluated, combined with P3HT, their performance in OPV devices. They found that F47, substituted with a carboxylic acid methyl ester (-CO_2CH_3) at the 4-position of one phenyl group and O-C_8H_{17}-substituted at the 3- and 4-locations of the other phenyl unit, showed a PCE as high as PCBM, while having a higher solubility than PCBM in a variety of organic solvents. Consistent with the differential pulse voltammetry (DPV) measurement conducted by Sukeguchi et al. [45], it can be concluded that the addition of electron donating groups leads to an increase of LUMO levels whereas the length, shape and number of these groups have a big influence on the solubility of the materials.

Lee et al. [46] synthesized the alkoxy-phenyl-substituted fulleropyrrolidine materials F48–F51 (Fig. 3.12), assessing the

Table 3.7 Photovoltaic performance of some fullerene acceptors

	V_{oc} (V)	J_{sc} (mA/cm^2)	FF (%)	PCE (%)	Reference
F32	0.60	10.2	45.0	2.80	[27]
F33	0.62	7.61	43.0	2.00	[27]
F34	0.58	8.00	45.0	2.10	[27]
PCBM	0.64	10.60	55.0	3.70	[27]
F35	0.676	6.72	56.8	2.58	[30]
PCBM	0.619	6.12	58.5	2.22	[30]
DPM-12	0.65	4.70	58.0	2.3	[43]
PCBM	0.55	–	–	–	[43]
DPM-6	0.69	8.47	50.0	2.60	[44]
PCBM	0.64	9.72	59.0	3.10	[44]
F36	0.61	5.41	59.0	1.93	[45]
F37	0.66	4.60	49.0	1.50	[45]
F38	0.60	4.43	39.0	1.05	[45]
F39	0.49	0.25	18.0	0.022	[45]
F41	0.54	3.13	59.0	1.00	[45]
F42	0.52	4.09	40.0	0.84	[45]
F43	0.35	0.69	21.0	0.051	[45]
F45	0.56	4.43	50.0	1.24	[45]
F46	0.48	4.20	43.0	0.86	[45]
F47	0.60	4.90	55.0	1.60	[45]
F48	0.60	1.21	29.0	0.21	[46]
F49	0.50	1.26	31.0	0.20	[46]
F50	0.50	2.40	34.0	0.40	[46]
F51	0.55	2.31	37.0	0.47	[46]
F52	0.50	0.98	20.0	0.37	[47]
F53	0.58	0.83	18.0	0.48	[47]
F54	0.71	0.62	27.0	0.56	[47]
F55	0.65	11.30	57.0	4.20	[48]
F56	0.64	2.32	32.60	0.48	[49]
F57	0.608	4.393	48.33	1.29	[50]
F58	0.60	7.60	35.0	1.60	[51]
C_{60}-Th-HX-a	0.43	0.73	44.0	0.14	[52]
C_{60}-Th-HX	0.58	6.58	64.0	2.44	[52]

impact of the increase of the length of the alkoxy-group from C_6 to C_{12}, C_{14} and finally C_{16}. All these molecules were readily soluble in common organic solvents (chloroform, toluene, etc.). Among these four materials, F51 gave the best photovoltaic performance when mixed with MEH-PPV. However, taking into account the still poor performance, also due to the choice of donor material and the use of bi-layer architecture, conclusions regarding the potential of this class of fullerene derivatives needs to be treated carefully. Camaioni et al. [47] synthesized the methanofullerene F52 as well as the phenyfulleropyrrolidines F53 and F54 before assessing their performance in OPV devices in combination with P3HT (structures shown in Fig. 3.12). All three compounds showed good solubility in common organic solvents and no substantial differences in photovoltaic performance were observed. While the performance of all the devices, initially reported, was rather small (<1%), preliminary results indicating that mild thermal annealing of a device consisting of an active layer with P3HT-F54 mixed in a 3:2 ratio resulted in power conversion efficiencies of up to 2.5%. More work appears to be necessary in order to assess fully the potential of compounds F52, F53, F54. Other non-PCBM-like mono-adduct derivatives F55-F58, C_{60}-Th-HX and C_{60}-Th-HX-a (the latter two being substituted thiophenefulleropyrrolidine), with structures shown in Fig. 3.12 and device data compiled Table 3.7 were also investigated [48–52]. Among them, F55, a substituted o-quinodimethane (dihydronaphthyl) adduct showed, combined with P3HT, the best photovoltaic performance Small changes such as inverting the position of two functional groups attached to the thiophene (C_{60}-Th-HX and C_{60}-Th-HX-a) were found to have in some cases a significant impact such as an increase of the PCE from 0.14% to 2.44%.

3.2.1.4 Non-PCBM-like bis- and multi-adducts

Following a similar trend as PCBM-like derivatives and having higher LUMO levels as their mono-equivalents, many other bis- and higher adducts have been designed and synthesized as shown in Figs. 3.13 and 3.14, targeting increased device V_{oc}. The photovoltaic performance based on P3HT as donor and these fullerene bis-adducts as acceptors are listed in Table 3.8. Niinomi et al. [53] synthesized several fullerene multiadducts bearing five substituted phenyl-rings and one hydrogen or methyl (F59–F63), six alkyl-groups (F64) or five butyl-phenyl and one metal sandwiched with cyclopentadiene (F65, F66). These materials have 50 π-electrons, so their LUMO levels are inherently higher than that of PCBM because of the contracted π-system. These kinds of materials can form ordered structures in the solid state which is helpful for improving electron mobility. While the parent compound F59 was found to be unstable because the hydrogen atom attached to the fullerene core is both reductively and oxidatively labile, the LUMO levels of F60–F66 and PCBM have been reported to be −3.42, −3.43, −3.42, −3.36, −3.32, −3.38, −3.39, and −3.80 eV, respectively. Please note that while several different values have been reported for the LUMO level of PCBM, comparison of the properties of newly synthesized molecules with those of PCBM provides usually all required information. F60–F66 have good solubility in common organic solvents (chlorobenzene, toluene, chloroform, etc.). Except the devices based on F62 and F64, devices have higher V_{oc} than when using PCBM due to their higher LUMO levels. F62- and F64-based devices are thought to show very low V_{oc} because F62 bears as many as ten non-conducting dodecyl chains, and F64 has a non-conical shape and therefore cannot stack with each other [53]. Varotto et al. [54] synthesized the 1,4-bis-adducts, F67 and F68 bearing in each case one phenyl and one tris-alkoxy-phenyl. The LUMO levels of F68 and PCBM are reported in this work to be −4.24 and −4.30 eV, respectively. These 1,4-diadducts have a lower symmetry than PCBM, thus they have better absorption in the visible region. Due to their higher LUMO levels, photovoltaic devices based on these derivatives as acceptors have ca. 70 mV higher V_{oc} than similar PCBM-based devices. Matsuo et al. also synthesized one silicon-containing 1,4-diadduct derivative, 1,4-bis-(dimethylphenylsilylmethyl)[60]fullerene (SIMEF) [55]. SIMEF has lower glass transition (T_g, 88 °C) and crystallization (T_c, 149 °C) temperatures than PCBM (122 °C and 195 °C), as well as a higher LUMO level (−3.74 eV) than PCBM, reported here to be −3.84 eV. Consistently, a device based on SIMEF was found to have a V_{oc} which is 0.15 V higher than that of corresponding PCBM devices.

Several cycloaddition bis-adducts have been reported. For instance, a dihydronaphthyl-bis-adduct ($NC_{60}BA$), also called bis-o-quinodimethane-C_{60} (bis-o-QDM-C_{60}) or o-xylenyl-bis-adduct (OXCBA), and corresponding compounds bearing additional substitutions on the phenyl ring have been described [56]. Evaluated with P3HT, compared to PCBM control devices, improved efficiencies have been observed. C_{60}-bis-indene (ICBA) is another [4 + 2]-cycloaddition-adduct which has been receiving significant attention due to encouraging device performance reported with P3HT as donor [56c,57]. While similar results have been obtained with the C_{70}-equivalent ($IC_{70}BA$) [58], the bis-adducts of 1,1′-biindene $BC_{60}BA$, Fig. 3.13) showed a—compared to PCBM—only the expected increase in V_{oc} but a pronounced drop in J_{sc} resulted in a lower overall performance [59].

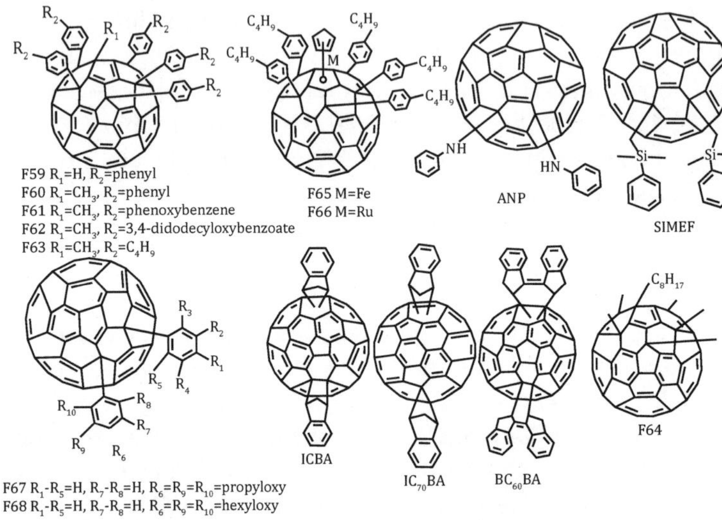

Figure 3.13 Chemical structures of non-PCBM-like bis- and multi-adducts.

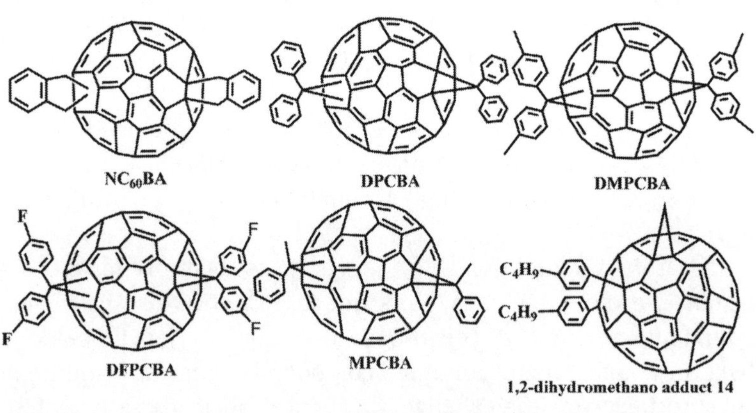

Figure 3.14 Chemical structures of some non-PCBM-like bis- and multi-adducts.

A group of diphenyl-methano-based C_{60} bis-adducts, di(4-methylphenyl)methano-C_{60} (DMPCBA), di(4-fluorophenyl) methano-C_{60}-bis-adduct(DFPCBA), diphenylmethano-C_{60}-bis-adduct

(DPCBA) and methylphenylmethano-C_{60}-bis-adduct (MPCBA) have been synthesized by cyclopropanation and evaluated in combination with P3HT for OPV applications [60]. An excellent solubility in common organic solvents was found and the electrochemical properties of these materials have been assessed by means of cyclic voltammetry. LUMO levels about 0.1 eV higher than that of PCBM were found for all molecules indicating that functionalization of the phenyl has nearly no effect on the electronic structure. As a result, an increased V_{oc} was observed in all cases but only DMPCBA enabled a PCE which was significantly higher than that of the P3HT:PCBM device.

Table 3.8 Photovoltaic performance of some fullerene acceptors

	V_{oc} (V)	J_{sc} (mA/cm^2)	FF (%)	PCE (%)	Reference
F59	0.25	0.20	43.0	0.02	[53]
F60	0.72	1.90	34.0	0.45	[53]
F61	0.76	2.80	51.0	1.08	[53]
F62	0.30	0.20	34.0	0.02	[53]
F63	0.63	3.80	38.0	0.91	[53]
F64	0.42	0.80	37.0	0.12	[53]
F65	0.73	2.50	38.0	0.68	[53]
F66	0.61	2.10	32.0	0.40	[53]
F67	0.61	6.16	62.0	2.34	[54]
F68	0.65	3.87	47.0	1.18	[54]
SIMEF	0.75	10.50	65.0	5.20	[55]
$NC_{60}BA$	0.82	9.88	67.0	5.37	[56]
ICBA	0.84	10.61	72.7	6.48	[57]
$IC_{70}BA$	0.84	9.73	69.0	5.64	[58]
$BC_{60}BA$	0.82	1.92	42.52	0.67	[59]
DPCBA	0.75	9.48	55.90	4.00	[60]
DMPCBA	0.87	9.05	65.5	5.20	[60]
DFPCBA	0.68	7.25	57.3	2.80	[60]
MPCBA	0.73	8.32	57.4	3.50	[60]
1,2-dihydromethano adduct 14	0.82	7.10	58.0	3.40	[61]

Other promising fullerene derivatives include 1,2-dihydromethano-functionalized 1,4-diorgano [60] fullerenes such as the 1,2-dihydromethano adduct 14 shown in Fig. 3.14 [61]. Such molecules are reported to be diastereomerically pure 56-π-electron fullerenes having a higher LUMO level than similar compounds after only one functionalization, i.e., without an additional dihydromethano-group.

Compared with PCBM-based devices and due to their increased LUMO levels, most fullerene bis-adducts allow for higher V_{oc}. However, care needs to be taken, as these bis-adducts are usually mixtures of different isomers with different LUMO levels, resulting in films with lower electron mobility compared to PCBM. Such energy level distribution is not only detrimental to charge transport, but possibly also to electron transfer. It remains therefore challenging to produce devices with higher V_{oc} compared to PCBM, but without an accompanying decrease in FF and J_{sc}.

3.2.1.5 Fullerene acceptor materials for low bandgap polymer solar cells

Due to its commercial availability, P3HT has been the donor of choice for most development efforts towards new solution-processed acceptor materials. However, P3HT has a relatively wide optical gap (2.0 eV) and its film mixed with PCBM can only absorb 30% of sunlight which puts intrinsic limitations on J_{sc}. As we know, the J_{sc} of photovoltaic devices is mainly determined by the capability of the active layer to absorb sunlight, the exciton diffusion efficiency in the donor phase, the exciton separation efficiency in the interface between donor phase and acceptor phase, free charge transport in both donor and acceptor phases, and the charge collection efficiency by two electrodes. Light absorption within the active layer is considered as a critical parameter and is mainly controlled by the optical gap of the donor material. Therefore, donor materials with lower bandgaps will be helpful for realizing higher J_{sc} and thus higher PCE. Based on the above considerations, many low bandgap polymer donor materials have been designed and synthesized (see Chapter 1). The absorption spectra of two of such low bandgap polymers both blended with $PC_{70}BM$ in comparison with a P3HT:$PC_{70}BM$ mixture is shown in Fig. 3.15. For getting high J_{sc}, the high-wavelength absorption edge of an ideal donor material should be in the black region in Fig. 3.15.

Besides J_{sc}, V_{oc} of P3HT:PCBM devices are also rather small (ca. 0.6 V) but as the V_{oc} of OPV devices is mainly determined by the difference between the HOMO of the donor material and the LUMO of the acceptor material, this problem can be overcome by the use of acceptor materials with higher LUMO levels (see above). However, further improvements of the PCE require the use of low bandgap/high performance donor materials. While many low bandgap polymers have been investigated, the choice of suitable material is also determined by the electron-acceptor material to be used. For instance, donor materials with bandgaps suitable for efficient absorption of solar light but HOMO levels similar to or even higher than that of PCBM will suffer from a low V_{oc}. The energetic diagram for a photovoltaic device using PSBTDT or PBDTT-C as donor combined with PCBM as acceptor material is given in Fig. 3.16. If suitable novel acceptors with higher LUMO levels than that of PCBM were used, the performance of devices using better absorbing low bandgap donor materials should be improved by both improved absorption and higher V_{oc}.

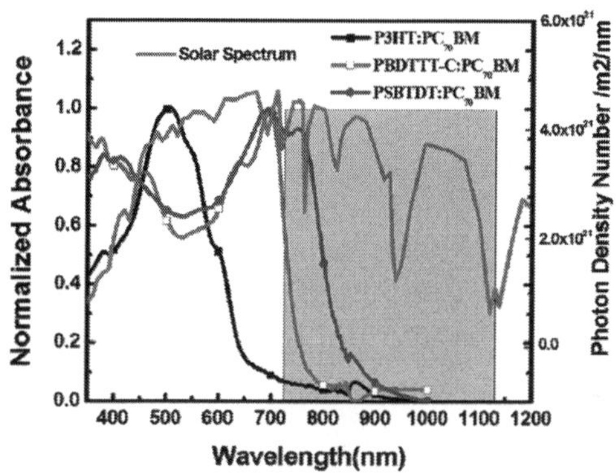

Figure 3.15 Absorption spectra of various active layers as well as the solar spectral flux.

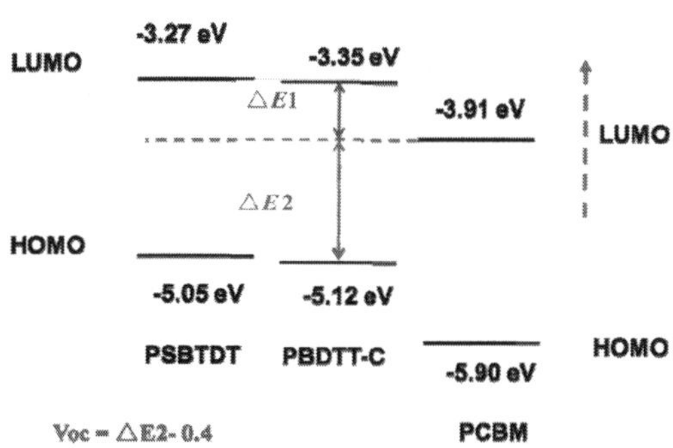

Figure 3.16 Energy diagram for a photovoltaic device.

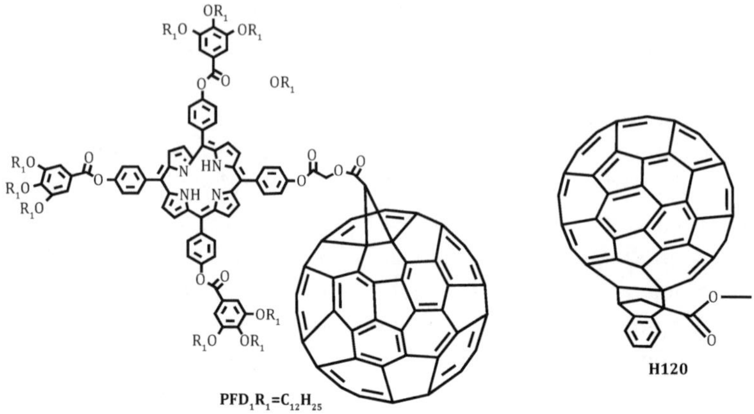

Figure 3.17 Structures of PFD1 and H120.

Still emerging, first studies addressing this topic have been conducted. For instance, Prof. A. Heeger's group at the University of California Santa Barbara fabricated the photovoltaic devices based on Si-PCPDTBT as donor and N-methyl-2-(3-hexylthiophen-2yl) pyrrolidine-C_{60} (C_{60}-TH-Hx) as acceptor [62]. The LUMO and HOMO energy levels of Si-PCPDTBT reported in this work are –3.31 eV and –4.8 eV, while for C_{60}TH-HX –3.8 eV and –5.3 eV are given. In the same work [62], LUMO and HOMO levels of –4.2 and 5.9 eV, respectively, are used for PCBM. Assessed in BHJ devices, the Si-PCPDTBT/C_{60}-TH-Hx combination resulted in a higher V_{oc} than the corresponding PCBM system but due to a lower J_{sc}, overall performances were very similar. In a collaborative effort involving Prof. Heeger's group and the laboratory of Prof. X. Gong at the University of Akron, PFD1, a porphyrin–fullerene dyad with a supramolecular "double-cable" structure (Fig. 3.17) was synthesized and tested as electron-acceptor in a BHJ device using again Si-PCPDTBT as donor material [63]. The porphyrin part in PFD1 is expected helping to absorb more sunlight and thus increase device short-circuit current while the alkyl-chains should lead to improved solubility of the acceptor. The LUMO of PFD1 is reported to be –3.96 eV. The J_{sc}, V_{oc}, FF and PCE of the devices based on Si-PCPDTBT and PFD1 were found to be 13.6 mA/cm^2, 0.66 V, 0.37 and 3.35%, while for the corresponding PCBM devices 11.5 mA/cm^2, 0.56 V, 0.56 and 4.05% have been measured, confirming the expected higher V_{oc} when using PFD1. He et al. [64], synthesized the novel fullerene acceptor H120, an indene adduct bearing a carboxylic acid methyl ester group, as shown in Fig. 3.17. It has good solubility (ca. 30 mg/ml in chloroform) in common device fabrication solvents and a 0.05 eV higher LUMO level than that of PCBM. OPV devices based on P3HT, PBDTTT-C, and PBDTTDPP as donors, and H120 as acceptor were fabricated. Their performances reached 4.2%, 6.0% and 6.2%, respectively.

3.2.2 Conclusions Regarding Fullerenic Acceptor Materials for High-Performance Solar Cells

To engineer best suited electron-acceptor materials, a significant number of parameters such as solubility, stability, energy levels, electron mobility and miscibility needs to be taken into account. Fortunately, the relative ease of controlled functionalization of fullerenes allows for the adjustment of most properties.

For instance, sufficient solubility in device fabrication solvents is necessary to realize good device morphology, necessary for high performance. Renz et al. [26] investigated the relationship between the solubility of fullerene derivatives and the J_{sc} of resulting devices. It was found that poor solubility leads to the formation of aggregates and thus small J_{sc}. Once a solubility of approx. 25 mg/ml in chlorobenzene was reached by the fullerene derivative, J_{sc}, reached a plateau of about 8–10 mA/cm^2. Thus the solubility of the fullerene derivative directly influences the blend morphology and displays an important parameter for efficient polymer–fullerene BHJ solar cell operation. Conducting a similar investigation, Troshin et al. [27] found that devices using fullerene derivatives with a solubility of 30–80 mg/ml in chlorobenzene have the best performance. However, as the relationship between the solubility and structure of functionalized fullerenes is still not clear, the systematic design of compounds with specific solubilities is not straightforward. Stability of fullerene materials is another very important characteristic. As many devices need to be annealed at high temperatures in order achieve their best possible performance, thermal stability of both donor and acceptor materials is required. In addition, materials should be stable relative to moisture taking into account possible shortcomings of encapsulation, but also device architectures in which buffer layers dissolved in water are deposited on the top of the active layer.

Finally, as described above for several fullerene bis-adducts, higher LUMO levels of the acceptor (but not higher than that of the donor material!) will lead to an increased V_{oc} and can be helpful for attaining improved overall performance assuming no or a very limited drop of J_{sc} and FF.

3.3 Inorganic Semiconductor Acceptor Materials

Inorganic semiconductors are another well-investigated class of materials used as acceptor materials in OPVs. In fact, dye-sensitized solar cells (DSSC) [65] can be considered similar in scope to the devices described in this section, as the photocurrent-producing interface is formed between an organic and an inorganic semiconductor (the inorganic semiconductor most often being TiO_2 or in some cases ZnO). When combined with organic donor materials, as discussed here, the inorganic acceptor is often nanostructured to produce a large donor/acceptor interface, and these acceptor nanostructures can be divided into four types according to their size and shape [66–68]: (1) zero-dimensional nanostructures restricted in all three dimensions, such as nanocrystal quantum dots, clusters, etc.; (2) one-dimensional nanostructures restricted in two dimensions, such as nanowires, nanorods, nanotubes and nanobelts; (3) two-dimensional nanostructures restricted in one dimension, such as nanoplates, nanodisks, ultrathin films, multilayers, superlattices, etc.; (4) three-dimensional nanostructures such as those used in DSSCs which are complicated three-dimensional nanostructures or voids composed of lower dimensional nanostructures, such as nanoflowers, four-arm rods, porous silica materials, etc.

Due to their high electron affinity and ability to be structured into nanoscale dimensions, many inorganic semiconductor nanomaterials can be used as acceptor materials, often mixed into a BHJ with a conjugated polymer donor. This type of solar cell is termed "organic/inorganic hybrid solar cell," or simply "hybrid solar cell." Various inorganic semiconductors, such as CdSe, CdS, TiO_2, ZnO, PbS and $CuInSe_2$, are commonly employed as acceptor materials in hybrid solar cells [69]. The potential advantages of hybrid solar cells stem from the fact that inorganic semiconductors possess high mobility, good stability, and the potential for near-infrared absorption in the case of inorganic semiconductor nanomaterials containing sulfur or selenium. The fact that hybrid cells can be solution-processed together with a conjugated polymer enables the same ease of fabrication as offered by polymer:fullerene cells.

Hybrid solar cells have been mainly fabricated by the following two methods: (1) Nanorod array structures of inorganic nanoparticles with regular morphology are grown on the substrate surface vertically, then a polymer is deposited to fill in the voids. TiO_2 [70], ZnO [71] and CdTe [72] arrays are the most reported in this class of hybrid solar cells. These regular array structures with tunable morphology are helpful for tuning the distance between acceptor materials and exciton dissociation, but the device performance was limited by the filling fractions of donor materials in the solid volume of the arrays. Recently, scientists have reported many methods to increase the amount of polymer materials in the arrays of nanoparticles [73]. The highest performance based on TiO_2 nanoparticles was 0.58% [74].

(2) The inorganic semiconductor nanoparticles and polymer were dissolved in the same solvent, and the active layer was obtained after spin coating. CdSe, CdTe, CdS, ZnO, TiO_2, $CuInS_2$, $CuInSe_2$, PbS and PbSe have been reported as acceptor materials in this kind of BHJ-structure hybrid solar cell. Among them, hybrid solar cells based on CdSe, ZnO and TiO_2 have been most often investigated. The energy levels and electron mobilities of these semiconductors are listed in Table 3.9.

Table 3.9 Properties of some inorganic semiconductors

	E_g (eV)	Type	Electron mobility ($cm^2/V \cdot s$)	Absorption coefficient (cm^{-1})
Si	1.11	I	1500	$10^{3,a}$
CdSe	1.74	D	720	$7 \times 10^{4,b}$
CdS	2.53	D	200	$4 \times 10^{4,b}$
CdTe	1.50	D	1000	$1.56 \times 10^{4,b}$
$CuInS_2$	1.55	D	200	$10^{5,b}$
$CuInSe_2$	1.02	D	100–1000	$5 \times 10^{4,c}$
TiO_2	3.2	I	1	
ZnO	3.3	D	205	$1.5 \times 10^{5,d}$
PbSe	0.82		2.5	

D, direct; I, indirect; [a] at 1000 nm; [b] at 500 nm; [c] at 800 nm; [d] at 340 nm.

In the next sections, we will discuss applications of different kinds of semiconductor nanomaterials in hybrid solar cells.

3.3.1 CdSe, CdTe, CdS and Related Materials

CdSe nanoparticles have a relatively high conduction (−3.7 eV) as well as valence band (−5.8 eV). Their energy levels are very close to the LUMO and HOMO energy levels of PCBM and therefore likely to be suitable as acceptor materials in hybrid solar cells. Greenham et al. [75] reported that 95% of MEH-PPV's PL was quenched after being mixed with CdSe nanoparticles. However, due to the low electron transporting ability of 5-nm sized CdSe nanoparticles without the possibility of forming continuous electron transporting pathways, the device performance was only 0.01%. In 1999, Huynh et al. [76] used CdSe nanorods as acceptor in a P3HT system. The device performance reached 1.7% when loading 90 weight % of CdSe nanorods having a diameter of 7 nm and a length of 60 nm [77]. These results indicate that a rod structure, in contrast to nanoparticles, may facilitate electron transport and that diameter and length of the rods are likely to have an influence on device performance. For instance, the absorption edge could be tuned from 650 nm to 720 nm by increasing the rod diameter from 3 nm to 7 nm. When the ratio of length to diameter (the aspect ratio) was increased from 1 to 10, the device performance was tripled. Sun et al. were the first to fabricate a photovoltaic device based on MDMO-PPV as donor and CdSe with a four-foot-like structure as acceptor. The PCE of such a device reached 1.8% with dichlorobenzene as solvent and could be further improved to 2.8% using a solvent having an even higher boiling point (trichlorobenzene) [78]. Compared with a device employing one-dimensional CdSe nanorods, devices based on four-foot-like CdSe have shown doubled performance [79]. By replacing the donor material in above devices with P3HT, they achieved a V_{oc} of 0.62 V, 8.79 mA/cm^2 of J_{sc} and 2.6% (PCE) [80]. Alivisatos' group also reported 2.6% device performance when using P3HT as donor and hyperbranched CdSe nanorods as acceptor [81]. Liu et al. [82] fabricated hybrid solar cells based on end-functional P3HT as donor but one-dimensional CdSe nanorods as acceptors using a chloroform–pyridine mixture as solvent. Pyridine was adsorbed on the surface of CdSe nanorods and P3HT was dissolved in chloroform, resulting in 0.57 V (V_{oc}), 5.8 mA/cm^2 (J_{sc}) and 1.4% (PCE), at a CdSe volume ration of 43.1%. With multi-armed CdS nanorods as acceptor and MEH-PPV as donor, Wang et al. [83] reached a PCE of 1.17%.

Zutz et al. [84] studied the influence of the Cd:Se precursor ratio during the synthesis of CdSe nanocrystals on the device performance, and found no obvious influence, the ability of the nanocrystals to accept electrons not being a limiting factor for higher efficiencies in hybrid solar cells. Radychev et al. investigated the origin of the impact of different nanocrystal surface modifications on the performance of CdSe/P3HT hybrid solar cells. They treated the CdSe quantum dots with pyridine and in some cases, in addition also with butylamine or octylamine. Compared with pyridine-treatment only, the use of butylamine improved the charge transport ability of CdSe quantum dots and removed hole traps which have been associated with the pyridine aromatic ring [85]. In order to avoid phase separation between polymer donor materials and inorganic acceptor materials, polymer donor materials and/or inorganic acceptors were treated with some ligands suitable to improve their miscibility. For example, Greaney et al. [86] showed that the V_{oc} of a device based on P3HT and CdSe can be improved by treating CdSe with tert-butylthiol. Consistent with this observation, the LUMO level of CdSe increased from −4.00 eV after the pyridine-treatment to −3.84 eV following the use of tert-butylthiol, resulting in an improvement of the PCE from 1% to 1.9%. However, the drawback of such ligand exchange approaches consists in reduced mobility of the inorganic acceptor materials.

Besides devices using P3HT as donor, many hybrid solar cells based on low bandgap polymers and CdSe have been reported. For instance, Dayal et al. fabricated photovoltaic devices based on PCPDTBT as donor material and four-foot-like CdSe nanocrystals as acceptor. Resulting devices show for V_{oc} 0.678 V, 10.1 mA/cm^2 for J_{sc} and a PCE of 3.19% [87]. Wang et al. [88] used APFO-3, a red polyfluorene copolymer, as donor and one-dimensional CdSe nanorods as acceptor to produce a photovoltaic device with 0.95 V of V_{oc}, 7.23 mA/cm^2 of J_{sc} and 2.4% of PCE.

While the morphology of CdTe nanocrystals is similar to that of CdSe nanocrystals and absorption better at longer wavelengths, the performance of CdTe-based hybrid cells was found to be poor. In order to gain a better understanding of this observation, Zhou et al. [89] designed and synthesized series of $CdSe_xTe_{1-x}$ nanocrystals (including $x = 0$ and $x = 1$) with four-arm rod structures ("tetrapod-shaped"), and determined energy levels of these materials using cyclic voltammetry. They found that the energy levels of nanocrystals relative to those of polymer donor materials have a big influence on the device performance. The conduction and valence bands of CdTe were found to be −3.06 and −5.01 eV, respectively, compared to −3.71 and −5.81 eV in the case of CdSe. The higher energy levels of CdTe lead to decreasing exciton dissociation efficiency between nanocrystals and MEH-PPV, the polymer used in this work, and therefore, a PCE of only 0.003%. Consistent with these results, Verma et al. [90] also fabricated devices using MEH-PPV as donor and CdTe nanocrystal as acceptor resulting in $V_{oc} = 0.33$ V, $J_{sc} = 0.48$ mA/cm^2, and PCE = 0.06%.

Using with PPV a more suitable donor and CdTe nanocrystals with a size of 2 nm as acceptor, Yu et al. [91] fabricated photovoltaic devices achieving 2.14% performance. A further improvement could be obtained with PSBTBT, a low bandgap polymer, and again CdTe, yielding $V_{oc} = 0.79$ V, $J_{sc} = 7.23$ mA/cm^2, FF = 0.56, and PCE = 3.2% [92].

Liao et al. [93] reported a new method for the preparation of the active layer. CdS single-crystal nanorods were grown *in situ* using P3HT as a soft template. For this purpose, cadmium acetate dihydrate and P3HT were allowed to react in mixtures between dichlorobenzene and dimethyl sulfoxide. One-dimensional CdS nanocrystals with various aspect ratios, depending on the reaction conditions, were obtained. It was demonstrated that increasing aspect ratios (up to 16) allowed for increased current densities, facilitating percolation pathways for electron transport and enabling power conversion efficiencies of up to 2.9%.

Lee et al. [94] fabricated hybrid solar cell devices by dipping nanocrystalline CdS films in aqueous solutions of the water-soluble polymers poly(2-ethynyl-*N*-aminopropylpyridiniumbromide) and poly(2-ethynyl-*N*-carboxypropylpyridiniumbromide), resulting in devices with a V_{oc} of 0.66 and 0.65 V, a J_{sc} of 6.99 and 7.42 mA/cm^2 as well as a PCE of 2.15% and 2.37%.

In summary, significant progress has been made using CdSe, CdTe, CdS, and related semiconductor nanomaterials as electron acceptors in hybrid solar cells, but their performance is still inferior to exclusively organic devices. For instance, phase separation between organic donor and inorganic acceptor phases while maintaining the electron transport ability of the latter remains challenging.

3.3.2 ZnO and ZnS

After the description of the synthesis of ZnO nanoparticles in the size range from 2 to 7 nm by addition of LiOH to an ethanolic zinc acetate solution or hydrolysis and condensation of zinc acetate dihydrate by potassium hydroxide in methanol [95], the use of such material in hybrid solar cells has been investigated by the groups of Janssen and Blom [96–99]. While ZnO nanoparticles have low conduction (−4.4 eV) and valence (−7.6 eV) bands, the wide bandgap (3.2 eV), leads to poor absorption of the sunlight. However, being low cost and environmentally friendly, ZnO could offer a new perspective towards "green electricity".

In their initial work from 2004, Janssen and co-workers chose MDMO-PPV as donor and 5 nm ZnO nanoparticles achieving—at white light illumination with intensity of 0.71 sun—a PCE of 1.6% [96]. When spin-coating diethylzinc (as a ZnO precursor) and MDMO-PPV from a common solvent followed by annealing, characterization by AFM showed a film in which crystalline ZnO was intimately mixed with MDMO-PPV. Device characterization gave a V_{oc} of 1.14 V, J_{sc} of 2.3 mA/cm^2, FF = 0.42 and PCE of 1.1%. The authors emphasized the more than 5-fold PCE improvement compared to similar TiO_2/polymer hybrid devices, due to the lower crystallization temperature of ZnO, compatible with the limits of the thermal stability of polymers [97].

According to Koster et al. [98], the lower device performance of the MDMO-PPV/ZnO system compared to MDMO-PPV/PCBM was induced by the lower hole mobility in the first case (5.5×10^{-10} m^2/V · s vs. 2.0×10^{-8} m^2/V · s in the latter case).

In 2006, Janssen and co-workers [99] also investigated P3HT/ZnO devices. Assessing the impact of the ZnO-P3HT ratio, they observed that performance reached a maximum at approximately 26 vol. % of ZnO. V_{oc}, J_{sc}, FF and PCE of a device were found to be 0.69 V, 1.71 mA/cm^2, 0.45 and 0.59%. After being annealed at 80 °C for 5 min, J_{sc}, FF, and PCE increased to 2.19 mA/cm^2, 0.55 and 0.92%. The improvements in short-circuit current and open-circuit voltage upon annealing are more pronounced for devices with lower amounts of ZnO. This is an indication that annealing mainly affects the P3HT phase. The improved chain ordering (stacking) of the P3HT chains caused by the thermal treatment is thought to result in enhanced hole transport. This improved charge transport is consistent with the increase in fill factor upon annealing.

Takanezawa et al. [100] have studied P3HT:PCBM/ZnO nanorod photovoltaic devices. ZnO seed-layers were prepared by spin-coating on the top of ITO either a zinc acetate/2-aminoethanol or a ZnO nanocrystal solution in ethanol. In the next step, ZnO nanorods were vertically grown on the surface of these ZnO seed layers using zinc nitrate hydrate and hexamethylenetetramine. Finally, P3HT:PCBM dissolved in chlorobenzene was spin-coated on top of the nanostructured ZnO substrates. The dependence of the photovoltaic performance on the ZnO nanorod length and the organic layer thickness was investigated. While without ZnO, the device gave 0.52 V of V_{oc}, 9.0 mA/cm^2 of J_{sc}, 38% for the fill factor (FF) and 1.8% of PCE, the presence of a 300 nm ZnO nanorod layer resulted in a significant increase of performance with V_{oc} = 0.57 V, J_{sc} = 9.6 mA/cm^2, FF = 50% and PCE = 2.7%. It is concluded that the ZnO nanorod array plays an important role in collecting photogenerated electrons and acts as a conducting path to the electrode. It should be emphasized that P3HT:PCBM/ZnO nanorod devices show also significant similarities with P3HT:PCBM BHJ cells using ZnO as interlayer (for instance electron transport layer), described in more detail in Chapter 4.

A significant number of other studies investigating the use of ZnO nanorods in P3HT-based hybrid cells have been conducted, but the device performance has been lower than 2%, mainly due to poor morphology and charge mobility [101].

Compared with ZnO, nanostructured ZnS was found to have a similar valence, but higher conduction band (–3.4 eV). The resulting bandgap of 4.0 eV leads to poor absorption of the solar light. In 2009, Bredol et al. [102] fabricated photovoltaic devices using P3HT as donor and ZnS as acceptor. Preparing an active layer with 50% wt. % ZnS and consistent with the high LUMO level (i.e., conduction band), a very high V_{oc} of 1.20 V has been reported. However, due to a low J_{sc}, a PCE of only 0.2% was reached. Enlargement of the active interface by improved control of the nanostructure, e.g., using other than spherical particle shapes like "nanowires," "nanorods," or "nanobelts," and also nanocomposites made from such materials are proposed to be used for improved performance.

3.3.3 TiO_2

Motivated by the high power efficiencies (~10%) of dye-sensitized TiO_2 solar cells, attributable for instance to the ultrafast charge transfer from the dye to the TiO_2, the high surface area of the TiO_2 film but also the efficient separation of opposite charges into different materials, the use of TiO_2 as electron acceptor material in polymer hybrid cells has been investigated [103–108].

Its conduction and valence bands were found to be –4.2 and –7.4 eV, both very close to those of ZnO. Comparing spin-coating a MEH-PPV layer on a TiO_2 film, the latter deposited before on an ITO-patterned glass substrate and the use of TiO_2/MEH-PPV solutions spun-cast onto an ITO-coated glass surface, a two order of magnitude higher photoconductivity has been observed in the first case [103].

In 2003, Janssen and coworkers [104] fabricated devices based on MDMO-PPV and TiO_2. For this purpose, they spin-coated a tetrahydrofuran (THF) solution containing MDMO-PPV and titanium (IV)isopropoxide ($Ti(OC_3H_7)_4$), forming TiO_2 by hydrolysis in the dark after deposition. V_{oc}, J_{sc}, and FF, were 0.52 V, 0.6 mA/cm^2, 0.42, respectively, with external quantum efficiencies (EQE) of up to 11%. Preparation of a similar MDMO-PPV/ZrO_2 film and comparison of light-induced electron spin resonance (LIESR) data allowed also for the unambiguous demonstration of photo-induced charge transfer between MDMO-PPV and TiO_2. Shim et al. [105] prepared via electro-spinning a TiO_2 nanowire architecture onto which MEH-PPV was spin-coated. Best devices, having aligned nanowires, resulted in a V_{oc} of 0.57 V, J_{sc} of 1.28 mA/cm^2, and PCE of 0.41%. Kim et al. [106] have investigated the effects of carboxylic groups on polythiophenes and the resulting photovoltaic performance of TiO_2-hybrid solar cells. Films of nanocrystalline TiO_2 were prepared by spin-coating a TiO_2 paste onto fluorine-doped tin oxide (SnO_2:F)-covered glass substrates. After sintering at 450 °C for 30 min followed by cooling to 120 °C, the coated substrates were dipped in a polythiophene solution for 1 day. It was found that carboxylic group containing polythiophenes show better PV performances than those containing ester groups due to their efficient adsorption on the surface of TiO_2. Depending on the polymer used, V_{oc}, J_{sc} and PCE of up to ≈0.46 V, 8.0 mA/cm^2 and 1.5% were observed.

In 2007, Shankar et al. [107] have investigated self-assembled hybrid polymer-TiO_2 nanotube array heterojunction solar cells. Films comprised of 4 μm long titanium dioxide nanotube arrays were fabricated by anodizing Ti foils in an ethylene glycol-based electrolyte. After cleaning and annealing, a carboxylated polythiophene derivative was self-assembled onto the TiO_2 nanotube arrays by immersing them in a solution of the polymer. The binding sites of the carboxylate moiety along the polymer chain provide multiple anchoring sites to the substrate, allowing for a stable rugged film. Resulting devices exhibited a PCE of 2.1%.

Ishwara et al. [108] have investigated the dependence of the open-circuit voltage, V_{oc}, of polymer/TiO_2 hybrid devices on the ionization potential of the polymer (IP). A bilayer structure consisting of flat layers of dense TiO_2 and polymer was chosen for this study in order to simplify analysis and to avoid sample variation due to different degrees of polymer penetration into TiO_2 nanostructure pores. Once corrected for differences in photocarrier generation by the polymers, the measured V_{oc} values vary linearly with the polymer IP, with a slope of 0.8 ± 0.1. This observation agrees with previous reports that the V_{oc} of purely organic donor–acceptor solar cells is limited by the energy difference between the HOMO level of the donor and the LUMO level of the acceptor, the latter being equivalent to the conduction band edge of TiO_2.

3.3.4 Future Prospects for Inorganic Nanoparticle Acceptor Materials

While significant progress has been made using inorganic nanoparticles as electron acceptor materials in hybrid PSCs, performance remains inferior to that of exclusively OPV devices. For instance, the J_{sc} of nearly all devices based on inorganic nanoparticle acceptors has been below 10 mA/cm^2, independent of the kind of polymeric donor, such as low bandgap or wide bandgap material, which was used in the devices.

Particularly, miscibility with polymer donor materials was found to be limited, making the formation of an efficient interpenetrating network challenging. However, treatment with ligands leading to the functionalization of the nanoparticles but also the use of modified polymers, e.g., bearing carboxylic acid group, were found to be possible pathways to address the miscibility issue at least partially.

Future work attempting to improve the morphology and therefore the charge transfer in inorganic/organic hybrid solar cells may include the use of small molecule donor materials but also of donor materials (polymers or small molecules) bearing functional groups suitable for better interaction with inorganic nanoparticles, e.g., via hydrogen bridges. Also, the use of other, e.g., more polar solvents such as alcohols or even water could be helpful for the formation of better active layer morphologies allowing for improved charge transport.

3.4 Summary

After the screening of a large range of materials including inorganic semiconductors, n-type small molecules and polymers, fullerene derivatives remain the dominant electron acceptor (n-phase) materials used within OPV devices. However, despite the investigation of a significant number of functionalization schemes, phenyl-C_{61}-butyric acid methyl ester ([60]PCBM) and its C_{70} equivalent, [70]PCBM, are still by far the most widely used compounds. Having otherwise very similar electronic and chemical properties as [60]PCBM, due to its absorption spectrum extended to longer wavelengths, the use of [70]PCBM, still significantly more expensive, has led in many cases to improved device performance, particularly when using low bandgap polymers. The long history of the development of new donor materials based on knowledge previously acquired in PCBM-based systems, coupled with the limited commercial availability of other fullerene derivatives, is likely to be a major reason for the continued PCBM dominance.

Nevertheless, taking into account the properties of many low bandgap polymers recently developed, further improvement of the performance of single junction but also tandem solar cells will require the development of fullerene derivatives with higher LUMO levels allowing for increased V_{oc} while ensuring efficient exciton separation and charge transport.

It was found that the solubility in the solvent used for active layer deposition, crystal properties, energy levels, absorption intensity and range, electron mobility, the numbers of isomers and the miscibility with the donor material have a very pronounced influence on the final device performance. Among these factors, the LUMO level of the acceptor, relative to the HOMO level of the donor, determines largely the V_{oc} of the final devices. The other factors were found to influence mainly J_{sc} and FF. The solubility, crystal properties and miscibility of a fullerene derivative with the donor material have significant influence on the active layer morphology, donor/acceptor phase separation and thus on J_{sc} and FF of the device. The absorption intensity and range will directly influence the number of photons absorbed by the active layer and thus affect device J_{sc} and FF. The electron mobility will influence the free charge transport efficiency as well as the frequency of exciton recombination, and therefore impact J_{sc} and FF. The number of isomers, particularly of multiple adducts, is thought to lead in many cases to a distribution of different energy, particularly, LUMO levels. Such energy distribution is likely to result in electron traps in the active layer and therefore reduced charge transport as well as electron mobility.

The design of better fullerene-based acceptor materials requires balancing all electronic, physical and chemical factors listed above. In addition, the possibility of a rather easy and high-yield synthesis at large scale including purification will be necessary for the further commercial development of selected materials and OPV in general.

While not convincingly demonstrated yet, it can be expected that the performance of single and tandem photovoltaic devices using state-of-the-art low bandgap polymers will be further improved by developing novel fullerene acceptors.

References

1. Rohlfing EA, Cox DM, Kaldor A (1984) Production and characterization of supersonic carbon cluster beams, *J Chem Phys*, **81**, 3322–3330.
2. Arbogast JW, Foote CS (1991) Photophysical properties of C_{70}, *J Am Chem Soc*, **113**, 8886–8889.
3. Xie QS, Cordero EP, Echegoyen L (1992) Electrochemical detection of C_{60}^{6-} and C_{70}^{6-}: Enhanced stability of fullerides in solution, *J Am Chem Soc*, **114**, 3978–3980.

4. Sariciftci NS, Smilowitz L, Heeger AJ, Wudl F (1992) Photoinduced electron transfer from a conducting polymer to buckminsterfullerene, *Science*, **258**, 1474–1476.

5. Sariciftci NS, Braun D, Zhang C, et al. (1993) Semiconducting polymer-buckminsterfullerene heterojunctions: Diodes, photodiodes and photovoltaic cells, *Appl Phys Lett*, **62**, 585–587.

6. Yu G, Gao J, Hummelen JC, Wudl F, Heeger AJ (1995) Polymer photovoltaic cells: Enhanced efficiencies via a network of internal donor-acceptor heterojunctions, *Science*, **270**, 1789–1791.

7. Halls JJM, Walsh CA, Greenham NC, et al. (1995) Efficient photodiodes from interpenetrating polymer networks, *Nature*, **376**, 498–500.

8. a) Zhou EJ, Cong JZ, Wei QS, Tajima K, Yang CH, Hashimoto K (2011) All-polymer solar cells from perylene diimide based copolymers: Material design and phase separation control, *Angew Chem Int Ed*, **50**, 2799–2803; b) Tan ZA, Zhou EJ, Zhan XW, et al. (2008) Efficient all-polymer solar cells based on blend of tris(thienylenevinylene)-substituted polythiophene and poly[perylene diimide-alt-bis(dithienothiophene)], *Appl Phys Lett*, **93**, 073309 (1–3); c) Fabiano S, Chen Z, Vahedi S, Facchetti A, Pignataro B, Loi MA (2011) Role of photoactive layer morphology in high fill factor all-polymer bulk heterojunction solar cells, *J Mater Chem*, **21**, 5891–5896; d) Zhan XW, Tan ZA, Zhou EJ, et al. (2009) Copolymers of perylene diimide with dithienothiophene and dithienopyrrole as electron-transport materials for all-polymer solar cells and field-effect transistors, *J Mater Chem*, **19**, 5794–5803.

9. a) Holcombe TW, Woo CH, Kavulak DFJ, Thompson BC, Frechet JMJ (2009) All-polymer photovoltaic devices of poly (3-(4-N-octyl)-phenylthiophene) from Grignard metathesis (GRIM) polymerization, *J Am Chem Soc*, **131**, 14160–14161; b) Cevik E, Ilicali D, Egbe DAM, Günes S (2012) Bulk heterojunction and inverted type solar cells using a CN-PPV derivative, *Sol Energy Mater Sol Cells*, **98**, 94–102; c) Mandoc MM, Veurman W, Koster LJA, et al. (2007) Charge transport in MDMO-PPV: PCNEPV all-polymer solar cells, *J Appl Phys*, **101**, 104512(1–5).

10. www.polyera.com

11. Schwenn PE, Gui K, Nardes AM, et al. (2011) A small molecule non-fullerene electron acceptor for organic solar cells, *Adv Energy Mater*, **1**, 73–81.

12. Zhou TL, Jia T, Kang BN, Li FH, Fahlman M, Wang Y (2011) Nitrile-substituted QA derivatives: New acceptor materials for solution-processable organic bulk heterojunction solar cells, *Adv Energy Mater*, **1**, 431–439.

13. Brunetti FG, Gong X, Tong M, Heeger AJ, Wudl F (2010) Strain and huckel aromaticity: Driving forces for a promising new generation of electron acceptors in organic electronics, *Angew Chem Int Ed*, **49**, 532–536.
14. Wang MF, Sun YM, Tong MH, et al. (2011) The N=S=N link as electron accepting moiety for stable, solution-processable conjugated oligomers, *J Polym Sci A Polym Chem*, **49**, 441–451.
15. Ren GQ, Ahmed E, Jenekhe SA (2011) Non-fullerene acceptor-based bulk heterojunction polymer solar cells: Engineering the nanomorphology via processing additives, *Adv Energy Mater*, **1**, 946–953.
16. Bloking JT, Han X, Higgs AT, et al. (2011) Solution-processed organic solar cells with power conversion efficiencies of 2.5% using benzothiadiazole/imide-based acceptors, *Chem Mater*, **23**, 5484–5490.
17. Anthony JE (2011) Small-molecule, nonfullerene acceptors for polymer bulk heterojunction organic photovoltaics, *Chem Mater*, **23**, 583–590.
18. Sivaraman N, Dhamodaran R, Kaliappan I, Srinivasan TG, Vasudeva Rao PR, Mathewsl CK (1992) Solubility of C_{60} in organic solvents, *J Org Chem*, **57**, 6077–6079.
19. Ruoff RS, Tse DS, Malhotra R, Lorents DC (1993) Solubility of C_{60} in a variety of solvents, *J Phys Chem*, **97**, 3379–3383.
20. Schinazi RF, McMillan A, Juodawlkis AS, et al. (1994) Anti-human immunodeficiency virus, toxicity in cell culture, and tolerance in mammals of a water-soluble fullerene, *Proc Electrochem Soc,* **94**(24), 689–696.
21. a) Hummelen JC, Knight BW, LePeq F, Wudl F, Yao J, Wilkins CL (1995) Preparation and characterization of fulleroid and methanofullerene derivatives, *J Org Chem*, **60**, 532–538; b) Gonzalez R, Hummelen JC, Wudl F (1995) The specific acid-catalyzed and photochemical isomerization of a robust fulleroid to a methanofullerene, *J Org Chem*, **60**, 2618–2620.
22. Wienk MM, Kroon JM, Verhees WJH, et al. (2003) Efficient methano[70] fullerene/MDMO-PPV bulk heterojunction photovoltaic cells, *Angew Chem Int Ed*, **42**, 3371–3375.
23. Zheng LP, Zhou QM, Deng XY, Yuan M, Yu G, Cao Y (2004) Methanofullerenes used as electron acceptors in polymer photovoltaic devices, *J Phys Chem B*, **108**, 11921–11926.
24. Seri M, Rossi E, Carofiglio T, et al. (2011) Efficient as-cast bulk-heterojunction solar cells based on a tert-butyl substituted methanofullerene acceptor, *J Mater Chem*, **21**, 18308–18316.

25. a) Yang C, Kim JY, Cho S, Lee JK, Heeger AJ, Wudl F (2008) Functionalized methanofullerenes used as n-type materials in bulk-heterojunction polymer solar cells and in field-effect transistors, *J Am Chem Soc*, **130**, 6444–6450; b) Kooistra FB, Knol J, Kastenberg F, et al. (2007) Increasing the open circuit voltage of bulk-heterojunction solar cells by raising the LUMO level of the acceptor, *Org Lett*, **9**, 551–554.

26. Renz JA, Troshin PA, Gobsch G, Razumov VF, Hoppe H (2008) Fullerene solubility–current density relationship in polymer solar cells, *Phys Stat Sol (RRL)*, **2**, 263–265.

27. Troshin PA, Hoppe H, Renz J, et al. (2009) Material solubility-photovoltaic performance relationship in the design of novel fullerene derivatives for bulk heterojunction solar cells, *Adv Funct Mater*, **19**, 779–788.

28. Zhao GJ, He YJ, Xu Z, et al. (2010) Effect of carbon chain length in the substituent of PCBM-like molecules on their photovoltaic properties, *Adv Funct Mater*, **20**, 1480–1487.

29. Yamada I, Pandey M, Hayashi Y, Shibata N, Soga T, Toru T (2011) Influence of new fullerene derivatives with fluorocarbon substituent on performance of polymer solar cells, *Phys Procedia*, **14**, 192–197.

30. Moriwaki K, Matsumoto F, Takao Y, Shimizu D, Ohno T (2010) Synthesis and properties of novel methanofullerenes having ethylthienyl and/or n-pentyl group for photovoltaic cells, *Tetrahedron*, **66**, 7316–7321.

31. Choi JH, Son KI, Kim T, Kim K, Ohkubo K, Fukuzumi KS (2010) Thienyl-substituted methanofullerene derivatives for organic photovoltaic cells, *J Mater Chem*, **20**, 475–482.

32. Chuang SC, Chiu CW, Chien SC, Chu CW, Chen FC (2011) 1-(3-methoxycarbonyl)propyl-2-selenyl-[6,6]-methanofullerene as a n-type material for organic solar cells, *Synth Met*, **161**, 1264–1269.

33. Zhang Y, Yip HL, Acton O, Hau SK, Huang F, Jen AKY (2009) A simple and effective way of achieving highly efficient and thermally stable bulk-heterojunction polymer solar cells using amorphous fullerene derivatives as electron acceptor, *Chem Mater*, **21**, 2598–2600.

34. Kim SO, Chuang DS, Cha H, et al. (2011) Thermally stable organic bulk heterojunction photovoltaic cells incorporating an amorphous fullerene derivative as an electron acceptor, *Sol Energy Mater Sol Cells*, **95**, 432–439.

35. Mikroyannidis JA, Kabanakis AN, Sharma SS, Sharma GD (2011) A simple and effective modification of PCBM for use as an electron acceptor in efficient bulk heterojunction solar cells, *Adv Funct Mater*, **21**, 746–755.

36. Park SH, Yang C, Cowan S, et al. (2009) Isomeric iminofullerenes as acceptors in bulk heterojunction organic solar cells, *J Mater Chem*, **19**, 5624–5628.
37. a) Ross RB, Cardona CM, Guldi DM, et al. (2009) Endohedral fullerenes for organic photovoltaic devices, *Nat Mater*, **8**, 208–212; b) Ross RB, Cardona CM, Swain FB, et al. (2009) Tuning conversion efficiency in metallo endohedral fullerene-based organic photovoltaic devices, *Adv Funct Mater*, **19**, 2332–2337.
38. a) Frost JM, Faist MA, Nelson J (2010) Energetic disorder in higher fullerene adducts: A quantum chemical and voltammetric study, *Adv Mater*, **22**, 4881–4884; b) Faist MA, Keivanidis PE, Foster S, et al. (2011) Effect of multiple adduct fullerenes on charge generation and transport in photovoltaic blends with poly(3-hexylthiophene-2,5-diyl), *J Polym Sci B*, **49**, 45–51; c) Schafferhans J, Deibel C, Dyakonov V (2011) Electronic trap states in methanofullerenes, *Adv Energy Mater*, **1**, 655–660.
39. Lenes M, Wetzelaer GAH, Kooistra FB, Veenstra SC, Hummelen JC, Blom PWM (2008) Fullerene bisadducts for enhanced open-circuit voltages and efficiencies in polymer solar cells, *Adv Mater*, **20**, 2116–2119.
40. Lenes M, Shelton SW, Sieval AB, Kronholm DF, Hummelen JC, Blom PWM (2009) Electrons trapping in higher adduct fullerene-based solar cells, *Adv Funct Mater*, **19**, 3002–3007.
41. He YJ, Peng B, Zhao GJ, Zou YP, Li YF (2011) Indene addition of [6,6]-phenyl-C_{61}-butyric acid methyl ester for high-performance acceptor in polymer solar cells, *J Phys Chem C*, **115**, 4340–4344.
42. Li CZ, Chien SC, Yip HL, et al. (2011) Facile synthesis of a 56π-electron 1,2-dihydromethano-[60]PCBM and its application for thermally stable polymer solar cells, *Chem Commun*, **47**, 10082–10084.
43. a) Riedel I, Martin N, Giacalone F, et al. (2004) Polymer solar cells with novel fullerene-based acceptor, *Thin Solid Films*, **451–452**, 43–47; b) Riedel I, Hauff EV, Parisi J, Martin N, Giacalone F, Dyakonov V (2005) Diphenylmethanofullerenes: New and efficient acceptors in bulk-heterojunction solar cells, *Adv Funct Mater*, **15**, 1979–1987.
44. Bolink HJ, Coronado E, Forment-Aliaga A, et al. (2011) Polymer solar cells based on diphenylmethanofullerenes with reduced sidechain length, *J Mater Chem*, **21**, 1382–1386.
45. Sukeguchi D, Singh SP, Reddy MR, et al. (2009) New diarylmethanofullerene derivatives and their properties for organic thin-film solar cells, *Beilstein J Org Chem*, **5**, 1–10.

46. Lee JK, Fujida K, Tsutsui T, Kim MR (2007) Synthesis and photovoltaic properties of soluble fulleropyrrolidine derivatives for organic solar cells, *Sol Energy Mater Sol Cells*, **91**, 892–896.

47. Camaioni N, Garlaschelli L, Geri A, Maggini M, Possamai G, Ridolfi G (2012) Solar cells based on poly(3-alkyl)thiophenes and [60]fullerene: A comparative study, *J Mater Chem*, **12**, 2065–2070.

48. Backer SA, Sivula K, Kavulak DF, Fréchet JMJ (2007) High efficiency organic photovoltaics incorporating a new family of soluble fullerene derivatives, *Chem Mater*, **19**, 2927–2929.

49. Mi DB, Kim JH, Yoon SC, Lee CJ, Lee JK, Hwang DH (2011) Synthesis and characterization of a novel fullerene derivative containing carbazole group for use in organic solar cells, *Synth Met*, **161**, 1330–1335.

50. Mi DB, Kim JH, Xu F, et al. (2011) Synthesis and characterization of a novel fullerene derivative for use in organic solar cells, *Sol Energy Mater Sol Cells*, **95**, 1182–1187.

51. Troshin PA, Romanova IP, Susarova DK, et al. (2010) The first phosphorus-containing fullerene derivative applied as an electron acceptor material in organic solar cells, *Mendeleev Commun*, **20**, 137–139.

52. Nam SY, Park EY, Kim TD, Cho S, Park JG, Lee KS (2011) Solution-processable fullerene derivatives for organic photovoltaics and n-type thin-film transistors, *Curr Appl Phys*, **11**, e44–e48.

53. Niinomi T, Matsuo Y, Hashiguchi M, Sato Y, Nakamura E (2009) Penta(organo)[60]fullerenes as acceptors for organic photovoltaic cells, *J Mater Chem*, **19**, 5804–5811.

54. Varotto A, Treat ND, Jo J, et al. (2011) 1, 4-Fullerene derivatives: Tuning the properties of the electron transporting layer in bulk-heterojunction solar cells, *Angew Chem Int Ed*, **50**, 5166–5169.

55. Matsuo Y, Sato Y, Niinomi T, Soga I, Tanaka H, Nakamura E (2009) Columnar structure in bulk heterojunction in solution-processable three-layered p-i-n organic photovoltaic devices using tetrabenzoporphyrin precursor and silylmethyl[60]fullerene, *J Am Chem Soc*, **131**, 16048–16050.

56. a) Meng XY, Zhang WQ, Tan ZA, et al. (2011) Dihydronaphthyl-based [60] fullerene bisadducts for efficient and stable polymer solar cells, *Chem Commun*, **48**, 425–427; b) Kim KH, Kang H, Nam SY, et al. (2011) Facile synthesis of o-xylenyl fullerene multiadducts for high open circuit voltage and efficient polymer solar cells, *Chem Mater*, **23**, 5090–5095; c) Voroshazi E, Vasseur K, Aernouts T, et al. (2011) Novel bis-C_{60} derivative compared to other fullerene bis-adducts in

high efficiency polymer photovoltaic cells, *J Mater Chem*, **21**, 17345–17352; d) Deng LL, Feng J, Sun LC, et al. (2012) Functionalized dihydronaphthyl-C_{60} derivatives as acceptors for efficient polymer solar cells with tunable photovoltaic properties, *Sol Energy Mater Sol Cells*, **104**, 113–120; e) Meng X, Xu Q, Zhang W, et al. (2012) Effects of alkoxy chain length in alkoxy-substituted dihydronaphthyl-based [60]fullerene bisadduct acceptors on their photovoltaic properties, *ACS Appl Mater Interfaces*, **4**, 5966–5973.

57. a) He YJ, Chen XY, Hou JH, Li YF (2010) Indene–C_{60} bisadduct: A new acceptor for high-performance polymer solar cells, *J Am Chem Soc*, **132**, 1377–1382; b) Zhao GJ, He YJ, Li YF (2010) 6.5% efficiency of polymer solar cells based on poly(3-hexylthiophene) and indene-C_{60} bisadduct by device optimization, *Adv Mater*, **22**, 4355–4358; c) Laird DW, Stegamat R, Daadi M, et al. (July 10, 2012) Organic photovoltatic devices comprising fullerenes and derivatives thereof, United States Patent US 8,217,260 B2.
58. He YJ, Zhao GJ, Peng B, Li YF (2010) High-yield synthesis and electrochemical and photovoltaic properties of indene-C_{70} bisadduct, *Adv Funct Mater*, **20**, 3383–3389.
59. He YJ, Chen XY, Zhao GJ, Hou JH, Li YF (2011) Biindene-C_{60} adducts for the application as acceptor in polymer solar cells with higher open-circuit-voltage, *Sol Energy Mater Sol Cells*, **95**, 899–903.
60. Cheng YJ, Liao MH, Chang CY, Kao WS, Wu CE, Hsu CS (2011) Di(4-methylphenyl)methano-C_{60} bis-adduct for efficient and stable organic photovoltaics with enhanced open-circuit voltage, *Chem Mater*, **23**, 4056–4062.
61. Zhang Y, Matsuo Y, Li CZ, Tanaka H, Nakamura E (2011) A scalable synthesis of methano[60]fullerene and congeners by the oxidative cyclopropanation reaction of silylmethylfullerene, *J Am Chem Soc*, **133**, 8086–8089.
62. Kim H, Seo JH, Park EY, et al. (2010) Increased open-circuit voltage in bulk-heterojunction solar cells using a C_{60} derivative, *Appl Phys Lett*, **97**, 193309–193312.
63. Wang CL, Zhang WB, Vanhorn RM, et al. (2011) A porphyrin-fullerene dyad with a supramolecular "double-cable" structure as a novel electron acceptor for bulk heterojunction polymer solar cells, *Adv Mater*, **23**, 2951–2956.
64. He YJ, You JB, Dou LT, et al. (2012) High performance low bandgap polymer solar cells with a non-conventional acceptor, *Chem Commun*, **48**, 7616–7618.

65. Hagfeldt A, Boschloo G, Sun LC, Kloo L, Pettersson H (2010) Dye-sensitized solar cells, *Chem Rev*, **110**, 6595–6663.

66. Jun YW, Choi JS, Cheon J (2006) Shape control of semiconductor and metal oxide nanocrystals through nonhydrolytic colloidal routes, *Angew Chem Int Ed,* **45**, 3414–3439.

67. Burda C, Chen XB, Narayanan R, El-Sayed MA (2005) Chemistry and properties of nanocrystals of different shapes, *Chem Rev*, **105**, 1025–1102.

68. Kumar S, Scholes GD (2008) Colloidal nanocrystal solar cells, *Microchim Acta*, **160**, 315–325.

69. Milliron DJ, Gur I, Alivisatos AP (2005) Hybrid organic—nanocrystal solar cells, *MRS Bull*, **30**, 41–44.

70. a) Ravirajan P, Haque SA, Durrant JR, Poplavskyy D, Bradley DDC, Nelson J (2004) Hybrid nanocrystalline TiO_2 solar cells with a fluorene–thiophene copolymer as a sensitizer and hole conductor, *J Appl Phys*, **95**, 1473–1480; b) Coakley KM, Liu YX, McGehee MD, Frindell KL, Stucky GD (2003) Infiltrating semiconducting polymers into self-assembled mesoporous titania films for photovoltaic applications, *Adv Funct Mater*, **13**, 301–306; c) Coakley KM, McGehee MD (2003) Photovoltaic cells made from conjugated polymers infiltrated into mesoporous titania, *Appl Phys Lett*, **83**, 3380–3382.

71. a) Vayssieres L, Keis K, Lindquist ST, Hagfeldt A (2001) Purpose-built anisotropic metal oxide material: 3D highly oriented microrod array of ZnO, *J Phys Chem B*, **105**, 3350–3352; b) Vayssieres L (2003) Growth of arrayed nanorods and nanowires of ZnO from aqueous solutions, *Adv Mater*, **15**, 464–466.

72. Kang YM, Park NG, Kim D (2005) Hybrid solar cells with vertically aligned CdTe nanorods and a conjugated polymer, *Appl Phys Lett*, **86**, 113101–113103.

73. a) Breeze AJ, Schlesinger Z, Carter SA, Brock PJ (2001) Charge transport in TiO_2/MEH-PPV polymer photovoltaics, *Phys Rev B*, **64**, 125205–125213; b) VanHal PA, Wienk MM, Kroon JM, et al. (2003) Photoinduced electron transfer and photovoltaic response of a MDMO-PPV: TiO_2 bulk-heterojunction, *Adv Mater*, **15**, 118–121.

74. Ravirajan P, Bradley DDC, Nelson J, et al. (2005) Efficient charge collection in hybrid polymer/TiO_2 solar cells using poly(ethylenedioxythiophene) polystyrene sulphonate as hole collector, *Appl Phys Lett*, **86**, 143101–143103.

75. Greenham NC, Peng XG, Alivisatos AP (1997) Charge separation and transport in conjugated polymer cadmium selenide nanocrystal

composites studied by photoluminescence quenching and photoconductivity, *Synth Met*, **84**, 545–546.

76. Huynh WU, Peng XG, Alivisatos AP (1999) CdSe nanocrystal rods/poly(3-hexylthiophene) composite photovoltaic devices, *Adv Mater*, **11**, 923–927.

77. a) Huynh WU, Dittmer JJ, Alivisatos AP (2002) Hybrid nanorod-polymer solar cells, *Science*, **295**, 2425–2427; b) Huynh WU, Dittmer JJ, Libby WC, Whiting GL, Alivisatos AP (2003) Controlling the morphology of nanocrystal-polymer composites for solar cells, *Adv Funct Mater*, **13**, 73–79.

78. Sun BQ, Marx E, Greenham NC (2003) Photovoltaic devices using blends of branched CdSe nanoparticles and conjugated polymers, *Nano Lett*, **3**, 961–963.

79. Sun BQ, Snaith HJ, Dhoot AS, Westenhoff S, Greenham NC (2005) Vertically segregated hybrid blends for photovoltaic devices with improved efficiency, *J Appl Phys*, **97**, 014914–014919.

80. Sun BQ, Greenham NC (2006) Improved efficiency of photovoltaics based on CdSe nanorods and poly(3-exylthiophene) nanofibers, *Phys Chem Chem Phys*, **8**, 3557–3560.

81. Gur I, Fromer NA, Chen CP, Kanaras AG, Alivisatos AP (2007) Hybrid solar cells with prescribed nanoscale morphologies based on hyperbranched semiconductor nanocrystals, *Nano Lett*, **7**, 409–414.

82. Liu J, Tanaka T, Sivula K, Alivisatos AP, Frechet JMJ (2004) Employing end-functional polythiophene to control the morphology of nanocrystal–polymer composites in hybrid solar cells, *J Am Chem Soc*, **126**, 6550–6551.

83. Wang L, Liu YS, Jiang X, Qin DH, Cao Y (2007) Enhancement of photovoltaic characteristics using a suitable solvent in hybrid polymer/multiarmed CdS nanorods solar cells, *J Phys Chem C*, **111**, 9538–9542.

84. Zutz F, Lokteva I, Radychev N, et al. (2009) Study of the influence of the Cd:Se precursor ratio during the synthesis of CdSe nanocrystals on the performance of CdSe/P3HT hybrid solar cells, *Phys Stat Sol A*, **206**, 2700–2708.

85. Radychev N, Lokteva I, Witt F, Kolny-Olesiak J, Borchert H, Parisi J (2011) Physical origin of the impact of different nanocrystal surface modifications on the performance of CdSe/P3HT hybrid solar cells, *J Phys Chem C*, **115**, 14111–14122.

86. Greaney MJ, Das M, Webber DH, Bradforth SE, Brutchey RL (2012) Improving open circuit potential in hybrid P3HT: CdSe bulk

heterojunction solar cells via colloidal tert-butylthiol ligand exchange, *ACS Nano*, **6**, 4222–4230.

87. Dayal S, Kopidakis N, Olson DC, Ginley DS, Rumbles G (2010) Photovoltaic devices with a low bandgap polymer and CdSe nanostructures exceeding 3% efficiency, *Nano Lett*, **10**, 239–242.

88. Wang P, Abrusci A, Wong HMP, Svensson M, Andersson MR, Greenham NC (2006) Photoinduced charge transfer and efficient solar energy conversion in a blend of a red polyfluorene copolymer with CdSe nanoparticles, *Nano Lett*, **6**, 1789–1793.

89. Zhou Y, Li YC, Zhong HZ, et al. (2006) Hybrid nanocrystal/polymer solar cells based on tetrapod-shaped $CdSe_xTe_{1-x}$ nanocrystals, *Nanotechnology*, **17**, 4041–4047.

90. Verma D, Rao AR, Dutta V (2009) Surfactant-free CdTe nanoparticles mixed MEH-PPV hybrid solar cell deposited by spin coating technique, *Sol Energy Mater Sol Cells*, **93**, 1482–1487.

91. a) Yu WL, Zhang H, Fan ZX, et al. (2011) Efficient polymer/nanocrystal hybrid solar cells fabricated from aqueous materials, *Energy Environ Sci*, **4**, 2831–2834; b) Yu WL, Zhang H, Tian HR, et al. (2012) Correlation between annealing-induced growth of nanocrystals and the performance of polymer: Nanocrystals hybrid solar cells, *J Phys Chem C*, **116**, 1322–1328.

92. Chen HC, Lai CW, Wu IC, et al. (2011) Enhanced performance and air stability of 3.2% hybrid solar cells: How the functional polymer and CdTe nanostructure boost the solar cell efficiency, *Adv Mater*, **23**, 5451–5455.

93. Liao HC, Chen SY, Liu DM (2009) In-situ growing CdS single-crystal nanorods via P3HT polymer as a soft template for enhancing photovoltaic performance, *Macromolecules*, **42**, 6558–6563.

94. Lee WJ, Mane RS, Min SK, Yoon TH, Han SH, Lee SH (2007) Nanocrystalline CdS-water-soluble conjugated polymers: High performance photoelectrochemical cells, *Appl Phys Lett*, **90**, 263503–263505.

95. a) Meulenkamp EA (1998) Synthesis and growth of ZnO nanoparticles. *J Phys Chem B*, **102**, 5566–5572; b) Pacholski C, Kornowski A, Weller H (2002) Self-assembly of ZnO: From nanodots to nanorods, *Angew Chem Int Ed*, **41**, 1188–1191.

96. Beek WJE, Wienk MM, Janssen RAJ (2004) Efficient hybrid solar cells from zinc oxide nanoparticles and a conjugated polymer, *Adv Mater*, **16**, 1009–1013.

97. Beek WJE, Slooff LH, Wienk MM, Kroon JM, Janssen RAJ (2005) Hybrid solar cells using a zinc oxide precursor and a conjugated polymer. *Adv Funct Mater*, **15**, 1703–1707.

98. Koster LJA, Strien WJV, Beek WJE, Blom PWM (2007) Device operation of conjugated polymer/zinc oxide bulk heterojunction solar cells, *Adv Funct Mater*, **17**, 1297–1302.

99. a) Beek WJE, Wienk MM, Janssen RAJ (2006) Hybrid solar cells from regioregular polythiophene and ZnO nanoparticles, *Adv Funct Mater*, **16**, 1112–1116; b) Quist PAC, Beek WJE, Wienk MM, Janssen RAJ, Savenije TJ, Siebbeles LDA (2006) Photogeneration and decay of charge carriers in hybrid bulk heterojunctions of ZnO nanoparticles and conjugated polymers, *J Phys Chem B*, **110**, 10315–10321.

100. Takanezawa K, Hirota K, Wei QS, Tajima K, Hashimoto K (2007) Efficient charge collection with ZnO nanorod array in hybrid photovoltaic devices, *J Phys Chem C*, **111**, 7218–7223.

101. a) Baeten L, Conings B, Boyen HG, et al. (2011) Towards efficient hybrid solar cells based on fully polymer infiltrated ZnO nanorod arrays, *Adv Mater*, **23**, 2802–2805; b) Oosterhout SD, Anton Koster LJA, Van Bavel SS, et al. (2011) Controlling the morphology and efficiency of hybrid ZnO: Polythiophene solar cells via side chain functionalization, *Adv Energy Mater*, **1**, 90–96; c) Sung YH, Liao WP, Chen DW, Wu CT, Chang GJ, Wu JJ (2012) Room-temperature tailoring of vertical ZnO nanoarchitecture morphology for efficient hybrid polymer solar cells, *Adv Funct Mater*, **22**, 3808–3814; d) Abdulalmohsin S, Cui JB (2012) Graphene-enriched P3HT and porphyrin-modified ZnO nanowire arrays for hybrid solar cell applications, *J Phys Chem C*, **116**, 9433–9438.

102. Bredol M, Matras K, Szatkowski A, Sanetra J, Schwab AP (2009) P3HT/ZnS: A new hybrid bulk heterojunction photovoltaic system with very high open circuit voltage, *Sol Energy Mater Sol Cells*, **93**, 662–666.

103. a) Arango AC, Carter SA, Brock PJ (1999) Charge transfer in photovoltaics consisting of interpenetrating networks of conjugated polymer and TiO_2 nanoparticles, *Appl Phys Lett*, **74**, 1698–1700; b) Carter SA, Scott JC, Brock PJ (1997) Enhanced luminance in polymer composite light emitting devices, *Appl Phys Lett*, **71**, 1145–1147.

104. VanHal PA, Wienk MM, Kroon JM, et al. (2003) Photoinduced electron transfer and photovoltaic response of a MDMO-PPV: TiO_2 bulk-heterojunction, *Adv Mater*, **15**, 118–121.

105. Shim HS, Na SI, Nam SH, et al. (2008) Efficient photovoltaic device fashioned of highly aligned multilayers of electrospun TiO_2 nanowire array with conjugated polymer, *Appl Phys Lett*, **92**, 183107–183110.

106. Kim YG, Walker J, Samuelson LA, Kumar J (2003) Efficient light harvesting polymers for nanocrystalline TiO_2 photovoltaic cells, *Nano Lett*, **3**, 523–525.

107. Shankar K, Mor GK, Prakasam HE, Varghese OK, Grimes GA (2007) Self-assembled hybrid polymer–TiO_2 nanotube array heterojunction solar cells, *Langmuir*, **23**, 12445–12449.

108. Ishwara T, Bradley DDC, Nelson J, et al. (2008) Influence of polymer ionization potential on the open-circuit voltage of hybrid polymer/TiO_2 solar cells, *Appl Phys Lett*, **92**, 053308–053310.

Chapter 4

Interfacial Layers

Riccardo Po,[a] Chiara Carbonera,[a] Andrea Bernardi,[a] and Nadia Camaioni[b]

[a]*Research Center for Non Conventional Energies—Istituto eni Donegani, eni SpA, Novara 28100, Italy*

[b]*Istituto per la Sintesi Organica e la Fotoreattività (CNR-ISOF), Consiglio Nazionale delle Ricerche, Bologna 40129, Italy*

Riccardo.po@eni.com

Interfacial layers are nowadays considered essential elements in organic photovoltaic devices. In this chapter, the different classes of interlayers will be reviewed after introducing the essential concepts related to the role they can play in the solar cells. The deposition methods of these interfacial layers, generally quite thin in comparison with the other layers forming the device structure, are briefly addressed at the end of the chapter.

4.1 Introduction

The essential components of any photovoltaic cell, including organic solar cells (OSCs), are the active layer, where charge carriers are photogenerated, and the two electrodes, at which

Organic Solar Cells: Fundamentals, Devices, and Upscaling
Edited by Barry P. Rand and Henning Richter

ISBN 978-981-4463-65-2 (Hardcover), 978-981-4463-66-9 (eBook)
www.panstanford.com

holes and electrons are collected. However, the typical structure of an organic cell, both in the conventional and in the inverted configuration, includes additional layers, interposed between the active layer and one or both electrodes (Fig. 4.1). These additional layers are referred to as either "buffer layers" or "interfacial layers" (or interlayers) and are essential to achieve highly efficient OSCs. Therefore, the suitable selection of the materials composing these layers, their deposition methods, and the optimization of their thickness are all factors that must be carefully considered. For a given active layer, significant improvements in the short-circuit current density (J_{SC}), open-circuit voltage (V_{OC}), and fill factor (FF) have been reported by choosing appropriate interlayers [1,2].

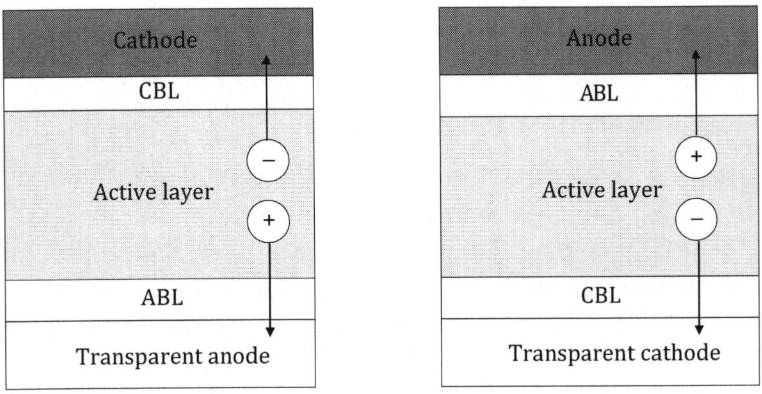

Figure 4.1 Architecture of conventional (left) and inverted (right) solar cells. ABL, anode buffer layer; CBL, cathode buffer layer.

A few reviews have been published recently, which deal with this subject in detail [3–5]. The aim of this chapter is to give the reader a brief overview of interfacial layers in OSCs and to provide a practical tool for the daily activity of researchers involved in the development of OSCs.

4.2 Charge Collection at Electrode Interfaces

One of the key points for a good performing OSC is undoubtedly represented by the quality of charge collection at the electrodes. Indeed, charges carriers, formed in the active layer upon exposure to sunlight, need to be effectively collected by the respective electrodes (holes at the anode and electrons at the cathode), in order to be exploited. The drift is assured by the intrinsic electric field generated by the difference in the work functions of the two asymmetric electrodes, but the overall charge collection process is not trivial, as several obstacles affect its efficiency.

The main obstacle is represented by the nonideal nature (i.e., nonohmic) of the contacts between the active layer and the electrodes. In real devices, barriers for charge extraction are present at one or both sides of the active layer, causing a diminished V_{OC} and affecting FF due to charge accumulation at the electrode–active layer interface. The poor energy-level alignment is a common feature of nonoptimized OSCs, originating from several sources (i.e., choice of the electrode material, nonconformal adhesion between adjacent layers, chemical reactions occurring at the electrode–active layer interface, etc.) and is responsible for nonoptimal performance. The use of suitable interlayers is an effective approach to improve energy-level alignment and avoid these aforementioned complications.

It is worth mentioning here the *integer charge model* developed for organic photovoltaic devices, which has been found to be valuable for most interfaces that are relevant to OSCs. This theory predicts Fermi-level pinning of a conducting interfacial layer to the organic semiconductor, until equilibrium is reached. No Fermi-level pinning takes place when spontaneous charge transfer between the two layers is not possible. A review of this subject is reported elsewhere [6–11].

To summarize, the optimization of device interfaces is of fundamental importance for the improvement of OSC performance. In general, interfacial layers inserted in between the electrodes and the active layer have largely proved to be effective in reducing deviation from ideal conditions, as will be discussed in more detail in the following sections. Examples of interfacial layers included in the structure of polymer solar cells (PSCs) will be mainly discussed in this chapter, but the same principles also apply to small-molecule organic solar cells (SMOSCs), fabricated both from solution and by thermal evaporation.

4.3 Roles of Interfacial Layers

In the previous paragraphs, the importance of the presence of interfacial layers in the architecture of OSCs has been underlined. In order to get a well-performing solar cell, it should be kept in mind that the appropriate choice of the material composing these layers (Table 4.1) is extremely critical. The right choice of the properties of the interlayer can in fact significantly contribute not only to the performance of OSCs but also to their stability (see Chapter 15). One of the clearest examples of the significant role of interfacial layers is given by inverted OSCs, largely more stable than conventional ones, in which the device polarity is inverted thanks to the appropriate choice of the anode buffer layer (ABL) and the cathode buffer layer (CBL). This demonstrates how the buffer layer engineering is as important as, for example, the choice and optimization of the cell active layer.

Table 4.1 Possible effects of interfacial layers

Effect classification	Details	Classes of materials or specific examples
Electrical effects	Electron blocking/hole transporting/suppression of recombination at the anode	High ionization potential (deep HOMO level) and high hole-mobility materials
	Hole blocking/electron transporting/suppression of recombination at the cathode	High electron affinity (deep LUMO level) and high electron-mobility materials
	Tuning of electrode work function	Materials with polar groups
	Device polarity inversion	Mainly TiO_x, and ZnO
	Decrease of contact resistance	poly(ethyleneglycol), fullerene derivatives
Exciton blocking effect	To avoid exciton quenching at electrodes	poly(aniline), tertiary aromatic amines, BCP
Optical effects	Enhancement of optical absorption	Reported for ZnO nanorods and other metal oxides
	Plasmonic effect	Metal nanoparticles (e.g., Ag, Au)
	Optical spacing effect	Reported for TiO_x, ZnO, MoO_3
Effects on device stability	UV blocking	Materials with strong optical absorption in the UV region
	Barrier to O_2, H_2O	Mainly metal oxides and metal salts
Mechanical effects	ITO (indium-doped tin oxide) surface planarization	Mainly reported for poly(3,4-ethylenedioxy-thiophene):poly(styrene-sulfonate) (PEDOT:PSS)
	Active layer protection during vacuum deposition of the top electrode	Mainly reported for MoO_3 and TiO_x

An extremely wide range of different effects have been reported for interfacial layers: electrical effects, optical effects, mechanical effects, exciton blocking effects, and even improvements of device stability, as mentioned earlier. A nonexhaustive schematic classification of the possible interlayer functions is shown in Table 4.1 and will be discussed in the following sections, with most examples related to PSCs, mainly based on the bulk-heterojunction (BHJ) approach, although the basic requirements of buffer layers do not significantly change for SMOSCs.

It should also be mentioned that the development of interfacial materials, tailored not only for conventional and inverted OSCs but also for tandem solar cells (see Chapter 6) as well as for ITO-free transparent electrodes (see Chapter 5), is an important research and technological topic that will probably be considered in a more consistent way in the near future.

4.3.1 Electrical Effects

As mentioned in Section 4.2, charge collection at the electrodes is one of the key points for producing efficient OSCs. Many factors playing a role in this process are reflected in the electrical characteristics of the cell.

Most of the commonly used materials to make electrical contacts of OSCs are able to collect both holes and electrons, that is, they show poor selectivity, leading to charge recombination at the electrodes. The typical example is ITO, widely used as the bottom transparent electrode, with a work function of around 4.7 eV, usually in-between the relevant energy levels of the active layer of a typical OSC. The selectivity issue is much more relevant in solar cells made with the BHJ structure, as opposed to a bilayer configuration, in which carriers of opposite sign are generated throughout the bulk of the cell, hence also close to both contacts. Electrode selectivity can be quite improved by inserting a buffer layer at the electrode–active layer interface with appropriate energy levels in order to collect just one type of the opposite charge carriers: an interlayer with a highest occupied molecular orbital (HOMO) level aligned with that of the electron-donor photoactive materials will select holes, whereas an interlayer with a lowest unoccupied molecular orbital (LUMO) level preferably aligned with that of the electron-acceptor photoactive materials will select electrons (Figure 4.2). Nevertheless, the proper position of the energy levels is not enough to assure an effective charge collection if the ability of the interlayer to transport the selected charge carriers is poor. This means that, in addition to the energy-level requirements, sufficient hole and electron mobility is needed for ABLs (also called hole-transporting layers) and CBLs (also called electron-transporting layers), respectively. Similar considerations apply to *excitons* that need to be dissociated before reaching an electrode interface.

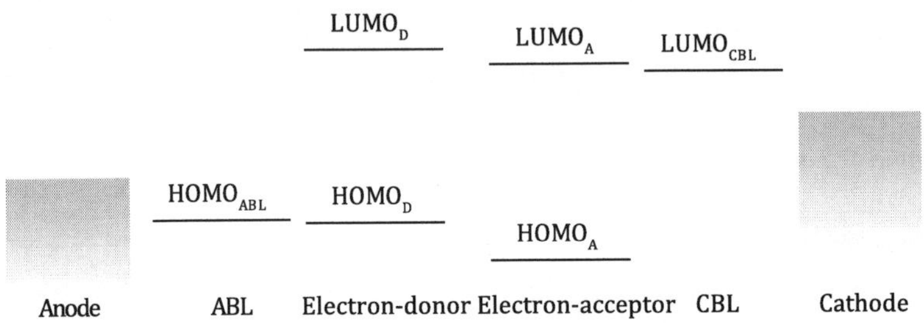

Figure 4.2 Ideal energy-level diagram for an OSC including buffer layers at both electrodes.

A further requirement for interlayers is to prevent the increase of the contact resistance. Otherwise, the overall series resistance of the device is increased and a detrimental effect for cell performance is observed. For this reason, high-resistivity buffer materials are usually effective only in very thin layers (see, for instance, Sections 4.4.1.4 and 4.4.3.2). Actually, the proper selection of the interlayer material, combined with the optimization of the thickness, can even decrease the contact resistance of the cell [12,13]. Finally, as already mentioned in the previous paragraph, the appropriate use of buffer layers allows the fabrication of solar cells with inverted polarity onto the same bottom transparent electrode used for the conventional architecture (usually ITO). In summary, the main electric functions of buffer layers are the tuning of the energy-level alignment at the electrode–active layer interfaces (in other words, improving their ohmic character), with a consequent enhanced selectivity, and the definition of cell polarity. Other minor functions can be mentioned, which can have an impact on the electrical behavior of the cell though cannot be strictly defined as "electrical effects": the control of the surface properties (e.g., to affect the morphology of the overlying active layer or to modify the surface energy to prevent interfacial dewetting and delamination) and the improvement of the stability of the electrode–active layer interface. For a deeper insight into this subject, the reader can refer to a recent review by Yip and Jen [4].

4.3.2 Optical Effects

It is quite evident that the overall efficiency of a solar cell strongly depends on the intensity of absorbed light and on how efficiently light is converted into free charge carriers.

The influence of buffer layers on the propagation of light through the cell structure has been considered, and it has been demonstrated that the right choice of buffer layers can significantly increase light absorption [14], by reducing parasitic reflection of light at the interfaces, and maximizing the portion of light that reaches the active layer. Indeed, a mismatch in the refractive index of two adjacent layers can cause a reflection of light, which leads to a loss for the solar cell. At the same time, the absorption of light should be maximized within the active layer, reducing the losses in the other layers composing the device structure. Therefore, buffer layers that absorb effectively in the visible spectrum at the same wavelength range of the photoactive layers, such as low-work-function transition metals, should be avoided when possible, whereas materials that are quite transparent in the visible light, such as TiO_x, ZnO and PEDOT:PSS, seem to be preferable.

Some specific device architectures can enhance light absorption and optimize the use of light within the active layer, leading to higher power-conversion efficiency (PCE). Transparent buffer layers can act as optical spacers by modifying the spatial distribution of light intensity within the device. The portion of light that passes through the active layer and is reflected back by the metallic electrode can interfere with the incident light leading to intensity peaks within the device. By changing the relative thickness of both the active layer and the transparent buffer layers, it is possible to move the position of these peaks in the regions of the cell in which absorbed photons have higher probability to be converted into charges. For more information about optical interference effects, see Chapter 8. TiO_x cathodic buffer layers have been recently used as optical spacers increasing the device efficiency in conventional solar cells [15], while a thermally evaporated MoO_3/ITO bilayer below a silver anode has been used in an inverted configuration [16].

An alternative way of increasing light absorption is the use of the plasmonic effect. The use of metallic nanoparticles, whose dimensions are comparable to the wavelength of incident light, generate a localized intensification of the electromagnetic field, thus a localized increase of light absorption. These metal nanoparticles can be either deposited on the surface of the electrode or of the buffer layer, or dispersed in the bulk of the buffer layer or of the active layer. In addition to the near-field absorption, plasmonic structures may also improve the far-field absorption by enhancing the scattering. This effect has been recently demonstrated in nanostructured electrodes [17,18].

Gold nanospheres 20–30 nm in diameter, dispersed in PEDOT:PSS, have been used for generating plasmonic effect and proved to increase the efficiency of solar cells [19]. On the contrary, it has been argued that the increased contact area between PEDOT:PSS and the active layer, due to the insertion of the nanospheres in the buffer layer, and their influence on the arrangement of the active layer could have similar or even higher effect on PCE than plasmonic effect [20].

4.3.3 Mechanical Effects

Although electrical and optical functions are the main purposes of buffer layers, in several cases, they also have other positive side effects. For instance, it is reported that PEDOT:PSS films are able to planarize the ITO electrode, whose rough surface would otherwise result in shorts through the thin active layer [21]. Buffer layers may also protect the active layer from damage arising from the vacuum deposition process used for the top electrode deposition [22].

4.3.4 Barrier Effects

Interlayer materials positively affect the device durability by blocking the diffusion of unwanted species, coming from electrode degradation during the device operation [23]. In addition, some metal salts–based buffer layers have been found to represent good barriers to gases, thus enhancing the device environmental stability (see Section 4.4.2.2).

Finally, buffer materials that are transparent to visible light but able to absorb UV radiation provide a certain degree of protection of the active layer against photodegradation. This effect has been exploited to improve the solar cell durability [24].

4.4 Classes of Materials Used as Interfacial Layers

In the following paragraphs, the materials used as interfacial layers are classified and discussed according to their chemical nature. The chemical properties responsible for the behavior of interfacial layers are underlined, and the resulting effect on solar cells, in terms of V_{OC}, J_{SC} and FF improvement, is highlighted.

The reader may refer to the original papers (digital object identifiers, or dois, are given) for more detailed information.

4.4.1 Polymeric Compounds

4.4.1.1 PEDOT

Historically speaking, PEDOT can be considered the first example of a buffer layer used in OSCs. It still remains the most widely used buffer layer in manifold ways. PEDOT refers to the semiconducting polymer poly(3,4-ethylenedioxythiophene). Several formulations of the polymer are known and employed, from the pure polymer synthesized in situ to hybrid inks, each one optimized to serve one or more particular functions (smoothening the underlying layer roughness, improving the anode selectivity by providing an ohmic contact with the donor material thanks to a relatively high work function of 5.0–5.1 eV). Since its very first use in solar cells, dating back to 1998 [25], or probably in a more conscious way in 2002 [26], the quality of buffer layers based on this p-type polymer has been greatly improved thanks to the work of both academic researchers and PEDOT producers.

As mentioned earlier, PEDOT is an organic semiconductor very versatile and relatively easy to use, and it has been used alone or in dispersions with poly(styrenesulfonic acid) (PSS) counterion as a transparent ABL. More information and an extensive description of PEDOT:PSS dispersions and their characteristics and uses can be found in a couple of reviews [5,27].

A different form of PEDOT, doped with chlorine ions, is obtained by oxidative chemical vapor deposition (oCVD) and up to now has mainly found application in SMOSCs [28].

The deposition process involves the evaporation/sublimation of a suitable oxidant (e.g., $FeCl_3$, $CuCl_2$, Br_2) in a vacuum chamber followed by an EDOT flux (Fig. 4.3). The film is grown on glass or plastic substrates at the desired thickness. Finally, the substrate is removed from the reaction chamber and rinsed with alcohols to remove the unreacted catalyst. Because the dopant is not a strong acid, such as PSS, and water is not used in the material synthesis, several drawbacks of PEDOT:PSS are avoided. On the contrary, the necessity of disposing of a suited deposition chamber makes the use of oCVD-PEDOT less practical.

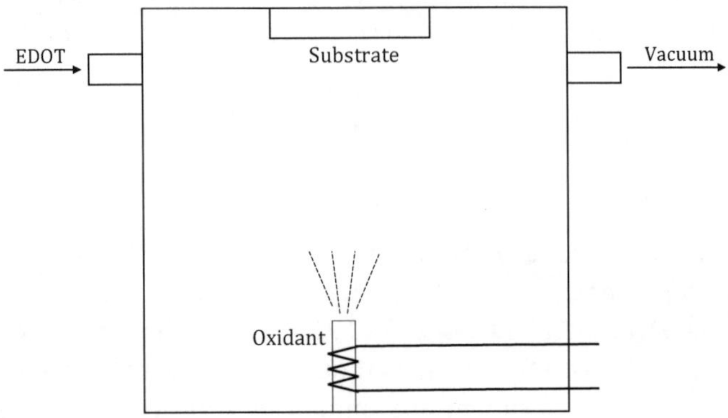

Figure 4.3 Scheme of an oCVD reactor.

Currently, almost all solar cells made with a *conventional* architecture include the PEDOT layer. Moreover, its commercial availability as "inks" of different formulations makes PEDOT:PSS compatible with the widely explored roll-to-roll processes for producing flexible devices (see Chapter 13).

Notwithstanding the interesting features of PEDOT, drawbacks are not missing, especially for PEDOT:PSS formulations. The acidity of these dispersions (typical pH between 1 and 2) causes the chemical degradation of the ITO electrode in the medium to long term [29], with migration of indium ions into the active layer. Furthermore, hygroscopicity of PEDOT:PSS may promote water absorption in unprotected devices, its migration to the cathode and the partial oxidation of the metal at the active layer/electrode interface [30]. The processes described contribute to accelerate the degradation of solar cell performance, even when encapsulated. When used as a buffer layer at the anode electrode (ABL) in *inverted* cells, PEDOT:PSS layer is sandwiched between the active blend and the top metal electrode, normally Ag or Au. Also in this case, the use of PEDOT:PSS ready-to-use inks is considered very interesting thanks to the advantage of wet deposition compared with alternative inorganic materials that are traditionally vacuum evaporated. A common problem to be faced, however, consists in the wettability and adhesion to the underlying active layer, which is commonly hydrophobic. Different approaches have been proposed in the literature to overcome this problem, mainly based on the addition of suitable surfactants or cosolvents able to reduce the surface energy of the underlying active layer [31].

Another drawback of PEDOT, recently outlined, is related to its mechanical failure in flexible devices. It has been demonstrated that the weak point of these devices, critical in the exploitation of their flexibility, is the interface between PEDOT:PSS and the active layer. Thermal annealing (with optimized time and temperature) has proved to help improvement, although V_2O_5, for instance, recently has shown superior properties [32].

Several ways have been attempted to improve the performances of PEDOT-based buffer layers. These consist both in chemical modifications (mainly additives, entries with "+" in Table 4.2, but also grafting functionalities or variation of the acidic form of PEDOT) and physical treatments, acting on the commercially available ink formulations or on the as-deposited buffer layer (e.g., spinning of a solvent on the top of the PEDOT:PSS film; entries with "&" in Table 4.2).

Typical additives used to improve the performance of PEDOT:PSS buffer layers (electrical conductivity, work function, and layer morphology, or, in some cases, stability) are monofunctional and polyfunctional alcohols.

Poly[3,4-(1-hydroxymethyl)ethylenedioxythiophene] (PHMEDOT, sometimes abbreviated as PEDTM) is a different form of PEDOT obtained through a hydroxymethyl functionalized monomer [33]. Another example is constituted by in situ electrochemically polymerized PEDOT with the addition of a thienylsilane molecular layer between ITO and PEDOT layers to provide a robust grafting [34]. The effect of inorganic additives such as NaOH has also been explored [35], as well as physical treatments, such as exposure to UV light [36]. Finally, there are a few cases of the use of modified oCVD-PEDOT used as buffer layer at the cathode electrode of SMOSCs [37].

Table 4.2 PEDOT and its modifications: examples of interlayers in conventional solar cells

Material	Buffer type	Article doi
PEDOT:PSS	ABL	10.1063/1.1739279
PEDOT:PSS & UV treatment	ABL	10.1016/j.orgel.2009.09.010
PEDOT nanorods	ABL	10.1149/1.3534201
PEDOT:PSS+DMSO	ABL	10.1016/j.solmat.2010.12.026
PEDOT:PSS+DMF	ABL	10.1016/j.solmat.2011.12.006
PEDOT:PSS+alcohols	ABL	10.1063/1.3600665
PEDOT:PSS & methanol	ABL	10.1016/j.solmat.2011.09.039
PEDOT:PSS & isopropanol	ABL	10.1016/j.solmat.2011.09.039
PEDOT:PSS & acetone	ABL	10.1016/j.solmat.2011.09.039
PEDOT:PSS+glycerol	ABL	10.1016/j.solmat.2007.04.001 10.1109/lpt.2008.2005422 10.1149/1.3196243
PEDOT:PSS+sorbitol	ABL	10.1016/j.cap.2011.03.061
PEDOT:PSS+NaOH	ABL	10.1016/j.solmat.2010.04.023 10.1021/jp2028615
Grafted PEDOT	ABL	10.1021/am800081k
PHMEDOT	ABL	10.1016/j.solmat.2009.03.007

4.4.1.2 Polymerized triphenylamines and carbazoles

The high hole mobility exhibited by carbazoles and tertiary aromatic amines (Fig. 4.4) [38], in addition to their shallow LUMO level that efficiently blocks electron leakage at ITO anode and charge recombination, makes them attractive candidates as anode buffer materials.

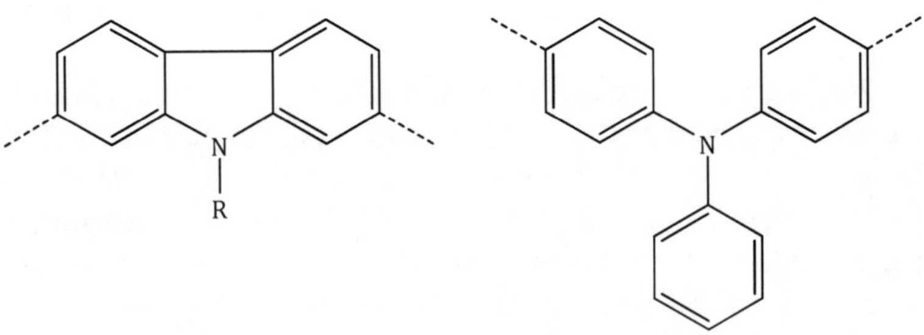

Figure 4.4 Structure of carbazole (left) and triphenylamine (right) monomer units.

In practice, two families of such aromatic amines have been used in PSCs as ABLs: (a) triarylamines or carbazoles cross-linked *via* radical reaction and (b) blends of a fluorene-triphenylamine copolymer with cross-linkable bis(trichlorosilyl) triarylamines (Table 4.3).

Table 4.3 Polymerized triphenylamines and carbazoles: examples of interlayers in conventional solar cells

Material	Buffer type	Article doi
Polymeric or X-linked aromatic tertiary amines	ABL	10.1063/1.2834697 10.1063/1.2715029 10.1021/am900634a
Doped X-linked triarylamines	ABL	10.1021/cm2024235
X-linked polycarbazole	ABL	10.1016/j.solmat.2011.05.044

The interlayer materials are deposited on ITO electrode by spin coating from organic solutions and, after thermal or photochemical treatment, an insoluble film is formed, which is not damaged by the following deposition of the active layer. The trichlorosilyl derivatives are also able to react with ITO, thus improving dramatically the adhesion to its surface. Improvements of J_{SC} and FF have been observed for the related solar cells; and in general, the PCE is much better than that measured for devices with bare ITO and similar or slightly better than those of devices with a PEDOT:PSS interlayer.

4.4.1.3 Conjugated cationic polymers

Conjugated polymers (CPs) with cationic or strongly polar groups in the side chains are readily soluble in solvents such as alcohols or water/alcohol mixtures. Since alcohols do not dissolve the commonly used active layers, they can be easily deposited also on top of them by coating or printing techniques. Indeed, when all-solution OSC devices are fabricated, orthogonal solvents are required for the different layers to prevent damaging the underlying layer. Examples of CPs are reported in Table 4.4.

The copolymers reported in the literature are based on fluorene, carbazole, triphenylamine, or thiophene units; ammonium salts, tertiary aliphatic amine, and phosphonate groups are the ionic/polar moieties in the lateral chains. Ionic and polar CPs are used as CBLs (thicknesses <10 nm; typically ~5 nm) both in conventional and inverted cells, and their main feature is the ability to generate strong dipoles at the active layer–electrode interface, decreasing the electrode work function. As a result, V_{OC} is often strongly increased [39] and, to a minor extent, FF as well, whereas J_{SC} is almost unaffected in most cases, although a simultaneous improvement of all the photovoltaic parameters has been also reported [1], leading to a record (at the time) device with PCE = 8.37%.

Table 4.4 Conjugated ionic and hydrophilic polymers: examples of interlayers in conventional (C) and inverted (I) solar cells

Material	Buffer type	Cell type	Article doi
Phosphonated triphenylamine polymer	CBL	I	10.1039/c1jm14970a
Phosphonated carbazole copolymers	CBL	I	10.1016/j.solmat.2011.09.030
Aminated carbazole copolymers	CBL	I	10.1016/j.solmat.2011.09.030
Cationic polythiophene	CBL	C	10.1039/c1jm12016f
Cationic fluorene copolymers	CBL	C	10.1039/b921775d 10.1063/1.3157278 10.1016/j.orgel.2009.02.009 10.1002/adfm.200902386 10.1021/ja2037673
Polar fluorene polymers	CBL	C	10.1016/j.solmat.2008.12.007 10.1039/b921775d 10.1002/adma.201101319 10.1002/adma.201103006
	CBL	I	10.1002/adma.201104579

4.4.1.4 Miscellaneous polymers

Apart from the categories discussed in the previous sections, several other polymeric materials have been used as interlayers.

Various forms of polyaniline (homopolymer nanotubes, sulfonated, grafted on polystyrenesulfonic acid) have been used by various groups as ABLs (Table 4.5), either alone or in combination with PEDOT:PSS in a multilayer structure, due to their good hole-transporting properties. Increased J_{SC} has been reported in all cases. Other CPs—that is, polythiophenes, proposed in combination with PEDOT:PSS as ABL, or naphthalenediimide copolymers as CBL—do not lead to meaningful improvement, if compared with more traditional interlayers.

Table 4.5 Other polymers: examples of interlayers for conventional (C) and inverted (I) solar cells

Material	Buffer type	Cell type	Article doi
Polyaniline	ABL	C	10.1016/j.orgel.2008.08.001
			10.1149/1.3071425
			10.1021/jp9083844
			10.1016/j.solmat.2009.08.01
			10.1007/s10008-005-0648-8
			10.1002/adfm.201100429
		I	10.1039/c1jm12019k
PEDOT:PSS/ polythiophenes	ABL	C	10.1063/1.3242006
			10.1021/am100076a
Poly(tetrafluoroethylene)	ABL	C	10.1063/1.2983742
Cationic polythiophene/ PEDOT:PSS	CBL	I	10.1002/adfm.201000304
			10.1002/aenm.201100714
PEO, PEG	CBL	C	10.1002/adma.200602597
			10.1039/b907773a
			10.1021/jp211089n
			10.1002/aenm.201100314
		I	10.1016/j.solmat.2008.11.002
X-linked naphthalenimide copolymers	CBL	I	10.1002/aenm.201100429
PDMS-PMMA	CBL	C	10.1016/j.orgel.2008.12.014

A limited number of non-CPs have also been explored as interlayers in PSCs, but with very small thicknesses (generally well below 5 nm) because of the insulating properties of the materials. A thin poly(tetrafluoroethylene) layer at the anode interface was found to induce a strong interfacial dipole reducing the barrier for hole extraction [40]. A pronounced enhancement of V_{OC}, and to a lesser extent of J_{SC} and FF, has been obtained. Improvements of V_{OC} and PCE have been also reported with oligo- or poly(ethylene oxide) CBLs in conventional BHJ solar cells. The deposition may be done by spin coating, coevaporation with the aluminum electrode or by spontaneous accumulation at the top interface when used as an additive to the blend solution for active layer deposition (see references in Table 4.5). Of course, the other deposition techniques reported in Section 4.5.1 can in principle also be applied, although there are not specific examples in the literature.

4.4.2 Low-Molecular-Weight Organic Compounds

4.4.2.1 Fullerene derivatives and carbonaceous materials

Tremolet de Villers et al. [41] demonstrated that the unfavorable vertical phase separation in conventional BHJ solar cells (i.e., accumulation of the donor toward the cathode electrode and of the acceptor toward the anode) can be circumvented by applying a thin evaporated layer of C_{60} as a CBL. Several different soluble fullerenes have also been used for the same purpose (Table 4.6), as well as polymeric structures of C_{60}, showing the advantage of being insoluble in common organic solvents, and therefore suitable as CBL in inverted solar cells [42].

Table 4.6 Fullerene derivatives, carbon nanotubes, graphene: examples of interlayers in conventional (C) and inverted (I) solar cells

Material	Buffer type	Cell type	Article doi
C_{60}	CBL	C	10.1016/j.orgel.2008.07.010 10.1021/jp907636f
PCBM	CBL	C	10.1021/jp9082163 10.1063/1.3677979
C_{60} derivatives	CBL	C	10.1002/aenm.201100522
C_{60} polymer	CBL	I	10.1039/c1jm10214a
Fluorinated fullerenes	CBL	C	10.1002/adma.200792876
C_{60}-capped PEG	CBL	C	10.1002/adma.201003996
$(Me_5Cp)_2Co$-doped C_{60}	CBL	I	10.1039/c1jm10214a
C nanoparticles + IL	CBL	C	10.1063/1.3237161
CNT	CBL	C	10.1021/nl070717l
	ABL	C	10.1016/j.orgel.2008.12.013 10.1016/j.solmat.2011.09.046 10.1021/nl070717l
GO_x	ABL	C	10.1002/adma.201102207 10.1021/jz201493d
GO_x + SWCNT	ABL	C	10.1002/aenm.201100466 10.1002/aenm.201100595
		I	10.1002/aenm.201100595

Due to their high electron affinity, fullerene derivatives are able to facilitate the transport of negative carriers toward the cathode. An increased FF is generally observed in solar cells including fullerene CBLs; less-evident improvements in J_{SC} are also attained in several cases. Fullerenes with fluorinated alkyl substituents form an ordered self-assembled monolayer [43] that has the additional effect to create an interfacial dipole, which decreases the cathode work function and raises the acceptor energy levels. Carboxylated fullerenes have been used for the interfacial functionalization of titanium and zinc oxide in inverted solar cells (see Section 4.4.4.1).

Carbon nanotubes and graphene are chemically closely related to fullerene, although they cannot be considered "low-molecular-weight" materials. Only few results have been reported on these materials as interfacial layers. They are wet-deposited from appropriate dispersions or, in a few cases, by transfer printing. Carbon nanotubes and graphene have been proposed both as anode and cathode modifiers, although they seem less effective as CBLs [44]. Enhancements in the efficiency of solar cell including this kind of ABLs have been reported when compared to bare-ITO devices, mainly due to an increase in FF; however, no meaningful improvements have been observed over PEDOT:PSS-based devices. The advantage of these ABLs over PEDOT:PSS could be an enhanced device stability, as reported by some authors [45].

4.4.2.2 Organic salts and metal complexes

Many of the metal salts and complexes used as interlayers in OSCs (Table 4.7) have been previously extensively used in organic light emitting diodes (OLEDs) as charge injection layers. Some materials, such as aluminum tris-8-hydroxyquinolinate, zinc bis[2-(2-benzothiazolyl)phenolato], and copper phthalocyanine (CuPc), exhibit an ambipolar transport behavior and are effective both as CBLs and ABLs, with improvements of V_{OC} and PCE for the related solar cells.

With some of these materials (i.e., lithium benzoate and titanium diisopropoxide bis(2,4-pentanedionate), a contribution to the decrease of the device resistance and to the increase of J_{SC} has been found.

Solar cells including interlayers based on some of the materials reported in Table 4.7 (i.e., zinc bis(2-(2-hydroxy-5-(p-dimethylamino-phenyl)phenyl) benzothiazolate) and lithium benzoate) are significantly improved from the point of view of the environmental stability, attributed to a barrier effect of the buffer material against the diffusion of air and moisture, due to the inherent crystallinity typical of metal salts.

Table 4.7 Metal complexes and organic salts: examples of interlayers in conventional (C) and inverted (I) solar cells

Material	Buffer type	Cell type	Article doi
Zinc 2-(2-hydroxyphenyl)	ABL	I	10.1039/c1jm13256c
benzothiazole complexes	CBL	C	10.1016/j.tsf.2011.02.049
Aluminum tris-8-	ABL	I	10.1039/c1jm13256c
hydroxyquinolinato	CBL	C	10.1016/j.tsf.2011.02.049
$CuPc(SO_3H)_4$	ABL	C	10.1016/j.solmat.2011.11.005
CuPc	ABL	C	10.1016/j.synthmet.2005.07.182
		I	10.1016/j.tsf.2004.10.018
	CBL	C	10.1016/j.renene.2009.11.044
Lithium benzoate	CBL	C	10.1016/j.solmat.2011.01.012
BPhen-Li	CBL	C	10.1016/j.orgel.2009.12.009
Titanium diisopropoxide	CBL	C	10.1063/1.2757125
bis (2,4-pentanedionate)		I	10.1002/adma.201104863

Depending on their solubility, these buffer layers are obtained through wet or vacuum deposition with a thickness <10 nm.

4.4.2.3 Miscellaneous

Most of the organic materials not included in the previous categories are electron–poor conjugated compounds (see *infra* for exceptions), and thus have been used as CBLs (Table 4.8). They are typically deposited as very thin layers, always <10 nm and typically around 5 nm or less, for which the most significant improvements are achieved for the related devices.

Bathocuproine (BCP, Fig. 4.5), which is very effective as an interlayer in small-molecule OSCs, offers only marginal advantages when used in PSCs. However, an interesting example involving BCP is reported by Jung et al. [46] in which the electrical and optical effects of BCP are exploited.

A multilayer, transparent bathocuproine/silver/molybdenum oxide [BCP (10 nm)/Ag (10 nm)/MoO_3 (25 nm)] electrode made by thermal evaporation has been proposed, exhibiting an optical transmittance over 75 percent and a low sheet resistance (8.6 Ω/sq). The different refractive indexes of the materials in the multilayer fulfill the optimum zero-reflection condition [47] (i.e., most of the incident light can transmit through the multilayer electrode and propagate into the organic layer, and only a small fraction of the light is reflected back at each interface in dielectric/metal/dielectric multilayers because of Fresnel reflection).

Table 4.8 Low-molecular-weight organic compounds: examples of interlayer in conventional (C) and inverted (I) solar cells

Material	Buffer type	Cell type	Article doi
$CH_3CH_2CH_2Si(OEt)_3$	ABL	C	10.1063/1.2778548
$CF_3CH_2CH_2SiCl_3$	ABL	C	10.1063/1.2778548
$H_2N(CH_2)_3Si(OEt)_3$	ABL	C	10.1063/1.2778548
Pentacene	CBL	C	10.1063/1.2998583
Perylene	CBL	I	10.1016/j.tsf.2004.10.018
Perylenetetracarboxylic dianhydride	CBL	C	10.1016/j.tsf.2011.10.163
Peryleneimides	CBL	I	10.1021/am201027j
Bathocuproine	CBL	C	10.1016/j.renene.2009.11.044 10.1021/jp902198y 10.1016/S0379-6779(02)01174-8 10.1021/cm8008122 10.1002/aenm.201100411
		I	10.1016/j.tsf.2004.10.018
Cationic quinacridone	CBL	C	10.1002/adfm.201101257
2-(4-biphenyl)-5-(4-t-butyl-phenyl)1,3,4-oxadiazole	CBL	C	10.1016/j.tsf.2011.02.049
Phosphorated spirobisfluorenes	CBL	C	10.1016/j.solmat.2012.01.036 10.1021/jp2036235
Pyronin B	CBL	C	10.1117/12.620957

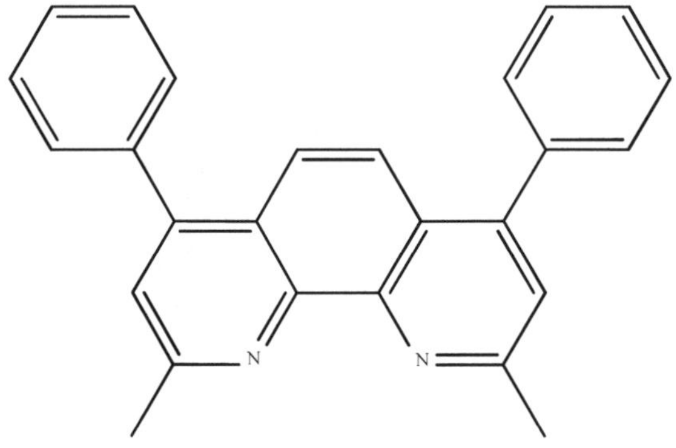

Figure 4.5 Chemical structure of bathocuproine.

Functional silanes represent a notable structural exception in this category, used to modify the ITO anode through a reaction with superficial –OH groups [48]. The silane with the trifluoromethyl substituent afforded a solar cell with increased V_{OC} and J_{SC} due to the control (increase) of the work function of the ITO electrode.

4.4.3 Inorganic Compounds

4.4.3.1 Metal oxides

A large variety of oxides of different metals (mainly transition metals) have been reported as efficient buffer layers for both conventional and inverted cells, namely high-work-function oxides (V_2O_5, VO_x, MoO_3, WO_3, WO_x, NiO, NiO_x, ReO_2, CrO_x, IrO_x, AgO_x, Cu_2O, In_2O_3) at the anode (Table 4.9) and low-work-function oxides (TiO_2, TiO_x, ZnO, ZnO:SnO_2, Al_2O_3, Al-doped ZnO, PbO, CaO) at the cathode (Table 4.10). Titanium and zinc oxide are currently widely used as CBLs, deposited over the ITO electrode of inverted solar cells, allowing the use of high-work-function metals (such as silver) as the top anode and thus enhancing the overall oxidative stability of the cell, due to the higher oxidation potential.

Table 4.9 Metal oxides: examples of anode buffer layers in conventional (C) and inverted (I) solar cells

Material	Cell type	Article doi
V_2O_5, VO_x	C	10.1063/1.2174093
		10.1016/j.orgel.2011.10.006
		10.1002/aenm.201100076
	I	10.1002/adma.201102142
		10.3390/ma4010169
		10.1002/adfm.201101402
		10.1088/0957-4484/19/25/255202
		10.1063/1.2918983
		10.1063/1.2212270
		10.1063/1.2972113
		10.1016/j.orgel.2009.08.011
MoO_3	C	10.1016/j.tsf.2009.07.044
		10.1063/1.2174093
		10.1063/1.3220064

(Continued)

Table 4.9 (*Continued*)

Material	Cell type	Article doi
		10.1016/j.orgel.2010.02.006
		10.1021/cm903495b
		10.1143/jjap.49.01ac02
		10.1016/j.solmat.2011.12.004
		10.1002/adfm.201102054
		10.1002/adma.201100038
		10.1002/adma.201104771
		10.1039/c2jm14911g
	I	10.1063/1.3196763
		10.1021/jp1003984
		10.1063/1.3026741
		10.1063/1.3039076
		10.1039/b921396a
		10.3390/ma3114915
		10.1039/c1ee01316e
		10.1063/1.3250176
		10.1016/j.solmat.2011.07.008
		10.1016/j.orgel.2011.06.015
WO_3, WO_x	C	10.1016/j.orgel.2009.03.016
		10.1016/j.solmat.2011.06.021
		10.1016/j.orgel.2012.01.033
	I	10.1063/1.3076134
		10.1021/jp201890q
ReO_2	C	10.1007/s11801-010-9290-5
NiO, NiO_x	C	10.1073/pnas.0711990105
		10.1016/j.solmat.2010.08.004
		10.1002/aenm.201100234
		10.1088/0022-3727/43/44/445101
		10.1016/j.solmat.2012.01.041
		10.1016/j.tsf.2011.10.059
		10.1016/j.solmat.2010.04.077
		10.1063/1.3673843
	I	10.1016/j.solmat.2011.11.011
		10.1016/j.solmat.2011.11.032
CrO_x	C	10.1016/j.tsf.2011.02.013
IrO_x	C	10.1016/j.solmat.2010.12.039
Cu_2O	C	10.1021/jp1013169
In_2O_3	C	10.1016/j.tsf.2011.12.044

The layer thickness of metal–oxide interlayers is typically in the range 5–30 nm. The most common technique for their deposition is sputtering or evaporation, but also atomic layer deposition (ALD) [49] and electron-beam evaporation have been reported. Direct spin coating of nanoparticle dispersions or sol-gel methods are especially useful for TiO_2 and ZnO, making the process particularly attractive for the all-solution fabrication of inverted devices in a roll-to-roll process [50]. Although sol-gel route generally provides a reliable method to obtain efficient buffer layers even on plastic substrates, the nanoparticle route is still challenging. ZnO nanoparticle dispersions have a limited shelf-life. The reported inverted cells generally show a lower PCE than the conventional cells, due to the high density of recombination centers (defects) represented by the dangling bonds in ZnO nanoparticles [51]. A mild UV-ozone treatment has been proposed [52] that can overcome this limitation. In addition, the effective sintering of nanoparticles is tricky, leading to unsatisfactory adhesion properties. Recently, MoO_3 inks have also been developed [53,54].

Table 4.10 Metal oxides: examples of CBLs in conventional (C) and inverted (I) solar cells

Material	Cell type	Article doi
TiO_2, TiO_x	C	10.1016/j.orgel.2009.12.022
		10.1002/adma.200501825
		10.1016/j.cap.2009.11.084
		10.1063/1.2730746
		10.1002/adma.200602653
		10.1063/1.2908035
		10.1063/1.2937844
		10.1063/1.3159622
		10.1063/1.3073710
		10.1063/1.3685718
	I	10.1063/1.2402890
		10.1063/1.3455108
		10.1016/j.solmat.2008.06.012
		10.1016/j.tsf.2006.11.096
		10.1088/0957-4484/19/25/255202
		10.1063/1.3628319
		10.1016/j.solmat.2011.08.004
		10.1063/1.3026741
		10.1016/j.orgel.2011.06.015
		10.1021/jp201890q
ZnO	C	10.1016/j.solmat.2007.12.004
		10.1039/c1jm13569d
		10.1021/jp203539k
		10.1021/jz200045x
	I	10.1016/j.orgel.2009.03.006
		10.1016/j.solmat.2008.11.033
		10.1063/1.2359579

(Continued)

Table 4.10 (*Continued*)

Material	Cell type	Article doi
		10.1021/jp071418n
		10.1063/1.2945281
		10.1002/adma.201004301
		10.1039/b921396a
		10.1016/j.solmat.2011.09.045
		10.1002/aenm.201100227
		10.1016/j.solmat.2011.11.026
		10.1016/j.solmat.2011.03.011
		10.1021/jp1003984
		10.1063/1.2972113
		10.1016/j.orgel.2009.08.011
		10.1063/1.3039076
		10.1039/b921396a
		10.3390/ma3114915
		10.1039/c1ee01316e
		10.1016/j.solmat.2011.11.032
		10.1021/jz200045x
ZnO/TiO_2	I	10.1021/jp2063589
		10.1016/j.solmat.2011.05.005
$ZnO:SnO_x$	I	10.1016/j.orgel.2012.01.011
Al_2O_3	I	10.1039/c0jm00662a
Al-doped ZnO	I	10.1016/j.orgel.2011.05.027
SiO_x	C	10.1063/1.1446988
PbO	I	10.1016/j.orgel.2011.07.023
CaO	C	10.1016/j.synthmet.2008.06.011

Depending on the material and the deposition technique, different morphologies can be achieved for metal–oxide interlayers, that is, compact layers, nanoparticles, nanorods, the latter being particularly effective for efficient charge collection and transport, with the widest variety of reported morphologies for the most investigated materials: titanium and zinc oxides. MoO_3 at the Ag anode in inverted cells, as well as TiO_2 and ZnO at the cathode in conventional cells, has been demonstrated to act also as optical spacers, further increasing cell efficiency [55]. Titanium dioxide also offers some protection against UV radiation, thus limiting the active layer degradation [24]. The filtering effect of TiO_2 is attributed to the presence of trap states, which causes light absorption below 390 nm.

Finally, zinc oxide, deposited from nanoparticle dispersions, is a very efficient recombination layer in tandem solar cells [56]; tandem cells and the role of the recombination layer are discussed in detail in Chapter 6.

4.4.3.2 Miscellaneous

Historically, lithium fluoride, transferred from OLEDs, has been one of the first CBLs [57] and is still often used in lab-scale devices.

However, the slow dissociation of LiF and diffusion of Li have been found to cause the degradation of solar cells, with a significant reduction of V_{OC} and J_{SC}. Thus, although convenient for lab-scale solar cells fabrication, LiF is unsuitable for stable devices and hence for scaling up the manufacturing process. Other several alkali metal fluorides have been tested (Table 4.11) with the effect of promoting the formation of a dipole at the interface and also the doping of the active layer [58].

Since the inorganic materials reported in Table 4.11 are mostly insulating, very thin layers (1 nm or less) are used. NaF behaves like LiF, but is effective at lower thicknesses [58].

Table 4.11 Inorganic compounds: examples of interlayers in conventional (C) and inverted (I) solar cells

Material	Buffer type	Cell type	Article doi
CuSCN	ABL	C	10.1246/cl.2007.762
Au np	ABL	C	10.1016/j.cplett.2008.01.013
LiF	CBL	C	10.1063/1.1345834
			10.1063/1.1446988
NaF	CBL	C	10.1063/1.2723077
KF	CBL	C	10.1063/1.2723077
CsF	CBL	C	10.1016/j.solmat.2009.01.005
Cs_2CO_3	CBL	C	10.1063/1.2937202
		I	10.1063/1.2918983
			10.1063/1.2212270
			10.1016/j.solmat.2011.07.008
CdS	CBL	C	10.1016/j.synthmet.2007.09.007
		I	10.1016/j.orgel.2011.09.007
$LiCoO_2$	CBL	C	10.1016/j.tsf.2007.12.036

An effective alternative to LiF is cesium carbonate, which greatly improves the FF and V_{OC} of solar cells. The effect arises from the thermal decomposition of carbonate to cesium oxide, or even metallic cesium, occurring during the vacuum evaporation of the Cs_2CO_3 layer or during the thermal annealing of the solution deposited salt solution, and the consequent formation of a low-work-function material [59]. In this case, contrary to LiF, the decomposition of the salt seems not to produce negative effects, although stability data are not reported in detail.

4.4.4 Hybrid Buffer Layers

4.4.4.1 Metal oxides + organic compounds

The interfacial modification of a metal oxide with a functional organic compound offers the possibility to finely tune the properties of the metal oxide (Table 4.12).

Table 4.12 Interfacially modified metal oxides: examples of interlayer in conventional (C) and inverted (I) solar cells

Materials	Buffer type	Cell type	Article doi
Organic acids/ZnO	CBL	C	10.1063/1.2919524 10.1002/adma.200703050
PEG tridecyl ether/TiO_x	CBL	I	10.1063/1.2885724
TiO_2/carboxylated C_{60}	CBL	I	10.1039/b808004f
TiO_2/aromatic acids	CBL	I	10.1039/b808004f
TiO_2/lauric acid	CBL	I	10.1039/b808004f
ZnO/PCBM	CBL	I	10.1016/j.solmat.2009.08.019
ZnO/X-linked C_{60} derivative	CBL	I	10.1021/ja100236b
ZnO/carboxylated C_{60}	CBL	I	10.1063/1.3028094 10.1021/am100238e 10.1063/1.3394679
AZO/phosphonated C_{60}	CBL	I	10.1002/aenm.201100668
AZO/hexylphosphonic acid	CBL	I	10.1002/aenm.201100668

The organic compound can be placed either between the electrode and the metal oxide or between the latter and the active layer, and can either electrostatically interact with the inorganic species or be chemically bound through a –COOH (more commonly) or a $-PO_3H_2$ group, forming a self-assembled monolayer. TiO_2 and ZnO are the metal oxides interfacially modified, thus these kinds of hybrid interfacial layers act as CBLs. Their electron selectivity can be further enhanced by using a fullerene derivative as an interfacial modifier [60,61].

The polarity of the organic molecule is important, and determines the direction of the dipole. For instance, benzoic acids with electron-rich substituents in *para* position (e.g., –OR) improve device performance, while electronegative substituents (e.g., –CN, $-CF_3$) have the opposite effect [62]. In the case of fullerene derivatives, the proper anchoring groups can be attached to the structure according to the known methods for fullerene functionalization [63,64]. Detailed examples of the chemistry involved in the synthesis of these functional fullerenes are reported, for instance, in Ref. [65].

The use of these buffer layers causes a significant increase of FF and a slight improvement of J_{SC}, while V_{OC} remains approximately unchanged (can decrease with unsuitable organic derivatives).

4.4.4.2 Other multilayer buffers made by miscellaneous materials

In this section, buffer layers with a multilayered structure, that is, combining two or more materials already discussed in the previous sections, are briefly summarized. The first layer indicated in each row of Table 4.13 is the one in contact with the electrode.

Table 4.13 Multilayered structures: examples of interlayers in conventional (C) solar cells

Materials	Buffer type	Article doi
PEDOT:PSS/pentacene	ABL	10.1021/jp902198y 10.1016/j.tsf.2010.12.252 10.1016/j.renene.2009.11.044
PEDOT:PSS/Al_2O_3/PEDOT:PSS	ABL	10.1063/1.2974087
AgO_x/PEDOT:PSS	ABL	10.1063/1.2830619
PEDOT:PSS/CuPc	ABL	10.1016/j.renene.2009.11.044
V_2O_5/CuPc, CuPc/V_2O_5	ABL	10.1143/jjap.49.04dk14
MoO_3/CuPc	ABL	10.1016/j.solmat.2011.06.027
GO_x/NiO_x	ABL	10.1016/j.solmat.2011.06.008
Au np/thiophene copolymer	ABL	10.1016/j.solmat.2011.06.043
Au clusters/BPhen:Li	CBL	10.1016/j.orgel.2009.12.009
LiF/Au	CBL	10.1063/1.2988190
LiF/$SnCl_2$Pc	CBL	10.1016/j.solmat.2011.05.038
LiF/C_{60}	CBL	10.1063/1.3297876
LiF/Bathocuproine	CBL	10.1016/j.solmat.2010.04.077
LiF/tetracyanoquinodimethane: Leucocrystal violet	CBL	10.1016/j.solmat.2011.02.014

Such combinations have been proposed to take advantage of the benefits of more materials, in particular, to favor charge transport through the alignment resulting from cascaded energy levels. Gold clusters or nanoparticles, used in a couple of cases both at the anode and at the cathode, can help in improving J_{SC}, thanks to an enhanced charge carrier collection process.

A further advantage provided by multilayer buffers consists in the active layer protection in conventional solar cells. For instance, several authors placed an additional layer between LiF CBL and the underlying active layer to avoid the negative effects of LiF decomposition (see Section 4.4.3.2).

4.4.4.3 Composite materials

The last class of buffer layers (Table 4.14) includes those based on truly hybrid—or composite—materials, prepared either by spin-coating from mixed solutions of different precursors or by vacuum coevaporation.

The properties and the effects on solar cell behavior are various, and the reader should refer to the original papers.

Table 4.14 Composite materials: examples of interlayers in conventional (C) and inverted (I) solar cells

Material	Buffer type	Cell type	Article doi
PEDOT:PSS+Au np	ABL	C	10.1039/c1jm12820e 10.1063/1.3174914
PEDOT:PSS+CNT	ABL	C	10.1063/1.3690056 10.1155/2011/392832
PEDOT:PSS+graphene	ABL	C	10.1166/jnn.2010.2107
Peryleneimides:F8BT	CBL	I	10.1021/am201027j
ZnO:PEG	CBL	C	10.1002/aenm.201100154
Cs_2CO_3:BPhen	CBL	I	10.1016/j.orgel.2012.01.009

4.5 Deposition Methods

Due to the large variety of materials used as buffer layers for OSCs, a wide variety of deposition techniques are used, each one with its own peculiarities, advantages, and limits. Here a nonexhaustive overview of the most commonly used methods is given.

The deposition processes can be divided into two classes: (i) wet processes, in which the film is obtained, usually in ambient condition, or at least in a controlled atmosphere, from a solution containing the material to be deposited or its precursors and (ii) vacuum processes that require the use of specific vacuum chambers and specific limits in process pressure.

4.5.1 Wet Deposition

One of the main potential advantages of OSCs is the possibility of using low- cost and low-energy-consuming techniques for their fabrication. These techniques are usually deposition processes in which materials are dissolved in solutions or inks that can be deposited in a controlled way onto a wide variety of substrates.

A further distinction should be made between wet processes suitable for the preparation of laboratory-scale devices (usually, with an area of up to a few square centimeters) and those that can be scaled up and easily integrated in roll-to-roll systems for high-throughput production of flexible solar modules. These techniques will be treated in Chapter 13 of this book.

Spin-coating deposition is undoubtedly the most widely used wet method for laboratory purposes. It is a low-cost and simple technique that allows good uniformity and control of the thickness as well as a good reproducibility, but it suffers from its limited scalability and low efficiency of material usage. Spin-coating has been used for all the materials that can be dissolved in a solvent (such as polymers, organic compounds, fullerene derivative, salts, carbon nanotubes), or which can be prepared by sol-gel method from suitable precursors (as in the case of metal oxides used as CBLs in inverted solar cells). An alternative deposition technique, sometimes used in laboratory, is dip-coating, consisting in the dipping of the substrate directly in the solution to be deposited. This technique has been sometimes used, for the deposition of self-assembled monolayers (SAMs) or carbon materials such as carbon nanotubes and graphene [44]. For CPs deposited onto a conductive substrate (e.g., ITO), electropolymerization can also be applied.

The above-mentioned techniques are not suitable for large-scale production of solar modules; however, alternative wet techniques can in principle be integrated in a continuous roll-to-roll process, and for some of them a complete scalability has already been demonstrated [66,67]. Among them, gravure [67] and flexographic printing [67], as well as slot die coating [50,68], have been experimented in continuous production processes on flexible substrates. It is also worth mentioning that spray coating [69] and doctor-blading techniques [70], though not yet integrated in continuous processes, have been successfully used for the deposition of metallic oxides used as CBLs in inverted cells.

Although wet deposition processes are highly desirable for low-cost production of large-scale solar modules, their drawbacks should be mentioned, such as (i) the complexity of ink formulation, which has to be optimized for each deposition technique, or (ii) the possibility for a solvent to partially dissolve the underlying layers, or even (iii) the difficulty of superposition of different materials due to a limited wettability of the underlying layer. As for point (i), because printing and coating techniques have strict requirements in terms of ink viscosity and surface tension, great attention must be paid when using additives, that must not interfere with the main purpose of the material interlayer.

Finally, some materials, such as several metal oxides, are often deposited by a sol-gel process. This consists in the use of a colloidal solution of a precursor of the final product (sol), which is then transformed in a network of particles of the desired material (gel). The sol can be easily deposited with a wet process, and then the evolution from the sol to the gel is usually obtained by a thermal treatment. Quite high temperature can be needed for this process, especially when a high level of order and crystallinity is wanted, but these temperatures could be incompatible with most common flexible substrates such as PET and PEN. Therefore, vacuum processes could be preferable in some cases, for the deposition of specific layers.

4.5.2 Vacuum Deposition

Though vacuum deposition techniques are in general complex, energy-consuming, and time-consuming, and so are in principle not preferred, they offer a very good control of the deposition process and do not suffer from problems of incompatibility of superposed layers. For this reason vacuum techniques are widely used for building multistacked structures.

Thermal evaporation has been widely used for the deposition of low-molecular-weight compounds, such as small molecules (CuPc, BCP, fullerenes, etc.) both in conventional and inverted cells, and in general for insoluble materials.

Sputtering [71] is a potential alternative to thermal evaporation, widely used, mainly for the deposition of metal oxides on ITO to be used as CBLs in inverted solar cells. A solid target is bombarded by very energetic particles, usually a plasma, and atoms or molecules are ejected from the target. They can diffuse in the reaction chamber and finally they can be deposited on the substrate. A well-controlled thin-film growth is obtained in this manner. Electron-beam deposition [71] is quite similar to sputtering but rarely used. With this technique, an electron beam is used instead of an ion plasma to strip away the atoms from the target.

A quite different technique that is gaining interest in the thin-film deposition field is ALD [72]. Usually two precursors of the material that has to be deposited are subsequently flushed into a reactive chamber in a gaseous phase. The first precursor reacts with the substrate giving origin to a conformal monoatomic coverage of the layer. Then the unreacted gas is pumped away and the second precursor is flushed in the reactor. It reacts with the first precursor and generates a single layer of the final material. Once again the unreacted gas is pumped away and the process is repeated several times, until the desired thickness has been obtained. Due to its self-limiting characteristics, conformal and pinhole-free films can be deposited layer by layer at atomic scale. For this reason, it allows an excellent control of the thickness, and it is suitable for the deposition of very uniform and thin films, or for conformal coating of layers with complex morphology. The drawback of the technique is that it is quite slow, so that full integration on a roll-to-roll production line could , in principle, not be easily realized. ALD has been already used in the deposition of some metal oxide buffer layers, namely MoO_3, TiO_x, ZnO and Al_2O_3 (Section 4.4.3.1) and CdS (Section 4.4.3.2).

4.6 Outlook

Achieving high-efficiency OSCs relies not only on the appropriate choice of the active materials and on the accurate control of the nanomorphology of the blend but also on the optimization of each layer and interface composing the device structure. The potential of materials with optimized electronic properties for photovoltaic application and formation of an optimal demixing of the donor/acceptor bicontinuous network could be overlooked in a nonoptimized cell structure. Indeed, a high efficiency cannot be achieved in the case of poor extraction of the photogenerated charge carriers and the main role of interfacial layers is precisely the reduction of losses associated with the extraction process. This justifies the great effort of the past years, devoted to the selection of appropriate interfacial layers able to effectively extract charge carriers. A lot of other positive effects have been reported in the literature, including mechanical and optical effects, all contributing to the optimization of device behavior, though the polarity inversion in inverted solar cells is undoubtedly of pivotal relevance.

As has been shown in this chapter, the classes of materials from which it is possible to draw are extremely varied as well as the possible ways in which these materials can be used (from self-assembled monolayers to thin individual layers, as additives or in the form of different kinds of nanoparticles). Some observed phenomena are not yet well understood; so much work still needs to be done to rationalize the available information and to build solid guidelines to refer to when looking for a specific effect. Besides performance, other relevant elements in view of an industrial implementation are (i) a constant behavior under reproducible process conditions and (ii) low cost. For a fully integrated process the deposition of the interlayers should be compatible with the deposition of the active layers; thus solution-processed interlayers are preferable for solution-processed cells, while vacuum-processed interlayers are preferable for vacuum-evaporated active layers. As far as solution-based deposition is concerned, not only the appropriate buffer material but also the ink formulation as a whole is critical; predictably, in the future great efforts ought to be devoted in developing optimized formulations and preventing possible harmful effects between their components.

In conclusion, the importance of the appropriate choice of interlayers between the active layer and the device contacts will become more and more critical as the efficiency of OSCs will increase, and so a sustained high-level research effort on this topic can be expected in the coming years.

References

1. He Z, Zhong C, Huang X, et al. (2011) Simultaneous enhancement of open-circuit voltage, short-circuit current density, and fill factor in polymer solar cells, *Adv Mater*, **23**, 4636–4643.

2. Gong X, Yang T, Wang M, et al. (2012) Inverted polymer solar cells with 8.4% efficiency by conjugated polyelectrolyte, *Energy Environ Sci*, **5**, 8208–8214.
3. Po R, Carbonera C, Bernardi A, Camaioni N (2011) The role of buffer layers in polymer solar cells, *Energy Environ Sci*, **4**, 285–310.
4. Yip H-L, Jen AKY (2012) Recent advances in solution-processed interfacial materials for efficient and stable polymer solar cells, *Energy Environ Sci*, **5**, 5994–6011.
5. Steim R, Kogler FR, Brabec CJ (2010) Interface materials for organic solar cells, *J Mater Chem*, **20**, 2499–2512.
6. Crispin A, Crispin X, Fahlman M, Berggren M, Salaneck WR (2006) Transition between energy level alignment regimes at a low band gap polymer-electrode interfaces, *Appl Phys Lett*, **89**, 213503.
7. Braun S, Salaneck WR, Fahlman M (2009) Energy-level alignment at organic/metal and organic/organic interfaces, *Adv Mater*, **21**, 1450–1472.
8. Osikowicz W, de Jong MP, Salaneck WR (2007) Formation of the interfacial dipole at organic-organic interfaces: C_{60}/polymer interfaces, *Adv Mater*, **19**, 4213–4217.
9. Lindstrom CD, Zhu XY (2006) Photoinduced electron transfer at molecule-metal interfaces, *Chem Rev*, **106**, 4281–4300.
10. Crispin BX, Geskin V, Crispin A, et al. (2002) Characterization of the interface dipole at organic-metal interfaces, *J Am Chem Soc*, **124**, 8131–8141.
11. Ishii H, Sugiyama K, Ito E, Seki K (1999) Energy level alignment and interfacial electronic structures at organic/metal and organic/organic interfaces, *Adv Mater*, **11**, 605–625.
12. Potscavage WJ Jr, Sharma A, Kippelen B (2009) Critical interfaces in organic solar cells and their influence on the open-circuit voltage, *ACC Chem Res*, **42**, 1758–1767.
13. Clarke TM, Durrant JR (2010) Charge photogeneration in organic solar cells, *Chem Rev*, **110**, 6736–6767.
14. Hadipour A, Cheyns D, Heremans P, Rand BP (2011) Electrode considerations for the optical enhancement of organic bulk heterojunction solar cells, *Adv Energy Mater*, **1**, 930–935.
15. Kim JY, Kim SH, Lee HH, et al. (2006) New architecture for high-efficiency polymer photovoltaic cells using solution-based titanium oxide as an optical spacer, *Adv Mater*, **18**, 572–576.

16. Chen F-C, Wu J-L, Hung Y (2010) Spatial redistribution of the optical field intensity in inverted polymer solar cells, *Appl Phys Lett*, **96**, 193304.

17. Luhman WA, Lee SH, Johnson TW, Holmes RJ, Oh S-H (2011) Self-assembled plasmonic electrodes for high-performance organic photovoltaic cells, *Appl Phys Lett*, **99**, 103306.

18. Niesen B, Rand BP, Van Dorpe P, et al. (2013) Plasmonic efficiency enhancement of high performance organic solar cells with a nanostructured rear electrode, *Adv Energy Mater*, **3**, 145–150.

19. Chen F-C, Wu J-L, Lee C-L, Hong Y, Kuo C-H, Huang MH (2009) Plasmonic-enhanced polymer photovoltaic devices incorporating solution-processable metal nanoparticles, *Appl Phys Lett*, **95**, 013305.

20. Fung DDS, Qiao L, Choy WCH, et al. (2011) Optical and electrical properties of efficiency enhanced polymer solar cells with Au nanoparticles in a PEDOT–PSS layer, *J Mater Chem*, **21**, 16349–16356.

21. Peumans P, Forrest SR (2001) Very-high-efficiency double-heterostructure copper phthalocyanine/C_{60} photovoltaic cells, *Appl Phys Lett*, **79**, 126–128.

22. Tanaka K, Kasahara M, Usui H (2000) Preparation of ITO electrode on the organic layer by sputtering, *Electron Commun Jpn 2*, **83**, 23–30.

23. Schlatmann AR, Wilms Floet D, Hilberer A, et al. (1996) Indium contamination from the indium–tin–oxide electrode in polymer light-emitting diodes, *Appl Phys Lett*, **69**, 1764–1766.

24. Sun H, Weickert J, Hesse HC, Schmidt-Mende L (2011) UV light protection through TiO_2 blocking layers for inverted organic solar cells, *Sol Energy Mater Sol Cells*, **95**, 3450–3454.

25. Roman LS, Mammo W, Pettersson LAA, Andersson MR, Inganäs O (1998) High quantum efficiency polythiophene, *Adv Mater*, **10**, 774–777.

26. Aernouts T, Geens W, Poortmans J, Heremans P, Borghs S, Mertens R (2002) Extraction of bulk and contact components of the series resistance in organic bulk donor-acceptor-heterojunctions, *Thin Solid Films*, **403–404**, 297–301.

27. Po R, Carbonera C, Bernardi A, Tinti F, Camaioni N (2012) Polymer- and carbon-based electrodes for polymer solar cells: Toward low-cost, continuous fabrication over large area, *Sol Energy Mater Sol Cells*, **100**, 97–114.

28. Bhattacharyya D, Howden RM, Borrelli DC, Gleason KK (2012) Vapor phase oxidative synthesis of conjugated polymers and applications, *J Polym Sci Part B*, **50**, 1329–1351.

29. Jørgensen M, Norrman K, Krebs FC (2008) Stability/degradation of polymer solar cells, *Sol Energy Mater Sol Cells*, **92**, 686–714.

30. Voroshazi E, Verreet B, Buri A, Müller R, Di Nuzzo D, Heremans P (2011) Influence of cathode oxidation via the hole extraction layer in polymer:fullerene solar cells, *Org Electron*, **12**, 736–744.

31. Glatthaar M, Niggemann M, Zimmermann B, et al. (2005) Organic solar cells using inverted layer sequence, *Thin Solid Films*, **491**, 298–300.

32. Dupont SR, Olivera M, Krebs FC, Dauskardt RH (2011) Interlayer adhesion in roll-to-roll processed flexible inverted polymer solar cells, *Sol Energy Mater Sol Cells*, **97**, 171–175.

33. Kim YS, Park JH, Lee S-H, Lee Y (2009) Polymer photovoltaic devices using highly conductive poly (3,4-ethylenedioxythiophene-methanol) electrode, *Sol Energy Mater Sol Cells*, **93**, 1398–1403.

34. Rider DA, Harris KD, Wang D, et al. (2009) Thienylsilane-modified indium tin oxide as an anodic interface in polymer/fullerene solar cells, *ACS Appl Mater Interfaces*, **1**, 279–288.

35. Kim H, Park J, Lee S, Ha C-S, Kim Y (2011) Effect of strong base addition to hole-collecting buffer layer in polymer solar cells, *Sol Energy Mater Sol Cells*, **95**, 349–351.

36. Lee HK, Kim J-K, Park OO (2009) Effects of UV light-irradiated buffer layer on the performance of polymer solar cells, *Org Electron*, **10**, 1641–1644.

37. Barr MC, Carbonera C, Po R, Bulovic V, Gleason KK (2012) Cathode buffer layers based on vacuum and solution deposited poly(3,4-ethylenedioxythiophene) for efficient inverted organic solar cells, *Appl Phys Lett*, **100**, 183301.

38. Yan H, Yoon M-H, Facchetti A, Marks TJ (2005) Organic field-effect transistors based on a crosslinkable polymer blend as the semiconducting layer, *Appl Phys Lett*, **87**, 183501.

39. Oh S-H, Na S-I, Jo J, Lim B, Vak D, Kim D-Y (2010) Water-soluble polyfluorenes as an interfacial layer leading to cathode-independent high performance of organic solar cells, *Adv Funct Mater*, **20**, 1977–1983.

40. Kang B, Tan LW, Silva SRP (2008) Fluoropolymer indium-tin-oxide buffer layers for improved power conversion in organic photovoltaics, *Appl Phys Lett*, **93**, 133302.

41. Tremolet de Villers B, Tassone CJ, Tolbert SH, Schwartz BJ (2009) Improving the reproducibility of P3HT:PCBM solar cells by controlling the PCBM/cathode interface, *J Phys Chem C*, **113**, 18978–18982.

42. Cho N, Yip H-L, Hau SK, et al. (2011) N-doping of thermally polymerizable fullerenes as an electron transporting layer for inverted polymer solar cells, *J Mater Chem*, **21**, 6956–6961.

43. Wei Q, Nishizawa T, Tajima K, Hashimoto K (2008) Self-organized buffer layers in organic solar cells, *Adv Mater*, **20**, 2211–2216.

44. Chaudhary S, Lu H, Muller AM, Bardeen CJ, Ozkan M (2007) Hierarchical placement and associated optoelectronic impact of carbon nanotubes in polymer-fullerene solar cells, *Nano Lett*, **7**, 1973–1979.

45. Yun J-M, Yeo J-S, Kim J, et al. (2011) Solution-processable reduced graphene oxide as a novel alternative to PEDOT:PSS hole transport layers for highly efficient and stable polymer solar cells, *Adv Mater*, **23**, 4923–4928.

46. Jung GH, Hong K, Dong WJ, Kim S, Lee J-L (2011) BCP/Ag/MoO_3 transparent cathodes for organic photovoltaics, *Adv Energy Mater*, **1**, 1023–1028.

47. Hashimoto K-Y (2000) *Surface acoustic wave devices in telecommunications: modelling and simulations*, springer, berlin.

48. Kim JS, Park JH, Lee JH, Jo J, Kim D-Y, Cho K (2007) Control of the electrode work function and active layer morphology via surface modification of indium tin oxide for high efficiency organic photovoltaics, *Appl Phys Lett*, **91**, 112111.

49. Zhou Y, Cheun H, Potscavage WJ, Fuentes-Hernandez C, Kim S-J, Kippelen B (2010) Inverted organic solar cells with ITO electrodes modified with an ultrathin Al_2O_3 buffer layer deposited by atomic layer deposition, *J Mater Chem*, **20**, 6189–6194.

50. Krebs FC, Tromholt T, Jørgensen M (2010) Upscaling of polymer solar cell fabrication using full roll-to-roll processing, *Nanoscale*, **2**, 873–886.

51. Fonoberov VA, Alim KA, Balandin AA, Xiu FX, Liu JL (2006) Photoluminescence investigation of the carrier recombination processes in ZnO quantum dots and nanocrystals, *Phys Rev B*, **73**, 165317.

52. Chen S, Small CS, Amb CM, et al. (2012) Inverted polymer solar cells with reduced interface recombination, *Adv Energy Mater*, **2**, 1333–1337.

53. Murase S, Yang Y (2012) Solution processed MoO_3 interfacial layer for organic photovoltaics prepared by a facile synthesis method, *Adv Mater*, **24**, 2459–2462.

54. Hammond SR, Meyer J, Widjonarko NE, et al. (2012) Low-temperature, solution-processed molybdenum oxide hole-collection layer for organic photovoltaics, *J Mater Chem*, **22**, 3249–3254.

55. Zhao DW, Liu P, Sun XW, Tan ST, Ke L, Kyaw AKK (2009) An inverted organic solar cell with an ultrathin Ca electron-transporting layer and MoO_3 hole-transporting layer, *Appl Phys Lett*, **95**, 153304.

56. Gilot J, Wienk MM, Janssen RAJ (2007) Double and triple junction polymer solar cells processed from solution, *Appl Phys Lett*, **90**, 143512.

57. Shaheen SE, Brabec CJ, Sariciftci NS, Padinger F, Fromherz T, Hummelen JC (2001) 2.5% efficient organic plastic solar cells, *Appl Phys Lett*, **78**, 841–843.

58. Ahlswede E, Hanisch J, Powalla M (2007) Comparative study of the influence of LiF, NaF, and KF on the performance of polymer bulk heterojunction solar cells, *Appl Phys Lett*, **90**, 163504.

59. Liao HH, Chen L-M, Xu Z, Li G, Yang Y (2008) Highly efficient inverted polymer solar cell by low temperature annealing of Cs_2CO_3 interlayer, *Appl Phys Lett*, **92**, 17303.

60. Huang J-S, Chou C-Y, Lin CF (2010) Enhancing performance of organic–inorganic hybrid solar cells using a fullerene interlayer from all-solution processing, *Sol Energy Mater Sol Cells*, **94**, 182–186.

61. Hau SK, Cheng Y-J, Yip H-L, Zhang Y, Ma H, Jen AKY (2010) Effect of chemical modification of fullerene-based self-assembled monolayers on the performance of inverted polymer solar cells, *ACS Appl Mater Interfaces*, **2**, 1892–1902.

62. Yip H-L, Hau SK, Baek NS, Ma H, Jen AKY (2008) Polymer solar cells that use self-assembled-monolayer-modified ZnO/metals as cathodes, *Adv Mater*, **20**, 2376–2382.

63. Prato M (1997) [60]Fullerene chemistry for materials science applications, *J Mater Chem*, **7**, 1097–1109.

64. Li C-Z, Yip H-L, Jen AKY (2012) Functional fullerenes for organic photovoltaics, *J Mater Chem*, **22**, 4161–4177.

65. Hau SK, Cheng YJ, Yip H-L, Zhang Y, Ma H, Jen AKJ (2010) Effect of chemical modification of fullerene-based self-assembled monolayers on the performance of inverted polymer solar cells, *ACS Appl Mater Interfaces*, **2**, 1892–1902.

66. Krebs FC (2009) Fabrication and processing of polymer solar cells: A review of printing and coating techniques, *Sol Energy Mater Sol Cells*, **93**, 394–412.

67. Kopola P, Aernouts T, Guillerez S, et al. (2010) High efficient plastic solar cells fabricated with a high-throughput gravure printing method, *Sol Energy Mater Sol Cells*, **94**, 1673–1680.

68. Espinosa N, Dam HF, Tanenbaum DM, Andreasen JW, Jørgensen M, Krebs FC (2011) Roll-to-roll processing of inverted polymer solar cells using hydrated vanadium(V)oxide as a PEDOT:PSS replacement, *Materials*, **4**, 169–182.

69. Kang YJ, Lim K, Jung S, et al. (2012) Spray-coated ZnO electron transport layer for air-stable inverted organic solar cells, *Sol Energy Mater Sol Cells*, **96**, 137–140.

70. Stubhan T, Salinas M, Ebel A, et al. (2012) Increasing the fill factor of inverted P3HT:PCBM solar cells through surface modification of Al-doped ZnO via phosphonic acid-anchored C_{60} SAMs, *Adv Energy Mater*, **2**, 532–535.

71. Martin PM (2010) *Handbook of deposition technologies for films and coatings*, 3rd ed, Elsevier, Oxford, ISBN: 978-0-8155-2031-3.

72. George SM (2010) Atomic layer deposition: An overview, *Chem Rev*, **110**, 111–131.

Chapter 5

Electrodes in Organic Photovoltaic Cells

Seunghyup Yoo,[a] Jung-Yong Lee,[b] Hoyeon Kim,[a] and Jaemin Lee[b]
[a]*Department of Electrical Engineering,*
Korea Advanced Institute of Science and Technology,
Daejeon 305-701, Republic of Korea
[b]*Graduate School of Energy, Environment, Water and Sustainability,*
Korea Advanced Institute of Science and Technology,
Daejeon 305-701, Republic of Korea
syoo@ee.kaist.ac.kr, jungyong.lee@kaist.ac.kr

This chapter explores technologies behind electrodes used in organic photovoltaics (OPVs) with an emphasis on transparent electrodes (TEs). In the first part, we describe important electrical and optical roles of electrodes, and discuss their general implications for designing efficient OPVs. Specific examples of various electrodes are then introduced, both those currently in use and those regarded as promising for realizing the potential benefits of OPVs.

5.1 Introduction

Electrodes collect photo-generated carriers and deliver them to external loads, and thus they are an indispensable element in any type

Organic Solar Cells: Fundamentals, Devices, and Upscaling
Edited by Barry P. Rand and Henning Richter

ISBN 978-981-4463-65-2 (Hardcover), 978-981-4463-66-9 (eBook)
www.panstanford.com

of solar cell. Compared to the established inorganic semiconductors like Si and GaAs, organic semiconductors have two major differences that influence the choice of electrodes or their configurations in solar cells: (i) the conductivity of organic semiconductors is relatively small [1]; and (ii) controlled carrier doping [2] is difficult, although not impossible, and cells are often fabricated with intrinsic semiconductors. The limited conductivity of organic semiconductors, together with the thin-film form factor of OPVs, makes it virtually impossible to adopt stand-alone grid-type electrodes, which are commonly used in bulk crystalline Si solar cells [3]. Instead, electrodes in OPVs should cover the whole device active region, and at least one electrode should be transparent or semitransparent to transmit incoming photons from the sun. Likewise, the difficulty in controlled carrier doping often limits the choice of an electrode material suitable for a given organic semiconductor, because the formation of an ohmic contact with a certain electrode material is not as readily achievable for intrinsic semiconductors as for doped semiconductors [2]. All of these factors make the right choice of electrode materials and configurations much more critical in OPVs than in inorganic bulk solar cells.

The purpose of this chapter is thus to elucidate the important roles of electrodes in OPVs, from both electrical and optical perspectives, and to provide the reader with the overall guideline for designing high-efficiency OPVs whose performance is balanced with affordability and reliability. Following the general descriptions, various electrode-related technologies reported to date are introduced, to discuss the practical issues that can occur when one tries to implement them in actual OPVs.

5.2 Electrodes in OPVs: Their Role and Importance

5.2.1 Overview

Organic solar cells typically consist of an anode, a cathode, and donor–acceptor heterojunctions which effectively function like pn-junctions in inorganic semiconductors. Once electrons and holes are generated by a successive process of photon absorption—creation of excitons—charge separation, they are selectively collected by the cathode and anode, respectively [4]. Cell configurations may be classified, first (depending on the location of the cathode with respect to a substrate) as "normal or conventional" (the cathode is away from the substrate) or "inverted" (the cathode is on the substrate side) [5,6]. This is mainly because most organic solar cells have been built to date with their anodes on the substrate side. Although not as common as the above classification, OPVs can also be divided into "bottom-illumination" and "top-illumination" OPVs [7]. Similar to the terminology of "bottom-emission" and "top-emission" used for organic light-emitting diodes (OLEDs) [8], a "bottom-illumination OPV" refers to an OPV in which the TE is located on a substrate so that incident light can be illuminated first through the substrate (as in a superstrate configuration for inorganic solar cells). Here the definition of "bottom" or "top" does not indicate the actual physical direction; it is rather based on the convention adopted in OLED development that defines the substrate side as "bottom." All four of the combinations of the above classifications are summarized in Fig. 5.1.

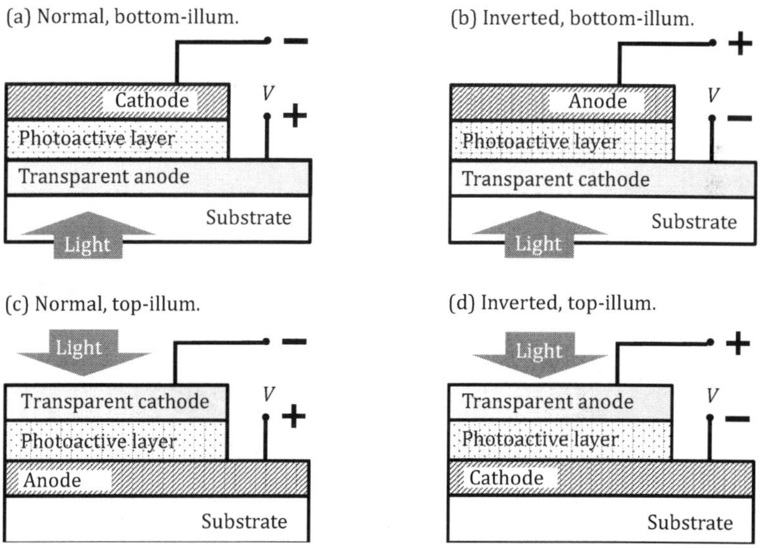

Figure 5.1 The classification of organic solar cells depending on the electrode configuration: (a) bottom-illumination normal; (b) bottom-illumination inverted; (c) top-illumination normal; (d) top-illumination inverted.

One should note that these different electrode configurations often carry a far more significant meaning than what one would imagine for a simple positional difference of electrodes. For example, bottom-illumination devices have most often been adopted, and thus relatively well-studied, yet they work only with transparent substrates. Top-illumination devices can work with various substrates regardless of substrate transparency [7], but securing top-TEs that do not damage underlying organic materials is often of concern [9,10]. Likewise, inverted devices are often regarded to be potentially better in terms of stability than normal devices because cathodes, which are more reactive than anodes in most cases, are buried underneath active and anode layers, although device development and lifetime investigations are still ongoing [6,11–13].

5.2.2 Effect of Sheet Resistance of Electrodes

5.2.2.1 Effect on the performance of individual cells

The resistance for conduction through a thin film (shown in Fig. 5.2), along the direction parallel to a substrate, depends not only on electrical resistivity (ρ) but also on film geometry, by Eq. (5.1):

$$R_{\text{film}} = \left(\frac{\rho}{d}\right)\frac{L_1}{L_2} = R_{\text{sheet}}\frac{L_1}{L_2} \tag{5.1}$$

in which R_{sheet} is the sheet resistance of a film defined by

$$R_{\text{sheet}} \equiv \frac{\rho}{d} \tag{5.2}$$

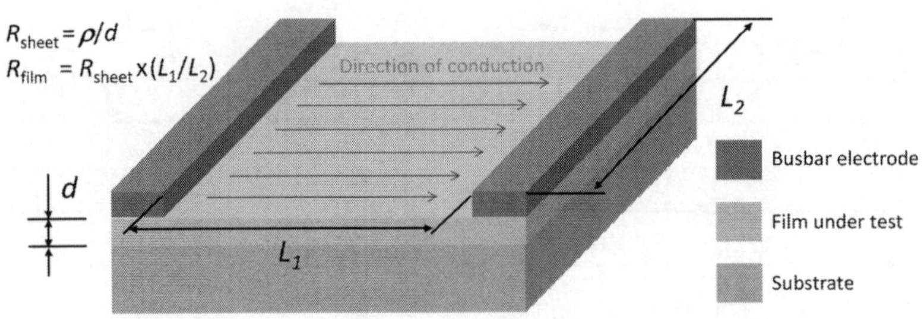

Figure 5.2 Sheet resistance R_{sheet} of a thin conducting film with a resistivity ρ and a thickness d.

To couple out the geometrical effect, R_{sheet}, or the resistance of a square-shaped thin film, ($L_1 = L_2$) is commonly adopted as a term to characterize the electrical property of thin conducting films. For example, indium tin oxide (ITO) films, which are commonly used as TEs in OPVs and flat-panel displays, exhibit R_{sheet} of 10 Ω/sq., defining the benchmark R_{sheet} to be met by other competing technologies.

The effect of R_{sheet} on the performance of solar cells may be best understood using an equivalent circuit model based on an ideal diode, a current source representing photo-generated current density (J_{ph}), and parasitic series (R_s) and shunt (R_p) resistances (Fig. 5.3) [14]. Among those parameters, the presence of non-zero R_s can be particularly detrimental to cell performance, and is directly related to R_{sheet} of TEs. This negative effect is aggravated when the photo-generated current is large, preventing extraction of all the power a PV cell can otherwise provide.

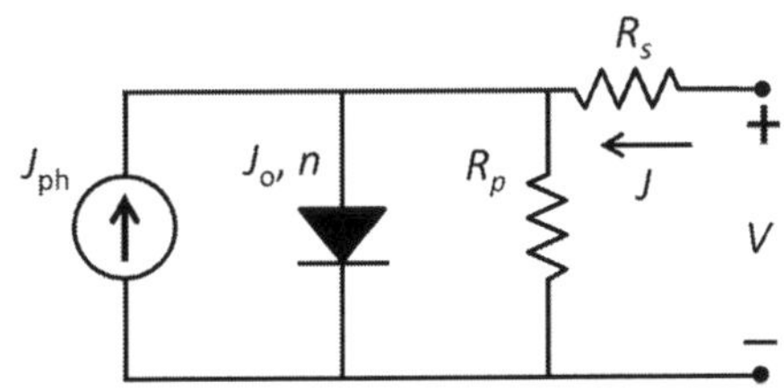

Figure 5.3 Equivalent circuit of a solar cell. J_0 and n refer to the reverse saturation current density and ideality factor of a diode, respectively.

Figure 5.4a and b shows current density (J)–voltage (V) characteristics of a model solar cell described by the equivalent circuit presented in Fig. 5.3 with its specific series resistance $R_S A$ varying from 0 to several hundreds of Ω cm^2 for J_{ph} of 1 and 10 mA/cm^2, respectively. (R_p assumed to be infinite. "A" refers to the active area of the solar cell.) One can easily confirm that the effect of R_s becomes significant when a cell operates with a high photocurrent density. This is mainly because the voltage drop across a device due to the resistance increases with the current flowing through the device. The effect of R_s on the fill factor (FF) of a solar cell is approximated with:

$$FF \approx FF_0(1 - r_S) \tag{5.3}$$

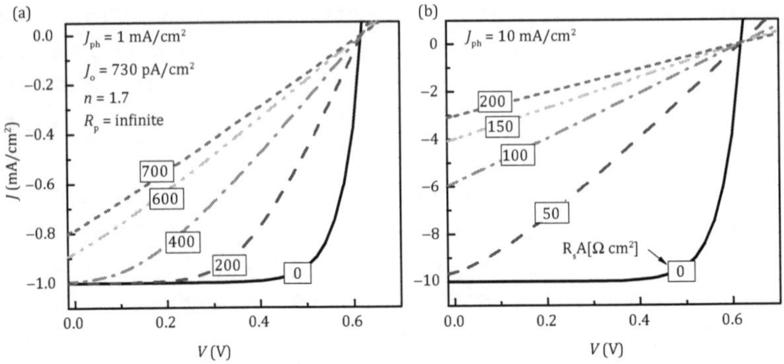

Figure 5.4 Effect of series resistance on the solar cell performance.

in which FF_0 is the fill factor in absence of a series resistance and r_S is the so-called "normalized series resistance" defined by

$$r_S = R_S/R_{CH} \equiv R_S J_{SC} A/V_{OC} = R_S I_{SC}/V_{OC} \tag{5.4}$$

where R_{CH} is the characteristic resistance of a given PV cell defined by V_{OC}/I_{SC} [14].

A study by Choi, Potscavage, and Kippelen indicates that R_s associated with the R_{sheet} of TEs is the parameter that most significantly limits overall performance as the size of devices becomes large [15]. Solar cells being essentially large-area devices, the importance of low R_{sheet} cannot therefore be overemphasized. In that regard, the authors stressed the beneficial role of auxiliary metal grids that are embedded within transparent conductors or deposited on their top surfaces to bring down the net R_{sheet} (see Section 5.3.2.2 for an example of auxiliary metal grids used together with transparent conducting polymers). One should be reminded that such a scheme will be meaningful only if the optical shadowing loss associated with it is smaller than the relative amount of electrical power that the proposed scheme can recover from the initial R_S-induced power loss. The proper electrical/optical design of metal grids is thus of a great importance.

5.2.2.2 Considerations for modules

In a module consisting of plural unit cells as shown in Fig. 5.5, the total series resistance will also scale with the number of cells participating in a module. Let's assume that the shunt resistance of individual cells is sufficiently large and that the connection scheme has been well laid out such that there are no connection-related leakage paths. Then one may consider only the series resistance that comes from that of individual cells and connections. For simplicity, let us assume that the connection-related resistance is negligible. Then, the total series resistance values ($R_S^{(N)}$) of a series-connected and a parallel-connected module are given by $N \times R_S^{(0)}$ and $R_S^{(0)}/N$, respectively, in which N is the number of unit cells and R_{S0} is the series resistance of an individual cell. It should be noted that it is the *normalized* resistance that counts when it comes to the effect of parasitic resistance on the performance of solar cells. With the aforementioned assumptions, the normalized series resistance of a module would not vary too much from that of an individual one; this is because the following scaling relations hold for series and parallel connected modules, respectively [16]:

$$\begin{cases} V_{OC}^{(N)} \approx N \times V_{OC}^{(0)};\ I_{SC}^{(N)} \approx I_{SC}^{(0)};\ R_S^{(N)} \approx N \times R_S^{(0)} & \text{(series)} \\ V_{OC}^{(N)} \approx V_{OC}^{(0)};\ I_{SC}^{(N)} \approx N \times I_{SC}^{()};\ R_S^{(N)} \approx R_S^{(0)}/N & \text{(parallel)} \end{cases} \quad (5.5)$$

For this reason, one does not have to worry about the increased R_S in a series-connected module except for those coming from interconnections; likewise, one may not expect any gain from a reduced R_S in a parallel-connected module.

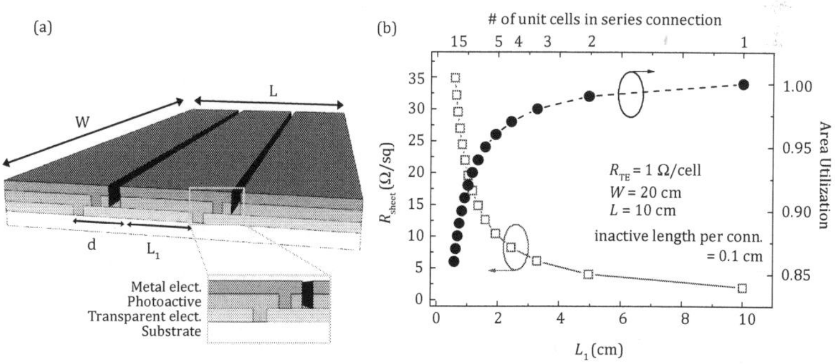

Figure 5.5 (a) Typical series-connected module design. (b) The effect of R_{sheet} of transparent conductors on L_1 and on the area utilization.

In a module, the negative effect of large R_{sheet} can be lessened to some degree by varying the aspect ratio of a unit cell constituting the module. That is, the actual resistance of an electrode per cell can be decreased with respect to its sheet resistance by the factor of (L/W) at the expense of the increased number of cells per module. However, one should remember that there is always an inactive area per each cell-to-cell connection in a module (see Fig. 5.5). Furthermore, one may not reduce the (L/W) ratio to an arbitrarily small value in consideration of the practical patterning resolution of conventional techniques used for OPV fabrication. Hence it will be critical to maintain the sheet resistance of TEs at the lowest possible minimum, or at least below a few tens of Ω/sq. If R_{sheet} is larger than that, auxiliary metal grids will be necessary to avoid resistance-induced degradation of cell/module performance [17]. For a more detailed treatment of this topic, the interested reader is referred to Chapter 14.

5.2.3 Optical Role of Electrodes in Organic Solar Cells: Effect on the Photocurrent Generation

The optical property of electrodes that is of pivotal importance to the operation of OPVs is the transmittance of TEs ($=T_{\text{TE}}$). In most cases, T_{TE} should be close to unity in the absorption band of the active layers in a given OPV so that most of the sunlight may penetrate into the active layers without a significant loss, for example, through absorption and/or reflection processes. The case can be subtle if TEs involve thin metal films. Because thin metal films are translucent (semi-reflective/semi-transparent), there can be an absorption enhancement within active layers due to a cavity resonance effect [18–20]. To accommodate those various situations, it is important to analyze the optical properties of OPVs "as a whole" rather than to look at the properties of TEs alone.

The overall optical properties of OPVs are best analyzed with transfer-matrix formalism (TMF) in the framework of thin-film wave optics as detailed in Chapter 8. For simplicity's sake, the participating layers of OPVs may be categorized into a thick substrate, an electrode 1, photoactive layers, and an electrode 2 as shown in Fig. 5.6a. Each electrode may be in a multilayer configuration containing buffer layers and optical capping layers, etc., in addition to actual electrode layers. Such a simplification, of treating a subset of multilayers as a single effective layer, is called "Smith's method" [21]. Under the TMF approach and Smith's method, the absorption within the active layer ($=A_a(\lambda)$) can be expressed for a light with wavelength of λ by [18]

$$A_a(\lambda)\alpha_a(\lambda)T_{sub}(\lambda)\left[T_1(\lambda)g_{cavity}(\lambda)\right]F_{int}(d_a,\lambda) \tag{5.6}$$

in which α_a is the absorption coefficient of the active layer, T_{sub} is the transmittance for a light incident from air into the substrate, and $g_{cavity}(\lambda)$ is the function quantifying the cavity-resonance enhancement defined by

$$g_{cavity}(\lambda)=\left|\frac{1}{1-\sqrt{R_1^{(-)}R_2^{(+)}}\,e^{-\alpha_a d_a}\exp(i\Delta\phi_{rt})}\right|^2_{\Delta\phi_{rt}=\frac{2\pi}{\lambda}(2n_a d_a)+\phi_1^{(-)}+\phi_2^{(+)}} \tag{5.7}$$

and $F_{int}(d_a, \lambda)$ is the term determined by the electric field distribution that is mainly governed by the interference effect between the light propagating toward electrode 2 and that reflected from it. In Eqs. (5.6) and (5.7), the subscript "a" indicates that the corresponding parameter is defined for the active layer; the superscript "+(−)" indicates that the corresponding parameter is for the light propagating to the right (left) in Fig. 5.6a. R, d, n, κ, and ϕ are reflectance, thickness, refractive index, extinction coefficient, and phase change upon reflection, respectively. If the internal reflectance by electrode 1 for the light coming from the active layer [$=R_1^{(-)}$] is close to zero, $g_{cavity}(\lambda)$ or the function that quantifies the cavity-resonance enhancement is close to unity and thus the cavity-resonance enhancement is negligible. This is what usually happens for conventional cells that typically employ TEs of ITO, PEDOT:PSS, or ITO/PEDOT:PSS. In cells with metallic TEs, however, $R_1^{(-)}$ is often no longer negligible and thus one should carefully monitor the competition between T_1 and g_{cavity}. Figure 5.6b presents $T_1 \times g_{cavity}$ of a cell fulfilling the cavity resonance condition as a function of $R_1^{(-)}$ obtained for various values of $\alpha_a d_a$ using $T_1 = 1 - R_1^{(-)} - A_1^{(-)}$ with $A_1^{(-)}$ assumed to be 10%. One can see that the cavity resonance effect can be significant for photoactive materials with low absorption while the resonance effect diminishes for photoactive materials with high absorption. Note also that the optimal $R_1^{(-)}$ increases as $\alpha_a d_a$ decreases.

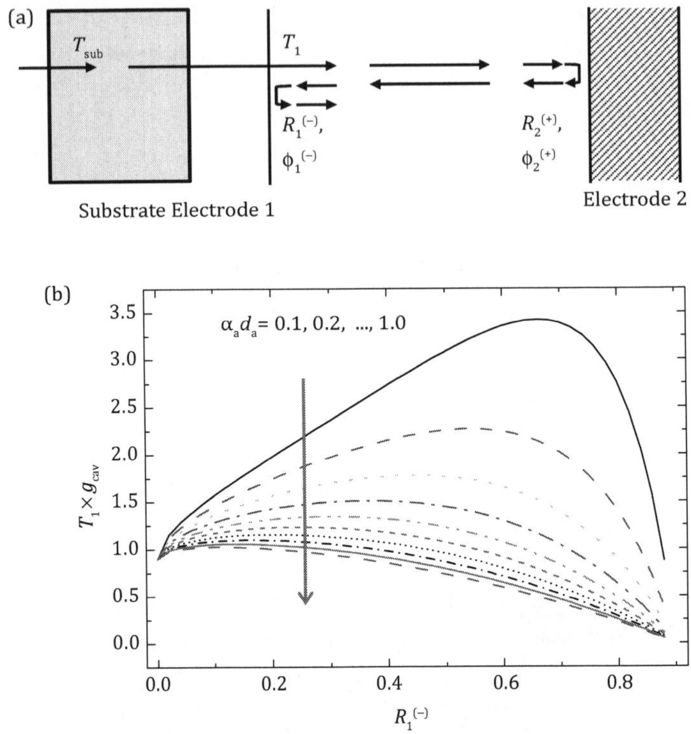

Figure 5.6 (a) Multilayer configuration of typical organic solar cells. Electrodes may consist of multiple layers including anodic or cathodic buffers and/or other optical in-coupling layers. Electrode 1 is assumed to be a TE. (b) $T_1 \times g_{cav}$ vs. $R_1^{(-)}$ for various values of αd and resonance-enhancement effect. Note that $T_1 = 1 - R_1^{(-)} - A_1^{(-)}$. For simplicity, $A_1^{(-)} = 0.1$, $R_2^{(+)} = 1.0$, and $\Delta\phi_{rt} = 2\,m\pi$($m$: integer) were assumed in the calculation.

While the above consideration holds for a specific wavelength, device optimization should eventually be done by maximizing the number of excitons ($N_{excitons}$) generated within active layers over the whole solar spectrum. In the case of bulk-heterojunction type OPVs, $N_{excitons}$ may be expressed by the overlap integral between $A_a(\lambda)$ and $N_{ph}(\lambda)$ or the photon flux under AM1.5G condition [22]:

$$N_{excitons} = \int N_{ph}(\lambda) A_a(\lambda) d\lambda \tag{5.8}$$

In the case of multilayer-type OPVs, the peak location of the optical field inside the active layer with respect to the location of major donor/acceptor heterojunctions also plays an important role and thus their optimization process should include such considerations as well. One may refer to Chapter 8 for more details.

5.2.4 Choosing the Right Transparent Electrode

5.2.4.1 Correlation between optical and electrical properties: figure of merit for transparent conductors

In the previous sections, the important roles of optical and electrical properties of electrodes in OPVs were discussed. In most cases, the performance-limiting electrode is the transparent one, as the materials constituting TEs have a relatively low conductivity, yielding a relatively large R_{sheet} at a given T_{TE}. It should be noted that the optical properties of a TE are closely related to its electrical properties, often in such a way that improving one compromises the other; thus the choice or optimization of an electrode should consider both of those properties in a balanced way. For such proposes, it would be useful to define a figure of merit (FOM) that scales with the *overall* performance of TEs.

The transmittance (T) of a free-standing TE with a thickness of d in air is related approximately to its sheet resistance by [23]

$$T \approx \left(1+\frac{Z_0}{2}\sigma_{opt}d\right)^{-2} = \left(1+\frac{188.5\Omega}{R_{sheet}\times\left(\sigma_{dc}/\sigma_{opt}\right)}\right)^{-2} \qquad (5.9)$$

in which σ_{opt} and σ_{dc} are the optical and dc conductivities, respectively, and Z_0 is the impedance of free space given by 377 Ω. Note that "$Z_0/2$" in the coefficient in front of ($\sigma_{opt}d$) in the first part of Eq. (5.9) is expressed in some literature [24] using $2\pi/c$, which is a CGS-equivalent of $Z_0/2$ in MKS units. Figure 5.7a and b clearly shows a trade-off between T and R_{sheet} of a conducting thin film and that T increases with the ratio of σ_{dc}/σ_{opt} for a given R_{sheet}. Because of the latter, the ratio of σ_{dc}/σ_{opt} has been accepted as a useful FOM for TEs. σ_{dc}/σ_{opt} of ITO, for example, is on the order of several hundreds.

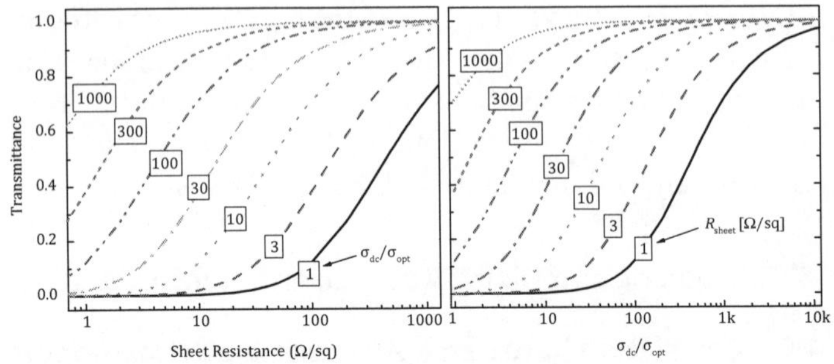

Figure 5.7 Transmittance (T) vs. sheet resistance [left]; T vs. σ_{dc}/σ_{opt} ratio [right] according to Eq. (5.9).

Other FOMs may also be used instead of σ_{dc}/σ_{opt}. A straightforward example of an alternative FOM might be the ratio of σ_{dc}/α proposed by Gordon [25], where α refers to the absorption coefficient of a TE. Widely adopted transparent conductors like ITO typically exhibit σ_{dc}/α ratios around 5 Ω^{-1}.

5.2.4.2 Other considerations

Other things to consider when choosing (or developing) the right (transparent) electrode for OPVs include not only the FOMs mentioned above but also the various aspects such as cost and processability [25,26]. It is of course challenging to find an electrode material fulfilling all the requirements simultaneously while still ensuring an acceptable performance level; one may thus need to make a strategic compromise depending on applications. The various aspects to consider are summarized as follows.

- *Spectral range of the transparency window: solar transmittance*

The FOM defined at a specific visible wavelength (typically at 550 nm) by σ_{dc}/σ_{opt} or σ_{dc}/α may suffice for general-purpose or display-oriented applications, but considerations for solar applications should take into account the broadband spectral range of the solar spectrum, or in general the spectrum incident on the solar cell (e.g. for indoor applications) [27]. To this end, the average transmittance weighted over the solar spectrum, called "solar transmittance (T_{solar})," is often used, and it is defined for typical terrestrial applications by [27,28]:

$$T_{\text{solar}} = \frac{\int T(\lambda) u_{\text{AM1.5G}}(\lambda) d\lambda}{\int u_{\text{AM1.5G}}(\lambda) d\lambda}, \tag{5.10}$$

where $u_{\text{AM1.5G}}(\lambda)$ is the spectral photon flux density of the AM1.5G spectrum.

- *Work function*

In OPVs, electrodes have to be chosen such that there is no energy barrier for collection of holes and electrons. Because most OPVs are made of intrinsic semiconductors, the work function of anodes [cathodes] needs to be matched to or smaller [larger] than the ionization potential [electron affinity] of donor [acceptor] compounds to ensure such barrier-less contacts. In addition, the work function of anodes should be larger than that of cathodes, in particular in M-I-M type OPV cells, to correctly define their polarity and the direction of built-in fields. In actual situations, such notions can face subtle issues when interface dipoles [29] are present or when Fermi-level pinning [30] occurs, and thus they should be used as a "first-order" design rule; careful engineering should be addressed in a case-by-case basis. Use of intentionally-doped organic semiconductors or anodic/cathodic buffer layers often lessens the requirements on the work function of electrodes, allowing for more degrees of freedom in choosing electrode materials, considering various other aspects.

- *Stability*

While organic semiconductors themselves are vulnerable to oxygen and humidity, electrodes can also be a weak link in terms of device stability. For example, low-work-function cathode materials such as Ca or buffers can be sensitive to the presence of O_2 and water vapor [31]; the migration of indium in ITO-based electrodes into active layers [32,33] and acidic corrosion due to common buffer/electrode layers PEDOT:PSS [34] can also be a cause of device degradations. Hence the choice of electrode materials and their ultimate engineering should embrace such possibilities so that one can avoid them or at least find a way to suppress such behaviors. For more information on stability and lifetime issues, please see Chapter 15.

- *Deposition temperature and possibility of deposition-induced damage*

The deposition temperature of conductor materials should be compatible with their substrates. For use with plastic substrates, for example, a deposition temperature should be smaller than 150–200 °C, with the exception of some heat-resistant polymers such as polyimides that can withstand a temperature as high as 350 °C [35]. Even for glass substrates, a deposition temperature higher than 500 °C should rather be avoided, to prevent deformation and the diffusion of sodium atoms present in common soda-lime glasses [25]. Restrictions become severe when it comes to a top electrode, because it needs to be deposited after and on top of the organic materials. Due to the relatively low glass-rubber transition temperature and decomposition temperature of organic materials, it is desirable for the deposition temperature of a top electrode to be as low as possible. Sputtering damage on organic materials is also known for OLED applications [36], and it can particularly be of concern in applications like semitransparent solar cells, where top electrodes could involve sputtered transparent conductive oxides (TCOs).

- *Processability/cost*

Organic photovoltaic cells are trying to position themselves as low-cost solar cells with a good performance–cost ratio. In realizing such a goal, a high electrode cost can potentially become a bottleneck. According to a study by Krebs and coworkers [26], the cost of TEs can be responsible for more than half of the total cost in OPVs. Furthermore, current ITO-based OPVs are often subject to a price hike due to ITO supply shortages, and heavy demands from the flat-panel display industry [37]. In addition to the cost of the electrode material itself, the overall cost will also depend on how it is processed. It would be best if electrodes can be deposited in a high-throughput process such as printing [38]. Large-area scalability should also be secured. Processability issues do not only include cost aspects and scalability, but also involve the ease of patterning and the resistance to chemicals and post-processing steps.

- *Mechanical properties: flexibility and adhesion strength*

The excellent flexibility of organic compounds used in OPVs is one of the important properties that can provide OPVs with distinctive features against other competing technologies [39]. Unfortunately, most TCOs are brittle and thus vulnerable to strain-induced failure [40]. This is one of the reasons why alternative materials like thin metal, conducting polymers, metal nano-wires, and graphene are being actively studied, even though they are currently less favored or established than TCOs like ITO, in terms of performance and processing [41]. There is also an effort to make TCOs themselves flexible, for example, in a hybrid approach combining thin amorphous (or less crystalline) TCO layers and thin metal films [42]. Another mechanical aspect to consider is adhesion strength to a substrate or to a layer underneath [25,43,44]; if adhesion strength is weak, electrodes may be subject to delamination during processing, operation, and bending [45].

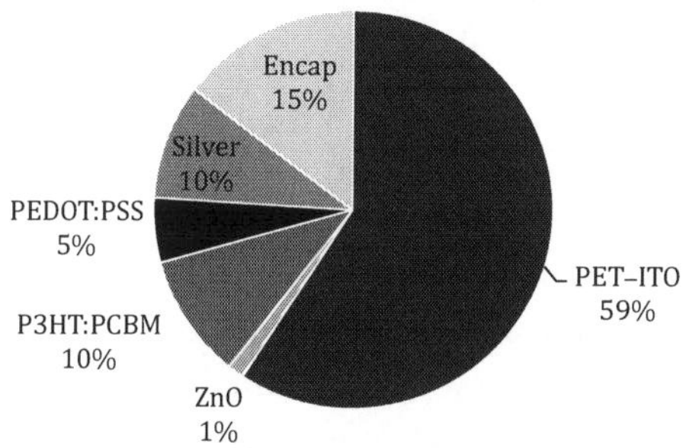

Figure 5.8 Relative material cost (with respect to 105.78 Euro per 1.7 m^2 panel) of various elements in OPVs on PET substrates studied by Medford et al. [26]. The cost shown here is only for the cost paid to purchase each element and does not include the processing cost or labor cost occurred thereafter. One should note that the price of ITO films could vary if it is directly prepared "in house" rather than purchased.

5.3 Examples of Electrodes in OPVs

5.3.1 Transparent Conductive Oxides

5.3.1.1 Overview

In a conductor, the transparent spectral window is determined by its plasma frequency (ω_p) given by [46]

$$\omega_p = \left[\frac{Ne^2}{m^* \varepsilon_0}\right]^{1/2} = \frac{2\pi c}{\lambda_p} \tag{5.11}$$

in which N is the carrier density and λ_p or plasma wavelength is the wavelength corresponding to ω_p. Carriers in a conductor cannot respond to photons with angular frequency [wavelength] higher [shorter] than ω_p [λ_p], and the conductor thus becomes transparent in the spectral region of $\lambda < \lambda_p$ while it is reflective for $\lambda > \lambda_p$ [25]. When m^* is assumed to be equal to electronic mass (m_e), $N < 1.1 \times 10^{21}$ cm^{-3} should be met to fulfill the condition of $\lambda_p > 1$ μm. Most common TCOs have $\lambda_p > 1$ μm, rendering them transparent in the visible and near IR range. For this reason, TCOs have been the default choice for TEs in many thin-film solar cells and flat-panel displays.

5.3.1.2 TCOs in OPVs

The most commonly used TCOs in PVs include (but are not limited to) ZnO:Al (AZO), SnO_2:F (FTO), and In_2O_3:Sn (ITO). As summarized in Table 5.1 [25], their typical resistivity values are on the order of 100–200 μΩ cm, which is 50–100 times larger than that of Ag or Cu. Nevertheless, such a level of resistivity is low enough to realize 10 Ω/sq when they are 100–200 nm thick. Due to their low absorption coefficient in the visible and NIR region, T of approx. 85–90% is readily achieved in such a thickness range.

Table 5.1 Characteristics of TCOs commonly used in PV applications

	ZnO:Al (AZO)	SnO_2:F (FTO)	In_2O_3:Sn (ITO)
λ_p (nm)	>1300	>1600	>1000
Resistivity (μΩ cm)	150	200	100
Work function (eV)	4.1	4.9	4.8
Abundance (ppm)	40 (ZnO)	40 (SnO_2)	0.1 (In_2O_3)
Etchant	Dilute acid or ammonium chloride	Zn + HCl or $CrCl_2$	Diluted HCl + HNO_3
Other notable characters	• High transparency • Low WF (may be used as a cathode)	• Low cost; • Thermally stable • High dep. temperature	• High conductivity • Easy, reproducible patterning • Low roughness
Major applications	a-Si cells	Dye-sensitized cells	Organic PVs, flat-panel displays

Source: [25].

While AZO and FTO are widely used in other thin-film PV cells and dye-sensitized solar cells [3,47,48], ITO, which is a far more expensive option than AZO and FTO, has been most popular in OPVs due to its excellent property, as well as the relatively well-established understanding, accrued from OLED research, of how to control its electronic properties and processing [49,50]. The relatively low roughness of ITO electrodes [25,51] has also been the reason for its popularity in OPVs, as the thickness of their active layers is often around 100 nm, making them vulnerable to electrical shorts. The work by Yang and Forrest shows that the rms roughness values of commercially obtained FTO (750 nm thick) and ITO (150 nm thick) electrodes are approximately 39 and 2.8 nm, respectively, demonstrating the challenge one faces when building OPVs on FTO-coated substrates [51]. While most of the cautions regarding electrical shorts have been taken for individual cells, the relatively large thickness of these low-cost TCOs can also be problematic as it could lead to shorts along the cell edges in series-connected modules.

In spite of the many advantages of ITO over other TCOs, the scarcity of indium and a heavy demand from the flat-panel industry have been a severe cost burden for OPVs targeting ultra-low-cost alternatives to conventional solar cells. In fact, significant fluctuation in the price of indium has recently been observed, illustrating those concerns are real and can indeed be serious [41]. Hence, a significant research effort is focused on finding a low-cost alternative to ITO in OPVs. Those low-cost TCOs may of course be regarded as reasonable candidates considering the maturity of the technology established in other thin-film PV technologies, and thus several efforts have been made to realize efficient OPVs with them. One should come up with solutions which can overcome the issues raised above and, at the same time, are scalable to reliable, large-area module fabrication.

5.3.1.3 Pending issues

(1) Applicability to flexible OPVs

The frequently raised issue about TCOs is the mechanical fragility of oxide-based materials, which could limit their application in flexible devices (see Fig. 5.9). Furthermore, TCOs on flexible substrates often show inferior electrical and optical properties compared to those on glass. This is mainly due to the high temperature required for deposition or post-processing of common TCO electrodes. Note that the service temperature of common plastic substrates is typically lower than 200 °C with a few exceptions. For example, the ITO electrodes, which are typically produced by RF sputtering process, should be annealed at a high temperature to ensure high T and low R_{sheet}. Solutions circumventing those issues are being developed and include a combinatorial approach using thin TCOs and metal (typically Ag) layers as a hybrid multilayer electrode.

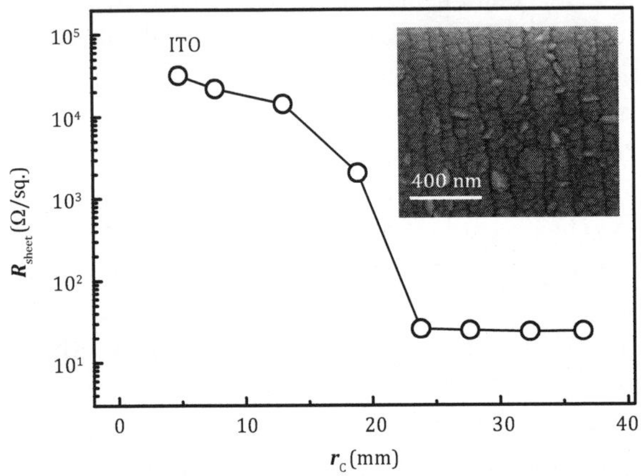

Figure 5.9 R_{sheet} of ITO electrodes on a plastic substrate vs. radius of curvature or r_c after repeated bending of 50 times at r_c. Inset: A top-view SEM image of an ITO electrode on a plastic substrate obtained after the repeated bending of 1000 times at r_c of 18 mm. (Bending was done under tensile stress.) [45] (Reprinted with permission from *Organic Electronics*. Copyright 2009. Elsevier B.V.)

(2) Controlled surface roughness for improved light harvesting

While the rough surface of TCOs is problematic in OPVs in most cases, controlled texturing of TCOs has been essential in improving the light absorption and thus the efficiency in many types of solar cells [52–57]. The texturing scatters the incident light at the TCO/active layer interface and thus increases the optical path length, for example, by a factor of $1/\cos\theta_{scat}$ for a normally incident ray having the internal scattering angle of θ_{scat}. Furthermore, it can make an incident ray go through multiple bounces within an active layer, effectively enhancing the optical path length. Both phenomena allow more photons to be absorbed within photoactive layers so that more charge carriers can be generated. A chemical etching process [54–57] has proven useful for controlled roughening of TCO surfaces as can be seen in Fig. 5.10a–d, in which the etching time of approximately 2–10 min showed the loss of the films thickness up to 1080 nm. The natural growth of roughened TCOs based on a chemical vapor deposition (CVD) method has also been proposed for simplicity, but at the expense of reduced *T* and conductivity [58] (Fig. 5.10e). To take full advantage of the TCO-texturing effect in OPVs, device structure or fabrication method should be carefully chosen so that the possibility of electrical shorts may be suppressed. For example, an organic deposition technology with a better conformal coverage will be preferred, if available, as can be seen in the report by Yang and Forrest, who utilized organic vapor-phase deposition (OVPD) instead of conventional vacuum thermal evaporation. Use of thick, transparent buffer layers, if available, may also be helpful [59].

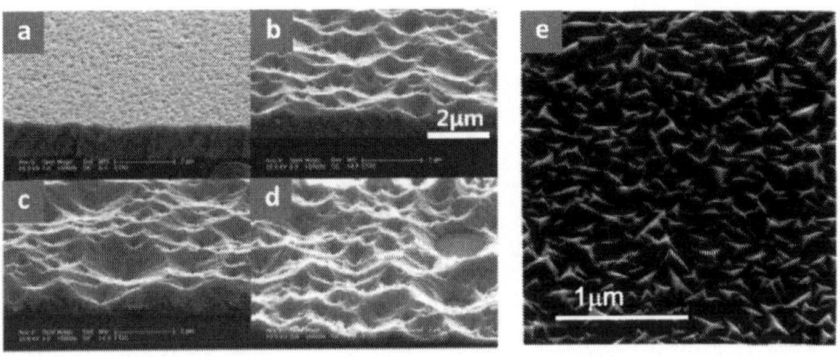

Figure 5.10 SEM images of textured ZnO films: (a) As-deposited ZnO, (b)–(d) textured ZnO films according to the etching time. (e) CVD-grown ZnO with naturally formed textures on the surface. Images in (a)–(d) are from Ref. [57] (Reprinted with permission from *Electrochem Sol State Lett*, The Electrochemical Society) and that (e) is from Ref. [58]. (Reprinted with permission from *JJ Appl Phys* Copyright 1998, The Japan Society of Applied Physics)

5.3.2 Conducting Polymers

5.3.2.1 Overview: working principles and potential advantages

Conducting polymers belong to a special class of polymeric materials along which electrical conduction can occur, differing from most of the other polymers that are electrically insulating

[60,61]. Conductive polymers eventually aim at combining the merit of electrically metal-like characteristics and the excellent processability/mechanical properties of polymeric materials [60]. They are also called "intrinsic conducting polymers" to emphasize their difference from a material system such as conductive epoxy, in which polymeric binders are used to provide a certain functionality but actual conduction occurs through high-conductivity fillers such as Ag or graphite [62]. Intrinsic conducting polymers ("conducting polymers" hereafter for simplicity's sake) conduct electrical carriers along their polymer chains based on π-conjugated systems, where π-electrons can be highly delocalized along the chain [60]. See Fig. 5.11 for the structure of common conducting polymers and note the π-conjugation present in those systems in such a way that double and single bonds alternate periodically throughout their respective systems.

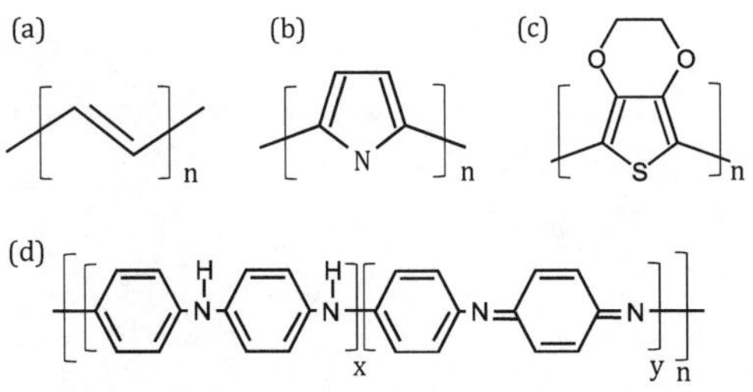

Figure 5.11 Chemical structure of a few exemplary conducting polymers: (a) trans-polyacetylene (b) polypyrrole (c) polyethylene dioxythiophene (PEDOT) (d) polyaniline [$x = 0$, $y = 1$ (pernigraniline); $x = 1$, $y = 0$ (leucoemeraldine); $x = y = 0.5$ (emeraldine)] [60,62,64].

The periodic repetition of a unit cell consisting of double and single bonds in conjugated polymers leads to the formation of a so-called π-band and π*-band that are essentially equivalent to valence and conduction bands in inorganic solids [60,63]. With the π-band all filled up, however, the inherent state of conducting polymers is either insulating or semiconducting. It is doping that finally endows conducting polymers with a metallic nature and high conductivity [60–62]. Dopants take out charges from the polymer main unit (or donate charges into it), yielding a defect state often referred to as quasiparticles such as charged solitons, polarons, and bipolarons (see Table 5.2 for definitions and examples).

Table 5.2 Various quasiparticles in conducting polymers

Name	Characteristics/ definition	Spin (S) charge (Q)	Example[a] for polyacetylene and polypyrrole
Soliton	Neutral radical	S = ½, Q = 0	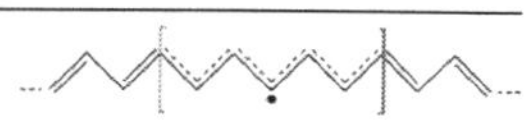
Charged soliton	Cation(+) or anion(−)	S = 0, Q = ±e	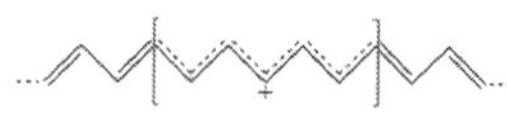
Polaron	Radical cation(+) or anion (−)	S = ½, Q = ±e	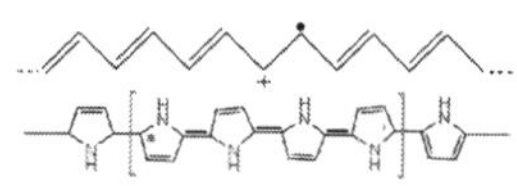
Bipolaron	Doubly charged cation or anion	S = 0, Q = ±2e	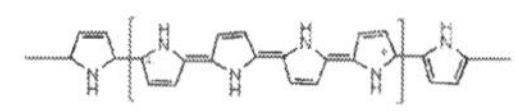

Source: [65,66].

[a] In the case of charged species, only a positive one is shown for simplicity. The dot indicates an unpaired electron.

In a trans-polyacetylene (trans-$(CH)_x$) shown in Fig. 5.11a, or the conducting polymer developed through the pioneering work of Heeger, MacDiarmid, Shirakawa, and coworkers, charge transfer reactions between main units and dopants create a plurality of charged solitons shown in Table 5.2, (i) either directly from a neutral defect state called "(neutral) soliton" that is rare but present when a synthesized $(CH)_x$ happens to contain odd number of carbon atoms, (ii) or by a double-step process involving (a) creation of charged polarons; and (b) creation of a charged soliton by combining two charged polarons as depicted in Fig. 5.12a. These charged solitons eventually form an energy band within the energy gap between the π-band and π^*-band [65,66]. Increasing the dopant concentration widens the bandwidth of the charged-soliton band until the energy gap is closed, resulting in a metallic state, as shown in Fig. 5.12a. By varying the dopant concentration, the conductivity of polyacetylene can be tuned in a very wide range from 10^{-5} to $>10^2$ S/cm [67]. Record conductivity as high as 10^5 S/cm had been reported for trans-$(CH)_x$ doped with iodine with the help of "stretching" that improved microscopic ordering [68].

In most other conducting polymers such as polythiophene and polypyrrole, which are poly(heterocycles), solitons are not energetically stable. In such a case, doping initially creates polarons but no charged solitons; when two polarons "meet" each other as doping concentration increases, bipolarons, which are essentially doubly charged cation or anion, are created. The energy band formed by the plurality of bipolarons is often responsible for conduction in poly(heterocycles). (Detailed description on the conduction mechanism in various conducting polymers is beyond the scope of this chapter. Readers may refer to Refs. [65,66] and references therein for further details.)

Despite the demonstration of impressively high electrical conductivity in doped conducting polymers in their early history, their practicality had often been questioned due to instability in ambient air and humidity [69,70] and due to poor processability. Unlike most of the other conducting polymers, those based on a complex of polyethylene dioxythiophene (PEDOT) and poly(styrenesulfonate) (PSS), commonly called "PEDOT:PSS," are easily processed from aqueous solution and were shown to display a relatively good environmental stability, providing unique yet practical opportunities in a wide range of applications [71,72] PEDOT is polymerized from a monomer of 3,4-ethylenedioxythiophene (EDOT) in the presence of PSS anions as the counter ions in water by so-called "oxidative polymerization" with oxidizing agents such as peroxodisulfate (e.g., $K_2S_2O_8$, $Na_2S_2O_8$) [71,73]. This polymerization process is unique in that it simultaneously performs the three key desired tasks: (i) polymerization, (ii) doping of PEDOT, and (iii) formation of stabilized PEDOT:PSS complexes in water. Note that dopants that help PEDOT to form the bipolaron band shown in Fig. 5.12b and to improve its conductivity are oxidants, not PSS. PSS works as counter-ions that stabilize the complex in water and should not be considered dopants. In fact, the PSS content in PEDOT:PSS complex is larger than that of PEDOT with PEDOT to PSS molar [weight] ratio, ranging from 1:1.9 [1:2.5] to 1:15.2 [1:20]. In addition, PSS is regarded as electrically inactive [71].

In addition to processability and environmental stability, the optical transparency of PEDOT:PSS in the visible spectral range, and its mechanical flexibility, make it highly attractive as a TE in organic devices. In the following section, detailed discussion and examples are thus provided, mainly for the cases where PEDOT:PSS films are adopted as TEs in OPVs.

5.3.2.2 Conducting polymers used in OPVs

Due to the advantages and promise mentioned in the previous section, PEDOT:PSS has been studied intensively as TEs in organic devices [74–82] as well as in some other applications where the high

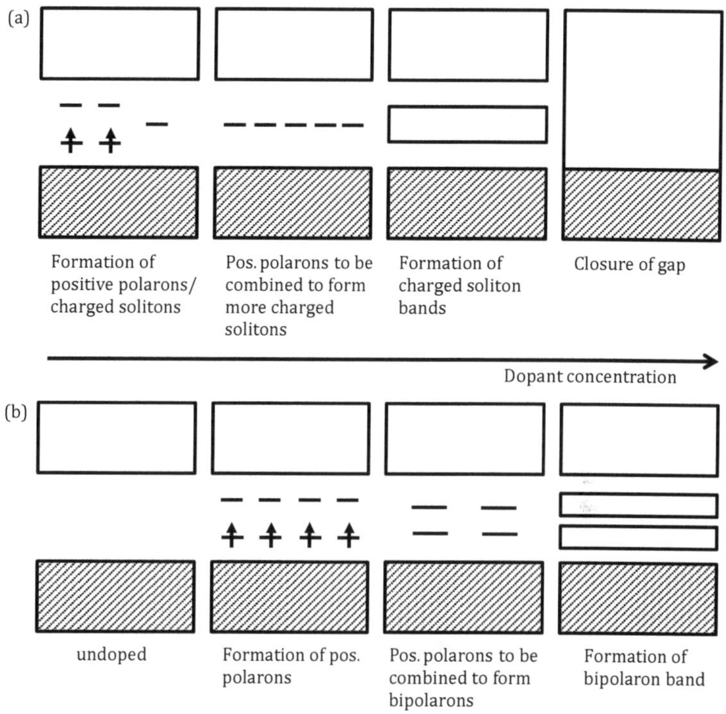

Figure 5.12 Evolution of the energy band structure according to dopant concentration in (a) trans-polyacetylene and (b) poly(heterocycles) such as polypyrrole and polythiophene. Examples shown here are for oxidizing (electron accepting) dopants [65,66].

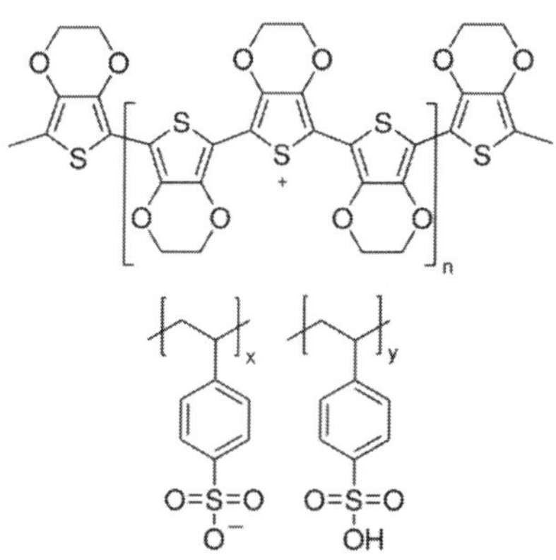

Figure 5.13 The chemical structure of poly(3,4-ethylene dioxythiophene)-poly(styrenesulfonate) (PEDOT:PSS).

conductivity of ITO or similar TCO is not necessarily required. One example of the latter may be found in the case where PEDOT:PSS was adopted as TEs for electroluminescent (EL) light sources used in a keypad of cellular phones [83]. The contoured shape of a keypad benefits from the conformal coating capability inherent to the solution processability of PEDOT:PSS, and the EL light source barely requires a large electrical current and thus high sheet resistance is tolerable.

In the case of OLEDs or OPVs, however, a certain level of R_{sheet} should be secured as discussed in previous sections. The standard grade PEDOT:PSS exhibits a conductivity of merely 1 S/cm, which is far smaller than that of conventional TCOs (~10^4 S/cm). In that respect, numerous studies have been devoted to improving its conductivity, as in some of the examples summarized in Table 5.3. While the conductivity of PEDOT:PSS can be influenced by several factors, such as the ratio between PEDOT and PSS, a small addition of solvent with high boiling point (b.p.) or a polar nature into an originally water-based PEDOT:PSS dispersion was consistently proven effective in improving the conductivity to a significant degree [76,84–88]. Upon addition of dimethyl sulfoxide (DMSO) or ethylene glycol (EG), for example, a certain grade of commercially available PEDOT:PSS can now exhibit conductivity as high as 1,000 S/cm [71,89]. Other types of treatment have also been reported recently, which involve the addition of fluoro-surfactant called Zonyl [90] or hexafluoroacetone (HFA) [91]; or immersion of solidified PEDOT:PSS films into methanol [92]. The proposed mechanisms of the conductivity enhancement due to addition of these solvents include (i) reorientation of polymers for better interconnection incurred by a high b.p.-solvent functioning as a plasticizer [85]; screening effect due to polar solvents hindering Coulomb electrostatic interaction between carriers in PEDOT and negatively charged counter-ions (PSS) [84]; (iii) washing of PSS during a film-forming stage [86]; (iv) conformational change into a case where linear or extended coil conformation is dominant [88]. Another notable method involves the so-called "vacuum vapor phase polymerization" (VPP), and the conductivity as high as 3400 S/cm has recently been reported for PEDOT films prepared by the VPP technique [93].

With the significantly enhanced conductivity, PEDOT:PSS may be considered suitable, although not perfect, as a stand-alone TE in organic solar cells. Vosgueritchian et al. improved the opto-electrical characteristics of PEDOT:PSS by adding a fluorosurfactant and demonstrated a sheet resistance as low as 46 Ω/sq at T ($\lambda = 550$ nm) of 82% [90]. The power conversion efficiency (PCE) of the P3HT: PCBM device applying the conductive polymer TE showed ca. 2.16% (device area = 0.05 cm^2), which is quite comparable to 2.22% of their control device. Xia et al. modified the PEDOT:PSS with HFA [91]. By increasing the number of treatments, they obtained a similar range of R_{sheet} and T. With the modified PEDOT:PSS solution, they obtained PCE of 3.56% with FF as large as 0.64 for P3HT:PCBM-based OPVs. These results are clearly different from those observed in most of the early seminal experiments where the cell performance was severely limited by the high sheet resistance of PEDOT:PSS. [74,75]

Table 5.3 Summary of PEDOT:PSS treatment results.

Method	Max. conductivity achieved (S/cm)	Relative enhancement	Reference
Adding DMSO, DMF, or THF	80	100	[84]
Adding EG	160	400	[88]
Adding DMSO	550	1250	[76]
Adding EG	634	634	[78]
Adding EG + post-solvent-treatment (dip in EG)	1418	1418	[78]
post-solvent-treatment (dip in EG)	1362	4540	[71]

To overcome the limitation of sheet resistance, whose effect becomes especially severe in large-area devices, auxiliary grids were often shown to be essential as briefly pointed out in Section 5.2.2.1. In particular, Galagan et al. demonstrated that PEDOT:PSS used in combination with Ag grids *embedded* within barrier layers can address problems of electrical shorts commonly found in such devices due to the thinness of PEDOT:PSS/organic active layers (see Fig. 5.14) [77].

Another approach that can ease the limitation of high R_{sheet} may be to use multiple spin-coating techniques. Once PEDOT:PSS gets solidified, its integrity is relatively well kept with additional PEDOT:PSS solution dispensed onto it, making multiple spin-coating processes possible to some degree. Shinar and coworkers demonstrated that efficient OLEDs can be realized based on a double-layer PEDOT:PSS [82], and Kim et al. demonstrated that the

T–R_{sheet} relationship of a multilayer PEDOT:PSS can be well described by Eq. (5.9) with σ_{dc}/σ_{opt} of approximately 36 [78] (see Fig. 5.15 for T and R_{sheet} as a function of the thickness of PEDOT:PSS layers prepared by multiple spin-coating technique). Because the benefits of the reduced R_{sheet} in a multi-layer PEDOT:PSS for the efficiency of OPVs can differ depending on various factors (e.g., device area, level of photocurrent/light intensity, etc.), it is not so trivial to predict how many layers are optimal for an OPV device; nevertheless, this multiple coating technique is expected to offer a significant degree of freedom in designing/engineering an OPV cell based on PEDOT:PSS electrodes.

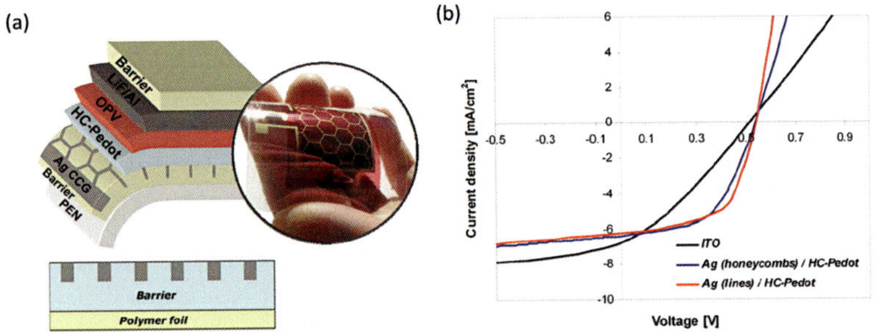

Figure 5.14 (a) Schematic diagram and photograph of an organic solar cell with embedded Ag grids and PEDOT:PSS electrodes (b) J–V characteristics of a cell shown in (a) compared with that of ITO-based reference cells [77] (Reprinted with permission from *Sol Energy Mater Sol Cells,* Copyright 2012. Elsevier B.V.)

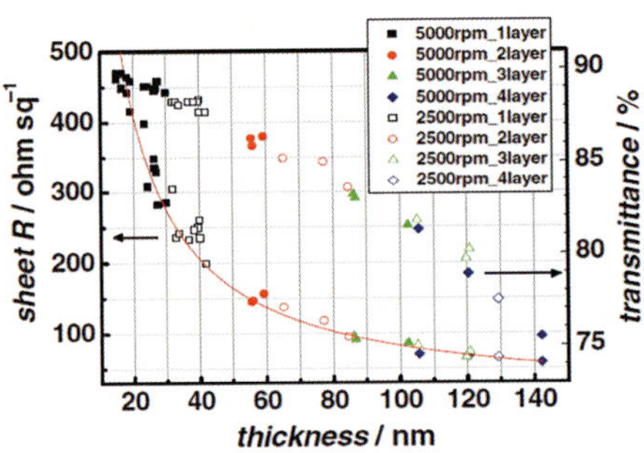

Figure 5.15 (a) T and R_{sheet} presented as a function of thickness prepared by a multiple spin-coating method. The trend curve shown by solid line is $R_{sheet} = \rho_0$/thickness with R_0 of 8.2×10^{-3} Ω-cm [78] (Reprinted with permission from *Adv Funct Mater,* Copyright 2011. Wiley).

The differentiating feature of PEDOT:PSS or similar conducting polymers may be found in the fact that it inherits the excellent mechanical properties of polymers. The recent report by Vosgueritchian et al. indeed shows that the PEDOT:PSS films prepared on poly(dimethylsiloxane) (PDMS) substrates can exhibit an excellent stretchability with no degradation in R_{sheet} even after 5000-time stretching cycles [90].

5.3.3 Thin Metallic Films

5.3.3.1 Overview

Typical metals have λ_p corresponding to the blue or UV region [94], and thus they are reflective or absorptive in most of the visible and IR region important in solar applications. Nevertheless, thin metal films can have a finite transmittance to a certain degree and their inherent high conductivity is such an attractive feature that one cannot overlook their potential importance as a transparent conductor. In particular, metals like Ag, Au, and Cu have low refractive indices (n) close to zero [72,95], allowing their thin films to have a relatively low absorption. Equation (5.12) shows below that the absorbed power per unit area (I_{abs}) in a thin non-magnetic medium with the thickness of d is proportional to the product of $n\kappa$ [95]:

$$I_{abs} = I_0\left(1-\exp(-\alpha d)\right) \approx \alpha d I_0 = d\left(\frac{4\pi\kappa}{\lambda}\right)\left(\frac{1}{2}nc\varepsilon_0|E_0|^2\right) \quad (5.12)$$

in which I_0 and E_0 are the optical irradiance and field just inside the medium of concern before light having the wavelength of λ propagates over a thickness of d, and κ and α are the extinction and absorption coefficients, respectively, of the metal. Nevertheless, one should note that low absorption (A) in a metallic film with low n does not necessarily mean high transmittance (T) because $T = 1 - R - A$, and because reflectance (R) can still be large. In this case, an anti-reflection (AR) coating approach involving additional dielectric layers can effectively suppress reflection at a certain spectral band, making the resultant multilayer unit transparent in that spectral range [95].

In this section, we will first describe a few key points to consider in real (non-idealized) thin metal films and then will introduce the concept of multilayer TEs based on this AR-coating approach so that one can take full advantage of what thin metal films can eventually offer for OPVs.

5.3.3.2 Practical considerations: morphological effect

The optical properties of thin metallic films in the context of transparent conductors in organic devices have been explored by several groups [10,96–99]. The key difference with respect to the optical properties of bulk or thick metal layers comes mainly from the morphological effect that complicates the situation with undesirable phenomena such as scattering [96] and plasmonic effects [100]. These non-ideal effects typically tend to increase A by making n [96,99] or κ [100,101] larger, and thus should be minimized or avoided, if possible.

When a thin metal film is formed on a certain layer, its microscopic morphology can differ significantly depending on various factors including surface properties (roughness, etc.) or chemical composition of the layer underneath, deposition conditions (rate, substrate temperature, etc.), and deposition method (thermal evaporation, sputtering, e-beam evaporation, etc.) [102]. Because of that, there is typically a certain critical thickness (d_c) below which a thin metal film is composed of a plurality of grains and thus their optical properties differ significantly from those of bulk or thick equivalents. The resultant optical constants of a film with $d < d_c$ should then be characterized in a case-by-case basis by spectroscopic ellipsometry in a test sample closely matching a real situation [102]. Figure 5.16 demonstrates that Ag grown on a bare, non-planarized glass substrate suffers from island-like morphologies [101]. One can easily confirm that such Ag films with poor morphologies have much lower transmittance than what is expected from optical simulations. This is mainly due to the enhanced absorption mediated by a surface plasmon effect.

In most cases, it is thus wise to reduce d_c by identifying the origin of such non-ideality or at least by finding a way to avoid or circumvent it. In the example shown in Fig. 5.17, deposition of a Cs_2CO_3 layer having a nominal thickness of 1 nm on organic layers greatly helps uniform Ag layers to nucleate and grow so that n of the Ag layers becomes significantly reduced, and thus the absorption-induced decrease in T is recovered to a significant degree [99].

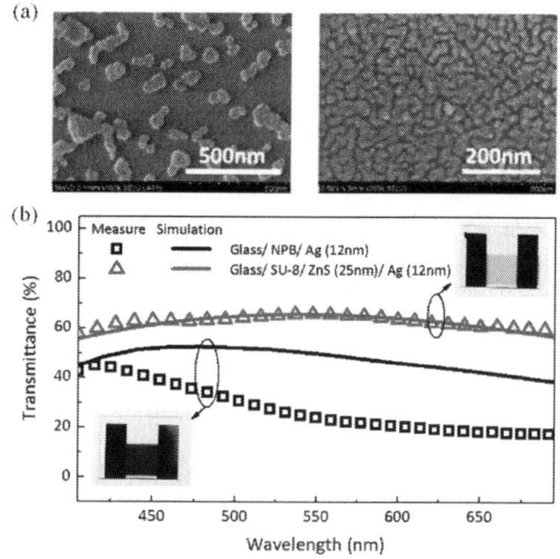

Figure 5.16 Difference in optical properties of Ag films: (a) SEM image of those with island-like (left) vs. uniform (right, still grainy) morphologies [45] (Reprinted with permission from *Org Electron*, Copyright 2009. Elsevier). (b) Optical simulation is based on (n, κ) obtained for thin-film Ag [101].

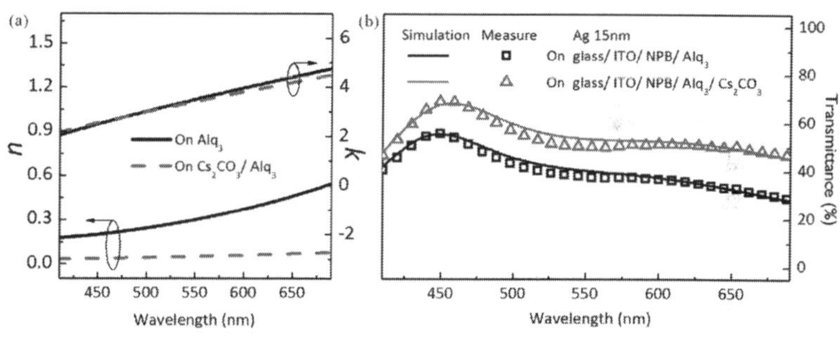

Figure 5.17 Example of Ag-layer engineering: effect of ultrathin Cs_2CO_3 layers on growth of Ag thin films [99].

Not only optical but also electrical properties depend sensitively on thin-film morphologies. That is, the resistivity of a metal thin film is typically larger than that of its bulk equivalent due to the surface and grain boundary effect, yielding electron scattering [98]. Such departure from the bulk properties becomes severe as the thickness of a metal film becomes thinner. Shtein and his coworkers showed that the thin-film resistivity (ρ_{tf}) of their thermally evaporated

Ag films and many other metallic films reported previously in the literature can be modeled by linearly combining the resistivity values based on (i) Fuchs–Sondheimer (F–S) model describing the surface effect in a thin metal film where the bulk mean free path can be comparable to the film thickness; and (ii) Mayadas–Shatzkes (M–S) models describing the effect of various scattering such as grain boundary scattering, background scattering, and external surface scattering [98,103,104]:

$$\rho_{\text{tf}} = \rho_{\text{F-S}} + \rho_{\text{M-S}} - \rho_{\text{bulk}} \tag{5.13}$$

where $\rho_{\text{F-S}}$ and $\rho_{\text{M-S}}$ are the resistivity of thin metal films under the F–S and M–S models (see Ref. [98] for details). With a few adjustable parameters such as surface reflection coefficient, grain boundary reflection coefficient, and grain size as a linear function of film thickness, the experimental data for resistivity were shown to be tightly fitted to the proposed model whose resultant ratio of $\rho_{\text{tf}}/\rho_{\text{bulk}}$ and its effect on R_{sheet} of a Ag layer are shown in Fig. 5.18a and b, respectively.

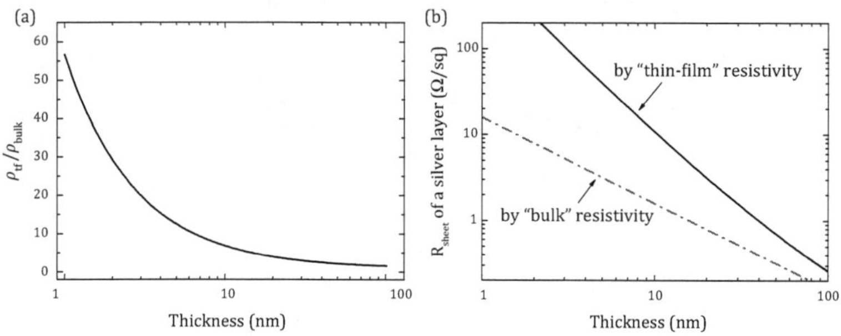

Figure 5.18 (a) The ratio of "thin-film" resistivity (ρ_{tf}) to "bulk" resistivity (ρ_{bulk}) vs the thickness of a metal film. (b) R_{sheet} of a silver layer vs. thickness obtained for the case of ρ_{tf} and ρ_{bulk}. Calculations in (a) and (b) are based on the fitting parameters obtained in Ref. [98].

From Figs. 5.18a and b, the departure from the ideal "bulk" case becomes aggravated rapidly as thickness decreases to below 10–20 nm, confirming the importance of finding a growth condition(s) for uniform, grain-less compact metal layers particularly for the thickness range of 10–20 nm or below.

5.3.3.3 Multilayer transparent electrodes based on dielectric–metal–dielectric structure

Even though a thin metal film is engineered to become highly uniform for the suppression of optical absorption, its reflectance can still be significant in the visible spectrum as discussed above, hindering many of the incident photons from being transmitted through an electrode into its active medium when used for a TE in OPVs. With a multilayer structure in a configuration of dielectric/metal/dielectric (DMD) layers or oxide/metal/oxide (OMO) layers, the reflectance from a thin metal film can also be tuned close to zero (or even to a higher value, if necessary) in the visible spectrum, provided that the metal layer is not too thick (typically a thickness less than a few tens of nm is desired for Ag, for example) and that dielectric layers consist of lossless, high-index materials (e.g., TiO_x, ZnS, ZnSe, Te_2O_3, MoO_x, WO_3, high-index organic materials, etc.) [18,19,105]. This is based on thin-film optics and works in essentially the same way as an anti-reflection or high-reflection coating [95]. For example, see Fig. 5.19a for the optical properties of ZnS/Ag/ZnS multilayers [96]. Also note in Fig. 5.19b that the high refractive index of dielectric layers is critical for achieving the highest transmittance from a DMD electrode with a metal layer having a given thickness [45,95].

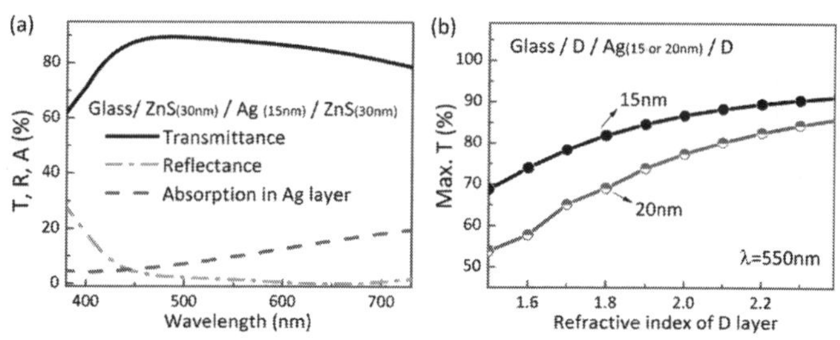

Figure 5.19 (a) Simulated *T*, *R*, *A* vs. λ of glass/ZnS (30 nm)/Ag(15 nm)/ZnS (30 nm). (b) Maximum achievable *T* calculated for *n* at λ = 550 nm of glass/D/Ag(15 or 20 nm)/D in which the thickness of dielectric layers D having the refractive index of *n* is optimized for maximal transmittance.

In a DMD multilayer electrode used in an organic device, the inner D layer or the dielectric layer interfacing between a metal layer and an organic layer plays an electrical role as well as an optical role, enabling the DMD electrode to have a proper polarity and/or functionality. For example, the inner D layer of WO_3 or MoO_x layers may function as a hole-selective buffer layer in OPVs [106,107].

As discussed in the previous section, the optical optimization should consider the whole device structure rather than simply looking at the transmittance of the TEs themselves [18]. Considering the presence of a finite reflectance from DMD electrodes, it is especially important to carefully monitor the competition between electrode transmittance and the resonance effect as discussed in Section 5.2.3. Efficiency corresponding to >90% of that of ITO-based control cells has been realized in OPVs employing $ZnS/Ag/WO_3$ electrodes reported by Han et al. [18] as well as in those employing $MoO_x/Ag/MoO_x$ electrodes by Sergeant et al. [19]. Given the loss of incoming photons due to the reflectance from these electrodes at substrate/electrode interface, a strategic optimization considering the interplay between T and resonance enhancement was a key to the successful demonstration of such comparable efficiencies [18,19]. Although not as efficient, OPVs with $ZnS/Cu/WO_3$ electrodes were also fabricated to exhibit efficiency corresponding to >85% of ITO-based control cells, being demonstrated as cells with a good cost-performance balance [108]. Although not termed as a DMD electrode or a multilayer TE, dielectric-capped metal thin-film electrodes [109] work in a similar way, and the optical structure of OPVs made thereof is essentially equivalent because organic layers directly interfacing with the metal electrode in such a system can be regarded as an inner dielectric layer of DMD electrodes.

Cells with these electrodes showed a mechanical flexibility far superior to those with ITO electrodes due partly to the malleable nature of metal layers and to the thinness of the layers involved (see Fig. 5.20). When these electrodes are used as a top electrode in semitransparent solar cells, the reflectance of metal electrodes typically results in asymmetric optical properties. Such asymmetric optical properties were shown to be useful in maximizing the efficiency for one illumination side with little expense to the overall cell transmittance [20]. Such a character can be useful in building-integrated PV (BIPV) applications provided that the efficiency-maximized side faces outside a building.

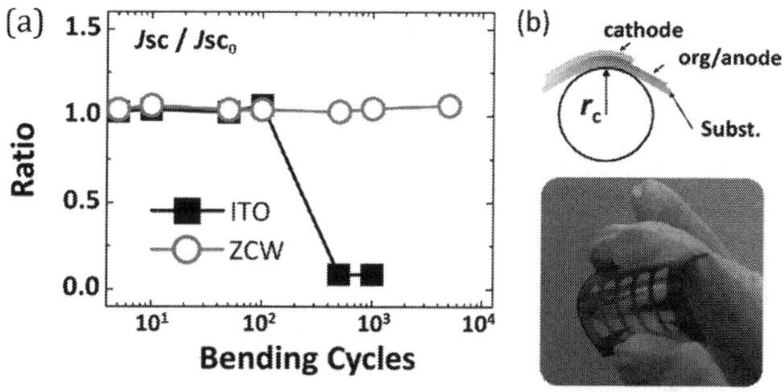

Figure 5.20 (a) Normalized short-circuit current density (J_{SC}) as a function of bending cycles [bending radius (r_C) = 1.83 cm]. (b) Test geometry used in the bending test in (a) and the photograph of ZnS–Cu–WO_3-based flexible organic solar cells [18,108]. (Reprinted with permission from *Sol Ener Mater Sol Cells*, Copyright 2012. Elsevier B.V.)

5.3.4 Nanowire Network

5.3.4.1 Overview

Solution-processable silver nanowires (AgNW) random network films may be considered as another promising alternative to ITO in demanding applications like OPVs and organic light emitting diodes (OLED) [110–116]. Rationales for this trend include a FOM for AgNW comparable to that of ITO, as reported by Coleman and colleagues [117,118]. Lee et al. proved that the opto-electrical property of the AgNW films could indeed rival that of ITO (Fig. 5.21) and reported that the AgNW films can replace ITO in organic solar cells for the first time [111].

Optically, randomly nanostructured AgNW films tend to scatter incident light, leading to a relatively large haze of 1–10%. Although such optical haze may yield adverse effects such as blurring in display applications, it can be advantageous in obtaining higher PCE for OPV by means of enhanced optical absorption, as reported by Peumans and Cui [110].

Mechanically, TEs consisting of AgNWs proved highly reliable on flexible substrates; devices utilizing AgNWs have been shown to be stable under mechanical stress [113,119,120]. Therefore, the AgNW films hold benefits as novel TEs for various flexible thin-film optoelectronic devices. However, the relatively rough surface morphology of the AgNW electrodes (see Fig. 5.21a) has posed a major problem when applied in thin-film optoelectronic devices, because the roughness tends to create shorts in the devices. Indeed, the roughness of deposited AgNW networks on a flat substrate is intrinsically large; the peak-to-peak distance can be more than double the diameter of the wires, because the random networks are constructed by stacking the wires. To address this challenge, several proof-of-concept strategies have been suggested for manipulating the rough surface topology of AgNW electrodes: for example, applying high pressure (~2.0×10^4 psi) to mechanically flatten the deposited AgNW as shown in Fig. 5.22a and b [121], laminating the AgNW into the conducting polymer PEDOT:PSS [119], and embedding the AgNWs from a glass substrate to polymer matrices such as polyvinyl alcohols and poly acrylates [120,122] and a combination thereof [119].

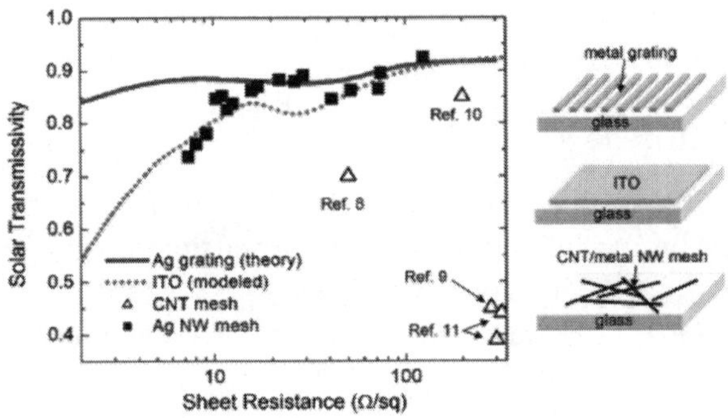

Figure 5.21 Solar transmissivity vs. sheet resistance of various transparent electrodes. The opto-electrical property of the AgNW films is comparable to that of ITO [111] (Reprinted with permission from *Nano Lett*, Copyright 2008, American Chemical Society).

Practically, it is necessary for an optoelectronic device electrode to have an appropriate work function. The work function of the AgNWs film can be easily adjusted, to be utilized interchangeably both as cathode and anode, by additional charge transporting layers, such as PEDOT:PSS, ZnO, TiO_x, etc., which can serve simultaneously as effective planarization layers [112,119,123].

Many solution-based methods have been proposed to deposit AgNW TEs, including doctor blade-coating [124,125], drop-casting [111], spin coating [112], and spray coating [115]. With these methods, the electrical and optical characteristics of AgNW films can be readily controlled. Lamination of the AgNWs on the organic layer was also suggested as a dry coating method [119,121]. An annealing step at approximately 200 °C is often required to firmly secure the connection between AgNWs, as it turns out that the contact resistances between wires were mainly limiting the electrical conductivity of the AgNW films. The insulating polymer surfactant wraps around the NWs, preventing efficient electrical conduction across the AgNWs. The phenomenon is closely related to the growth of the AgNWs. Refer to the next section for the detailed growth mechanism of the AgNWs.

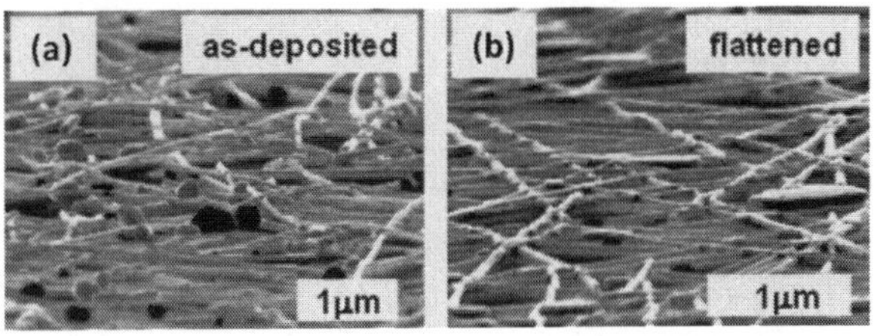

Figure 5.22 SEM images of (a) as-deposited AgNW networks and (b) flattened AgNW networks. Due to the rough surface of AgNW networks, the flattening process is required for thin-film devices [121] (Reprinted with permission from *Nano Lett*, Copyright 2010, American Chemical Society).

5.3.4.2 Fabrication of AgNWs

In order to utilize them as TEs, a significant effort has been made to fabricate nanowires from various metals; copper, gold, and silver have been intensively studied [110,126–132]. Among the metals, silver has become the center of interest owing to its high conductivity that is retained even in the nanowire structure [135]. A few methods to fabricate AgNWs include the template method [137], vaporization method [138], and solution-based synthesis [136]. Although template and vaporization methods can produce high-quality nanowires, they need complex processes to obtain nanowires, leading to high production cost and long growth time.

Recently, the polyol process, a solution-based synthesis method, has received much attention due to its simplicity and easy scalability at low production cost [110,135,139]. In the polyol process, EG and glycerol are often used as solvent. The solvent is mixed with polyvinylpyrrolidone (PVP), a surfactant, and heated for a certain amount of time. As silver nitrate is added into the mixture, the silver ions are reduced by the solvent and form silver nanoparticle seeds. Constant provision of silver ions accelerates Oswald ripening and relatively larger silver particles grow continuously, while the smaller particles are dissolved. Among the particles, 5-fold multi-twinned particles that have a pentagonal cross-section shape become starting points for NW growth, as illustrated in Fig. 5.23 [136].

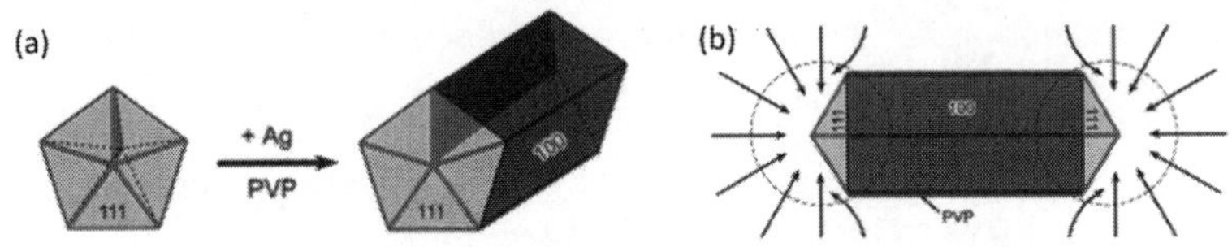

Figure 5.23 (a) Schematic of growth of AgNWs and (b) selective reaction of PVP with [111] facets of a AgNWs [136] (Reprinted with permission from *Nano Lett*, Copyright 2003, American Chemical Society).

Silver ions are reduced and deposited on the ten [111] facets to reduce the surface energy. As the particle grows into rod-like shapes, it exposes [100] facets at the five sides along the longitudinal axis. The [100] facets are stabilized by interacting with the oxygen atom of pyrrolidone, leaving only [111] facets unstabilized. As a result, the rods keep growing into nanowires. In principle, the reaction continues until the nanoparticles smaller than 5 nm disappear completely [135]. The AgNWs synthesized by this method can have various dimensions according to the synthesis conditions [135,139]. The length and the diameter of the AgNWs usually range a few micrometers to 100 μm and tens of nanometers, respectively [112,135,136,140].

5.3.4.3 Example of nanowire network used in OPVs

The AgNWs network films have been regarded as attractive transparent conductors for OPVs, which require high-performance opto-electrical TEs. Lee et al. fabricated small molecular OPVs on AgNW electrodes, demonstrating similar power conversion efficiency (PCE) as the control device relying on ITO, as shown in Fig. 5.24, although the PCE was much less than that of the state-of-the-art OPVs [111].

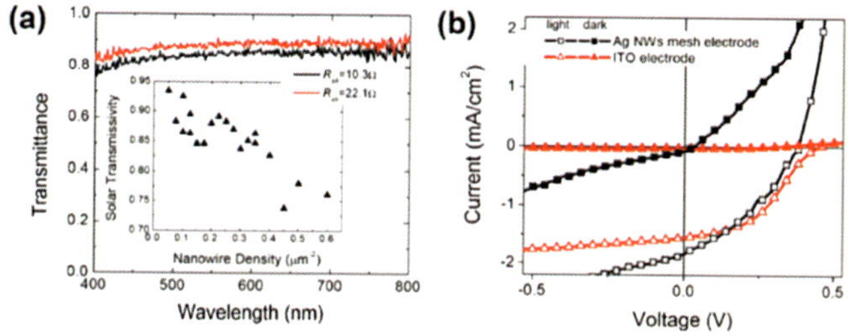

Figure 5.24 (a) Transmittance spectra of AgNW films for two different sheet resistances of 10.3 and 22.1 Ω/sq, respectively. The inset shows the solar transmissivity according to NW density. (b) Current density–voltage characteristics of a small molecule-based OPV on the AgNW and ITO electrodes [111] (Reprinted with permission from *Nano Lett,* Copyright 2008 American Chemical Society).

Gaynor et al. demonstrated all solution-processed inverted OPVs, using P3HT:PCBM as an active layer, to yield PCE of 2.5%. They laminated the AgNWs on the PEDOT:PSS layer as the top electrode. The suggested lamination process can be potentially compatible with a roll-to-roll manufacturing process [141]. Lee et al. reported semitransparent small molecular weight (CuPc/C_{60}) OPV by laminating the AgNW mesh as a top transparent conducting cathode layer. The semitransparent OPV exhibited a PCE of 0.63%, which was 57% of that of a corresponding opaque device with a conventional metal cathode [121]. Yang et al. demonstrated AgNW-based OPVs with a PCE of ca. 3% using an active layer (thickness: ~300 nm) based on a blend of polybenzo[1,2-*b*:4,5-*b*′]dithiophene-5,6-difluoro-4,7-dithien-2-yl-2,1,3-benzothiadiazole (PBnDT-DTffBT) and PCBM on glass substrates [113].

Due to the intrinsically rough AgNW films as mentioned above, many approaches were attempted to make the surface of the AgNW network film flatter. Gaynor et al. suggested embedding the AgNWs into the PEDOT:PSS layer to produce a TCE having root mean square roughness of 11.9 nm, yielding a successful ITO-free P3HT:PCBM-based OPV with a PCE of 4.2%, comparable to that on ITO (Fig. 5.25a) [119]. It should be noted that they had to use a relatively thicker PEDOT:PSS (125 nm) layer than usual in order to create a flat electrode surface. Zhu et al. have also planarized the rough AgNW film with a thick PEDOT:PSS (200 nm) layer, resulting in a PCE of 3.4% in P3HT:PCBM-based OPVs [142]. Before spin-coating the PEDOT:PSS layer, the AgNWs were treated with sol–gel TiO_2 to enhance the adhesion of the AgNWs to the substrates, because otherwise, the PEDOT:PSS could peel off the pristine AgNWs that are weakly bound to the substrate [142].

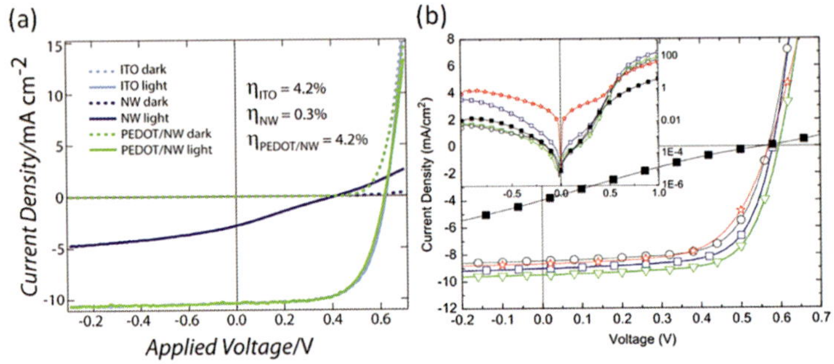

Figure 5.25 The effect of planarization in AgNW-based OPVs: *J–V* characteristics under 1 Sun illumination (AM 1.5G) of P3HT:PCBM-based devices with electrodes of (a) ITO, AgNW, or AgNW/PEDOT:PSS [119] and (Reprinted with permission from *Adv Mater,* Copyright 2011. Wiley) (b) ITO/ZnO (black filled squares), AgNW/ZnO (red stars), AgNW/Ag/ZnO (green triangles), or AgNW/colloidal ZnO (black circles) [125]. Inset in (b): Dark *J–V* characteristics in semilog scale (Reprinted with permission from *Sol Energy Mater Sol, Cells* Copyright 2012. Elsevier B.V.).

However, thick PEDOT:PSS can be optically absorptive, and it may incur long-term stability issues. In an effort to find an alternative to the PEDOT:PSS, metal oxides such as TiO_x and ZnO with large optical bandgaps have been proposed as planarization layers. Leem et al. have fabricated inverted P3HT:PCBM-based OPVs, where the AgNW layer was treated with a thick TiO_x (200 nm) layer, to exhibit performance comparable to that of the control OPV on ITO (PCE: 3.5%) [112]. Morgenstern et al. and Ajuria et al. have treated the AgNW networks with ZnO nanoparticles (NPs) to successfully fabricate a P3HT:PCBM-based OPV, as illustrated in Fig. 5.25b. However, they had to deposit thick active layers of 400 and 600 nm, respectively, presumably due to shorting problems [123,125]. Gaynor et al. reported a flexible P3HT:PCBM-based OPV on PET with a PCE of 3.8% [119]. Yang et al. showed that the DTffBT:PCBM-based OPV fabricated on PET retained an efficiency of 2.3% (originally, 2.5%) after 10 bending cycles at a bending radius of 7.2 mm [113].

Recently, the AgNW-embedded polymer substrate was suggested as one of the easy and simple ways to attain both the flexibility and the reduced roughness of the electrode platforms [122]. By embedding the nanowires into the polymer composite, a smooth surface could be obtained (R_{rms}: 1–5 nm), thus increasing the feasibility of the AgNW films in thin film organic solar cells. Similarly, Yu et al. fabricated P3HT:PCBM-based OPVs on ITO and AgNW-embedded polymer substrates, as demonstrated in Fig. 5.26 [143]. The OPV showed almost similar PCE of 3.3%, demonstrating that embedded AgNWs can be an excellent choice as flexible electrodes for OPV devices. The OPV also showed excellent mechanical flexibility inherited from both the flexible organic materials and the AgNWs, as discussed earlier; after bending at a 5-mm radius of curvature, the PCE remained unchanged.

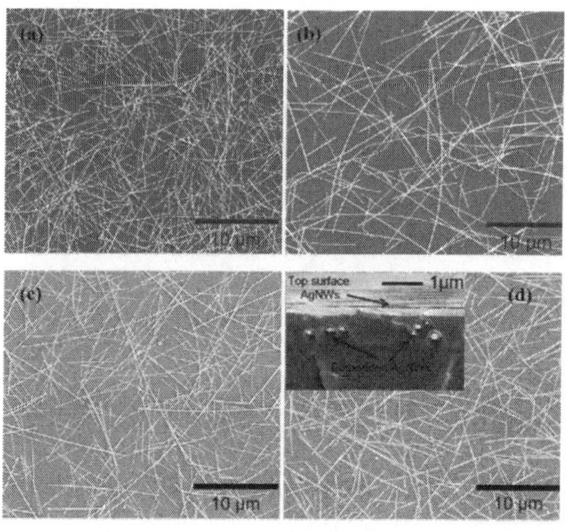

Figure 5.26 (a)–(c) SEM images of AgNW electrodes with various lengths on glass. (d) AgNWs embedded in polymer composite. The inset shows a cross-section of the composite electrode; the AgNWs are well embedded [143] (Reprinted with permission from *Adv Mater*, Copyright 2011. Wiley).

5.3.5 Carbon-Based Nanomaterials: CNTs and Graphene

5.3.5.1 Overview

In addition to the TE technologies mentioned in the previous sections, those based on carbon-based nanomaterials such as carbon nanotubes (CNTs) and graphene, shown in Fig. 5.27, are also under intense investigation [144]. Those who are involved in this field expect the superior mechanical property and conductivity of these materials may open up a new opportunity in TE markets. This section briefly summarizes the basic mechanism of CNTs and graphene and their potential implications as TEs in OPVs.

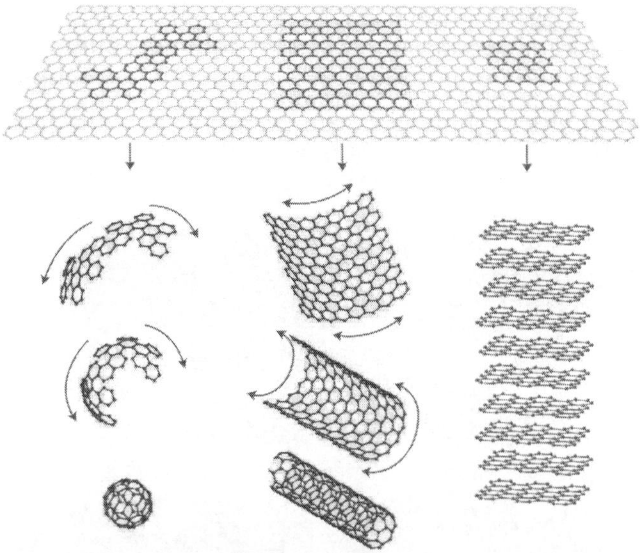

Figure 5.27 Illustrative diagram for construction of carbon-based nanomaterials: 0D buckyballs (C_{60}), 1D CNTs, and 2D graphenes [144] (Reprinted with permission from *Nat Mater*, Copyright 2007. Nature Publishing Group).

5.3.5.2 Carbon nanotubes (CNTs) for transparent electrodes in OPVs

CNTs have been studied since the early 1990s and TE technologies based thereon are regarded to be at a relatively mature developmental stage when compared to other emerging technologies [41,145,146].

The intrinsic property of individual CNTs is known to have a carrier mobility exceeding 10^5 cm^2/V sec, yielding conductivity as high as 2×10^5 S/cm. When they are made into thin films, however, the conductivity gets reduced to typically less than mid- 10^3 S/cm, due to interconnection-related retardation in transport properties [41]. The σ_{dc}/σ_{opt} ratio of CNT films reported in the literature appears to be scattered but mostly falls in the range of 5–50 [147]; for a film with R_{sheet} of 20 Ω/sq., σ_{dc}/σ_{opt} ratio of 5 and 50 corresponds to T of 12% and 71%, respectively. Likewise, for a film with T of 85%, the corresponding R_{sheet} values are 445 and 44.5 Ω/sq for σ_{dc}/σ_{opt} ratio of 5 and 50, respectively. Due to such limitations in the conductivity of actual CNT films, their immediate use appears to still be limited to applications like touch panels, which have a less stringent requirement for R_{sheet} than displays or solar cells. Results reported for OPVs employing CNT TEs [148–153] are also consistent with such considerations. As shown in Table 5.4, most of the OPVs with CNTs suffered from low FF values, which is regarded to stem mainly from the high series resistance, owing to the relatively large R_{sheet} of CNT films. Vulnerability to electrical shorts has also been of concern, and it often required buffer layers like PEDOT:PSS capable of planarization [148–150]. Nevertheless, recent reports by Barnes et al. appear to indicate that the 3D feature of CNT films may be useful in extracting carriers by providing interpenetrating networks at an active layer/electrode interface, allowing for relatively high photocurrent to be generated in spite of their low transmittance [152].

Table 5.4 Performance of OPVs with CNT transparent electrodes.

Anode	Cell performance J_{sc} (mA/cm^2) V_{oc} (V) FF Eff. (%)	TE spec. [T, R_{sheet}(Ω/sq)]	Reference
SWNT/PEDOT	6.5/0.500/0.30/0.99	0.4/200	[148]
SWNT/PEDOT	9.2/0.560/0.29/1.5	0.7/50	[149]
ITO/MWNT/PEDOT	11.0/~0.440/~0.36/1.7	–	[150]
SWNT	11.4/0.540/0.55/3.4	0.55/24	[152]
SWNT (top elec.)	6.8/0.480/0.42/1.5	0.58/360	[153]

5.3.5.3 Graphene for transparent electrodes in OPVs

Graphene is a two-dimensional sheet of carbon-based material in which carbons are arranged in a hexagonal lattice with sp^2 bonding, as can be seen in Fig. 5.27 [144]. Its unique properties had been expected but it was only after Geim, Novoselov and coworkers discovered a few-layer graphene by mechanical exfoliation of graphite in 2004 that its actual existence was proven experimentally [154]. Since then, several synthetic routes for graphene have been developed, which may be categorized into (i) mechanical exfoliation; (ii) chemical exfoliation; (iii) CVD; (iv) thermal decomposition of SiC; (v) reduction of graphene oxide (GO) [144].

Graphene has a unique cone-shape band structure that results in zero-effective mass near the band edge, enabling very high carrier mobility, up to 1.5×10^4 cm^2/V sec, and high conductivity [144]. Sheet resistance and transmittance of a "defect-free" N-layer graphene are predicted to scale as $R_{sheet} \sim 64/N$ [Ω/sq] and $T \sim 100–2.3\ N$ [%], respectively [41,155]. The R_{sheet} and T expected for an *ideal* four-layer graphene would be, for example, ~16 Ω/sq and 91%, respectively, corresponding to σ_{dc}/σ_{opt} ratio of ~240. With such potentially superior optical/electrical properties, graphene is being regarded as an attractive candidate for alternative TE technologies. Relatively easy patterning that can be done via plasma processing as well as the inherently planar, low-roughness nature [156] of graphene may also be regarded as important advantages from the practical perspectives of TE technologies.

It should be noted that σ_{dc} and thus the σ_{dc}/σ_{opt} ratio of graphene vary significantly depending on the fabrication method used; this is because the fabrication technique impacts the size or density of grains, purity, and multilayer morphologies, all of which can influence carrier transport properties. The actual σ_{dc}/σ_{opt} ratio is thus typically much smaller than the values expected for idealized graphene layers. For example, the solution-processing-based GO-reduction method, regarded as a method that may realize the low-cost potential of graphene electrodes, often results in R_{sheet} that is much higher (typically on the order of kΩ/sq) than those of ideal graphene layers [157]. On the other hand, a CVD process has resulted in graphene TEs with a relatively low R_{sheet}, as recently demonstrated by Hong and his coworkers [158]. Combined with HNO_3-based p-doping and a multilayer stacking method based on a multiple transfer process, the authors were able to obtain four-layer graphene exhibiting R_{sheet} as low as 30 Ω/sq at T of 90%. In addition, they laid out a scheme to prepare graphene layers on large substrates by a roll-to-roll (R2R) dry transfer process [see Fig. 5.28 for a basic process flow commonly adopted for CVD-grown graphene and their proposed R2R transfer scheme].

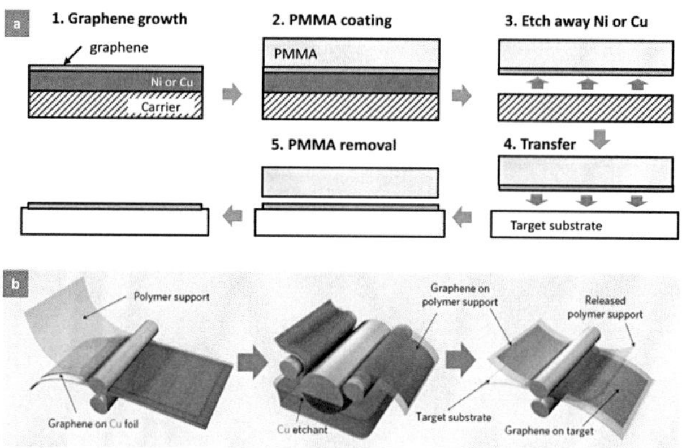

Figure 5.28 (a) Process flow of graphene growth (by CVD) and transfer to a target substrate. (b) R2R transfer process scheme proposed by Bae et al. [158] (Reprinted with permission from, *Nat Nanotech,* Copyright 2010. Nature Publishing Group).

Nevertheless, obtaining such high-performance graphene TEs in a reasonably practical scale still appears to be a highly challenging task. Long-term instability of some doping mechanisms [159,160] is also of concern. Organic solar cells with graphene TEs reported in some of the seminal works had *J–V* characteristics that were severely limited by the high R_{sheet} (>~5 kΩ/sq) of the graphene electrodes prepared by GO-reduction method or a bottom-up route based on spin-coating of hexadodecyl-substituted superphenalene and subsequent heating at 1100 °C under Ar [see Table 5.5 for summary]. Recent works using multilayer CVD-grown graphene TEs displayed much improved performance with FF larger than 40%, raising hope for a further boost in the near future (see Fig. 5.29).

Table 5.5 Performance of OPVs with graphene electrodes.

Anode/active [method]	Cell performance J_{sc} (mA/cm^2) V_{oc} (V) FF Eff. (%)	TE spec. [T, R_{sheet}(Ω/sq)]	Ref.
G/P3HT:PCBM [B.U.[a]]	0.36[b]/0.380/0.25/0.29	0.85/18k	[161]
G/CuPc/C_{60} [GO_R]	2.10[c]/0.480/0.34/0.40	0.80/5k	[162]
G/P/CuPc/C_{60} [CVD]	4.73/0.480/0.52[d]/1.2	0.89/3.5k	[156]
G/P/P3HT:PCBM [CVD]	9.03/0.600/0.48/2.6	0.87/610	[163]
G/M/CuPc/C_{60} [CVD]	3.41/0.540/0.40/0.75	0.92/250	[164]

[a] G=graphene, P=PEDOT:PSS, B.U.=bottom-up approach, M=MoO_x.
[b] Measured under 1.2 sun.
[c] Measured under 0.85 sun.
[d] Very small device area (0.75 mm^2).

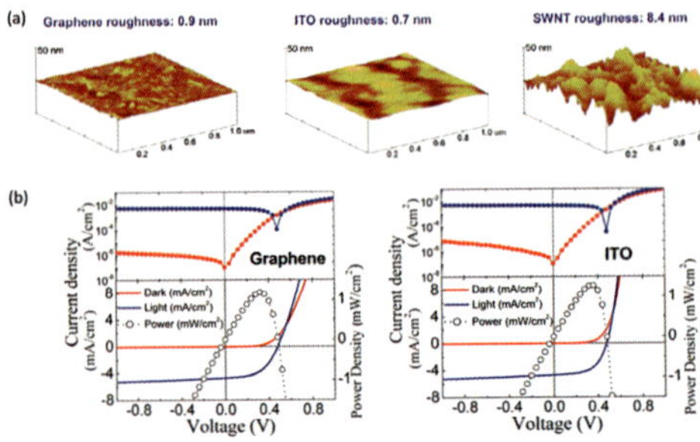

Figure 5.29 (a) Surface roughness of graphene, ITO, and CNT electrodes. (b) Performance example of OPVs using graphene as transparent electrodes [156] (Reprinted with permission from *ACS Nano*, Copyright (2010) American Chemical Society).

5.4 Summary and Outlook

Electrode technologies relevant to OPV cells were briefly reviewed from both electrical and optical perspectives. As OPVs are inherently thin-film, large area devices, TEs were noted as an essential element that allows simultaneously for light transmission and carrier collection. Various TE technologies and their applications in OPVs were introduced for those which could potentially be an alternative to ITO electrodes, which have been dominant in most OPVs. Their current status may be summarized as shown in Table 5.6.

Table 5.6 Summary of transparent electrode technologies.[a]

Electrode [Ref]	T (%)[b]	R_{sheet} (Ω/sq)	σ_{dc}/σ_{opt} ratio[c]	Advantages	Issues
ITO [165]	83	8.5	227	Well-established process/ mechanically robust/easy patternability	Indium price fluctuation/ poor flexibility
FTO [166]	83	13	149	Relatively low cost among TCOs	Surface roughness
AZO [167]	88	10	285	Low work function (if needed)/ relatively low cost	Weak to corrosive chemical environment/ relatively low mechanical hardness

(*Continued*)

Table 5.6 (*Continued*)

Electrode [Ref]	T (%)[b]	R_{sheet} (Ω/sq)	σ_{dc}/σ_{opt} ratio[c]	Advantages	Issues
Thin metal [98] (Ag, 12 nm)	45	6	64	High sheet conductance/ flexibility possible	High reflectance (low transmittance). Need to find the uniform growth condition for thin metal
D/M/D[d]					
$ZnS/Ag/WO_3$ [18]	85	14	159	Optical tuning possible by layer structure variation/high conductance/ flexibility	Optical design can be non-trivial/Spectral dependence of T/trilayer structure may increase the overall fabrication cost
$MoO_x/Ag/MoO_x$ [19]	78	6	229		
Ag-NWs [111]	80	10	159	High sheet conductance/ flexibility possible/ solution-processable	Difficult to control surface roughness
PEDOT:PSS [78] (vapor-phase growth [93])	86 (85)	150 (58)	16 (38)	High flexibility/ solution-processable/ low roughness	Relatively low conductivity/ condition to enhance conductivity still under development/ relatively poor stability
CNTs [149]	70	50	19	High flexibility	Limited sheet conductance
Graphene [158]	90	30	116	High flexibility/ relatively good sheet conductance	Related processes are not yet mature and still under development

[a] Based on actual experimental data reported in the literature or obtained from specifications of commercially available products; ultimately achievable values may be higher in some TE technologies. Some of the data other than those from commercial products may not necessarily be regarded typical; actual (T, R_{sheet}) values can depend sensitively on various conditions such as processing and compositions.

[b] Value near the wavelength of 550 nm. In case of ITO and FTO, values are for commercially available products having R_{sheet} around 8–15 Ω/sq.

[c] Calculated from the single (T, R_{sheet})-combination quoted in the table. Note that σ_{dc}/σ_{opt} ratio changes rather sensitively, and the value quoted here may differ from the value estimated from a trend line obtained by fitting (T, R_{sheet}) data measured over several thickness values. In some cases like DMD electrodes and Ag-NWs, σ_{dc}/σ_{opt} ratios are *effective* values, rather than intrinsic material parameters.

[d] Dielectric–metal–dielectric electrodes.

It is likely that ITO or a similar TCO technology will continue to play a major role, considering its technological maturity and performance level that other TE technologies still struggle to match; nevertheless, some emerging TE technologies are also expected to find their use in applications that prefer ultralow-cost fabrication and/or flexibility. Until their performance level becomes on par with that of ITO-based technologies and their processing conditions/methods are fully established, hybrid approaches—like conducting polymer/metal grid combination—may serve as an interim solution.

References

1. Kearns D, Tollin R, Calvin GMJ (1960) Electric properties of organic solids. II Effects of added electron acceptor on metal-free phthalocyanine, *Chem Phys,* **32**, 1020–1025.
2. Walzer K, Maennig B, Pfeiffer M, Leo K (2007) Highly efficient organic devices based on electrically doped transport layers, *Chem Rev,* **107**, 1233–1271.
3. Green M (1982) *Solar cells: Operating principles, technology, and system applications*, Prentice-Hall, Englewood Cliffs, NJ.
4. Heremans P, Cheyns D, Rand BP (2009) Strategies for increasing the efficiency of heterojunction organic solar cells: Material selection and device architecture, *Acc Chem Res,* **42**, 1740–1747.
5. Li G, Chu C-W, Shrotriya V, Huang J, Yang Y (2006) Efficient inverted polymer solar cells, *Appl Phys Lett,* **88**, art. no. 253503.
6. Yip H-L, Jen AK-Y (2012) Recent advances in solution-processed interfacial materials for efficient and stable polymer solar cells, *Energ Environ Sci,* **5**, 5994–6011.
7. Barr MC, Howden RM, Lunt RR, Bulović V, Gleason KK (2012) Top-illuminated organic photovoltaics on a variety of opaque substrates with vapor-printed poly(3,4-ethylenedioxythiophene) top electrodes and MoO_3 buffer layer, *Adv Energ Mater,* **2**, 1404–1409.
8. Han S, Feng X, Lu ZH, Johnson D, Wood R (2003) Transparent-cathode for top-emission organic light-emitting diodes, *Appl Phys Lett,* **82**, 2715–2717.
9. Hau SK, Yip H-L, Zou J, Jen AK-Y (2009) Indium tin oxide-free semi-transparent inverted polymer solar cells using conducting polymer as both bottom and top electrodes, *Org Electron,* **10**, 1401–1407.

10. Meiss J, Riede MK, Leo K (2009) Optimizing the morphology of metal multilayer films for indium tin oxide (ITO)-free inverted organic solar cells, *J Appl Phys,* **105**, art. no. 063108.

11. Ma H, Yip H-L, Huang F, Jen AK-Y (2010) Interface engineering for organic electronics, *Adv Funct Mater,* **20**, 1371–1388.

12. Chen S, Small CE, Amb CM, et al. (2012) Inverted polymer solar cells with reduced interface recombination, *Adv Energ Mater,* **2**, 1333–1337.

13. He Z, Zhong C, Su S, Xu M, Wu H, Cao Y (2012) Enhanced power-conversion efficiency in polymer solar cells using an inverted device structure, *Nat Photon*, **6**, 593–597.

14. Yoo S, Domercq B, Kippelen B (2005) Intensity-dependent equivalent circuit parameters of organic solar cells based on pentacene and C_{60}, *J Appl Phys,* **97**, art. no. 103706.

15. Choi S, Potscavage WJ, Kippelen B (2009) Area-scaling of organic solar cells, *J Appl Phys,* **106**, art. no. 054507.

16. Yoo S, Potscavage WJ, Domercq B, Kim J, Holt J, Kippelen B (2006) Integrated organic photovoltaic modules with a scalable voltage output, *Appl Phys Lett,* **89**, art. no. 233516.

17. Manceau M, Angmo D, Jørgensen M, Krebs, FC (2011) ITO-free flexible polymer solar cells: From small model devices to roll-to-roll processed large modules, *Org Electron,* **12**, 566–574.

18. Han S, Lim S, Kim H, Cho H, Yoo S (2010) Versatile multilayer transparent electrodes for ITO-free and flexible organic solar cells, *IEEE J Sel Topics Quant Electron,* **16**, 1656–1664.

19. Sergeant NP, Hadipour A, Niesen B, et al. (2012) Design of transparent anodes for resonant cavity enhanced light harvesting in organic solar cells, *Adv Mater,* **24**, 728–731.

20. Han D, Kim H, Lee S, Seo M, Yoo S (2010) Realization of efficient semitransparent organic photovoltaic cells with metallic top electrodes: Utilizing the tunable absorption asymmetry, *Opt Express,* **18**, A513–A521.

21. Macleod HA (2001) *Thin Film Optical Filters,* Taylor & Francis, New York.

22. Pettersson LAA, Roman LS, Inganäs O (2009) Modeling photocurrent action spectra of photovoltaic devices based on organic thin films, *J Appl Phys,* **86**, 487–496.

23. Glover RE, Tinkham M (1957) Conductivity of superconducting films for photon energies between 0.3 and 40 kTc, *Phys Rev,* **108**, 243–256.

24. Hu L, Hecht DS, Grüner G (2004) Percolation in transparent and conducting carbon nanotube networks, *Nano Lett*, **4**, 2513–2517.
25. Gordon RG (2000) Criteria for choosing transparent conductors, *MRS Bull*, **25**, 52–57.
26. Medford AJ, Lilliedal MR, Jørgensen M, et al. (2010) Grid-connected polymer solar panels: Initial considerations of cost, lifetime, and practicality, *Opt Express*, **18**, A272–A285.
27. Valkonen E, Karlson B, Ribbing C-G (1984) Solar optical properties of thin films of Cu, Ag, Au, Cr, Fe, Co, Ni, and Al, *Sol Energy*, **32**, 211–222.
28. Lee JY, Connor ST, Cui Y, Peumans P (2008) Solution-processed metal nanowire mesh transparent electrodes, *Nano Lett*, **8**, 689–692.
29. Crispin X, Geskin V, Crispin A, et al. (2002) Characterization of the interface dipole at organic/metal interfaces, *J Am Chem Soc*, **124**, 8131–8141.
30. Tengstedt C, Osikowicz W, Salaneck WR, Parker ID, Hsu C-H, Fahiman M (2006) Fermi-level pinning at conjugated polymer interfaces, *Appl Phys Lett*, **88**, 053502.
31. Paetzolda R, Winnacker A, Henseler D, Cesari V, Heuser K (2003) Permeation rate measurements by electrical analysis of calcium corrosion, *Rev Sci Instrum*, **74**, 5147–5150.
32. Luo Y, Aziz H, Popovic ZD, Xu G (2007) Degradation mechanisms in organic light-emitting devices: Metal migration model versus unstable tris(8-hydroxyquinoline) aluminum cationic model, *J Appl Phys*, **101**, art. no. 034510.
33. Yambem SD, Liao K-S, Alley NJ, Curran SA (2012) Stable organic photovoltaics using Ag thin film anodes, *J Mater Chem*, **22**, 6894–6898.
34. Suh Y, Lu N, Lee SH, et al. (2012) Degradation of a thin Ag layer induced by poly(3,4-ethylenedioxythiophene):Polystyrene sulfonate in a transmission electron microscopy specimen of an inverted polymer solar cell, *ACS Appl Mater Interfaces*, **4**, 5118–5124.
35. Yan M, Kim TW, Erlat AG, et al. (2005) A transparent, high barrier, and high heat substrate for organic electronics, *Proc IEEE*, **93**, 1468–1477.
36. Parthasarathy G, Burrows PE, Khalfin V, Kozlov VG, Forrest SR (1998) A metal-free cathode for organic semiconductor devices, *Appl Phys Lett*, **72**, 2138–2140.
37. Chipman, A (2007) A commodity no more, *Nature*, **449**, 131.
38. Parashkov R, Becker E, Riedl T, Johannes HH, Kowalsky W (2005) Large area electronics using printing methods, *Proc IEEE*, **93**, 1321–1329.

39. Forrest SR (2004) The path to ubiquitous and low-cost organic electronic appliances on plastic, *Nature*, **428**, 911–918.

40. Wang GF, Tao XM, Wang RX (2008) Flexible organic light-emitting diodes with a polymeric nanocomposite anode, *Nanotechnology*, **19**, art. no. 145201.

41. Hecht DS, Hu L, Irvin G (2011) Emerging transparent electrodes based on thin films of carbon nanotubes, graphene, and metallic nanostructures, *Adv Funct Mater*, **23**, 1482–1513.

42. Park Y-S, Choi K-H, Kim H-K (2009) Room temperature flexible and transparent ITO/Ag/ITO electrode grown on flexible PES substrate by continuous roll-to-roll sputtering for flexible organic solar photovoltaics, *J Phys D Appl Phys*, **42**, art. no. 235109.

43. Jansen KW, Delahoy AE (2003) A laboratory technique for the evaluation of electrochemical transparent conductive oxide delamination from glass substrates, *Thin Solid Films*, **423**, 153–160.

44. Lee S-H, Yang C-W, Park J-W (2012) The effect of the surface physical properties of polymer substrates on the adhesion and cracking of transparent conductive oxide (TCO) coatings, *Surf Coat Tech*, **207**, 24–33.

45. Cho H, Yun C, Park J-W, Yoo S (2009) Highly flexible organic light-emitting diodes based on ZnS/Ag/WO_3 multilayer transparent electrodes, *Org Electron*, **10**, 1163–1169.

46. Kittel C (1991) *Introduction to Solid State Physics*, 6th ed. Wiley & Sons, New York.

47. O'Reagan B, Grätzel M (1991) A low-cost, high-efficiency dye-sensitized solar cells based on colloidal TiO_2 films, *Nature*, **353**, 737–740.

48. Hüpkes J, Müller J, Rech B (2007) Texture etched ZnO:Al for silicon thin film solar cells, in *Transparent conductive oxide: Basics and applications in thin film solar cells* (Ellmer K, Klein A, Rech B, eds) Springer, Berlin.

49. Mason MG, Hung LS, Tang CW, Lee ST, Wong KW, Wang M (1999) Characterization of treated indium–tin–oxide surfaces used in electroluminescent devices, *J Appl Phys*, **86**, 1688–1692.

50. Kim JS, Friend RH, Caciallia F (1999) Surface energy and polarity of treated indium–tin–oxide anodes for polymer light-emitting diodes studied by contact-angle measurements, *J Appl Phys*, **86**, 2774–2778.

51. Yang F, Forrest SR (2006) Organic solar cells using transparent SnO_2:F anodes, *Adv Mater*, **18**, 2018–2022.

52. Yablonovitch E (1982) Statistical ray optics, *J Opt Soc Am,* **72**, 899–907.

53. Campbell P, Green MA (1987) Light trapping properties of pyramidally textured surfaces, *J Appl Phys,* **62**, 243–249.

54. Kluth O, Rech B, Houben L, et al. (1999) Texture etched ZnO:Al coated glass substrates for silicon based thin film solar cells, *Thin Solid Films,* **351**, 247–253.

55. Müller J, Kluth O, Wieder S, et al. (2001) Development of highly efficient thin film silicon solar cells on texture-etched zinc oxide-coated glass substrates, *Sol Energy Mater Sol Cells,* **66**, 275–281.

56. Müller J, Rech B, Springer J, Vanecek M (2004) TCO and light trapping in silicon thin film solar cells, *Sol Energy,* **77**, 917–930.

57. Lee JM, Yun SJ, Kim JK, Lim JW (2011) Texturing of Ga-doped ZnO transparent electrode for a-Si:H thin film solar cells, *Electrochem Sol State Lett,* **14**, B124–B126.

58. Sang BS, Yamada A, Konagai M (1998) Textured ZnO thin films for solar cells grown by a two-step process with the atomic layer deposition technique, *J Appl Phys Part 2-Lett,* **37**, L206–L208.

59. Youn HS, Yang M (2010) Solution-processed polymer light-emitting diodes utilizing a ZnO/organic ionic interlayer with Al, *Appl Phys Lett,* **97**, art. no. 3526308.

60. Heeger AJ (2001) Semiconducting and metallic polymers: The fourth generation of polymeric materials (nobel lecture), *Angew Chem Int Ed,* **40**, 2591–2611.

61. Shirakawa H (2001) The discovery of polyacetylene film: The dawning of an era of conducting polymers (nobel lecture), *Angew Chem Int Ed,* **40**, 2574–2580.

62. MacDiarmid AG (2001) "Synthetic metals": A novel role for organic polymers (nobel lecture), *Angew Chem Int Ed,* **40**, 2581–2590.

63. Andre JM, Delhalle J, Brédas JL (1991) *Quantum chemistry aided design of organic polymers: An introduction to the quantum chemistry of polymers and its applications,* World Scientific Publishing Co., Singapore.

64. Feast WJ, Friend RH (1996) Synthesis, processing and material properties of conjugated polymers, *Polymer,* **37**, 5017–5047.

65. Salaneck WR, Friend RH, Brédas JL (1999) Electronic structure of conjugated polymers: Consequences of electron lattice coupling, *Phys Reports,* **319**, 231–251.

66. Chandrasekhar P (1999) *Conducting polymers, fundamentals and applications: A practical approach,* Springer, Berlin.

67. Chiang CK, Fincher Jr CR, Park YW, et al. (1977) Electrical conductivity in doped polyacetylene, *Phys Rev Lett,* **39**, 1098–1101.

68. Naarmann H, Theophilou N (1987) New process for the production of metal-like, stable polyacetylene, *Synth Met,* **22,** 1–8.

69. Billingham NC, Calvert PD (1989) Electrically conducting polymers: A polymer science viewpoint, *Adv Polym Sci,* **90**, 1–104.

70. Muenstead H (1988) Aging of electrically conducting organic materials, *Polymer,* **29**, 296–302.

71. Elschner A, Kirchneyer S, Lövenich W, Merker U, Reuter K (2011) *PEDOT: Principles and applications of an intrinsically conductive polymer,* CRC Press, Boca Raton, FL.

72. Groenendaal LB, Jonas F, Freitag D, Pielartzik H, Reynolds JR (2000) Poly(3,4-ethylenedioxythiophene) and its derivatives: Past, present, and future, *Adv Mater,* **12**, 481–494.

73. Heywang G, Jonas F (1992) Poly(alkylenedioxythiophene)s—new, very stable conducting polymers, *Adv Mater,* **4**, 116–118.

74. Arias AC, Granström M, Thomas DS, Petritsch K, Friend RH (1999) Doped conducting-polymer–semiconducting-polymer interfaces: Their use in organic photovoltaic devices, *Phys Rev B,* **60**, 1854–1860.

75. Zhang F, Johansson M, Andersson MR, Hummelen JC, Inganäs O (2002) Polymer photovoltaic cells with conducting polymer anodes, *Adv Mater,* **14**, 662–665.

76. Na S-I, Kim S-S, Jo J, Kim D-Y (2008) Efficient and flexible ITO-free organic solar cells using highly conductive polymer anodes, *Adv Mater,* **20**, 4061–4067.

77. Galagan Y, Coenen EWC, Sabik S, et al. (2012) Evaluation of ink-jet printed current collecting grids and busbars for ITO-free organic solar cells, *Sol Energy Mater Sol Cells,* **104**, 32–38.

78. Kim YH, Sachse C, Machala ML, May C, Müller-Meskamp L, Leo K (2011) Highly conductive PEDOT:PSS electrode with optimized solvent and thermal post-treatment for ITO-free organic solar cells, *Adv Funct Mater,* **21**, 1076–1081.

79. Fehse K, Walzer K, Leo K, Lövenich W, Elschner A (2007) Highly conductive polymer anodes as replacements for inorganic materials in high-efficiency organic light-emitting diodes, *Adv Mater,* **19**, 441–444.

80. Carter SA, Angelopoulos M, Karg S, Brock PJ, Scott JC (1997) Polymeric anodes for improved polymer light-emitting diode performance, *Appl Phys Lett,* **70**, 2067–2069.

81. Koh T-W, Choi J-M, Lee S, Yoo S (2010) Optical outcoupling enhancement in organic light-emitting diodes: Highly conductive polymer as a low-index layer on microstructured ITO electrodes, *Adv Mater,* **22**, 1849–1853.

82. Cai M, Ye A, Xiao T, et al. (2012) Extremely efficient indium–tin-oxide-free green phosphorescent organic light-emitting diodes, *Adv Mater,* **24**, 4337–4342.

83. Web resource. http://www.agfagraphics.com/sp/global/en/internet/main/solutions/orgacon_electronic_materials/applications_orgacon/el_lamps/index.jsp (retrieved November 2012).

84. Kim JY, Jung JH, Lee DE, Joo J (2002) Enhancement of electrical conductivity of poly(3,4-ethylenedioxythiophene)/poly(4-styrene-sulfonate) by a change of solvents, *Synth Met,* **126**, 311–316.

85. Pettersson LAA, Ghosh S, Inganäs O (2002) Optical anisotropy in thin films of poly(3,4-ethylenedioxythiophene)–poly(4-styrenesulfonate), *Org Electron,* **3**, 143–148.

86. Jönsson SKM, Birgerson J, Crispin X, et al. (2003) The effects of solvents on the morphology and sheet resistance in poly(3,4-ethylenedioxythiophene)–polystyrene sulfonic acid (PEDOT–PSS) films, *Synth Met,* **139**, 1–10.

87. Wang T, Qi Y, Xu J, Hu X, Chen P (2005) Effects of poly(ethylene glycol) on electrical conductivity of poly(3,4-ethylenedioxythiophene)–poly(styrenesulfonic acid) film, *Appl Surf Sci,* **250**, 188–194.

88. Ouyang J, Chu C-W, Chen F-C, Xu Q, Yang Y (2005) High-conductivity poly(3,4-ethylenedioxythiophene)–poly(styrenesulfonate) film and its application in polymer optoelectronic devices, *Adv Funct Mater,* **15**, 203–208.

89. Web resource. http://clevios.com/en/conductivepolymers/pedot-pss-conductive-polymers.aspx (retrieved November 2012).

90. Vosgueritchian V, Lipomi DJ, Bao Z (2012) Highly conductive and transparent PEDOT:PSS films with a fluorosurfactant for stretchable and flexible transparent electrodes, *Adv Funct Mater,* **22**, 421–428.

91. Xia Y, Sun K, Ouyang J (2012) Solution-processed metallic conducting polymer films as transparent electrode of optoelectronic devices, *Adv Mater,* **24**, 2436–2440.

92. Alemu D, Wei H-U, Hod K-C, Chu C-W (2012) Highly conductive PEDOT:PSS electrode by simple film treatment with methanol for ITO-free polymer solar cells, *Energy Environ Sci,* Available Online (DOI: 10.1039/c2ee22595f).

93. Fabretto MV, Evans DR, Mueller M, et al. (2012) Polymeric material with metal-like conductivity for next generation organic electronic devices, *Chem Mater*, **24**, 3998–4003.

94. Simmons J, Potter KS (2000) *Optical Materials,* Academic Press, San Diego, CA.

95. Macleod HA (2010) *Thin-Film Optical Filters,* 4th ed. CRC Press, Boca Raton, FL.

96. Liu X, Cai X, Qiao J, Mao J, Jiang N (2003) The design of ZnS/Ag/ZnS transparent conducive multilayer films, *Thin Solid Films,* **441**, 200–206.

97. Pang H, Yuan Y, Zhou Y, et al. (2007) ZnS/Ag/ZnS coating as transparent anode for organic light emitting diodes, *J Luminescence,* **122–123**, 587–589.

98. O'Connor B, Haughn C, An K-H, Pipe KP, Shtein M (2008) Transparent and conductive electrodes based on unpatterned, thin metal films, *Appl Phys Lett,* **93**, art. no. 223304.

99. Cho H, Choi J-M, Yoo S (2011) Highly transparent organic light-emitting diodes with a metallic top electrode: The dual role of a Cs_2CO_3 layer, *Opt Express,* **19**, 1113–1121.

100. Rand BP, Peumans P, Forrest SR (2004) Long-range absorption enhancement in organic tandem thin-film solar cells containing silver nanocluster, *J Appl Phys,* **96**, 7519–7526.

101. Cho H, Yun C, Yoo S (2010) Multilayer transparent electrode for organic light-emitting diodes: Tuning its optical characteristics, *Opt Express,* **18**, 3404–3414.

102. Kaiser N (2002) Review of the fundamentals of thin-film growth, *Appl Opt,* **41**, 3053–3060.

103. Horvath G, Bankuti J (1988) Resistivity increase in thin conducting films considering the size effect, *Phys Stat Sol (a),* **110**, 549–554.

104. Tellier CR, Tosser AJ, Boutrit C (1977) The Mayadas-Shatzkes conduction model treated as a Fuchs-Sondheimer model, *Thin Solid Films,* **44**, 201–208.

105. Tao C, Xie G, Liu C, et al. (2009) Semitransparent inverted polymer solar cells with MoO_3/Ag/MoO_3 as transparent electrode, *Appl Phys Lett,* **95**, art. no. 053303.

106. Shrotriya V, Li G, Yao Y, Chu C-W, Yang Y (2006) Transition metal oxides as the buffer layer for polymer photovoltaic cells, *Appl Phys Lett,* **88**, art. no. 073508.

107. Han S, Shin WS, Seo M, Gupta D, Moon S-J, Yoo S (2009) Improving performance of organic solar cells using amorphous tungsten oxides as an interfacial buffer layer on transparent anodes, *Org Electron,* **10**, 791–797.

108. Lim S, Han D, Kim H, Lee S, Yoo S (2012) Cu-based multilayer transparent electrodes: A low-cost alternative to ITO electrodes in organic solar cells, *Sol Energy Mater Sol Cells,* **101**, 170–175.

109. Meiss J, Allinger N, Riede MK, Leo K (2008) Improved light harvesting in tin-doped indum oxide (ITO)-free inverted bulk-heterojunction organic solar cells using capping layers, *Appl Phys Lett,* **93**, art. no. 103311.

110. Hu LB, Kim HS, Lee JY, Peumans P, Cui Y (2010) Scalable coating and properties of transparent, flexible, silver nanowire electrodes, *ACS Nano,* **4**, 2955–2963.

111. Lee JY, Connor ST, Cui Y, Peumans P (2008) Solution-processed metal nanowire mesh transparent electrodes, *Nano Lett,* **8**, 689–692.

112. Leem DS, Edwards A, Faist M, Nelson J, Bradley DDC, de Mello JC (2011) Efficient organic solar cells with solution-processed silver nanowire electrodes, *Adv Mater,* **23**, 4371–4375.

113. Yang LQ, Zhang T, Zhou HX, Price SC, Wiley BJ, You W (2011) Solution-processed flexible polymer solar cells with silver nanowire electrodes, *ACS Appl Mater Interfaces,* **3**, 4075–4084.

114. Liu CH, Yu X (2011) Silver nanowire-based transparent, flexible, and conductive thin film, *Nanoscale Res Lett,* **6**, art. no. 75.

115. Scardaci V, Coull R, Lyons PE, Rickard D, Coleman JN (2011) Spray deposition of highly transparent, low-resistance networks of silver nanowires over large areas, *Small,* **7**, 2621–2628.

116. Tokuno T, Nogi M, Karakawa M, et al. (2011) Fabrication of silver nanowire transparent electrodes at room temperature, *Nano Res,* **4**, 1215–1222.

117. De S, King PJ, Lyons PE, Khan U, Coleman JN (2010) Size effects and the problem with percolation in nanostructured transparent conductors, *ACS Nano,* **4**, 7064–7072.

118. De S, Coleman JN (2011) The effects of percolation in nanostructured transparent conductors, *MRS Bull,* **36**, 774–781.

119. Gaynor W, Burkhard GF, McGehee MD, Peumans P (2011) Smooth nanowire/polymer composite transparent electrodes, *Adv Mater*, **23**, 2905–2910.

120. Yu ZB, Zhang QW, Li L, et al. (2011) Highly flexible silver nanowire electrodes for shape-memory polymer light-emitting diodes, *Adv Mater*, **23**, 664–668.

121. Lee JY, Connor ST, Cui Y, Peumans P (2010) Semitransparent organic photovoltaic cells with laminated top electrode, *Nano Lett*, **10**, 1276–1279.

122. Zeng XY, Zhang QK, Yu RM, Lu CZ (2010) A new transparent conductor: Silver nanowire film buried at the surface of a transparent polymer, *Adv Mater*, **22**, 4484–4488.

123. Morgenstern FSF, Kabra D, Massip S, et al. (2011) Ag-nanowire films coated with ZnO nanoparticles as a transparent electrode for solar cells, *Appl Phys Lett*, **99**, art. no. 183307.

124. Krantz J, Richter M, Spallek S, Spiecker E, Brabec CJ (2011) Solution-processed metallic nanowire electrodes as indium tin oxide replacement for thin-film solar cells, *Adv Funct Mater*, **21**, 4784–4787.

125. Ajuria J, Ugarte I, Cambarau W, Etxebarria I, Tena-Zaera R, Pacios R (2012) Insights on the working principles of flexible and efficient ITO-free organic solar cells based on solution processed Ag nanowire electrodes, *Sol Energy Mater Sol Cells*, **102**, 148–152.

126. Link S, El-Sayed MA (1999) Spectral properties and relaxation dynamics of surface plasmon electronic oscillations in gold and silver nanodots and nanorods, *J Phys Chem B*, **103**, 8410–8426.

127. Ah CS, Do Hong S, Jang DJ (2001) Preparation of $Au_{core}Ag_{shell}$ nanorods and characterization of their surface plasmon resonances, *J Phys Chem B*, **105**, 7871–7873.

128. Zhang KW, Meng LJ, Li J, Liu WL, Tang Y, Zhong JX (2008) Structure and thermal stability of gold nanowire encapsulated in carbon nanotube, *Acta Phys Sinica*, **57**, 4347–4355.

129. Lu WS, Shi W, Wang W, Jiang L (2008) One-step synthesis of gold nanowire network capped by diacetylene molecules under ultrasonic irradiation, *Coll Surf A-Physicochem Eng Aspects*, **317**, 239–246.

130. Messing ME, Hillerich K, Johansson J, Deppert K, Dick KA (2009) The use of gold for fabrication of nanowire structures, *Gold Bull*, **42**, 172–181.

131. Messing ME, Hillerich K, Bolinsson J, et al. (2010) A comparative study of the effect of gold seed particle preparation method on nanowire growth, *Nano Res*, **3**, 506–519.

132. Wu H, Hu LB, Rowell MW, et al. (2010) Electrospun metal nanofiber webs as high-performance transparent electrode, *Nano Lett,* **10**, 4242–4248.

133. Zaraska L, Sulka GD, Jaskula M (2012) Fabrication of free-standing copper foils covered with highly-ordered copper nanowire arrays, *Appl Surf Sci,* **258**, 7781–7786.

134. Sun YG, Gates B, Mayers B, Xia YN (2002) Crystalline silver nanowires by soft solution processing, *Nano Lett,* **2**, 165–168.

135. Sun YG, Yin YD, Mayers BT, Herricks T, Xia YN (2002) Uniform silver nanowires synthesis by reducing $AgNO_3$ with ethylene glycol in the presence of seeds and poly(vinyl pyrrolidone), *Chem Mater,* **14**, 4736–4745.

136. Sun YG, Mayers B, Herricks T, Xia YN (2003) Polyol synthesis of uniform silver nanowires: A plausible growth mechanism and the supporting evidence, *Nano Lett,* **3**, 955–960.

137. Riveros G, Green S, Cortes A, Gomez H, Marotti RE, Dalchiele EA (2006). Silver nanowire arrays electrochemically grown into nanoporous anodic alumina templates, *Nanotechnology,* **17**, 561–570.

138. Kim EU, Baeg KJ, Noh YY, et al. (2009) Templated assembly of metal nanoparticles in nanoimprinted patterns for metal nanowire fabrication, *Nanotechnology,* **20**, art. no. 355302.

139. Wiley B, Sun YG, Mayers B, Xia YN (2005) Shape-controlled synthesis of metal nanostructures: The case of silver, *Chem Eur J,* **11**, 454–463.

140. Lee P, Lee J, Lee H, et al. (2012) Highly stretchable and highly conductive metal electrode by very long metal nanowire percolation network, *Adv Mater,* **24**, 3326–3332.

141. Gaynor W, Lee JY, Peumans P (2010) Fully solution-processed inverted polymer solar cells with laminated nanowire electrodes, *ACS Nano,* **4**, 30–34.

142. Zhu R, Chung CH, Cha KC, et al. (2011) Fused silver nanowires with metal oxide nanoparticles and organic polymers for highly transparent conductors, *ACS Nano,* **5**, 9877–9882.

143. Yu ZB, Li L, Zhang QW, Hu WL, Pei QB (2011) Silver nanowire-polymer composite electrodes for efficient polymer solar cells, *Adv Mater,* **23**, 4453–4457.

144. Geim AK, Novoselov KS (2007) The rise of graphene, *Nat Mater,* **6**, 183–191.

145. Wu Z, Chen Z, Du X, et al. (2004) Transparent, conductive carbon nanotube films, *Science,* **305**, 1273–1276.

146. Hu L, Hecht DS, Gruner G (2004) Percolation in transparent and conducting carbon nanotube networks, *Nano Lett,* **4**, 2513–2517.

147. Hecht DS, Heintz AM, Lee R, et al. (2011) High conductivity transparent carbon nanotube films deposited from superacid, *Nanotechnology,* **22**, art. no. 075201.

148. Pasquier AD, Unalan AE, Kanwal A, Miller S, Chhowalla M (2005) Conducting and transparent single-wall carbon nanotube electrodes for polymer-fullerene solar cells, *Appl Phys Lett,* **87**, art. no. 203511.

149. Van De Lagemaat J, Barnes TM, et al. (2006) Organic solar cells with carbon nanotubes replacing In_2O_3:Sn as the transparent electrode, *Appl Phys Lett,* **88**, art. no. 233503.

150. Ulbricht R, Lee SB, Jiang X, et al. (2007) Transparent carbon nanotube sheets as 3-D charge collectors in organic solar cells, *Sol Energy Mater Sol Cells,* **91**, 416–419.

151. Ago H, Petritsch K, Shaffer MSP, Windle AH, Friend RH (1999) Composites of carbon nanotubes and conjugated polymers for photovoltaic devices, *Adv Mater,* **11**, 1281–1285.

152. Barnes TM, Bergeson JD, Tenent RC, et al. (2010) Carbon nanotube network electrodes enabling efficient organic solar cells without a hole transport layer, *Appl Phys Lett,* **96**, art. no. 243309.

153. Kim YH, Sachse C, Zakhidov AA, et al. (2012) Combined alternative electrodes for semi-transparent and ITO-free small molecule organic solar cells, *Org Electron,* **13**, 2422–2428.

154. Novoselov KS, Geim AK, Morozov SV, et al. (2004) Electric field in atomically thin carbon films, *Science,* **306**, 666–669.

155. Mak KF, Sfeir MY, Wu Y, Lui CH, Misewich JA, Heinz TF (2008) Measurement of the optical conductivity of graphene, *Phys Rev Lett,* **101**, art. no. 196405.

156. Gomez De Arco L, Zhang Y, et al. (2010) Continuous, highly flexible, and transparent graphene films by chemical vapor deposition for organic photovoltaics, *ACS Nano,* **4**, 2865–2873.

157. Eda G, Fanchini G, Chhowalla M (2008) Large-area ultrathin films of reduced graphene oxide as a transparent and flexible electronic material, *Nature Nanotech,* **3**, 270–274.

158. Bae S, Kim H, Lee Y, et al. (2010) Roll-to-roll production of 30-inch graphene films for transparent electrodes, *Nature Nanotech,* **5**, 574–578.

159. Jackson R, Domercq B, Jain R, Kippelen B, Graham S (2008) Stability of doped transparent carbon nanotube electrodes, *Adv Funct Mater,* **18**, 2548–2554.

160. Park H, Rowehl JA, Kim KK, Bulovic V, Kong J (2010) Doped graphene electrodes for organic solar cells, *Nanotechnology,* **21**, art. no. 505204.

161. Wang X, Zhi L, Tsao N, Tomović Ž, Li J, Müllen K (2008) Transparent carbon films as electrodes in organic solar cells, *Angew Chem Int Ed,* **47**, 2990–2992.

162. Wu J, Becerril HA, Bao Z, Liu Z, Chen Y, Peumans P (2008) Organic solar cells with solution-processed graphene transparent electrodes, *Appl Phys Lett,* **92**, art. no. 263302.

163. Choe M, Lee BH, Jo G, et al. (2010) Efficient bulk-heterojunction photovoltaic cells with transparent multi-layer graphene electrodes, *Org Electron,* **11**, 1864–1869.

164. Park H, Brown PR, Bulović V, Kong J (2012) Graphene as transparent conducting electrodes in organic photovoltaics: Studies in graphene morphology, hole transporting layers, and counter electrodes, *Nano Lett,* **12**, 133–140.

165. Web resource. Retrieved from http://www.pgo-online.com

166. Web resource (Pilkington Global Glass Handbook 2012) Retrieved from http://www.pilkington.com

167. Park S-M, Ikegami T, Ebihara K (2006) Growth of transparent conductive Al-doped ZnO films and device applications, *J Appl Phys,* **45**, 8453–8456.

Chapter 6

Tandem and Multijunction Organic Solar Cells

Jan Gilot[a,b] and René A. J. Janssen[c]
[a]*Holst Centre, TNO, Eindhoven, 5656 AE, The Netherlands*
[b]*TMC, Eindhoven, 5657 DC, The Netherlands*
[c]*Molecular Materials and Nanosystems, Eindhoven University of Technology, Eindhoven, 5612 ZA, The Netherlands*
jan.gilot@tno.nl

The best way to increase the efficiency of solar cells is by making better use of the energy of the absorbed photons. By stacking more photoactive layers with complementary absorption spectra, the ultimate achievable efficiency increases dramatically. Organic tandem solar cells have evolved strongly in the past years to reach 10–12% power conversion efficiency in 2013. The development of the intermediate layers is crucial for tandem cells where energy-level alignment, conductivity, and compatible processing with the photoactive layers are major concerns. The optimization and characterization of tandem cells are more complicated than that of single junction cells because of the interdependence of the subcells. This chapter provides an overview of the recent literature of multijunction organic solar cells and discusses the options for further improving these novel devices.

Organic Solar Cells: Fundamentals, Devices, and Upscaling
Edited by Barry P. Rand and Henning Richter

ISBN 978-981-4463-65-2 (Hardcover), 978-981-4463-66-9 (eBook)
www.panstanford.com

6.1 Introduction of Tandem Cells

The success for commercialization of organic solar cells will depend on improving three key parameters: efficiency, lifetime, and cost. In this chapter, we present an appropriate way to enhance the efficiency of polymer solar cells by using a multiple junction configuration in which two or more cells with different absorption spectra, i.e., different bandgaps, are stacked.

The efficiency of an organic solar cell strongly depends on the optical bandgap (E_g) of the donor and acceptor materials used ($E_g = \min(E_{g,donor}, E_{g,acceptor})$). By lowering E_g a larger part of the solar spectrum can be absorbed, which under ideal conditions results in an increase of the short-circuit current density (J_{sc}) and decrease of the open-circuit voltage (V_{oc}). As a consequence, the power conversion efficiency, which is proportional to the product of J_{sc} and V_{oc}, maximizes at some intermediate value of E_g.

This tradeoff between J_{sc} and V_{oc} is intimately related to the fact that photons with energies smaller than E_g cannot be absorbed and photons with larger energies will lose their excess energy via thermal equilibration. In the detailed balance limit, first described by Shockley and Queisser,[1] the power conversion efficiency (η) of a single bandgap solar cell under non-concentrated air mass 1.5 global (AM1.5G) sunlight reaches a maximum value of ~33% for E_g = 1.1–1.4 eV. The conversion losses can be significantly reduced by converting different parts of the solar spectrum in separate subcells that are tailored to a specific wavelength range. In a tandem solar cell, thermalization losses of high-energy photons are reduced by conversion in the subcell with a wide E_g and transmission losses are lowered by absorption of the low energy photons in the subcell with a small E_g.

De Vos determined that in the detailed balance limit the maximum efficiency of a solar cell under non-concentrated sunlight, modeled as black-body irradiation, rises to 42% for a tandem cell with E_g's of 1.9 and 1.0 eV, and to 49% for triple junction solar cells with E_g's of 2.3, 1.4, and 0.8 eV.[2] Brown and Green calculated the detailed balance limit for classical, two and three junction series constrained tandem devices to be 45% and 51%.[3] To establish design rules for new materials and device configurations, various empirical and fundamental approaches to the limiting efficiency have been developed for organic solar cells, where an additional loss is introduced by the necessity of the donor/acceptor concept.[4] Empirical models that predict 10–11% efficiencies for single-junction solar cells[5,6] suggest that the efficiency of organic tandem cells will increase the achievable performance by 40% to about 15%, for E_g's of 1.6 and 1.3 eV compared to that of a single junction at $E_g = 1.4$ eV.[7] Using a similar methodology, the model can be extended for triple-junction solar cells. For subcells with a bandgap of 1.8, 1.4, and 1.15 eV the performance can reach up to 17.5% assuming an FF of 65%, an internal quantum efficiency (IQE) of 85%, and an external quantum efficiency (EQE) of 65%.[8] This is an improvement of 65% compared to single-junction cells.

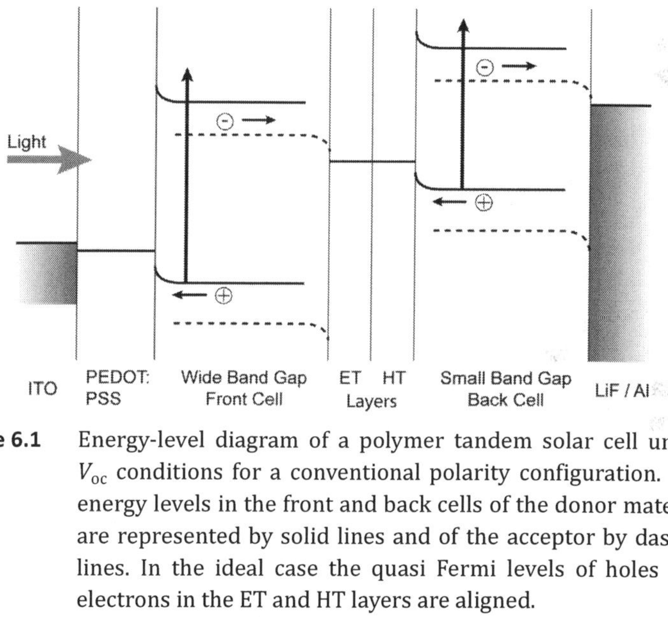

Figure 6.1 Energy-level diagram of a polymer tandem solar cell under V_{oc} conditions for a conventional polarity configuration. The energy levels in the front and back cells of the donor material are represented by solid lines and of the acceptor by dashed lines. In the ideal case the quasi Fermi levels of holes and electrons in the ET and HT layers are aligned.

The commercially most interesting configuration of an organic tandem solar cell is a two-terminal series connection, which can be constructed by stacking the two photoactive layers using an intermediate contact layer. Under operation, Kirchhoff's law dictates that equal currents flow through the two subcells and the voltage of the tandem cell is a summation of the voltages of the subcells. In most modern organic tandem cells the intermediate contact consists of an electron-transporting (ET) layer, collecting electrons from one subcell, and a hole-transporting (HT) layer, collecting holes from the other subcell. The energy levels of the ET and HT layer should align with $LUMO_{acceptor}$ and $HOMO_{donor}$, respectively. To avoid voltage losses, the levels of the ET and HT layers must align at the ET/HT interface. The energy-level diagram of a polymer-tandem solar cell under V_{oc} conditions is shown in Fig. 6.1 for a conventional polarity configuration of the solar cell. Further requirements for the intermediate contact will be discussed in detail in Section 6.3.

6.2 History and Current Status of Tandem Solar Cells

The first organic tandem solar cells were reported in 1990 and consisted of two evaporated small-molecule bilayer cells stacked on top of each other. The tandem provided a voltage that was nearly twice that of the single-junction cell (0.78 vs. 0.44 V).[9] By the end of 2012, i.e., more than 20 years since the first report, organic tandem cells provide power conversion efficiencies exceeding 10% for evaporated and solution-processed organic materials.[10,11] Here a brief summary of the history of tandem solar cells and the recent most efficient solar cells are presented. For a more extensive overview, Hadipour et al., Ameri et al., and Sista et al. have published comprehensive reviews on the recent published progress in small-molecule and polymer-tandem solar cells.[12–14]

6.2.1 History of Tandem Solar Cells

In 1990, Hiramoto et al. reported on the first organic tandem solar cell stacking two identical small-molecule bilayer subcells (Me-PTC/H_2Pc).[9] To provide an effective intermediate contact for recombination of electrons and holes generated in the adjacent subcells, a 2 nm gold layer was evaporated between the subcells. In 2002, Yakimov and Forrest presented an identical approach with comparable small molecules (CuPc/PTCBI) stacking two, three, and five subcells resulting in a V_{oc} that was almost the summation of the V_{oc}'s of the subcells.[15] Ultrathin (~0.5 nm) layers of Ag clusters were placed between each subcell to serve as recombination centers. In 2004, Xue et al. improved the device efficiency up to 5.7% by using a hybrid planar-mixed heterojunction where a blend layer composed of co-evaporated CuPc donor and C_{60} acceptor molecules was sandwiched between homogeneous CuPc and C_{60} layers.[16] The same year, Drechsel et al. and Männig et al. introduced the p–i–n type heterojunction architecture with doped ET and HT layers differing from the materials in the photoactive layer.[17,18] Tuning the thickness of the ET and HT layers enabled the optimization of the optical absorption in the photoactive layers using the ET and HT layers as an optical spacer. Because of the interference between incident light and light reflected from the metal back electrode, the electric field component of the light maximizes at specific distances from the electrode. These distances depend on wavelength. By inserting a transparent conductive material between the two photoactive layers it is possible to tune the position of the photoactive layers such that they coincide with the anti-nodes of the standing wave of the optical electrical field and thereby maximize the optical absorption. The extra transparent layer acts effectively as an optical spacer, keeping the two photoactive layers at optimized positions.

Also in 2004, Triyana et al. attempted to optimize a triple-junction solar cell by modifying the type of metal clusters together with the order of donor and acceptor and by fine-tuning the thickness of the middle subcell.[19,20] In 2007, Cheyns et al. studied different metals as efficient recombination centers of holes and electrons at the contact between the two subcells.[21] Recently, Timmreck et al. have shown that a high doping level in the ET and HT layer allows for omitting the metal clusters to obtain efficient conversion contacts.[22]

The first step toward polymer-tandem solar cells was reported in 2006 by Dennler et al. who stacked two different subcells with different absorption spectra.[23] The first subcell was a solution-processed diffused bilayer P3HT:[60]PCBM and the second subcell an evaporated ZnPc:C_{60} bulk heterojunction. The intermediate contact consisted of evaporated ET and HT layers based on C_{60} and ZnPc with a 1 nm Au recombination layer in between, resulting in a 2.3% efficiency. The V_{oc} of the tandem was the addition of the V_{oc}'s of the subcells. Similar tandem configurations, using P3HT:[60] PCBM and CuPc:C_{60}, were reported by Colsmann et al., Janssen et al., and Zhao et al.[24–26] In 2012, Lassiter et al. combined solution-processed and evaporated small molecules in a 6.6% efficient tandem cell.[27]

The first tandem cell composed of two polymer subcells was presented by Kawano et al. in 2005 by stacking two MDMO-PPV:[60]PCBM subcells with an ITO (sputtered)/PEDOT:PSS (spin cast) intermediate layer.[28] Sakai et al. deposited MoO_3 on a sputtered ITO layer to separate a P3HT:[60]PCBM subcell from a P3HT:[70]PCBM subcell, resulting in a 5.1% polymer-tandem cell.[29] The triple cell reported by Zhao et al. stacking three P3HT:[60]PCBM subcells separated by Al/MoO_3 resulted in a V_{oc} of 1.73 V.[30]

The polymer-tandem solar cell reported by Hadipour et al. in 2006 consisted of subcells with complementary absorption spectra.[31] The intermediate contact was a combination of evaporated metals. Later, Hadipour et al. introduced an embedded spacer in the stack, resulting in a four-terminal device allowing for arbitrary series or parallel connection of the two subcells.[32] By varying the thickness of the spacer layer, the transmission of the front cell and spacer was tuned to maximize the absorption in the back cell.

The first organic tandem cells with solution-processed photoactive and intermediate layers were published by Gilot et al. in 2007.[33] The combination of ZnO nanoparticles from acetone (ET layer) and pH-neutral PEDOT:PSS (HT layer) formed a closed intermediate layer by which tandem- and triple-junction cells were made with MDMO-PPV:[60]PCBM and P3HT:[60]PCBM as photoactive layers. Figure 6.2 shows a cross section TEM image of the ZnO/PEDOT:PSS intermediate ET/HT layer with other photoactive layers from later results.[34] Later that year, Kim et al. presented a fully solution-processed tandem cell of subcells with complementary absorption spectra with a power conversion efficiency of 6.5%.[35] PCPDTBT:[60]PCBM and P3HT:[70]PCBM were combined in a tandem cell separated by a TiO_x/PEDOT:PSS intermediate layer. This power conversion efficiency value was later corrected by Sista et al. to 5.8% because of the high conductivity of the PEDOT:PSS used.[36] Parasitic current generation in illuminated bordering areas of the active area contribute to the total current of the subcell and thereby leading to an overestimation. This is easily avoided by masking the bordering areas with a non-transparent material or by removing the photoactive layers precisely around the active area.

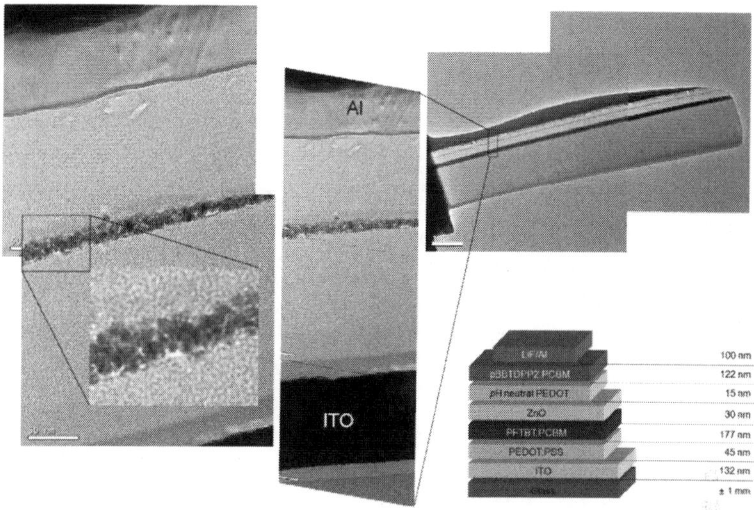

Figure 6.2 TEM cross section of a tandem solar cell with a ZnO/PEDOT:PSS intermediate electrode. This image corresponds to solar cells described in Ref. [34].

6.2.2 High-Efficiency Tandem Solar Cells

Since the publication of an efficiency of 5.7% for evaporated and 6.5% for solution-processed organic tandem solar cells in 2004 and 2007, respectively, large steps have been made toward the record efficiencies by the end of 2012.

The record efficiency of a regular configuration in solution-processed tandem solar cells was first improved by Yang et al. in 2011 to 7.0% followed by Gevaerts et al. reaching 7.0% as well in 2012.[37,38]

The solution-processed tandem solar cell with an inverted configuration was introduced in 2011 by Chou et al. with a partly evaporated intermediate layer of MoO_3/Al (HT layer) and a solution-processed precursor ZnO (ET layer) resulting in an efficiency of 5.1%.[39] One year later Kouijzer et al. showed a fully solution-processed intermediate layer from PEDOT:PSS/ZnO with an efficiency of 5.8%.[40]

The continuing progress on the more ideal small bandgap polymer led to improvements in the efficiency of both the regular and inverted structure. With PEDOT:PSS/ZnO as intermediate layer, the record efficiency gradually increased from 8.6% to 9.5% with PBDTT-DPP possessing either sulfur or selenium atoms on the thiophene moiety.[41,42] By using an even more ideal small bandgap polymer (PDTP-DFBT), a certified record efficiency of 10.6% was published.[11] In the same way of using a more suitable small bandgap polymer in the back cell, the 7.0% of the regular configuration was improved to 8.9% for a tandem cell and 9.6% for a triple junction cell.[43] Even with a combination of two identical absorbing layers of PDTP-DFBT, a 10.2% efficient solar cell was constructed.[44]

These record efficiencies are not limited to an intermediate layer based on ZnO and PEDOT:PSS. Zhou et al. showed that ethoxylated polyethylenimine (PEIE) could be used to modify the work function of highly conductive PEDOT:PSS selectively on one side, creating a work function contrast of 1.3 eV between the two interfaces of the intermediate layer with both photoactive layers.[45] Efficiencies up to 8.2% were reached, with the highest FF achieved so far for a tandem cell of 0.72.

Also for evaporated tandem solar cells the use of subcells with complementary absorption spectra resulted in higher efficiencies. In 2010, Cheyns et al. showed that the combination of a green absorbing and near-IR absorbing small molecule resulted in an efficiency of 5.2%, which is an improvement of 40% compared to the individual single-junction cells.[46] One year later, Riede et al. published a 6.1% tandem cell where the two complementary absorbing subcells were separated by a conductive spacing layer to optimize the current generation in both subcells.[47] Early 2013, Heliatek GmbH, a German company, announced that it achieved an efficiency of 12% for an organic multijunction solar cell of 1.1 cm^2 active area with evaporated layers.[10]

This recent progress shows the potential of organic tandem solar cells to achieve high efficiencies, exceeding the performance of single-junction cells.

6.3 Intermediate Layers

An organic tandem solar cell comprises two subcells that are separated from each other by an intermediate layer. This intermediate layer must fulfill several functions. It physically separates the two subcells and prevents—in the case of solution processing—interaction of the solvents of the newly deposited layer with the already present layer. The intermediate layer itself is, in the case of solution processing, deposited from orthogonal solvents (see Section 6.3.1) to avoid damaging the first subcell. The intermediate layer also connects the two subcells electrically by recombining the electrons extracted from one subcell on one side with the holes collected from the second subcell on the other side. The work functions on both sides of the intermediate layer are tuned to create an open-circuit voltage in each subcell comparable to a single junction cell with corresponding photoactive layer. This intermediate layer also functions as an optical spacer in the tandem cell. Preferentially, it is completely transparent for the light absorbed in the subcells. Modifying the thickness of either of the photoactive or intermediate layers redistributes the optical electric field inside the solar cell, enabling an extra tool for optimizing tandem solar cells. We will elaborate on all these properties in this section.

6.3.1 Materials

The intermediate layer consists of at least two different layers to fulfill the different properties on both sides: an ET layer and a HT layer. In the case of solution-processed tandem cells, the choice of solvent is crucial. The solvent should be innocuous to the photoactive layer and at the same time should wet the surface sufficiently so that the dissolved material adheres well to the photoactive layer after deposition. Common solvents for photoactive layer blends as chloroform, chlorobenzene, toluene, 1,2-dichlorobenzene, 1,2,4-trichlorobenzene, and o-xylene are unsuitable for this purpose as the underlying photoactive layer will be dissolved upon deposition of the ET layer. The solubility, especially of the PCBM, turns out to be the largest obstacle. Even solvents that are considered as poor or non-solvents for PCBM (cyclohexane, cyclohexanone, and diisopropylether) can effectively remove all PCBM out of the photoactive layer despite the very low solubility of PCBM. Suitable solvents that can be used on top of most photoactive layers without interfering with the layer are acetone, methanol, ethanol, isopropanol, ethanolamine, and 2-ethoxyethanol (this list is non-exhaustive). An alternative technique to enable multilayer solution processing is to alter the solubility of the layer once it has been formed by using thermocleavable materials as proposed by Hagemann et al.[48]

Metal oxides, small organic molecules, and polymers have all been applied as ET and HT layers. For solution processing, polymer (PEDOT:PSS) and metal oxide (NiO, MoO_3, V_2O_5, WO_3, ITO) based HT layers are widely used, while Cs_2CO_3, ZnO, TiO_2, and calcium niobate nanosheets[49] or doped variants of the latter two are commonly applied as ET layer. The focus for the deposition of these layers is from solution; however, evaporation or sputtering in vacuum can also be applied. More recently, the use of ethoxylated polyethylenimine (PEIE) was demonstrated as a first solution-processable organic material that can be used to modify the work function of PEDOT:PSS and create an effective intermediate contact and thereby omitting the use of a separate ET layer.[45,50] Evaporated small molecule-based organic tandem solar cells incorporate inorganic interlayers, such as thin metal (Ag, Au) or metal oxide layers (MoO_3), and electrically- doped organic layers (C_{60}, BCP, DiNPB, BPhen, etc.). Doping of the latter occurs by metals (Au, Yb) or metal oxides, or by dedicated organic dopant materials.

6.3.2 Aligning of Energy Levels of ET and HT Layers in the Intermediate Layer

The key for success of tandem solar cells in a series connection is an intermediate contact layer that does not introduce any voltage losses. However, often the voltage of the tandem cell is not the exact sum of the voltages of the two single junction cells. This can be attributed to a non-ohmic contact between the ET and HT layers. If the charge-carrier density in one of the transporting layers is insufficient for compensating the diffused charges from the other transporting layer, the energy levels of the electrons and holes do not align (Fig. 6.3a) and an additional energy (or voltage) is required for charges to pass the junction in the dark, increasing the voltage onset of the dark current. Under illumination the p–n junction at the ET/HT layer interface causes a small voltage drop compared to the sum of the V_{oc}'s of the subcells. Combined, these effects result in an s-shaped current density to voltage (*J*–*V*) curve (Fig. 6.3c).

A first approach to align the energy levels of the ET and HT layers is by evaporating thin metal (0.5–2 nm of Au or Ag) clusters as recombination centers between ET and HT layers.[15,51] This has been used in solution-processed and evaporated tandem cells. However, these metal clusters cause absorption and reflection of light, which influence the current generation in both subcells and reduce the power conversion efficiency of the tandem cell. The efficiency of the recombination contact also depends on the type of metal cluster.[22] Possible explanations are the doping of the transporting layer by metal diffusion or the interface dipoles originating at the interface between metal and organics resulting in a built-in voltage at the interface.

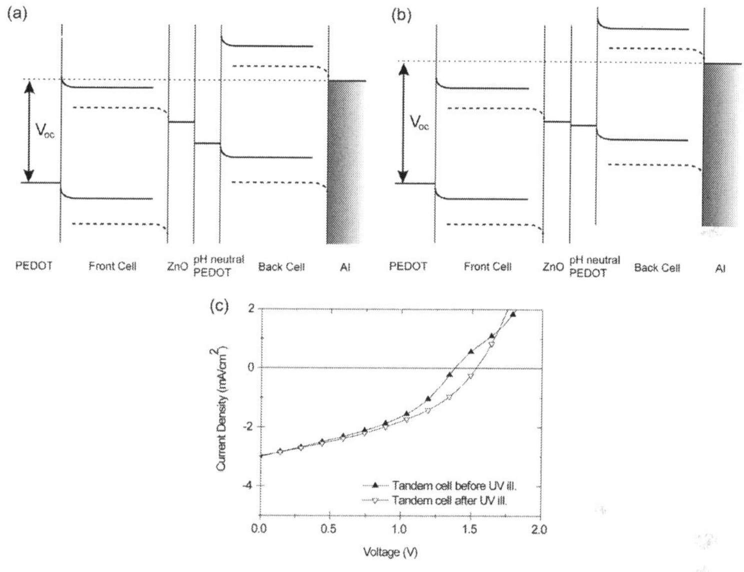

Figure 6.3 (a, b) The energy band diagram of a tandem cell with ZnO/pH neutral PEDOT as intermediate contact before (a) and after (b) UV illumination. The band bending of ZnO is modified upon exposure to UV light. (c) At low charge-carrier density in ZnO ET layers (before UV illumination) of the ZnO/pH-neutral PEDOT intermediate electrode an s-shaped *J–V* curve is obtained, at higher charge density (after UV illumination) the s-shape is no longer present.

A second way to overcome the voltage loss induced in the intermediate layer is by highly doping the ET and HT layers.[22] Via co-evaporation of specific dopants in the ET and HT layer, the inorganic tandem solar cell method of tunnel diodes is mimicked. High doping concentrations induce high charge-carrier concentrations on both sides of the recombination contact leading to a high recombination rate at the contact. Hereby, the energy levels of the ET and HT layers effectively align at the junction giving an ideal intermediate layer where the V_{oc} of the tandem cell is the addition of the V_{oc}'s of the subcells of a tandem solar cell.

A third method is commonly applied in solution-processed tandem solar cells. By inserting a metal oxide/PEDOT intermediate layer, an ohmic contact can be obtained by sufficient doping of the two materials. PEDOT is already heavily doped (certainly the highly conductive variant) and a metal oxide, ZnO or TiO_x, can very easily be doped by exposure to UV irradiation (Fig. 6.3b).[33,36]

To explain the beneficial effect of UV light on the metal oxide/PEDOT contact, it is important to consider that the electrical properties of metal oxides are strongly dependent on the precise stoichiometry of the metal oxide. Trapped positively charged defects are compensated by electrons in the conduction band that have reacted with molecular oxygen from the ambient to form bound O_2^- at the surface.[52,53] Generation of an electron/hole pair by exposure to UV light in the absence of O_2 results in recombination of the photogenerated hole with the electron trapped by O_2^- at the surface of the metal oxide, inducing desorption of neutral O_2 (Fig. 6.4a) and leaving the mobile electron in the conduction band.[54,55] As a consequence, the concentration of mobile electrons in the nanoparticle is enhanced resulting in a high conductivity of the metal oxide layer.[56,57] This process is reversed upon exposure to air.

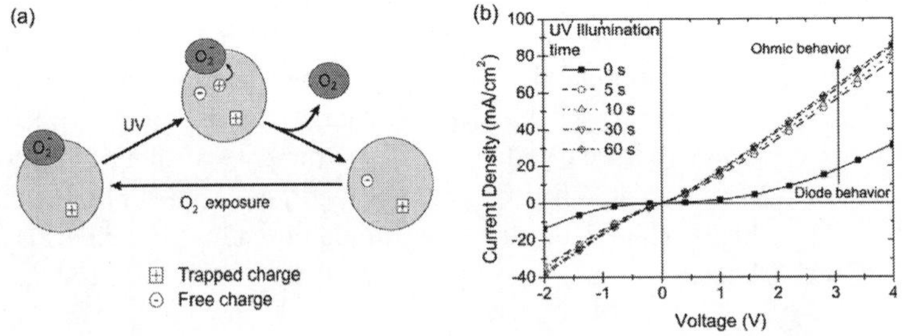

Figure 6.4 (a) Schematic representation of the changes induced by UV illumination of ZnO nanoparticles. A photogenerated hole induces desorption of O_2 resulting in a mobile electron. This process is reversed upon exposure to O_2. (b) J–V characteristics of an ITO/ZnO/pH neutral PEDOT/LiF/Al device before and after UV illumination.

A study of the electrical characteristics of an ITO/metal oxide/PEDOT/LiF/Al device clarifies the influence of UV illumination on the metal oxide/PEDOT contact. The *J–V* characteristic in dark of the pristine diode shows some rectification (Schottky contact). This rectification can be explained by the presence of an injection barrier for electrons from the metal oxide into PEDOT. Upon exposure to UV light, the junction becomes ohmic (Fig. 6.4b).[58] The increase in current density (for $V \neq 0$) saturates after 30 s of UV illumination.

The formation of an ohmic contact between the n-type metal oxide and the p-type PEDOT upon photodoping implies a considerable amount of charge transfer between the metal oxide and the pH-neutral PEDOT. Using ellipsometry, Lakhwani et al. observed significant changes in the optical constants of ZnO and PEDOT upon UV illumination measured on a glass substrate covered with ZnO/PEDOT.[59] A reversible decrease of the extinction coefficient was noticed in the band of the PEDOT centered at ~900 nm. This band is assigned to the oxidized form of PEDOT and its decrease provides an indication that some reduction of oxidized PEDOT segments occurred by transfer of electrons from the photodoped ZnO.

The carrier density in ZnO after photodoping was estimated to be $10^{19}/cm^3$, determined from ellipsometry measurements of the Moss–Burstein shift and Kelvin probe measurements.[59] This doping level is comparable to the results of other experimental studies of doped ZnO (In, Al).[60,61] Thus the exposure to UV irradiation increases the mobile carrier concentration and affords electron transfer and hence an Ohmic contact at ZnO/PEDOT interface layers.

6.3.3 The Use of PEDOT as HT layer in Solution-Processed Tandem Solar Cells

Solution-processed tandem solar cells often comprise an intermediate layer of a metal oxide in combination with PEDOT. In the case of using ZnO as the metal oxide, the standard dispersion of PEDOT:PSS in water dissolves the ZnO layer due to its acidic nature (pH $\approx$ 2). This causes a processing issue for tandem cells with the normal polarity configuration, where ZnO and then PEDOT are deposited sequentially. To solve these problem, a PEDOT dispersion at neutral pH (pH $\approx$ 7), which is innocuous toward the ZnO layer, is commonly employed.[33] Neutralizing PEDOT reduces its work function and thereby increases the difference with the oxidation potential of the polymer to a level which is too large for an ohmic contact.[62] As a result, V_{oc} is no longer determined by the offset between the highest occupied molecular orbital of the donor ($HOMO_{donor}$) and the lowest unoccupied molecular orbital of the acceptor ($LUMO_{acceptor}$), but by the pH-neutral PEDOT work function/$LUMO_{acceptor}$ offset. As a result, the ZnO/PEDOT intermediate layer is only suited in combination with back cells based on a donor with an ionization potential of –4.8 ~4.9 eV, such as P3HT.

To enable the use of donor polymers with a higher ionization potential than P3HT and hence a higher open-circuit voltage in the back cell, an interfacial layer is required on top of the pH-neutral PEDOT. Spin casting a Nafion perfluorinated resin solution from ethanol can restore the ohmic contact between the donor polymer of the back cell and the pH-neutral PEDOT.[62] A double junction with two identical polymers with an ionization potential of –5.4 eV (PF10TBT) has an open-circuit voltage virtually as high as the single-junction device.[62]

6.3.4 Triple- and Multiple-Junction Solar Cells Leading to High Open-Circuit Voltages

Tandem solar cells consisting of two subcells with a high open-circuit voltage have reached voltages approaching 2 V. The combination of two solution-processed PF10TBT:[60]PCBM subcells with a ZnO/PEDOT/Nafion intermediate contact as described in Section 6.3.3 produced a V_{oc} of 1.92 V,[62] while an evaporated merocyanine tandem solar cell with an intermediate layer of Al/MoO_3 reached a V_{oc} of 2.07 V.[63] These devices combined two identical subcells to show a proof of principle to achieve high voltages in tandem cells rather than to pursue the best power conversion efficiency.

The use of an evaporated intermediate layer greatly facilitates the fabrication of triple- and multiple-junction solar cells that provide V_{oc}'s exceeding that of the tandem cell. Stacking three subcells with a moderate open-circuit voltage, like P3HT, with an intermediate contact of Al/MoO_3 resulted already in an overall V_{oc} of 1.73 V.[30] Four ZnPc/C_{60} subcells on top of each other separated by a LiF/Au/MoO_x intermediate electrode showed a V_{oc} of 2.07 V and five CuPc/PTCBI with Ag recombination centers in between lead to a V_{oc} of

1.73 V.[15,74] Further increase of the V_{oc} of the multiple junction cell has been reported for triple-junction cells of TAPC:C_{60} connected by Bphen/Ag/HAT-CN/MoO_x intermediate layer having a V_{oc} of 2.71 V.[65] Recently, Zou et al. reported ClAlPc/C_{60} multijunction cells connected using a BCP/Ag/MoO_3 intermediate contact with V_{oc}'s of 3.50 V for a five-fold and 5.89 V for a ten-fold junction.[66]

Multiple junction solar cells are achievable with solution-processed intermediate contacts as well. For the ZnO/pH-neutral PEDOT intermediate layer multiple junction solar cells with three and six photoactive layers were processed in a similar fashion as the tandem cell. A P3HT:[60]PCBM photoactive layer was preceded by, respectively, two and five MDMO-PPV:[60]PCBM photoactive layers. The layer thicknesses are increased from front to back cell to permit sufficient light transmission to the back cell in order to maintain a decent current. V_{oc}'s of 1.71 and 1.84 V were obtained for the completed device before UV illumination, but these increased significantly to 2.19 and 3.58 V, respectively, upon applying UV light (Fig. 6.5).

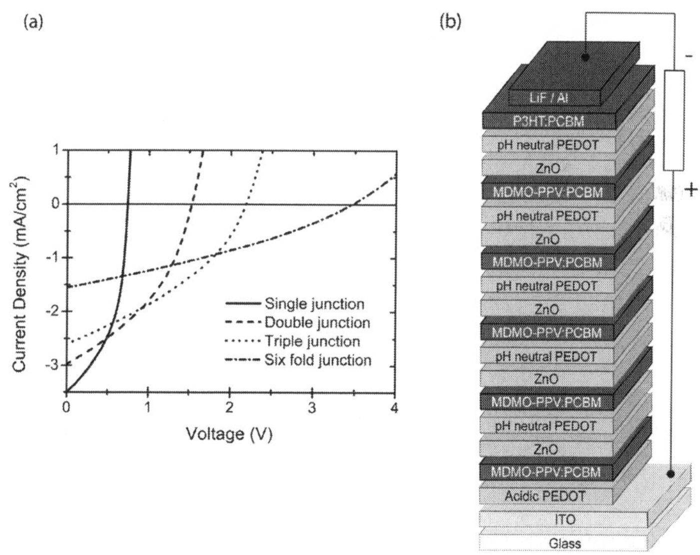

Figure 6.5 (a) J–V characteristics after UV illumination of a single-junction P3HT:[60]PCBM cell with additional MDMO-PPV:[60]PCBM layers forming a double, triple, and six-fold junction device. (b) Device layout of a six-fold junction based on MDMO-PPV:[60]PCBM and P3HT:[60]PCBM active layers.

The V_{oc} of the multiple-junction cells deviate significantly from the estimated V_{oc} obtained via addition of the voltages of the single-junction cells. This is a consequence of the lower work function of pH-neutral PEDOT and the related loss in V_{oc}, as explained in Section 6.3.3 for all layers except the first one deposited on acidic PEDOT. Also the reduced light intensity reaching the photoactive layers closest to the reflective electrode results in a decrease of the actual V_{oc}.[67]

6.4 Optimization

The optical and electrical interaction of two subcells complicates the optimization of performance in a tandem cell. Finding the optimal thickness of the layers is key to achieve high efficiencies and exceed the performance of a single-junction cell. Different approaches are used to optimize solution-processed and evaporated tandem cells. The thickness of the photoactive layer in evaporated organic solar cells is limited because of excitonic effects, low charge-carrier mobilities, and lifetimes, while for solution-processed tandem cells the conductivity and thereby the thickness of the electron and hole transport layers are the limiting factors. In this section, we will highlight the various methods that have been applied to extract the highest performance from a specific tandem cell.

Absorption in the photoactive layers is generally evaluated by optical modeling using the transfer matrix formalism (for a more complete treatment, the reader is referred to Chapter 8). This formalism allows determining the reflection, transmission, absorption, and the amplitude of the electromagnetic field within a whole stack of thin layers (thin compared to the coherence length of the light). The stack is treated as a one-dimensional system in which the amplitude of the electromagnetic field vector is calculated by taking into account the phase information. The formalism considers incoming and outgoing electromagnetic waves propagating in opposite direction within the multilayer system. Changeovers between two adjacent layers result in partial transmission and reflection at the interface. Propagation within a layer leads to a decay of the amplitude in the case of absorption. These effects can be written in terms of the optical constants n and k, describing both the forward and backward propagation of the electric field vector. By these calculations, the amplitude and phase of the electric field are known within the system at any point and the local absorption can be calculated. Multiplying the integrated absorption for each wavelength over the whole photoactive layer thickness with the AM1.5G solar spectrum results in the total number of photons absorbed in that photoactive layer for the specified wavelength. The maximum current density can be calculated assuming that each absorbed photon gives rise to a collected electron (i.e., the IQE is 100%).[68]

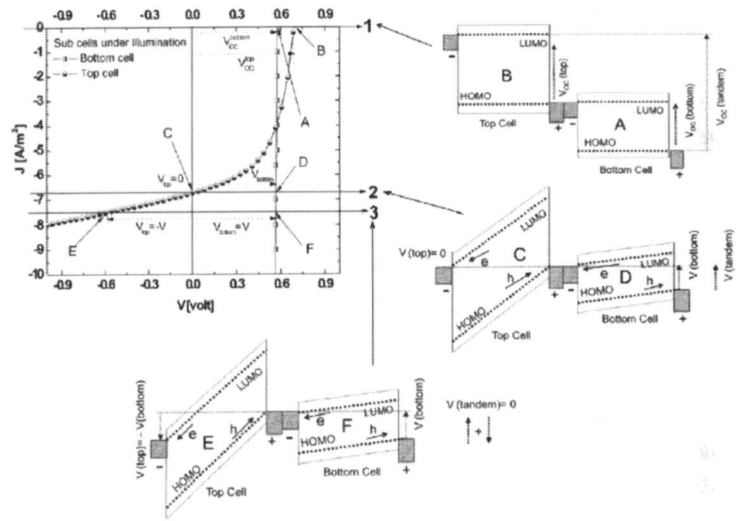

Figure 6.6 A close-up of the current density-voltage characteristic of two subcells under illumination where one subcell delivers higher open-circuit voltages, while its photocurrent is much lower than the other subcell. The energy-level diagrams for different conditions during a *J*–*V* measurement explain the electrical interaction between two subcells: open-circuit voltage (1), short-circuit conditions of one subcell (2), and short-circuit conditions of tandem cell. Reproduced from Ref. [68], with permission from Elsevier.

The electrical performance, interaction, and mutual influence of individual subcells within a tandem configuration are often unknown and more difficult to access. By Kirchhoff's law, a constant current will flow through two interconnected subcells in series and the voltages generated by the subcells will add up. These equations are valid in every point of the J–V curve of a tandem cell.[69] This implies that when one subcell generates more current than the other, charges (holes or electrons) of the first subcell arriving at the interface are not compensated for by opposite charges from the second subcell. As a result, the excess of holes/electrons will positively/negatively charge the intermediate electrode of the subcell respectively. This charging reduces the effective voltage across the excess current-generating subcell, and thus also the extracted current from this subcell. On the other hand, the additional holes/electrons in the intermediate contact provide a stronger voltage drop across the current-limiting subcell (this subcell is more reversed biased) and, therefore, a higher current flows through this subcell. Steady state is reached when the lowered current in the excess-generating subcell is equal to the enhanced current of the current-limiting subcell (Fig. 6.6).

6.4.1 Optimization of Solution-Processed Tandem Cells

Owing to the large number of layers in tandem solar cells, optimization of their performance is challenging. For solution-processed solar cells, the layers most sensitive to variation of the thickness in terms of efficiency are the photoactive layers. Increasing layer thickness results in an oscillatory pattern in current density because of interference effects (optical) and mostly in a decrease of FF due to space charge effects and recombination in the photoactive layer (electrical). In a stacked tandem cell, the subcells influence one another optically and electrically. The combined analysis of the optical absorption and electrical characteristics of the individual single junction subcells is essential to identify the optimum device layout of the corresponding tandem cell.

Based on the optical constants and thicknesses of all layers, one can determine the IQE via optical modeling as the ratio between the experimental short-circuit current density (J_{sc}) and the calculated $J_{sc,max}$ (IQE = 100%) for a series of single-junction cells with varying layer thickness. The IQEs can be combined with optical modeling of the complete stack of layers in the tandem cell to predict the J_{sc} that would be generated by each individual subcell for a given combination of photoactive layer thicknesses (Fig. 6.7). The bold line in Fig. 6.7 indicates the intersection between both surfaces and defines the conditions for current matching between front and back cell.

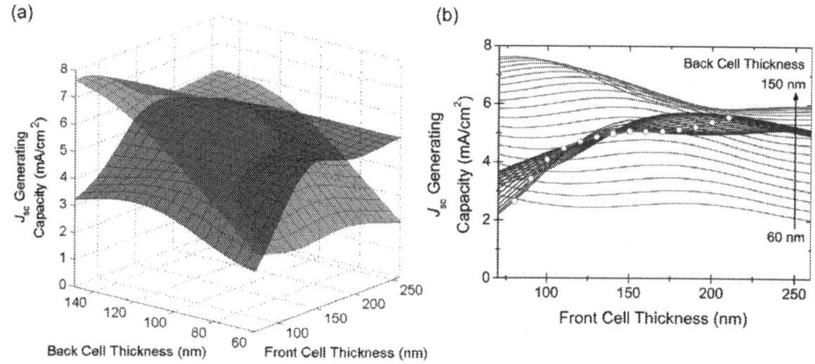

Figure 6.7 (a) 3D and (b) 2D plot of J_{SC}-generating capacities of the front (wine) and back cell (green) vs. the front and back cell layer thicknesses. The red line indicates the intersection of both surfaces indicating matching current densities. Reproduced from Ref. [34], with permission from Wiley VCH.

Assuming that the intermediate contact does not introduce any electrical losses, the *J*–*V* curve of the tandem cell can be calculated. In a series configuration, the voltage of the tandem cell equals the sum of the voltages of the subcells for each value of current density as described before. For the analysis of the electrical interactions of the subcells in the tandem cell, *J*–*V* curves of representative single junction cells can be used with thicknesses covering the whole range of interest. After normalization, these curves must be interpolated to have a complete picture of the evolution of the *J*–*V* characteristics (V_{OC} and FF) for varying layer thickness of the photoactive layer. After multiplication with the calculated J_{SC}-generating capacity, the resulting *J*–*V* curves of front and back cell must be added up according to Kirchhoff's law for series connection. The sum represents the *J*–*V* curve of a tandem cell with the concerned subcell layer thicknesses and the corresponding parameters (J_{SC}, V_{OC}, FF, and η) can be extracted (Fig. 6.8).

This was theoretically and experimentally tested with a tandem cell of a PFTBT:[60]PCBM* front cell, a pBBTDPP2:[60]PCBM back

* Please note that reference [34] is using the term PFTBT when referring to PF10TBT.

cell, and a ZnO/pH-neutral PEDOT:PSS intermediate electrode where the J_{sc}-generating capacity and the FF were not matched.[34] By mathematically constructing the J–V curves of the tandem cells from representative single-junction cell characteristics and taking into account the electronic and optical interactions between the subcells, maximum efficiencies of about 4.8% can be reached for configurations with a 150–200 nm front cell and a 120–150 nm back cell (Fig. 6.8d). Notably, the J_{sc} of the tandem cell surpasses the lowest J_{sc}-generating capacity of the two subcells. At short-circuit conditions for the tandem cell, the back cell operates in forward bias, whereas the front cell experiences an equally large but opposite reverse bias, which assists carrier collection in this subcell (Fig. 6.9a). Especially for organic solar cells, which generally exhibit a field-assisted collection, this can be a large effect. The significance of the bias often depends on the FF of the current-limiting subcell and more specifically on the shunt resistance. The FF of the tandem cell is primarily dictated by the FF of the current-limiting subcell.

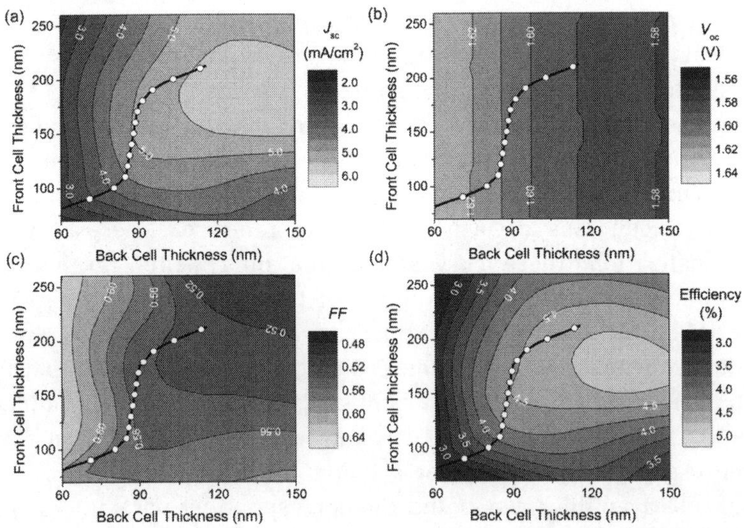

Figure 6.8 Contour plot of predicted J_{sc}, V_{oc}, FF, and η of tandem cells calculated by mathematically constructing J–V curves from representative single-junction characteristics. The isoline of matching J_{sc}-generating capacities of the subcells is represented by the bold line.

In contrast to existing views, matching the photocurrents of the subcells is not a leading design criterion for optimum performance, because the short-circuit current of a polymer tandem cell can exceed that of the current-limiting subcell. For highly efficient subcells, the gain in current will be marginal.

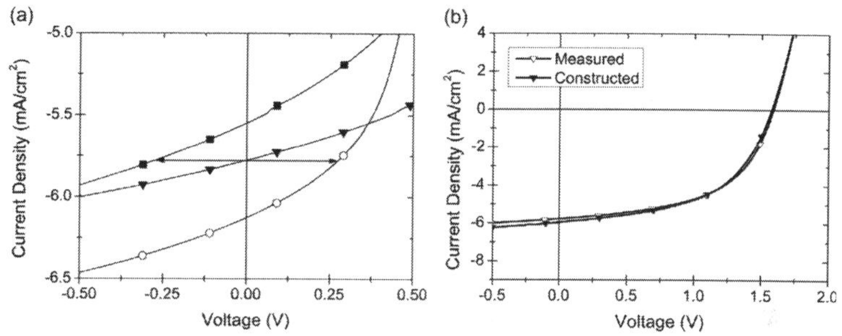

Figure 6.9 (a) An enlarged area of the J–V curves with arrows indicating the forward and reverse electrical bias over the two subcells showing that the short-circuit current density of the tandem cell can exceed the one of the current-limiting subcell. ■ is the current-limiting subcell, ○ is the excess current-generating subcell, ▼ is the tandem cell. (b) The measured and constructed (from single-junction reference cells) J–V curves coincide nicely.

For experimental verification, the J–V curves of two single-junction reference cells, identical to the individual subcells of the tandem, can be measured under a light intensity that provides the J_{SC} equal to the J_{SC}-generating capacity calculated for the subcell inside the tandem under AM1.5G illumination. Combining these two J–V curves leads to a constructed J–V curve of the tandem that is in excellent agreement with the measured J–V curve of the real tandem (Fig. 6.9b), demonstrating that the behavior of a tandem solar cell can be predicted using the single-junction reference cells. Moreover, the excellent correspondence indicates that the ZnO/pH-neutral PEDOT intermediate contact does not introduce significant losses.

The approach described here is an empirical approach for the electrical interconnection making use of physical reference cells for the J–V curves. Another option is to combine optical modeling with a numerical electronic device model.[70,71] The J–V curves are simulated taking into account drift and diffusion of charge carriers, a temperature- and field-dependent generation rate of free carriers, bimolecular recombination, and the effect of space charge on the electric field in the device. As described by Moet et al., the dissociation rate of bound pairs into free carriers is described by Onsager–Braun theory, using an initial electron–hole separation distance with a Gaussian distribution in energy and a bound pair decay rate to parameterize the dependence of the dissociation probability on electric field and temperature.[72,73] Recombination of free carriers is assumed to be of the Langevin type, with a recombination constant dominated by the slowest carrier, and results in a bound pair that can either dissociate again or decay to the ground state.[74] Other relevant parameters are the electron and hole mobilities, the spatially averaged relative dielectric constant and the effective energy gap between the highest occupied molecular orbital (HOMO) energy of the donor and the lowest unoccupied molecular orbital (LUMO) energy of the acceptor. The exciton generation profiles obtained by optical modeling are used as an input for the electronic model. The crosscheck with experimental data gives satisfying results (Fig. 6.10).

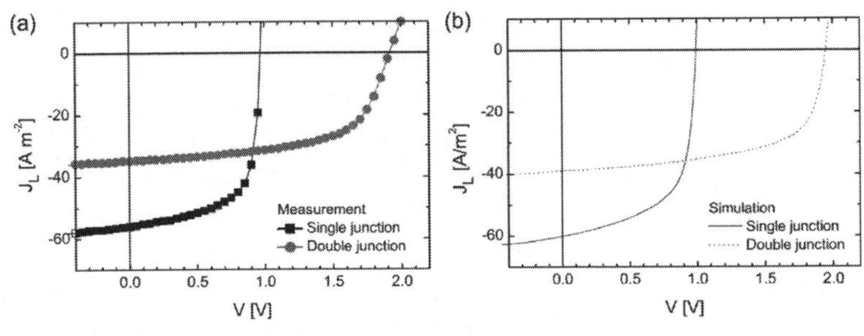

Figure 6.10 Comparison between measured (a) and calculated (b) double-junction organic solar cells with two PF10TBT:[60]PCBM photoactive layers. The *J–V* curve has been predicted by optoelectrical modeling. Reproduced from Ref. [62], with permission from AIP.

6.4.2 Optimization of Evaporated Tandem Solar Cells

Low charge-carrier mobilities are often found in evaporated small-molecule layers and reduce the FF for too thick films. Therefore, the subcell thickness can generally not be used as a parameter to optimize the performance. Fortunately, in p–i–n tandem solar cells the transport layers are highly conductive, which opens the opportunity to place thin photoactive layers in the maximum of the optical electric field by adjusting their layer thickness, provided the ET and HT layers are sufficiently optically transparent. It is even more beneficial to position the red absorbing material in the front cell because of interference effects. Scheuppel et al. and Riede et al. have elegantly shown that by adjusting the thickness of the p-doped HT layer, which acts as an optical spacer, the current density and hence the efficiency of the tandem solar cell follows an interference pattern (Fig. 6.11).[47,75] The highest efficiency is then obtained either in the first interference maximum or in the second when the subcells are too thick. This interference pattern is studied by optical simulations.

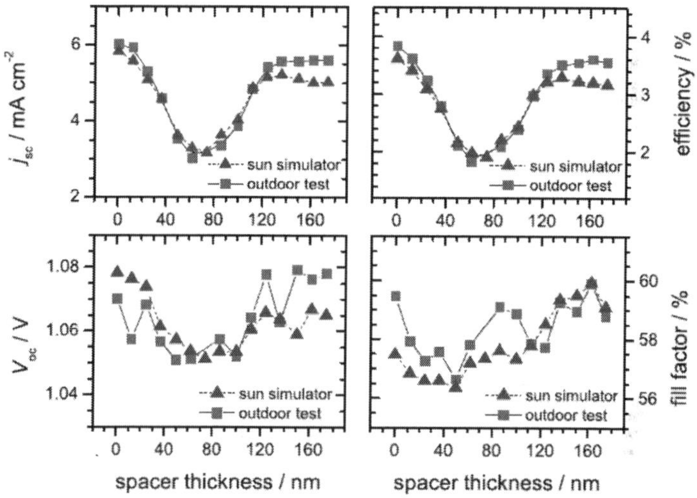

Figure 6.11 Adjusting the thickness of the p-doped HT layer in the intermediate electrode of a small molecule evaporated tandem cell results in a variation of the current–voltage parameters enabling to tune the performance of the tandem cell, here measured under solar simulator and outdoor. Reproduced from Ref. [75], with permission from AIP.

6.5 Characterization of Organic Tandem Solar Cells

The power conversion efficiency and spectral response are the most relevant parameters that characterize solar cells. Accurate characterization for single-junction organic cells has been described by Kroon et al. and Shrotriya et al.[76,77] In this section it will be shown that the measurement of the power conversion efficiency and the external quantum efficiency of tandem cells implies further challenges.

6.5.1 Determination of Power Conversion Efficiency

Accurate determination of the power conversion efficiency is imperative for the credibility of the organic solar cell community. An accurate efficiency measurement of a tandem cell is much more complicated than for a single-junction cell. The efficiency η is defined as follows:

$$\eta = \frac{J_{sc} \cdot V_{oc} \cdot \mathrm{FF}}{P_{inc}} \tag{6.1}$$

where P_{inc} is the power density of the incident light. A solar simulator only approximates the standardized air mass 1.5 global (AM1.5G) solar spectrum and power density. V_{oc} and FF are not significantly influenced by the discrepancy between both spectra and can be determined from the *J–V* characteristics under the solar simulator. The J_{sc} on the other hand strongly depends on the incident spectrum and on the intensity. Therefore, the response of both subcells under a solar simulator is different than when illuminated with AM1.5G solar light and thereby influencing the outcome of the measurement in a beneficial or adverse way. A solar simulator spectral mismatch factor *M* corrects for the approximated match of the solar spectrum and the difference in spectral response between the tested and calibrated cell. In the case of tandem solar cells, *M* of both subcells must be identical for the solar simulator used.

M is computed via

$$M = \frac{\int E_R(\lambda) \cdot S_R(\lambda) d\lambda}{\int E_S(\lambda) \cdot S_R(\lambda) d\lambda} \cdot \frac{\int E_S(\lambda) \cdot S_T(\lambda) d\lambda}{\int E_R(\lambda) \cdot S_T(\lambda) d\lambda} \tag{6.2}$$

with $E_S(\lambda)$ the solar simulator spectrum, $E_R(\lambda)$ the AM1.5G solar spectrum, and $S_T(\lambda)$ and $S_R(\lambda)$ the spectral response of the test cell and reference cell, respectively.

A way to tune the spectrum of the solar simulator toward a mismatch factor matching of both subcells in the tandem cell has been established at NREL.[41] The simulator used nine separate wavelength bands of light to build a spectrum so that the ratio of current for the front cell under the reference spectrum and the simulator spectrum is the same as the ratio of current for the rear cell under the reference spectrum and the simulator spectrum. In this way, each device junction behaved the same under the simulator spectrum as it would have behaved under the reference spectrum.

For the determination of M, the spectral response (S) of the individual subcells is required. The spectral response is related to the external quantum efficiency (EQE) via $\text{EQE}(\lambda) = S(\lambda)\ (hc/q\lambda)$, where q is the elementary charge, λ the wavelength, h the Planck constant, and c the speed of light. EQE measurements of single-junction solar cells are carried out with monochromatic light. However, in tandem solar cells, the response to monochromatic light represents the response of the current-limiting subcell as the two subcells are connected in series. To determine the EQE for the individual subcells, bias illumination is required.[78] The spectrum and intensity of bias illumination is chosen such that the bias light is predominantly absorbed by one of the two subcells, resulting in charge generation in that subcell that largely exceeds the charge generation in the other subcell. As a consequence, this other subcell becomes current limiting for the tandem cell over the entire wavelength range and the EQE of the tandem cell represents the EQE of the current-limiting subcell.

6.5.2 Determination of External Quantum Efficiency

For a more precise characterization of the tandem cell, one wants to know the generated current density of the two subcells. This current density is determined by integrating the external quantum efficiency of the subcell with the AM1.5G solar spectrum.

Accurate determination of the EQE of the subcells of an organic tandem cell involves quite a few particularities because of field-assisted current generation and the non-linear light intensity dependence of organic solar cells.[79] These processes complicate the measurements in organic tandem solar cells compared to inorganic tandem solar cells and make that a forward bias voltage is required to create short-circuit conditions for the addressed subcell and flood light on top of the modulated bias light to operate the subcell at

1 sun conditions. Ideally this flood light would be AM1.5G simulated solar light to proportionally illuminate both subcells, but this is often practically not feasible and halogen-based white light is used.

By optically biasing the tandem cell, the much more strongly absorbing subcell will be operating close to its open-circuit voltage (V_{oc}). Therefore, at short-circuit conditions for the tandem cell, the current-limiting subcell must be operating under reverse bias (□ in Fig. 6.12a). Polymer solar cells generally have a lower effective shunt resistance than silicon-based solar cells, producing sloped *J–V* curves under reverse bias and can display field-assisted charge collection. This results in a higher current density in the current-limiting subcell and to an overestimation of the EQE when the tandem cell is operating at short-circuit conditions as described in Section 6.4.1.

To compensate for this overestimation, a forward electrical bias is required. Applying a forward bias voltage on a solar cell moves the *J–V* curve of the entire cell toward more negative voltage. In a tandem cell, the optically biased subcell is continuously operating close to V_{oc} regardless the voltage over the tandem cell. Therefore, applying a forward bias voltage on the tandem cell moves the *J–V* curve of the tandem and the current-limiting subcell to more negative voltages enabling the measurement of the current-limiting subcell operating at short-circuit conditions (Fig. 6.12b).

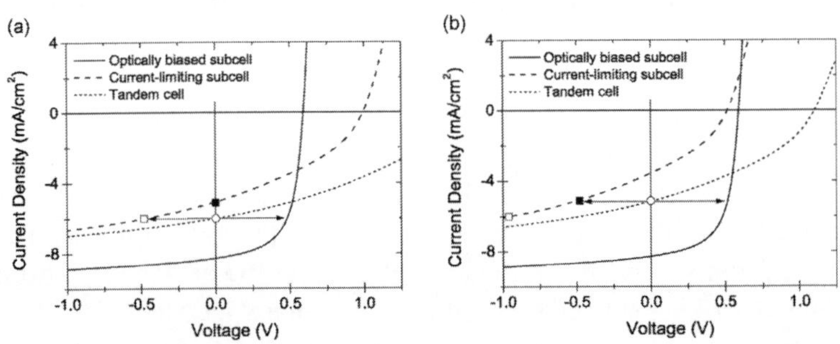

Figure 6.12 (a) Influence of the bias voltage of the optically biased subcell on the current-limiting subcell. ■ is the target value while □ is the obtained value when measuring on the tandem cell under short-circuit conditions. (b) The effect of a forward bias voltage on the subcells of a tandem solar cell. This enables the measurement of the current-limiting subcell at short-circuit conditions in contrast to (a). Reproduced from Ref. [79] with permission from Wiley VCH.

A lock-in technique with modulated monochromatic light is required when measuring the response of a subcell using flood light to operate this subcell at 1 sun condition. When the short-circuit current density depends non-linearly on the light intensity (I), the $S(\lambda)$ measured with the lock-in detection can deviate from the real $S(\lambda)$ because the differential spectral responsivity ($DS(\lambda)$) will be measured instead of the spectral responsivity $S(\lambda)$. When the photocurrent is sub-linear with light intensity, e.g., due to light-intensity-dependent losses,[80] the differential spectral responsivity $DS(\lambda)$ measured with the lock-in technique is not equal to the spectral responsivity $S(\lambda)$, but off by the second term on the right in Eq. (6.3).

$$DS = S + I\frac{\mathrm{d}S}{\mathrm{d}I} \tag{6.3}$$

When the photocurrent is sub-linear with light intensity ($\mathrm{d}S/\mathrm{d}I < 0$), $DS(\lambda)$ is less than $S(\lambda)$. The sub-linear light intensity dependence of some organic solar cells is attributed to an increased carrier concentration under illumination, which increases recombination and hinders carrier transport due to space-charge buildup in the device. Wehenkel et al. have developed a method to correct this error via a scaling factor.[81] How to correct for this error in the measurement for tandem cells is not clear yet; however, it certainly plays a roll.

6.6 Alternative Constructions of Tandem Solar Cells

So far this chapter focused on organic tandem solar cells with series-connected subcells aiming for a high efficiency. In this section organic tandem cells with subcells connected in parallel are presented as well as tandem cells combining organic and inorganic subcells. Alternative constructions avoiding processing difficulties of the multiple layer tandem solar cells are shown with stacked, folded, and laminated individual subcells.

6.6.1 Tandem Cells Connected in Parallel

Parallel interconnection can be achieved by three- or four-terminal tandem cells. In a three-terminal tandem cell, one subcell is stacked on top of a subcell with an opposite polarity configuration. They are separated by a highly conductive electrode, mostly a thin metal layer forming an intermediate contact with the following structure: ET layer/metal/ET layer or HT layer/metal/HT layer. Power conversion efficiencies of almost 5% have been reached with a PEDOT:PSS/Au/V_2O_5 or a MoO_3/Au/MoO_3 intermediate contact.[82,83] Four-terminal constructions have been made either with a transparent embedded optical spacer (e.g., poly(trifluoroethylene) (PTrFE))[32] or by stacking two independent, discrete single-junction cells where the front one has a semi-transparent electrode.[84,85]

6.6.2 Stacked, Folded, and Laminated Discrete Cells

Discrete semi-transparent cells can be used to tune the transmittance by stacking two or more subcells and constructing semi-transparent tandem and multiple junction cells.[85,86] For evaporated tandem cells a power conversion efficiency of almost 5% was achieved with a transmittance of 24% in the visible range.[87]

Independent cells folded in a V-shaped geometry, so that unabsorbed photons are reflected to the other subcell, can be interconnected both in series or in parallel (Fig. 6.13a).[88,89] Three major benefits of this technique are an enhanced path length upon tilting, light trapping upon multiple reflections, and enabling complementary absorption spectra on both sides of the V-shaped geometry upon reflection of non-absorbed light.

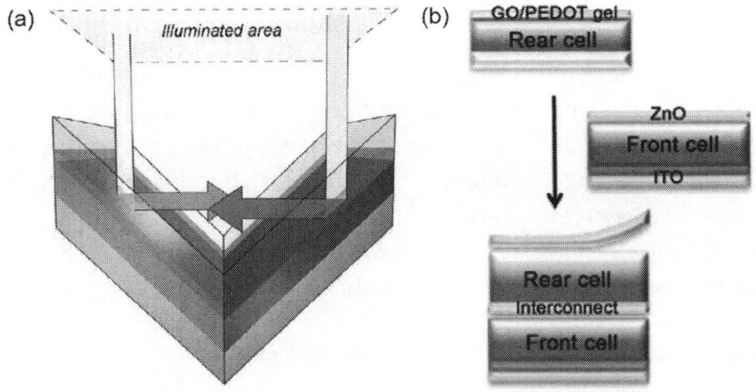

Figure 6.13 Sketches of (a) folded and (b) laminated tandem cells. (a) Reproduced from Ref. [88] with permission from AIP. (b) Reproduced from Ref. [90] with permission from ACS.

Two subcells can also be processed separately and laminated on top of each other afterward (Fig. 6.13b). Graphene oxide/PEDOT gel or graphene oxide/SWCNT is applied as a sticky intermediate contact while the top cell is transferred via a polydimethylsiloxane (PDMS) stamp on the bottom cell.[90,91]

All these technologies with independent cells avoid the processing difficulties of the intermediate electrode, but will be difficult to accomplish on large areas.

6.6.3 Combination of Inorganic and Organic Subcells

The discrete absorption bands of organic materials can complement with the dip in absorption of other PV technologies. The combination of a dye-sensitized solar cell based on an indoline dye with ZnPc:C_{60} subcell resulted in a power conversion efficiency of 4% where ZnPc:C_{60} complemented the absorption in the red region of the spectrum.[92] PCPDTBT and PBDTTT were introduced for the same beneficial contribution to a thin film amorphous silicon solar cell to result in tandem cells with power conversion efficiencies of, respectively, 1.8% and 5.7% coming from either 5.7% and 4.1% efficiency for the thin film amorphous silicon solar cells.[93,94] Organic solar cells can even boost the efficiency of copper–indium–gallium–selenide (CIGS) solar cells over 20% by absorbing more in the blue region of the spectrum, which can be relatively 10% more than just CIGS.[95] To avoid the problem of current matching this is preferably achieved with a four-terminal tandem solar cell.

6.7 Conclusion

Tandem and by extension multiple-junction solar cells possess the potential of increasing the efficiency of solar cells significantly. Since 1990, organic solar cells joined the exploration of this important and exciting field of research. Over the years the power conversion efficiency of tandem organic solar cells increased to over 10% for both solution-processed and evaporated photoactive layers. The improvements can largely be attributed to progress on high-performance materials for the photoactive layers and device configurations. Especially for tandems, the development and understanding of the intermediate layer separating the two photoactive layers physically, but interconnecting them electrically at the same time, has been crucial for progress. The energy-level alignment of ET and HT layer in the intermediate layer, which prevents a voltage loss, can be achieved via metal cluster insertion, highly doping of the transport layers or UV illumination depending on the type of transport layer used.

Optimization and characterization of organic tandem solar cells is more demanding than for single-junction organic solar cells or inorganic tandem solar cells. The electrical and optical interplay between the two subcells connected in series complicates the calculation of the ideal layer thicknesses of all layers to achieve the highest power conversion efficiency. In solution-processed cells, the thickness of the photoactive layers is tunable while for evaporated solar cells the thickness of the intermediate electrode serves as an optical spacer to position the two photoactive layers in the maximum of the optical electrical field for their absorption spectrum. The combination of optical and electrical modeling or a more empirical approach with single-junction reference cells gives insight into the optimal range of layer thicknesses. Measuring accurately the efficiency under a solar simulator asks for adjustments to the simulator to ensure the proportional treatment of both subcells during the measurement. Also for the determination of the external quantum efficiency, extra attention is required for the non-linear light intensity dependence and field-assisted current generation inherent to organic solar cells.

Besides series connection which is the most favorable for large area processing, other configurations have been developed like parallel interconnection, folded or stacked constructions, and even combinations with other photovoltaic technologies.

The improvement in efficiency makes tandem and multiple-junction solar cells extremely interesting to make organic photovoltaics a cost-competitive technology to silicon solar cells. The development of both higher performing polymers and small molecules and better controlled transporting layers will boost the performance of tandem solar cells even further in the future, likely beyond 15%. The role of the multiple junction devices with more than one absorbing layer herein will be crucial.

References

1. Shockley W, Queisser HJ (1961) Detailed balance limit of efficiency of p-n junction solar cells, *J Appl Phys*, **32**, 510–519.
2. De Vos A (1980) Detailed balance limit of the efficiency of tandem solar cells, *J Phys D Appl Phys*, **13**, 839–846.
3. Brown AS, Green MA (2002) Limiting efficiency for current-constrained two-terminal tandem cell stacks, *Prog Photovolt*, **10**, 299–307.
4. Janssen RAJ, Nelson J (2013) Factors limiting device efficiency in organic photovoltaics, *Adv Mater*, **25**, 1847–1858.
5. Scharber MC, Mühlbacher D, Koppe M, et al. (2006) Design rules for donors in bulk-heterojunction solar cells—towards 10% energy-conversion efficiency, *Adv Mater*, **18**, 789–794.
6. Koster LJA, Mihailetchi VD, Blom PWM (2006) Ultimate efficiency of polymer/fullerene bulk heterojunction solar cells, *Appl Phys Lett*, **88**, 093511.
7. Dennler G, Scharber M, Ameri T, et al. (2008) Design rules for donors in bulk-heterojunction tandem solar cells—towards 15% energy-conversion efficiency, *Adv Mater*, **20**, 579–583.
8. Van Eersel H, Gilot J, Wienk MM, Janssen RAJ (2010) Unpublished results.
9. Hiramoto M, Suezaki M, Yokoyama M (1990) Effect of thin gold interstitial-layer on the photovoltaic properties of tandem organic solar cells, *Chem Lett*, 327–330.
10. Press release: http://www.heliatek.com/newscenter/latest_news/neuer-weltrekord-fur-organische-solarzellen-heliatek-behauptet-sich-mit-12-zelleffizienz-als-technologiefuhrer/?lang=en
11. You J, Dou L, Yoshimura K, et al. (2013) A polymer tandem solar cell with 10.6% power conversion efficiency, *Nat Commun*, **4**, 1446.
12. Hadipour A, de Boer B, Blom PWM (2008) Organic tandem and multi-junction solar cells, *Adv Funct Mater*, **18**, 169–182.
13. Ameri T, Dennler G, Lungenschmied C, Brabec CJ (2009) Organic tandem solar cells: A review, *Energy Environ Sci*, **2**, 347–363.
14. Sista S, Hong Z, Chenx L-M, Yang Y (2011) Tandem polymer photovoltaic cells—current status, challenges and future outlook, *Energy Environ Sci*, **4**, 1606–1620.
15. Yakimov A, Forrest SR (2002) High photovoltage multiple-heterojunction organic solar cells incorporating interfacial metallic nanoclusters, *Appl Phys Lett*, **80**, 1667–1669.

16. Xue J, Uchida S, Rand BP, Forrest SR (2004) Asymmetric tandem organic photovoltaic cells with hybrid planar-mixed molecular heterojunctions, *Appl Phys Lett*, **85**, 5757–5759.

17. Drechsel J, Männig B, Kozlowski F, et al. (2004) High efficiency organic solar cells based on single or multiple PIN structures, *Thin Solid Films*, **451–452**, 515–517.

18. Männig B, Drechsel J, Gebeyehu D, et al. (2004) Organic p-i-n solar cells, *Appl Phys A-Mater Sci Process*, **79**, 1–14.

19. Triyana K, Yasuda T, Fujita K, Tsutsui T (2004) Effects of different materials used for internal floating electrode on the photovoltaic properties of tandem type organic solar cell, *Jpn J Appl Phys*, **43**, 2352–2356.

20. Triyana K, Yasuda T, Fujita K, Tsutsui T (2005) Tandem-type organic solar cells by stacking different heterojunction materials, *Thin Solid Films*, **477**, 198–202.

21. Cheyns D, Gommans H, Odijk M, Poortmans J, Heremans P (2007) Stacked organic solar cells based on pentacene and C_{60}, *Sol Energy Mater Sol Cells*, **91**, 399–404.

22. Timmreck R, Olthof S, Leo K, Riede MK (2010) Highly doped layers as efficient electron–hole recombination contacts for tandem organic solar cells, *J Appl Phys*, **108**, 033108.

23. Dennler G, Prall H-J, Koeppe R, Egginger M, Autengruber R, Sariciftci NS (2006) Enhanced spectral coverage in tandem organic solar cells, *Appl Phys Lett*, **89**, 073502.

24. Colsmann A, Junge J, Kayser C, Lemmer U (2006) Organic tandem solar cells comprising polymer and small-molecule subcells. *Appl Phys Lett*, **89**, 203506.

25. Zhao DW, Sun XW, Jiang CY, Kyaw AKK, Lo GQ, Kwong DL (2008) Efficient tandem organic solar cells with an Al/MoO_3 intermediate layer, *Appl Phys Lett*, **93**, 083305.

26. Janssen AGF, Riedl T, Hamwi S, Johannes H-H, Kowalsky W (2007) Highly efficient organic tandem solar cells using improved connecting architecture, *Appl Phys Lett*, **91**, 073519.

27. Lassiter BE, Zimmerman JD, Panda A, Xiao X, Forrest SR (2012) Tandem organic photovoltaics using both solution and vacuum deposited small molecules, *Appl Phys Lett*, **101**, 063303.

28. Kawano K, Ito N, Nishimori T, Sakai J (2006) Open circuit voltage of stacked bulk heterojunction organic solar cells, *Appl Phys Lett*, **88**, 073514.

29. Sakai J, Kawano K, Yamanari T, et al. (2010) Efficient organic photovoltaic tandem cells with novel transparent conductive oxide interlayer and poly(3-hexylthiophene): Fullerene active layers, *Sol Energy Mater Sol Cells*, **94**, 376–380.
30. Zhao DW, Sun XW, Jiang CY, Kyaw AKK, Lo GQ, Kwong DL (2009) An efficient triple-tandem polymer solar cell, *IEEE Electron Dev Lett*, **30**, 490–492.
31. Hadipour A, de Boer B, Wildeman J, et al. (2006) Solution-processed organic tandem solar cells, *Adv Funct Mater*, **16**, 1897–1903.
32. Hadipour A, de Boer B, Blom PWM (2007) Solution-processed organic tandem solar cells with embedded optical spacers, *J Appl Phys*, **102**, 074506.
33. Gilot J, Wienk MM, Janssen RAJ (2007) Double and triple junction polymer solar cells processed from solution, *Appl Phys Lett*, **90**, 143512.
34. Gilot J, Wienk MM, Janssen RAJ (2010) Optimizing polymer tandem solar cells, *Adv Mater*, **22**, E67–E71.
35. Kim JY, Lee K, Coates NE, et al. (2007) Efficient tandem polymer solar cells fabricated by all-solution processing, *Science*, **317**, 222–225.
36. Sista S, Park M-H, Hong Z, et al. (2009) Highly efficient tandem polymer photovoltaic cells, *Adv Mater*, **22**, 380–383.
37. Yang J, Zhu R, Hong Z, et al. (2011) A robust inter-connecting layer for achieving high performance tandem polymer solar cells, *Adv Mater*, **23**, 3465–3470.
38. Gevaerts VS, Furlan A, Wienk MM, Turbiez M, Janssen RAJ (2012) Solution processed polymer tandem solar cell using efficient small and wide bandgap polymer:fullerene blends, *Adv Mater*, **24**, 2130–2134.
39. Chou C-H, Kwan WL, Hong Z, Chen L-M, Yang Y (2011) A metal-oxide interconnection layer for polymer tandem solar cells with an inverted architecture, *Adv Mater*, **23**, 1282–1286.
40. Kouijzer S, Esiner S, Frijters CH, Turbiez M, Wienk MM, Janssen RAJ (2012) Efficient inverted tandem polymer solar cells with a solution-processed recombination layer, *Adv Energy Mater*, **2**, 945–949.
41. Dou L, You J, Yang J, et al. (2012). Tandem polymer solar cells featuring a spectrally matched low-bandgap polymer, *Nature Phot*, **6**, 180–185.
42. Dou L, Chang W-H, Gao J, Chen C-C, You J, Yang Y (2012) A selenium-substituted low-bandgap polymer with versatile photovoltaic applications, *Adv Mater*, **25**, 825–831.

43. Li W, Furlan A, Hendriks KH, Wienk MM, Janssen RAJ (2013) Efficient tandem and triple-junction polymer solar cells, *J Am Chem Soc*, **135**, 5529–5532.
44. You J, Chen C-C, Hong Z, et al. (2013) 10.2% Power conversion efficiency polymer tandem solar cells consisting of two identical sub-cells, *Adv Mater*, **25**, 3973–3978.
45. Zhou Y, Fuentes-Hernandez C, Shim JW, Khan TM, Kippelen B (2012) High performance polymeric charge recombination layer for organic tandem solar cells, *Energy Environ Sci*, **5**, 9827–9832.
46. Cheyns D, Rand BP, Heremans P (2010) Organic tandem solar cells with complementary absorbing layers and a high open-circuit voltage, *Appl Phys Lett*, **97**, 033301.
47. Riede M, Uhrich C, Widmer J, et al. (2011) Efficient organic tandem solar cells based on small molecules, *Adv Funct Mater*, **21**, 3019–3028.
48. Hagemann O, Bjerring M, Nielsen NC, Krebs FC (2008) All solution processed tandem polymer solar cells based on thermocleavable materials, *Sol Energy Mater Sol Cells*, **92**, 1327–1335.
49. Chang L, Holmes MA, Waller M, Osterloh FE, Moulé AJ (2012) Calcium niobate nanosheets as a novel electron transport material for solution processed multi-junction polymer solar cells, *J Mater Chem*, **22**, 20443–20450.
50. Shim JW, Zhou Y, Fuentes-Hernandez C, et al. (2012) Studies of the optimization of recombination layers for inverted tandem polymer solar cells, *Sol Energy Mater Sol Cells*, **107**, 51–55.
51. Ajuria J, Etxebarria I, Cambarau W, et al. (2010) Inverted ITO-free organic solar cells based on p and n semiconducting oxides. New designs for integration in tandem cells, top or bottom detecting devices, and photovoltaic windows, *Energy Environ Sci*, **4**, 453–458.
52. Sharma P, Sreenivas K, Rao KV (2003) Analysis of ultraviolet photoconductivity in ZnO films prepared by unbalanced magnetron sputtering, *J Appl Phys*, **93**, 3963–3970.
53. van Dijken A, Meulenkamp EA, Vanmaekelbergh D, Meijerink A (2000) Identification of the transition responsible for the visible emission in ZnO using quantum size effects, *J Lumin*, **90**, 123–128.
54. Beek WJE, Wienk MM, Kemerink M, Yang X, Janssen RAJ (2005) Hybrid zinc oxide conjugated polymer bulk heterojunction solar cells, *J Phys Chem B*, **109**, 9505–9516.
55. Arnold MS, Avouris P, Pan ZW, Wang ZL (2003) Field-effect transistors based on single semiconducting oxide nanobelts, *J Phys Chem B*, **107**, 659–663.

56. Studenikin SA, Golego N, Cocivera M (2000) Carrier mobility and density contributions to photoconductivity transients in polycrystalline ZnO films, *J Appl Phys*, **87**, 2413–2421.

57. Verbakel F, Meskers SCJ, Janssen RAJ (2007) Electronic memory effects in diodes of zinc oxide nanoparticles in a matrix of polystyrene or poly(3-hexylthiophene), *J Appl Phys*, **102**, 083701.

58. Kong J, Lee J, Kim G, Kang H, Choia Y, Lee K (2012) Building mechanism for a high open-circuit voltage in an all-solution-processed tandem polymer solar cell, *Phys Chem Chem Phys*, **14**, 10547–10555.

59. Lakhwani G, Roijmans R, Kronemeijer AJ, Gilot J, Janssen RAJ, Meskers SCJ (2010) Probing charge carrier density in a layer of photodoped ZnO nanoparticles by spectroscopic ellipsometry, *J Phys Chem C*, **114**, 14804–14810.

60. Kim KJ, Park YR (2001) Large and abrupt optical bandgap variation in In-doped ZnO, *Appl Phys Lett*, **78**, 475–477.

61. Lu JG, Fujita S, Kawaharamura T, et al. (2007) Carrier concentration dependence of bandgap shift in n-type ZnO:Al films, *J Appl Phys*, **101**, 083705.

62. Moet DJD, de Bruyn P, Blom PWM (2010) High work function transparent middle electrode for organic tandem solar cells, *Appl Phys Lett*, **96**, 153504.

63. Steinmann V, Kronenberg NM, Lenze MR, et al. (2011) A simple merocyanine tandem solar cell with extraordinarily high open-circuit voltage, *Appl Phys Lett*, **99**, 193306.

64. Taima T, Yamanari T, Sakai J, Yoshida Y (2010) Tandem organic photovoltaic cells based on low-molecular-weight semiconductors, *J Appl Phys*, **49**, 01AC04.

65. Zhang M, Wang H, Tang CW (2012) Tandem photovoltaic cells based on low-concentration donor doped C_{60}, *Org Electron*, **12**, 249–251.

66. Zou Y, Deng Z, Potscavage WJ, Hirade M, Zheng Y, Adachi C (2012) Very high open-circuit voltage of 5.89 V in organic solar cells with 10-fold-tandem structure, *Appl Phys Lett*, **100**, 243302.

67. Koster LJA, Mihailetchi VD, Ramaker R, Blom PWM (2005) Origin of the light intensity dependence of the short-circuit current of polymer/fullerene solar cells, *Appl Phys Lett*, **86**, 123509.

68. Dennler G, Forberich K, Ameri T, et al. (2007) Design of efficient organic tandem cells: On the interplay between molecular absorption and layer sequence, *J Appl Phys*, **102**, 123109.

69. Hadipour A, de Boer B, Blom PWM (2008) Device operation of organic tandem solar cells, *Org Electron*, **9**, 617–624.

70. Moet DJD, de Bruyn P, Kotlarski JD, Blom PWM (2010) Enhanced efficiency in double junction polymer:fullerene solar cells, *Org Electron*, **11**, 1821–1827.

71. Nam YM, Huh J, Jo WH (2011) A computational study on optimal design for organic tandem solar cells, *Sol Energy Mater Sol Cells*, **95**, 1095–1101.

72. Braun CL (1984) Electric field assisted dissociation of charge transfer states as a mechanism of photocarrier production, *J Chem Phys*, **80**, 4157–4161.

73. Mihailetchi VD, Koster LJA, Hummelen JC, Blom PWM (2004) Photocurrent generation in polymer–fullerene bulk heterojunctions, *Phys Rev Lett*, **93**, 216601.

74. Koster LJA, Mihailetchi VD, Blom PWM (2006) Bimolecular recombination in polymer/fullerene bulk heterojunction solar cells, *Appl Phys Lett*, **88**, 052104.

75. Schueppel R, Timmreck R, Allinger N, et al. (2010) Controlled current matching in small molecule organic tandem solar cells using doped spacer layers, *J Appl Phys*, **107**, 044503.

76. Kroon JM, Wienk MM, Verhees WJH, Hummelen JC (2002) Accurate efficiency determination and stability studies of conjugated polymer/fullerene solar cells, *Thin Solid Films*, **403–404**, 223–228.

77. Shrotriya V, Li G, Yao Y, Moriarty T, Emery K, Yang Y (2006) Accurate measurement and characterization of organic solar cells, *Adv Funct Mater*, **16**, 2016–2023.

78. Burdick J, Glatfelter T (1986) Spectral response and I-V measurements of tandem amorphous-silicon alloy solar cells, *Sol Cells*, **18**, 301–314.

79. Gilot J, Wienk MM, Janssen RAJ (2010) Measuring the external quantum efficiency of two-terminal polymer tandem solar cells, *Adv Funct Mater*, **20**, 3904–3911.

80. Koster LJA, Kemerink M, Wienk MM, Maturová K, Janssen RAJ (2011) Quantifying bimolecular recombination losses in organic bulk heterojunction solar cells, *Adv Mater*, **23**, 1670–1674.

81. Wehenkel DJ, Hendriks KH, Wienk MM, Janssen RAJ (2012) The effect of bias light on the spectral responsivity of organic solar cells, *Org Electron*, **13**, 3284–3290.

82. Sista S, Hong Z, Park M-H, Xu Z, Yang Y (2010) High-efficiency polymer tandem solar cells with three-terminal structure, *Adv Mater*, **22**, E77–E80.

83. Guo X, Liu F, Yue W, Xie Z, Geng Y, Wang L (2009) Efficient tandem polymer photovoltaic cells with two subcells in parallel connection, *Org Electron*, **10**, 1174–1177.

84. Shrotriya V, Wu EH-E, Li G, Yao Y, Yang Y (2006) Efficient light harvesting in multiple-device stacked structure for polymer solar cells, *Appl Phys Lett*, **88**, 064104.

85. Tang Z, George Z, Ma Z, et al. (2012) Semi-transparent tandem organic solar cells with 90% internal quantum efficiency, *Adv Energy Mater*, **2**, 1467–1476.

86. Bailey-Salzman RF, Rand BP, Forrest SR (2006) Semitransparent organic photovoltaic cells, *Appl Phys Lett*, **88**, 233502.

87. Meiss J, Menke T, Leo K, et al. (2011) Highly efficient semitransparent tandem organic solar cells with complementary absorber materials, *Appl Phys Lett*, **99**, 043301.

88. Tvingstedt K, Andersson V, Zhang F, Inganäs O (2007) Folded reflective tandem polymer solar cell doubles efficiency, *Appl Phys Lett*, **91**, 123514.

89. Rim S-B, Zhao S, Scully SR, McGehee MD, Peumans P (2007) An effective light trapping configuration for thin-film solar cells, *Appl Phys Lett*, **91**, 243501.

90. Tung VC, Kim J, Cote LJ, Huang J (2011) Sticky interconnect for solution-processed tandem solar cells, *J Am Chem Soc*, **133**, 9262–9265.

91. Tung VC, Kim J, Huang J (2012) Graphene oxide:single-walled carbon nanotube-based interfacial layer for all-solution-processed multijunction solar cells in both regular and inverted geometries, *Adv Energy Mater*, **2**, 299–303.

92. Bruder I, Karlsson M, Eickemeyer F, et al. (2009) Efficient organic tandem cell combining a solid state dye-sensitized and a vacuum deposited bulk hetero junction solar cell, *Sol Energy Mater Sol Cells*, **93**, 1896–1899.

93. Kim T, Jeon JH, Han S, et al. (2011) Organic-inorganic hybrid tandem multijunction photovoltaics with extended spectral response, *Appl Phys Lett*, **98**, 183503.

94. Seo JH, Kim D-H, Kwon S-H, et al. (2012) High efficiency inorganic/organic hybrid tandem solar cells, *Adv Mater*, **24**, 4523–4527.

95. Beiley ZM, McGehee MD (2012) Modeling low cost hybrid tandem photovoltaics with the potential for efficiencies exceeding 20%, *Energy Environ Sci*, **5**, 9173–9179.

Part II

Characterization, Modeling, and Fundamental Insights

Chapter 7

Bulk Heterojunction Morphology Control and Characterization

Tao Wang[a,b] and David G. Lidzey[a]

[a]*Department of Physics and Astronomy, The University of Sheffield, Hicks Building, Hounsfield Road, Sheffield S3 7RH, United Kingdom*

[b]*School of Materials Science and Engineering, Wuhan University of Technology, Wuhan, 430070, China*

t.wang@hotmail.com, d.g.lidzey@sheffield.ac.uk

In this chapter, we review the results of a number of experimental studies that elucidate the nanoscale morphology of the active layer of a polymer : fullerene bulk heterojunction (BHJ) solar cell. In particular, we discuss the growing understanding regarding the role of miscibility that drives phase separation between conjugated polymers and fullerene derivatives. We show that BHJ structures containing phase-separated domains having a characteristic distribution of sizes can be probed using a range of morphological and optical techniques. We then review the methods used to probe and control film morphology, vertical stratification, and structural evolution through modification of film deposition protocol and postdeposition treatments. Finally, we correlate changes in BHJ nanoscale morphology with the optoelectronic properties of organic photovoltaic devices.

Organic Solar Cells: Fundamentals, Devices, and Upscaling

Edited by Barry P. Rand and Henning Richter

ISBN 978-981-4463-65-2 (Hardcover), 978-981-4463-66-9 (eBook)

www.panstanford.com

7.1 Introduction

A bulk heterojunction (BHJ) solar cell is a photovoltaic device in which the active semiconductor layer is composed of an intimate blend of materials that can either donate or receive electrons, with such electronic properties deriving from different electron affinities and ionization potentials.[1–4] The function of the BHJ is to absorb sunlight, dissociate the resultant excited states (excitons) into electrons and holes, and then to provide a series of conduction paths by which such charge carriers can be extracted. Organic materials have been widely explored as functional semiconductors for BHJ solar cell applications, as such materials potentially combine low cost of synthesis, ease of thin film deposition, and mechanical flexibility. In this chapter, we review the nanoscale morphology of the active layer in a BHJ as characterized by a number of different techniques. We explore typical nanostructures found within polymer–fullerene BHJ films, and relate this to methods used to process and deposit such films. Our central objective is to provide an understanding of the routes by which molecular nano- and mesostructure defines the resultant performance of organic photovoltaic (OPV) devices.

At the heart of an organic BHJ device is a composite thin film that comprises a conjugated polymer that has been blended or otherwise combined with a fullerene derivative that can efficiently accept an excited-state electron. A range of electron acceptors have been explored for this application (for more information, refer Chapter 3), including the fullerene derivatives $PC_{61}BM$, $PC_{71}BM$, ICBA, etc.[5] To create the active layer of an organic BHJ device, a conjugated polymer and the electron-accepting fullerene are usually codissolved into a single solvent (or a mixture of solvents), and then cast into a thin film. As the film dries, the polymer and the fullerene undergo partial phase separation, forming a series of heterojunctions that are distributed throughout the bulk of the film. Following the absorption of light, excitons are generated that can subsequently undergo diffusion to a heterojunction where they can subsequently dissociate into an electron and a hole. The nanoscale morphology of the active layer is known to affect the photon-to-electron generation process, i.e., exciton generation, diffusion, separation, and collection, and thus plays a critical role in determining the power conversion efficiency of an OPV device.[6–9] Exploring the nanoscale morphology within a polymer–fullerene OPV has proved a rich area of study, with researchers exploring the structural order of the materials (e.g., whether they are crystalline or amorphous), the miscibility, purity, and size of the phase-separated domains, and the relative distribution of materials in a direction normal to the plane of the film. There are a number of good articles that have reviewed work exploring the morphology of the active layer of a BHJ solar cell to which an interested reader is directed.[10–15]

The molecular structure and energy gap of a conjugated polymer can be tuned by chemical synthesis (for more information, see Chapter 1), resulting in a much wider range of materials for photovoltaic applications than is possible with inorganic counterparts.[16,17] The morphology of a polymer:fullerene photovoltaic blend can be affected by a number of parameters, including the interaction parameter between the components, the solubility of the materials within a particular solvent, the rate at which the solvent is permitted to evaporate, the film deposition temperature, the method used to deposit the film, etc. Rather than providing a summary of all the different morphologies that can be found within an OPV, this chapter concentrates on a number of photovoltaic blends that we believe are generally representative of the structure of a BHJ. We then discuss the different morphologies found within such films together with factors that influence their structure. In particular, we discuss the organization of a polymer and a fullerene in a BHJ at a molecular level, and address issues such as material miscibility, the size of the phase-separated domains, and the vertical distribution of material within a thin film. We also discuss the morphology of a BHJ in an as-cast state, together with the structures formed upon thermal and solvent annealing. Finally, we discuss the use of time-resolved techniques to study the evolution of nanoscale morphology in a polymer–fullerene blend film *during* the solvent casting process and as it is exposed to posttreatment processes. Our aim is to provide a general overview of the types of structure and treatments used to modify structure within an OPV that we hope will be of interest to researchers exploring entirely new material systems.

7.2 Organization of Polymers and Fullerenes at the Molecular Level

Compared to crystalline molecular materials, conjugated polymers are at best only semicrystalline and are usually characterized by coexisting crystalline and amorphous domains. Grazing incidence wide-angle X-ray scattering (GIWAXS) has been widely used to characterize nanostructure in OPV thin films. Because the crystal structures of many conjugated polymers are not well known, techniques such as two-dimensional (2D) GIWAXS permit parameters such as coherence length and ordering and orientation with respect to the substrate plane to be characterized.[18]

Poly(3-hexylthiophene) (P3HT) is probably the most extensively studied electron-donating OPV material. P3HT is a semicrystalline conjugated polymer and—under suitable processing conditions—films are formed having crystalline domains that have a range of sizes and spatial distributions. This has provided an interesting test bed to correlate molecular and mesoscopic structure within a BHJ with performance of an OPV device. Figure 7.1a shows a 2D GIWAXS image of a pure P3HT film cast from chlorobenzene (CB). The (100) peak at $q_z = 0.37$ Å^{-1} is the first-order diffraction peak of the lamellar spacing of P3HT crystallites separated by alkyl side chains. A lamellar spacing of $d = 1.68$ nm can be determined using $d = 2\pi/q_z$. The (200) and (300) peaks that we identify in the figure are the second- and third-order diffraction of the lamellar ordering. The (010) diffraction peak observed in the in-plane direction at $q = 1.68$ Å^{-1} is associated with π-stacking of the conjugated molecular backbones separated by around 0.37 nm. The 2D GIWAXS pattern of P3HT therefore indicates an "edge-on" molecular orientation as shown in Fig. 7.1c. Here, the P3HT side chains sit perpendicular to the plane of the substrate, with the π-stacking direction being parallel to the substrate plane. In OPV applications, a P3HT polymer is usually selected that has a high degree of regio-regularity. This generally results in films in which the molecules adopt an "edge-on" orientation. However, in films that are spin-cast from P3HT having a low degree of regio-regularity, molecules generally adopt a "face-on" orientation in which the plane of the conjugated backbone is positioned parallel to the substrate.[19] It is also evident that processing conditions, such as choice of solvent, interaction between the polymer and the substrate, can also impact on the relative orientation of P3HT molecules in an OPV film.[20,21] For both packing configurations, any radial broadening (evidenced by the diffraction spots appearing as being "arc-like") implies an increase in the orientation distribution of polymer domains.

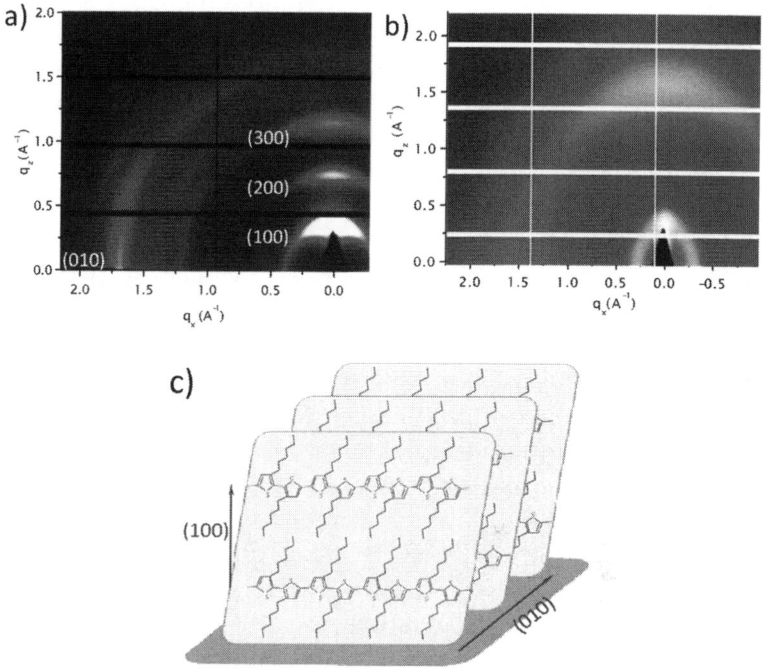

Figure 7.1 Two-dimensional GIWAXS images of (a) P3HT and (b) PCDTBT thin films. (c) A schematic of "edge-on" orientation of P3HT crystallites.

Many new so-called "donor–acceptor" conjugated copolymers are, however, more amorphous in nature. Such donor–acceptor copolymers contain both electron-rich (donating) and electron-deficient (accepting) motifs within each monomer unit, with the monomer then polymerized to form a conjugated polymer.[22] Although they display very different phase behavior compared to semicrystalline P3HT, the power conversion efficiency of BHJ OPVs based on copolymers having a low degree of structural order can be high. One example of such a material is the polymer poly[N-9′-heptadecanyl-2,7-carbazole-alt-5,5-(4′,7′-di-2-thienyl-2′,1′,3′-benzothiadiazole)] (PCDTBT).[23] Figure 7.1b shows a 2D GIWAXS image of a PCDTBT thin film prepared by spin-casting from CB. Compared with P3HT, PCDTBT has a simpler diffraction pattern. Rather than undergoing partial crystallization, PCDTBT molecules simply undergo π-stacking when spin-cast from solution. As shown in Fig. 7.1b, π-stacking (in an as-cast PCDTBT thin film) is evidenced by a scattering "crescent" observed at $q = 1.57$ Å^{-1} in the out-of-plane direction. This indicates that the conjugated backbones predominantly adopt a "face-on" π-stacked orientation with respect to the silicon substrate. The inner diffraction ring observed at $q_z = 0.31$ Å^{-1} is attributed to first-order scattering from side chain ordering, and is associated with the distance (approximately 20 Å) between lamella of PCDTBT backbones separated by two alkyl side chains.[24,25]

To improve the solubility and therefore solution processability of low energy gap conjugated polymers, it is usually necessary to attach alkyl side chains along the molecular backbone. Such side chains can, however, affect the energy gap of the polymer, its miscibility with fullerenes, and also the propensity for the molecular backbone to undergo structural ordering. Recent work has demonstrated that the chemical structure of the side chains can affect the π–π stacking distance and the stacking coherence length, with both parameters having a substantial effect on OPV device efficiency. For example, Yiu and coworkers found that the structure of an alkyl side chain has a pronounced effect on molecular packing in PDPP2FT polymers.[26] Compared with branched side chains, linear alkyl side chains have been found to reduce the π–π stacking distance between conjugated backbones and also increase the coherence length of both π–π stacking and lamella order. It was concluded that branched side chains are not coplanar to the polymer backbone; however, linear side chains are able to maintain backbone planarity and promote longer range structural order. Other work has studied the effect of the structure of alkyl side chains in the PTB family of copolymers. It was found that linear octyl chains on the BDT unit in a PTB copolymer reduces the π–π stacking distance compared to a branched ethylhexyl side chain.[27] As a result of reduced π–π stacking distance, there is improved charge transport and reduced density of defects (that act as charge traps), and consequently improved device efficiency.

In a spin-cast P3HT:PCBM blend that has not been subject to further treatment, the electron acceptor PCBM is in a largely amorphous form. For this reason, a GIWAXS spectra of a P3HT:PCBM blend typically has a broad symmetric ring at q = 1.41 Å^{-1} (see Fig. 7.2a), which is characteristic of PCBM. However, at a higher loading ratio (e.g., at a relative weight fraction of 1:4 in a PCDTBT:$PC_{71}BM$ blend), Bragg scattering from $PC_{71}BM$ is characterized by three broad, symmetric rings at q = 0.67, 1.36, and 2.0 Å^{-1}, as can be seen in Fig. 7.2b. The broad, symmetric nature of the diffraction rings indicates that the nanoscale $PC_{71}BM$ domains that exist are oriented randomly with respect to the film substrate. Upon thermal annealing, microscale, needle-like PCBM crystal clusters can often be observed in such blends as a result of cold crystallization of PCBM. Using aperture-adjusted and focused field-emission transmission electron microscopy (FE-TEM), Kim et al.[28] deduced that PCBM domains in a (1:1) P3HT:PCBM blend film are characterized by a distorted, asymmetric body-centered cubic (bcc) crystal nanostructure. A schematic illustration of the molecular packing of PCBM molecules in such a domain is shown in Fig. 7.3c.

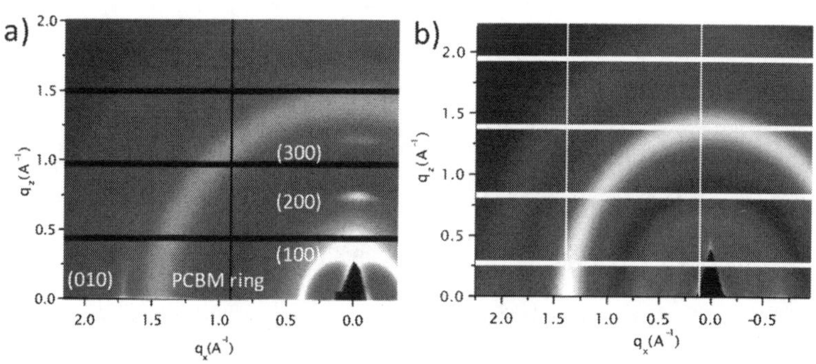

Figure 7.2 Two-dimensional GIWAXS images of (a) P3HT:PCBM (1:0.8) and (b) PCDTBT:$PC_{71}BM$ (1:4) blend films.

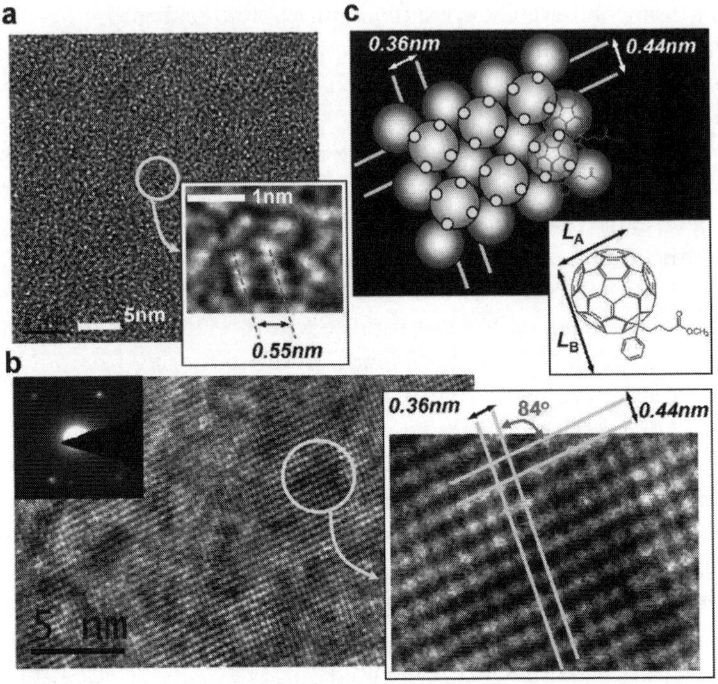

Figure 7.3 Aperture-adjusted and focused FE-TEM images of P3HT:PCBM (1:1) blend film annealed at 140 °C for 2 h. (a) FE-TEM image detailing a P3HT-enriched region. (b) FE-TEM image detailing a PCBM-enriched region (inset: electron diffraction pattern). (c) Schematic illustration of the arrangement of PCBM molecules that lead to a distorted bcc crystal structure: The apparent dimensions of a single PCBM molecule are given by the length scales L_A (0.71 nm) and L_B (~1.1 nm) as shown. (Reprinted from Kim et al., *ACS Nano,* 3, 2557 (2009) with permission from American Chemical Society).

7.3 Organization of Polymers and Fullerenes during Solvent Casting

After casting from solution into a thin film, solvent loss occurs from a polymer : fullerene photovoltaic blend due to the process of evaporation. This process triggers a partial demixing of the blend components in order to minimize the free energy of the system. During this process, a number of self-organization processes occur, including crystallization, self-assembly, and π–π stacking. The nanoscale structures formed in the photovoltaic blend thin film after solution casting will depend upon the miscibility of the components, the film drying kinetics, and the surface energy of the substrate. The nanoscale morphology formed in such a process is highly dependent on processing conditions and—for a single material set—a range of different morphological structures can be generated. Such structures have a significant effect on eventual BHJ device efficiency. Such an observation highlights the necessity of unraveling the mechanisms by which nanostructure develops during the film solution-deposition process. Although direct (real-space) morphological tracking of such structures has not proved possible due to a combination of rapid structural changes and other technical limitations, optical probes such as X-ray scattering and spectroscopic ellipsometry (SE) have been used to determine structure evolution during the solvent casting process in polymer : fullerene photovoltaic blends.

The evolution of structure in P3HT:PCBM blend films during solution casting has been investigated using both SE[29] and X-ray scattering techniques.[30] SE can determine the thickness evolution of a film from a wet state to a solid state, a quantity from which the solid concentration of the blend in the drying film can be calculated. The complex refractive index of a film (i.e., n and k) can also be extracted via modeling of the SE data, with structure changes being deduced via changes in optical properties. A three-stage solvent loss process for P3HT:PCBM blend films during casting has been identified from the evolution of k spectra of P3HT:PCBM blend by Wang et al.[29] In Fig. 7.4a, we show the evolution of the extinction coefficient spectrum (k) of a P3HT:PCBM blend as it dries after being cast from a CB solution. In Stage I, the structure of the k-spectrum closely resembles that of dilute P3HT in solution, indicating weak inter-chain interactions. During Stage II, the intensity of the k spectrum increases rapidly and the spectrum red shifts with vibronic peaks indicating the emerging presence of crystalline P3HT. Such changes in the k spectrum result from the fact that P3HT molecules in a crystalline state have extended electronic conjugation length along the molecular backbone and a reduced degree of inhomogeneous broadening. In Stage III, the k spectrum resembles that of a solid film, with only small increases in optical absorption observed. Figure 7.4b shows the thickness of the P3HT:PCBM film thickness as a function of time after casting, with the onset of the three stages in Fig. 7.4a identified accordingly. It can be seen that in Stage I there is rapid solvent evaporation and linear thinning of the film. Stage II is characterized by a reduced rate of film thinning, and commences once the solid content of P3HT:PCBM in the drying film reaches 50% and ends once the solid content reaches ~90% by volume. This stage is marked by rapid P3HT crystallization. Stage III corresponds to a further reduction in solvent evaporation rate and a slowing of the rate of crystallization. The evolution of *k* spectra in different P3HT:PCBM blends cast from different solvents follows the same trends as illustrated in Fig. 7.4c.

The three-stage drying process of a P3HT:PCBM blend revealed from SE measurements is also evidenced by time-

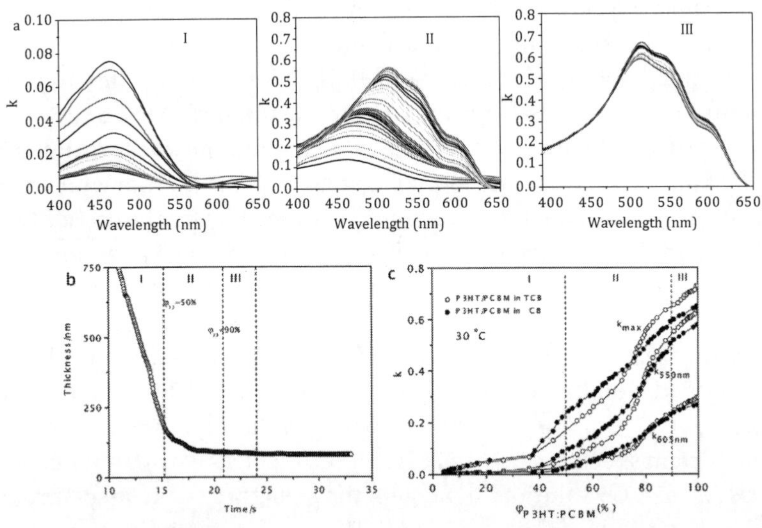

Figure 7.4 (a) Evolution of optical extinction coefficient (*k*) and (b) thickness during the drying process of a P3HT:PCBM (1:0.8 wt%) thin film after bar coating. Spectra in part (a) are split over time intervals of 1 s (Stage I), 0.1 s (Stage II), and 1 s (Stage III). (c) Evolution of *k* at three different wavelengths as a function of volume fraction of P3HT:PCBM blends in the drying film. Stages I–III in part (a) correspond to those in part (b) and (c) (Adapted from Wang et al., *Soft Matter,* 6, 4128 (2010). Copyright ©The Royal Society of Chemistry).

resolved GIWAXS measurements. The intensity changes of the (100) peak of P3HT in a P3HT:PCBM blend during the drying process are shown in Fig. 7.5a. It can be seen that the P3HT diffraction peak appears once the film has reached a certain solid content as a result of solvent evaporation. Again, it appears that the rate of crystallization is initially fast and then slows. A "self-annealing" process within P3HT crystallites is also evidenced, where the *d*-spacing between P3HT lamella reduces during the crystallization process.

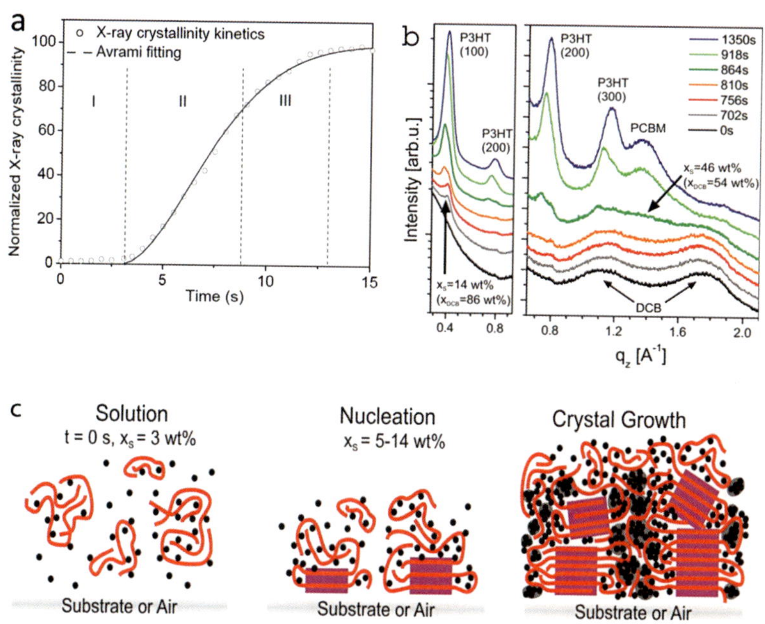

Figure 7.5 Development of nanoscale crystallinity in a P3HT:PCBM blend thin film during casting. Part (a) shows the (normalized) crystallinity of P3HT (inferred from the intensity of Bragg scatter from the (100) plane) over the course of film drying. (Reprinted from Wang et al., *Soft Matter,* 6, 4128 (2010) with permission from The Royal Society of Chemistry) Part (b) shows the evolution of X-ray scattering from a P3HT:PCBM blend film recorded as it dries. In part (c), a schematic of the three stages of morphology evolution in P3HT:PCBM blend films is shown. (Reprinted from Schmidt-Hansberg et al., *ACS Nano,* 5, 8579 (2010) with permission from American Chemistry Society).

Schmidt-Hansberg et al.[30] have tracked the changes of X-ray diffraction peaks of a P3HT:PCBM blend over a much broader q range. A series of GIWAXS diffraction curves recorded from a P3HT:PCBM thin film as it undergoes drying are shown in Fig. 7.5b. It was found that on casting a P3HT:PCBM (1:0.8 wt%) blend from dichlorobenzene (DCB), Bragg scatter from P3HT (100) planes was observed when the film contained a solid content of 14%. Interestingly, this technique suggests that P3HT crystallization occurs at a much lower solid concentration than was suggested by SE measurements.[29] The exact reason for this discrepancy is not clear. One possible reason is that polymer aggregation or self-assembly has already occurred in solution prior to film deposition.[31] Notably, Bragg scatter from PCBM was found to occur at a much higher solid content of approximately 50%. This suggests that PCBM aggregation results from the self-organization of P3HT, with the planarization and stacking of adjacent P3HT chains displacing PCBM molecules into amorphous regions of the polymer matrix. This would increase the local concentration of the fullerene to the level necessary to drive aggregation.

Such results suggest that self-organization of the polymer and fullerene during solution casting does not occur spontaneously. This is illustrated in Fig. 7.5c, where we show a schematic of structure organization during the solidification of a P3HT:PCBM film after solution casting. At early times, the solid content within the film is still below that required to drive self-organization of either the polymer or the fullerene. Above a certain solid content threshold, the polymer molecules begin to crystallize, an effect that is correlated with a significant strengthening and a red shift of their optical absorption. Phase separation occurs at a later stage, during which nanoscale domains of PCBM form. It has also been suggested that the evolution of P3HT domain orientation is mediated by the substrate, with a relatively broader distribution of orientations adopted by domains located within the "bulk" of the film.[32]

The exact sequence of polymer crystallization and fullerene aggregation can be affected by the different relative solubility of the polymer and fullerene components in the solution. This effect has been demonstrated for a number of photovoltaic blends dissolved in different solvent mixtures. The evaporation rate of the solvent and the selective solubility of the solvent to polymer and fullerene components will clearly drive morphology changes during solvent casting, an effect identified by Russell and coworkers[33] in studies of PCPDTBT:PCBM blend films. Here, it was found that if an additive was added to a solution that has a higher vapor pressure and a selective solubility for PCBM, the PCPDTBT first crystallizes into fibrils, a process that is followed by the aggregation of PCBM in the amorphous PCPDTBT domains. This effect results from the solvent becoming less favorable for the PCPDTBT while completely solubilizing the PCBM during the film drying. Studies on the photovoltaic blend pDPP:PCBM dissolved in a DCB:CF (dichlorobenzene:chloroform) solvent mixture identified a similar effect.[34] Here, the pDPP dissolves well in the low boiling point solvent CF, but not in high boiling point solvents such as CB and DCB. After solution casting from a DCB:CF solution, the fast evaporation of CF results in the crystallization of pDPP into fibrils. The PCBM component, however, remains dissolved in the slowly evaporating DCB solvent. Consequently, the pDPP crystallization was observed at a much earlier stage than the aggregation of PCBM as observed by in situ grazing incidence X-ray diffraction. In this case, differences in the solvent vapor pressures and in the solubility of the components are seen to result in a blend morphology having a range of length scales.

7.4 Miscibility of the Polymer and Fullerene

The morphology of a BHJ active layer is often simplistically described in terms of a bicontinuous network of domains, composed of regions of pure polymer and pure fullerene, with both sets of domains formed as a result of microscale phase separation. It is now known, however, that this description is oversimplistic, as such structures do not in general contain pure phases separated by distinct interfaces.[35,36] Rather, in a P3HT and PCBM blend, a degree of miscibility of PCBM in the amorphous phase of the polymer has been demonstrated by a bilayer diffusion study using dynamic secondary ion mass spectrometry (DSIMS) and cross section scanning electron microscopy.[35] This is illustrated in Fig. 7.6a–c, where it can be seen that PCBM diffuses into an adjacent P3HT layer upon annealing at 100 °C, with the interface between the two layers completely disappearing after thermal annealing at 150 °C for 5 min. DSIMS measurements—summarized in Fig. 7.6d—are able to provide a quantitative measurement of PCBM concentration as a function of depth into the film through a ^{2}H label. This reveals that the efficiency of diffusion of PCBM into the P3HT layer is as expected temperature dependent. Chen et al.[36] have also shown that amorphous P3HT and PCBM are highly miscible and that PCBM can penetrate into a P3HT layer through amorphous P3HT regions and then form a BHJ-like structure within a few seconds of annealing at 150 °C.

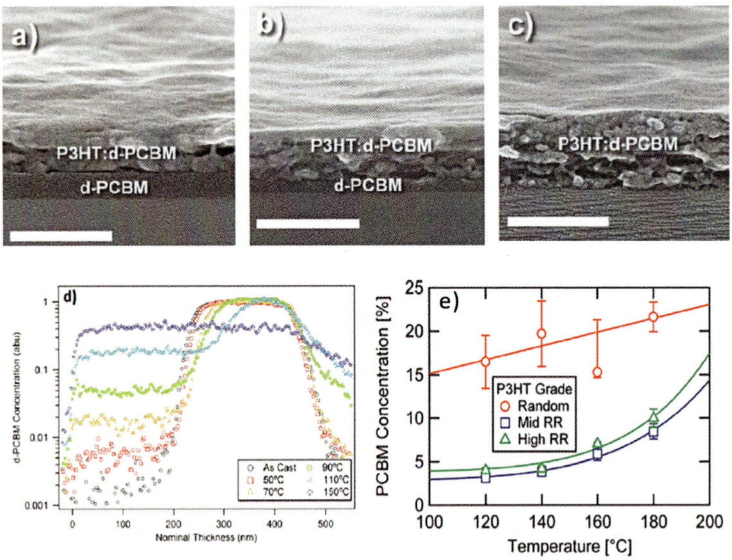

Figure 7.6 Cross-sectional scanning electron microscopy (SEM) images of a P3HT/d-PCBM bilayer on a silicon wafer annealed at (a) 70 °C, (b) 110 °C, and (c) 150 °C for 5 min. The scale bar represents a distance of 200 nm. (d) DSIMS profiles of ^{2}H in bilayer samples of P3HT and d-PCBM annealed for 5 min at different temperatures. (Reprinted from Treat et al., *Adv Energy Mater,* 1, 82 (2011) with permission from Wiley-VCH) (e) Miscibility phase diagram for P3HT mixed with PCBM. (Reprinted from Collins et al., *J Phys Chem Lett,* 1, 3160 (2010) with permission from American Chemistry Society).

Collin et al.[37] have deduced the miscibility phase diagram for the P3HT:PCBM system at different temperatures (from 100 °C to 200 °C) from measurements of near-edge X-ray absorption fine structure (NEXAFS), as shown in Fig. 7.6e. It can be seen that the miscibility of PCBM with P3HT increases with increasing temperature. This miscibility is also significantly greater in a regio-random P3HT:PCBM blend compared with blends containing P3HT having an intermediate or a high degree of regio-regularity. Dadmun and coworkers[38,39] have employed neutron scattering to explore the miscibility of P3HT:PCBM by studying the diffusion of PCBM in bilayer films upon thermal annealing and have concluded that the miscibility limit of PCBM in P3HT is 22 vol %, a result in accord with NEXAFS studies. Ruderer et al.[40] have shown that within a 1:1 ratio P3HT:PCBM blend, there is in fact 18 vol % PCBM dispersed in the amorphous P3HT phase. Jamieson et al.[41] reached a similar conclusion by estimating the exciton diffusion length from the PL quenching in polymer–fullerene blends. Here blends of PCBM with the polymers PCPDTBT, PCDTBT, pBTTT, and F8DTBT were studied, and it was shown that the exciton diffusion length in such blends is less than 1 nm, a result indicative of mixing between the polymer and PCBM on a length scale less than 1 nm.

X-ray scattering has revealed that in selected blend systems (e.g., pBTTT:PCBM) the intercalation of fullerenes between polymer chains occurs at a molecular level. McCulloch et al. first proposed the intercalation of PCBM between the side chains of a polyterthiophene (C6-TT) polymer.[42] McGehee and coworkers then used GI-WAXS, photoluminescence quenching, and charge mobility studies to investigate intercalation in various polymer:fullerene blend systems.[43,44] Figure 7.7 illustrates the intercalation of PCBM into the crystalline polymer pBTTT.[43] GI-WAXS measurements indicated an increase in the lamellar packing distance in pBTTT:PCBM blend from ca. 21 Å in a sample of pure pBTTT polymer to ca. 30 Å in a 1:4 polymer:fullerene blend. This increase in length scale originated from an intercalation of the fullerene into the polymer crystal, forming a highly ordered bimolecular crystal.[43] Further support for this idea comes from X-ray scattering studies that also suggest fullerene intercalation into semicrystalline pTT and PQT polymers.[43] In contrast however, X-ray studies did not evidence an expanded polymer *d*-spacing in a P3HT:PCBM blend, suggesting that in this system, intercalation does not occur as the free volume between polymer side chains is not sufficiently large for the fullerenes to undergo intercalation.

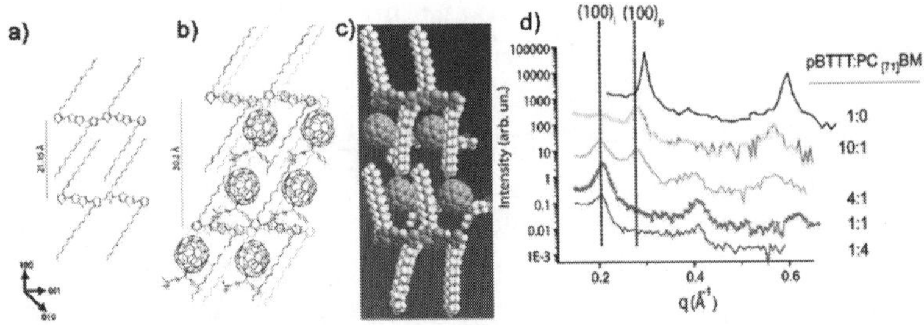

Figure 7.7 Schematic of $PC_{71}BM$ intercalation in the crystal lattice of pBTTT. (a) The interdigitation of the side chains in pristine pBTTT crystals. (b) $PC_{71}BM$ intercalation in a pBTTT:$PC_{71}BM$ blend. (c) A space-filling schematic model of $PC_{71}BM$ intercalation in a pBTTT:$PC_{71}BM$ blend made using ChemDraw3D. (d) X-ray scattering spectra of annealed pBTTT:$PC_{71}BM$ films having varying weight ratios, demonstrating the intercalation of the $PC_{71}BM$ into the pBTTT lattice. Here, the lamellar spacing increases from ca. 21 Å in the pristine pBTTT polymer to ca. 30 Å in a 1:4 pBTTT:$PC_{71}BM$ blend. (Reprinted from Mayer et al., *Adv Funct Mater*, 19, 1173 (2009) with permission from Wiley-VCH).

In systems where intercalation does occur, the proximity of PCBM to the polymer backbone results in efficient exciton quenching; however, the PCBM molecules that are intercalated are not "electronically connected" and do not therefore provide an effective percolation pathway for electron transport. The larger polymer molecules, however, are still assumed to act as efficient hole–transport materials even after intercalation. As a result of this, a higher PCBM loading in a blend is required in order to completely "fill up" spaces between the polymer molecules and drive aggregation of excess PCBM into pure PCBM domains that act as continuous electron transport pathways. Indeed, empirical studies have shown that the optimum polymer:fullerene blend ratio is close to 1:1 in a nonintercalated blend, but is significantly higher (around 1:4) in an intercalated blend. It is also thought that intercalation may change the conformation of a polymer backbone from being coil-like in the absence of PCBM to more rod-like in the presence of PCBM, thus increasing hole mobility.[45]

Further work has explored the efficiency of intercalation by varying various parameters such as polymer side chain spacing, side chain branching, and fullerene size. For example, it has been found that the fullerene derivative bis$PC_{71}BM$—which is larger because of extra side chain—cannot intercalate into pBTTT.[44] In other work, Sista et al.[46] explored benzodithiophene polymers having different spacing between the side chains along the molecular backbone. They found that a higher PCBM loading was required to optimize device efficiency when the spacing between the side chains increased. Interestingly, however, their X-ray diffraction study indicated that the lamellar spacing in the blend is identical to that of the pristine polymer. In this case, it was concluded that PCBM was able to intercalate between the polymer side chains without influencing the solid-state packing of the polymer. It is clear, however, that using X-ray scattering to determine PCBM intercalation in a conjugated polymer through changes in lamellar spacing can be problematic, as many conjugated polymers exhibit a low degree of crystallinity. For example, the conjugated polymers PCDTBT and MDMO-PPV are largely amorphous in an as-cast state and thus the determination of changes in lamellar spacing is difficult as the 2D X-ray scattering pattern does not contain sharp diffraction peaks. Nevertheless, it is highly likely that molecular mixing occurs in both of these blends, as optimized device efficiency is reached at a polymer : fullerene blend ratio of 1:4, with such blends displaying enhanced hole mobility, efficient PL quenching, with no obvious phase separation visible using an optical microscope.[24,47]

While photovoltaic blends that undergo intercalation can result in more efficient PL quenching (a characteristic that is generally beneficial to the performance of BHJ OPVs), this is not uniformly true. For example, recent work by Miller et al.[48] found that ICBA fullerenes improve the efficiency of solar cells only when they do not intercalate between the polymer side chains. Here, it was suggested that the poor electron transport in the molecularly mixed domains may account for the reduced device performance of blends in which the fullerene underwent intercalation.

7.5 Phase Separation and Domain Size

A number of studies have suggested that at least three phases exist in the active layer of a P3HT:PCBM OPV device: pure crystalline P3HT, pure PCBM, and one or more mixed phases of P3HT and PCBM. This morphology is also common in other semicrystalline or amorphous polymer blends incorporating PCBM. In extreme cases—for example, using highly crystalline polymers such as pBTTT—the phase in which there is intimate mixing can be a cocrystal, whereas in largely amorphous polymers such as PCDTBT and MDMO-PPV, the intimately mixed phase is best described as an amorphous solid solution. Figure 7.8 illustrates a schematic of the various different phases believed to exist in P3HT:PCBM and PCDTBT:PCBM blends. The size of the domains comprising the relatively pure components depends on a number of factors including the miscibility between polymer and fullerene, the relative solubility of the two components in the solution, the blending ratio, and the film drying kinetics after solution casting.

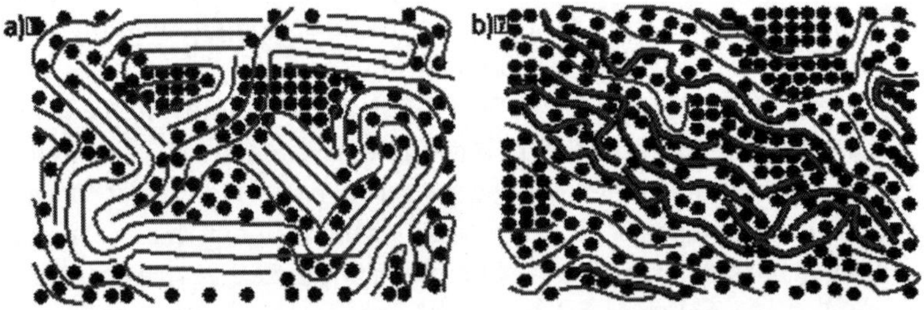

Figure 7.8 A schematic of the different phases in a blend of (a) a semicrystalline polymer (such as P3HT) and PCBM and (b) an amorphous polymer (such as PCDTBT) and PCBM.

A number of different microscopy techniques have been employed to characterize domain size in partially phase-separated polymer : fullerene BHJ films. Defocused TEM has been used to characterize the bicontinuous network in cross section through a P3HT:PCBM film.[49] However, the observation of polymer : fullerene phase separation and the length scale probed by this method has been recently challenged by Mackay et al.[50] who showed that a pure amorphous polystyrene film also shows similar structure when probed using defocused TEM. While defocusing can potentially enhance the inherent contrast between the two phases, it can also lead to spatial filtering arising from the contrast transfer function. Here, additional analysis must be performed to prevent artifacts.[51] Loos et al. have investigated nanoscale morphology in a P3HT:PCBM blend using bright-field TEM[52] (see Fig. 7.9a and b) and have also constructed a three-dimensional map of nanoscale organization in a P3HT:PCBM blend using electron tomography (see Fig. 7.9c).[53] Electron tomography is only sensitive to the crystalline P3HT component in the blend and has therefore been used to map the distribution of semicrystalline P3HT nanowires having a diameter of 15 nm. Differences in the plasmon peak energies between P3HT and PCBM permit energy-filtered TEM (EFTEM) to be used to image P3HT nanowires, revealing structures having a diameter of around 10 nm and a length of around 25 nm.[54] Helium ion microscopy (HeIM) has also been employed to examine a P3HT:PCBM blend.[55] Here, a network of slightly elongated PCBM domains having a spatial periodicity of (20 ± 4) nm and a length of (12 ± 8) nm were identified in a spin-cast P3HT:PCBM blend having a 1:1 blend ratio.

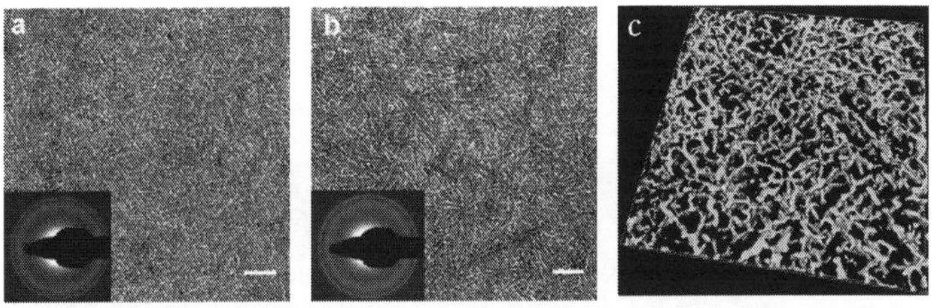

Figure 7.9 Bright-field TEM images of 50-nm-thick P3HT:PCBM films that are (a) as-cast and (b) thermally annealed. The scale bar in each case is 200 nm. The insets are the corresponding electron diffraction patterns. (Reprinted from van Bavel et al., *Macromolecules,* 42, 7396 (2009) with permission from American Chemistry Society) (c) A 3D electron tomography image of a thermally annealed blend film. (Reprinted from van Bavel et al., *Nano Lett,* 9, 507 (2009) with permission from American Chemistry Society).

Scattering techniques can also reveal domain size in BHJ blends. Grazing incidence small-angle X-ray scattering (GISAXS) measurements using resonant soft X-ray scattering have identified a length scale for phase separation of around 20 nm in an annealed P3HT:PCBM blend film.[56] This is in good agreement with transmission small-angle X-ray scattering (SAXS) measurements that have determined a separation length scale of 25 nm in both drop-cast and thermal-annealed P3HT:PCBM films.[51] Ruderer et al.[40] have estimated the PCBM domain size in P3HT:PCBM blends at various PCBM weight fractions using grazing incidence small-angle neutron scattering (GISANS). Figure 7.10a shows the experimental (left column) and simulated (right column) 2D GISANS patterns of a P3HT:PCBM BHJ film cast at 1:1 blend ratio. Figure 7.10b and c shows the horizontal (b) and vertical (c) line cuts through the GISANS data (symbols). The fitted data (plotted using solid lines in Fig. 7.10b and c) to GISANS data reveal the existence of PCBM domains having different length scales. Here, it was found that the distance between PCBM domains that—in Figure 7.10d—are defined as the structure size is around 20 nm for a broad range of PCBM weight fractions. Much larger PCBM domains with a radius (defined as domain radius in Fig. 7.10d) of 100–200 nm are detectable; however, these only comprise a very small fraction of the total. Smaller PCBM domains having a radius (defined as object radius in Fig. 7.10d) between 3 and 10 nm dominate, with a maximum domain radius of 10 nm identified at a PCBM content of 50 wt%.

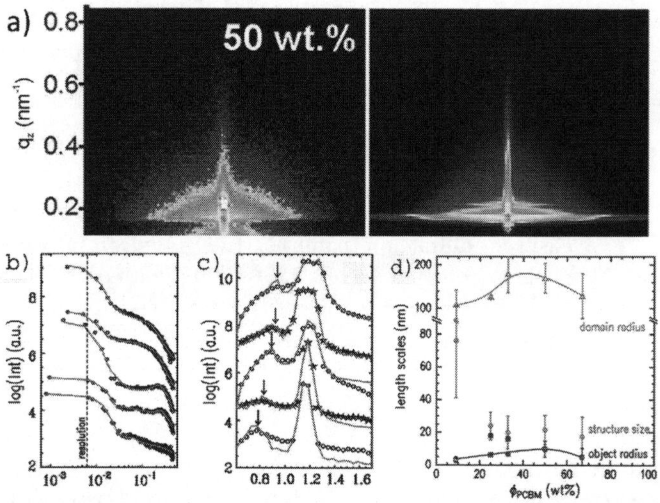

Figure 7.10 (a) Two-dimensional GISANS data (left column) and corresponding pattern from simulation using a computer program called IsGISAXS[51] (right column) of a P3HT:PCBM BHJ film at 1:1 ratio. Horizontal (b) and vertical (c) line cuts of the GISANS data (symbols) and corresponding fits (solid lines) of P3HT:PCBM films at various PCBM concentrations. (d) Distance (structure size), small (object radius) and large (domain radius) PCBM domains from fits to GISANS data at various PCBM content in P3HT:PCBM blend. (Reprinted from Ruderer et al., *J Phys Chem Lett*, **3**, 683 (2012) with permission from American Chemical Society).

7.6 Impact of Solvent Additives on Morphology

The primary casting solvent can have a dramatic impact on the morphology of an organic BHJ and consequently affects the PCE of an OPV device. This is exemplified in OPV devices based on a PCDTBT:$PC_{71}BM$ blend.[23] Here, the addition of a few percent of a solvent "additive"—usually to the primary solvent—has been demonstrated as an effective method to improve device efficiency.[58] To act as an effective additive, it is necessary that the additive is miscible with the polymer and fullerene components, and has a higher boiling point relative to that of the primary solvent, as well as a preferential solubility with the PCBM.[59,60] This effect has been pronounced in low bandgap donor–acceptor conjugated polymers such as PTB7[61] and PCPDTBT.[60,62,63] Solvent additives can affect the nanoscale morphology that is developed during the film drying process. When properly employed, such additives can result in a reduction in domain size, the formation of a smooth film surface with low roughness, and the generation of self-organization having a higher degree of structural order.

Characterization using TEM has shown that solvent additives can significantly reduce the size of PCBM domain when it is blended with the donor–acceptor polymers PTB7,[61] PBTTPD,[64] PDPPTPT,[65] PSi, and PGe.[66] This is illustrated in the TEM image shown in Fig. 7.11, where domains having a size of 150 nm are observed in a PBTTPD:$PC_{71}BM$ film that is cast without an additive, with the domain size reduced to ca. 30 nm on addition of the additive diiodohexane (DIH). A more uniform morphology was also observed on addition of 3 vol% of diiodooctane (DIO) to PTB7/$PC_{71}BM$ blend films.[61] Transmission SAXS has revealed that the additive DIO can selectively dissolve PCBM aggregates that form in a PTB7 blend solution (using a CB solvent), allowing the formation of a finer domain size and a significant improvement in device efficiency.[67]

Solution-processing additives can also result in an increased degree of crystallinity of the polymer phase. This has been shown to occur in a PBTTPD:$PC_{71}BM$ blend upon the addition of DIH to the primary solvent CF as revealed using GISAXS/GIWAXS.[64] The polymer PCPDTBT has also been found to crystallize through the presence of solvent additives.[33,59,68,69] Using time-resolved GIWAXS, Rogers et al.[68] observed a lower nucleation barrier for PCPDTBT crystallization on addition of octanedithiol (ODT) to the primary CB solvent, thereby encouraging crystallite growth over an extended timescale. This is illustrated in Fig. 7.12, where 2D GIWAXS scattering patterns from a PCDPTBT:$PC_{71}BM$ 1:3 wt% blend film cast from CB or CB/ODT after 78 min after spin-casting are shown. Films cast from CB/ODT solutions show pronounced Bragg scatter from the (100) plane of semicrystalline PCPDTBT. As demonstrated in an earlier study, PCPDTBT crystallization requires the PCBM component to remain in solution (as facilitated by the additive) for a sufficiently long timescale such that polymer crystallization is not disrupted. Russell and coworkers concluded that the presence of the additive leads to a multilength scale morphology, comprising domains of pure, crystalline PCPDTBT fibrils, and a PCBM-rich mixture containing amorphous PCPDTBT.[33]

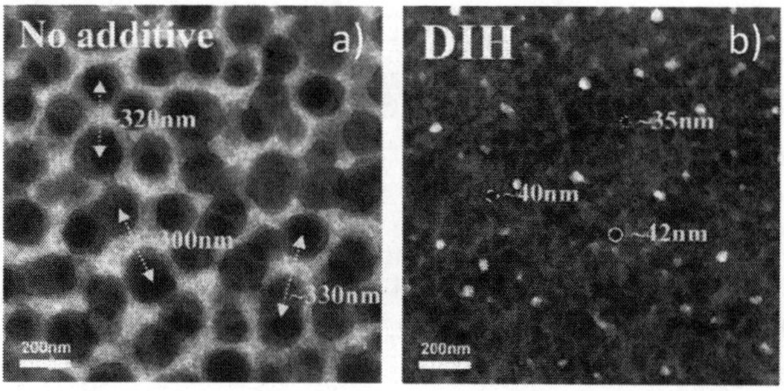

Figure 7.11 TEM images of PBTTPD/$PC_{71}BM$ films prepared in the (a) absence and (b) presence of the additive DIH. (Reprinted from Su et al., *Adv Mater,* 23, 3315 (2011) with permission from Wiley-VCH).

In some material systems, the film morphology formed via the presence of additives cannot be replicated via thermal annealing, as polymer self-organization can be inhibited by the simultaneous thermally activated diffusion of PCBM molecules. However, in other material systems, the presence of a solvent additive is less effective in improving OPV efficiency. For example, the addition of DIO to a PCBM blend with a *N*-alkylthieno[3,4-c]pyrrole-4,6-dione (TPD)-based polymer with straight chain side groups showed a marginal effect on efficiency improvement.[70] This was explained by the fact that this polymer already forms ordered structures in the primary solvent. While the addition of 3 vol% of DIO to PDTSTPD/ $PC_{71}BM$ in a CB solution results in a large increase of PCE, such improvements are less pronounced when a DIO additive is used with a DCB primary casting solvent.[71] A similar conclusion has also been reached when using a DIO additive in a DCB-cast PTB7/ $PC_{71}BM$ blend device.[61]

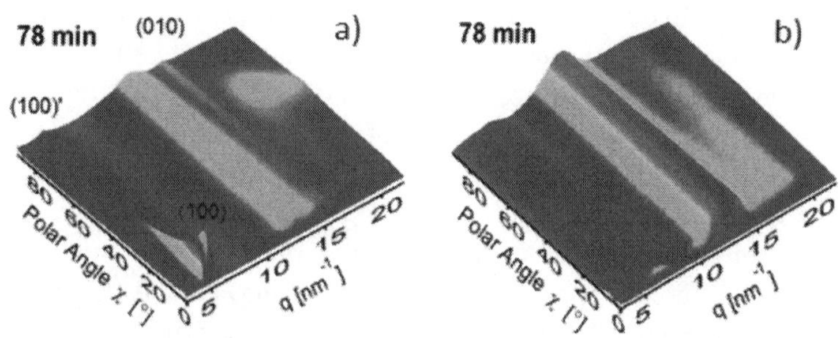

Figure 7.12 Three-dimensional sector plots of PCPDTBT:$PC_{71}BM$ blends cast from CB containing (a) 3% or (b) 0% ODT after 78 min of spin coating, showing the scattered intensity vs polar angle χ and scattering vector q. (Reprinted from Rogers, et al., *J Am Chem Soc*, 134, 2884 (2012) with permission from American Chemistry Society).

7.7 Post-Deposition Processes and Techniques

7.7.1 Impact of Thermal Annealing on Nanoscale Morphology and Optoelectronic Properties

7.7.1.1 Two contrasting effects on structural order and device performance

In semicrystalline polymers such as P3HT, the fast vitrification that occurs during spin-casting usually limits the degree of self-ordering as well as the formation of an effective bicontinuous network suitable for charge extraction. Thermal annealing at elevated temperatures can, however, provide a driving force for the molecules to reorganize or proceed towards a state of dynamic thermal equilibrium (although such an equilibrium state may never actually be reached). Importantly, thermal annealing can be used to improve both the relative crystallinity and degree of phase separation in a polymer : fullerene blend. When applied correctly, the efficiency of photocurrent generation can also be improved by thermal annealing as a result of improved long-wavelength absorption, increased charge carrier mobility, and thus reduced charge carrier recombination.

Thermal annealing has proved to be effective at improving the efficiency of devices made from P3HT:PCBM[72] and PSBTBT:PCBM[73,74] blends. Here, the minimum temperature required for structural modification to occur in a blend film is thought to be coincident with the glass transition temperature T_g of the photovoltaic blend film.[75] This defines a temperature beyond which molecules within a film have sufficient thermally activated motion to undergo reorganization. Below T_g, a film is defined as "glassy" and is thus relatively rigid. In the absence of an external driving force, structural changes within a glassy film are energetically unfavorable. In blend films incorporating semicrystalline materials, thermal annealing may change the average inter-chain spacing between polymer chains, their orientation with respect to the sample substrate, and the size of domains. For small molecules such as PCBM, thermal annealing can lead to the formation of interconnected nanoscale aggregates together with a diffusion of material toward the film surface, potentially improving the efficiency of electron transport and extraction from a regular (top cathode-based) OPV. Micron-scale PCBM crystallization may also occur if a sample is subjected to an elevated annealing temperature and/or prolonged annealing times. Phase separation at this length scale is commonly described as "excessive" or "coarse," as it greatly exceeds the diffusion length of photo-generated excitons.

For some material systems, thermal annealing has been found to be detrimental to the formation of structural order within a BHJ and can therefore reduce OPV efficiency. For example, PCDTBT is a largely amorphous polymer that—in an as-cast film—is characterized by weak order resulting from π–π stacking. Recent work[76] has found that PCDTBT molecules form lamellar structures with an "edge-on" orientation after thermal treatment at a temperature above the bulk T_g (ca. 130 °C) of PCDTBT. Using both GIWAXS and theoretical calculations, Lu et al.[76] argued that lamella crystallinity in PCDTBT takes the form of bilayer ordering, where a pair of backbones are positioned "side-by-side" with the alkyl side chains pointing outward. The molecular structure in PCDTBT films upon thermal treatment above the bulk T_g is thus a combination of "face-on" π–π stacking and "edge-on" lamella ordering. Wang et al.[24] found, however, that the overall degree of π–π stacking order in an annealed PCDTBT film is reduced (as shown in Fig. 7.13a), an effect correlated with reduced charge carrier mobility (as shown in Fig. 7.13b) in an annealed film. Such effects are consistent with device studies that show that PCDTBT OPV efficiency is reduced upon thermal annealing around the bulk T_g, suggesting that lamellar ordering confers reduced charge mobility in PCDTBT devices.

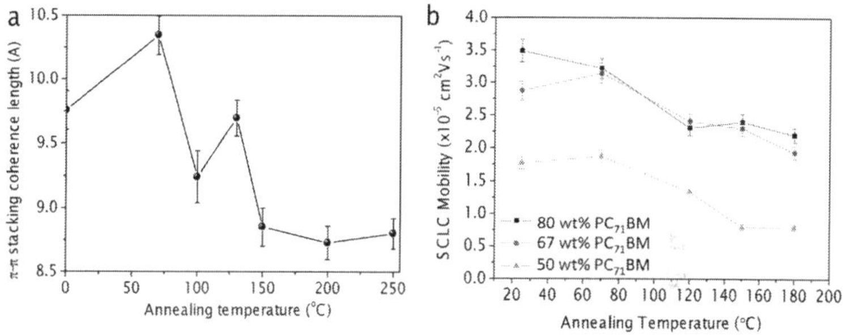

Figure 7.13 (a) The π–π stacking coherence length of PCDTBT in a thin film as a function of annealing temperature as measured by GIWAXS. (b) SCLC hole-carrier mobility of a PCDTBT:$PC_{71}BM$ thin film device as a function of annealing temperature (annealing time of 30 min). (Reprinted from Wang et al., *Adv Funct Mater*, 22, 1399 (2012) with permission from Wiley-VCH).

Other work on the OPV blend of poly(thienothiophene benzodithiophene) (PTB1) and PCBM has demonstrated that thermal annealing also reduces device efficiency.[77] Here, GIWAXS reveals that π–π stacked polymer backbone planes are oriented parallel to the substrate surface, with such π–π stacking being more ordered after thermal annealing. However, large PCBM domains having a size 10–200 nm are formed in the annealed film and decrease the effective interfacial area between PTB1 and PCBM. This overcoarse phase separation subsequently reduces the charge separation and charge collection efficiencies.

7.7.1.2 Time-resolved morphology characterization during thermal annealing

In situ studies of the thermal annealing process can be used to identify the sequence in which donor and acceptor materials undergo phase separation and self-organization, and the rates at which these processes occur. This information cannot be deduced by studying final film morphology alone. The advantage of using such in situ studies is that they can—in principle—identify the presence of transitional morphologies, permitting thermal annealing protocols to be designed that can drive the formation of entirely new structures from such transitional states.

Figure 7.14 shows the changes in P3HT crystallinity that occur during thermal annealing as determined using GIWAXS.[78] It can be seen that P3HT crystallization is a relatively fast process, occurring over a period of a few minutes at an annealing temperature of 140 °C. Such studies also reveal that the rate of P3HT crystallization in a P3HT:PCBM blend reduces upon the increasing PCBM content in the blend.

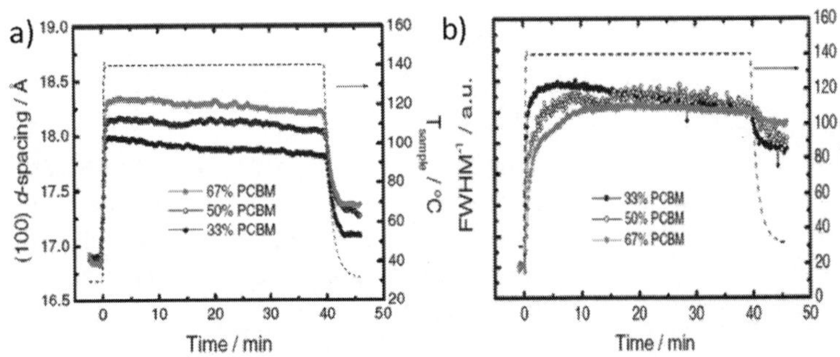

Figure 7.14 (a) Evolution of the *d*-spacing of P3HT extracted from the (100) diffraction peak in out-of-plane line profiles at three different blend concentrations. (b) Evolution of reciprocal FWHM during thermal annealing of P3HT:PCBM blend films having different composition, as determined from the P3HT (100) peak in the out-of-plane direction. The temperature of the sample during annealing is also shown using dashed lines. (Reprinted from Agostinelli, T. et al., *Adv Funct Mater,* 21, 1701 (2011) with permission from Wiley-VCH).

Empirical studies reveal that the optimum annealing temperature used when thermal annealing a P3HT:PCBM blend for OPV applications is generally in the range 140–150 °C, a temperature that greatly exceeds the glass transition temperature (T_g) of the polymer in the blend (which lies between 20 °C and 60 °C[79]). Note that variations in polymer blend T_g are dependent on a number of factors, including the regio-regularity of the polymer and its molecular weight, and on the concentration of PCBM in the blend.[75,79,80] At a temperature above the polymer T_g, the P3HT molecules have a relatively high chain mobility that permits them to undergo structural reorganization, as well as encouraging the diffusion and aggregation of PCBM molecules within the polymer matrix.

At present, it is not clear whether the growth of P3HT domains is isotropic or whether it occurs prior to the aggregation of PCBM. Resolving this issue is hampered by the use of different experimental conditions in each study. Lilliu et al.[81] identified an anisotropic rate of crystallite growth along the (100) and (010) planes, the latter evolving at a rate that is approximately an order of magnitude slower. Furthermore, the growth of PCBM aggregates was observed to occur at a slower rate than the crystallization of P3HT as evidenced by (100) scattering. These results support the hypothesis advanced by Agostinelli et al.[78] who proposed phase separation during thermal annealing is driven by P3HT crystallization, with the delayed aggregation of PCBM molecules occurring over relatively longer time scales. However, a separate study by Wu et al.[82] that combined GIWAXS and GISAXS showed that morphology evolution was a competition between P3HT crystallization and aggregation of PCBM. It was shown that PCBM aggregation (as measured by GISAXS) occurs faster than P3HT self-organization (as measured by GIWAXS), and the growth rate of polymer domains is roughly equal along orthogonal directions. More work is clearly necessary to address these discrepancies, as the order at which blend components self-organize will have a clear impact on a rational determination of optimum isothermal annealing times during the postdeposition thermal processing of a device.

SE has also been used to study structural changes during thermal annealing of P3HT:PCBM[83] and PCDTBT:PCBM[24] blends. Figure 7.15a shows the change in thickness that occurs in a P3HT:PCBM 1:0.7 wt% blend thin film as it is heated, isothermally annealed, and then cooled back to room temperature. Figure 7.15b shows changes in the optical absorption (extinction coefficient, k) spectrum at three selected wavelengths, again as the film is heated, annealed, and cooled. From changes in film thickness and optical absorption spectra, a number of physical processes can be identified. These include solvent loss once the film temperature exceeds T_g of the blend, volume/structure relaxation during isothermal annealing followed by depletion/coarse phase separation of PCBM from the bulk to film surface. The increase in optical extinction coefficient during isothermal annealing and cooling can be clearly observed in Fig. 7.15b and occurs due to crystallization and reduced free volume of P3HT molecules. It is worth noting that fluctuations in the absorption coefficient during the heating process around T_g result from the evaporation of solvent that was originally trapped in the film during the solution spin-casting process. It can be seen that during the final cooling cycle, the absorption coefficient increases markedly due to P3HT crystallization. In fact, control over this process is important, as returning an OPV device relatively slowly to room temperature can result in a ~5% relative improvement in device efficiency compared to a device which has been rapidly

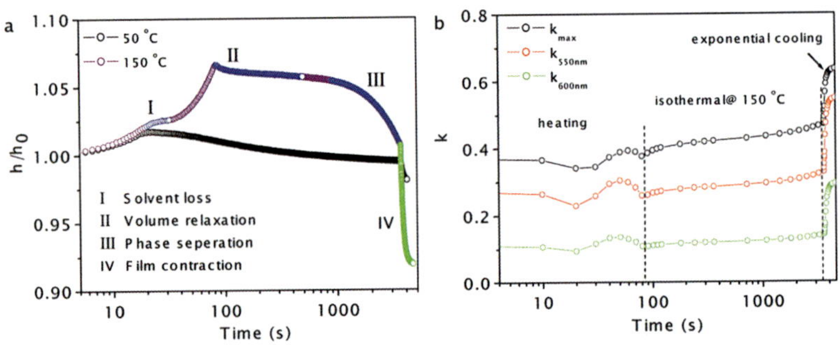

Figure 7.15 (a) The change in film thickness of a P3HT:PCBM blend during heating, isothermal annealing, and cooling. The physical processes identified are highlighted using different colors. (b) The evolution of k_{max}, $k_{550\ nm}$, and $k_{600\ nm}$ of the k spectra of a P3HT:PCBM blend film during heating, isothermal annealing, and cooling. (Reprinted from Wang et al., *Adv Funct Mater*, 21, 1383 (2011) with permission from Wiley-VCH).

cooled.[83]

During isothermal annealing of a P3HT:PCBM blend, a slow contraction of film volume has been ascribed to a volume relaxation process in which the free volume of the polymer chains reduces. A similar process has also been observed upon thermally annealing PCDTBT. This volume relaxation process can result from a change in polymer morphology from an amorphous to a crystalline state, as well as from one amorphous state to another more densely packed amorphous state. The latter process is more applicable to volume relaxation in a noncrystallizable polymer or when the annealing temperature is below the T_g of the film.

At the latter stages of thermal annealing, film thinning results from coarse phase separation between the blend components. Here, the phase-separated PCBM component in the bulk of the film migrates toward the film surface and forms larger aggregates or needle-like crystals, a process that both reduces the average film thickness and increases surface roughness. Annealing at higher temperatures (below the melt temperature of the blend) increases the rate of volume relaxation and phase separation. It has been found that coarse phase separation of $PC_{71}BM$ in a PCDTBT:$PC_{71}BM$ blend only occurs at a temperature ≥155 °C. $PC_{71}BM$ has been identified to undergo cold-crystallization process when heated to 155 °C,[24,84] thus suggesting that this process drives the coarse phase separation of $PC_{71}BM$ in a PCDTBT:$PC_{71}BM$ blend. Such coarse phase separation is deleterious to OPV efficiency, with the PCE of PCDTBT:$PC_{71}BM$ OPVs fabricated under optimum processing conditions—often only requiring a "low-temperature" (<100 °C) thermal annealing—reducing from over 6% to around 4% when the device is annealed above 150 °C.[25]

7.7.2 Impact of Solvent Annealing on Nanoscale Morphology and Optoelectronic Properties

The term "solvent vapor annealing" is used to describe a process in which either a dry or a semidry thin film is placed in a solvent vapor atmosphere. The exposure to solvent vapor plasticizes the BHJ film, increasing molecular mobility and thus promoting self-organization.

Solvent vapor annealing has been found to improve the performance of P3HT:PCBM solar cells.[85,86] The time scale for solvent vapor annealing to achieve optimized device performance depends on the type of solvent used and its solvent vapor pressure. In a P3HT:PCBM blend, solvent vapor annealing times that have been explored range from several minutes to more than an hour. Using Raman spectroscopy and photoluminescence measurements, Miller et al.[86] observed a rapid reduction in disorder in the P3HT phase within the first minute of solvent vapor annealing. Other work by Campoy-Quiles et al.[87] on solvent vapor annealing of a P3HT:PCBM blend evidenced an evolution of optical density and peak absorption wavelength that occurred over a period of several minutes. As can be seen in Fig. 7.16, a simultaneous red shift and enhancement in absorption intensity resulting from P3HT crystallization occurs during solvent vapor annealing. Interestingly, Jo et al. found that solvent vapor annealing times of more than 1 hour were required to improve device efficiency to a point that could be created via thermal annealing alone.[88] It has been shown[88] that solvent vapor annealing can improve the crystallinity of P3HT without inducing a high degree of PCBM aggregation. Tang et al.[89] reported a multistep approach to control the evolution of P3HT and PCBM components within a blend film by subjecting the film to different solvent vapors. In this work, PCBM was shown to form aggregates having sizes of ca. 30 nm when the blend film was exposed to tetrahydrofuran (THF) vapor. A second step was then applied using carbon disulfide (CS_2) vapor that reduced the size of the PCBM aggregates to ca. 20 nm. This process was shown to substantially increase the crystallinity of

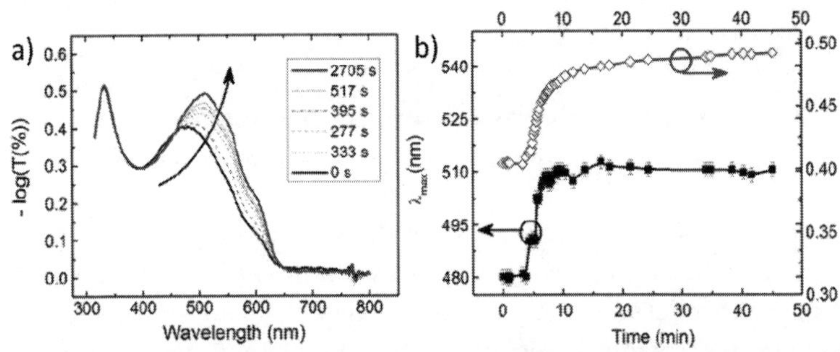

Figure 7.16 (a) Optical density of a P3HT:PCBM film as a function of solvent annealing time. (b) Spectral position of the absorption maximum (squares) and magnitude (diamonds). (Reprinted from Campoy-Quiles et al., *Thin Solid Films*, 519, 2678 (2011) Copyright © Elsevier).

P3HT as confirmed by X-ray scattering.

Verploegen et al. studied changes in the nanostructure in P3HT, PCBM, and a P3HT:PCBM blend during solvent vapor annealing.[90] Here, GIWAXS was used to monitor the real-time evolution of crystallinity upon exposure to various solvents including CF, CB,

DCB, THF, toluene, hexane, etc. It was found that vapor annealing using CF and THF solvents both result in a swelling of the P3HT lamellar layer, a decrease in the π–π stacking distance, an increase in the crystalline order of the P3HT component, and phase separation of P3HT and PCBM domains. Solvent vapor annealing using CB, DCB, and toluene resulted in similar effects. In contrast, however, vapor annealing using hexane—a solvent that has a low miscibility with P3HT and PCBM—resulted in a swelling of the P3HT lamellar layer without any impact on the P3HT backbone conformation or the crystallization of PCBM. Figure 7.17 shows the changes in the layer spacing and full-width half-maximum linewidth (FWHM) of Bragg scatter from P3HT lamella as well as the distance and Bragg scatter linewidth of π–π stacking in a pure P3HT film during vapor annealing using CF, THF, and hexane. It can be seen in Fig. 7.17a that the P3HT lamellar layer spacing increases by around 3% upon exposure to CF, THF, and hexane vapors. After this initial exposure, there is then a general reduction in layer spacing; however, on removal of the vapor the layer spacing is larger than before the exposure process. This indicates that solvent vapor annealing induces a swelling and rearrangement of P3HT alkyl side chains. The decrease of lamella-spacing linewidth shown in Fig. 7.17b indicates that there is an increase in domain size as a result of increased P3HT crystallinity.

Although solvent vapor increases the P3HT lamellar layer spacing, the π–π stacking distance and linewidth both reduce upon the introduction of CF and THF vapor and undergo further reductions when the solvent is removed, as shown in Fig. 7.17c and d. A reduced π–π stacking distance is clearly beneficial for solar cell device efficiency as it will improve hole charge mobility. It can be seen that there is negligible change in π–π stacking distance and linewidth on exposure to a hexane vapor annealing due to the low miscibility between P3HT and hexane. Hexane appears therefore to only swell the alkyl side chains, without swelling the conjugated-polymer backbone. Whilst exposure to a CF vapor does not induce the crystallization of PCBM, THF vapor annealing results in PCBM crystallization, a result consistent with previous observations by Tang et al.[89] Verploegen et al. suggested that THF might act as a nucleation site for PCBM

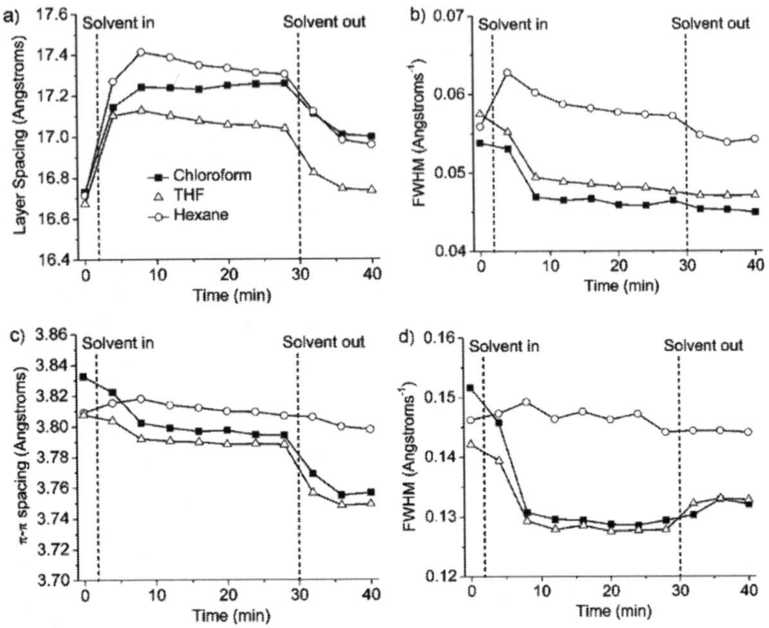

Figure 7.17 (a) Layer spacing, (b) layer (FWHM) linewidth, (c) π–π stacking distance, and (d) π–π stacking linewidth of P3HT lamella in a pure P3HT film upon exposure to solvent vapors of CF, THF, and hexane. The lamellar layer spacing and linewidth were determined from the in-plane (100) diffraction peak of P3HT, with the π–π stacking distance and linewidth determined from the π–π stacking diffraction peak in out-of-plane direction. (Reprinted from Verploegen et al., *Chem Mater,* 24, 3923 (2012) with permission from American Chemical Society).

crystallization or otherwise direct the growth of small existing crystallites within the bulk of the film.

While the crystalline form of P3HT is usually a lamellar-type structure with a large aspect ratio of crystalline domains, recent work has indicated that the morphology of P3HT crystalline domains can be tuned by solvent vapor annealing using CS_2, which permits a control of the nucleation density during the recrystallization process.[91,92] As shown in Fig. 7.18, long-aspect-ratio P3HT crystalline domains can form at high nucleation density,

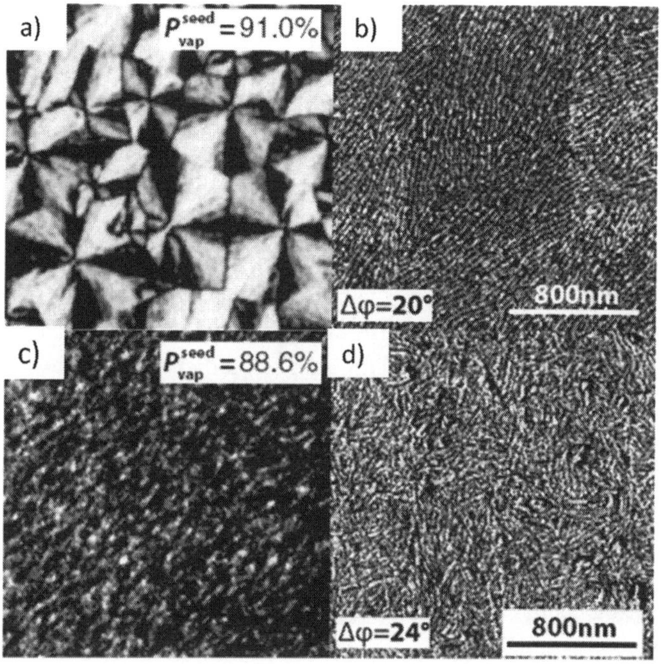

Figure 7.18 Polarized optical microscopy (POM) (a) and AFM (b) images of the P3HT crystal morphology after recrystallization at low nucleation density during CS_2 vapor annealing. The AFM phase image shows the alignment of crystalline lamellae within the bright domains visible in the POM image. POM (c) and AFM (d) images of P3HT crystal morphology after recrystallization at high nucleation density during CS_2 vapor annealing. POM images shown in parts (a) and (c) are 100 μm × 100 μm (Reprinted from Crossland et al., *Adv Funct Mater,* 21, 518 (2011) with permission from Wiley-VCH).

whereas the P3HT lamellae can emanate radially from the nucleation center to form micron-scale spherulites at low nucleation density.

7.7.3 Comparison between Thermal and Solvent Annealing

Compared with thermal annealing, solvent vapor annealing is a relatively low-energy technique that can be used to modify film nanostructure. A significant advantage of solvent annealing is its applicability to BHJ devices that are fabricated on flexible, plastic substrates that would otherwise degrade or warp when subjected to thermal annealing at elevated temperature.

There are similar effects that occur in structural evolution in a P3HT:PCBM blend upon both solvent vapor annealing and thermal annealing; however, notable differences are apparent. It is found that solvent vapor annealing does not induce a reorientation of the P3HT lamellar layer and the π–π stacking diffraction peak remains significantly stronger in the out-of-plane direction than that in the in-plane direction after solvent vapor annealing.[90] In contrast, it has been found that the out-of-plane π–π stacking peak disappeared and the in-plane π–π stacking peak increased upon thermal annealing above the P3HT melting temperature.[84] The π–π stacking peak in the out-of-plane direction originates from π–π stacks having their stacking direction perpendicular to the substrate, a molecular orientation that is more favorable for charge transport in a solar cell device. GISAXS measurements also show that solvent vapor annealing using CF and THF has a moderate effect on nanoscale phase separation in a P3HT:PCBM blend; however, thermal annealing induces more abrupt phase separation, resulting in larger P3HT and PCBM domains.

7.8 Vertical Component Distribution

Due to surface and interface effects, vertical stratification of the electron donors and acceptors within a BHJ can occur during the film deposition process. Such vertical stratification can affect the charge transport and extraction processes and consequently OPV device efficiency. For example, the enrichment of a hole-transporting polymer at the cathode interface can block electron extraction, whereas an enriched layer of electron-transporting fullerenes at the anode interface can block hole extraction. A number of techniques have been used to probe the depth profile of the components in a blend including neutron reflectivity (NR),[93] X-ray reflectivity (XRR),[94] SE,[95,96] and DSIMS.[97] Other techniques have been used to characterize the composition at or near the surface of a film, including NEXAFS[98] and X-ray photoelectron spectroscopy (XPS).[99] Such techniques provide valuable information about the surface of a film; however, they are insensitive to bulk composition and the composition near the buried interface with the underlying substrate.[18] To investigate bulk and buried interfacial regions using surface-sensitive techniques clearly requires the use of processing techniques that can progressively and accurately remove material from the film surface (e.g., ion sputtering, etching, or delamination); however, such processes need to be used with care as they can themselves generate structures (artifacts) that are not present in an unprocessed film.

By modeling specular neutron scattering the depth profile of a conjugated polymer and a fullerene in a BHJ can be probed because of the high contrast in their neutron scattering length density (SLD). Work (see Fig. 7.19a) by Parnell et al.[93] showed that in a CB spin-cast P3HT:PCBM thin film, the surface layer is relatively depleted in PCBM, with a PCBM-rich layer positioned at the substrate–film interface. Ellipsometry modeling (see Fig. 7.19b) by Germack et al.[96] on a P3HT:PCBM blend revealed a similar depth profile. Such a PCBM depletion layer at the film interface (which is effectively a P3HT-rich surface layer) has also been reported using a range of surface analysis techniques including NEXAFS, XPS, and DSIMS, regardless of the substrate on which the film was deposited.[98–100] In contrast however, Mackay et al.[50] reported a higher PCBM concentration both at the substrate and near the air interface, with thermal annealing producing a qualitatively similar profile-albeit with slightly different interfacial concentrations. It has been proposed that the enrichment of blend components near the substrate interface is significantly affected by the surface energy of the substrate.[95,101]

NR studies by Parnell et al. indicated that PCBM molecules migrate toward the film surface on thermal annealing and thereby reduce the volume of the P3HT surface-rich layer. This thermal treatment did not apparently affect the PCBM concentration at the substrate interface. Device work on P3HT:PCBM has shown that thermal annealing can significantly improve OPV efficiency, via increased crystallinity (improving light absorption) and phase separation (improving charge dissociation and transport). It is thus tempting to conclude that the PCBM-depleted surface of an as-cast P3HT:PCBM thin film is partly responsible for its relatively poor performance when incorporated in an OPV device, and that thermal annealing helps improve electron extraction via increasing the PCBM concentration close to the device cathode. In other work, Dadmun et al.[39] used NR to study annealed P3HT–PCBM bilayers, and reported that PCBM selectively segregates to

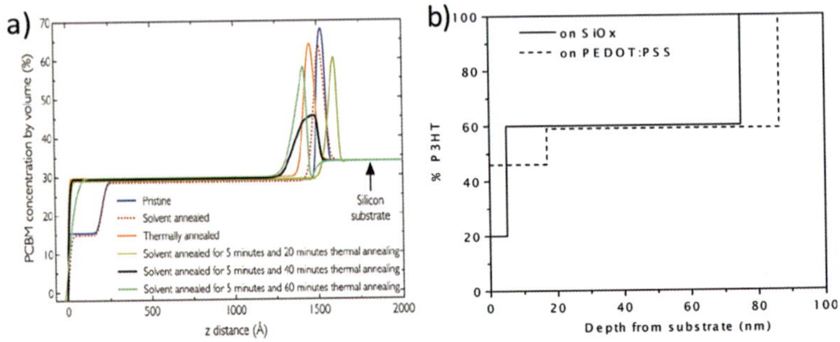

Figure 7.19 (a) Vertical PCBM distribution in a series of P3HT:PCBM blends as probed by NR. (Reprinted from Parnell et al., *Adv. Mater.*, 22, 2444 (2010) with permission from Wiley-VCH) (b) Vertical P3HT distribution in a P3HT:PCBM blend deposited on SiO_x and PEDOT:PSS substrates as extracted from ellipsometry fitting. (Reprinted from Germack et al., *Macromolecules*, 43, 3828 (2010) with permission from American Chemical Society).

the air interface, and concluded that such diffusion resulted from entropic reasons as the system proceeded toward an equilibrium state.

Other work has addressed vertical stratification in films composed of fullerene derivatives blended with other donor–acceptor conjugated polymers. Felicissimo et al.[102] used XPS to study a polyfluorene F8DTBT film blended with PCBM, with an enriched polymer layer also detected at the film surface. Anselmo et al.[103] used DSIMS and NEXAFS to study the vertical profile of donor–acceptor polyfluorene copolymers blended with PCBM and found that there is also a polymer-rich layer near the film surface that becomes more concentrated as the polar nature of the polymer is increased. Lu et al.[74] reported that in an as-spin-cast blend film of PSBTBT:PCBM, there is a PSBTBT-enriched layer at the film surface (see Fig. 7.20a). Recent NR[104] on a (1:4) PCDTBT:PCBM blend that was spin-cast from CF on PEDOT:PSS showed that the as-cast film is characterized by a PCBM-enriched layer at the film surface and a relative depletion of PCBM toward the anode interface (see Fig. 7.20b). This distribution of

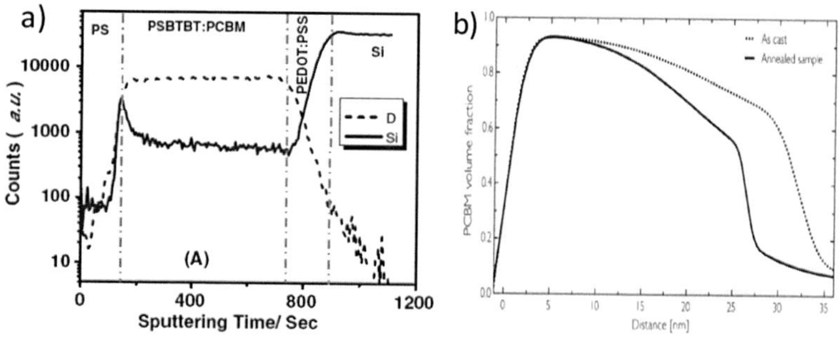

Figure 7.20 (a) DSIMS profile of an as-spun PSBTBT:PCBM blend. (Reprinted from Lu et al., *Adv. Energy Mater.*, 1, 870 (2011) with permission from Wiley-VCH) (b) PCBM vertical distribution as extracted from neutron scattering. In both cases the film is positioned on a PEDOT:PSS layer. (Reprinted from Staniec et al., *Adv Energy Mater*, 1, 499 (2011) with permission from Wiley-VCH).

donor and acceptor materials will clearly facilitate charge extraction in OPV having a cathode deposited onto the film surface.

Wang et al.[105] have studied vertical stratification of $PC_{71}BM$ when blended in a range of different polycarbazole donor–acceptor copolymers (namely PCDTBT, P1, and P2) and deposited by spray coating on a PEDOT:PSS thin film. As shown in Fig. 7.21, a region in which the $PC_{71}BM$ concentration is relatively depleted—that is, its concentration is lower than its 80 wt% composition in the original polymer:fullerene blend. This $PC_{71}BM$-depleted region exists at the PEDOT:PSS interface for all blend films studied, with vertical stratification apparently dependent on the substrate temperature applied during the film drying process. For example, at substrate temperature of 40 °C, the $PC_{71}BM$-depleted region in a PCDTBT: $PC_{71}BM$ blend had a width of 17 nm, a roughness of 6 nm, and an average $PC_{71}BM$ concentration of 65 wt%. When the substrate deposition temperature was increased to 60 °C, the $PC_{71}BM$ concentration in the depleted region increased to 78 wt%, with the width of the region reducing to 13 nm (having a roughness of 7 nm). This depleted region was almost completely suppressed when the substrate deposition temperature was increased to 80 °C, a temperature that most likely significantly accelerates the film drying

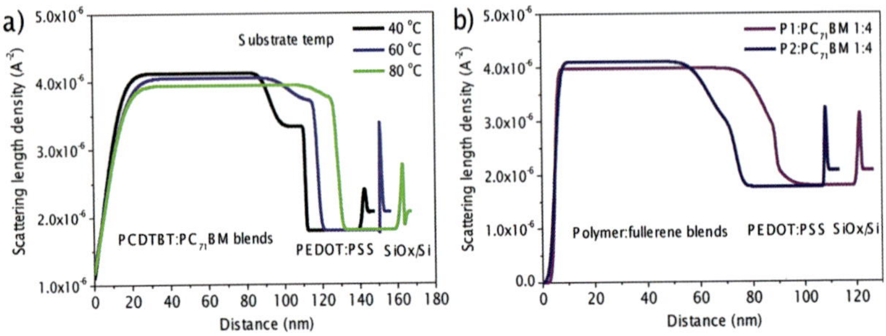

Figure 7.21 (a) SLD profiles of PCDTBT:$PC_{71}BM$ (1:4) blends deposited by spray-coating at three different substrate temperatures from a CB solution. (b) SLD profiles of P1:$PC_{71}BM$ and P2:$PC_{71}BM$ blends deposited by spray-coating at a substrate temperature of 40 °C from a CB solution. The thicknesses of the blend films are slightly different for all samples. (Reprinted from Wang et al., *Adv Energy Mater,* **3**, 505–512 (2013) with permission from Wiley-VCH).

dynamics and thereby suppresses the vertical stratification of $PC_{71}BM$.

A number of studies have explored vertical segregation upon thermal annealing in a BHJ when coated by a cathode layer. Using XPS, Orimo et al.[106] found that the degree of surface segregation upon annealing in a P3HT:PCBM blend is dependent on whether the film has a free surface or been capped with an aluminum cathode. These authors found that in a blend film that is annealed before a cathode is deposited, there was actually a P3HT-rich layer positioned at the film surface. However, if the annealing process was performed after the deposition of a cathode, the film–cathode interface was instead rich in PCBM. This effect is similar to that reported by Lu et al.[74] who showed that thermal annealing promotes the segregation of PSBTBT to a free surface in a blend film without the presence of the cathode layer; however, PCBM enrichment occurs toward the cathode interface if annealing occurs when the cathode layer is present (Fig. 7.22). Mauger et al.[107] recently investigated the vertical composition of PCBM in P3HT:PCBM film capped with a metal film using NR. It was found that surface energy of the substrate and an additive added to the casting solution both affect the PCBM concentration at the anode and cathode interfaces. It was argued that low-work-function metals donate electrons to the PCBM through interfacial charge transfer and exohedral doping,

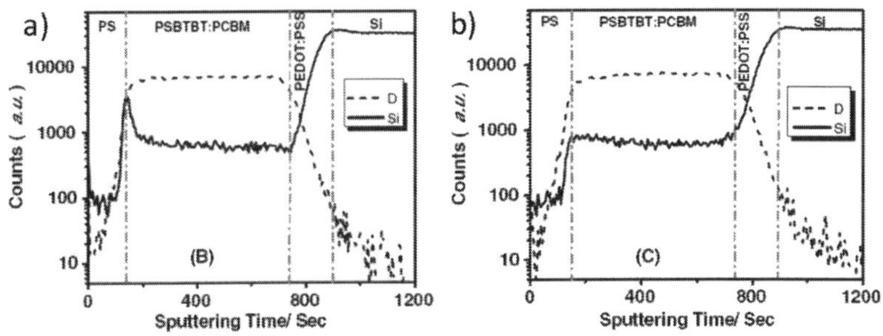

Figure 7.22 DSIMS profile of PSBTBT:PCBM blend films annealed at 150 °C for 1 min (a) before and (b) after the deposition of the Al cathode. (Reprinted from Lu et al., *Adv Energy Mater,* 1, 870 (2011) with permission from Wiley-VCH).

and that the presence of charge on the PCBM affects its vertical stratification during the thermal annealing process.

Clearly, more work is needed to determine the intrinsic mechanisms that control vertical stratification during a film casting process and any subsequent posttreatment processes, as well as its relevance to device performance. We note that Germack et al. found that the performance of P3HT:PCBM OPV devices having differences in the degree of vertical segregation at the interface with different hole transport layers were in fact relatively similar.[96] A similar conclusion was also reached in other work[108] in which it was shown that OPV device performance is insensitive to the precise details of interfacial segregation, provided that the interface between the active layer and the electrode is not composed of a pure layer of an organic semiconductor that hinders charge extraction.

7.9 A Brief Summary of Characterization Techniques

The development of the morphology of polymer : fullerene photovoltaic BHJ blend involves a number of processes, including molecular packing, ordering and orientation, mixing and separation, and the vertical segregation of donors and acceptors within the active layer. The length scale associated with structure formed by such processes ranges from sub nanometer to tens of nanometers. For example, the *d*-spacing, correlation length, ordering, and orientation that occur

Table 7.1 A summary of techniques used to probe the nanostructure of BHJ solar cells

Methods	**Relevant structures characterized**	**Regions probed within a thin film**	**Notes**
GI-WAXS	Molecular packing, ordering, and orientation	Depth resolution possible using angular-dependent measurements	Risk of beam damage to materials
GI-SAXS	Phase separation, domain size		Risk of beam damage to materials
GI-SANS	Phase separation, domain size		Long measurement time
SEM	Phase separation, domain size	Near-surface	Risk of beam damage to materials, low resolution
TEM	Phase separation, domain size	Bulk	Risk of beam damage to materials, defocused TEM can bring artifacts due to low electron density contrast
HeIM	Phase separation, domain size	Near-surface	Risk of beam damage to materials
Three-dimensional electron tomography	Phase separation, domain size	Bulk	Risk of beam damage to materials, probing crystalline phase only
NR	Vertical component distribution	Depth profile	Contrast between SLD is high, but requires long measurement time due to low neutron beam intensity
XRR	Vertical component distribution	Depth profile	Requires short measurement time because of high X-ray beam intensity, but the contrast between electron density is low
DSIMS	Vertical component distribution	Depth profile	Different evaporating rate of polymer and fullerene can potentially generate artifacts
SE	Complex refractive index, vertical component distribution	Depth profile	Depth profiles are constrained by mathematical models, e.g., linear, parabolic, exponential, etc.
XPS NEXAFS	Composition near the surface region	Near-surface	Require ion sputtering, etching, or delamination to remove materials to establish depth profile, can potentially generate artifacts

in the polymer and fullerene phases have a typical length scale of Angstroms to several nanometers. This length scale can in fact be probed using a number of techniques, wide-angle X-ray scattering (WAXS) being one such example.[109] A range of morphological (e.g., electron microscopy) and scattering (e.g., SAXS) techniques have been developed to probe the domain structures that result from phase separation. Resonant soft X-ray scattering has recently been demonstrated to provide information on the purity of polymer and fullerene domains and the nature of their interfaces (i.e., whether they are diffuse or sharp).[110] NR, XRR, DSIMS, NEXAFS spectroscopy, and XPS are able to probe depth-dependent composition. In Table 7.1, we summarize the various characterization techniques that have been discussed in this chapter. We outline the specific structures that can be used to probe in a BHJ and also list known issues that should be considered when using such characterization techniques in order to avoid sample modification and the introduction of artifacts.

7.10 Conclusions

We have reviewed the morphologies that can exist in a number of polymer : fullerene BHJ solar cells and the routes by which such structures can emerge. We have discussed a range of techniques by which molecular nanostructure can be characterized. We show that the nanostructure that exists within a polymer : fullerene BHJ solar cell is highly dependent on the choice of donor and acceptor materials, together with the processing conditions used both to deposit the film and then to subsequently improve PV cell efficiency. Effective strategies for BHJ nanostructure increasingly favor molecular-scale mixing of electron donor and acceptors in the active layer film resulting in effective charge carrier generation, as well as developing an interpenetrating network of each component toward the device electrodes to ensure efficient charge carrier transport. BHJ films should also ideally possess a vertical distribution of donor and acceptor components, with a relatively higher concentration of electron donors (acceptors) positioned toward the anode (cathode) to enhance charge extraction and reduce leakage currents and recombination.

We believe the eventual commercialization of BHJ organic PV will require a number of challenges to be solved—the most pressing of which are currently the improvement of device operational lifetime and the increase in device efficiency. Solving such issues will require a combination of techniques, including the synthesis of improved materials, the design of appropriate device architectures, the development of deposition technologies suitable for scale-up, and device modularization. We hope, however, that we have shown that control over BHJ morphology is a vital issue that must be understood as it will underpin the eventual successful development of this technology.

References

1. Brabec CJ, Sariciftci NS, Hummelen JC (2001) Plastic solar cells, *Adv Funct Mater,* **11**, 15–26.
2. Blom PWM, Mihailetchi VD, Koster LJA (2007) Device physics of polymer: fullerene bulk heterojunction solar cells, *Adv Mater,* **19**, 1551–1566.
3. Dennler G, Scharber MC, Brabec CJ (2009) Polymer-fullerene bulk heterojunction solar cells, *Adv Mater,* **21**, 1323–1338.
4. Hoppe H, Sariciftci NS (2004) Organic solar cells: An overview, *J Mater Res,* **19**, 1924–1945.
5. Li CZ, Yip HL, Jen AKY (2012) Functional fullerenes for organic photovoltaics, *J Mater Chem,* **22**, 4161–4177.
6. Mihailetchi VD, Xie HX, de Boer B, Koster LJA, Blom PWM (2006) Charge transport and photocurrent generation in poly(3-hexylthiophene):methanofullerene bulk-heterojunction solar cells, *Adv Funct Mater,* **16**, 699–708.
7. Clarke TM, Ballantyne AM, Nelson J, Bradley DDC, Durrant JR (2008) Free energy control of charge photogeneration in polythiophene/fullerene solar cells: The influence of thermal annealing on P3HT/PCBM blends, *Adv Funct Mater,* **18**, 4029–4035.
8. Marsh RA, Hodgkiss JM, Albert-Seifried S, Friend RH (2010) Effect of annealing on P3HT:PCBM charge transfer and nanoscale morphology probed by ultrafast spectroscopy, *Nano Lett,* **10**, 923–930.
9. Etzold F, Howard IA, Forler N, et al. (2012) The effect of solvent additives on morphology and excited-state dynamics in PCPDTBT:PCBM photovoltaic blends, *J Am Chem Soc,* **134**, 10569–10583.

10. Chen LM, Hong Z, Li G, Yang Y (2009) Recent progress in polymer solar cells: Manipulation of polymer : fullerene morphology and the formation of efficient inverted polymer solar cells, *Adv Mater,* **21**, 1434–1449.
11. Brady MA, Su GM, Chabinyc ML (2011) Recent progress in the morphology of bulk heterojunction photovoltaics, *Soft Matter,* **7**, 11065–11077.
12. Ruderer MA, Müller-Buschbaum P (2011) Morphology of polymer-based bulk heterojunction films for organic photovoltaics, *Soft Matter,* **7**, 5482–5493.
13. Liu F, Gu Y, Jung JW, Jo WH, Russell TP (2012) On the morphology of polymer-based photovoltaics, *J Polym Sci Part B: Polym Phys,* **50**, 1018–1044.
14. Nagarjuna G, Venkataraman D (2012) Strategies for controlling the active layer morphologies in OPVs, *J Polym Sci Part B: Polym Phys,* **50**, 1045–1056.
15. Mayukh M, Jung IH, He F, Yu L (2012) Incremental optimization in donor polymers for bulk heterojunction organic solar cells exhibiting high performance, *J Polym Sci Part B: Polym Phys,* **50**, 1057–1070.
16. Zhou HX, Yang LQ, You W (2012) Rational design of high performance conjugated polymers for organic solar cells, *Macromolecules,* **45**, 607–632.
17. Duan CH, Huang F, Cao Y (2012) Recent development of push-pull conjugated polymers for bulk-heterojunction photovoltaics: Rational design and fine tailoring of molecular structures, *J Mater Chem,* **22**, 10416–10434.
18. Delongchamp DM, Kline RJ, Herzing A (2012) Nanoscale structure measurements for polymer-fullerene photovoltaics, *Energy Environ Sci,* **5**, 5980–5993.
19. Sirringhaus H, Brown PJ, Friend RH, et al. (1999) Two-dimensional charge transport in self-organized, high-mobility conjugated polymer, *Nature,* **401**, 685–688.
20. Yang H, LeFevre SW, Ryu CY, Bao Z (2007) Solubility-driven thin film structures of regioregularpoly(3-hexyl thiophene) using volatile solvents, *Appl Phys Lett,* **90**, 172116.
21. Kim DH, Jang Y, Park YD, Cho K (2006) Layered molecular ordering of self-organized poly(3-hexylthiophene) thin films on hydrophobized surfaces, *Macromolecules, 39*, 5843–5847.
22. Bian L, Zhu E, Tang J, Tang W, Zhang F (2012) Recent progress in the design of narrow bandgap conjugated polymers for high-efficiency organic solar cells, *Prog Polym Sci,* **37**, 1292–1331.

23. Park SH, Roy A, Beaupré S, et al. (2009) Bulk heterojunction solar cells with internal quantum efficiency approaching 100%, *Nat Photon,* **3**, 297–302.

24. Wang T, Pearson AJ, Dunbar ADF, et al. (2012) Correlating structure with function in thermally annealed PCDTBT:$PC_{70}BM$ photovoltaic blends, *Adv Funct Mater,* **22**, 1399–1408.

25. Beiley ZM, Hoke ET, Noriega R, et al. (2011) Morphology-dependent trap formation in high performance polymer bulk heterojunction solar cells, *Adv Energy Mater,* **1**, 954–962.

26. Yiu AT, Beaujuge PM, Lee OP, Woo CH, Toney MF, Frechet JMJ (2012) Side-chain tenability of furan-containing low band-gap polymers provides control of structural order in efficient solar cells, *J Am Chem Soc,* **134**, 2180–2185.

27. Szarko JM, Guo J, Liang Y, et al. (2010) When function follows form: Effects of donor copolymer side chains on film morphology and BHJ solar cell performance, *Adv Mater,* **22**, 5468–5472.

28. Kim Y, Nelson J, Zhang T, et al. (2009) Distorted asymmetric cubic nanostructure of soluble fullerene crystals in efficient polymer: fullerene solar cells, *ACS Nano*, **3**, 2557–2562.

29. Wang T, Dunbar ADF, Staniec PA, et al. (2010) The development of nanoscale morphology in polymer: fullerene photovoltaic blends during solvent casting, *Soft Matter,* **6**, 4128–4134.

30. Schmidt-Hansberg B, Sanyal M, Klein MFG, et al. (2011) Moving through the phase diagram: Morphology formation in solution cast polymer-fullerene blend films for organic solar cells, *ACS Nano,* **5**, 8579–8590.

31. Koppe M, Brabec CJ, Heiml S, et al. (2009) Influence of molecular weight distribution on the gelation of P3HT and its impact on the photovoltaic performance, *Macromolecules,* **42**, 4661–4666.

32. Sanyal M, Schmidt-Hansberg B, Klein MFG, et al. (2011) Effect of photovoltaic polymer/fullerene blend composition ratio on microstructure evolution during film solidification investigated in real time by X-ray diffraction, *Macromolecules,* **44**, 3795–3800.

33. Gu Y, Wang C, Russell TP (2012) Multi-length-scale morphologies in PCPDTBT/PCBM bulk-heterojunction solar cells, *Adv Energy Mater,* **2**, 683–690.

34. Liu F, Gu Y, Wang C, et al. (2012) Efficient polymer solar cells based on a low bandgap semi-crystalline DPP polymer-PCBM blends, *Adv Mater,* **24**, 3947–3951.

35. Treat ND, Brady MA, Smith G, et al. (2011) Interdiffusion of PCBM and P3HT reveals miscibility in a photovoltaically active blend, *Adv Energy Mater,* **1**, 82–89.

36. Chen D, Liu F, Wang C, Nakahara A, Russell TP (2011) Bulk heterojunction photovoltaic active layers via bilayer interdiffusion, *Nano Lett,* **11**, 2071–2078.

37. Collins BA, Gann E, Guignard L, He X, McNeill CR, Ade H (2010) Molecular miscibility of polymer-fullerene blends, *J Phys Chem Lett,* **1**, 3160–3166.

38. Yin W, Dadmun M (2011) A new model for the morphology of P3HT/PCBM organic photovoltaics from small-angle neutron scattering: Rivers and streams, *ACS Nano,* **5**, 4756–4768.

39. Chen H, Hegde R, Browning J, Dadmun MD (2012) The miscibility and depth profile of PCBM in P3HT: Thermodynamic information to improve organic photovoltaics, *Phys Chem Chem Phys,* **14**, 5635–5641.

40. Ruderer MA, Meier R, Porcar L, Cubitt R, Müller-Buschbaum P (2012) Phase separation and molecular intermixing in polymer–fullerene bulk heterojunction thin films, *J Phys Chem Lett,* **3**, 683–688.

41. Jamieson FC, Domingo EB, McCarthy-Ward T, Heeney M, Stingelin N, Durrant JR (2012) Fullerene crystallisation as a key driver of charge separation in polymer/fullerene bulk heterojunction solar cells, *Chem Sci,* **3**, 485–492.

42. Koppe M, Scharber M, Brabec C, Duffy W, Heeney M, McCulloch I (2007) Polyterthiophenes as donors for polymer solar cells, *Adv Funct Mater,* **17**, 1371–1376.

43. Mayer AC, Toney MF, Scully SR, et al. (2009) Bimolecular crystals of fullerenes in conjugated polymers and the implications of molecular mixing for solar cells, *Adv Funct Mater,* **19**, 1173–1179.

44. Cates NC, Gysel R, Beiley Z, et al. (2009) Tuning the properties of polymer bulk heterojunction solar cells by adjusting fullerene size to control intercalation, *Nano Lett,* **9**, 4153–4157.

45. Brabec CJ, Heeney M, McCulloch I, Nelson J (2011) Influence of blend microstructure on bulk heterojunction organic photovoltaic performance, *Chem Soc Rev,* **40**, 1185–1199.

46. Sista P, Xue B, Wilson M, et al. (2012) Influence of the alkyl substituents spacing on the solar cell performance of benzodithiophene semiconducting polymers, *Macromolecules,* **45**, 772–780.

47. Cates NC, Gysel R, Dahl JEP, Sellinger A, McGehee MD (2010) Effects of intercalation on the hole mobility of amorphous semiconducting polymer blends, *Chem Mater*, **22**, 3543–3548.
48. Miller NC, Sweetnam S, Hoke ET, et al. (2012) Molecular packing and solar cell performance in blends of polymers with a bisadduct fullerene, *Nano Lett*, **12**, 1566–1570.
49. Moon JS, Lee JK, Cho S, Byun J, Heeger AJ (2009) "Columnlike" structure of the cross-sectional morphology of bulk heterojunction materials, *Nano Lett*, **9**, 230–234.
50. Kiel JW, Kirby BJ, Majkrzak CF, Maranville BB, Mackay ME (2010) Nanoparticle concentration profile in polymer-based solar cells, *Soft Matter*, **6**, 641–646.
51. Parnell AJ, Cadby AJ, Mykhaylyk OO, et al. (2011) Nanoscale phase separation of P3HT PCBM thick films as measured by small-angle X-ray scattering. *Macromolecules*, **44**, 6503–6508.
52. van Bavel SS, Sourty E de With G, Frolic K, Loos J (2009) Relation between photoactive layer thickness, 3D morphology, and device performance in P3HT/PCBM bulk-heterojunction solar cells, *Macromolecules*, **42**, 7396–7403.
53. van Bavel SS, Sourty E de With G, Loos J (2009) Three-dimensional nanoscale organization of bulk heterojunction polymer solar cells, *Nano Lett*, **9**, 507–513.
54. Herzing AA, Richter LJ, Anderson IM (2010) 3D nanoscale characterization of thin-film organic photovoltaic device structures via spectroscopic contrast in the TEM, *J Phys Chem C*, **114**, 17501–17508.
55. Pearson AJ, Boden SA, Bagnall DM, Lidzey DG, Rodenburg C (2011) Imaging the bulk nanoscale morphology of organic solar cell blends using helium ion microscopy, *Nano Lett*, **11**, 4275–4281.
56. Swaraj S, Wang C, Yan H, et al. (2010) Nanomorphology of bulk heterojunction photovoltaic thin films probed with resonant soft X-ray scattering, *Nano Lett*, **10**, 2863–2869.
57. Lazzari R (2002) IsGISAXS: A program for grazing-incidence small-angle X-ray scattering analysis of supported islands, *J Appl Crystallogr*, **35**, 406–421.
58. Yao Y, Hou J, Xu Z, Li G, Yang Y (2008) Effects of solvent mixtures on the nanoscale phase separation in polymer solar cells, *Adv Funct Mater*, **18**, 1783–1789.
59. Roger JT, Schmidt K, Toney MF, Kramer EJ, Bazan GC (2011) Structural order in bulk heterojunction films prepared with solvent additives, *Adv Mater*, **23**, 2284–2288.

60. Lee JK, Ma WL, Brabec CJ, et al. (2008) Processing additives for improved efficiency from bulk heterojunction solar cells, *J Am Chem Soc*, **130**, 3619–3623.

61. Liang Y, Zheng X, Xia J, et al. (2010) For the bright future—bulk heterojunction polymer solar cells with power conversion efficiency of 7.4%, *Adv Mater*, **22**, E135–E138.

62. Peet J, Kim JY, Coates NE, et al. (2007) Efficiency enhancement in low-band gap polymer solar cells by processing with alkane dithiols, *Nat Mater*, **6**, 497–500.

63. Hoven CV, Dang X-D, Coffin RC, Peet J, Nguyen T-Q, Bazan GC (2010) Improved performance of polymer bulk heterojunction solar cells through the reduction of phase separation via solvent additives, *Adv Mater*, **22**, E63–E66.

64. Su MS, Kuo CY, Yuan MC, Jeng US, Su CJ, Wei KH (2011) Improving device efficiency of polymer/fullerene bulk heterojunction solar cells through enhanced crystallinity and reduced grain boundaries induced by solvent additives, *Adv Mater*, **23**, 3315–3319.

65. Bijleveld JC, Gevaerts VS, Nuzzo DD, et al. (2010) Efficient solar cells based on an easily accessible diketopyrrolopyrrole polymer, *Adv Mater*, **22**, E242–E246.

66. Amb CM, Chen S, Graham KR, et al. (2011) Dithienogermole as a fused electron donor in bulk heterojunction solar cells, *J Am Chem Soc*, **133**, 10062–10065.

67. Lou SJ, Szarko JM, Xu T, Yu L, Marks TJ, Chen LX (2011) Effects of additives on the morphology of solution phase aggregates formed by active layer components of high-efficiency organic solar cells, *J Am Chem Soc*, **133**, 20661–20663.

68. Rogers JT, Schmidt K, Toney MF, Bazan GC, Kramer EJ (2012) Time-resolved structural evolution of additive-processed bulk heterojunction solar cells, *J Am Chem Soc*, **134**, 2884–2887.

69. Agostinelli T, Ferenczi TAM, Pires E, et al. (2011) The role of alkane dithiols in controlling polymer crystallization in small band gap polymer: fullerene solar cells, *J Polym Sci Part B: Polym Phys*, **23**, 2284–2288.

70. Piliego C, Holcombe TW, Douglas JD, Woo CH, Beaujuge PM, Frechet JMJ (2010) Synthetic control of structural order in N-alkylthieno[3,4-c]pyrrole-4,6-dione-based polymers for efficient solar cells, *J Am Chem Soc*, **132**, 7595–7597.

71. Chu TY, Lu J, Beaupre IS, et al. (2011) Bulk heterojunction solar cells using thieno[3,4-c]pyrrole-4,6-dione and dithieno[3,2-b:2′,3′-d]silole copolymer with a power conversion efficiency of 7.3%. *J Am Chem Soc*, **133**, 4250–4253.

72. Ma W, Yang C, Gong X, Lee K, Heeger AJ (2005) Thermally stable, efficient polymer solar cells with nanoscale control of the interpenetrating network morphology, *Adv Funct Mater,* **15**, 1617–1622.

73. Chen HY, Hou J, Hayden AE, Yang H, Houk KN, Yang Y (2010) Silicon atom substitution enhances interchain packing in a thiophene-based polymer system, *Adv Mater,* **22**, 371–375.

74. Lu HY, Akgun B, Rusell TP (2011) Morphological characterization of a low-band gap crystalline polymer:PCBM bulk heterojunction solar cells, *Adv Energy Mater,* **1**, 870–878.

75. Pearson AJ, Wang T, Jones RAL, et al. (2012) Rationalizing phase transitions with thermal annealing temperatures for P3HT:PCBM organic photovoltaic devices, *Macromolecules,* **45**, 1499–1508.

76. Lu XH, Hlaing H, Germack DS, et al. (2012) Bilayer order in a polycarbazole-conjugated polymer, *Nat Commun,* **3**, 795.

77. Guo JC, Liang YY, Szarko J, et al. (2010) Structure, dynamics, and power conversion efficiency correlations in a new low band gap polymer: PCBM solar cell, *J Phys Chem B,* **114**, 742–748.

78. Agostinelli T, Lilliu S, Labram JG, et al. (2011) Real-time investigation of crystallization and phase-segregation dynamics in P3HT:PCBM solar cells during thermal annealing, *Adv Funct Mater,* **21**, 1701–1708.

79. Hopkinson PE, Staniec PA, Pearson AJ, et al. (2011) A phase diagram of the P3HT:PCBM organic photovoltaic system: Implications for device processing and performance, *Macromolecules,* **44**, 2908–2917.

80. Zhao J, Swinnen A, Van Assche G, Manca J, Vanderzande D, Van Mele B (2009) Phase diagram of P3HT/PCBM blends and its implication for the stability of morphology, *J Phys Chem B,* **113**, 1587–1591.

81. Lilliu S, Agostinelli T, Pires E, Hampton M, Nelson J, Macdonald JE (2011) Dynamics of crystallization and disorder during annealing of P3HT/PCBM bulk heterojunctions, *Macromolecules,* **44**, 2725–2734.

82. Wu W-R, Jeng U-S, Su C-J, et al. (2011) Competition between fullerene aggregation and poly(3-hexylthiophene) crystallization upon annealing of bulk heterojunction solar cells, *ACS Nano,* **5**, 6233–6243.

83. Wang T, Pearson AJ, Jones RAL, Lidzey DG (2011) Evolution of structure, optoelectronic properties and device performance of polythiophene:fullerene solar cells during thermal annealing, *Adv Funct Mater,* **21**, 1383–1390.

84. Verploegen E, Mondal R, Bettinger CJ, Sok S, Toney MF, Bao Z (2010) Effects of thermal annealing upon the morphology of polymer–fullerene blends, *Adv Funct Mater,* **20**, 3519–3529.

85. Zhao Y, Xie Z, Qu Y, Geng Y, Wang L (2007) Solvent vapor treatment induced performance enhancement of poly(3 hexylthiophene): methanofullerene bulk-heterojunction photovoltaic cells, *Appl Phys Lett,* **90**, 043504.

86. Miller S, Fanchini G, Lin Y-Y, et al. (2008) Investigation of nanoscale morphological changes in organic photovoltaics during solvent vapor annealing, *J Mater Chem,* **18**, 306–312.

87. Campoy-Quiles M, Schmidt M, Nassyrov D, et al. (2011) Real-time studies during coating and post-deposition annealing in organic semiconductors. *Thin Solid Films,* **519**, 2678–2681.

88. Jo J, Kim SS, Na SI, Yu BK, Kim DY (2009) Time-dependent morphology evolution by annealing processes on polymer : fullerene blend solar cells, *Adv Funct Mater,* **19**, 866–874.

89. Tang H, Lu G, Li L, Li J, Wang Y, Yang X (2010) Precise construction of PCBM aggregates for polymer solar cells via multi-step controlled solvent vapor annealing, *J Mater Chem,* **20**, 683–688.

90. Verploegan E, Miller CE, Schmidt K, Bao Z, Toney MF (2012) Manipulating the morphology of P3HT-PCBM bulk heterojunction blends with solvent vapor annealing, *Chem Mater,* **24**, 3923–3931.

91. Crossland EJW, Rahimi K, Reiter G, Steiner U, Ludwigs S (2011) Systematic control of nucleation density in poly(3-hexylthiophene) thin films, *Adv Funct Mater,* **21**, 518–524.

92. Crossland EJW, Tremel K, Fischer F, et al. (2012) Anisotropic charge transport in spherulitic poly (3-hexylthiophene) films, *Adv Mater,* **24**, 839–844.

93. Parnell AJ, Dunbar ADF, Pearson AJ, et al. (2010) Depletion of PCBM at the cathode interface in P3HT/PCBM thin films as quantified via neutron reflectivity measurements, *Adv Mater,* **22**, 2444–2447.

94. Liu HJ, Jeng U, Yamada NL, et al. (2011) Surface and interface porosity of polymer/fullerene-derivative thin films revealed by contrast variation of neutron and X-ray reflectivity, *Soft Matter,* **7**, 9276–9282.

95. Campoy-Quiles M, Ferenczi T, Agostinelli T, et al. (2008) Morphology evolution via self-organization and lateral and vertical diffusion in polymer : fullerene solar cell blends, *Nat Mater,* **7**, 158–164.

96. Germack DS, Chan CK, Kline RJ, et al. (2010) Interfacial segregation in polymer/fullerene blend films for photovoltaic devices, *Macromolecules,* **43**, 3828–3836.

97. Chen DA, Nakahara A, Wei DG, Nordlund D, Rusell TP (2011) P3HT/PCBM bulk heterojunction organic photovoltaics: Correlating efficiency and morphology, *Nano Lett,* **11**, 561–567.

98. Tillack AF, Noone KM, MacLeod BA, et al. (2011) Surface characterization of polythiophene:fullerene blends on different electrodes using near edge X-ray absorption fine structure, *ACS Appl Mater Interf*, **3**, 726–732.

99. Zheng X, Chen LM, Yang G, et al. (2009) Vertical phase separation in poly(3-hexylthiophene): fullerene derivative blends and its advantage for inverted structure solar cells, *Adv Funct Mater*, **19**, 1227–1234.

100. Xue B, Vaughan B, Poh CH, et al. (2010) Vertical stratification and interfacial structure in P3HT:PCBM organic solar cells, *J Phys Chem C*, **114**, 15797–15805.

101. Germack DS, Chan CK, Hamadani BH, et al. (2009) Substrate-dependent interface composition and charge transport in films for organic photovoltaics, *Appl Phys Lett*, **94**, 233303.

102. Felicissimo MP, Jarzab D, Gorgoi M, et al. (2009) Determination of vertical phase separation in a polyfluorenecopolymer: fullerene derivative solar cell blend by X-ray photoelectron spectroscopy, *J Mater Chem*, **19**, 4899–4901.

103. Anselmo AS, Lindgren L, Rysz J, et al. (2011) Tuning the vertical phase separation in polyfluorene:fullerene blend films by polymer functionalization, *Chem Mater*, **23**, 2295–2302.

104. Staniec PA, Parnell AJ, Dunbar ADF, et al. (2011) The nanoscale morphology of a PCDTBT:PCBM photovoltaic blend, *Adv Energy Mater*, **1**, 499–504.

105. Wang T, Scarratt NW, Yi H, et al. (2013) Fabricating high performance, donor-acceptor copolymer solar cells by spray-coating in air, *Adv Energy Mater*, **3**, 505–512.

106. Orimo A, Masuda K, Honda S, et al. (2010) Surface segregation at the aluminum interface of poly(3-hexylthiophene)/fullerene solar cells, *Appl Phys Lett*, **96**, 043305.

107. Mauger SA, Chang L, Friedrich S, et al. (2013) Self-assembly of selective interfaces in organic photovoltaics, *Adv Funct Mater*, **23**, 1935–1946.

108. Wang H, Gomez ED, Kim J, et al. (2011) Device characteristics of bulk-heterojunction polymer solar cells are independent of interfacial segregation of active layers, *Chem Mater*, **23**, 2020–2023.

109. Rivnay J, Mannsfeld SCB, Miller CE, Salleo A, Toney MF (2012) Quantitative determination of organic semiconductor microstructure from the molecular to device scale, *Chem Rev*, **112**, 5488–5519.

110. Ye L, Zhang S, Ma W, et al. (2012) From binary to ternary solvent: Morphology fine-tuning of D/A blends in PDPP3T-based polymer solar cells, *Adv Mater*, **24**, 6335–6341.

Chapter 8

Optical Modeling and Light Management in Organic Photovoltaic Devices

Olle Inganäs, Zheng Tang, Jonas Bergqvist, and Kristofer Tvingstedt
Biomolecular and Organic Electronics, Linköping University,
Center of Organic Electronics, Department of Physics, Chemistry and Biology,
Linköping University, Linköping, Sweden
ois@ifm.liu.se

We review light management for thin film organic solar cells, including modeling of the optical properties of active semiconductors, device modeling with various models, optimization of device geometries, and different avenues for improving the absorption of solar light through light management in such devices.

8.1 Introduction

The design and optimization of organic photovoltaic (OPV) devices requires control over all the different processes of optical absorption, charge generation, and charge extraction and is a daunting task. The first stages in these processes are concerned with optical incoupling and optical absorption and form the topics of this chapter. Clearly, the design of photovoltaic devices requires

Organic Solar Cells: Fundamentals, Devices, and Upscaling
Edited by Barry P. Rand and Henning Richter

ISBN 978-981-4463-65-2 (Hardcover), 978-981-4463-66-9 (eBook)
www.panstanford.com

considerable elements of device geometry optimization for generating large photocurrents, the process in which the incoming photon is converted into an electron in a photovoltaic device. This optimization can be limited to that of finding optimal thicknesses of the different layers building a planar thin film multilayer photovoltaic device, but all the possibilities of forming more complex geometries for enhancing incoupling of light, often subsumed under the term light management, considerably extend these geometries. In this chapter, we attempt to cover the basic elements of electromagnetic modeling of devices, both planar and nonplanar structures, and also describe various ways of modifying geometries for enhanced optical incoupling in the photoactive layer of thin film devices. We describe methods to obtain the necessary data for such modeling, by the use of spectroscopic ellipsometry measurements of individual materials, and modeling of the optical properties of the materials building the device, in the form of dielectric functions within the optical wavelength range (optical materials modeling). The ultimate goal of optical modeling is to allow the optical design of devices that assures maximum photon energy dissipation inside the photoactive layer of the organic solar cell, and that minimizes losses such as reflectance and parasitic absorption in the electrodes of the device. The number of photons dissipating in the active layer hence determines the maximum photocurrent. With an internal quantum efficiency of 100%, the fraction of solar photons dissipating in the active layer equals the external quantum efficiency. Subsequent charge generation and transport, determining the real values of internal quantum efficiency, are here left outside of the treatment, but always all the different conditions for light incoupling and management must be designed taking these issues in mind.

The final device must be designed for high power conversion efficiency, and combining multiple materials with different bandgaps can help to reach this target. Tandem solar cells are therefore also treated from the point of view of optical optimization. Sometimes the request is not that of optimizing photocurrent or power, but rather to extract electrical energy from a semitransparent device, a case that we also describe through modeling and examples.

We exclude the treatment of concentrating optics-feeding photovoltaic devices, as the use of OPV materials in this application is at present rather irrelevant, due to the low mobility in the thin film OPV active layers. This will not make it attractive to enhance the photocurrent density by concentration of light, as ohmic losses will be very severe.

It is rather the possibility of coating large areas with active materials by printing methods that define the technical avenue of OPVs, in nonconcentrating devices. The coating processes are accessible to optical investigation, and we describe the recent development of optical imaging of reflectance during solution processing of OPV materials, a metrology tool for process control of solution-processed OPVs.

Some of these topics have been covered in previous books [1,2], PhD theses [3–8], and papers [9,10].

8.2 Optical Modeling of Materials

The required information for optical simulation of absorption in a thin film multilayer device is the dielectric function of the different materials in all the layers. Some of these materials are standardized and with dielectric functions available in common databases, such as for elemental metals used as cathodes; others are yet unknown and associated with a new material, typically the case for novel polymer : fullerene blends formed from compounds coming out of fresh synthesis. Spectroscopic ellipsometry is the method used to experimentally collect sufficient data to build an optical materials model (Fig. 8.1). This is often done at the early phase of evaluating a new material, to better know the upper limits to photocurrent generation, and to evaluate optimal geometries. For multilayers prepared by vacuum evaporation of organic thin films, the rate of deposition, substrate temperature, and other deposition conditions are parameters of importance for the resulting material, the effects of which can be followed by spectroscopic ellipsometry [11]. For materials deposited from solutions, preparation of thin films of the organic semiconductor blend on a representative surface by spin coating gives the samples. These are often chosen concurrently with device evaluation, as the conditions for forming the thin film (spin coating conditions, solvents, cosolvents, blend stoichiometry, etc.) all contribute in forming the final material. It is desired to form the thin film with no defects, such as particles, spin ghosts (reduced thickness due to particle hindering flow), etc., that make the surface inhomogeneous even to the eye. A flat and uniform film with no defects is the desired object for making spectroscopic ellipsometry measurements at a number of different angles (Fig. 8.1). The ellipsometric angles delta and psi are the outcome of these measurements, for the wavelength range covered, and need to be used in conjunction with the creation of a model for the thin film material. Alternative methods to construct the dielectric function rely on optical transmission and reflection spectra from thin films and devices. This is feasible, but less convenient than the use of ellipsometric data for establishing the dielectric function.

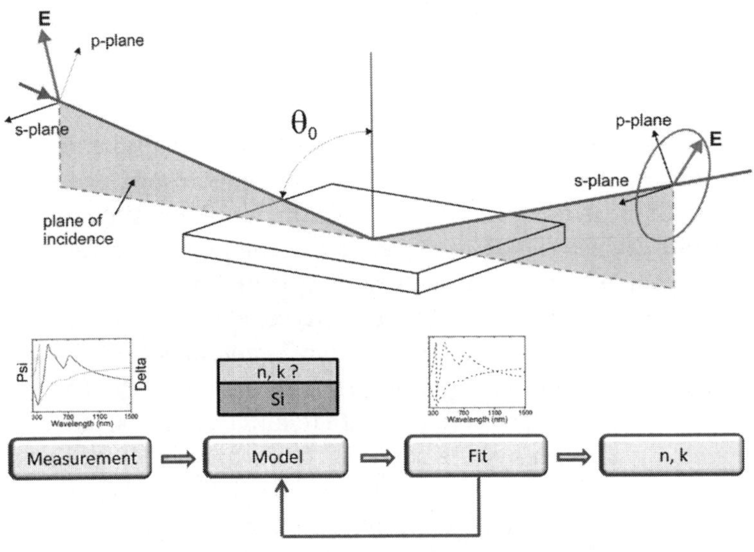

Figure 8.1 Spectroscopic ellipsometry of thin organic films—mode of analysis. An incoming ray of polarized light is reflected from a thin organic film residing on a solid substrate, here Si. The state of polarization of reflected light is measured and represented as two angles, delta and psi. The wavelength and the incoming angle of light are varied in spectroscopic ellipsometry.

The creation of a model starts with measurements on a ~100-nm-thick organic thin film (comparable to the thickness used in OPV devices) on a silicon substrate, as the dielectric function of silicon generally gives a high contrast to that of organics. Ellipsometry measures the change of polarization for a light beam reflected on a sample. The output is generally the two ellipsometric angles psi and delta, which constitute a polar description of the ratio between the reflectance for light for the two polarizations, r_p and r_s. The desired optical property used for modeling is the dielectric function $\varepsilon = \varepsilon_1 + i\varepsilon_2$ (where ε_1 represents the dispersive component and ε_2 represents the absorptive component) and the thickness d of the film. As discussed later in this chapter, accurate thickness determination of organic thin films can be complicated. Consequently there are three unknown parameters in ε_1, ε_2, and the thickness (d) of the thin film, but only two experimental parameters. This is generally overcome by using the fact that organic semiconductors are transparent below the bandgap. This enables modeling the transparent ($\varepsilon_2 = 0$) region with the mathematical Cauchy dispersion model. With a known thickness d and a model of ε_1 in the transparent region, extending the model above the bandgap can be done by using a wavelength-by-wavelength approach, where a numerical inversion of the experimental data is used for determining ε_1 and ε_2 of the full wavelength region measured. This will however include measurement noise, and it is not guaranteed to produce a physically meaningful model of ε_1 and ε_2 because Kramers–Kronig (KK) consistency is not enforced. This is why polynomials such as b-splines [12] can be used to smoothen the spectra. Another approach is to use physical oscillator models to model the dielectric function when the thickness is known. Here critical point [13] or Tauc–Lorentz [14] models are suitable for bandgap materials, which can be fitted to match a previously measured absorption spectrum. An advantage of these oscillator models is the possibility to do accurate bandgap determination. These oscillator models also guarantee a KK consistent model of ε. Representative dielectric functions ($\varepsilon = \varepsilon_1 + i\varepsilon_2$) and optical constants (n, k) for electrodes (metallic, ITO, PEDOT(PSS)) and photovoltaic active materials are shown in Fig. 8.2.

It would be desirable to be able to construct the dielectric function of a blend between two materials by computation alone, starting from known dielectric functions of the materials blended. This is generally possible with the help of mean field models relevant for electromagnetics (Bruggemann, Maxwell–Wagner), where the structural elements in the blend are smaller than the wavelength of light. Our attempts to do this in some blends of alternating polyfluorenes and fullerenes failed, and we found that the Wiener limits which evaluate the validity of this procedure could not be obtained [15].[a] Others have claimed success in modeling material [16–18] but do not show that the Wiener limits are fulfilled, which is a test of self-consistency in the mean field model. As the possibility of varying composition gradients are many, including also pure top and bottom layers in blends, it is often possible to find simulations that fit data, but to establish uniqueness in these solutions is very difficult, and not yet shown. This barrier to simulations of new materials in silico therefore takes us back to the laboratory and the laborious combination of blend stoichiometry, processing, and thickness, in order to take the data that enable the creation of a suitable dielectric function.

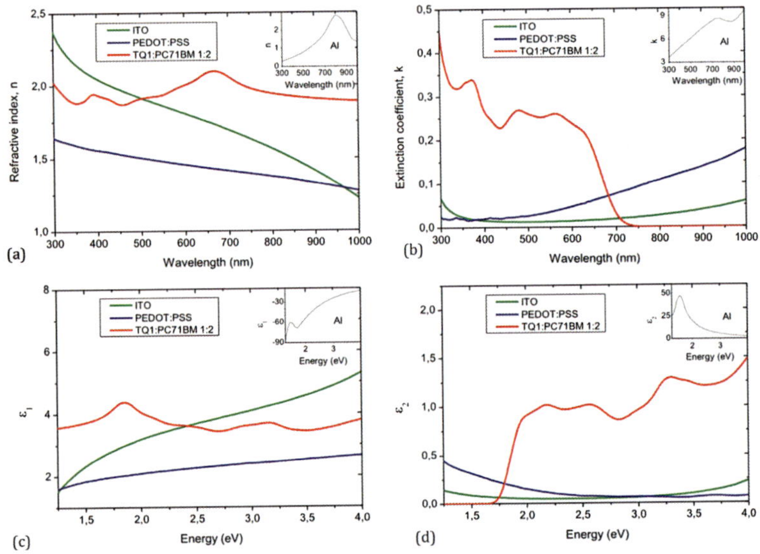

Figure 8.2 Dielectric functions versus energy and optical constants versus wavelength, for ITO, PEDOT:PSS, Al, and TQ1:$PC_{71}BM$ (1:2), typical materials used in OPV devices.

To simulate the dielectric function of a polymer/fullerene blend of course requires knowledge of the contributing elements. As long as these might be domains of pure crystalline polymer, crystalline

[a]The Wiener limits evaluate whether a mean field model can at all be applied to describe the dielectric function of a composite material based on the dielectric function of the contributing phases.

polymer with intercalated fullerenes of different concentration, amorphous polymer with fullerenes in different concentration [19,20], it is rather expected that the simple addition of dielectric functions for pure materials will not be possible. The complex phase diagrams of these blends indicate that a two-phase system will be a rare event, and the number of phases can be large. The fact that the spin-coated film is not in thermodynamic equilibrium, thus enforcing the need for solvent or thermal annealing to reach desirable properties in some blends, also indicates that the dielectric functions of blends reflect the nanostructure in the material. Now that the nanostructure of polymer/fullerene blends can be imaged by electron tomography [21,22], this would be the suitable input for a materials model. However, as electron tomography so far does not allow the identification of the composition of the electron-rich and the electron-poor phases in polymer/fullerene blends giving rise to electron scattering, it would be fruitless to use these geometries as inputs for dielectric materials models. For blends of organic and inorganic metal oxides, the contrast may allow better reconstructions [23] and pave the way for materials modeling. In addition, the creation of a new electronic state in contact between the polymer donor and fullerene acceptor, the charge transfer state [24], is yet another element which contributes (albeit weakly) to the dielectric function in the blend, and which needs to be considered in materials modeling.

The more ambitious project, to be able to predict the dielectric function for a polymer from quantum chemical calculations, was tested on some low bandgap alternating polyfluorenes, but useful predictions could only be obtained with semiempirical methods [25]; Density functional theory predictions were inadequate to predict optical properties. However, recent progress in quantum chemistry has led to a better situation, and evaluation of low-bandgap polymers using quantum chemistry methods has seen advances [26,27]. The very ambitious project to identify optical transitions in organic heterojunctions by quantum chemistry is still in an early phase.

8.3 Optical Modeling of Devices

The device geometry in OPVs is typically a multilayer stack consisting of thin layers of electrodes, absorber/photocurrent generator layers, buffer layers, p-type contact, or n-type contact layers to create selective contacts for charge extraction, exciton blocking layers, etc., deposited on a common transparent substrate in a sandwich fashion. The layer stack may include several different absorber layers, for tandem solar cells, and in this case we will also find current recombination layers for matching photocurrents from the different absorber layers. The electrodes may be transparent, semitransparent, or totally reflecting. The substrate can easily be 1000 times thicker than the total multilayer structure. The thin film layers can be 5–100 nm thick, individually, and in a number that in a tandem solar cell easily reaches 6–8. Our choice of device geometries include standard (substrate/transparent anode/active layer/reflecting cathode), inverted (substrate/transparent cathode/active layer/reflecting anode), reversed (substrate/metallic cathode/active layer/transparent anode), and semitransparent (substrate/transparent cathode/active layer/transparent anode) device stacks as illustrated in Fig. 8.3.

For each of these layers, the appropriate dielectric function is required as input for calculations of photon dissipation in the different layers of the device. The calculations can take many forms, and among the methods used we find the transfer matrix method (TMM) [1,2,28,29] (also named scattering matrix method), the rigorous coupled wave analysis method [30], the transmission line model [31], and numerical methods based on finite element method (FEM) simulations [32–34]. In some cases both thin and thick film optics are found, and the combined treatment of coherent and incoherent light is necessary [35,36].

Among these, we will focus mainly on TMM and FEM methods, where the TMM is well established in the field, extensively used, and also implemented in standard simulation programs for photovoltaic devices.[b]

8.3.1 The Transfer Matrix Method

The TMM is well established for analysis of multilayer optics [37] and the relevance of the method to OPV devices was first demonstrated by Pettersson [28,38]. The importance of optical input had already

[b]A commercial software SETFOS (http://www.fluxim.com/) incorporates the transfer matrix model.

been appreciated [39], but not with the full electromagnetic treatment. Subsequent work on optical modeling of devices [40,41] followed the same lines, which we will describe here.

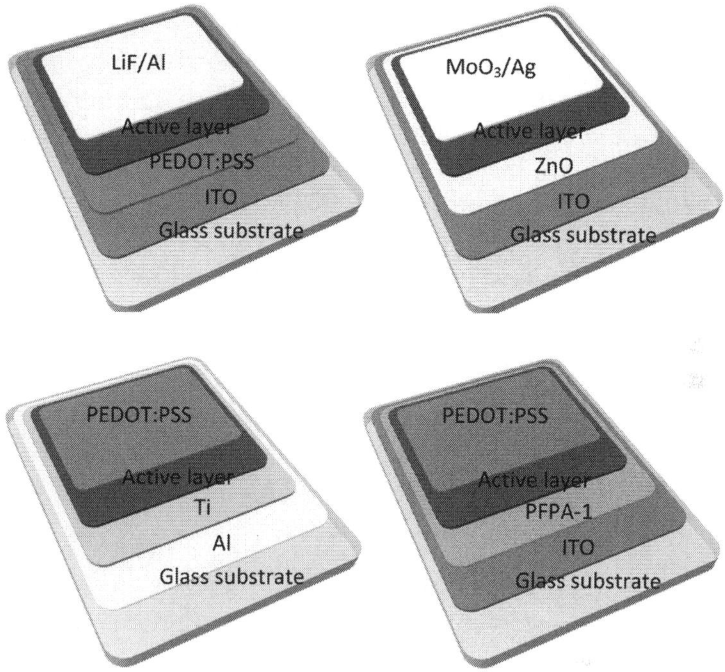

Figure 8.3 Geometry of OPV devices: the standard with a metal cathode deposited by evaporation on top of the active layer, which is contacted from below with a transparent electrode; the inverted device with a modified ITO as transparent bottom cathode and a high-work-function metal electrode as top anode; the reversed with a metal cathode as substrate, the active layer deposited on top of this and a transparent electrode deposited on top; and the semitransparent geometry with two transparent electrodes on top and on bottom of the active layer. The choice of metals, transparent electrodes, and buffer layers may vary considerably; here we present materials combinations relevant for the simulations later presented.

We consider a one-dimensional model for our multilayer devices, where sharp boundaries separate the different layers with different dielectric functions (Fig. 8.4). We assume the layers to be optically flat, to show linear optical behavior, and consider incoming radiation. Several of the active and passive materials in OPV show anisotropy, which has to be included for analyzing nonnormal incidence. It would be easy to assume normal incoming radiation, as this will remove the necessity to handle anisotropic properties, as long as all the layers are isotropic or uniaxial anisotropic. But this will also exclude a number of well-established strategies to enhance incoupling of light to the active layer, based on patterning of the layers. It also removes the influence of wave-guiding, an important aspect of OPVs, which we will return to. We assume coherence of the radiation throughout the multilayer stack, but take care to allow also thick substrates and intermediate layers, where the assumption of coherence does not hold, and where simple addition of electromagnetic amplitudes is not possible.

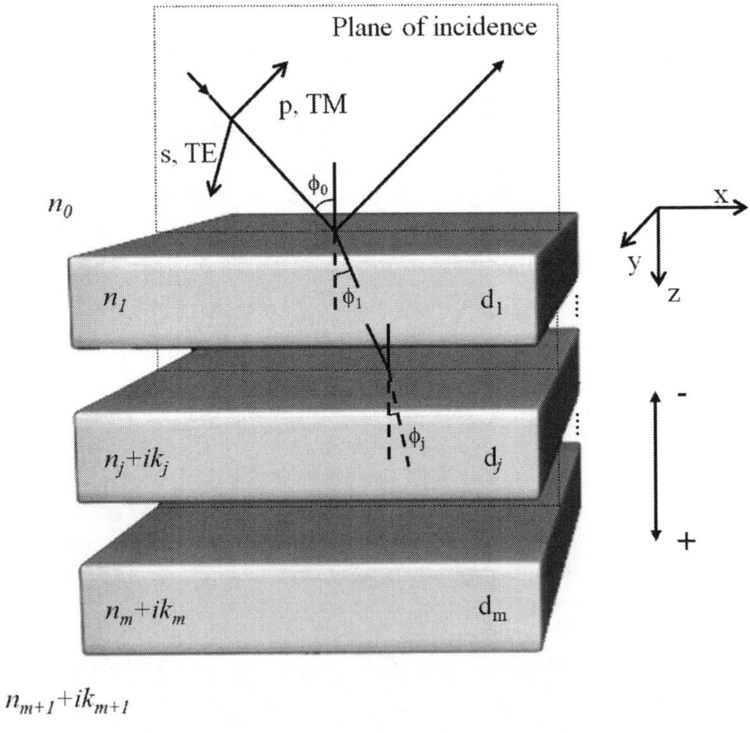

Figure 8.4 Sketch of electromagnetic analysis mode.

We will first consider the case of a monochromatic wave impinging on the stack, and later extend to polychromatic and solar light illumination. We start by decomposing the optic electrical field E into an incident (E^+) and a reflected (E^-) component:

$$E = E^+ + E^- \quad \text{for } s \text{ or } p \tag{8.1}$$

For a flat isotropic, or uniaxial anisotropic, layer with optical axis normal to the plane, E, E^+, and E^- do not depend on x and y. Therefore, we write

$$E(z) = E^+(z) + E^-(z), \tag{8.2}$$

one equation for each polarization state, s and p. The indexation with s and p will be omitted in most cases beneath, where normal incidence is applied, but is of importance for non-normal incidence. We relate two nearby points, z_1 and z_2 by assuming linearity

$$\begin{aligned} E^+(z_1) &= M_{11}E^+(z_2) + M_{12}E^-(z_2) \\ E^-(z_1) &= M_{21}E^+(z_2) + M_{22}E^-(z_2) \end{aligned} \tag{8.3}$$

Introducing the generalized field vectors $(E^+, E^-)_z^T$ and the scattering matrix, M, consisting of the numbers M_{ij}, $i, j = 1, 2$, we compound these equations into a single vector–matrix relation:

$$\begin{pmatrix} E^+ \\ E^- \end{pmatrix}_{z_1} = M \begin{pmatrix} E^+ \\ E^- \end{pmatrix}_{z_2}$$

If z_1 and z_2 are found in the same layer, and no interfaces are found in between we write

$$\begin{pmatrix} E^+ \\ E^- \end{pmatrix}_{z_1} = L \begin{pmatrix} E^+ \\ E^- \end{pmatrix}_{z_2}$$

but if z_1 and z_2 are on opposite sides of an interface, with z_1 inside one and z_2 inside the next layer we have

$$\begin{pmatrix} E^+ \\ E^- \end{pmatrix}_{z_1} = I \begin{pmatrix} E^+ \\ E^- \end{pmatrix}_{z_2}$$

L is denoted *layer matrix* and I *interface matrix*. For layer j we write L_j and for the interface between layer i and j, I_{ij}. Thus, for two points z_0 in layer 0 and z_{m+1} in layer m+1,

$$M = I_{01}L_1I_{12}L_2I_{23}L_3 \cdots L_m I_{m,m+1} = \left(\prod_{v=1}^{m} I_{v-1,v}L_v \right) I_{m,m+1} \tag{8.4}$$

$(E^+,E^-)_0^T$ is the same as $(E_{in},E_{refl})^T$ where in indicates incoming and refl the reflected field amplitude, and for $(E^+,E^-)_{m+1}^T$ we put $(E_{trans},0)^T$.

The Fresnel complex transmission and reflection coefficients then relate the amplitudes as

$$t = \frac{E_{m+1}^+}{E_0^+} \tag{8.5}$$

$$r = \frac{E_0^-}{E_0^+} \tag{8.6}$$

Applying Eq. (8.3) $\begin{cases} E_0^+ = M_{11}E_{m+1}^+ + 0 \\ E_0^- = M_{21}E_{m+1}^+ + 0 \end{cases}$

gives

$$\begin{cases} t = \dfrac{1}{M_{11}} \\ r = \dfrac{M_{21}}{M_{11}} \end{cases} \tag{8.7}$$

It is now possible to determine I_{ij}. The relation $\begin{pmatrix} E^+ \\ E^- \end{pmatrix}_i = \begin{pmatrix} I_{11} & I_{12} \\ I_{21} & I_{22} \end{pmatrix} \begin{pmatrix} E^+ \\ E^- \end{pmatrix}_j$ is true for all $(E^+, E^-)^T$ in particular for the situation in Fig. 8.5b: $E_j^+ = t_{ij}E_i^+, E_j^- = 0$ and $E_i^- = r_{ij}E_i^+$ which gives I_{11} and I_{21} and the situation in Fig. 8.5c: $E_j^+ = r_{ji}E_j^-, E_i^+ = 0$ and $E_i^- = t_{ji}E_j^-$. Fresnel formulae give $r_{ij} = -r_{ji}$ and $t_{ij} = \frac{1-r_{ji}^2}{t_{ji}}$ from which it is possible to derive the complete I_{ij} matrix:

$$I_{ij} = \frac{1}{t_{ij}} \begin{pmatrix} 1 & r_{ij} \\ r_{ij} & 1 \end{pmatrix} \tag{8.8}$$

I_{ij} can be expressed in the indices of refraction and angle of incidence, via Fresnel formulae for r_{ij} and t_{ij}.

We now turn to L_j.

Passing through layer (see Fig. 8.5d) j of thickness d and index of refraction n and angle of refraction φ, we find the phase difference β accumulated as the wave traverses layer j

$$\beta_j = \frac{2\pi d}{\lambda} n_j \cos\phi_1 = \xi_j d$$

where $\xi_j = \frac{2\pi}{\lambda} n_j \cos\phi_1$. Note that both n_j and φ_j are complex quantities, in general.

Then $E^+_{d=0} = e^{i\beta} E^+_d$ and $E^-_{d=0} = e^{-i\beta} E^-_d$ which gives

$$L_j = \begin{pmatrix} e^{i\beta_j} & 0 \\ 0 & e^{-i\beta_j} \end{pmatrix} \tag{8.9}$$

With L_j and I_{ij} we can analyze the E-field in layer j.

The environment around layer j is partitioned into incident system indicated with ′, and reflected system indicated with ″ (Fig. 8.5e),

$$M = M'_j \, L_j M''_j$$

where

$$M'_j = \left(\prod_{\nu=1}^{j-1} I_{\nu-1,\nu} L_\nu\right) I_{j-1,j}$$

$$M''_j = \left(\prod_{\nu=j+1}^{m} I_{\nu-1,\nu} L_\nu\right) I_{m,m+1}$$

The incident system and reflected system have their own reflection and transmission coefficients:

$$t'_j = \frac{1}{M'_{j,11}} \quad t''_j = \frac{1}{M''_{j,11}}$$
$$r'_j = \frac{M'_{j,21}}{M'_{j,11}} \quad r''_j = \frac{M''_{j,21}}{M''_{j,11}} \tag{8.10}$$

derived in the same manner as Eq. (8.7).

We introduce transfer coefficients that couple between layer j and the incoming light.

$$t^+_j = \frac{E^+_j}{E^+_0} \tag{8.11}$$

$$t^-_j = \frac{E^-_j}{E^+_0} \tag{8.12}$$

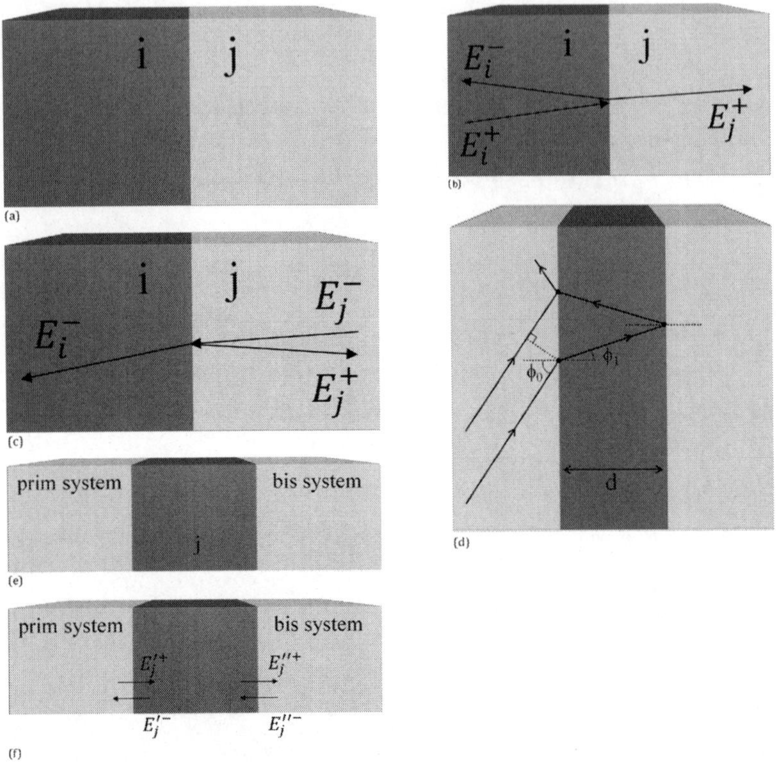

Figure 8.5. Details of electromagnetic terminology. (a) Derivation of the *I*-matrix, *i* and *j* are two adjacent layers in the stack. (b) A ray from the left, which gives reflection and refraction. (c) A ray from the right, which is reflected and refracted. (d) Deriving β. A ray passes through a medium of thickness *d*. (e) A certain layer *j* in the stack. (f) Output and input to layer *j* in the form of electrical field amplitudes. The arrows indicate the direction of propagation, not the field amplitudes.

With these transfer coefficients, closed expressions can be derived. For the incident system

$$\begin{pmatrix} E_0^+ \\ E_0^- \end{pmatrix} = \begin{pmatrix} M'_{j,11} & M'_{j,12} \\ M'_{j,21} & M'_{j,22} \end{pmatrix} \begin{pmatrix} E_j'^+ \\ E_j'^- \end{pmatrix}$$

which gives

$$E_0^+ = M'_{j,11} E_j'^+ + M'_{j,12} E_j'^- \tag{8.13}$$

The reflected system

$$\begin{pmatrix} E_j''^{+} \\ E_j''^{-} \end{pmatrix} = \begin{pmatrix} M_{j,11}'' & M_{j,12}'' \\ M_{j,21}'' & M_{j,22}'' \end{pmatrix} \begin{pmatrix} E_{m+1}'^{+} \\ 0 \end{pmatrix}$$

contains related elements such as $r_j'' = \dfrac{M_{j,21}''}{M_{j,11}''} = \dfrac{E_j''^{-}}{E_j''^{+}}$.

Applying Eq. (8.9)

gives $\begin{pmatrix} E_j'^{+} \\ E_j'^{-} \end{pmatrix} = \begin{pmatrix} e^{i\beta_j} & 0 \\ 0 & e^{-i\beta_j} \end{pmatrix} \begin{pmatrix} E_j''^{+} \\ E_j''^{-} \end{pmatrix}$, i.e.,

$$r_j'' = \frac{E_j''^{-}}{E_j''^{+}} = \frac{E_j'^{-} j}{E_j'^{+}} \frac{e^{i\beta_j}}{e^{-i\beta j}} = r_j' e^{i2\beta_j}.$$

Divide Eq. (8.13) with $E_j'^{+}$ and introduce $\rho_j' = -\dfrac{M_{j,12}'}{M_{j,11}'}$. Then

$$t_j^{+} = \frac{t_j'}{1 - \rho_j' r_j'' e^{i2\beta_j}} \tag{8.14}$$

and by dividing Eq. (8.13) with $E_j'^{-}$ it is easy to show that

$$t_j^{-} = t_j^{+} r_j'' e^{i2\beta_j} \tag{8.15}$$

We can now express the E-field inside layer j. From $E_j(z) = E_j^{+}(z) + E_j^{-}(z)$ we use Eqs. (8.11), (8.12), (8.14), and (8.15) and finally collecting all coefficients in terms of matrix elements,

$$E_j(z) = \frac{M_{j,11}'' e^{i\xi_j(z-d_j)} + M_{j,21}'' e^{i\xi_j(d_j - z)}}{M_{j,11}' M_{j,11}'' e^{-i\xi_j d_j} + M_{j,12}' M_{j,11}'' e^{i\xi_j d_j}} E_0^{+} \tag{8.16}$$

Here z belongs to layer j.

The Poynting theorem (1,18) can be interpreted as a statement of conservation of energy,

$$Q(z) = \langle -\nabla \cdot \boldsymbol{S} \rangle \tag{8.17}$$

where Q is the (time average) energy flow dissipation per time unit at the point z, $\langle\,\rangle$ indicates time average, and $\boldsymbol{S}$ is the Poynting vector,

$\boldsymbol{S} = \boldsymbol{E} \times \boldsymbol{H}$

Because time-averaging of a product of $\tilde{\boldsymbol{A}} = \boldsymbol{A}_0 e^{i(kx-wt)}$ and $\tilde{\boldsymbol{B}} = \boldsymbol{B}_0 e^{i(kx-wt)}$ can (19) be performed as $\frac{1}{2}\mathrm{Re}(\tilde{\boldsymbol{A}}\tilde{\boldsymbol{B}}^*)$ and

$-\nabla \cdot \boldsymbol{S} = \left(\boldsymbol{H}\frac{\partial \boldsymbol{B}}{\partial t} + \boldsymbol{E}\frac{\partial \boldsymbol{D}}{\partial t}\right)$ using Maxwell equations, one can show that

$$Q(z) = \frac{1}{2} c\varepsilon_0 \alpha n \,|\, \boldsymbol{E}(z) \,|^2 \tag{8.18}$$

Q has the unit $W/(m^2 \cdot nm)$ where it is indicated that z, the distance, is measured in nm. The speed of light is $c = 3.00 \times 10^8$ m/s, ε_0 is permittivity of vacuum, 8.85×10^{-12} F/m, n is the real index of refraction, α is the absorption coefficient, $\alpha = 4\pi k / \lambda$ with $\tilde{n} = n + ik$, and λ is the vacuum wavelength; and $\boldsymbol{E}(z)$ is the total electrical optical field at the point z. The factor 1/2 is due to averaging of the rapid frequency variation, ca 10^{14} Hz, in the optical field.

Q is a discontinuous function of z, and Q at the points corresponding to the interfaces is not well defined.

We will use the integrated dissipation Q from an interval of z,

$$\int_{z \in \text{interval}} Q(z)dz$$

when generating predictions or comparing our calculations to experiments.

For a given layer, j, in the stack, the absorptance A_j is related to Q as

$$A_j = \frac{1}{S_0} \int_{z \in \text{layer } j} Q(z)dz \tag{8.19}$$

where S_0 is the irradiance from air for a given wavelength. These As will be indexed as $A_{\text{active layer}}$ and A_{contacts} as explained later.

$\boldsymbol{E}(z)$ is dependent on the incoming field to the stack. This is affected by the substrate, provided incident light passes through the substrate before coupling into the active layer. There is a loss due to reflection at the air–substrate interface as well as multiple reflection at the substrate–air interface (Fig. 8.6). Reflection occurs in the air–glass interface with some of the light reflected and some transmitted to the stack. At the stack, incoming light is partially reflected or transmitted. The reflected part goes back toward the glass–air interface where some parts exit the substrate and reenter into air, adding with the light just one time reflected, while some is reflected back to the stack.

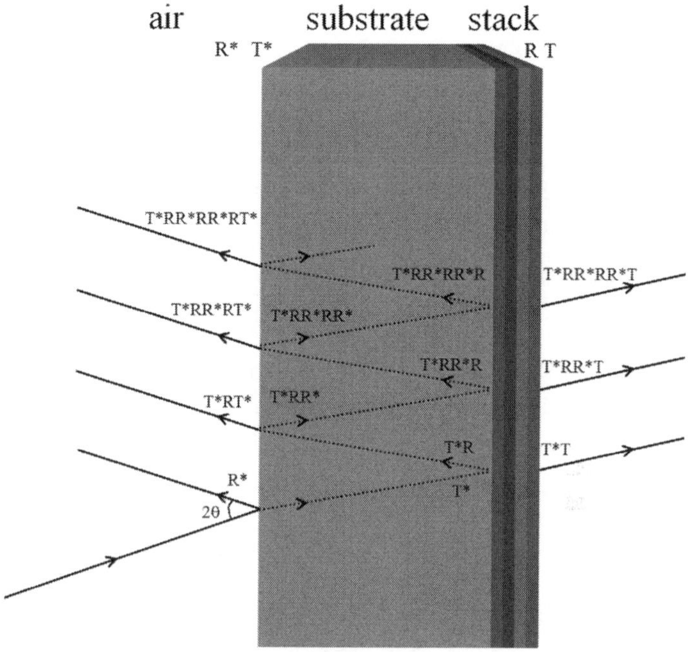

Figure 8.6 Optics in the structure. *R* are reflectances and *T* transmittances. Star (*)-marked quantities refer to the glass substrate, unmarked to the stack. The derivation was done assuming a finite *T* although it is later assumed that *T* is zero due to the thick Al layer. The angle of incidence is θ. Normal incidence is used throughout the study.

Using the Poynting vector the irradiance in air, B, is

$$B = \frac{1}{2} c\varepsilon_0 \,|\, \boldsymbol{E}_0 \,|^2$$

For the optical description of a photovoltaic device we divide the device into two optical elements; first, the optical stack consisting of the multilayer of thin films; transparent electrodes, organic semiconductor blend, buffer layers, metallic electrodes; second, the substrate. Typically, the thicknesses for the stack layers are much less than the irradiated wavelength ($\approx$0.5 μm), and the thickness of the substrate is of the order of millimeters. This makes it necessary to add light coherently inside the stack and incoherently, using irradiances, when including the substrate.

We let the air–substrate interface be characterized by a reflectance R^* (Fig. 8.6). The substrate has a transmittance, T^*. The reflectance and transmittance for the stack are R and T, respectively. All these quantities can be calculated from the complex indices of refraction of the materials and the different layer thicknesses, through Fresnel equations. For normal incidence from a material i to j, the reflectance is

$$reflectance = \frac{(n_j - n_i)^2 + (k_j - k_i)^2}{(n_j + n_i)^2 + (k_j + k_i)^2} \tag{8.20}$$

The total transmittance for the whole structure (Fig. 8.6) is calculated by adding energy (irradiance) quantities. Then,

$$T_{\text{tot}} = T^* \left(\sum_{i=0}^{\infty} (RR^*)^i \right) T = T^* \frac{1}{1 - RR^*} T = \frac{T^* T}{1 - RR^*}$$

For the total reflectance,

$$R_{\text{tot}} = R^* + T^* \left(\sum_{i=0}^{\infty} (R^* R)^i \right) RT^* = R^* + T^* \frac{1}{1 - RR^*} RT^*.$$

With no absorption in the substrate, i.e., $1 = R^* + T^*$, we obtain

$$T_{\text{tot}} = \frac{(1 - R^*)\mathrm{T}}{1 - RR^*} \tag{8.21}$$

and

$$R_{\text{tot}} = R^* + \frac{(1 - R^*)(1 - R^*)}{1 - RR^*} R = \frac{R^* + R - 2R^* R}{1 - RR^*} \tag{8.22}$$

The irradiance to the optical stack is

$$T^* \left(\sum_{i=0}^{\infty} (R^* R)^i \right) B = T^* \frac{1}{1 - RR^*} B = \frac{1 - R^*}{1 - RR^*} B$$

Writing this as $\frac{1}{2} c\varepsilon_0 n_g \left|\boldsymbol{E}_{o,g}\right|^2$, where the subscript g indicates a quantity in the substrate, reaches

$$\left|\boldsymbol{E}_{o,g}\right|^2 = \frac{2(1 - R^*)B}{\varepsilon_0 c n_g (1 - RR^*)}$$

This is the square modulus amplitude that has to be used as input to the Q calculation. Compared to a stack in air the expression is modified with the factor $\frac{(1-R^*)}{n_g(1-RR^*)}$.

Under the assumption of zero stack transmittance, fully transmitting substrate, and no scattering effects,

$$1 = R_{\text{tot}} + \sum_j A_j \tag{8.23}$$

8.3.2 Optical Electrical Field Distribution from Simulations

The possibility to calculate the distribution of the electrical field from the incoming light is essential, as this information cannot be obtained with experimental methods. By solving the matrix problem to give this information, we can calculate the profile of the electrical field throughout the device stack, for any incident wavelength for which we know the dielectric function. In Fig. 8.7, we see such an example, where continuity of the E-field at interfaces is visible, and we note that the thickness of the device is comparable to the spatial variation of the time-independent field.

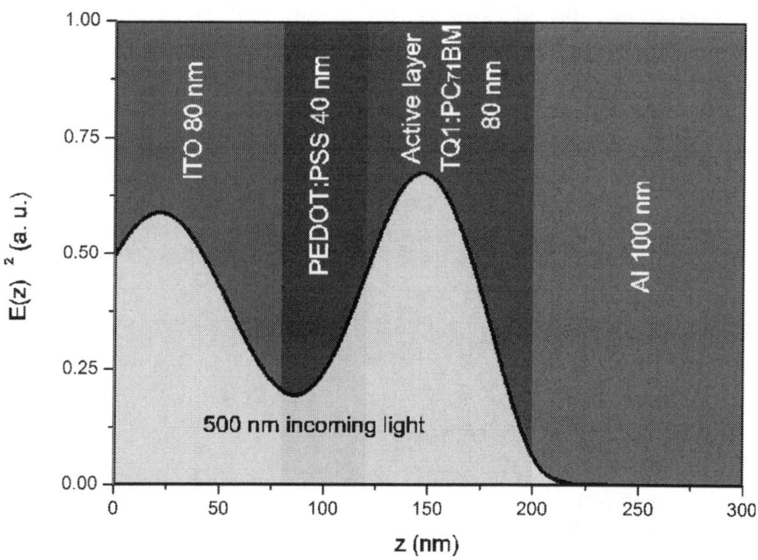

Figure 8.7 Spatial field distribution for a standard solar cell based on TQ1:PC_{71}BM.

8.3.2.1 Light Redistribution in Standard and Nonstandard Geometries

The optical model now described can be used to illustrate the distribution of the optical power dissipation, for the different layers as well as for different wavelengths [29]. For each wavelength and device geometry, the incoming photon flux can be distributed into reflection, transmission, absorption in active layer, and absorption in electrodes, where electrodes may also include buffer layers of various kinds. As absorptances are additive, the device absorptance $A_{tot} = \Sigma_j A_j$ and we single out absorptance in the active layer A_a.

It is of importance to be able to predict the wavelength-dependent incoupling efficiency, that is, the fraction of photons absorbed in that layer in the device.

An illustrative demonstration of this is found in Fig. 8.8 where a polymer blend of the alternating donor–acceptor polymer TQ1 [42] and $PCBM_{71}$ is used in contact with ITO/PEDOT(PSS) as anode and Al as cathode. The dielectric function for the blend (Fig. 8.2a), and those for contacts, and the chosen geometry of the device (Fig. 8.3) can be combined to calculate the power dissipation at different wavelengths across the solar spectrum. The absorption of the polymer blend extends to 750 nm, and then reflection and dissipation in the electrodes accounts for all of the incoming energy flux. This calculation is done for one specific geometry, but we can easily vary the thicknesses within the model, to predict optical power dissipation in varying geometries.

8.3.2.2 Quantum Efficiencies of Charge Generation

The external quantum efficiency η_{EQE} is experimentally determined, and is defined as the number of electrons generated in the photovoltaic device (often at short circuit) divided by the number of incoming photons to the structure. It is measured for monochromatic light.

This function is related to η_{IQE}, the internal quantum efficiency, a quantity cleared from reflection:

$$\eta_{IQE} = \text{(number of electrons at short circuit)/(number of photons absorbed in active layer)} \tag{8.24}$$

Not all photons are absorbed and therefore $\eta_{IQE} \geq \eta_{EQE}$. For nontransmissive devices $T_{tot} = 0$, we obtain $1 = R_{tot} + A_{tot}$ and using additivity of A one gets $1 = R_{tot} + A_w + A_a$, where A_w is the absorptance in parts of the device that do not generate photocurrent (electrodes, buffer layers).

Thus,

$$\eta_{IQE} = \frac{\eta_{EQE}}{A_a} \tag{8.25}$$

For a transmissive device $T_{tot} > 0$, we obtain $1 = R_{tot} + A_{tot} + T_{tot}$. For devices made from multiple-absorbing and charge-generating layers, such as in tandem cells and in bilayer cells, efficiencies of charge generation can be defined with respect to these multiple layers in the layer-wise internal quantum efficiency [29].

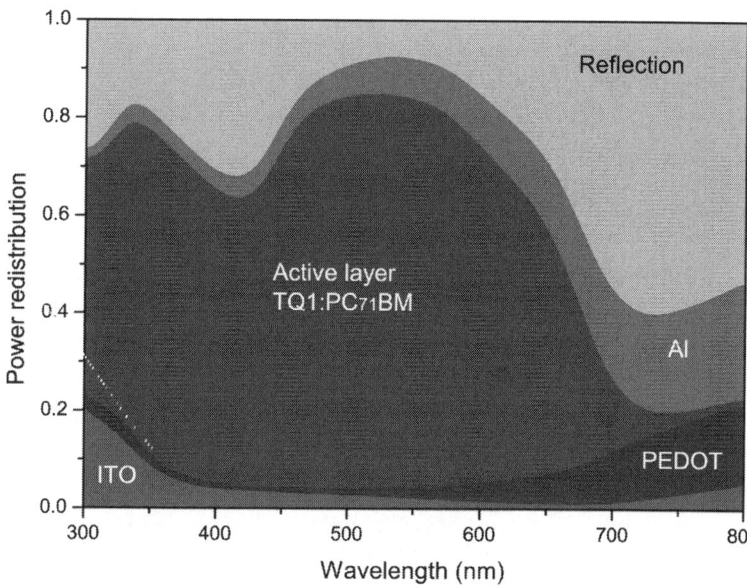

Figure 8.8 Optical power dissipation profile (redistribution plot) for a standard solar cell (ITO 80 nm/PEDOT:PSS 40 nm/TQ1:$PC_{71}BM$ 80 nm/Al 100 nm).

Determination of absorption in the active layer can be challenging since it cannot be independently measured, due to optical interference. Therefore, TMM is often employed in determining the absorption of different layers in the solar cell, as well as reflectance and transmission through the device.

This mode of calculation needs to predict both reflectance and transmission of the device, which can be directly compared to experiments. This is one of the points of confrontation between theory and experiments, and success is not always there, as noted in early studies [15]. Alternative routes for combining materials and device data have recently been developed, and improve upon previous modeling efforts. TMM modeling results for semitransparent solar cells are presented in Fig. 8.9.

There are in principle two routes available to determine absolute values of internal quantum efficiency for thin film OPVs, depending on what set of measurement tools are available and mostly trusted. The first one is based on TMM modeling to obtain the modeled absorptance in all different cell layers and then divide the measured EQE with the modeled active layer absorptance.

$$\eta_{\text{IQE}}(V,d,\lambda)=\frac{\eta_{\text{EQE}}(V,d,\lambda)}{A_{\text{a}}} \tag{8.26}$$

Values for dielectric functions of new materials are likely only available via ellipsometry. It is very important to have correct values of n and k for the active layer as well as for the employed electrodes, to get trusted values of A_{a}. Accurate knowledge of thickness of all layers is further required. Small errors in thickness will affect the modeled photon dissipation in the different layers. As both ellipsometry and thickness determination of solution-processed bulk heterojunction films can often introduce error, deviations can often persist when comparing calculational and experimental data.

The second route relies on measured cell reflectance (R) (transmittance (T)) and less on ellipsometry. Here one measures R (and T if present) and then subtract only the modeled parasitic absorption in the electrodes. The rest must be active layer absorptance. Again then divide measured EQE with this "*measured and modeled*" active layer absorptance.

$$\eta_{\text{IQE}}(V,d,\lambda)=\frac{\eta_{\text{EQE}}(V,d,\lambda)}{1-(R+T+A_{\text{w}})} \tag{8.27}$$

The benefit of this second approach is that very accurate values of n and k for the active layer blend are not as important, particularly if thick active layers are used. Only using Beer-Lambert's law for

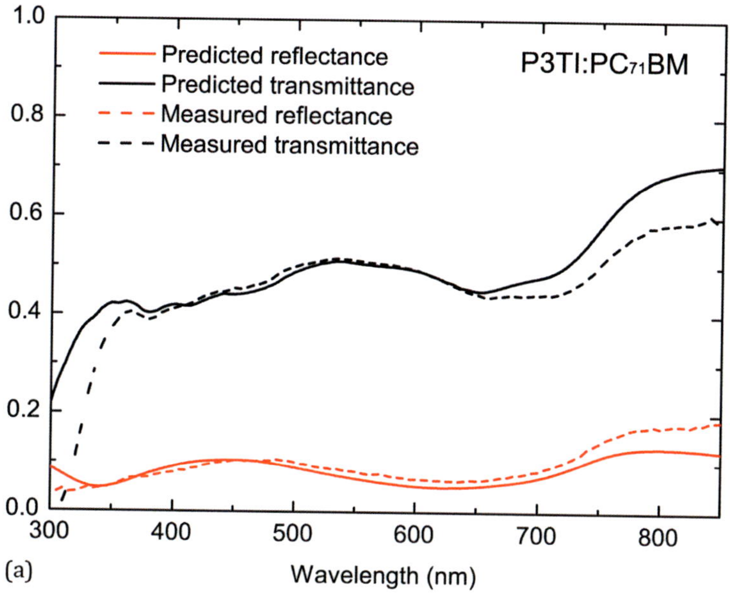

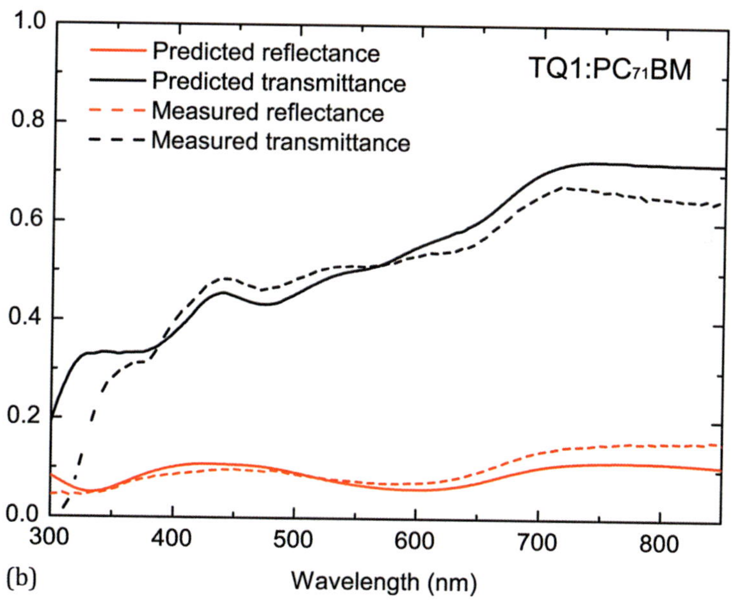

Figure 8.9 Predicted reflectance and transmittance compared with experiments for semitransparent solar cells based on (a) P3TI:$PC_{71}BM$ and (b) TQ1:$PC_{71}BM$ (ITO 80 nm/PEDOT:PSS 40 nm/active layer 80 nm/Al 100 nm).

an active layer film on glass substrate, with measured absorption to determine extinction coefficient k in combination with an *assumed* refractive index n of a set value around 2, still gives very high accuracy in the model [43]. Values of n and k for the electrode materials are still required. These are to some extent available in literature and databases and a laboratory with a decent reflectance accessory tool can in principle determine the internal quantum efficiency this way.

Combinations of the two above approaches are however recommended for the most accurate IQE determination. Here one includes both measured device reflectance $R(\lambda)$ (and T if present) and full ellipsometric data for all layers. The best ways to determine whether the TMM model is good or bad is to add the TMM modeled value of complete cell absorptance (all layers) to the measured absolute cell reflectance (R). In a nontransparent solar cell these values should add to 1 (at all wavelengths). Frequently, however, the first trial modeled reflectance (1-modeled cell dissipation) does not accurately follow the spectral shape of the measured cell reflectance. Then, rely on the device reflectance measurements, and correct the values used as inset in the TMM model, either thickness or dielectric functions. As thickness measurements for spin-coated bulk heterojunction films are not very accurate, adapting the active layer thickness in the model is very justified, simple and the first thing that should be tried, as long as it does not diverge too much from measured values. If no fits are obtained, new dielectric functions might need to be measured and implemented. If good fits are obtained between modeled and measured device reflectance, a fairly successful IQE determination is expected.

We have used the second method described above to calculate the internal quantum efficiency in semitransparent solar cells based on two low bandgap materials (Fig. 8.10). The internal quantum efficiency obtained from either TQ1:$PC_{71}BM$ or P3TI:$PC_{71}BM$ solar cell is fairly constant over the absorbing wavelength region of the polymer, indicating a photon energy independent charge carrier generation and transport in the device. The internal quantum efficiency for long wavelengths, where the active layer does not strongly absorb, is less accurate. When the active layer absorption approaches zero, determination of internal quantum efficiency becomes impossible since absorption appears in the denominator. Accurate determination of IQE could only be performed for the wavelengths where the active layer strongly absorbs.

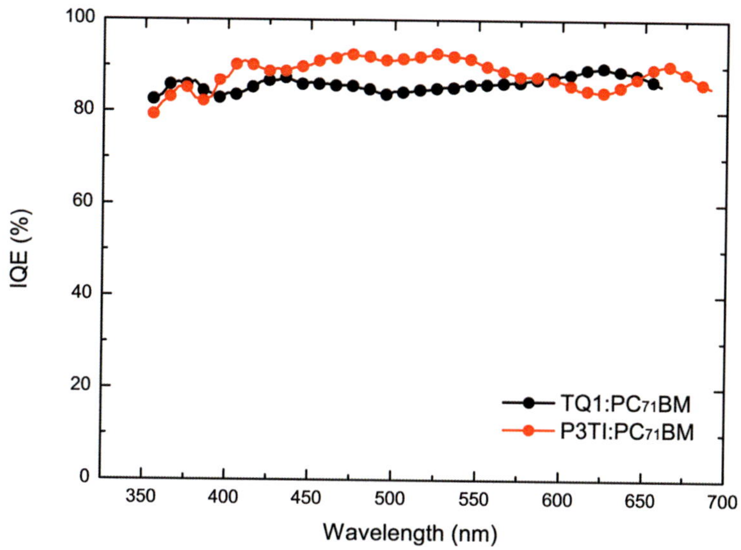

Figure 8.10 Internal quantum efficiency in semitransparent solar cells based on P3TI:$PC_{71}BM$ and TQ1:$PC_{71}BM$ from optical modeling (modified ITO 80 nm/active layer 80 nm/PEDOT:PSS 110 nm).

8.3.2.3 Spatial Distribution of Power Dissipation

The one-dimensional model also allows predictions of quantities not accessible to experiments, such as the internal distribution of optical power dissipation. We use the z-dependence of the power dissipation to demonstrate this in Fig. 8.11, where the same material and device is used. Plotted is the power dissipation profile within the active layer for three different wavelengths: above, at the edge, and below the bandgap. We do not observe an exponential decrease (Beer–Lambert-like) of power density in these thin layers where interference is strong; reflection in the metal electrode leads to interference and local extrema of power dissipation. Where the active material has almost no absorption, we only note dissipation in the electrodes. It is also clear that the power dissipation is a noncontinuous function, with steps at interfaces, due to different dielectric properties of the layers in the solar cell stack.

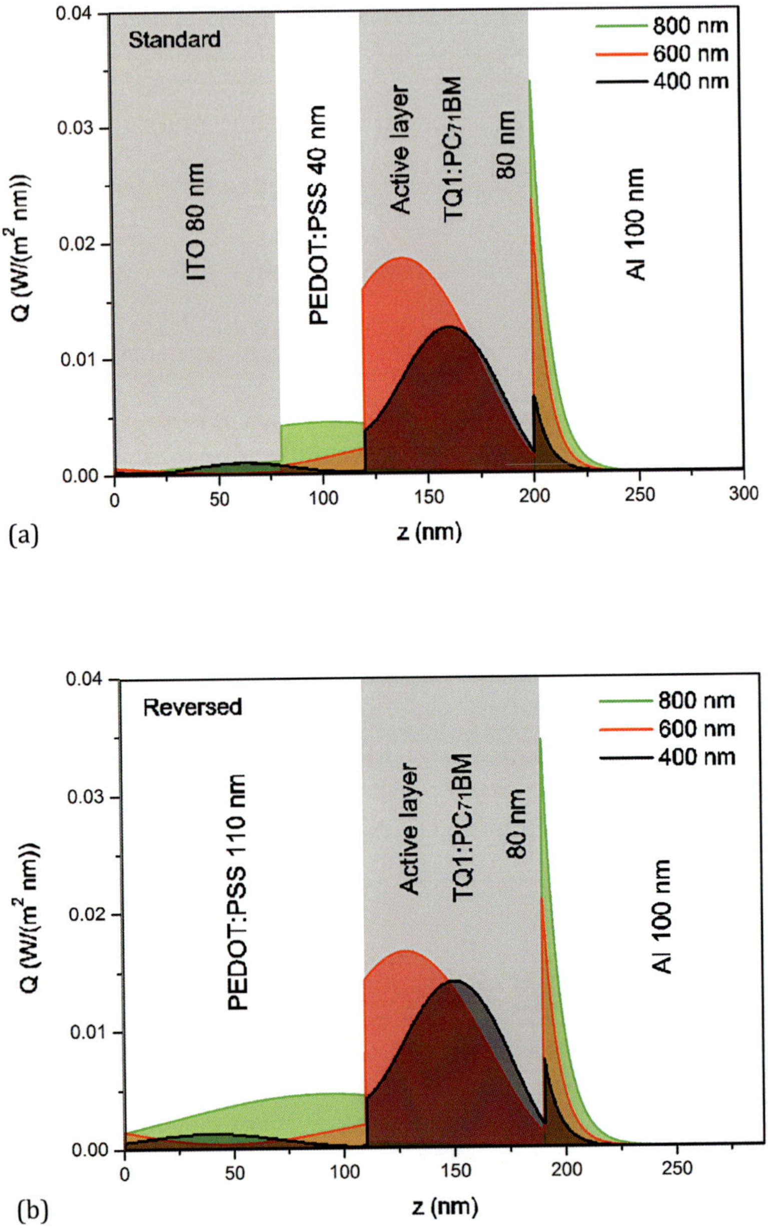

Figure 8.11 (Continued).

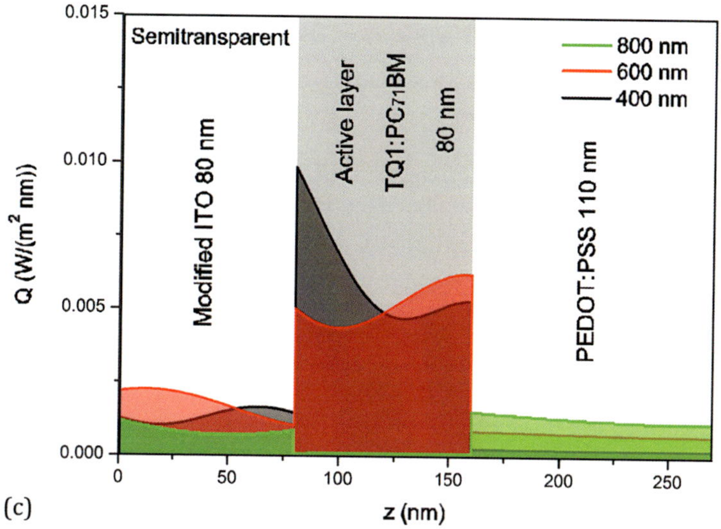

(c)

Figure 8.11 Spatial distribution of power dissipation for incoming light of wavelengths 400, 600, 800 nm for (a) standard, (b) reversed, (c) semitransparent solar cells based on TQ1:PC_{71}BM.

When varying the geometry of the device, we can improve on the optical absorptance in the active layer. Increasing thickness of the active layer causes more integrated dissipation in this layer, and thus increasing potential photocurrent (Fig. 8.12). As evident in Fig. 8.12 there might appear local maxima of power dissipation due to the interference, a powerful tool for studies of charge generation and transport.

8.3.3 Integrated Power Dissipation for Polychromatic Illumination

We often use the maximum photocurrent that can be obtained from the active layer when illuminated by light, by assuming an internal quantum efficiency of 100% for charge generation from the optical power dissipation. We could choose any illumination spectrum, but we will of course mainly choose a relevant solar spectrum as input in these calculations. The simplest approximation is to treat the sun as a blackbody radiator. Modulating the peak position to 515 nm and the peak value (1.5 W/m^2) to resemble the Air Mass 1.5 standard, we reach

$$B = \frac{2hc^2}{\lambda^5} \frac{1}{e^{\frac{hc}{\lambda kT}} - 1} \tag{8.28}$$

with $T = 5630$ K (equivalent solar blackbody temperature) and the Planck constant $h = 6.63 \times 10^{-34}$ Js. λ is the vacuum wavelength. As this spectrum is not valid at the Earth surface, due to absorption in the atmosphere, a more appropriate choice is the tabulated solar spectra found in databases. For instance, the ASTMG173 AM1.5 spectrum available from NREL gives a solar illumination of 1000 W/m^2 in the full range (280–4000 nm).

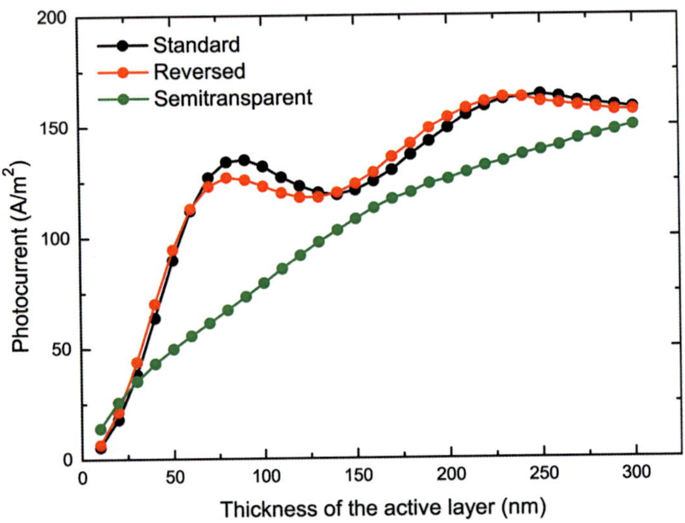

Figure 8.12 Maximal photocurrent plots for standard, reversed, and semitransparent cells based on TQ1:$PC_{71}BM$.

As we are working with quantities measured in real units we can also calculate how much of the total incoming light t is absorbed in the active layers, a distant upper limit of the power conversion efficiency. This is done with the integral of $Q(z)$ within wavelengths of 300–800 nm for TQ1/$PC_{71}BM$ in a standard geometry device (Fig. 8.13a), reversed geometry device (Fig. 8.13b), and a semitransparent device (Fig. 8.13c) under AM1.5 solar illumination.

To calculate the photocurrent generated from this flux of absorbed photons, we must consider the distribution over wavelengths, and apply the principle of one photon absorbed gives one electron, thus assuming a 100% internal quantum efficiency for all wavelengths of excitation. This estimated upper limit to photocurrent is found in Fig. 8.12 for the $TQ1/PC_{71}BM$ system, for varying active layer thickness in a standard device geometry, reversed geometry with metal cathode, and semitransparent device geometry.

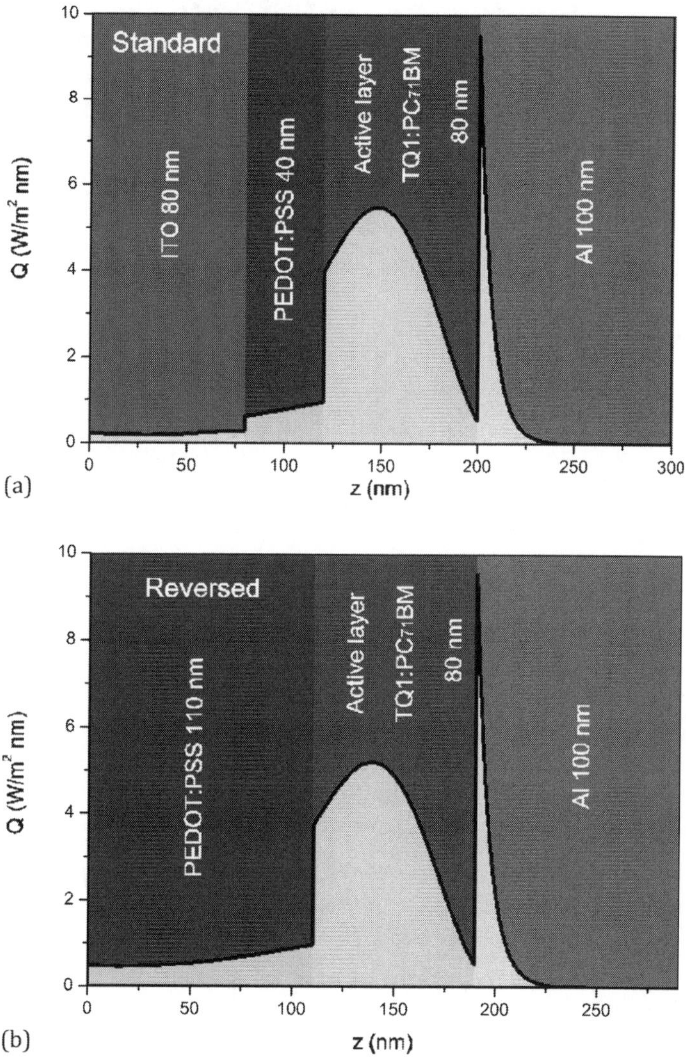

(a) (b)

Figure 8.13 (Continued).

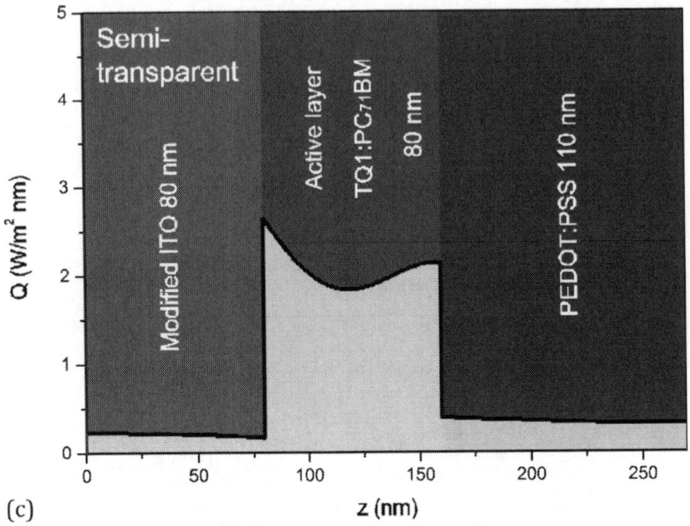

(c)

Figure 8.13 $Q(z)$ plots for (a) standard, (b) reversed, (c) semitransparent solar cells based on TQ1:PC_{71}BM.

For the standard and reversed solar cells with a reflective back electrode, the maximum photocurrent varies with active layer thickness in a nonmonotonic function, with local maxima due to interference [44–47]. The existence of local maxima for power dissipation is general, but the variation with thickness varies depending on an active material. The optical analysis of low bandgap materials indicated a great need for absorbing layers with a more than linear increase of extinction coefficient for increasing wavelength [48], thus explaining the problem of alternating polyfluorene polymers where increasing spectral coverage generally led to lower power conversion efficiency [49–55].

Manipulation of this interference-dominated limit to photocurrents, by introduction of optical spacers [38], has been proposed to optimize photocurrent, as discussed below.

8.3.4 Coherence and Incoherence Mixed

In the case of planar multilayer stacks on optically thick substrates, or with interlayers of thicknesses much larger than the wavelength, optical coherence is not expected across these thick layers. This requires incoherence and coherence in a combined treatment. For the simplest case of a thick dielectric transparent substrate, this is already included in Ref. [28] and a unifying treatment was presented by Centurioni [35]. The case is highly relevant for tandem solar cells with thick spacers between the two active layers [36], while thin film tandems are better treated under coherence conditions.

8.3.4.1 Anisotropy in Multilayers

The fact that the wave vector and the Poynting vector are not in general parallel inside anisotropic media causes some problems. For treating the problem of nonnormal light impinging on uniaxial media, it is necessary to introduce a complex angle, to handle these problems. The derivation of the proper form of the Poynting vector and the energy dissipation is actually still a matter of debate, with relevance also for treatment of light propagation in metamaterials [56,57].

Even the very realistic introduction of oblique incidence to optically anisotropic layers creates considerable modeling difficulties. Our early treatment of isotropic layers [28] avoids this situation, but for most devices one or more layers will be optically anisotropic [25,58–62]. In particular, the ubiquitous contact material PEDOT:PSS was found to be optically anisotropic [63], thus demonstrating the need for a treatment of this case also. In general, this can lead to situations where the electromagnetic wave becomes inhomogeneous, and where the Poynting vector is not parallel with the wave vector, and the standard expressions for power dissipation becomes inappropriate. The TMM does not give an easy solution for this case, as we noted, but this was something that was not understood in some TMM modeling [64,65] but later more widely acknowledged [66]. A recent presentation [67] restates the possibility to model isotropic layers also at oblique incidence, but avoids the difficulty inherent in modeling anisotropy. A general treatment of the problem of stratified anisotropic media is found in Ref. [37] with a more recent treatment by Schubert [68], but this takes us out of the TMM and onto four-dimensional models for light propagation.

8.4 Light Management in OPVs

Coupling light into photovoltaic devices of standard materials often includes large optical scattering elements, of wavelength size. As this is not easy to accommodate into thin film OPV devices, where the thickness is smaller than the wavelength, thin and flat geometries will be the standard geometry. These may be combined with transparent or reflective electrodes in several device geometries.

8.4.1 One Transparent Electrode and One Reflective

Where there is one transparent electrode and one reflective, we may either have a standard geometry (substrate/transparent electrode/active layer/metal contact) or inverted geometry, where the transparent electrode has been converted to the cathode and the reflective metal electrode is anode. Most inverted structures use ITO modified by surface layers, as found for instance in Refs. [69–78], or modified PEDOT [79,80]. An alternative form of inverted, or reversed, geometry uses a top illuminated (substrate/metallic reflector/active layer/transparent electrode) [81–84], and the transparent electrode is often variants of PEDOT (see Chapter 4 for a more complete treatment of buffer layers and formation of inverted cells). These devices all share a common optical structure, and the electromagnetic mode structure has similarities, but also differences due to the different thicknesses of the different electrode material.

Optical optimization of devices in these geometries has been reported, for the standard geometry with bilayer [58] or bulk heterojunctions [15,46,85]. Optical limitation for low bandgap materials was analyzed [48,85]. For the inverted devices with a modified ITO cathode, optical modeling [86,87] did show differences compared to standard geometries. New materials and new device architectures necessitate new calculations [88].

8.4.1.1 Optical Spacers

An extensive literature has developed on the topic of optical spacers, which were first used in simple bilayer devices [38], where the optical field can be designed for a maximum close to a donor/acceptor junction, thus maximizing conditions for absorbing light and generating charges. In this case, the spacer was also an absorbing material. Dielectric layers were also introduced in small molecule/molecule bilayer solar cells, in the form of an exciton blocking layer [89,90]. Transparent optical spacers which act to redistribute the optical power dissipation within the device was later done with spacer layers located at the cathode, in standard geometry devices [91]. The problem here is, however, that of designing a spacer thickness valid for all wavelengths. Claims of optical spacers did not come also with demonstration that photocurrent was actually varying with wavelength in the way that interference would predict; in fact, when investigated, the improvement of photocurrent generation in OPVs with optical spacers did not show this wavelength variation, thus disproving the concept. Other causes of the success of optical spacers can thus be plausible, and the use of TiO_2 as a spacer layer is one example where electrical properties contribute to the improved photocurrent. In systematic and comparative electromagnetic simulations using TM and FEM methods we demonstrated [92] that the inclusion of a dielectric optical spacer at the cathode certainly can affect the optical power dissipation within the active layer, but that the optical spacer cannot contribute to enhanced power dissipation if the optically optimal thickness of the active layer was already chosen; in this case only lower dissipation can be obtained with an optical spacer. However, if the thickness is not chosen as optimal, the inclusion of an optical spacer can redistribute and improve optical dissipation [84], as demonstrated for three different polymer/fullerene blends. And it is often the case that optimal thickness from the viewpoint of power dissipation is thicker than what is desirable for the electrical performance of the active material, leading to lower electrical fields, longer paths for charge transport, and low power conversion efficiency. For small molecule/molecule blends prepared by vacuum evaporation, this is often the case, and optical spacers can thus enable improvements. These thin dielectrics were often used for a combination of exciton blocking and as optical spacers, combined in a single layer.

8.4.2 Transparent Devices

The transparent electrode necessary for optical incoupling in OPV can be made from many materials and with different processing methods. However, for semitransparent solar cells, both anode and cathode must be transparent. Standard geometries of the OPV use metal anodes or cathodes, which work both as electrodes and optical reflectors. With semitransparent electrodes substituted for this metal electrode, see-through solar cells can be built. Semitransparent electrodes are also necessary as the recombination layer in stacked tandem solar cells [93], but do not have to carry much current along the electrode, and thus are designed for different parameters.

The transparent bottom electrodes can be modified ITO that act as either anodes [94–97] or cathodes [72,98–100], the same as those used to construct solar cells with a top reflective electrode. Conducting polymers can also be used as transparent bottom electrodes, which are usually constructed from different forms of PEDOT [81,101–103] suitable for anodes. For cathodes PEDOT is also possible, in different chemical forms [104,105].

Substitution of top reflective electrodes has been done by lamination of metal nanowire mesh [106], deposition of ultrathin metals [107,108], or transparent metal oxides on top of active layers [97–99,109,110]. Metal oxides can be used as anodes, but can also be converted to work as cathodes. PEDOT processed from aqueous solution can also be used as a top anode, once the problem of wettability of PEDOT on the hydrophobic active layers is solved [81]. Bilayer PEDOT comprising the commercial PEDOT material CPP with good wettability, and PEDOT PH1000 with high conductivity,[c] has been demonstrated to work as top transparent anode [72,111], and recently PEDOT PH1000 mixed with a surfactant has also been used as a top transparent anode to construct solar cells [83].

The development of the transparent electrodes enables the rapid improvements in quality of the semitransparent solar cells based on organic materials. Good transparency, color perception, and rendering properties could also be realized by employing Al-doped ZnO top cathode and low bandgap active layer to construct semitransparent solar cells [97]. Meiss et al. [112] have shown that the semitransparent solar cells can be more transparent in the visible wavelength region using infrared absorbing materials as the photoactive layer. With the help of special transparent cathodes based on silver nanowires, PCEs of 4% and transmission of 60% over the visible spectrum has been reported [113]. Efficient tandem solar cells can be constructed by simply stacking semitransparent solar cells on top of each other [114] or by having semitransparent solar cells on top of a standard/reversed solar cell with a back-reflecting electrode [115].

---

[c]PEDOT CPP and PEDOT PH1000 are brand names of PEDOT:PSS dispersions from commercial suppliers.

Recently, semitransparent solar cells with IQE over 90% and transmittance as high as 50% have been achieved via proper electrode surface modification [114].

Semitransparent solar cells create new opportunities to combine with optical elements. With a far-field metallic reflector on the back, light transmitted through the device is once more injected in the photovoltaic device, and current is improved. With two or more semitransparent devices stacked on top of each other, separated by the thickness of the substrate and thus without coherence between waves in the two devices, photocurrent can be improved considerably. In this geometry, loss of energy in the semitransparent electrodes is reduced, as the metallic reflector is the dominant loss element. More photocurrent can thus be collected by stacking a number of semitransparent devices than what is obtained in the standard geometry with a single device [114].

8.4.3 Optical Cavities

As the active layers in the multilayer stack, of thicknesses below half a wavelength of light, are confined by reflective electrodes, the microcavity optics treated in quantum electrodynamics becomes relevant. These aspects have long been incorporated in the analysis of light emission from organic devices [116], through their impact on spectral and angular distributions. The moderate cavity effects found in OPVs with ITO and metal electrodes are all included in the TMM simulations. However, introducing thin metal layers as transparent electrodes strongly enhances the reflectivity of one electrode, and moves the devices closer to microcavity optics, as the light transmitted through the transparent metal can be reused in multiple reflections. Use of transparent metal anodes or cathodes also includes dielectric layers sandwiching the thin metal, a necessity to establish enough of optical input in the active layer residing between electrodes. Important developments of these transparent metal electrodes have been shown by introducing dielectric/metal/dielectric electrodes into OPVs [117–120], which were later extensively used in top contact electrodes [107,121–130]. The intuitively expected problem of high reflectivity of the semitransparent electrode can be offset due to the confinement of light in the microcavity [131].

A more recent development in light confinement is so far only theoretical. Incoupling of light into very thin films of low-index photovoltaic materials in between two high-index materials compresses light into the photovoltaic material. By introducing textured surfaces, a considerable fraction of solar light can in this way be injected into extremely thin films, in the nanophotonic regime. The predictions from this theory, which is based on the wave treatment of light, go considerably beyond the classical Yablonovitch limit [132] for light trapping in solar cells. In this theory, the optical path enhancement is limited by $4n^2$, where n is the optical index of the photovoltaic material. This prediction of the ergodic limit is based on ray optics, and relevant for layers thicker than the wavelength of light. For thicknesses below, wave optics is necessary, and considerable improvements of light trapping should be possible. Calculations demonstrating this effect have been presented [131,133,134], but not yet experimentally verified.

8.5 Out of Planarity—Structured OPVs

8.5.1 Microstructured Interfaces in OPV

As the planar geometries of standard OPVs are not necessarily the optimal geometry for high performance, the very limiting assumptions of the TM methods become restrictive. For folded geometries (Fig. 8.14) [135,136], where multiple reflections of light with different interference effects occur for different incoming angles, this becomes a real drawback. The combination of structures of many micrometer size and layer thicknesses of 10–100 nm makes this an awkward combination. In some cases the combination of ray-tracing with TM methods is possible, but sometimes these become obscure. Another approach is to do the full solution of Maxwell's equation in a numerical solution. We have used this method extensively, in particular for folded cell geometries [32,33]. These folded cells offer considerable light trapping opportunities, and can be made on very different dimensions, from a few wavelengths of light to millimeters and centimeters. We have used a commercial software module to deduce the power dissipation properties in folded cells of various materials, and also applied this treatment to tandem solar cells.

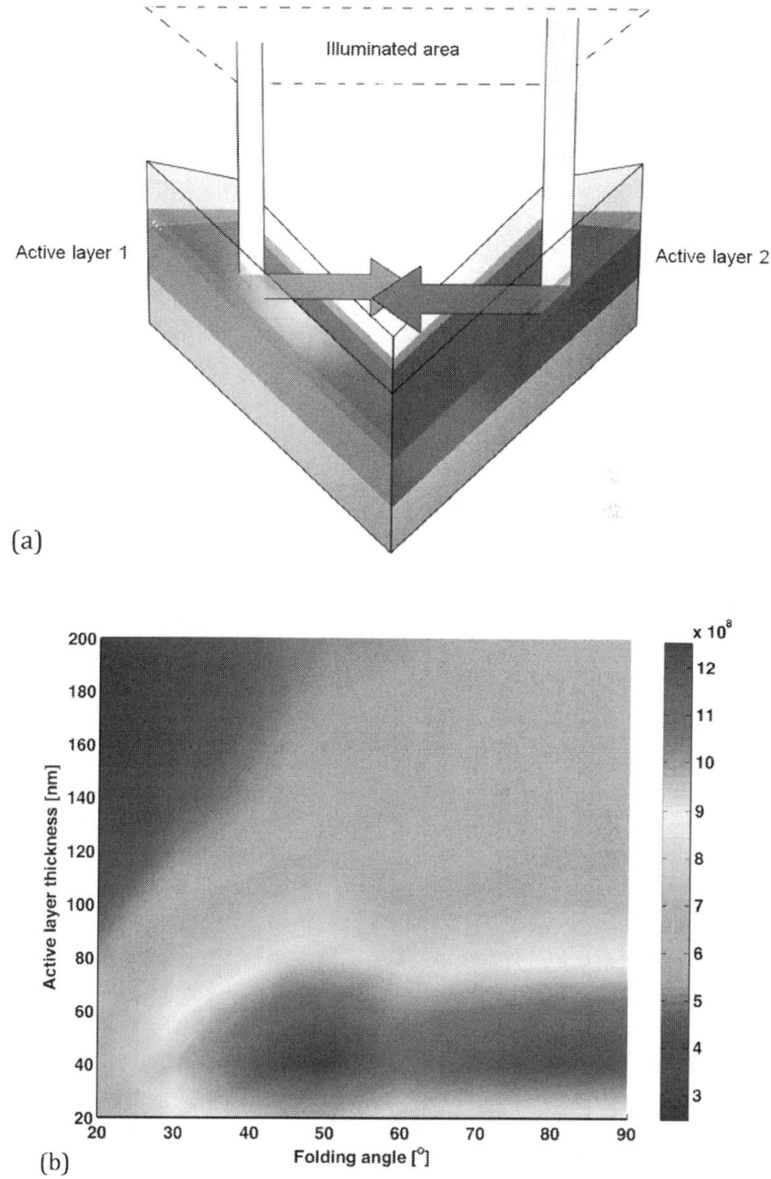

Figure 8.14 (a) Folded devices can improve on light trapping. (b) Calculated photocurrent generation of a planar or folded cell (for the material APFO-3:$PC_{60}BM$), here expressed as the photocurrent per unit volume (A/m^3), where two local maxima can be observed. Figure reproduced from Ref. [8] with permission from B. Viktor Andersson.

The FEM method allows geometry optimization of folded solar cells, with varying active layer thickness and varying folding angle [33]. We can also compare the difference of power dissipation obtained by suppressing the optical anisotropy in the FEM calculations and find these to be small.

As an example, consider the power dissipation in a folded cell with two different materials, of higher and lower bandgap and with some degree of optical complementarity, where minima of absorption in one material coincides with maxima of absorption in the other [32]. Using FEMs and the (anisotropic) dielectric functions of the polymer blends APFO-3 and APFO-Green5 we can solve for the power dissipation within the folded structure (Fig. 8.15), and note that for the chosen thicknesses of the active layers, the power dissipation in the lower bandgap materials is higher. As these devices can conceivably be connected in serial or in parallel mode, we can also evaluate the current balancing in the serial mode connection through these calculations. One way of presenting this information is through contour plots of (potential) photocurrent from the two materials in the two devices, as a function of thickness of the active layers, folded into a defined geometry (Fig. 8.16). Here, we note that the combination of two materials with current balancing follows a trajectory of thickness combinations. We can also perform these calculations for planar tandem stacks, where no light trapping is found, and evaluate the attractiveness of these alternative geometries [32].

8.5.2 Nano- and Microstructured Interfaces in OPV

The routes to light incoupling by modification of boundaries between layers in the multilayer devices have been pursued for a considerable time. Imprinting of diffraction gratings in the metal electrode/active material interface enables light incoupling for a narrow range of wavelengths [137] and proper polarization, but is not a helpful avenue for collecting more solar light; for this very reason only a fraction of the solar spectrum is affected. The use of broadband light incouplers has taken different forms [138]; a recent contribution shows how spontaneous buckling of interfaces due to stretching/compression of elastomeric substrates can help to collect more light in thin film solar cells [139].

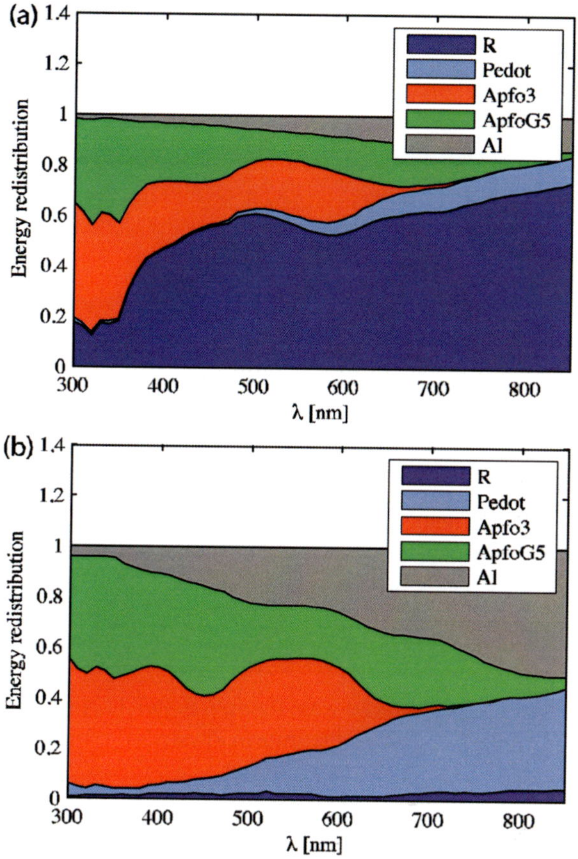

Figure 8.15 Power dissipation in folded tandem. Reproduced from figure 4 in Ref. [32] with permission from the American Institute of Physics.

Light traps are ubiquitous in inorganic photovoltaics, and can be found in many forms also in organics. Only moderate concentration of light is expected to be useful, and therefore micropatterned optical elements are most relevant. These come in geometries of Winston apertures [89] and microlenses [140], and more.

With the recent development of semitransparent solar cells comes the opportunity to incorporate light trap in the optical far field, and thus not compromising the electrical functions of the solar cell. We recently demonstrated a light trap with a micropatterned metal reflector at the back [141], with the reflector at an angle to incoming light, in order to make some of the reflected light trapped inside the device due to total internal reflection at interfaces. This considerably improved photocurrent, compared to the planar mirror.

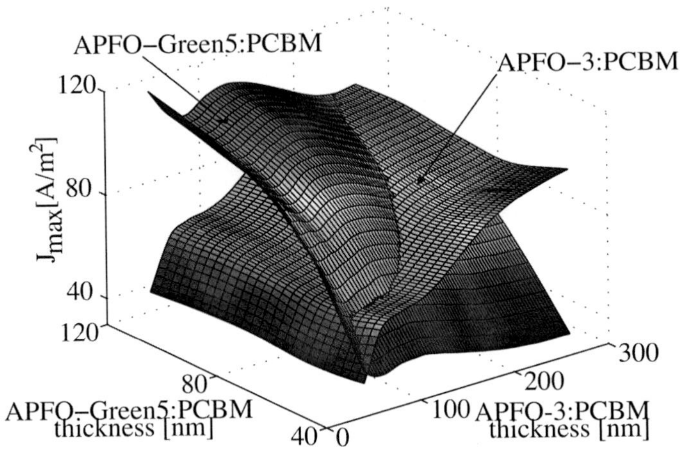

Figure 8.16 Balancing photocurrents from a folded tandem cell. Potential photocurrent from the two layers must be identical for series connection; this crossing of the two surfaces is shown in the graph. Reproduced from figure 5 in Ref. [32] with permission from the American Institute of Physics.

8.5.3 Plasmonics

Many paths to gain enhanced optical input have been attempted, and great expectations have been given to the use of nanoplasmonic effects in improving light input in photovoltaic devices. This area has recently been reviewed [142–146] in the broader context of photovoltaics, both inorganic and organic, and plasmonic effects have been demonstrated in both classes of devices. The electromagnetic plasmon excitations at a metal/dielectric interface are relevant for both light absorption and light emission [147]. How to improve upon light incoupling using these phenomena has been less clear. Plasmonics comes in two geometries particularly relevant to OPVs, in nanoparticle plasmonic phenomena, as reported in tandem solar cells with recombination layers formed from metal nanoparticles [148] and in thin film plasmonic phenomena. The plasmons at a dielectric/metal interface as exploited in a thin film geometry with coplanar metal and dielectric can be directly excited by patterning the metal interface, thereby enabling the wave-vector conditions for plasmon excitation.

Studies of inorganic thin film devices have largely been driving this field [149], but many of the theoretical simulations indicate that organic materials are better able to fit the required properties in order to use plasmons to improve light trapping. Still, the competition between energy dissipation in the metal used for plasmonic phenomena and the semiconductor is critical. A criterion to evaluate this distribution has been suggested [145].

8.5.3.1 Plasmons at Planar Metal/Dielectric Interfaces

An early study [150] of plasmonic phenomena in an OPV used imprinted metal/dielectric interfaces, and designed these using electromagnetic calculations. Very distinct plasmon peaks were observed in the reflection spectra of the devices, as predicted by theory [150], but this energy was not used to generate photocurrent in an efficient manner. Simulations indeed demonstrated that the power dissipation was found inside the metal layer, and thus not at all contributing to photocurrent generation, only to heating of the electrode. Detailed electromagnetic simulations have contributed to clarify the role of far-field light scattering, near-field electromagnetic field enhancement, and coupling to wave-guided modes in the plethora of phenomena present in thin film OPVs. The possibility of improvement of optical absorption in the active layer is shown in reports [151]. Optimization of plasmonic metallic gratings for enhancing absorption in the active layer thus could lead to improvements of absorptance by 16% [152,153], and with multiperiod grating improvements might reach 20%, from simulations. Similar numbers come from others, also considering the different TE and TM modes [154]. Positioning gratings at both sides of the active layer also improves absorption by similar numbers [155]. The polarization-dependent effects of nanowires, causing large differences for TE and TM polarization, can be suppressed by choosing hexagonal nanoparticle lattice geometries [156].

Experimental demonstrations of such effects are also reported, but however rarely contributing to improvement of overall performance of OPVs. However, some studies have given quite different results [157], and formation of more charge carriers has been observed with spectroscopic methods [158].

Early demonstrations of nanohole silver electrodes showed the presence of plasmonic phenomena [159]. Related simulations and experiments are in agreement [160], and the improvement of the plasmonic structure, here a nanovoid metallic top contact, is very considerable but the absolute performance not impressive. Further demonstrations of the nanovoid top contact were reported [161], and in great detail resolve the conflicting effects of plasmonic reflector geometry.

8.5.3.2 Plasmons at Nanoparticle Metal/Dielectric Interfaces

Nanoparticle plasmons are found associated with metallic nanoparticles, typically made from silver or gold, but potentially from other metals. Here the radius of curvature is small, and the plasmon excitations of the particle are given by the geometry, coupled to the electromagnetic environment. Nanoparticle effects on the electromagnetic phenomena include both classical light scattering, in the form of Mie scattering in the far field, and more pronounced plasmonic phenomena in the near field. These can be very strong, and form the basis for using plasmonics in sensing, where single molecule sensitivity is possible due to the very intense electromagnetic fields created in the nearest distance to the nanoparticle. To resolve the contributions from near-field plasmonics and from far-field light scattering is not easy. Early studies point to the use of nanoparticle plasmons in generating photocurrent in OPVs through improved absorption [162,163] but also when nanoparticles were used as the recombination layer in an early organic tandem solar cell [148].

Studies based on numerical solution of the electromagnetic problem of absorption in the presence of plasmonic metallic nanoparticles indicate that the location of the nanoparticles within the device can have a considerable impact on the improvement, or lack of such [164]. Whether this simulation is done for a material [165] or for a material in a device geometry [166] may make a considerable effect on the outcome, as the coupling of plasmonic effects, light scattering, interference, and waveguiding is very complex. The coating on nanoparticles also has a major impact on the degree of improvement, as this is a near-field effect, and a coating will effectively reduce the volume in which this improvement of absorption in the active layer occurs.

Plasmonic nanoparticles might be located in the active layer, in the electrodes, or as separate buffer layers in between electrodes and contact layers; in bilayer and tandem devices they may also be located at the interface between different active layers. Where Au nanoparticles were located in a lattice, and extending into the active layer, considerable contributions of the particle plasmons to active layer absorption were identified [166]. When added into the active layers, the effects of nanoparticles are manifold, and not limited to plasmonics; charge trapping, recombination centers, excitation quenching sites, modification of nanostructure, and order in active material may contribute to the sequence of optical and electrical events during formation of photocurrent. Considering that the introduction of the nanoparticles often occurs with inclusion of stabilizing surfactants or surface layers, which may or may not be part of the resulting active material, it is expected that many different phenomena can be obtained. Alternative routes include synthesis of metal nanoparticles from silver salts inside the active material [167].

It is reported that the improved performance of a low-bandgap polymer/fullerene blend is due to light scattering, due to the introduction of silver nanoparticles in the active material [168]. The degradation of device performance was reported for another polymer/fullerene blend with nanoparticles located at the buffer layer contact to the blend [169]. It is also reported that inclusion of gold nanoparticles in the blend has a deleterious impact on the performance of P3HT:PCBM blends [170]. Inclusion of gold nanoparticles in the active material can improve absorption, as others have reported [171], but the end result is also due to several effects of nanoparticles on the charge collection. Nanoparticle-based electrodes have been evaluated [172,173]. Addition of nanoparticles in all the different layers has also been reported to somewhat improve performance, but contributed to from a number of mechanisms [174]. Others argue that the effects of nanoparticles are not optical, but electrical [175]. In tandem solar cells, where the nanoparticles may be included in the recombination layer between top and bottom absorbers of the tandem, early reports [148,163] and recent reports [176] demonstrate distinct effects due to improved absorption in the layers. The assignment of these to light scattering in the far-field, or to near-field modification, may still be debated.

Recent theoretical and experimental work show how particle plasmons can be combined with thin film organic devices [177,178] where enhancement of light incoupling can easily be offset by the quenching of excited states at the metal nanoparticle. In nanostructured electrodes incorporating designer plasmons, the effects seen are often related to light scattering of the particles. Though the effects of plasmons in enhanced absorption are present, and can be used to improve the energy dissipation in the active material at the edge of the absorption, the improvements are minor and confined to a narrow wavelength interval [161].

Dual-plasmonic and light-scattering nanostructures have recently been demonstrated, where plasmonic gratings at one electrode and nanoparticles are combined to improve the PCE of the polymer/fullerene blend [179].

Alternative uses of plasmonic scatterers include a recent demonstration of light scattering for improved absorption in the active layer, that locate the plasmonic structure outside the electrical pathway of the OPV [180]. Furthermore, optical antennas have been proposed as suitable for thin film organics [134].

Given the considerable number of new theoretical models and concepts in plasmonics [134], this will be an active field in the coming years.

8.6 The Solar Day and the Solar Year

What is desired from photovoltaic devices and modules is electrical energy at low cost, and the prevalent metrics of evaluating the efficiency of modules is therefore not always the wisest choice. When considering the light management issues connected to this metric, all the factors of solar light come into play; the daily variation of solar intensity, spectrum, and angle of incidence; the distribution between direct and indirect (scattered) light; the yearly variation of angles, intensity, and spectrum. Not all of these factors can easily be put into standardized metrics, and none of them are available through the simple metric of power conversion efficiency.

For planar organic solar cells, with great similarity to multilayer thin film inorganic solar cells (a-Si : H, CIGS, CdTe, etc.), an important difference is the lower refractive index, typically ca. 2 in the active materials. This enables improved optical incoupling from the ambient, and can bring a considerable improvement compared to thick film inorganics. Analysis of the incoupling in OPVs [64] verifies that the optimal angle for solar light incoupling is not at normal incidence, but rather at glancing incidence. Incorporating this physics into models of optical incoupling into planar and nonplanar tandem solar cells [32,33], also including tandem solar cells, the outcome of light incoupling can be compared. For normal incidence, the multiple reflections in the folded solar cells prolong the path of photon absorption, and improved photocurrent is obtained per incident solar light, albeit at the price of larger amounts of active materials and substrate in the device. It can be shown that the collected optical power per volume of active material actually has two local maxima, as a function of fold angle in folded cells with very thin active layer (Fig. 8.14b) [8].

With time-varying spectra representing the ascent, zenith, and descent of the sun at a location in Central Europe, it is possible to evaluate the improved incoupling over the solar day and the solar year [34]. These simulations account for the spectral, angular, and intensity variations of light incoming to a solar cell fixed in position at this location. The angular dependence of light incoupling here makes a sizable impact on photocurrent. Also here the folded cells considerably improve on light trapping and collected optical power.

For most optical simulations, diffuse and indirect light is not included. The fact that the angular variation of optical incoupling is beneficial for OPVs compared to inorganic PV, and the prevalence of indirect light under Northern climates, contributes to improved optical incoupling over the solar year, thus improving the number of kilowatthours generated per nominal installed peak power capacity. This is one of three factors that much improve energy collection from OPVs.

Due to the poor electrical transport properties of active materials and higher recombination at higher charge carrier density, lower light intensity leads to higher power conversion efficiency. This downside of device performance under standard metrics therefore somewhat improves the energy-capturing ability of OPVs, compared to inorganic devices. The poor electrical transport is also temperature activated, and therefore OPVs improve in photocurrent under moderate heating, as found in solar modules, while inorganic

PV decreases due to loss of photovoltage. Taken together, these three factors can improve the energy generated from OPVs with 30% compared to silicon devices [181].

8.7 Optical Imaging of Processing of Polymer Photovoltaic Modules

Light management can also be used as a tool for control of the printing process for OPVs. The advantage of production in organic solar cells coming from liquids relies on printing the active materials in solvents, and during printing removing the solvent and cosolvents that carry donor and acceptor into the thin film. As the concentration of materials is 1–2%, we must coat a liquid layer of 10 μm to arrive at a final thickness of 100–200 nm. This should be done at great speed but with nanostructure forming over many square meters per second in a reproducible and controllable manner, which means the removal of solvent and formation of the blend are of interest, and preferably possible to correlate to the outcome in the form of nanostructure and photocurrent generation. Removal of the solvent can be followed by interference imaging methods (Fig. 8.17). The phase boundaries of the drying liquid film are found at a substrate (which can be chosen as reflective metal) and at the top liquid/gas interface, where solvent is evaporating. Interference of light reflected in the top and bottom surface can be used to precisely measure the rate of evaporation of solvents and cosolvents [182], which was also shown to correlate with the formation of crystalline phases in the blend [183]. This was done pointwise, with one laser beam and one photodiode probing the specular reflected light from a drying film on a substrate. Extensions to more laser beams [184] or imaging of larger areas, as well as to microscopic imaging [185] considerably extend the power of the method. We can now follow the variations of solvent removal from the liquid film, across a surface, both for spin-coating [186] and for blade-coating methods [187]. The blade-coating conditions are similar to those expected for large-scale printing of areal photovoltaic devices, and it might eventually be feasible to use the imaging method for process control. We show in Fig. 8.17 an image of the materials deposition from blade coating a polymer/fullerene solution and with the inset showing the variation of reflected light intensity as a function of time, for one chosen pixel. Variations of evaporation rates across the surface can thereby be imaged.

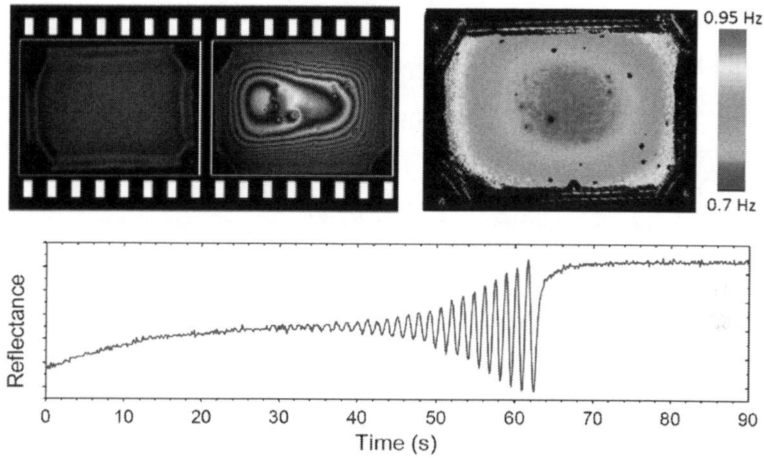

Figure 8.17 Two reflectance image snapshots of the drying process of a spin-coated thin film (upper left). The graph shows the interference effects on the reflectance as the wet film is thinning. A mapping of the average frequency of the visible interference fringes (upper right), proportional to the evaporation rate of the solvent, shows that the edges have a higher evaporation rate than the central parts.

8.8 Summary

In summary, we have reviewed the state of the art for optical modeling of OPV materials and devices. Classical electromagnetic theory is sufficient to handle the optical aspects of thin film OPVs, and is an important tool in the further development of devices and processing of OPV materials, as well as indicating required properties for the design of new materials. The current race to incorporate photonic and plasmonic incoupling structures in OPV devices may enable improvements in the light management into these thin film devices, to improve photocurrents and power conversion efficiency.

Acknowledgments

We acknowledge critical reading of the manuscript from Wolfgang Tress, Feng Gao, Viktor Andersson, and Scott Mauger. Research in OPVs at Linköping is funded by the Swedish Energy Agency.

References

1. Inganäs O, Roman LS (2003) Organic photodiodes: From diodes to blends, *in Organic Photovoltaics,* (Brabec CJ, Dyakonov V, Parisi J, Sariciftci NS, eds), Springer, Berlin.
2. Persson N-K, Inganäs O (2005) Simulations of optical processes in organic photovoltaic devices, *in Organic Photovoltaics: Mechanisms, Materials and Devices,* (Sun S-S, Sariciftci NS, eds), CRC Press, Boca Raton, Florida, USA.
3. Pettersson LAA (1999) Optical modelling and characterization of layers and multilayer structures. PhD dissertation (Linköping University, Linköping).
4. Roman LS (2000) Organic photodiodes: Optical, electrical and morphological aspects of polymer based devices. PhD dissertation (Linköping University, Linköping).
5. Nyberg T (2002) Nano and micro patterned organic devices: From neural interfaces to optoelectronic devices. PhD dissertation (Linköping University, Linköping).
6. Persson N-K (2005) Optical modelling of conjugated polymers—from materials to devices. PhD dissertation (Linköping University, Linköping).
7. Tvingstedt K (2008) Light trapping and alternative electrodes for organic photovoltaic devices. PhD dissertation (Linköping University, Linköping).
8. Andersson BV (2012) Electron tomography and optical modelling for organic solar cells. PhD dissertation (Linköping University, Linköping).
9. Niggemann M, Riede M, Gombert A, Leo K (2008) Light trapping in organic solar cells, *Phys Stat Sol A-Appl Mater Sci,* **205**(12), 2862–2874.
10. Ko DH, et al. (2011) Light-trapping nano-structures in organic photovoltaic cells, *J Mater Chem,* **21**(41), 16293–16303.
11. Wynands D, *et al.* (2012) Spectroscopic ellipsometry characterization of vacuum-deposited organic films for the application in organic solar cells, *Org Electron,* **13**(5), 885–893.

12. Johs B, Hale JS (2008) Dielectric function representation by B-splines, *Phys Stat Sol A-Appl Mater Sci*, **205**(4), 715–719.
13. Campoy-Quiles M, *et al.* (2005) Ellipsometric characterization of the optical constants of polyfluorene gain media, *Adv Func Mater*, **15**(6), 925–933.
14. Jellison GE, Modine FA (1996) Parameterization of the optical functions of amorphous materials in the interband region, *Appl Phys Lett*, **69**(3), 371–373.
15. Persson NK, Arwin H, Inganas O (2005) Optical optimization of polyfluorene-fullerene blend photodiodes, *J Appl Phys*, **97**(3), 034503.
16. Campoy-Quiles M, et al. (2008) Morphology evolution via self-organization and lateral and vertical diffusion in polymer:fullerene solar cell blends, *Nat Mater*, **7**(2), 158–164.
17. Madsen MV, *et al.* (2011) Ellipsometry as a nondestructive depth profiling tool for roll-to-roll manufactured flexible solar cells, *J Phys Chem C*, **115**(21), 10817–10822.
18. Engmann S, Turkovic V, Gobsch G, Hoppe H (2011) Ellipsometric investigation of the shape of nanodomains in polymer/fullerene films, *Adv Energy Mater*, **1**(4), 684–689.
19. Muller C, et al. (2008) Binary organic photovoltaic blends: A simple rationale for optimum compositions, *Adv Mater*, **20**(18), 3510–3515.
20. Muller C, et al. (2011) Phase behaviour of liquid-crystalline polymer/fullerene organic photovoltaic blends: Thermal stability and miscibility, *J Mater Chem*, **21**(29), 10676–10684.
21. Andersson BV, Herland A, Masich S, Inganas O (2009) Imaging of the 3D nanostructure of a polymer solar cell by electron tomography, *Nano Lett*, **9**(2), 853–855.
22. van Bavel SS, Sourty With G, Loos J (2009) Three-dimensional nanoscale organization of bulk heterojunction polymer solar cells, *Nano Lett*, **9**(2), 507–513.
23. Oosterhout SD, et al. (2009) The effect of three-dimensional morphology on the efficiency of hybrid polymer solar cells, *Nat Mater*, **8**(10), 818–824.
24. Vandewal K, Tvingstedt K, Gadisa A, Inganas O, Manca JV (2009) On the origin of the open-circuit voltage of polymer-fullerene solar cells, *Nat Mater*, **8**(11), 904–909.
25. Persson NK, Sun M, Kjellberg P, Pullerits T, Inganas O (2005) Optical properties of low band gap alternating copolyfluorenes for photovoltaic devices, *J Chem Phys*, **123**(20), 204718.

26. Mondal R, *et al.* (2009) Molecular design for improved photovoltaic efficiency: Band gap and absorption coefficient engineering, *J Mater Chem*, **19**(39), 7195–7197.
27. Risko C, McGehee MD, Bredas JL (2011) A quantum-chemical perspective into low optical-gap polymers for highly-efficient organic solar cells, *Chem Sci*, **2**(7), 1200–1218.
28. Pettersson LAA, Roman LS, Inganas O (1999) Modeling photocurrent action spectra of photovoltaic devices based on organic thin films, *J Appl Phys*, **86**(1), 487–496.
29. Pettersson LAA, Roman LS, Inganas O (2001) Quantum efficiency of exciton-to-charge generation in organic photovoltaic devices, *J Appl Phys*, **89**(10), 5564–5569.
30. Niggemann M, Glatthaar M, Gombert A, Hinsch A, Wittwer V (2004) Diffraction gratings and buried nano-electrodes—architectures for organic solar cells, *Thin Solid Films*, **451**, 619–623.
31. Stathopoulos NA, *et al.* (2011) A transmission line model for the optical simulation of multilayer structures and its application for oblique illumination of an organic solar cell with anisotropic extinction coefficient, *J Appl Phys*, **110**(11), 114506.
32. Andersson BV, Persson NK, Inganas O (2008) Comparative study of organic thin film tandem solar cells in alternative geometries, *J Appl Phys*, **104**(12), 124508.
33. Andersson V, Tvingstedt K, Inganas O (2008) Optical modeling of a folded organic solar cell, *J Appl Phys*, **103**(9), 094520.
34. Andersson BV, Wuerfel U, Inganas O (2011) Full day modelling of V-shaped organic solar cell, *Sol Energy*, **85**(6), 1257–1263.
35. Centurioni E (2005) Generalized matrix method for calculation of internal light energy flux in mixed coherent and incoherent multilayers, *Appl Opt*, **44**(35), 7532–7539.
36. Persson NK, Inganas O (2006) Organic tandem solar cells—modelling and predictions, *Sol Energy Mater Sol Cells*, **90**(20), 3491–3507.
37. Azzam RMA, Bashara NM (1977) *Ellipsometry and polarised light*, North-Holland Pub. Co., Amsterdam.
38. Roman LS, Mammo W, Pettersson LAA, Andersson MR, Inganas O (1998) High quantum efficiency polythiophene/C-60 photodiodes, *Adv Mater*, **10**(10), 774–777.
39. Halls JJM, Pichler K, Friend RH, Moratti SC, Holmes AB (1996) Exciton diffusion and dissociation in a poly(p-phenylenevinylene)/C_{60} heterojunction photovoltaic cell, *Appl Phys Lett*, **68**(22), 3120–3122.

40. Rostalski J, Meissner D (2000) Photocurrent spectroscopy for the investigation of charge carrier generation and transport mechanisms in organic p/n-junction solar cells, *Sol Energy Mater Sol Cells,* **63**(1), 37–47.
41. Stubinger T, Brutting W (2001) Exciton diffusion and optical interference in organic donor-acceptor photovoltaic cells, *J Appl Phys,* **90**(7), 3632–3641.
42. Wang EG, et al. (2010) An easily synthesized blue polymer for high-performance polymer solar cells, *Adv Mater,* **22**(46), 5240–5244.
43. Burkhard GF, Hoke ET, McGehee MD (2010) Accounting for interference, scattering, and electrode absorption to make accurate internal quantum efficiency measurements in organic and other thin solar cells, *Adv Mater,* **22**(30), 3293–3297.
44. Hoppe H, Arnold N, Sariciftci NS, Meissner D (2003) Modeling the optical absorption within conjugated polymer/fullerene-based bulk-heterojunction organic solar cells, *Sol Energy Mater Sol Cells,* **80**(1), 105–113.
45. Lacic S, Inganas O (2005) Modeling electrical transport in blend heterojunction organic solar cells, *J Appl Phys,* **97**(12), 124901.
46. Moule AJ, Bonekamp JB, Meerholz K (2006) The effect of active layer thickness and composition on the performance of bulk-heterojunction solar cells, *J Appl Phys,* **100**(9), 094503.
47. Sylvester-Hvid KO, *et al.* (2007) Analyzing poly(3-hexyl-thiophene): 1-(3-methoxy-carbonyl)propyl-1-phenyl-(6,6)C-61 bulk-heterojunction solar cells by UV-visible spectroscopy and optical simulations, *J Appl Phys,* **102**(5), 054502.
48. Persson NK, Wang XJ, Inganas O (2007) Optical limitations in thin-film low-band-gap polymer/fullerene bulk heterojunction devices, *Appl Phys Lett,* **91**(8), 083503.
49. Wang XJ, *et al.* (2004) Infrared photocurrent spectral response from plastic solar cell with low-band-gap polyfluorene and fullerene derivative, *Appl Phys Lett,* **85**(21), 5081–5083.
50. Zhang FL, *et al.* (2005) Polymer solar cells based on a low-bandgap fluorene copolymer and a fullerene derivative with photocurrent extended to 850 nm, *Adv Func Mater,* **15**(5), 745–750.
51. Wang XJ, *et al.* (2005) Enhanced photocurrent spectral response in low-bandgap polyfluorene and C_{70}-derivative-based solar cells, *Adv Func Mater,* **15**(10), 1665–1670.
52. Wang XJ, *et al.* (2006) Polymer solar cells with low-bandgap polymers blended with C_{70}-derivative give photocurrent at 1 μm, *Thin Solid Films,* **511**, 576–580.

53. Perzon E, Wang XJ, Admassie S, Inganas O, Andersson MR (2006) An alternating low band-gap polyfluorene for optoelectronic devices, *Polymer*, **47**(12), 4261–4268.
54. Zhang F, et al. (2006) Low band-gap alternating fluorene copolymer/ methanofullerene heterojunctions in efficient near infrared polymer solar cells, *Adv Mater*, **18**, 2169–2173.
55. Inganas O, *et al.* (2004) Low bandgap alternating polyfluorene copolymers in plastic photodiodes and solar cells, *App Phys A-Mater Sci Processing*, **79**(1), 31–35.
56. Markel VA (2008) Correct definition of the Poynting vector in electrically and magnetically polarizable medium reveals that negative refraction is impossible, *Opt Express*, **16**(23), 19152–19168.
57. Costa JT, Silveirinha MG, Alu A (2011) Poynting vector in negative-index metamaterials, *Phys Rev B*, **83**(16).
58. Persson NK, Schubert M, Inganas O (2004) Optical modelling of a layered photovoltaic device with a polyfluorene derivative/fullerene as the active layer, *Sol Energy Mater Sol Cells*, **83**(2–3), 169–186.
59. Zhokhavets U, Gobsch G, Hoppe H, Sariciftci NS (2004) Anisotropic optical properties of thin poly(3-octylthiophene)-films as a function of preparation conditions, *Synt Met*, **143**(1), 113–117.
60. Zhokhavets U, *et al.* (2003) Anisotropic optical properties of conjugated polymer and polymer/fullerene films, *Thin Solid Films* **444**(1–2), 215–220.
61. Campoy-Quiles M, Etchegoin PG, Bradley DDC (2005) On the optical anisotropy of conjugated polymer thin films, *Phys Rev B*, **72**(4), 045209.
62. Campoy-Quiles M, *et al.* (2008) On the determination of anisotropy in polymer thin films: A comparative study of optical techniques, *Phys Stat Sol C-Curr Topic Sol Stat Phys*, **5**, 1270–1273.
63. Pettersson LAA, Ghosh S, Inganas O (2002) Optical anisotropy in thin films of poly(3,4-ethylenedioxythiophene)-poly(4-styrenesulfonate), *Org Electron*, **3**(3-4), 143–148.
64. Dennler G, *et al.* (2007) Angle dependence of external and internal quantum efficiencies in bulk-heterojunction organic solar cells, *J Appl Phys*, **102**(5), 054516.
65. Cheyns D, *et al.* (2008) The angular response of ultrathin film organic solar cells, *Appl Phys Lett*, **92**(24), 243310.
66. Meyer A, Ade H (2009) The effect of angle of incidence on the optical field distribution within thin film organic solar cells, *J Appl Phys*, 106(11), 113101.

67. Kim J, Jung S, Jeong I (2012) Optical modeling for polarization-dependent optical power dissipation of thin-film organic solar cells at oblique incidence, *J Opt Soc Korea,* **16**(1), 6–12.

68. Schubert M (1996) Polarization-dependent optical parameters of arbitrarily anisotropic homogeneous layered systems, *Phys Rev B,* **53**(8), 4265–4274.

69. Li G, Chu CW, Shrotriya V, Huang J, Yang Y (2006) Efficient inverted polymer solar cells, *Appl Phys Lett,* **88**(25), 253503.

70. Liao HH, Chen LM, Xu Z, Li G, Yang Y (2008) Highly efficient inverted polymer solar cell by low temperature annealing of Cs_2CO_3 interlayer, *Appl Phys Lett,* **92**(17), 173303.

71. Krebs FC (2008) Air stable polymer photovoltaics based on a process free from vacuum steps and fullerenes, *Sol Energy Mater Sol Cells,* **92**(7), 715–726.

72. Zhou YH, et al. (2009) Inverted and transparent polymer solar cells prepared with vacuum-free processing, *Sol Energy Mater Sol Cells,* **93**(4), 497–500.

73. Chen LM, Hong ZR, Li G, Yang Y (2009) Recent progress in polymer solar cells: Manipulation of polymer:fullerene morphology and the formation of efficient inverted polymer solar cells, *Adv Mate*r, **21**(14–15), 1434–1449.

74. Krebs FC (2009) All solution roll-to-roll processed polymer solar cells free from indium-tin-oxide and vacuum coating steps, *Org Electron,* **10**(5), 761–768.

75. Zhou YH, *et al.* (2010) Indium tin oxide-free and metal-free semitransparent organic solar cells, *Appl Phys Lett,* **97**(15), 153304.

76. Sun YM, Seo JH, Takacs CJ, Seifter J, Heeger AJ (2011) Inverted polymer solar cells integrated with a low-temperature-annealed sol-gel-derived ZnO film as an electron transport layer, *Adv Mater,* **23**(14), 1679–1683.

77. Larsen-Olsen TT, Bundgaard E, Sylvester-Hvid KO, Krebs FC (2011) A solution process for inverted tandem solar cells, *Org Electron,* **12**(2), 364–371.

78. Zhou YH, *et al.* (2012) A universal method to produce low-work function electrodes for organic electronics, *Science,* **336**(6079), 327–332.

79. Lindell L, *et al.* (2006) Transparent, plastic, low-work-function poly(3,4-ethylenedioxythiophene) electrodes, *Chem Mater,* **18**(18), 4246–4252.

80. Osikowicz W, *et al.* (2004) Transparent low-work-function indium tin oxide electrode obtained by molecular scale interface engineering, *Appl Phys Lett,* **85**(9), 1616–1618.

81. Nyberg T (2004) An alternative method to build organic photodiodes, *Synt Met,* **140**(2–3), 281–286.
82. Manceau M, Angmo D, Jorgensen M, Krebs FC (2011) ITO-free flexible polymer solar cells: From small model devices to roll-to-roll processed large modules, *Org Electron,* **12**(4), 566–574.
83. Tang Z, et al. (2012) Interlayer for modified cathode in highly efficient inverted ito-free organic solar cells, *Adv Mater,* **24**(4), 554–558.
84. Zimmermann B, *et al.* (2006) Organic solar cells with inverted layer sequence incorporating optical spacers—simulation and experiment, art. no. 61970G-1, *Photonics for solar energy systems,* Proceedings of the Society of Photo-Optical Instrumentation Engineers (Spie), ed Gombert A, **6197**, 1–1.
85. Moule AJ, Meerholz K (2007) Minimizing optical losses in bulk heterojunction polymer solar cells, *Appl Phys B-Lasers Opt,* **86**(4), 721–727.
86. Ameri T, *et al.* (2008) Realization, characterization, and optical modeling of inverted bulk-heterojunction organic solar cells, *J Appl Phys,* **103**(8), 084506.
87. Ameri T, *et al.* (2010) Fabrication, optical modeling, and color characterization of semitransparent bulk-heterojunction organic solar cells in an inverted structure, *Adv Func Mater,* **20**(10), 1592–1598.
88. Albrecht S, *et al.* (2012) Light management in PCPDTBT:$PC_{70}BM$ solar cells: A comparison of standard and inverted device structures, *Org Electron,* **13**(4), 615–622.
89. Peumans P, Bulovic V, Forrest SR (2000) Efficient photon harvesting at high optical intensities in ultrathin organic double-heterostructure photovoltaic diodes, *Appl Phys Lett,* **76**(19), 2650–2652.
90. Peumans P, Forrest SR (2001) Very-high-efficiency double-heterostructure copper phthalocyanine/C_{60} photovoltaic cells, *Appl Phys Lett,* **79**(1), 126.
91. Kim JY, et al. (2006) New architecture for high-efficiency polymer photovoltaic cells using solution-based titanium oxide as an optical spacer, *Adv Mater,* **18**(5), 572–576.
92. Andersson BV, Huang DM, Moule AJ, Inganas O (2009) An optical spacer is no panacea for light collection in organic solar cells, *Appl Phys Lett,* **94**(4), 043302.
93. Shrotriya V, Wu EHE, Li G, Yao Y, Yang Y (2006) Efficient light harvesting in multiple-device stacked structure for polymer solar cells, *Appl Phys Lett,* **88**(6), 064104.

94. Chen FC, Wu JL, Hsieh KH, Chen WC, Lee SW (2008) Polymer photovoltaic devices with highly transparent cathodes, *Org Electron,* **9**(6), 1132–1135.

95. Koeppe R, *et al.* (2009) Organic solar cells with semitransparent metal back contacts for power window applications, *ChemSusChem,* **2**(4), 309–313.

96. Nickel F, *et al.* (2010) Cathodes comprising highly conductive poly (3,4-ethylenedioxythiophene):poly(styrenesulfonate) for semi-transparent polymer solar cells, *Org Electron,* **11**(4), 535–538.

97. Colsmann A, et al. (2011) Efficient semi-transparent organic solar cells with good transparency color perception and rendering properties, *Adv Energy Mater,* **1**(4), 599–603.

98. Huang JS, Li G, Yang Y (2008) A semi-transparent plastic solar cell fabricated by a lamination process, *Adv Mater,* **20**(3), 415–419.

99. Schmidt H, *et al.* (2009) Efficient semitransparent inverted organic solar cells with indium tin oxide top electrode, *Appl Phys Lett,* **94**(24), 243302.

100. Xia XY, et al. (2010) Infrared-transparent polymer solar cells, *J Mater Chem,* **20**(39), 8478–8482.

101. Zhang FL, Johansson M, Andersson MR, Hummelen JC, Inganas O (2002) Polymer photovoltaic cells with conducting polymer anodes, *Adv Mater,* **14**(9), 662–665.

102. Winther-Jensen B, Krebs FC (2006) High-conductivity large-area semi-transparent electrodes for polymer photovoltaics by silk screen printing and vapour-phase deposition, *Sol Energy Mater Sol Cells,* **90**(2), 123–132.

103. Admassie S, et al. (2006) A polymer photodiode using vapour-phase polymerized PEDOT as an anode, *Sol Energy Mater Sol Cells,* **90**(2), 133–141.

104. Jakobsson FLE, *et al.* (2006) Towards all-plastic flexible light emitting diodes, *Chem Phys Lett,* **433**(1–3), 110–114.

105. Gadisa A, *et al.* (2006) Transparent polymer cathode for organic photovoltaic devices, *Synt Met,* **156**(16–17), 1102–1107.

106. Lee JY, Connor ST, Cui Y, Peumans P (2010) Semitransparent organic photovoltaic cells with laminated top electrode, *Nano Lett,* **10**(4), 1276–1279.

107. Meiss J, Riede MK, Leo K (2009) Optimizing the morphology of metal multilayer films for indium tin oxide (ITO)-free inverted organic solar cells, *J Appl Phys,* **105**(6), 063108.

108. Meiss J, *et al.* (2011) Highly efficient semitransparent tandem organic solar cells with complementary absorber materials, *Appl Phys Lett,* **99**(4), 043301.

109. Bailey-Salzman RF, Rand BP, Forrest SR (2006) Semitransparent organic photovoltaic cells, *Appl Phys Lett,* **88**(23), 233502.

110. Lunt RR, Bulovic V (2011) Transparent, near-infrared organic photovoltaic solar cells for window and energy-scavenging applications, *Appl Phys Lett,* **98**(11), 113305.

111. Zhou Y, *et al.* (2010) Indium tin oxide-free and metal-free semitransparent organic solar cells, *Appl Phys Lett,* **97**(15), 153304.

112. Meiss J, Holzmueller F, Gresser R, Leo K, Riede M (2011) Near-infrared absorbing semitransparent organic solar cells, *Appl Phys Lett,* **99**(19), 193307.

113. Chen CC, *et al.* (2012) Visibly transparent polymer solar cells produced by solution processing, *ACS Nano,* **6**(8), 7185–7190.

114. Tang Z, et al. (2012) Semi-transparent tandem organic solar cells with 90% internal quantum efficiency, *Adv Energy Mater,* **2**(12), 1467–1476.

115. Ajuria J, *et al.* (2011) Inverted ITO-free organic solar cells based on p and n semiconducting oxides. New designs for integration in tandem cells, top or bottom detecting devices, and photovoltaic windows, *Energy Environ Sci,* **4**(2), 453–458.

116. Granlund T, Pettersson LAA, Anderson MR, Inganas O (1997) Interference phenomenon determines the color in an organic light emitting diode, *J Appl Phys,* **81**(12), 8097–8104.

117. O'Connor B, An KH, Pipe KP, Zhao Y, Shtein M (2006) Enhanced optical field intensity distribution in organic photovoltaic devices using external coatings, *Appl Phys Lett,* **89**(23), 233502.

118. Oyamada T, Sugawara Y, Terao Y, Sasabe H, Adachi C (2007) Top light-harvesting organic solar cell using ultrathin Ag/MgAg layer as anode, *Jap J Appl Phys 1, Review Papers* 46(4A), 1734–1735.

119. O'Connor B, Haughn C, An KH, Pipe KP, Shtein M (2008) Transparent and conductive electrodes based on unpatterned, thin metal films, *Appl Phys Lett,* **93**(22), 223304.

120. Winkler T, *et al.* (2011) Efficient large area semitransparent organic solar cells based on highly transparent and conductive ZTO/Ag/ZTO multilayer top electrodes, *Org Electron,* **12**(10), 1612–1618.

121. Meiss J, Allinger N, Riede MK, Leo K (2008) Improved light harvesting in tin-doped indum oxide, (ITO)-free inverted bulk-heterojunction organic solar cells using capping layers, *Appl Phys Lett,* **93**(10), 103311.

122. Meiss J, *et al.* (2009) Efficient semitransparent small-molecule organic solar cells, *Appl Phys Lett,* **95**(21), 213306.

123. Meiss J, Riede MK, Leo K (2009) Towards efficient tin-doped indium oxide (ITO)-free inverted organic solar cells using metal cathodes, *Appl Phys Lett,* **94**(1), 013303.

124. Long YB (2009) Improving optical performance of inverted organic solar cells by microcavity effect, *Appl Phys Lett,* **95**(19), 193301.

125. Meiss J, Furno M, Pfuetzner S, Leo K, Riede M (2010) Selective absorption enhancement in organic solar cells using light incoupling layers, *J Appl Phys,* **107**(5), 053117.

126. Lee J, Kim SY, Kim C, Kim JJ (2010) Enhancement of the short circuit current in organic photovoltaic devices with microcavity structures, *Appl Phys Lett,* **97**(8), 083306.

127. Ghosh DS, Betancur R, Chen TL, Pruneri V, Martorell J (2011) Semi-transparent metal electrode of Cu-Ni as a replacement of an ITO in organic photovoltaic cells, *Sol Energy Mater Sol Cells,* **95**(4), 1228–1231.

128. Jung GH, Hong K, Dong WJ, Kim S, Lee JL (2011) BCP/Ag/MoO_3 transparent cathodes for organic photovoltaics, *Adv Energy Mater,* **1**(6), 1023–1028.

129. Sergeant NP, et al. (2012) Design of transparent anodes for resonant cavity enhanced light harvesting in organic solar cells, *Adv Mater,* **24**(6), 728–732.

130. Lin HW, et al. (2012) Device engineering for highly efficient top-illuminated organic solar cells with microcavity structures, *Adv Mater,* **24**(17), 2269–2272.

131. Green MA (2011) Enhanced evanescent mode light trapping in organic solar cells and other low index materials, *Prog Photovol Res Appl,* **19**, 473–477.

132. Yablonovitch E (1982) Statistical ray optics, *J Opt Soc Am,* **72**(7), 899–907.

133. Yu ZF, Raman A, Fan SH (2010) Fundamental limit of nanophotonic light trapping in solar cells, *Proc Natl Acad Sci USA,* **107**(41), 17491–17496.

134. Callahan DM, Munday JN, Atwater HA (2012) Solar cell light trapping beyond the ray optic limit, *Nano Lett,* **12**(1), 214–218.

135. Tvingstedt K, Andersson V, Zhang F, Inganas O (2007) Folded reflective tandem polymer solar cell doubles efficiency, *Appl Phys Lett,* **91**(12), 123514.

136. Rim SB, Zhao S, Scully SR, McGehee MD, Peumans P (2007) An effective light trapping configuration for thin-film solar cells, *Appl Phys Lett,* **91**(24), 243501.

137. Roman LS, et al. (2000) Trapping light in polymer photodiodes with soft embossed gratings, *Adv Mater*, **12**(3), 189–195.
138. Niggemann M, *et al.* (2006) Functional microprism substrate for organic solar cells, *Thin Solid Films*, **511**, 628–633.
139. Kim JB, *et al.* (2012) Wrinkles and deep folds as photonic structures in photovoltaics, *Nat Photonics*, **6**(5), 327–332.
140. Tvingstedt K, Dal Zilio S, Inganas O, Tormen M (2008) Trapping light with micro lenses in thin film organic photovoltaic cells, *Opt Express*, **16**(26), 21608–21615.
141. Tvingstedt K, Tang Z, Inganas O (2012) Light trapping with total internal reflection and transparent electrodes in organic photovoltaic devices, *Appl Phys Lett*, **101**(16), 163902.
142. Atwater HA, Polman A (2010) Plasmonics for improved photovoltaic devices, *Nat Mater*, **9**(3), 205–213.
143. Ferry VE, Munday JN, Atwater HA (2010) Design considerations for plasmonic photovoltaics, *Adv Mater*, **22**(43), 4794–4808.
144. Mallick SB, Sergeant NP, Agrawal M, Lee JY, Peumans P (2011) Coherent light trapping in thin-film photovoltaics, *MRS Bull*, **36**(6), 453–460.
145. Green MA, Pillai S (2012) Harnessing plasmonics for solar cells, *Nat Photonics*, **6**(3), 130–132.
146. Hagglund C, Apell SP (2012) Plasmonic near-field absorbers for ultrathin solar cells, *J Phys Chem Lett*, **3**(10), 1275–1285.
147. Barnes WL, Dereux A, Ebbesen TW (2003) Surface plasmon subwavelength optics, *Nature*, **424**(6950), 824–830.
148. Rand BP, Peumans P, Forrest SR (2004) Long-range absorption enhancement in organic tandem thin-film solar cells containing silver nanoclusters, *J Appl Phys*, **96**(12), 7519–7526.
149. Rockstuhl C, Fahr S, Lederer F (2008) Absorption enhancement in solar cells by localized plasmon polaritons, *J Appl Phys*, **104**(12), 123102.
150. Tvingstedt K, Persson NK, Inganas O, Rahachou A, Zozoulenko IV (2007) Surface plasmon increase absorption in polymer photovoltaic cells, *Appl Phys Lett*, **91**(11), 113514.
151. Min CJ, *et al.* (2010) Enhancement of optical absorption in thin-film organic solar cells through the excitation of plasmonic modes in metallic gratings, *Appl Phys Lett*, **96**(13), 133302.
152. Abass A, Le KQ, Alu A, Burgelman M, Maes B (2012) Dual-interface gratings for broadband absorption enhancement in thin-film solar cells, *Phys Rev B*, **85**(11), 115449.

153. Abass A, Shen HH, Bienstman P, Maes B (2011) Angle insensitive enhancement of organic solar cells using metallic gratings, *J Appl Phys,* **109**(2), 023111.

154. Sefunc MA, Okyay AK, Demir HV (2011) Plasmonic backcontact grating for P3HT:PCBM organic solar cells enabling strong optical absorption increased in all polarizations, *Opt Express,* **19**(15), 14200–14209.

155. Shen HH, Maes B (2011) Combined plasmonic gratings in organic solar cells, *Opt Express,* **19**(23), A1202–A1210.

156. Zhu JF, *et al.* (2011) Plasmonic effects for light concentration in organic photovoltaic thin films induced by hexagonal periodic metallic nanospheres, *Appl Phys Lett,* **98**(15), 151110.

157. Chen XH, Zhao CC, Rothberg L, Ng MK (2008) Plasmon enhancement of bulk heterojunction organic photovoltaic devices by electrode modification, *Appl Phys Lett,* **93**(12), 123302.

158. Kulkarni AP, Noone KM, Munechika K, Guyer SR, Ginger DS (2010) Plasmon-enhanced charge carrier generation in organic photovoltaic films using silver nanoprisms, *Nano Lett,* **10**(4), 1501–1505.

159. Reilly TH, van de Lagemaat J, Tenent RC, Morfa AJ, Rowlen KL (2008) Surface-plasmon enhanced transparent electrodes in organic photovoltaics, *Appl Phys Lett,* **92**(24), 243304.

160. Lal NN, et al. (2011) Enhancing solar cells with localized plasmons in nanovoids, *Opt Express,* **19**(12), 11256–11263.

161. Niesen B, *et al.* (2012) Near-field interactions between metal nanoparticle surface plasmons and molecular excitons in thin-films. Part II: Emission, *J Phys Chem C,* **116**(45), 24215–24223.

162. Stenzel O, Stendal A, Voigtsberger K, Vonborczyskowski C (1995) Enhancement of the photovoltaic conversion efficiency of copper phthalocyanine thin-film devices by incorporation of metal-clusters, *Sol Energy Mater Sol Cells,* **37**(3–4), 337–348.

163. Westphalen M, Kreibig U, Rostalski J, Luth H, Meissner D (2000) Metal cluster enhanced organic solar cells, *Sol Energy Mater Sol Cells,* **61**(1), 97–105.

164. Shen HH, Bienstman P, Maes B (2009) Plasmonic absorption enhancement in organic solar cells with thin active layers, *J Appl Phys,* **106**(7), 073109.

165. Vedraine S, et al. (2011) Intrinsic absorption of plasmonic structures for organic solar cells, *Sol Energy Mater Sol Cells,* **95**, S57–S64.

166. Diukman I, Tzabari L, Berkovitch N, Tessler N, Orenstein M (2011) Controlling absorption enhancement in organic photovoltaic cells by patterning Au nano disks within the active layer, *Opt Express,* **19**(1), A64–A71.

167. Naidu BVK, et al. (2008) Novel hybrid polymer photovoltaics made by generating silver nanoparticles in polymer:fullerene bulk-heterojunction structures, *Sol Energy Mater Sol Cells,* **92**(4), 397–401.

168. Wang DH, et al. (2011) Enhanced power conversion efficiency in PCDTBT/PC_{70}BM bulk heterojunction photovoltaic devices with embedded silver nanoparticle clusters, *Adv Energy Mater,* **1**(5), 766–770.

169. Liu HJ, Goh WP, Leung MY, Li YN, Norsten TB (2012) Effect of nanoparticle stabilizing ligands and ligand-capped gold nanoparticles in polymer solar cells, *Sol Energy Mater Sol Cells,* **96**(1), 302–306.

170. Topp K, *et al.* (2010) Impact of the incorporation of Au nanoparticles into polymer/fullerene solar cells, *J Phys Chem A* **114**(11), 3981–3989.

171. Wang CCD, et al. (2012) Optical and electrical effects of gold nanoparticles in the active layer of polymer solar cells, *J Mater Chem,* **22**(3), 1206–1211.

172. Kim SS, Na SI, Jo J, Kim DY, Nah YC (2008) Plasmon enhanced performance of organic solar cells using electrodeposited Ag nanoparticles, *Appl Phys Lett,* **93**(7), 073307.

173. Chen FC, *et al.* (2009) Plasmonic-enhanced polymer photovoltaic devices incorporating solution-processable metal nanoparticles, *Appl Phys Lett,* **95**(1), 013305.

174. Xie FX, Choy WCH, Wang CCD, Sha WEI, Fung DDS (2011) Improving the efficiency of polymer solar cells by incorporating gold nanoparticles into all polymer layers, *Appl Phys Lett,* **99**(15), 153304.

175. Fung DDS, et al. (2011) Optical and electrical properties of efficiency enhanced polymer solar cells with Au nanoparticles in a PEDOT-PSS layer, *J Mater Chem,* **21**(41), 16349–16356.

176. Yang J, *et al.* (2011) Plasmonic polymer tandem solar cell, *ACS Nano,* **5**(8), 6210–6217.

177. Niesen B, *et al.* (2012) Near-field interactions between metal nanoparticle surface plasmons and molecular excitons in thin-films. Part I: Absorption, *J Phys Chem C,* **116**(45), 24206–24214.

178. Niesen B, et al. (2013) Plasmonic efficiency enhancement of high performance organic solar cells with a nanostructured rear electrode, *Adv Energy Mater,* **3**(2), 145–150.
179. You JB, et al. (2012) Surface plasmon and scattering-enhanced low-bandgap polymer solar cell by a metal grating back electrode, *Adv Energy Mater,* **2**(10), 1203–1207.
180. Dissanayake D, Roberts B, Ku PC (2011) Plasmonic backscattering enhanced inverted photovoltaics, *Appl Phys Lett,* **99**(11), 113306.
181. Leo K, *et al.* (2013) *Energies* in press.
182. Heriot SY, Jones RAL (2005) An interfacial instability in a transient wetting layer leads to lateral phase separation in thin spin-cast polymer-blend films, *Nat Mater,* **4**(10), 782–786.
183. Schmidt-Hansberg B, *et al.* (2011) Moving through the phase diagram: Morphology formation in solution cast polymer-fullerene blend films for organic solar cells, *ACS Nano,* **5**(11), 8579–8590.
184. Schmidt-Hansberg B, *et al.* (2011) Spatially resolved drying kinetics of multi-component solution cast films for organic electronics, *Chem Eng Proc,* **50**(5–6), 509–515.
185. Ebbens S, *et al.* (2011) In situ imaging and height reconstruction of phase separation processes in polymer blends during spin coating, *ACS Nano,* **5**(6), 5124–5131.
186. Hou LT, *et al.* (2011) Lateral phase separation gradients in spin-coated thin films of high-performance polymer:fullerene photovoltaic blends, *Adv Funct Mater,* **21**(16), 3169–3175.
187. Bergqvist J, Mauger SA, Tvingstedt K, Arwin H, Inganas O (2013) In situ reflectance imaging of organic thin film formation from solution deposition, *Sol Energy Mater Sol Cells,* **114**, 89–98.

Chapter 9

Spectroscopy of Charge-Carrier Dynamics—From Generation to Collection

Sarah R. Cowan[a] **and Natalie Banerji**[b]

[a]*National Center for Photovoltaics, National Renewable Energy Laboratory, Colorado 80401–3305, USA*

[b]*Ecole Polytechnique Fédérale de Lausanne, Institute of Chemical Sciences & Engineering, EPFL SB ISIC GR-MO), CH-1015 Lausanne, Switzerland*

sarah.cowan@nrel.gov, natalie.banerji@epfl.ch

A fundamental understanding of the physical processes, which contribute to efficient mobile carrier generation and transport in organic solar cells, remains under debate. We explore the role that spectroscopic and electrical techniques have played in understanding the processes relevant to photogeneration of charge carriers and the interplay between recombination and collection of photogenerated charge carriers. An inexhaustive survey of the way in which sets of measurement techniques complement each other or have presented conflicting sets of data on the same materials systems, and continue to shape popular perception of photovoltaic current generation in bulk heterojunction (BHJ) systems is presented.

Organic Solar Cells: Fundamentals, Devices, and Upscaling

Edited by Barry P. Rand and Henning Richter

ISBN 978-981-4463-65-2 (Hardcover), 978-981-4463-66-9 (eBook)

www.panstanford.com

9.1 Introduction

The functioning of any solar cell is based on the absorption of light by an active material sandwiched between two electrodes, followed by the formation of mobile charge carriers, which integrate an external circuit. Organic photovoltaics (small molecule and polymer based) rely on the concept, first introduced in 1989, that charge carriers are generated by photoinduced charge separation (CS) at the heterojunction between an electron donor and acceptor.[1] The efficiency of CS in bilayer devices is however limited, because excitons can recombine before reaching the D/A interface. A breakthrough was achieved in 1995 with the discovery that D and A can be blended to form a nanoscale BHJ active layer material.[2]

Progress in the fabrication of organic BHJ solar cells continues to proceed at a rapid pace. While record power conversion efficiencies (PCEs) in the field were limited for a decade below 5%, the last 5 years has seen the synthesis of a vast new class of molecules: donor–acceptor molecules, in which the push–pull structure of a single molecule stabilizes the intramolecular charge transfer process allowing for a more delocalized primary excitation and highly efficient charge transfer to the intercalated electron transport molecular matrix. This new class of molecules, which has demonstrated record single-cell efficiencies (7 to >10%) in polymer (many conjugated units) and small-molecule (single monomer unit) form, has opened a new chapter of progress in organic solar cell PCE and stability, and in physical understanding of morphology, charge transport, and recombination processes.

Here, the role that spectroscopic and electrical techniques have played in understanding the physical processes relevant to photogeneration of charge carriers and the interplay between recombination and collection of photogenerated charge carriers will be detailed. Along the way, we present an inexhaustive survey of the way in which sets of measurement techniques complement each other or have presented conflicting sets of data on the same materials systems, and continue to shape popular perception of photovoltaic current generation in BHJ systems.

For an introduction to a discussion on the use of spectroscopy to understand the physical operation of an organic solar cell, we briefly outline some basic concepts and terms that have been comprehensively described in other sources.[3,4] We start at the beginning: In 1995, two research groups first published blended heterojunctions exhibiting photovoltaic behavior: polymer: polymer[5] and polymer:fullerene[2] BHJs. The two components mixed in solution and cast into a thin film formed a complex internal structure. The BHJ in organic BHJ solar cells is composed of two intimately mixed components: an absorber (donor) material, usually p-type and primarily responsible for light absorption and hole transport, and an acceptor material, usually n-type and with a high electron affinity, and primarily responsible for facilitating an efficient charge transfer process and efficient electron transport. The nomenclature, donor and acceptor, derives from each material's role in the charge transfer process at the p-n junction interface, such that in a single charge transfer process, the "donor" material donates an electron across the interface and the acceptor material "accepts" the electron. The interface between the donor and acceptor materials makes up the disordered p-n junction.

While it has been shown to be possible to draw many parallels between crystalline p-n junction solar cells (mainly made of inorganic materials) and the disordered BHJ solar cell, a few critical physical differences make BHJ solar cells a class unto themselves, to wit: lack of long-range crystalline atomic or molecular structure in the p-and n-type regions results in a significantly higher binding energy for the primary exciton and a significantly lower carrier mobility. Many of the materials used in BHJ solar cells, detailed at length elsewhere in this book, have been shown to exhibit long-range order when grown in pristine (unmixed) films under ideal conditions[6] and short-range order when intermixed in BHJ solar cells.[7] However, even in systems that show short-range order, solar cell performance remains characterized by the physics of disordered semiconductors.

The theoretical framework of crystalline semiconductors, based on the periodicity of an ordered atomic lattice, cannot apply to polymers and small molecules because while polymers and small molecules may be locally ordered around a rigid conjugated backbone, the molecules are only weakly linked via van der Waals forces. In such weak bonding conditions, the electronic wave functions of the molecules in a thin film do not significantly overlap. Hence, the bulk properties of an organic thin film are, to first order, quite similar to properties of individual molecules, leading to several significant differences between crystalline semiconductors and the disordered molecular semiconductors used in organic BHJ solar cells, as summarized in Table 9.1. Weak intermolecular interactions result in novel mechanisms for the formation of defects, an exciton binding energy above the thermal voltage ($V_T = kT/e \sim 25$ meV), the localization of charge carriers, and lower carrier mobility for electrons and holes.

Descriptions of the formation of a bandgap via a carbon chain with alternating single and double bonds, commonly referred to as a conjugated backbone, and of charge transport in systems like the one outlined above have been described in detail elsewhere. We refer you to books[3,4] or the literature for further details.

Table 9.1 Significant differences between crystalline semiconductors and disordered molecular semiconductors in physical processes relevant to photovoltaic applications

Property or Process	→	Crystalline Semiconductor	Disordered Molecular Semiconductor
Band structure	→	Atomic lattice periodity (see Bloch's theorem)	Intramolecular periodicity (1st order) and intermolecular interactions (2nd order)
Defects	→	Disruptions of the atomic lattice—vacancies and interstitial defects	Intramolecular conformation, molecular packing, electrically active impurities
Defect states	→	Discrete gap states	Band tails, energetically broadened gap states

(Continued)

Table 9.1 (*Continued*)

Property or Process	→	Crystalline Semiconductor	Disordered Molecular Semiconductor
Photogeneration	→	Strong (covalent/ ionic) bonding between lattice atoms → $E_b << kT$ (25 meV)	Weak intermolecular bonding → material-dependent E_b, 40 meV < E_b < 1 eV
Carrier transport	→	Delocalized charge carriers, high-mobility transport	Intramolecularly localized charge carriers, lossy intermolecular "hopping" transport
Primary transport mechanism in a solar cell architecture	→	Delocalized charge screen field, diffusion transport	Localized charge does not efficiently screen internal field, drift transport
Theoretical framework	→	Classic band theory	Band theory modified for disordered materials, wide array of molecularly based models

Here, we directly address the results of these phenomena, i.e., photogeneration of charge carriers and how the interplay between recombination and collection of photogenerated carriers directly contributes to photovoltaic performance.

9.2 Controversy and Uncertainty Regarding the Physical Concepts in a BHJ Solar Cell

While organic semiconductors have been shown to form efficient devices, including commercialized light-emitting diodes, and thin film transistors and solar cells nearing the market, a fundamental understanding of the physical processes that contribute to efficient mobile carrier generation and transport remains under debate. Popular perceptions of the physical processes, which contribute to efficient photovoltaic current generation in BHJ solar cells, are as follows.

Photons incident upon a "plastic" solar cell initiate a series of optical and electrical processes ultimately resulting in the generation of charge in the external circuit. A photon absorbed by a polymer chain produces an excitation along the conjugated backbone.[4] Figure 9.1 outlines the time evolution of the photoexcitation and resulting charge carriers induced in a BHJ thin film in response to a single ultrafast pulse of light. Processes that have been observed by experiment are identified along with their rough time scale, and the quasiparticle/particle evolution highlights the commonly used physical picture of current generation in BHJ solar cells. This includes (1) absorption of a photon resulting in a primary photoexcitation, (2) relaxation of the primary excitation to a singlet exciton, (3) exciton diffusion to a donor–acceptor interface, (4) charge transfer to create mobile electrons and holes via the charge transfer state, (5) charge transport to the electrodes and charge collection, (6) residual presence of deep-level trapped carriers. Studies of loss processes for each of these steps, charge generation, charge transport, and charge collection, produce conflicting evidence on the role of excitons and charge transfer (CT) states, in particular, in *efficient* photovoltaic conversion.

Debate about the nature and evolution of the primary photoexcitation in conjugated polymers is ongoing.[4,8–12] There are two main schools of thought. On the one hand, conjugated polymers have been treated as one-dimensional semiconductors, where mobile electrons and holes are directly photogenerated in energy bands and eventually localize into a bound exciton state.[4,8,13] On the other hand, evidence exists for a more molecular approach, where the primary photoexcitation is already a bound exciton.[9,14,15] The limit between the two cases depends to some extent on both disorder and on the time scale at which the photoexcitation is observed. Thus, disorder in the conjugated polymer chain determines the delocalization of the excited state and hence its more semiconductor-like or more molecular-like behavior. In highly ordered polydiacetylene quantum wires, quantum coherence reaching tens of micrometers has been observed (macroscopic delocalization of the excited state).[16] In most cases, disorder in the dissolved or solid-state materials prevents the delocalization of the excited state over the entire chain and breaks it into much smaller, more molecular-like chromophores of variable size. Those can, however, couple together at short time delays (<200 fs), giving rise to collective delocalized states.[17,18] The primary photoexcitation observed directly after light absorption is therefore much more delocalized than the excited state species (exciton) formed following a series of complex relaxation processes.[3,4,17–33]

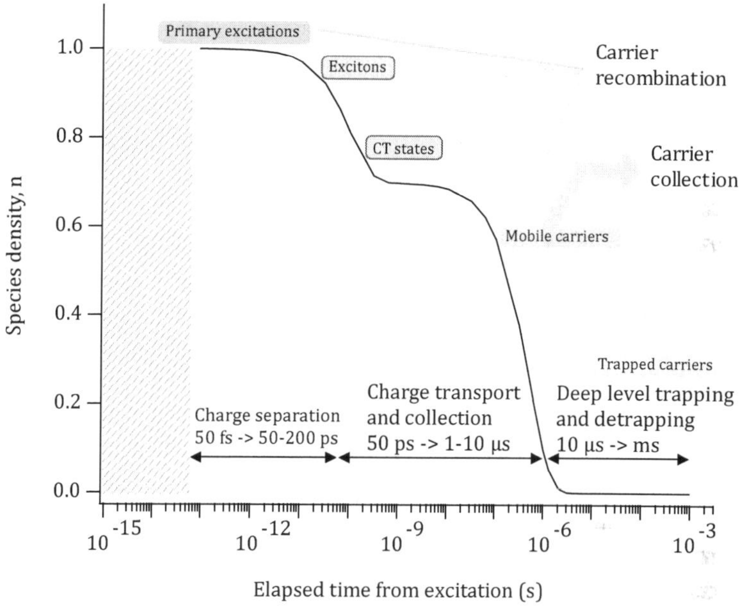

Figure 9.1 Species density temporal evolution in a BHJ thin film in response to a single ultrafast pulse of light with line width ~10 fs.

In the BHJ, CS at the donor–acceptor interface splits the excitation into a domain-segregated electron and hole. Many time-resolved studies have shown that the CS to the fullerene materials is nearly quantitative and ultrafast (<100 fs for many optimized systems).[5,34–43] Based on relaxation time scales found during time-resolved fluorescence studies on pristine polymers, the ultrafast CS can compete with primary excitation relaxation processes such as excited-state localization and exciton diffusion.[2,19,20]

Moreover, CS on the order of 100 fs (or less) limits exciton diffusion to the 0.1–0.2 nm length scale, according to published exciton diffusion coefficients.[6,44,45] However, nanoscale phase separation[7,46–51] is known to be necessary for efficient BHJ photovoltaics,[3,4,52] so that there is a discrepancy between the distance that an exciton *can* diffuse and the distance that it *needs to* diffuse in order to ensure ultrafast CS at the fullerene interface. We will discuss alternative mechanisms in which the primary photoexcitation, which is distinctively different from the bound exciton since it is more delocalized, reaches an interface and undergoes CS.

A body of research puts forward that all generation of free charge carriers passes through intermediate charge transfer (CT) states, in which electrons and holes are still coulombically bound. They are responsible for most geminate recombination events—a note: geminate recombination refers to recombination of charge-separated carriers that originally stem from the same photoexcitation. CT states, distinct electronic states existing at the interface between donor and acceptor materials and formed by the dissociation of the photoexcitation, are known in literature as a loss mechanism for polymer : polymer and polymer : small-molecule blends. CT states can be optically or electrically populated, and emission from these states can be measured via photoluminescence or electroluminescence, respectively. However, CT state emission has been measured, in contactless photo- and contacted electroluminescence experiments, to be orders of magnitude lower for blends, which produce efficient solar cells than for lossy BHJ blends. In other words, CT states seem unlikely to intervene in charge dissociation for efficient BHJ solar cells. Electroluminescence experiments on their own do not rule out the possibility that some or all charges pass through these states. However, the observation of free charge carriers on the hundreds of femtosecond time scale in efficient BHJ blends makes the probability that a majority of carriers pass through a bottleneck electronic state much less likely. It has recently been put forward that dissociation of the excited state at the D/A interface leads directly to free charges via delocalized states, so that trapping in the relaxed CT state is only a minor loss process.[53] The exact scope and magnitude of the role CT states play in the operation of an efficient BHJ solar cell remains a topic of intense debate and research.

A large fraction of the incident light absorbed in the BHJ in optimized devices is converted to photoinduced charge carriers that are collected as current in an external circuit—a process shown to be greater than 90% efficient in at least one of the donor–acceptor systems. Photogenerated current is reduced by internal recombination loss prior to charge collection at the electrodes. Competition between carrier sweep-out by the internal field and the loss of photogenerated carriers by recombination is one of the key issues to overcome for high-efficiency devices. Due to the relatively low charge-carrier mobility in common BHJ materials, the PCE does not scale linearly with the thickness of the active layer in polymer BHJ solar cells. As the active layer thickness increases, the path length of photogenerated electrons and holes to the electrodes increases. Thin devices can be made with almost 100% conversion efficiency of absorbed photons to electrons collected in the external circuit.[4,54] However, increasing the thickness of the active layer in these devices typically does not increase overall efficiency.[4,8–12,55,56] The solar cell active layer thickness, which maximizes PCE, is usually smaller than necessary for optimal light harvesting. The short-circuit current increases moderately (nonlinearly) with increasing active layer thicknesses,[57] but often, the open-circuit voltage is reduced and the fill factor drops dramatically. While a larger fraction of the incident photons are absorbed in thick film devices, the resulting photogenerated charge carriers are not efficiently extracted into the external circuit.

Recombination of free charge carriers, which unlike CS processes is dependent on the total thickness of the film, becomes the dominant loss mechanism as the active layer thickness increases, limiting short-circuit current, open-circuit voltage, and fill factor. Unfortunately, as researchers seek to increase the efficiency of polymer photovoltaics and companies begin the process of up-scaling this technology, increasing the thickness of the active layer of the BHJ becomes necessary due to the minimum film thicknesses that can be reproducibly fabricated from scalable printing processes (roll-to-roll printing, etc.).[55,58]

Recombination mechanisms are best classified by rate, as this is what principally distinguishes them experimentally. Geminate recombination refers to recombination of species that originate from the same absorbed photon and is thus a monomolecular process. This encompasses the recombination of the primary photo-excitation, of the exciton before the quasi-particle charge separates and of bound electron–hole pairs in the CT state. Geminate recombination rate is predicted to be proportional to the population of bound electron–hole pairs (in the exciton or CT state), X, by decay rate, k_f:

$$R_0 = k_f X \tag{9.1}$$

Trap-based recombination describes recombination through a trap state located within the energy gap of a material. In inorganic materials, where charge transport properties originate from the ordering of atoms in a lattice, trap states originate from interruptions in the ordered lattice. In organic materials, charge transport properties arise from inter- and intramolecular ordering. Trap states can include electrically active impurities, intramolecular energetically localized states, disordered energetic states at the interface between two molecules, or at the interface between two materials. In the case of BHJ, where majority charge-carrier populations are spatially separated by BHJ phase separation, recombination likely occurs in traps at the interface between the two phase-separated regions. The rate at which this trap is filled is assumed to be much faster than the recombination rate, such that an immobile population of trapped charge exists with which mobile carriers of the opposite sign may recombine. Equation (9.2) describes the trap-based recombination rate through a density of single-occupancy electron traps, n_t, where $n_{e,t}$ is the density of occupied traps, (n_t - $n_{e,t}$) is the density of unoccupied states available for electron occupation, and $k_{t,e(h)}$ is the rate of electron and hole capture:

$$R_1 = k_{t,e} n_e (n_t - n_{e,t}) + k_{t,h} n_h n_{e,t} \tag{9.2}$$

Bimolecular recombination describes interband recombination of an electron and hole (separated or in an excited state) which do not originate from the same photon. In the case of the BHJ, where majority charge-carrier populations are spatially separated by BHJ phase separation, recombination likely occurs at the interface between the two phase-separated regions. Because this recombination process depends on the spatial localization of a mobile electron and hole, the recombination rate is proportional to the mobile electron–hole product minus contribution from the intrinsic carrier density of both charge species (usually negligible).

The decay rate for bimolecular recombination in organic BHJ solar cells has been found to be related to the Langevin recombination rate, $\gamma = e\mu_{\text{eff}}/\varepsilon_0\varepsilon_{\text{eff}}$,

$$R_2 = \gamma(n_e n_h - n_i^2) \tag{9.3}$$

In an Auger recombination event, an excited state decaying to the ground state transfers its kinetic energy to a neighboring excited state, visualized in Fig. 9.2d as an electron–electron interaction with some decay rate, $k_{A,e(h)}$. The kinetically excited state usually relaxes back to its initial excited state via phonon interactions.

$$R_3 = k_{A,e}n_e(n_e n_h - n_i^2) + k_{A,h}n_h(n_e n_h - n_i^2) \tag{9.4}$$

Because of Auger recombination's third-order dependence on charge density, high charge densities are needed to observe Auger recombination, and it has not yet been observed in organic BHJ solar cells.

In the simplest cases of these recombination mechanisms, where the recombination rate is relatively low and $n_e \sim n_h$, $n_e >> n_i$, and $n_t >> n_{e,t}$, recombination rate equations can be simplified as summarized in Fig. 9.2.

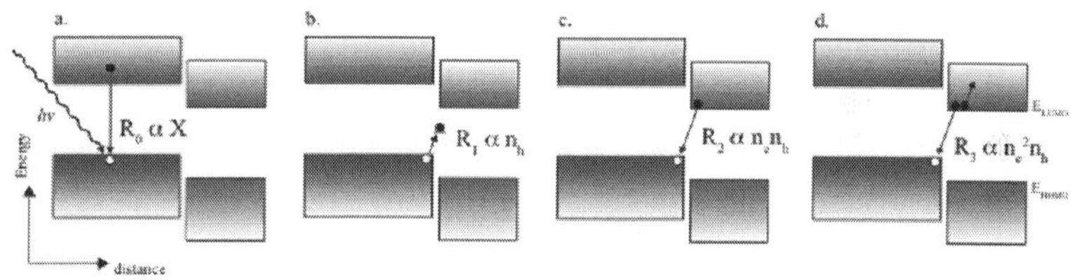

Figure 9.2 Recombination mechanisms by rate: (a) *Geminate* recombination rate proportional to the bound electron–hole density; (b) *Trap-based* recombination rate in the simplest case proportional to the density of mobile holes; (c) *Bimolecular* (interband) recombination rate proportional to the mobile electron and hole densities; (d) *Auger* recombination (transition energy transferred to neighboring particle) rate proportional to mobile holes, and square of mobile electrons.

Charge recombination has been studied via a variety of experimental methods, including time-of-flight (TOF),[59] steady-state current–voltage,[60–63] impedance spectroscopy,[64–66] photoinduced charge extraction by linearly increasing voltage (photo-CELIV),[67] double injection currents,[68] and transient absorption and transient photovoltage measurements.[69,70] At applied voltages above the maximum power point and at the open-circuit condition, bimolecular recombination reduces the current density and limits the fill factor, thereby decreasing PCE.[60,70] Determining primary recombination mechanisms and their physical origins is therefore essential to the future of organic BHJ solar cells.

In addition to processes internal to the BHJ materials, extraction of charge into an external circuit depends on one's ability to contact organic materials and draw charge out of the device. Metals including aluminum, silver, and gold are commonly used to contact the devices, as well as transparent conducting oxides, conducting polymers, carbon nanotubes, and graphene. Contact interlayers have been shown to be necessary as a buffer layer between the anode and cathode and the active layer of the device, the BHJ. These interlayers are thought to be necessary for a number of reasons. First, interlayers are necessary to "break the symmetry" of the device, or to create selectivity for primarily electron collection at the electron contact and primarily hole collection at the hole contact. Interlayers play a similar role to prevent injection of the opposite carrier. Common interlayer materials, to date, include semiconducting polymers like PEDOT:PSS, thin layers of metals with high electron affinity such as calcium and lithium fluoride, polymer polyelectrolytes, self-assembled monolayers, and transparent metal oxides including zinc oxide, molybdenum oxide, nickel oxide, and titanium oxide. Many of the metal oxides are additionally promising because of their stabilizing role in extending device lifetimes from seconds or minutes to months and years. Researchers are in the process of developing a rubric for ideal contacts, though this problem is much more difficult than it may seem (for further information, see Chapter 4).

Improvements in PCE in the field of organic photovoltaics are usually the result of entirely new materials, each with new requirements for contact layers and interlayers. Nevertheless, we will discuss here the materials parameters, which current research has shown to be critical to charge extraction: work function and energy band alignment, charge selectivity, interface defect density, transparency, and interface chemical interactions, and explore case studies in which new BHJ materials have necessitated specific contacts.

9.3 Overview of Spectroscopic and Electrical Methods for Probing Internal Processes

A wide variety of optical and electrical characterization techniques have been utilized to evaluate internal processes of organic photovoltaic devices. Many photovoltaic device characterization techniques are all-optical, the benefits of which include simple sample preparation—i.e., test structures requiring no metallic contacts. These all-optical characterization techniques probe physical processes such as photoabsorption (excitation), the evolution of the primary excitation, CS, and recombination of the excited state, CT state, and separate charges. They can be steady state or probe time evolution, or transient behavior. Still more organic photovoltaic characterization techniques are charge based, and require more detailed sample preparation, from the active layer of the device between metal electrodes to the fabrication of full device structures. Many of these techniques are listed in Table 9.2, categorized by the primary excitation source (photons or electrons) and the probed quantity (photons or electrons). Many of these techniques can be implemented in steady state or transient configurations to probe various windows of the time evolution of excitation and charge-carrier formation by changing the excitation source, detector, and electronics. Additionally, choice of energy of incoming photons or electrons will determine the process being probed. Note, however, that we do not discuss techniques for the chemical or physical characterization of these materials and devices. Such descriptions can be found elsewhere in this text.

9.3.1 Absorption/Transmission Spectroscopy

Absorption spectroscopy, or transmission spectroscopy, measures optical absorption and absorption coefficients, and can reveal a great deal more information concerning material sample crystallinity, effective optical bandgap, and some impurities. In such measurements, a broad-band light source is passed through a monochromator, such that single wavelengths of light pass through the sample in progression. Wavelength-dependent photons transmitted through the sample are measured by a detector, as illustrated in Fig. 9.3. Light incident on the sample can be absorbed, reflected, or scattered, reducing transmission. It is possible to separate and measure the relative quantities of photons reflected or transmitted by placing the source at a small angle off the orthogonal axis and measuring small angle reflectance. However, in most cases, reflection and scattering are assumed to be a relatively small effect reducing transmission, when compared to bulk absorption. For thin or largely transparent films, care must be taken in measurement as this assumption does not always hold. In order to properly measure the absorption of highly transparent or thin films, it is recommended to measure transmission and reflection.

Table 9.2 List of the primary spectroscopies used to evaluate the physical processes relevant to photogeneration and collection in organic BHJ solar cells

Photons In/ Photons Out	**Photons In/ Electrons Out**	**Electrons In/ Photons Out**	**Electrons In/ Electrons Out**
Photoluminescence	Photoemission spectroscopy	Electroluminescence	Impedance/ admittance spectroscopy
Fluorescence spectroscopy	UPS	Inverse photoelectron spectroscopy (IPES)	Double injection current
Transient absorption spectroscopy	XPS		C-V measurements
Vibrational spectroscopy	Photoconductivity		Kelvin probe
Transient microwave conductivity	Time-of-flight		Conductive AFM
Photoinduced absorption	Transient photocurrent		
Steady-state absorption	Transient photovoltage		
	Photo-CELIV		
	TDCF, TDDP		
	EQE/IQE		
	Photoconductive AFM		

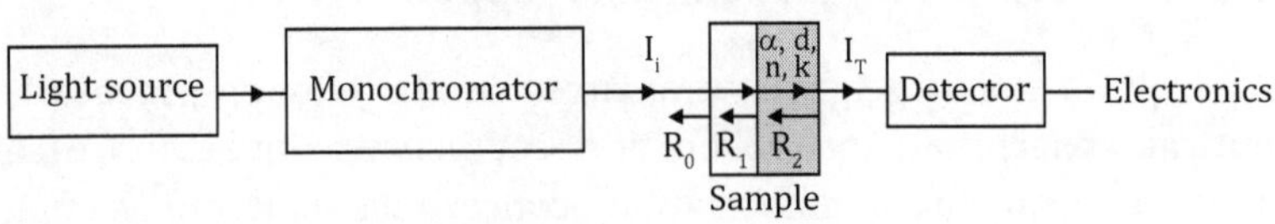

Figure 9.3 Transmission spectroscopy schematic.

Photons normally incident on a sample, I_i, undergo a series of physical processes that reduce the number of photons transmitted,

I_t. Transmittance, T, is defined as the ratio between the incident and transmitted light intensity:

$$T = I_t/I_i = 1 - A - R - S, \quad (9.5)$$

where this total fraction of incident light may be reduced by absorbance, A, reflectance, R, and scattering, S. At each interface, a fraction of the photons are reflected and transmitted, based on the difference in refractive index, n, between the two materials in intimate contact and the angle of incident light, θ, according to Snell's law. Within a material, light may be absorbed, based on the extinction coefficient, k, related to the absorption coefficient, α, and the length of the light path through the material, d. The absorption and extinction coefficients are related by the relation: $\alpha = 4\pi k/\lambda$, where λ is the wavelength of incident light. Refractive index and extinction coefficients are simply related to the complex dielectric constant of a material:

$$\tilde{\varepsilon} = \varepsilon_1 + i\varepsilon_2; \quad \varepsilon_1 = n^2 - k^2; \quad \varepsilon_2 = 2nk \quad (9.6)$$

For photon energies below their bandgap, materials are largely transparent. Light incident on a material below the bandgap is transmitted, or the material is "transparent" to photons at these wavelengths, or $\alpha(\lambda) \approx 0$. For materials with bandgap larger than the visible frequencies, such as glass, indium tin oxide, and others, the materials themselves are largely transparent to the human eye. At photon energies near the bandgap, E_G, light begins to be strongly absorbed by the material. The wavelength width of the absorption band depends on parameters of the material—specifically the width and structure (e.g., density of states) of the excited state energy band. The optical bandgap can be estimated from the wavelength-dependent absorption, but as illustrated in Fig. 9.4, amorphous organic materials often have moderately increasing absorption near the band edge and, in the case of [6,6]-phenyl-C_{61}-butyric acid methyl ester (PCBM) illustrated here, multiple absorption edges. Furthermore, in Fig. 9.4a, a plot of absorptance versus wavelength and absorptance versus photon energy inset (absorptance data extracted from Ref. [71]), demonstrates that linear extrapolation to zero absorption does not accurately predict the optical bandgap of a material. Extrapolation in this case, Fig. 9.4a, predicts lowest unoccupied molecular orbital (LUMO) edges at 3.3 and 4.1 eV, underestimating the observed energy levels by over 10% in the case of the primary absorption edge.

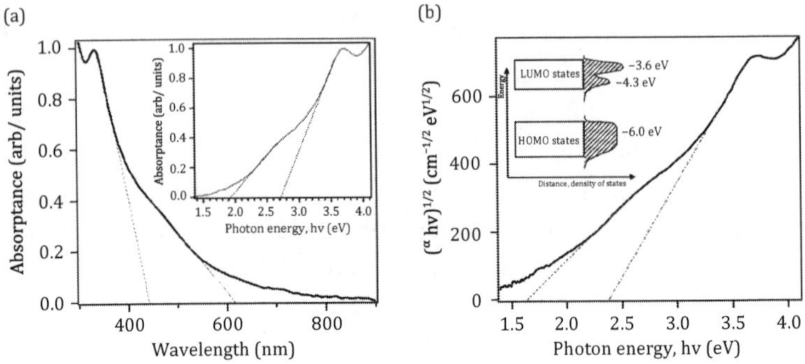

Figure 9.4 Illustrating extraction of an optical bandgap from absorption data (a) via absorption plotted vs. wavelength and photon energy, and (b) via Tauc plot for PCBM thin films. Inset: HOMO/LUMO energy band diagram inferred from Tauc plot for PCBM.

Tauc plots are commonly utilized to extract the optical energy gap from absorption data. Classic analysis of the structure of the band edge of amorphous materials[72] from 1966 from Tauc et al. has been utilized to analyze oxides, polymers, and small molecules with sufficient disorder. Tauc analysis cannot be strictly applied to conjugated semiconductors, where Bloch's theorem does not hold—strong intramolecular bonding does not extend to a full crystal structure. Efforts have been made over the past 30 years to determine the relationship between absorptivity and the energy gap to amorphous semiconductors, under the following assumptions: selection rules, based on lattice spacing, k, are relaxed, and probability of a band-to-band transition (photon absorption) shows reduced dependence on photon energy at the band edge. Common analysis of amorphous semiconductors assumes a parabolic density of state at the band edge, such that the relationship between absorptivity and energy gap can be described: $\alpha \propto (h\omega - E_g)^2/h\omega$.[73] Tauc plots may provide a better estimate of the optical energetic gap than simple absorption vs. wavelength, as illustrated in Fig. 9.4b. Here, spectrally resolved absorptance data for PCBM are used to estimate the absorption coefficient via Beer–Lambert Law: $T_y = 1{-}A_y = e^{-\alpha t}$. Absorption coefficient may also be measured via variable angle spectroscopic ellipsometry. Assuming a highest occupied molecular orbital (HOMO) energy level of –6.0 eV, the primary and secondary energy transition in the LUMO level can be estimated as –3.6 and –4.3 eV, respectively, significantly closer to the widely reported primary and secondary LUMO energy levels for PCBM.

9.3.2 Photoluminescence (PL)

Optical measurements can be generally described by the active interaction of light in the test structure—light can be absorbed, reflected, transmitted, and/or emitted. In photoluminescence measurements, an incident light source—usually higher energy than the primary absorption transition—is made incident on a photoactive film. Light of energy lower than the energy gap of the semiconductor may be used to probe midgap states and impurities. The light emitted by the sample by fluorescence or phosphorescence is then detected. (See Fig. 9.5 for a generalized example of steady-state photoluminescence experiment schematics.) Under excitation energy $E > E_G$, absorbed light creates a photoexcitation. The photoexcitation decays via a series of relaxations to the band edge, from where it can recombine to the ground state nonradiatively or radiatively (i.e., by emitting a photon).

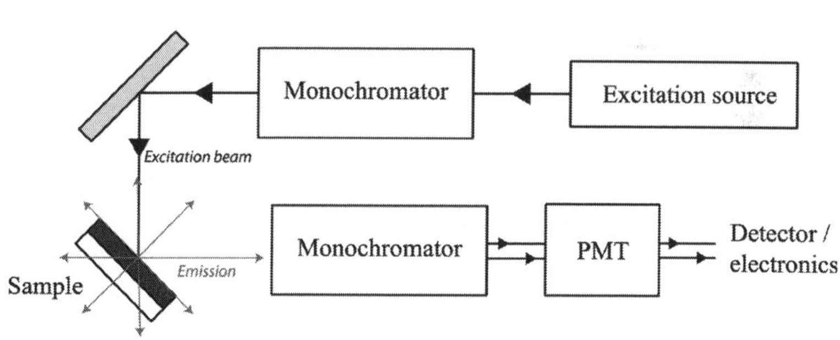

Figure 9.5 Steady-state photoluminescence experiment schematic.

Steady-state photoluminescence maps the emission spectrum due to excitation and is sometimes described as the fluorescence emission spectrum. The energy shift, and shape and structure of the photoluminescence reveal the energetic structure and disorder of the probed material in the excited state. Fluorescence excitation spectra are obtained when photoluminescence is detected at a given wavelength, while the spectrum of the excitation is scanned. By using pulsed laser experiments, time-resolved photoluminescence can be recorded. The time resolution depends on the duration of the laser pulses and typically spans the widest range of any other experiment—from femtoseconds (10^{-15} s) to steady state. A technique called time-correlated single-photon counting is often used to record nanosecond fluorescence dynamics, while fluorescence up-conversion (a gated experiment) is the most common to obtain the femtosecond evolution of the emission. Finally, fluorescence typically occurs from neutral excited states or from bound CT states, so that emission techniques are not appropriate to investigate free charge carriers.

9.3.3 Transient and Photoinduced Absorption Spectroscopy

All-optical spectroscopic techniques provide a particular advantage in that they allow the probing of absorption and charge generation processes without the complications of electrodes (i.e., recombination at the electrodes or charge injection, for instance) or an externally applied field. Time-resolved transient absorption and quasi-steady state photoinduced absorption are pump-probe techniques that measure the absorption spectrum of a sample that has previously been excited. Typically, a visible pump beam (above the bandgap) is absorbed by the sample, which generates a photoexcitation that subsequently evolves in time. For a BHJ material, this includes the processes of exciton relaxation, charge generation, and recombination described in Section 9.2. A probe beam (this can be in the visible or infrared range) then measures the absorption of the excited species and charges. Pulsed femtosecond and nanosecond lasers are used to visualize the excited-state processes in time. A motorized delay stage typically adjusts the arrival of the probe pulse with respect to the pump pulse, so that the evolution of the excited state can be followed. Photoinduced absorption (see Fig. 9.6) is widely used because it utilizes less expensive components, for example, a pulsed LED as the light source rather than a laser. Photoinduced absorption experiments are run in near-steady-state conditions, so that charge carriers that survive on the long time scale are probed. Sometimes, utilizing microsecond/millisecond time evolution with the help of a chopper gives insight into charge-separated carrier recombination mechanisms.

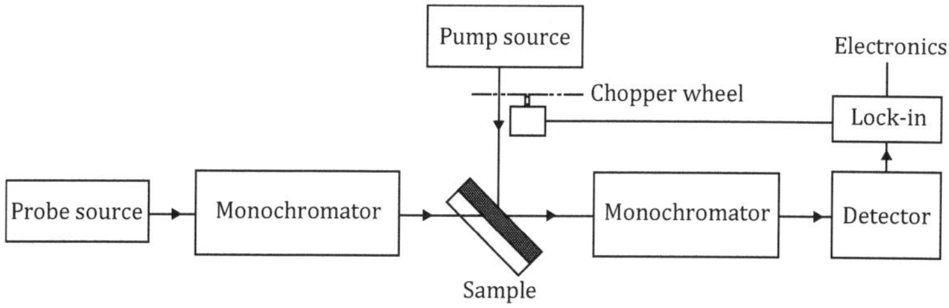

Figure 9.6 Photoinduced absorption experiment schematic.

9.3.4 Transient Microwave Conductivity

Transient microwave conductivity (TRMC) is also a pump-probe technique measuring the intrinsic quantum yield of the photogeneration of excitons and mobile carriers and the sum of electron and hole mobility at microwave frequencies.

In transient microwave conductivity measurements, a laser acts as a light source for absorption and charge photogeneration in a photoactive material (Fig. 9.7). A continuous (DC) microwave source acts as the probe beam, where the sample, mounted in a microwave cavity, is probed at the frequency of the cavity (GHz). A microwave detector measures the change in power of the incident and transmitted microwave beams, such that measured microwave photoconductance

$$\Delta G = -\frac{1}{K}\frac{\Delta P(t)}{P}, \tag{9.7}$$

where K is an experimentally determined cavity resonance calibration factor, and P is the microwave beam power.

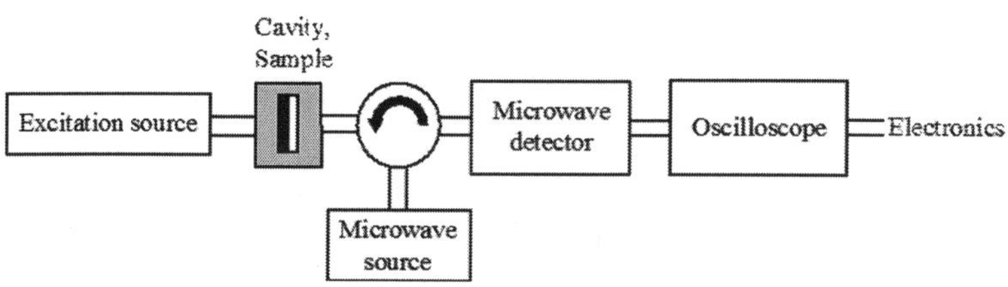

Figure 9.7 Transient microwave conductivity experiment schematic.

Physically, a continuous microwave beam of ~9 GHz provides a very low energy perturbation—over 500× less energy than kT at room temperature. The pulsed excitation source stimulates a band-to-band transition of carriers into excited states. Due to the continuum of excited states near the band edge present in organic semiconductors, these free carriers absorb some of the low-energy microwave pulse. However, as the excitation decays to the ground state, less of the probe beam is absorbed. Measuring the change in microwave photoconductance allows the experimental determination of the carrier mobility/yield product, $\phi\Sigma\mu$, by the relation

$$\Delta G = \beta e I_o F_A [\phi\Sigma\mu] \tag{9.8}$$

where β is a geometric factor of the resonant cavity shape, F_A is the fraction of light attenuated at the pump beam wavelength, I_o is the incident photon flux, e is the fundamental unit of charge, ϕ is the intrinsic quantum yield of mobile carrier formation, and $\Sigma\mu$ is the sum of electron and hole mobilities at the microwave frequency in the material. Practically, it can be rather difficult to deconvolve the carrier mobility/yield product, as measurement of the mobility at microwave frequencies is nontrivial. However, the technique can reveal relative fractions of mobile carrier formation and provide a useful tool for the contactless evaluation of new polymers and donor/acceptor blends for suitability in solar cell applications.

We note here that a related pump-probe technique uses terahertz (THz) instead of microwave probing. This technique is mostly sensitive to charge carriers, which are resonant at THz frequencies, while neutral excited states only yield a weak signal due to changes in the microscopic conductivity of the sample. Fourier transformation of the THz electric field transmitted through the sample in the presence of visible excitation yields the photoinduced spectrum in the THz range. The real and imaginary part of the conductivity (on the microscopic length scale of a THz oscillation) can be extracted. This yields information about the extent of binding of the charges, i.e., electrons and holes in a CT state can be distinguished from free ones.

9.3.5 Photoelectron Spectroscopies

Photoelectron spectroscopies allow the direct probing of electronic states—either binding energies of electrons in atoms or binding energies of energetic states in a bulk material, which can be related to the Fermi level. The most commonly used photoemission spectroscopies are X-ray photoelectron spectroscopy (XPS), ultraviolet photoelectron spectroscopy (UPS), and inverse photoelectron spectroscopy (IPES). XPS primarily analyzes chemical composition and chemical bonding by injecting photons with sufficient energy to eject core shell electrons from individual atoms, and will not be discussed here as this measurement technique is not directly used in the determination of charge generation and recombination processes. UPS probes the electronic ground state of bulk materials by using photons with enough energy to eject electrons from the ground state. The energy of these ejected electrons is measured via an electron energy analyzer as depicted in Fig. 9.8. The binding energy of these electrons with respect to the vacuum energy provide a map of the ground states, often referred to as valence states for inorganic materials, and measure the Fermi level for a material under low photoexcitation conditions.

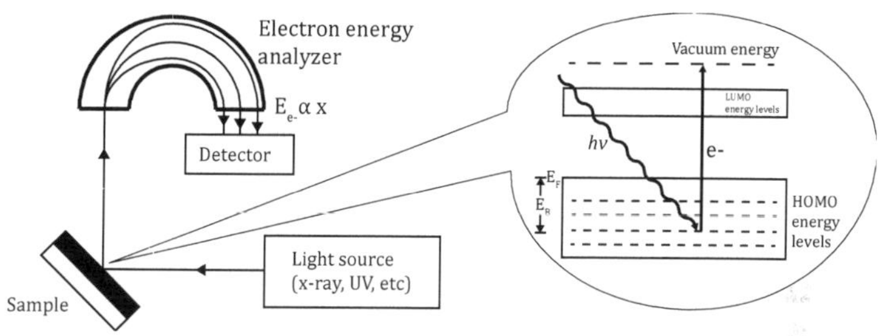

Figure 9.8 Ultraviolet photoemission spectroscopy experiment schematic.

IPES uses an inverse technique from UPS: electrons are injected into unoccupied excited states in a bulk material, and the photoemission of these excited electrons decaying into unoccupied ground states is measured. Hence, IPES is a complementary technique to UPS in that it maps out excited state energy levels above the Fermi level with respect to the vacuum level. Used together, the two techniques can provide a full picture of electronic states and can determine an electronic bandgap in semiconductor materials. Both UPS and IPES are primarily surface-sensitive techniques: while the penetration depth for high-energy electron and photon beams is relatively high (up to several microns), mean free path of emitted electrons or photons is less than 10 Å.[74] Most of the electrons or photons reaching the detector have been emitted from the first 2–3 nm of the bulk material's surface. One of the most powerful applications of these techniques is to measure the energy-level changes as monolayers of materials are deposited in situ in order to measure interfacial energetics.

9.3.6 Transient Photocurrent/Photovoltage Measurements

All-optical techniques are useful tools for probing materials properties and assessing photovoltaic processes from primary excitation to charge separation or, in Fig. 9.1, from $\sim 10^{-15}$ s after ultrafast excitation to $\sim 10^{-9}$ s in state-of-the-art modern organic solar cells. Electrical measurement techniques are best suited to probing carrier transport and collection at times after the creation of separated charge, and can often be conducted on fully functioning device structures, making it straightforward to correlate transient electrical measurements to device performance. Electrical measurement techniques can be roughly categorized as either steady state or transient. Accurate steady-state solar cell measurement, while not a trivial task to do accurately, has been well documented, and can be referred to V. Shrotriya and colleagues.[75] Here, we look in detail at transient techniques, with their ability to resolve electrical properties of the solar cell, including charge recombination and collection, carrier mobility and impurity density, and many other details of photovoltaic operation on time scales from $\sim 10^{-9}$ s after ultrafast excitation to $\sim 10^{-3}$ s in state-of-the-art modern organic solar cells.

Figure 9.9 outlines a general schematic for transient photocurrent experiments via a pulsed light source. Here, line width of the light source pulse, as well as the internal resistance and capacitance of the sample and the circuit determine the RC time constant of the photo-response. In the simplest circuit, a solar cell is connected directly to an oscilloscope. The high internal resistance of the oscilloscope applies a voltage across the device, biasing it toward open-circuit conditions. The transient photovoltage technique has typically been used to study photoresponse in organic BHJ solar cells near open circuit, where recombination processes dominate device performance. Inserting a low-resistance resistor in the circuit in parallel to the oscilloscope ensures that the high impedance of the oscilloscope is not dropped across the photovoltaic device. This technique is referred to as transient photocurrent, and generally probes photoresponse near the short-circuit condition, where charge is efficiently collected from the device. Finally, a voltage source may be added to control bias to the test cell and probe voltage-dependent recombination and collection dynamics. These techniques are myriad, and range from the simple—pulsed excitation source (pump) and steady-state DC voltage bias (probe) (bias-dependent transient photocurrent)—to the complex: photo-charge extraction by linearly increasing voltage (photo-CELIV), time-delayed dual pulse (TDDP), and time-delayed collection field (TDCF) measurements. Control of the extraction field used in photo-CELIV and control of photoexcited carrier dwell time used in TDDP and TDCF have allowed a detailed picture of the time-, voltage bias-, and carrier density-dependent recombination and charge collection in organic BHJ solar cells, and will continue to be excellent diagnostic tools as state-of-the-art polymers and devices evolve.

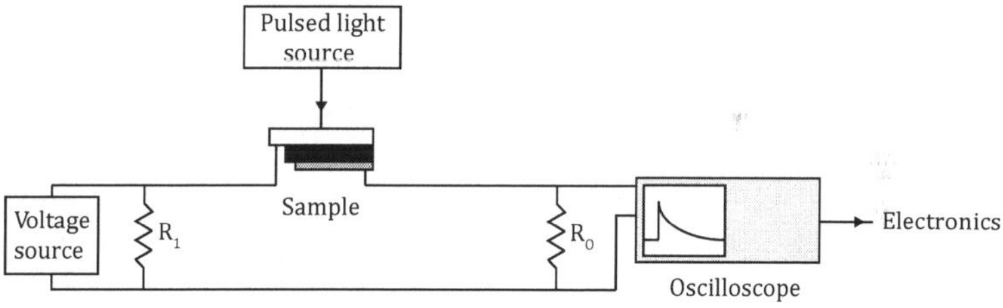

Figure 9.9 General schematic for transient photocurrent experiments.

9.3.7 Internal/External Quantum Efficiency Measurement

Internal, and external, quantum efficiency measurements measure a fully functioning solar cell's ability to convert absorbed, or incident, wavelength-dependent photogenerated carriers, respectively, into current collected in an external circuit. Inexpensive internal/external quantum efficiency measurement systems have become commercially available in the past 5 years, making these powerful wavelength-dependent measurements a common diagnostic tool in device laboratories.

In quantum efficiency measurements, broadband lamp light passes through a monochromator and chopper. This modulated light is incident on a solar cell device, with the entire photon flux under-filling the device. Photoexcitation from the light source creates current collected in an external circuit. Dark diode current (DC signal) and photocurrent (AC signal) are filtered by a lock-in amplifier, such that only current created by photoexcitation is recorded. Signal from the test cell is compared to measured photocurrent at identical wavelength and light intensity, or photon flux, such that EQE of the test cell is $EQE_T = I_T/I_R\ EQE_R$. The reference cell, calibrated to the solar flux, correlates measured current to the fraction of carriers collected in response to incident photons. The reference cell must have a known and well-characterized response to the solar spectrum, be stable over time, and have a high spectral response over the spectral response of the test cell. For instance, the commonly used Si reference cell has low responsivity between 300 and 400 nm, a region where PCBM and $PC_{70}BM$ show nonnegligible spectral responsivity. In these measurements, care must be taken to minimize noise in the quantum efficiency measurement in the 300–400 nm wavelength range. Internal quantum efficiency measurements simply factor out light absorption, $A = 1 - T - R - S$, by measuring light transmission, reflection, and scattering: IQE = EQE/A. Figure 9.10 illustrates an inclusive schematic for quantum efficiency measurement, assuming negligible light scattering and transmission. It is necessary to employ a spherically shaped photomultiplier tube array (integrating sphere) to accurately measure absorption if angular reflection, transmission, and/or angular scattering cannot be assumed to be negligible.

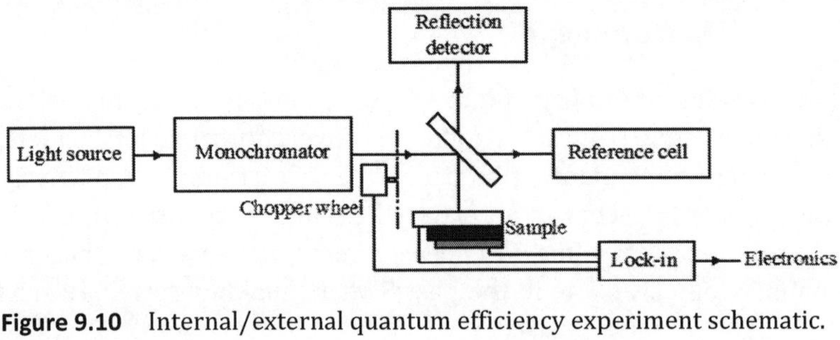

Figure 9.10 Internal/external quantum efficiency experiment schematic.

External quantum efficiency (EQE) measurements, sometimes referred to as incident photon conversion efficiency (IPCE), measure a number of loss processes in a device. Say, for simplicity, that loss processes are roughly divided into three categories: absorption, charge separation, and charge collection. Then we can predict external quantum efficiency, η_{EQE}, from the efficiency of these loss processes:

$$\eta_{EQE} = \eta_{Abs}\eta_{CS}\eta_{CC} \tag{9.9}$$

where η_{Abs} is absorption efficiency, η_{CS} is charge separation efficiency, and η_{CC} is charge collection efficiency. External quantum efficiencies from 60% to 80% are commonly reported in current literature within absorption wavelengths for the solar cell and for devices with BHJ film thicknesses of 60–120 nm. A 60% EQE would be reported if each process was 85% efficient; 70% EQE would be reported if each step were 90% efficient, and 80% EQE is expected if each step, from absorption to charge separation to charge collection, is 95% efficient! However, reported external quantum efficiency remains level or drops significantly as BHJ film thickness is increased, while the absorption efficiency of thicker films increases. This infers that the CS efficiency and/or charge collection efficiency must drop significantly as absorption efficiency increases, inferring that OPV PCEs remain limited by a device's ability to separate and extract charge from thicker BHJ films, or by the low mobility of the donor and acceptor materials themselves.

The following sections contain case studies on topics of intensive research in photoexcitation and CS (Section 9.4), the role of CT states in charge transfer and recombination (Section 9.5), the role of the internal electric field in charge collection (Section 9.6), bias- and charge density-dependent carrier recombination (Section 9.7), and the role of contacts in charge-carrier collection (Section 9.8).

9.4 Case Study: Photoexcitation and Charge Separation

Figure 9.11 visualizes the commonly used physical picture of current generation in BHJ solar cells. This includes (1) absorption of a photon, (2) relaxation of the primary excitation to a singlet exciton state, (3) exciton diffusion to a donor–acceptor interface, (4) CS (and possibly dissociation of an interfacial CT state) to create electrons and holes, and (5) charge transport to the electrodes. The exciton binding energy (E_b) in semiconducting polymers is intermediate between that of highly delocalized Wannier–Mott excitons in inorganic semiconductors (usually less than the thermal voltage, $V_T = kT/e \sim 25$ mV at room temperature, due to electric field screening in the material[4]) and localized Frenkel excitons in molecular crystals. Contrasting values for E_b for conjugated polymers ranging from 0.05 eV (in poly(phenylene vinylenes))[4] to ~0.4 eV[77–79] have been reported. According to the above-mentioned mechanism, diffusion and separation of tightly bound excitons limit the photoinduced CS at the heterojunction interfaces between the donor and acceptor phases, necessary for efficient current generation.

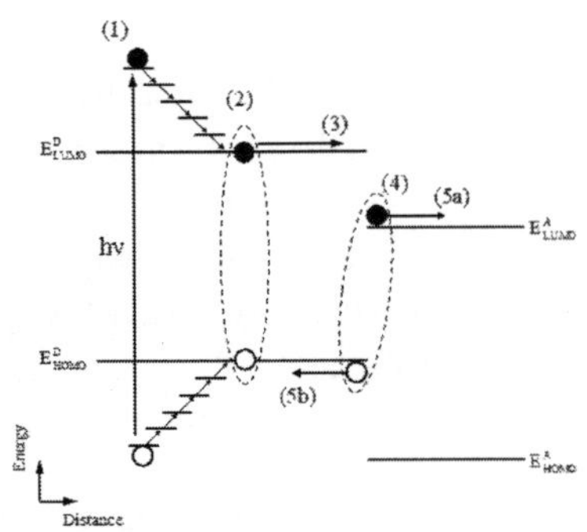

Figure 9.11 Commonly accepted outline of the physical processes converting absorbed photons to collected current: The photoexcitation (1) thermalizes to the band edge to form a bound exciton (2). This exciton diffuses (3) to a BHJ interface and charge transfers (4). Mobile electrons (5a) and holes (5b) are transported to the cathode and anode, respectively. However, the time scale of these processes is not consistent—exciton diffusion is too slow to account for ultrafast charge transfer. Reproduced with permission from Ref. [76]. Copyright 2012 by Wiley-VCH Verlag GmbH & Co. KGaA.

Let us examine the mechanism depicted in Fig. 9.11 in more detail. CS in the poly(2-methoxy-5-(3′-7′-dimethyloctyloxy)-1,4-phenylenevinylene) (MDMO-PPV): PCBM composite has been directly measured to occur within 45 fs after photoexcitation.[37]

Some references exist that point to much slower charge separation, with a distribution of time constants that is correlated to the distribution of exciton distances from the interface.[80] Nevertheless, most femtosecond transient absorption studies carried out on polymer : fullerene blends with optimized composition and morphology evidence the appearance of a large fraction of charges faster than the 100–200 fs time resolution of the experiments.[34–39,43,81]

Admitting that the CS indeed occurs on the 100 fs time scale, this implies that exciton diffusion to a fullerene interface also occurs in <100 fs. According to the exciton diffusion coefficients reported in literature for poly[2-methoxy-5-(2′-ethyl-hexyloxy)-*p*-phenylene vinylene] (MEH-PPV) and poly-3(hexylthiophene) (P3HT) (3×10^{-3} and 1.8×10^{-3} $cm^2\ s^{-1}$, respectively), excitons can diffuse by only 0.1–0.2 nm in 100 fs.[44,45] This is clearly much less than the typical length scale of about 10–20 nm in optimized BHJs, estimated by structural characterization in optimized morphologies.[46–51] Furthermore, point-like exciton diffusion by the hopping mechanism has been measured on numerous occasions to take 0.5–1 ps for a single hopping step,[21–27,30–32,82,83] which is again too slow to explain how excitons reach a fullerene interface in <100 fs. In agreement with this, an ultrafast transient absorption study of poly[2,7-(9,9-dioctylfluorene)-alt-5,5-(4,7′-di-2-thienyl-2′,1′,3-benzothiadiazole)] (APFO3) films with various concentrations of PCBM revealed that exciton hopping before CS only occurs in films with very low fullerene concentration and poor morphology, but does not play a role for favorable blends having ultrafast quantitative CS within 200 fs.[81]

One could argue that the 10 nm length scale for phase separation in the BHJ is largely overestimated, and that the donor and acceptor are in reality so intimately mixed that every excitation in the polymer occurs so close to a fullerene that no diffusion is necessary prior to charge separation. Ultrafast charge separation in an intimately mixed phase has indeed been observed in several polymer : fullerene blends. Against the intermixed phase being the sole reason for ultrafast CS speaks the experimental observation that PCE in very intimately mixed polymer : fullerene blends is often low, because charge transport is inefficient. For example, for the polymer poly(2,5-bis(3-alkylthiophen-2-yl)thieno[3,2-b]thiophene) (PBTTT), the fullerene intercalates into the crystalline polymer phase, but very low PCE was observed with the intimately mixed blend in spite of complete fluorescence quenching.[84] Good solar cell performance for this polymer could only be achieved with a high fullerene concentration leading to phase separation and pure fullerene domains for electron transport. Also, as-cast films of P3HT:PCBM have been shown to display intimate mixing and have a PCE of less than 1%.[85] This can be increased to 4–4.5% by thermal annealing, which increases the phase separation and leads to more crystalline polymer domains of about 10 nm. A recent publication demonstrates that the CS in annealed P3HT:PCBM blends might be slower (around 3 ps) due to the increased domain size compared to nonannealed films, where charges are formed within hundreds of femtoseconds.[86] The authors nevertheless point out that even in the annealed case the CS rate is still orders of magnitude higher than if it was truly limited by point-like exciton migration. They conclude that excitons must be highly spatially delocalized in the ordered P3HT domains, so that they can sample distances of the order of 10 nm on a very fast time scale—see Fig. 9.12a.

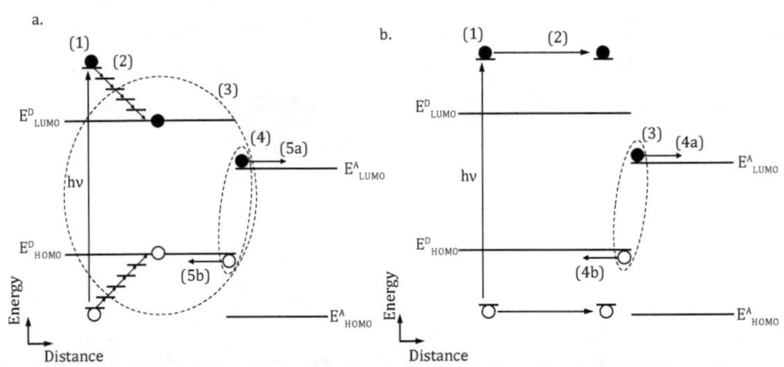

Figure 9.12 Two proposed mechanisms for ultrafast CS at a fullerene interface following polymer photoexcitation in organic BHJ solar cells: (a) The primary photoexcitation (1) thermalizes to the band edges (2) to form a bound exciton. This delocalized singlet exciton (3) can sample spatial distances of 1–10 nm, resulting in ultrafast CS (4) and mobile carrier generation (5a–b). (b) The highly delocalized primary photoexcitation (1)–(2) charge separates (3) before thermalization and dynamic localization in less than 100 fs and mobile carriers (4a–b) are formed in the two phases. Reproduced with permission from Ref. [76]. Copyright 2012 by Wiley-VCH Verlag GmbH & Co. KGaA.

It is likely a crude oversimplification to assign the excited state of conjugated polymers to a unique species, such as an exciton, without considering its temporal evolution. Banerji et al. show in two recently published studies that complex relaxation processes occur in the excited state of conjugated polymers, spanning time scales from femtoseconds to hundreds of picoseconds.[19,20] Even the early excited state measured with the typical 100–200 fs time resolution of ultrafast laser experiments is already considerably different and more localized than the one directly following excitation. In BHJ blends, ultrafast CS at the donor–acceptor interface and relaxation in the pure polymer phase are therefore competing processes.

Banerji et al. investigated pristine polymers poly[*N*-9″-heptadecanyl-2,7-carbazole-*alt*-5,5-(4′,7′-di-2-thienyl-2′,1′,3′-benzothiadiazole)] (PCDTBT) and P3HT by femtosecond-resolved fluorescence up-conversion spectroscopy.[19,20] The results with both polymers point to the formation of an initially delocalized photoexcitation, which localizes on the 100 fs time scale. This is followed by successive exciton migration steps (hopping by the Förster mechanism) to lower energy-localized states, which exist as a result of disorder (0.5–100 ps). In parallel, there is also slow conformational relaxation of the polymer backbone (up to ~200 ps in thin films). The ultrafast dynamic localization of the primary photoexcitation, evidenced in the fluorescence up-conversion study by a <200 fs red-shift of the emission spectrum and loss of polarization anisotropy, has been independently reported by several groups using, for example, photon-echo experiments or ultrafast transient absorption, and fluorescence depolarization experiments.[17,18,21,22,28,29,87–89] A delocalization of the primary photoexcitation over at least 11 nm has been reported for MEH-PPV.[29]

The origin of the existence and the decrease of short-lived delocalization in conjugated polymers is highly complex and originates from a number of entangled processes. Three main contributions to the dynamic localization have been identified by experimental[17,18,21,22,28,29,87–89] and theoretical[90–92] studies:

- Electronic relaxation to a more localized lowest energy state through a manifold of delocalized states formed by coupling of the chromophoric units.

- Self-trapping of the electronic states caused by strong coupling to nuclear modes, i.e., localization caused by structural distortion on the polymer chain.
- Coherent excitation energy transfer between chromophore units in the intermediate coupling regime.

The precise interplay and relative importance of the three processes is still a matter of some controversy. It depends to some extent on the nature of the investigated polymer (e.g., polythiophene or poly(p-phenylenevinylene), strength of coupling to nuclear modes, torsional freedom, chemical conjugation breaks) and on environmental effects such as disorder, medium (solution or film), and temperature.

Let us now come back to the discrepancy between the distance that a polymer exciton *can* diffuse in 100 fs (0.1–0.2 nm) and the distance that the photoexcitation *needs to* cover (~10 nm) in order to reach a fullerene interface in the BHJ to undergo ultrafast (<100 fs) charge separation. Based on a comparison between the time scales for the various relaxation processes occurring in the excited polymer and the time scale of charge separation, the <100 fs CS might very well occur from the delocalized primary photoexcitation, rather than from the relaxed exciton. It certainly occurs before the relative slow Förster exciton hopping, although this is commonly considered to be the mechanism that brings the exciton to an interface. Therefore, an alternative mechanism was proposed by Banerji et al. that is consistent with ultrafast CS during relaxation in the polymer, before localization of the primary excitation, as outlined in Fig. 9.12b.

Given the larger spatial extent that it samples, it seems very plausible that the primary photoexcitation can reach the interface faster than a relaxed exciton, which would account for the ultrafast CS in BHJ blends. In this case, the excited state does not really diffuse, but is transported by a more quantum-mechanical process involving delocalization. Intrachain coherent excitation energy transfer was indeed observed by Collini and Scholes in dissolved MEH-PPV on the 100 fs time scale.[93] Here, excited-state population and coherences move in concert between chromophores coupled in the intermediate regime. This was experimentally demonstrated by a specifically designed experiment (two-time anisotropy decay, TTDA) combined with two-dimensional electronic spectroscopy.

Surprisingly long electronic coherence (up to 250 fs) was observed at room temperature.

Moreover, the binding energy of the electron and hole contributions in the short-lived excited state might be reduced by the initial delocalization, and there is excess energy in the system, which can help charge separation, possible even directly to free charges without the intermediate CT state. A very recent study by Lanzani and coworkers directly evidenced the dissociation of unrelaxed neutral excited states in a low bandgap polymer at the fullerene interface.[94] Free (as opposed to CT bound) charge-carrier generation within tens of femtoseconds via delocalized states was enhanced by excitation into higher singlet states (more excess energy). Direct free charge-carrier generation will be discussed more in the next section.

In summary, Förster-type diffusion of a localized and relaxed exciton is unlikely to be involved in the CS mechanism in polymer : fullerene BHJ blends, because it is too slow. The much debated concepts of exciton diffusion length and exciton binding energy therefore may be less important for CS in "plastic" solar cells than has been commonly assumed. Delocalization of the primary photoexcitation and transport assisted by quantum effects offer a plausible explanation to the observation that provides the foundation for the entire field of BHJ solar cells: ultrafast photoinduced charge transfer.

9.5 Case Study: Role of the Charge Transfer State in Charge Dissociation and Recombination

Photoexcitation and CS at the D/A interface are widely accepted to be field-independent processes in pristine polymers, small molecules, and polymer : polymer or polymer : small-molecule blends. The primary excitation and bound electron–hole pair are charge neutral, which screens the effect of an electric field on the quasi-particle. During the CS process, the excitation is transferred across an interface between donor and acceptor regimes. There exists evidence for two competing viewpoints as to what happens next. (1) In such close proximity, the electron and hole are believed to still be correlated and coulombically bound (intermediate CT state).

The bound state must overcome the binding energy to create fully separated ("free") charge carriers. In such a case, the energy required to overcome the Coulomb potential binding these two charge carriers together was first described by Onsager in 1938,[95] expanded on by Braun in 1984[96] and by Tachiya in 1988,[97] and used to generate a full electrical model by Koster, Blom, and colleagues.[98] (2) CS directly creates fully separated and unbound charge carriers. The relevance of the elegant Onsager–Braun model has not been borne out in efficient BHJ blends. While it has been extensively shown to be possible to measure photoluminescence, and to optically and electrically pump directly into the CT state, photoluminescence from the most efficient systems is orders of magnitude lower than similar measurements on highly lossy systems. Device experiments by Heeger et al.[99] and Powell et al.[100] have shown that it is not necessary to have a significant (>0.1 eV) band offset between the donor and acceptor LUMO levels, thought to be necessary to split the CT state,[101] and transient photoconductivity experiments by Street et al., Neher et al., and Laquai and colleagues have shown that free carrier generation is field independent[102–104] in efficient BHJ blend devices and thin films. Recent experimental results[104] from Etzold, Laquai et al. suggest that the two processes (intermediate CT state and direct free charge generation) coexist and compete in BHJ solar cells, but that in the most efficient devices, only a negligible amount of charge passes through the coulombically bound CT state (a path for recombination on the pico/nanosecond timescale) and the majority of charge separates into free carriers independent of the CT states and electric field.

Classic Onsager theory describes the interaction between an excited electron and its "parent"—in the simplest case assumed to be a single ionized atom in a weak electrolyte solution. The excitation and its parent are assumed to be bound until enough energy is received to overcome the coulombic attraction of the negative and positively charged separate particles, or until the charge recombines—what Onsager names "initial" recombination—as visualized in Fig. 9.13. Today, geminate recombination is a term more commonly used to refer to the loss process involving the charges formed from the same primary excitation. In the Onsager theory of geminate, or "initial" recombination, excited charge and their parent species diffuse via Brownian motion in the Coulomb

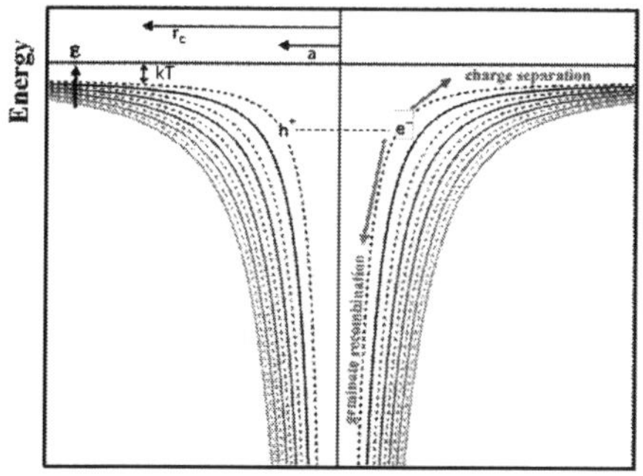

Figure 9.13 Coulombic capture and emission of an electron–hole pair via Onsager theory. The potential energy versus separation distance > kT must be overcome for exciton dissociation.

potential resulting in a probability for either recombination or full charge separation. The minimum Coulomb radius, r_C, sometimes called the Onsager radius, required for CS sufficient enough to result in dissociation via heat fluctuations in the material (i.e., the thermal energy, $V_T = kT/e$) is defined as

$$r_C = \frac{e^2}{4\pi\varepsilon kT} \tag{9.10}$$

where e is the fundamental unit of charge, k is the Boltzmann constant, T is temperature, and $\varepsilon = \varepsilon_o\varepsilon_r$ is the dielectric constant, ε_o is the permittivity of vacuum, and ε_r is the permittivity of the surrounding material. In other words, Onsager hypothesized a relationship between the capture radius, r_C, and the thermal voltage, kT, below which ($r_C < kT$) charges are bound by the Coulomb potential, and above which ($r_C > kT$) charges dissociate, as plotted in Fig. 9.14. Estimates for the capture radius of a polymer with a relative permittivity of 3 and silicon with a relative permittivity of ~ 11.7 are highlighted at room temperature conditions. Silicon is estimated to have a significantly lower capture radius than a polymer, with a difference of r_C ~ 3 nm versus r_C ~ 20 nm. The capture radius estimates the CS distance required for dissociation; however, this estimate is not expected to be entirely accurate for organic molecular materials, nor does experimental evidence bear out a literal interpretation of this relationship.[105] Braun attributes the lower exciton binding energies measured in molecular materials to a lower nearest-neighbor spacing than the exciton radius and the internal energy of the exciton itself.[96] When an electron and hole are closer than the capture radius, r_C, the probability of the charge pair recombining is estimated by Onsager for ions in an electrolyte and adapted to molecular systems by Braun[96] and Tachiya[97] to include electric field dependence and electron mean free path variations due to scattering and a finite reaction radius. Tachiya, for instance, points out that Onsager assumes recombination occurs at a zero reaction radius, an unrealistic assumption, but one that allows a relatively simple statement of the escape probability, ϕ, as a function of the capture radius, r_C, and the separation distance, r, and a weak electric field:

$$\phi = \exp\left(-\frac{r_C}{r}\right)\left(1+\frac{er_C}{2kT}E\right) \tag{9.11}$$

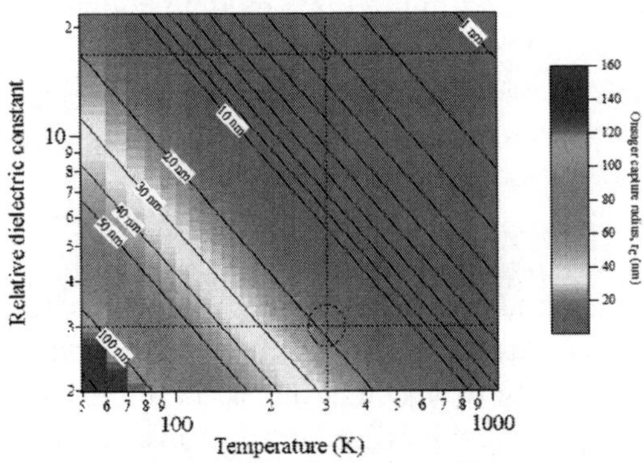

Figure 9.14 Dependence of the Onsager capture radius on temperature and relative permittivity. Approximate Onsager capture radius at room temperature is marked for silicon (ε_r ~ 11.7) and polymers (ε_r ~ 3).

The equation predicts linear dependence of dissociation probability with electric field, and a slope of $b = e^3E/8\pi\varepsilon k^2T^2$. Subsequent extensions to Onsager's original theory are increasingly complex. For instance, Braun's equation for field-dependent dissociation probability additionally depends on spatially averaged mobility, $<\mu>$, reduced exciton mass, m_X, $\hbar$, the order 1 Bessel function, J_1,

$$\phi(r,T,E) = \frac{k_D(E)}{(k_F = k_D(E))},$$
$$k_D(E) = \nu \exp(-\Delta E/kT) J_1[2\sqrt{2}(-b)^{1/2}]/\sqrt{2}(-b)^{1/2}, \tag{9.12}$$

where k_F and k_D are the recombination and dissociation rate constants, and the exponential prefactor $\nu = \langle \mu \rangle e/\varepsilon \left(m_x kT/3\pi\hbar^2\right)^{3/2}$. Braun's field-dependent dissociation probability can be simplified in the case of low electric field to show the same behavior as Onsager's result: linear dependence on electric field and a slope of b.

In practice, observing field dependence of free carrier dissociation in organic BHJ systems is a good deal more complex. There are, of course, two motivations for understanding mechanisms of charge dissociation in these organic materials: fundamental understanding of these materials in general, and a specific understanding of the mechanisms in efficient systems to guide development of new materials. Keep in mind also that it may not be possible to group all polymers together, mechanistically. Significant changes in chemical structure between different generations of polymers may cause varying degrees of primary excitation delocalization, for instance, favoring one mechanism over the other. Alone, thus far, among the solar materials, progress in organic photovoltaics has been the result of the synthesis of entirely new materials at each milestone, necessitating a study of the mechanistic behavior of each new class of materials. Here, we examine recent evidence both for and against electric-field-assisted charge separation, and a newly developing paradigm—that primary excitations have two paths to free carriers and the relative ratio of placement of charge either (1) into bound CT states with finite and field-dependent dissociation rates plays a large role in determining overall photovoltaic efficiency or (2) directly into delocalized free carriers.

Transient absorption spectroscopy (TAS) is a commonly implemented version of the all-optical pump-probe experiment diagrammed in Fig. 9.6. Specifically, TAS measures the wavelength-dependent time evolution in transmission or absorption of a probe light pulse due to an initial fast light pulse, usually temporally resolved to the femto- to nanosecond timescale. In 2010, Marsh, Hodgkiss, and Friend reported the direct measurement of electric field-assisted CS in polymer: fullerene solar cells via TAS.[106] Their work was not the first based on this spectroscopic method to make a positive or negative claim on field-dependent carrier dissociation. The research by Veldman et al. in 2008[107] and other groups supports Marsh et al.'s results, while Guo et al. in 2008,[108] and Jamieson et al.[109] and Howard et al.[110] in 2010 see evidence for field-independent charge dissociation, the result of ultrafast charge separation, from intensity- or field-dependent TAS data.

Let us focus directly on the work of Jamieson et al.[109] and Marsh et al.[106] who take similar data—bias-dependent transient absorption—on two different systems: PCPDTBT:$PC_{70}BM$ and P3HT:PCBM, respectively. Transient absorption experiments are described in more detail in Section 9.3.3. Data from these two papers are plotted in Fig. 9.15. Figure 9.15a,b measures the change in optical density from applied fields of –4 V (strongly reverse bias) to above open-circuit voltage (+1 V) and under pump fluences of 8 and 85 μJ/cm^2 for the system C-PCPDTBT:$PC_{70}BM$ modified with the octane dithiol (ODT) processing additive. Elsewhere in the 2010 work, similar measurements are performed on C-PCPDTBT:$PC_{70}BM$ without ODT and Si-PCPDTBT:$PC_{70}BM$. Neither the magnitude of the initial signal nor the shape changes for all three systems under increasing external bias, indicating that neither the ultrafast generation and CT dynamics ($t <$ 20 ns) preceding the resolution of the experimental setup nor the nano/microsecond CS and decay dynamics ($t >$ 20 ns) are field-dependent.

Strikingly similar data taken by Marsh and colleagues[106] for a different materials system and pump fluence, but similar voltage bias and time scale, show markedly different results. Figure 9.15c,d plots voltage bias-dependent photoinduced absorption data on the 1 ns $> t >$ 100 μs time scale for P3HT:PCBM blends, both untreated (pristine) and thermally annealed. Both devices show field dependence of the decay in transient absorption under a pump fluence of 12 μJ/cm^2. At low applied fields (reverse bias), TAS lifetimes are longer than under high applied fields (open circuit), corresponding to higher carrier collection in reverse bias and at short-circuit conditions. The voltage dependence of Marsh's TAS data validates the application of Onsager–Braun theory to the solar cells they measure and, due to the voltage dependence of carrier generation, suggest proof of a coulombically bound, excitonic CT state, whereas the lack of voltage dependence observed in the PCPDTBT:$PC_{70}BM$ system shows no evidence of a bound excitonic CT state.

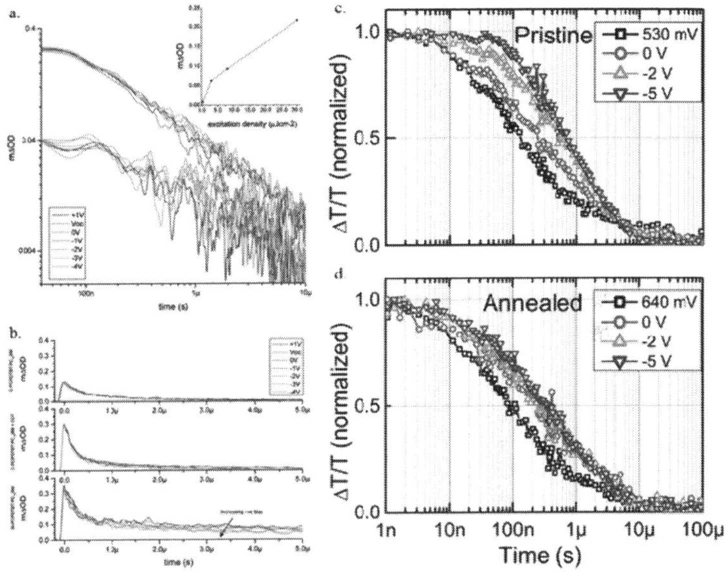

Figure 9.15 Transient absorption spectroscopic data from (a–b) Jamieson et al. for a PCPDTBT:$PC_{70}BM$ solar cell under (a) variable incident light intensity and (b) voltage bias. Copyright 2010 by the American Chemical Society. Reproduced with permission from Ref. [109]; and from (c-d) Marsh et al. for a P3HT:PCBM solar cell under (c) variable incident light intensity and (d) voltage bias. Reproduced with permission from Ref. [106]. Copyright 2010 by Wiley-VCH Verlag GmbH & Co. KGaA.

The reader must weigh the relative importance of the differences between the two experiments—two different materials systems and pump fluences. The pump fluences are less than an order of magnitude different between the two experiments, resulting in what is likely to be an insignificant difference in carriers in terms of the standard models for photogeneration. Additionally, Jamieson et al. in Fig. 9.15a,b show TAS spectra from two different pump fluences. The lower pump fluence signal is of lower magnitude and signal-to-noise ratio, but the two sets of voltage-bias dependent data show no voltage dependence in the time scales. Hence, it is likely that two materials systems (and perhaps even the same materials system under different fabrication conditions, though not those shown in these two studies) can suggest two different mechanisms of operation.

We have reviewed recent evidence both for and against electric-field-assisted charge separation, and now move on to examine a newly developing paradigm—that primary excitations have two paths to free carriers and the relative ratio of placement of charge either (1) directly into delocalized free carriers or (2) into bound CT states, where finite and field-dependent dissociation rates play a large role in determining overall photovoltaic efficiency. Two exceptional experiments were published recently: Etzold and colleagues track the temporal evolution of photoexcitation in efficient photovoltaic systems from 100 fs to 1 ms—10 orders of magnitude—in order to provide the most comprehensive look at the evolution of photoexcitation in the BHJ yet presented.[104] Bakulin et al. utilize an electro-optical pump-push experiment on functional organic photodiodes to determine the relative role of CT states in carrier dissociation. The study concludes that CT states play a relatively minor role in efficient OPVs and photoexcitations primarily dissociate via hot delocalized band states.[53]

Etzold et al. present the temporal evolution of photoexcitation in two efficient photovoltaic systems—P3HT:PCBM and PCDTBT:PCBM—from 100 fs to 1 ms.[104] Standard photoinduced absorption experiments allow the determination of transient absorption, with a new addition—a two-part delay stage for short-delay (short time scale—pico/nanoseconds) and long-delay (long time scale nano/microseconds) measurements. Qualitatively, transient absorption data present a story consistent with previously presented work: light absorption creates primary excitations, the majority of which charge separate on a time scale less than the temporal resolution of the measurement system ($t < 100$ fs). A small fraction of the charges relaxes toward the band edge (relaxation from 2.1 to 1.9 eV) and decay back to the ground state on a time scale longer than previously observed—hypothesized to be longer because transient absorption is extended in this experiment to time scales not observed in previous experiments and includes full spectral data at each time step—from 1.75 to 2.17 eV. Intensity-dependent decay dynamics reveal nongeminate recombination in the slow-decay (long-delay) regime.

Etzold and colleagues present an analytical model to quantify these processes, summarized by a system of differential equations relating ground state (GS) population with bound charge-carrier (CT state) population and spatially separated carrier (SSC) population via transfer and decay processes—(1) geminate decay of CT states back to the ground state with rate $k_{CT \to GS}$, (2) nongeminate decay of SSCs back to the ground state with rate γ and order λ:

$$\mathrm{CT}(t) = N_0(1-f)\exp(-k_{\mathrm{CT}\to\mathrm{GS}}t), \tag{9.13a}$$

$$\mathrm{SSC}(t) = \left(\lambda\gamma t + (N_0 f)^{-\lambda}\right)^{-1/\lambda}, \tag{9.13b}$$

$$\mathrm{GS}(t) = N_0(1-f)\left(1-\exp(-k_{\mathrm{CT}\to\mathrm{GS}}t)\right) + N_0 f - \left(\lambda\gamma t + (N_0 f)^{-\lambda}\right)^{-1/\lambda}, \tag{9.13c}$$

where N_0 is the initial photoexcitation density and f is the fraction of carriers which recombine nongeminately. These time-dependent equations allow quantitative mechanistic analysis of transient absorption data, and result in Etzold et al.'s conclusion—89% of carriers in the PCDTBT:PCBM blend film charge separate on the ultrafast time scale, and only 11% of carriers pass through CT states to geminately recombine. The fitted ratios are 85% and 15%, respectively, for an optimized P3HT:PCBM system. Etzold et al. therefore assert that 11% of the total carrier population is too small to contribute in an observed voltage-bias dependence of charge generation in efficient systems. They report 32% geminate recombination in as-cast (versus optimized) P3HT:PCBM films, and it follows that even less efficient systems may show even higher ratios of geminate recombination in blend films. Nevertheless, evidence presented by Etzold concludes that CT states play an insignificant role in charge dissociation for optimized systems. Etzold's data and discussion do not suggest a mechanism for ultrafast CT—now one of the most significant questions in the study of organic photovoltaics.

Fortunately for our discussion here, Bakulin and colleagues have begun to answer this question in a study published in 2012 in

Science.[53] Photo- and electroluminescence have been commonly used to probe the character of CT states in organic BHJ blends. However, these studies have overemphasized the role of CT states in efficient systems, where CT state photoluminescence has been shown to be decreased by orders of magnitude below inefficient BHJ systems. Bakulin and colleagues in the Cavendish Laboratory, University of Cambridge, utilize an electro-optical "pump-push" experiment on functional organic photodiodes to elucidate the ultrafast CT mechanism. A pump beam with visible wavelength (i.e., energy above the bandgap of the chosen organic material) creates excited states, which evolve over time into separated carriers and CT-bound carriers.

For all practical purposes, the carriers bound into relaxed CT states are lost to the system, as the time scale for CT state recombination strongly competes with CT state separation. The separated charge carriers are therefore formed by an alternative route, via higher lying delocalized CT_n states. Those hot states are very short-lived (a few hundreds of femtoseconds) and can either dissociate to the free charges (main route in efficient systems) or be trapped in the relaxed CT_0 state. The "pump-push" experiment delivers a second "push" pulse, an IR beam intending to directly excite the CT_0 to CT_n transition. A significant increase in photocurrent is observed as a result of the "push" pulse, suggesting that the excess energy assists the separation of CT state dissociation into free carriers within 200 fs. Additionally, the spectral signature of these "pushed" IR-excited CT states is similar to that of the carriers separated by the initial visible pump pulse, giving the CT state carriers a second chance to dissociate. The authors hypothesize that the IR push pulse optically repopulates delocalized states formed after primary excitation dissociation, and allows the study of these delocalized states. The implications of these data are that relaxed CT states play a role in decreasing the efficiency of poor OPV devices. Those poor systems end up with a larger fraction of charge in the CT_0 state, so that the "push" pulse can significantly help to untrap those charges (an external field is in this case also expected to have an effect). On the other hand, there is more direct generation of separate charges in efficient systems, so that the "push" pulse (or a field) cannot help to increase the photocurrent. The role of delocalized neutral and charged states in BHJ materials will undoubtedly be the subject of intense research in the coming years.

9.6 Case Study: Role of the Internal Field in Charge Collection

In a BHJ photodiode under illumination, excess charges are created in the polymer and fullerene domains due to photoexcitation and subsequent charge transfer, as discussed in the previous section. Mobile (free) carriers must then be generated and collected at opposite electrodes prior to recombination, as temporally described in Fig. 9.1, in order to contribute to measured current. Recombination lifetimes of the charge carriers in the donor and acceptor domains limit the efficient extraction of these carriers to an external circuit. While recombination can be minimized in optimized BHJ thin films (BHJ thickness <100 nm),[54] recombination increases as the BHJ film thickness is increased,[55,56] likely due to the low carrier mobility in the disordered donor and acceptor materials.

The strong internal field present within BHJ solar cells at short-circuit conditions has been stipulated to affect the generation of free charge carriers,[43,106,111,112] their loss due to various recombination processes,[70,113] and "sweep-out"—the process by which charge carriers escape recombination to be collected at the electrodes.[114–116] Results from TAS have often been interpreted according to a model in which CT from the polymer to the fullerene leads to the formation of a coulomb bound interfacial charge pair (also called an exciton, polaron pair,[117] or intermediate charge transfer state (CT state), Fig. 9.1).[43,111,112,118] The charge pair either undergoes geminate recombination or dissociates into free charge carriers that can ultimately be collected at the electrodes. The presence of an internal electric field at the short-circuit condition has long been considered necessary for the separation of the strongly bound charge pair. Although this seems to be confirmed by the observed electric field dependence of free charge generation,[106] some ultrafast transient absorption experiments have shown the generation of free carriers on the order of picoseconds, too fast to be interpreted as field-dependent charge separation.[108,110] As the meaningful interpretation of transient absorption results is made difficult by high-intensity excitation conditions, broad/overlapping spectral signatures, and complexity in assigning the bands, alternative techniques have been applied. Kniepert and colleagues use a new experimental method to study photocurrent extraction—TDCF.[114] TDCF measurements resolve photocurrent extraction on the nanosecond to microsecond timescale using an optical excitation pulse and time-delayed electric field extraction. Generated charge density is found to be independent of the applied electric field during illumination, indicating that CS is not a field-dependent process. Similarly, more investigations will be discussed in Section 9.7, which show that recombination in BHJ solar cells is bimolecular or trap-based, rather than geminate. Therefore, the possibility that CT from the polymer to the fullerene directly yields free charge carriers (without intermediate bound state that leads to geminate recombination) should be seriously considered.

While the role of the internal field in free charge generation is questionable, it certainly influences the collection efficiency of charges at the electrodes. Recent experimental results suggest that under high electric field conditions (short-circuit current, J_{sc}, and reverse bias), mobile carriers are efficiently swept out of the BHJ via drift transport on the order of ~100 ns.[114,116] This is again consistent with evidence of fast generation of free charge carriers,[19,108,112] because efficient charge collection must occur before recombination can significantly reduce the free charge-carrier population. Results from a variety of transient photocurrent techniques on functional polymer solar cells are consistent with carrier transport at a constant drift velocity at short circuit.[114,116,119] Here, we explore evidence for the idea that efficient charge collection at the short-circuit condition and in reverse bias in polymer solar cells is due to free carriers transported via drift in a strong built-in electric field—a process known in photodiodes as sweep-out[116,120]—and the role of the internal field in charge collection at the short-circuit condition. The sweep-out competes efficiently with non geminate recombination, so that charges can be collected.

While sweep-out is a well-known phenomenon in photodiodes, it has only recently been shown to play a large role in efficient charge collection in polymer solar cells in reverse bias and at the short-circuit condition, where the internal electric field can be high. Charge sweep-out is a classical concept, and when applied to BHJ materials require optimized charge extraction characteristics within the BHJ for charge to drift. Inefficient BHJ morphologies or inefficient charge collection at the contacts can lead to highly nonuniform spatially distributed charge densities, sometimes called space charge, where the electric field is locally screened and diffusion may play a larger role in transport. In disordered films, regions of localized charge are likely sites for increased recombination. Within a classical picture of charge transport, charge carriers with average carrier mobility, μ, drifting in an electric field, E, will transit a given distance, d, in a characteristic time, τ_{sw}—the sweep-out time. Here, we imagine a carrier traversing the BHJ, and define d equal to the BHJ thickness. The drift velocity, v, of these carriers is classically given as $v = \mu E = d\tau_{sw}$. The electric field across a given distance can be estimated as the internal voltage, V_{int}, per unit length, as long as the internal voltage is uniform across the distance. While photoconductive atomic force microscopy measurements reveal morphology-dependent nanoscale variations in the electric field,[121] we assume that these variations produce a nearly uniform average field on the macroscale. Hence, the sweep-out time is defined as $\tau_{sw} = d^2/2\mu V_{int}$, inversely proportional to the internal voltage. A geometric factor (here, 2) is inserted to account for uniform charge generation across the BHJ layer. Charge carriers must travel on average through only one-half the length of the BHJ to be collected.

Transient photocurrent experiments by Cowan et al. show this to be a good model of charge extraction over a wide range of internal voltages in BHJ solar cells composed of PCDTBT:$PC_{71}BM$ ([6,6]-phenyl-C_{71}-butyric acid methyl ester) and P3HT:PCBM.[116] Experiments were performed on solar cell architectures by pulsing a laser with 10 ns pulse duration and 1 Hz repetition rate. Current was measured by the voltage drop across a small resistor in series with the cell. Representative bias-dependent data are plotted in the inset of Fig. 9.16, where V is the externally applied voltage. Internal voltage is determined by the equation $V_{int} = V_{BI} - V$, where V_{BI} is the built-in voltage and sometimes called the flatband voltage elsewhere. V_{BI} is estimated from the voltage at which the transient photocurrent as a function of voltage changes sign.

Carrier decay times are extracted from the transient photocurrent decay data, and are shown in Fig. 9.16 for the two BHJ systems and two temperatures studied. The sweep-out time is found to be proportional to the inverse internal voltage from reverse bias (low inverse internal voltages) to approaching the maximum power point. At short circuit and room temperature, charge is extracted from both BHJ systems in <1 μs. Transient photoconductance data in Fig. 9.16 imply the existence of a well-defined macroscopic internal field; carrier sweep-out is proportional to the magnitude of this internal field and is limited by the carrier mobility.

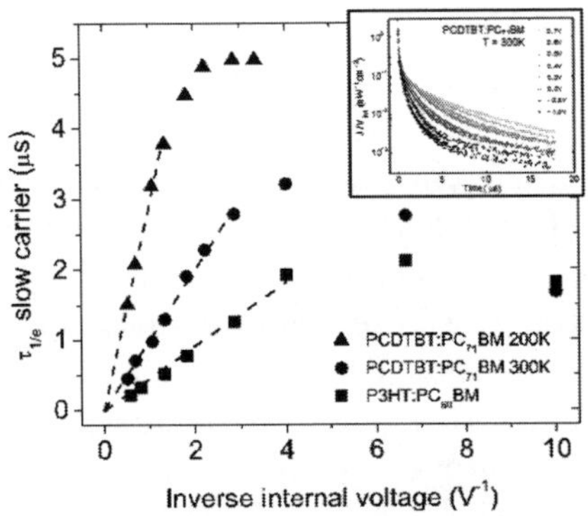

Figure 9.16 Estimated transit time of slow carriers vs. the inverse internal voltage ($1/V_{int}$). In the sweep-out limit (high internal voltage), data are fit to the drift mobility equation, $\tau_{sw} = d^2/2\mu V_{int}$. Inset: transient photoconductance vs. time for the PCDTBT:PC$_{71}$BM cell (T = 300K). Voltage applied to the cell is scanned from −1.0V to 0.7V. Reproduced with permission from Ref. [116]. Copyright 2011 by the American Physical Society.

Other experiments have qualitatively observed drift transport under high-field conditions, observed as linear charge extraction.[114,119] At external voltages near open circuit, where the internal field approaches zero, the transit time can no longer be described solely by drift transport—nongeminate recombination plays a large role and recombination is bias dependent.

These conclusions concerning the bias dependence of nongeminate charge recombination were recently confirmed by Mauer et al., who developed a novel experimental technique, a TDDP current–voltage measurement, which directly probed the dwell time of charge carriers in P3HT:PCBM devices.[115] Quasi-steady-state current–voltage curves were recorded as a function of time delay, intensity, and bias upon irradiation with two pulsed lasers, and the results were extrapolated to solar illumination. Nongeminate recombination, identified specifically as bimolecular recombination in many recent reports, is measured to be bias dependent. Low recombination is found at the short-circuit condition, in agreement with high current collection in the extraction regime. Current loss due to bimolecular recombination increases with applied bias, and at the low internal fields present at open-circuit condition, a large fraction of charge bimolecularly recombines. These new results and experimental method will be discussed in more detail in Section 9.7.

At the short-circuit condition, Mauer et al. report very small bimolecular recombination loss under 1 sun illumination conditions,[115] a result in agreement with differential current measurements performed by Koster and colleagues.[65] Relatively small current loss due to bimolecular recombination (2%–17% depending on sample preparation in these experiments) at the short-circuit condition has thus far been overlooked[61,122] because of efficient carrier collection in this regime. In the efficient charge collection regime, short-circuit current, J_{sc}, is thought to be dominated by charge-carrier sweep-out via drift transport in the high internal electric field,[116] as described by the expression for drift current, $J_n = e\mu n_e E J_{n_e} = e\mu n_e E$, where e is the elementary unit of charge, n_e is the density of electrons (or holes assuming $n_e \approx n_h$), and E is the internal electric field. The expression for drift current predicts that short-circuit current is linearly proportional to the density of photogenerated carriers (which has been shown to increase proportional to incident light intensity[113]). Due to high levels of charge collection, small amounts (<10%) of bimolecular current loss cause only small deviations from the intensity-dependent linearity of J_{sc}. Hence, the dependence of steady-state short-circuit current on incident light intensity, I, $J_{sc} \propto I^{\alpha}$, where the physical interpretation of the fitting parameter, α, is under debate, is not a high-precision measure of recombination mechanism as sometimes thought.[123] In the common interpretation of the fitting parameter, α, fit values of $\alpha = 1$ are attributed to first-order loss processes, and the values of $\alpha = 0.5$ to second-order (bimolecular) loss processes.[123] In recent literature, space charge effects were found to predict $\alpha = 0.75$.[122] Koster and colleagues improve upon these models, and present an analytically derived expression for the fraction of carriers lost to bimolecular recombination, η_{BR}, in relation to the steady-state fitting parameter, α: $\eta_{BR} = \alpha^{-1} - 1$. It is worthwhile noting that when $\alpha = 0.5$, the current, J_{sc}, is reduced by one order of magnitude, rather than completely recombining. However, it is still true that in this regime, current loss is dominated by bimolecular recombination. This new expression for bimolecular recombination loss efficiency, as well as differential current density measurements, should prove to be a useful tool in the determination of current loss in the extraction regime.

We conclude that there is interesting recent evidence to suggest the presence of an electric field drop across the BHJ layer, as a result of the built-in electric field and applied bias voltage. High charge-carrier collection at the short-circuit condition and in reverse bias results from drift transport of photogenerated charge within this electric field. Carrier extraction may be described by drift transport of charge to the contacts over a wide range of relevant applied bias voltages. Additionally, bimolecular (nongeminate) recombination can account for current loss at the short-circuit condition. Finally, bimolecular recombination is found to be bias dependent, a concept we will explore in further detail in the next section.

9.7 Case Study: Bias- and Charge Density-Dependent Charge-Carrier Recombination

Although research into the role of exciton recombination and geminate recombination via coulomb bound interfacial charge pairs (see Sections 9.4 and 9.5) is ongoing,[104,106,108,110,112] an interesting new hypothesis has emerged: that bimolecular recombination accounts for the majority of recombination from short circuit to open circuit,[65,114,115] and that bimolecular recombination is bias- and charge density-dependent.[60,70,114,115] This new hypothesis is notably interesting within the context of previous work, because it presupposes highly efficient charge formation and separation,[104] and that recombination through traps, either from material impurities[60,63] or unoptimized morphology,[124] does not appear to be a major loss mechanism in optimized cells.

Recombination of photogenerated charge carriers in polymer BHJ solar cells reduces the short-circuit current, the fill factor, and the open-circuit voltage.[61] Identifying the mechanisms of recombination is, therefore, fundamentally important for increasing the PCE. Bimolecular recombination, the inter-domain band-to-band transition, as sketched in Fig. 9.2c, involves two mobile and independent charge carriers, and hence the recombination rate, R, can be considered a second-order process with respect to the mobile charge-carrier density ($R \propto n_e^2$). In a device with trap-free electron and hole transport and a high density of photogenerated charge carriers, mobile carriers are most likely to recombine nonradiatively via bimolecular recombination. Trap-assisted recombination can be characterized by recombination of mobile charge with a trapped carrier (see Fig. 9.2b) and hence can be considered a first-order process with respect to the density of mobile charge ($R \propto n_h$, in the case of an electron trap). The presence of impurities and defects that function as traps enhances trap-assisted recombination, as will be described below. In disordered donor : acceptor blends,[125,126] it may be necessary to go beyond Shockley–Read–Hall (SRH) recombination theory, based on traps with discrete energy levels; for example, SRH recombination may occur through a distribution of states in the disordered band tails and through trap states with energy levels influenced by their local environment in the BHJ matrix.

Light intensity-dependent current–voltage measurements on polymer BHJ cells made from a variety of different semiconducting polymers and fullerenes show that the kinetics are voltage dependent[61,70,114,115] and evolve from charge-carrier sweep-out in the extraction regime (reverse bias and short circuit) to a carrier loss regime around open circuit. Bimolecular recombination dominates the charge kinetics as a result of increasing, bias-dependent charge-carrier density in the cell.[61,70] Bimolecular recombination is almost universally identified as the dominant recombination mechanism in optimized organic BHJ solar cells at open circuit.[61,62,70,127] However, due to an overwhelming background of charge-carrier sweep-out, a study of recombination at the short-circuit condition has been experimentally challenging until recently, as discussed in the previous section. Two new experimental methods have been implemented to study recombination across the whole *J–V* curve, and have identified bimolecular recombination as the dominant current loss mechanism across the pertinent range of bias voltages.

Using TDCF experiments,[114] Kniepert et al. probe charge collection and recombination dynamics via a "pump-probe" analogue: a diode-pumped, Q-switched Nd:YAG with 8 ns pulse width generates charge within a P3HT:PCBM BHJ solar cell under a bias voltage. After a time delay, allowing for charge to dynamically interact within the device, a high extraction bias ($V = -5$ V) is applied, pulling charge out of the device. Integrated photocurrent transients recorded upon extraction with variable time delay, t_d, from $t_d = 100$ ns to 10 μs allow analysis of the evolution of charge extraction and recombination within the device. There are two important observations to make from the data with respect to bimolecular recombination: First, at short circuit, the applied pre-bias voltage provides enough extraction field to efficiently sweep out most of the excited charge population. Given a delay time of 10 μs (the longest delay time probed), ~15% of the initially excited charge recombines before collection. As pre-bias is increased, the total charge population decreases as delay time is increased. This pre-bias and delay time-dependent reduction in charge density is well-reproduced by incorporating Langevin recombination[59,128] as the loss mechanism. Bimolecular (Langevin-type) recombination increases from ~17% at short circuit to ~58% at a pre-bias of 0.55 V, just slightly above V_{OC}.

Mauer and colleagues come to similar conclusions using another unique experimental method: TDDP current–voltage measurements.[115] TDDP probes P3HT:PCBM BHJ solar cell devices using two pulsed lasers with total fluence energy comparable to solar illumination. By varying the time delay between the two pulsed lasers, the interaction time of two charge populations is controlled. When the delay between pulses is long, two separate charge populations transit the BHJ layer, resulting in an equivalent collected current equal to twice the current collected from one pulse. A delay time of 9 μs is necessary to collect the total current, or in other words, most charge is extracted after excitation (100 fs pulse length is reported) in ~10 μs. As the time delay between pulses is decreased, the complementary charge populations are both present in the BHJ layer simultaneously. In the case of low delay times, these interacting charge populations recombine at higher rate than the individual charge populations. Additionally, applied bias is again seen to reduce the dwell time of charge when reverse bias is applied. As bias increases toward open circuit, charge is extracted more slowly in agreement with results from Cowan et al.[116] Longer dwell times, in addition to increased charge density, result in increased current loss. The measured charge density- and bias-dependent recombination rates are again well reproduced via a Langevin-type bimolecular recombination model.

In 1903, Langevin described recombination rate, R, of ions in a gas via the correlation between population densities of positively and negatively charged ions, $n_{e(h)}$, the dielectric constant of the medium, $\varepsilon_r\varepsilon_0$, and the mobility of the carrier species, $R = en_e n_h(\mu_e+\mu_h)/\varepsilon_r\varepsilon_0$[128]. The Langevin recombination model has been widely shown to apply to organic semiconductors, though in the case of the BHJ, the recombination rate has been measured to be reduced—hypothesized to be a function of the reduced correlation between electron and hole populations due to spatial separation of the populations in the BHJ phases. Recombination based on the correlation between electrons and holes is also widely referred to in organic photovoltaics as nongeminate recombination or, less often, intensity-dependent recombination.

Representative data in Fig. 9.17 show bias-dependent bimolecular recombination under steady-state conditions.[115] Bimolecular recombination increases with decreasing internal field, and this experiment implies that the increased recombination may be due to increased dwell times and higher charge densities. Additionally, in concert with recent results from other groups, these results indicate that the effect of bimolecular recombination should not be neglected at either short circuit or open circuit.

We move on to a discussion of recent evidence for charge density dependence of bimolecular recombination. An increasing number of reports published to date detail the charge density dependence of bimolecular recombination. Bimolecular recombination will occur more frequently, where populations of electrons and holes intersect, as the energetic transition occurs without mediation through a localized state, which could increase the probability of opposite charge species becoming correlated. Charge extraction and transient photovoltage experiments by Shuttle et al. have offered the most direct look into the BHJ charge density.[69,70,113,129] Additionally, varying the intensity of light incident on a BHJ solar cell has long been assumed to linearly increase the charge density.[113] In a recent paper, in 2010, Shuttle and colleagues used charge extraction techniques to probe the bias dependence of the BHJ charge density, as shown in Fig. 9.18.[113] Increasing charge density proportional to applied bias can be attributed to two effects: (a) increasing externally applied bias decreases the internal field for charge extraction, localizing photogenerated charge carriers. Additionally, (b) the potential barrier for charge injection from the contacts decreases via the applied voltage. Beyond the diode's turn-on voltage, a significant density of charge is injected from the contacts. Charge extraction experiments measure both mobile and trapped charge populations in the active layer, but due to the transient nature of the experiments, only observe effects from the photogenerated current.

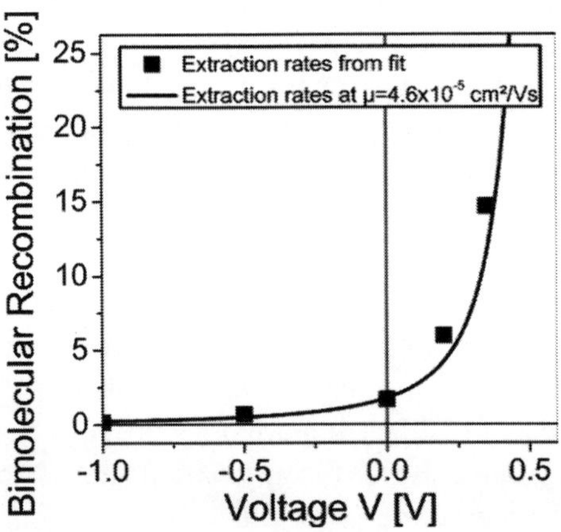

Figure 9.17 The bimolecular recombination rate, as determined from TDDP experiments, is found to increase nonlinearly with increasing bias in a P3HT:PCBM BHJ solar cell. Increasing recombination rate data (black squares) can be reproduced using the drift–diffusion model incorporating Langevin recombination (black line), and using an average mobility as seen in the graph. Reproduced with permission from Ref. [115]. Copyright 2011 by the *J Phys Chem Lett.*

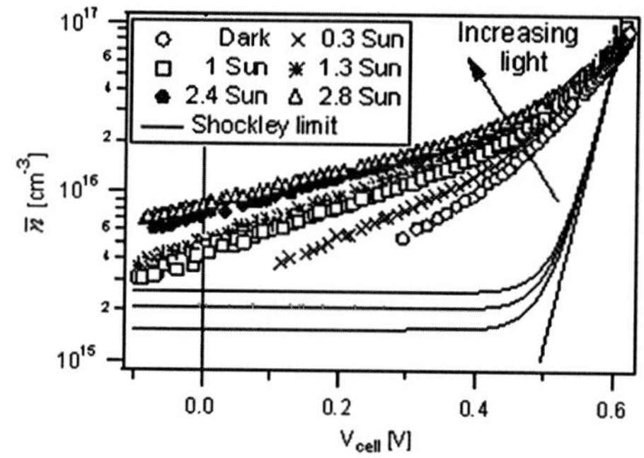

Figure 9.18 Bias dependence of the charge density of a P3HT:PCBM BHJ solar cell is plotted with varying incident light intensity. Charge density, as determined from charge extraction experiments, is found to increase more than an order of magnitude from short circuit to open circuit. Measured charge density dependence varies significantly from the Shockley limit (black lines). Reproduced with permission from Ref. [113]. Copyright 2010 by PNAS.

One may estimate the difference in charge density between short circuit and open circuit in a relatively optimized device with dynamics dominated by drift current sweep-out at short circuit and bimolecular recombination at open circuit.[61] At short circuit, the photocurrent $J_{sc} = J_{photo} = edGP_C(V = 0)J_{sc} = J_{photo} = edGP_C(V = 0)$ where d is the thickness of the BHJ, G is the photon flux absorbed by the solar cell per unit volume, and P_C is the charge collection probability. P_C approaches 1 at short circuit in the most efficient systems. If the current at short circuit is determined by drift, then $J_{sc} = 2e\mu n_{sc}E = edG$, where n_{sc} is the electron (or hole) charge density at short circuit. This equation can be rewritten in terms of the time scale for sweep-out, τ_{sw}: $G = n_{sc}/\tau_{sw}$. If we assume that all the photogenerated charge recombines bimolecularly at open circuit, $G = R(V_{oc}) = \gamma n_{oc}$, then we can relate charge density at open circuit, n_{oc}, to n_{sc}: $n_{sc}/\tau_{sw} = \gamma n_{oc}^2$. Using $n_{oc}/n_{sc} \sim 25$ from Fig. 9.18 for a P3HT:PCBM BHJ,[113] and $\tau_{sw} \sim 1$ μs[116] for a similarly prepared P3HT:PCBM BHJ solar cell, the bimolecular recombination coefficient γ can be calculated entirely from data to be $\gamma = 4 \times 10^{-13}$ cm^3/s. The bimolecular recombination coefficient has been reported to be in the range from 10^{-12} cm^3/s $> \gamma > 10^{-14}$ cm^3/s implying that the ratio of charge density at open circuit to short circuit may be in the range

$10 > n_{oc}/n_{sc} > 1000$ for polymer solar cells constructed of other materials or via other fabrication processes. Increasing charge densities can be correlated with increased bimolecular recombination, as indicated by TDDP current–voltage measurements, although further experiments will be necessary to determine the relationship between charge density and recombination rate and strength.

One such study investigates the charge density and trap density dependence of the recombination mechanism in polymer BHJ solar cells.[60] We note that while bimolecular recombination has been identified as the dominant recombination mechanism in optimized BHJ solar cells, investigations into the recombination mechanisms of unoptimized devices (both in material purity and morphology) will remain relevant as long as the optimization of devices using newly synthesized materials continues to play a role in forward progress of the field. In order to probe the recombination mechanism, small concentrations of impurity were introduced into PCDTBT:PCBM BHJ solar cells. It was found that impurity levels of even 1 part in 1,000 altered the electronic properties of the BHJ devices, reducing the open-circuit voltage, the short-circuit current, and the fill factor. Steady-state current–voltage studies showed a dramatic increase in the trap-assisted recombination rate when $PC_{84}BM$ ([6,6]-phenyl-C_{84}-butyric acid methyl ester) was introduced as a trap site into the polymer BHJ solar cells.[60] This trap density-dependent recombination was modeled as a sum of bimolecular and Shockley–Read–Hall recombination processes; the latter was dramatically enhanced by the addition of the $PC_{84}BM$ traps. Results of trapping and recombination discussed here are confirmed in the source literature,[60,66] via transient photocurrent and transient absorption and impedance spectroscopy.

The trap-assisted recombination rate (proportional to the trap density) and the generation rate (proportional to incident light intensity) control the cross-over from bimolecular to trap-assisted recombination, as plotted (both data and model) in Fig. 9.19. Cross sections of Fig. 9.19 at a single trap density reproduce the commonly plotted intensity dependence of the open-circuit voltage.[61,62] Above the cross-over point at low trap densities and high light intensities, the open-circuit voltage is relatively unaffected by the density of traps within the BHJ. Below the cross-over point at high trap densities, such as those potentially found in BHJ materials with impurities left over from the synthesis and device processing, recombination through traps becomes the dominant loss mechanism and results in a significant reduction in the open-circuit voltage at operating illumination levels.

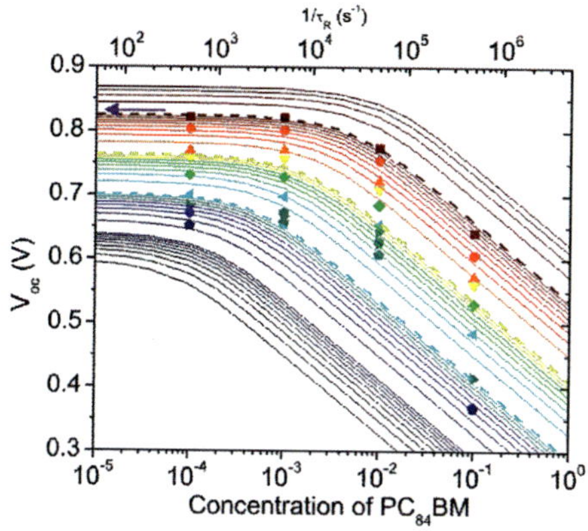

Figure 9.19 Intensity/charge density-dependent V_{oc} plotted as a function of $PC_{84}BM$ trap concentration (data points, lower x-axis) in a PCDTBT:PCBM BHJ solar cell. Bimolecular/SRH recombination model fits to the data (lines, upper x-axis): the trap-assisted recombination rate plotted vs. the generation rate of photoexcitations with color from blue to brown indicating an increase in generation rate. Arrow indicates the V_{oc} at 1 sun for the trap-free PCDTBT:PCBM solar cell devices. Dotted lines correspond to logarithmic increases in generation rate (G(brown) = 10^{21} cm^{-3} s^{-1}, G(yellow) = 10^{20} cm^{-3} s^{-1}, G(blue) = 10^{19} cm^{-3} s^{-1}). Reproduced with permission from Ref. [60]. Copyright 2011 by Wiley-VCH Verlag GmbH & Co. KGaA.

The dominant recombination mechanism is both incident light intensity dependent (proportional to charge density) and trap density dependent (inversely proportional to the free carrier lifetime). Here, we offer an interpretation of this data. At high incident light intensity, the higher density of charge carriers increases the probability of bimolecular recombination (an electron and a hole must find each other in the BHJ material). At low light intensity, charge carriers are typically farther apart and traps are more likely to be empty, making them viable routes for recombination. Increasing the trap density increases the probability of trap-assisted recombination. In practice, trap states are likely to result from impurities, polymerization defects, and poor phase separation (unoptimized formation of the BHJ material). Further research into the physical mechanisms of charge recombination may not only lead to more highly efficient organic BHJ solar cells but also decrease the time necessary to evaluate newly synthesized materials.

9.8 Case Study: Role of the Contacts in Charge Collection

Finally, a report on charge collection and recombination in organic photovoltaics would not be complete without a few sentences regarding the role of contacts and contact interlayers in collection, recombination, and overall device performance. BHJ active layer films rely on semiconducting and metallic contacts to extract charge into an external circuit.[130–133] Ultimately, the contacts play a larger role in organic BHJ solar cells than they do in inorganic (high-mobility) systems due to the nature of low carrier mobility charge transport. Metallic and semiconducting contacts set up a drift electric field (materials used in organic BHJ solar cells are assumed to not have sufficient carrier populations or mobility to compensate the applied field) for efficient charge collection.[116,134,135] In the thin BHJ layers used for the most internally efficient devices, there is likely to be a strong, uniform drift field across the entire device thickness in equilibrium. At steady state, an opposing field due to carrier populations and/or voltage bias can screen the built-in field. Contact work function has been shown to be critically important in determining device performance.[136–140] Furthermore, contacts must block carriers of the opposite sign from being extracted, or injected, into the active layer.[141] Hence, transport layers (for a full treatment, see Chapter 4), including PEDOT:PSS, Ca, ZnO, TiO_x, MoO_3, WO_3, NiO, and others, are usually implemented, based on work function and selectivity, and to improve fill factor and all the other performance parameters.

However, additional factors must be considered. For example, contacts must have favorable chemical interactions with the active layer. For example, in recent years, PEDOT:PSS and calcium have been shown to be a significant breakdown mechanism in lifetime studies as calcium oxidizes[142] and the mobile PSS dopant migrates into the active layer.[143] Unfavorable chemical interactions such as bonding or CT at the interface have been shown to significantly reduce performance by creating energy barriers to charge extraction at the contact: active layer interface.[144] Metal oxides have been shown to stabilize the devices, increasing device lifetime on the order of years.[143,145,146]

Research on the topic is ongoing to determine the relative importance of contact material properties in determining charge extraction and overall device performance: (1) work function, (2) surface energy/wetting properties, (3) stability, (4) carrier mobility, (5) interfacial chemical interactions such as bonding and charge transfer, and (6) blocking characteristics/selectivity.

9.9 Summary

Here, the role that spectroscopic and electrical techniques have played in understanding the physical processes relevant to photogeneration of charge carriers and the interplay between recombination and collection of photogenerated charge carriers is detailed. Along the way, we presented an inexhaustive survey of the way in which sets of measurement techniques complement each other or have presented conflicting sets of data on the same materials systems, and continue to shape the scientific perception of photovoltaic current generation in BHJ systems. The results of these phenomena, i.e., photogeneration of charge carriers and how the interplay between recombination and collection of photogenerated carriers directly contributes to photovoltaic performance, will be topics of intense research as OPV efficiencies continue to increase beyond 10%.

Acknowledgments

S. R. Cowan would like to acknowledge many fruitful discussions and work with Prof. Alan J. Heeger, as well as discussions with Dr. Anshuman Roy, Dr. Robert Street, Dr. Wei Lin Leong, Dr. Bradley A. MacLeod, and Dr. Dana C. Olson. Dr. Cowan's writing was supported by the Energy Efficiency and Renewable Energy Postdoctoral

Fellowship, and her work as a SunShot Postdoctoral Fellow. N. Banerji would like to acknowledge the Swiss National Science Foundation for funding of her research through the Ambizione Fellowship PZ00P2_136853.

References and Citations

1. Tang CW (1986) Two-layer organic photovoltaic cell, *Appl Phys Lett,* **48**, 183–185.
2. Yu G, Gao J, Hummelen JC, Wudl F, Heeger AJ (1995) Polymer photovoltaic cells: Enhanced efficiencies via a network of internal donor-acceptor heterojunctions, *Science,* **270**, 1789–1791.
3. Hadziioannou G, Malliaras GG (eds) (2006) *Semiconducting polymers,* 2nd ed, John Wiley & Sons.
4. Heeger A, Sariciftci NS, Namdas EB (2010) *Semiconducting and metallic polymers,* Oxford University Press: New York.
5. Halls JJM, Walsh CA, Greenham NC, et al. (1995) Efficient photodiodes from interpenetrating polymer networks, *Nature,* **376**, 498–500.
6. Liu J, Sun Y, Gao X, et al. (2011) Oriented poly(3-hexylthiophene) nanofibril with the π–π stacking growth direction by solvent directional evaporation, *Langmuir,* **27**, 4212–4219.
7. Mayer AC, Toney MF, Scully SR, et al. (2009) Bimolecular crystals of fullerenes in conjugated polymers and the implications of molecular mixing for solar cells, *Adv Funct Mater,* **19**, 1173–1179.
8. Heeger AJ (2010) Semiconducting polymers: The third generation, *Chem Soc Rev,* **39**, 2354–2371.
9. Scheblykin IG, Yartsev A, Pullerits T, Gulbinas V, Sundström V (2007) Excited state and charge photogeneration dynamics in conjugated polymers, *J Phys Chem B,* **111**, 6303–6321.
10. Barford W (2005) *Electronic and optical properties of conjugated polymers,* Oxford University Press: New York.
11. Lanzani G, Cerullo G, Polli D, Gambetta A, Zavelani-Rossi M, Gadermaier C (2004) Photophysics of conjugated polymers: The contribution of ultrafast spectroscopy, *Phys Stat Sol (a),* **201**, 1116–1131.
12. Sariciftci NS (1997) *Primary photoexcitations in conjugated polymers: Molecular excitation versus semiconductor band model,* World Scientific: River Edge, NJ.
13. Heeger AJ, Schrieffer JR, Su WP (1988) Solitons in conducting polymers, *Rev Mod Phys,* **60**, 781–850.

14. Bässler H, Arkhipov VI, Emelianova EV, et al. (2003) Excitons in π conjugated polymers, *Synth Met,* **135**, 377–382.

15. Bässler H (1996) Opto-electronic properties of conjugated polymers, *Macromol Symp,* **104**, 269–284.

16. Dubin F, Melet R, Barisien T, et al. (2005) Macroscopic coherence of a single exciton state in an organic quantum wire, *Nat Phys,* **2**, 32–35.

17. Yang X, Dykstra TE, Scholes GD (2005) Photon-echo studies of collective absorption and dynamic localization of excitation in conjugated polymers and oligomers, *Phys Rev B,* **71**, 045203.

18. Dykstra T, Kovalevskij V, Yang X, Scholes G (2005) Excited state dynamics of a conformationally disordered conjugated polymer: A comparison of solutions and film, *Chem Phys,* **318**, 21–32.

19. Banerji N, Cowan S, Leclerc M, Vauthey E, Heeger AJ (2010) Exciton formation, relaxation, and decay in PCDTBT, *J Am Chem Soc,* **132**, 17459–17470.

20. Banerji N, Cowan S, Vauthey E, Heeger AJ (2011) Ultrafast relaxation of the poly(3-hexylthiophene) emission spectrum, *J Phys Chem C,* **115**, 9726–9739.

21. Grage M, Pullerits T, Ruseckas A, Theander M (2001) Conformational disorder of a substituted polythiophene in solution revealed by excitation transfer, *Chem Phys,* **339**, 96–102.

22. Grage M, Zaushitsyn Y, Yartsev A, Chachisvilis M (2003) Ultrafast excitation transfer and trapping in a thin polymer film, *Phys Rev B,* **67**, 205207.

23. Hayes G, Samuel I, Phillips R (1995) Exciton dynamics in electroluminescent polymers studied by femtosecond time-resolved photoluminescence spectroscopy, *Phys Rev B,* **52**, 11569–11572.

24. Hayes G, Samuel I, Phillips R (1997) Ultrafast dynamics of photoexcitations in conjugated polymers, *Synth Met,* **84**, 889–890.

25. Kennedy S, Garro N, Phillips R (2001) Time-resolved site-selective spectroscopy of poly(p-phenylene vinylene), *Phys Rev B,* **64**, 115206.

26. Kersting R, Lemmer U, Mahrt R, et al. (1993) Femtosecond energy relaxation in π-conjugated polymers, *Phys Rev Lett,* **70**, 3820–3823.

27. Kersting R, Mollay B, Rusch M, Wenisch J, Leising G, Kauffmann HF (1997) Femtosecond site-selective probing of energy relaxing excitons in poly(phenylenevinylene): Luminescence dynamics and lifetime spectra, *J Chem Phys,* **106**, 2850–2864.

28. Milota F, Sperling J, Tortschanoff A, Szöcs V, Kuna L, Kauffmann H (2004) Probing phase relaxation by measuring fluorescence interference:

Polarization beating and electron–phonon coupling in conjugated polymers, *J Lumin*, **108**, 205–209.

29. Ruseckas A, Wood P, Samuel IDW, et al. (2005) Ultrafast depolarization of the fluorescence in a conjugated polymer, *Phys Rev B*, **72**, 115214.
30. Sperling J, Milota F, Tortschanoff A, et al. (2002) Femtosecond excitation tuning and site energy memory of population transfer in poly (p-phenylenevinylene): Gated luminescence experiments and simulation, *J Chem Phys*, **117**, 10877.
31. Sperling J, Milota F, Kauffmann H (2005) Excitonic relaxation in a conjugated polymer: Initial coherence beyond energy-dispersive population transfer, *Opt Spectrosc*, **98**, 729–739.
32. Warmuth C, Tortschanoff A, Brunner K, Mollay B, Kauffmann H (1998) Excitonic fs-luminescence rise in poly(phenylenevinylene), PPV, *J Lumin*, **76**, 498–501.
33. Wells NP, Boudouris BW, Hillmyer MA, Blank DA (2007) Intramolecular exciton relaxation and migration dynamics in poly(3-hexylthiophene), *J Phys Chem C*, **111**, 15404–15414.
34. Jespersen KG, Zhang FL, Gadisa A, Sundström V, Yartsev A, Inganäs O (2006) Charge formation and transport in bulk-heterojunction solar cells based on alternating polyfluorene copolymers blended with fullerenes, *Org Electron*, **7**, 235–242.
35. Zhang FL, Jespersen KG, Bjorstrom C, et al. (2006) Influence of solvent mixing on the morphology and performance of solar cells based on polyfluorene copolymer/fullerene blends, *Adv Funct Mater*, **16**, 667–674.
36. Bakulin AA, Hummelen JC, Pshenichnikov MS, van Loosdrecht PHM (2010) Ultrafast hole-transfer dynamics in polymer/PCBM bulk heterojunctions, *Adv Funct Mater*, **20**, 1653–1660.
37. Brabec C, Zerza G, Cerullo G, et al. (2001) Tracing photoinduced electron transfer process in conjugated polymer/fullerene bulk heterojunctions in real time, *Chem Phys Lett*, **340**, 232–236.
38. Hwang I-W, Kim JY, Cho S, et al. (2008) Bulk heterojunction materials composed of poly(2,5-bis(3-tetradecylthiophen-2-yl) thieno[3,2-*b*]thiophene): Ultrafast electron transfer and carrier recombination, *J Phys Chem C*, **112**, 7853–7857.
39. Hwang I-W, Soci C, Moses D, et al. (2007) Ultrafast electron transfer and decay dynamics in a small bandgap bulk heterojunction material, *Adv Mater*, **19**, 2307–2312.

40. Kraabel B, Hummelen JC, Vacar D, et al. (1996) Subpicosecond photoinduced electron transfer from conjugated polymers to functionalized fullerenes, *J Chem Phys,* **104**, 4267–4273.

41. Morita S, Zakhidov AA, Yoshino K (1996) Doping effect of buckminsterfullerene in conducting polymer—change of absorption-spectrum and quenching of luminescence, *Solid State Commun,* **82**, 249–252.

42. Sariciftci NS, Smilowitz L, Heeger AJ, Wudl F (1992) Photoinduced electron-transfer from a conducting polymer to buckminsterfullerene, *Science,* **258**, 1474–1476.

43. Tong M, Coates NE, Moses D, Heeger AJ, Beaupré S, Leclerc M (2010) Charge-carrier photogeneration and decay dynamics in the poly(2,7-carbazole) copolymer PCDTBT and in bulk heterojunction composites with $PC_{70}BM$, *Phys Rev B,* **81**, 125210.

44. Shaw PE, Ruseckas A, Samuel IDW (2008) Exciton diffusion measurements in poly(3-hexylthiophene), *Adv Mater,* **20**, 3516–3520.

45. Lewis A, Ruseckas A, Gaudin O, Webster G, Burn P, Samuel I (2006) Singlet exciton diffusion in MEH-PPV films studied by exciton-exciton annihilation, *Org Electron,* **7**, 452–456.

46. Dante M, Garcia A, Nguyen T-Q (2009) Three-dimensional nanoscale organization of highly efficient low band-gap conjugated polymer bulk heterojunction solar cells, *J Phys Chem C,* **113**, 1596–1600.

47. Erb T, Zhokhavets U, Gobsch G, et al. (2005) Correlation between structural and optical properties of composite polymer/fullerene films for organic solar cells, *Adv Funct Mater,* **15**, 1193–1196.

48. Moon JS, Takacs CJ, Cho S, et al. (2010) Effect of processing additive on the nanomorphology of a bulk heterojunction material, *Nano Lett,* **10**, 4005–4008.

49. Parnell AJ, Cadby AJ, Mykhaylyk OO, et al. (2011) Nanoscale phase separation of P3HT PCBM thick films as measured by small-angle X-ray scattering, *Macromolecules,* **44**, 6503–6508.

50. Walker B, Tomayo AB, Dang X-D, et al. (2009) Nanoscale phase separation and high photovoltaic efficiency in solution-processed, small-molecule bulk heterojunction solar cells, *Adv Funct Mater,* **19**, 3063–3069.

51. Chen D, Nakahara A, Wei D, Nordlund D (2011) P3HT/PCBM bulk heterojunction organic photovoltaics: Correlating efficiency and morphology, *Nano Lett,* **11**, 561–567.

52. Campoy-Quiles MM, Ferenczi TT, Agostinelli TT, et al. (2008) Morphology evolution via self-organization and lateral and vertical diffusion in polymer:fullerene solar cell blends, *Nat Mater,* **7**, 158–164.
53. Bakulin AA, Rao A, Pavelyev VG, et al. (2012) The role of driving energy and delocalized states for CS in organic semiconductors, *Science,* **335**, 1340–1344.
54. Park SH, Roy A, Beaupré S, et al. (2009) Bulk heterojunction solar cells with internal quantum efficiency approaching 100%, *Nat Photonics,* **3**, 297–302.
55. Sun Y, Takacs CJ, Cowan SR, et al. (2011) Efficient, air-stable bulk heterojunction polymer solar cells using MoO_x as the anode interfacial layer, *Adv Mater,* **23**, 2226–2230.
56. Lenes M, Koster LJA, Mihailetchi VD, Blom PWM (2006) Thickness dependence of the efficiency of polymer:fullerene bulk heterojunction solar cells, *Appl Phys Lett,* **88**, 243502.
57. Moule AJ, Meerholz K (2007) Minimizing optical losses in bulk heterojunction polymer solar cells, *Appl Phys B,* **86**, 721–727.
58. Peet J, Wen L, Byrne P, et al. (2011) Bulk heterojunction solar cells with thick active layers and high fill factors enabled by a bithiophene-co-thiazolothiazole push-pull copolymer, *Appl Phys Lett,* **98**, 043301.
59. Pivrikas A, Juška G, Mozer AJ, et al. (2005) Bimolecular recombination coefficient as a sensitive testing parameter for low-mobility solar-cell materials, *Phys Rev Lett,* **94**, 176806.
60. Cowan SR, Leong WL, Banerji N, Dennler G, Heeger AJ (2011) Identifying a threshold impurity level for organic solar cells: Enhanced first-order recombination via well-defined $PC_{84}BM$ traps in organic bulk heterojunction solar cells, *Adv Funct Mater,* **21**, 3083–3092.
61. Cowan S, Roy A, Heeger A (2010) Recombination in polymer-fullerene bulk heterojunction solar cells, *Phys Rev B,* **82**, 245207.
62. Koster LJA, Mihailetchi VD, Ramaker R, Blom PWM (2005). Light intensity dependence of open-circuit voltage of polymer:fullerene solar cells, *Appl Phys Lett,* **86**, 123509.
63. Mandoc MM, Kooistra FB, Hummelen JC, De Boer B, Blom PWM (2007) Effect of traps on the performance of bulk heterojunction organic solar cells, *Appl Phys Lett,* **91**, 263505.
64. Garcia-Belmonte G, Bisquert J (2010) Open-circuit voltage limit caused by recombination through tail states in bulk heterojunction polymer-fullerene solar cells, *Appl Phys Lett,* **96**, 113301.

65. Koster LJA, Kemerink M, Wienk MM, Maturová K, Janssen RAJ (2011) Quantifying bimolecular recombination losses in organic bulk heterojunction solar cells, *Adv Mater,* **23**, 1670–1674.

66. Leong WL, Cowan SR, Heeger AJ (2011) Differential resistance analysis of charge-carrier losses in organic bulk heterojunction solar cells: Observing the transition from bimolecular to trap-assisted recombination and quantifying the order of recombination, *Adv Energy Mater,* **1**, 517–522.

67. Mozer AJ, Dennler G, Sariciftci NS, et al. (2005) Time-dependent mobility and recombination of the photoinduced charge-carriers in conjugated polymer/fullerene bulk heterojunction solar cells, *Phys Rev B,* **72**, 035217.

68. Juška G, Arlauskas K, Sliaužys G, et al. (2005) Double injection as a technique to study charge-carrier transport and recombination in bulk-heterojunction solar cells, *Appl Phys Lett,* **87**, 222110.

69. Shuttle CG, Maurano A, Hamilton R, O'Regan BC, De Mello JC, Durrant JR (2008) Charge extraction analysis of charge-carrier densities in a polythiophene/fullerene solar cell: Analysis of the origin of the device dark current, *Appl Phys Lett,* **93**, 183501.

70. Shuttle C, O'Regan B, Ballantyne A, Nelson J, Bradley D, Durrant J (2008) Bimolecular recombination losses in polythiophene:fullerene solar cells, *Phys Rev B,* **78**, 113201.

71. Cook S, Ohkita H, Kim Y, Benson-Smith JJ, Bradley DDC, Durrant JR (2007) A photophysical study of PCBM thin films, *Chem Phys Lett,* **445**, 276–280.

72. Tauc J, Grigorovici R, Vancu A (1966) Optical properties and electronic structure of amorphous germanium, *Phys Stat Sol (b),* **15**, 627–637.

73. Morigaki K (1999) *Physics of amorphous semiconductors,* Imperial College Press: London.

74. Turner DW, Baker AD, Brundle CR (1970) *Molecular photoelectron spectroscopy,* Wiley-Interscience: London.

75. Shrotriya V, Li G, Yao Y, Moriarty T, Emery K, Yang Y (2006) Accurate measurement and characterization of organic solar cells, *Adv Funct Mater,* **16**, 2016–2023.

76. Cowan SR, Banerji N, Leong WL, Heeger AJ (2012) Charge formation, recombination, and sweep-out dynamics in organic solar cells, *Adv Funct Mater,* **22**, 1116–1128.

77. Alvarado SF, Seidler PF, Lidzey DG, Bradley DDC (1998) Direct determination of the exciton binding energy of conjugated polymers using a scanning tunneling microscope, *Phys Rev Lett,* **81**, 1082–1085.
78. Conwell EM, Mizes HA (1995) Photogeneration of polaron pairs in conducting polymers, *Phys Rev B,* **51**, 6953–6958.
79. Deibel C, Mack D, Gorenflot J, et al. (2010) Energetics of excited states in the conjugated polymer poly(3-hexylthiophene), *Phys Rev B,* **81**, 085202.
80. Barbour L, Pensack R, Hegadorn M, Arzhantsev S, Asbury J (2008) Excitation transport and CS in an organic photovoltaic material: Watching excitations diffuse to interfaces, *J Phys Chem C,* **112**, 3926–3934.
81. De S, Kesti T, Maiti M, et al. (2008) Exciton dynamics in alternating polyfluorene/fullerene blends, *Chem Phys,* **350**, 14–22.
82. Meskers S, Hübner J, Oestreich M, Bässler H (2001) Dispersive relaxation dynamics of photoexcitations in a polyfluorene film involving energy transfer: Experiment and Monte Carlo simulations, *J Phys Chem B,* **105**, 9139–9149.
83. Pina J, Seixas de Melo J, Burrows H, et al. (2009) Photophysical and spectroscopic investigations on (oligo)thiophene-arylene step-ladder copolymers. The interplay of conformational relaxation and on-chain energy transfer, *J Phys Chem B,* **113**, 15928–15936.
84. Cates NC, Gysel R, Beiley Z, et al. (2009) Tuning the properties of polymer bulk heterojunction solar cells by adjusting fullerene size to control intercalation, *Nano Lett,* **9**, 4153–4157.
85. Ma WL, Yang CY, Gong X, Lee K, Heeger AJ (2005) Thermally stable, efficient polymer solar cells with nanoscale control of the interpenetrating network morphology, *Adv Funct Mater,* **15**, 1617–1622.
86. Marsh RA, Hodgkiss JM, Albert-Seifried S, Friend RH (2010) Effect of annealing on P3HT:PCBM charge transfer and nanoscale morphology probed by ultrafast spectroscopy, *Nano Lett,* **10**, 923–930.
87. Lee YH, Yabushita A, Hsu C-S, et al. (2010) Ultrafast relaxation dynamics of photoexcitations in poly(3-hexylthiophene) for the determination of the defect concentration, *Chem Phys Lett,* **498**, 71–76.
88. Du J, Wang Z, Feng W, Yoshino K, Kobayashi T (2008) Simultaneous measurement of electronic and vibrational dynamics to clarify a geometrical relaxation process in a conjugated polymer, *Phys Rev B,* **77**, 195205.

89. Wells NP, Blank DA (2008) Correlated exciton relaxation in poly(3-hexylthiophene), *Phys Rev Lett,* **100**, 086403.

90. Barford W, Boczarow I, Wharram T (2011) Ultrafast dynamical localization of photoexcited states in conformationally disordered poly(p-phenylenevinylene), *J Phys Chem A,* **115**, 9111–9119.

91. Tozer OR, Barford W (2012) Exciton dynamics in disordered poly(p-phenylenevinylene). 1. Ultrafast interconversion and dynamical localization, *J Phys Chem A,* **116**, 10310–10318.

92. Walter M, Lupton J, Becker K, Feldmann J, Gaefke G, Höger S (2007) Simultaneous Raman and fluorescence spectroscopy of single conjugated polymer chains, *Phys Rev Lett,* **98**, 137401.

93. Collini E, Scholes GD (2009) Coherent intrachain energy migration in a conjugated polymer at room temperature, *Science,* **323**, 369–373.

94. Grancini G, Maiuri M, Fazzi D, et al. (2013) Hot exciton dissociation in polymer solar cells, *Nat Mater,* **12**, 29–33.

95. Onsager L (1938) Initial recombination of ions, *Phys Rev,* **54**, 554–557.

96. Braun CL (1984) Electric field assisted dissociation of CT states as a mechanism of photocarrier production, *J Chem Phys,* **80**, 4157–4161.

97. Tachiya M (1988) Breakdown of the Onsager theory of geminate ion recombination, *J Chem Phys,* **89**, 6929–6935.

98. Koster L, Smits E, Mihailetchi V, Blom P (2005) Device model for the operation of polymer/fullerene bulk heterojunction solar cells, *Phys Rev B,* **72**, 085205.

99. Gong X, Tong M, Brunetti FG, et al. (2011) Bulk heterojunction solar cells with large open-circuit voltage: Electron transfer with small donor-acceptor energy offset, *Adv Mater,* **23**, 2272–2277.

100. Schwenn PE, Gui K, Zhang Y, Burn PL, Meredith P, Powell BJ (2012) Kinetics of CT processes in organic solar cells: Implications for the design of acceptor molecules, *Org Electron,* **13**, 2538–2545.

101. Rand B, Burk D, Forrest S (2007) Offset energies at organic semiconductor heterojunctions and their influence on the open-circuit voltage of thin-film solar cells, *Phys Rev B,* **75**, 115327.

102. Street RA, Cowan S, Heeger AJ (2010) Experimental test for geminate recombination applied to organic solar cells, *Phys Rev B,* **82**, 121301.

103. Albrecht S, Schindler W, Kurpiers J, et al. (2012) On the field dependence of free charge-carrier generation and recombination in

blends of PCPDTBT/PC_{70}BM: Influence of solvent additives, *J Phys Chem Lett,* **3**, 640–645.

104. Etzold F, Howard IA, Mauer R, et al. (2011) Ultrafast exciton dissociation followed by nongeminate charge recombination in PCDTBT:PCBM photovoltaic blends, *J Am Chem Soc,* **133**, 9469–9479.

105. Karl N, Ziegler J (1975) Generation and transport of charge-carriers in the charge-transfer complex anthracene-pyromellitic-dianhydride, *Chem Phys Lett,* **32**, 438–442.

106. Marsh RA, Hodgkiss JM, Friend RH (2010) Direct measurement of electric field: Assisted CS in polymer:fullerene photovoltaic diodes, *Adv Mater,* **22**, 3672–3676.

107. Veldman D, İpek O, Meskers SCJ, et al. (2008) Compositional and electric field dependence of the dissociation of CT excitons in alternating polyfluorene copolymer/fullerene blends, *J Am Chem Soc,* **130**, 7721–7735.

108. Guo J, Ohkita H, Benten H, Ito S (2010) Charge generation and recombination dynamics in poly(3-hexylthiophene)/fullerene blend films with different regioregularities and morphologies, *J Am Chem Soc,* **132**, 6154–6164.

109. Jamieson FC, Agostinelli T, Azimi H, Nelson J, Durrant JR (2010) Field-independent charge photogeneration in PCPDTBT/PC_{70}BM solar cells, *J Phys Chem Lett,* **1**, 3306–3310.

110. Howard IA, Mauer R, Meister M, Laquai F (2010) Effect of morphology on ultrafast free carrier generation in polythiophene:fullerene organic solar cells, *J Am Chem Soc,* **132**, 14866–14876.

111. Hwang I-W, Moses D, Heeger AJ (2008) Photoinduced carrier generation in P3HT/PCBM bulk heterojunction materials, *J Phys Chem C,* **112**, 4350–4354.

112. Keivanidis PE, Clarke TM, Lilliu S, et al. (2010) Dependence of CS efficiency on film microstructure in poly(3-hexylthiophene-2,5-diyl):[6,6]-phenyl-C_{61} butyric acid methyl ester blend films, *J Phys Chem Lett,* **1**, 734–738.

113. Shuttle C, Hamilton R, O'Regan B, Nelson J, Durrant J (2010) Charge-density-based analysis of the current–voltage response of polythiophene/fullerene photovoltaic devices, *Proc Natl Acad Sci,* **107**, 16448.

114. Kniepert J, Schubert M, Blakesley JC, Neher D (2011) Photogeneration and recombination in P3HT/PCBM solar cells

probed by time-delayed collection field experiments, *J Phys Chem Lett,* **2**, 700–705.

115. Mauer R, Howard I (2011) Effect of external bias on nongeminate recombination in polythiophene:methanofullerene organic solar cells, *J Phys Chem Lett,* **2**, 1736–1741.

116. Cowan SR, Street RA, Cho S, Heeger AJ (2011) Transient photoconductivity in polymer bulk heterojunction solar cells: Competition between sweep-out and recombination, *Phys Rev B,* **83**, 035205.

117. Deibel CC, Strobel TT, Dyakonov VV (2009) Origin of the efficient polaron-pair dissociation in polymer-fullerene blends, *Phys Rev Lett,* **103**, 036402.

118. De S, Pascher T, Maiti M, et al. (2007) Charge recombination in alternating polyfluorene copolymer/fullerene blends. *J Am Chem Soc,* **129**, 8466–8472.

119. Christ NS, Kettlitz SW, Valouch S, et al. (2009) Nanosecond response of organic solar cells and photodetectors, *J Phys Chem,* **105**, 104513.

120. Moses D, Wang J, Yu G, Heeger A (1998) Temperature-independent photoconductivity in thin films of semiconducting polymers: Photocarrier sweep-out prior to deep trapping, *Phys Rev Lett,* **80**, 2685–2688.

121. Reid OG, Xin H, Jenekhe SA, Ginger DS (2010) Nanostructure determines the intensity-dependence of open-circuit voltage in plastic solar cells, *J Phys Chem,* **108**, 084320.

122. Koster LJA, Mihailetchi VD, Xie H, Blom PWM (2005) Origin of the light intensity dependence of the short-circuit current of polymer/fullerene solar cells, *Appl Phys Lett,* **87**, 203502.

123. Rose A (1963) *Concepts in photoconductivity and allied problems,* Wiley-Interscience: New York.

124. Mandoc MM, Veurman W, Koster LJA, De Boer B, Blom PWM (2007) Origin of the reduced fill factor and photocurrent in MDMO-PPV:PCNEPV all-polymer solar cells, *Adv Funct Mater,* **17**, 2167–2173.

125. Bisquert J, Garcia-Belmonte G (2011) On voltage, photovoltage and photocurrent in bulk heterojunction organic solar cells, *J Phys Chem Lett,* **2**, 1950–1964.

126. Street RA (2011) Localized state distribution and its effect on recombination in organic solar cells, *Phys Rev B,* **84**, 75208.

127. Credgington D, Hamilton R, Atienzar P, Nelson J, Durrant JR (2011) Non-geminate recombination as the primary determinant of open-circuit voltage in polythiophene:fullerene blend solar cells: An analysis of the influence of device processing conditions, *Adv Funct Mater,* **21**, 2744–2753.

128. Langevin P (1903) Recombinaison Et Mobilités Des Ions Dans Les Gaz, *Ann Chim Phys,* 433.

129. Shuttle CG, O'Regan BC, Ballantyne AM, et al. (2008) Experimental determination of the rate law for charge-carrier decay in a polythiophene:fullerene solar cell, *Appl Phys Lett,* **92**, 093311.

130. Braun S, Salaneck WR, Fahlman M (2009) Energy-level alignment at organic/metal and organic/organic interfaces, *Adv Mater,* **21**, 1450–1472.

131. Hwang J, Wan A, Kahn A (2009) Energetics of metal–organic interfaces: New experiments and assessment of the field, *Mater Sci Eng R: Rep,* **64**, 1–31.

132. Heimel G, Romaner L, Zojer E, Brédas J-L (2008) The interface energetics of self-assembled monolayers on metals, *Acc Chem Res,* **41**, 721–729.

133. Garcia-Belmonte G, Boix PP, Bisquert J, et al. (2010) Influence of the intermediate density-of-states occupancy on open-circuit voltage of bulk heterojunction solar cells with different fullerene acceptors, *J Phys Chem Lett,* **1**, 2566–2571.

134. Barker JA, Ramsdale CM, Greenham NC (2003) Modeling the current-voltage characteristics of bilayer polymer photovoltaic devices, *Phys Rev B,* **67**, 75205.

135. Mihailetchi V, Koster L, Hummelen J, Blom P (2004) Photocurrent generation in polymer-fullerene bulk heterojunctions, *Phys Rev Lett,* **93**, 216601.

136. Brabec C, Cravino A, Meissner D, et al. (2001) Origin of the open-circuit voltage of plastic solar cells, *Adv Funct Mater,* **11**, 374–380.

137. MacLeod BA, Horwitz NE, Ratcliff EL, et al. (2012) Built-in potential in conjugated polymer diodes with changing anode work function: Interfacial states and deviation from the Schottky–Mott limit, *J Phys Chem Lett,* **3**, 1202–1207.

138. Greiner MT (2011) Universal energy-level alignment of molecules on metal oxides, *Nat Mater,* **11**, 76–81.

139. Guerrero A, Marchesi LF, Boix PP, et al. (2012) How the charge-neutrality level of interface states controls energy level alignment

in cathode contacts of organic bulk-heterojunction solar cells, *ACS Nano,* **6**, 3453–3460.

140. Potscavage WJ, Sharma A, Kippelen B (2009) Critical interfaces in organic solar cells and their influence on the open-circuit voltage, *Acc Chem Res,* **42**, 1758–1767.
141. Ratcliff EL, Zacher B, Armstrong NR (2011) Selective interlayers and contacts in organic photovoltaic cells, *J Phys Chem Lett,* **2**, 1337–1350.
142. Cros S, De Bettignies R, Berson S, et al. (2011) Definition of encapsulation barrier requirements: A method applied to organic solar cells, *Sol Energy Mater Sol Cells,* **95**, S65–S69.
143. Lloyd MT, Peters CH, Garcia A, et al. (2011) Influence of the hole-transport layer on the initial behavior and lifetime of inverted organic photovoltaics, *Sol Energy Mater Sol Cells,* **95**, 1382–1388.
144. Garcia A, Welch GC, Ratcliff EL, Ginley DS, Bazan GC, Olson DC (2012) Improvement of interfacial contacts for new small-molecule bulk-heterojunction organic photovoltaics, *Adv Mater,* **24**, 5368–5373.
145. Peters CH, Sachs Quitana IT, Kastrop JP, Beaupré S, Leclerc M, McGehee MD (2011) High efficiency polymer solar cells with long operating lifetimes, *Adv Energy Mater,* **1**, 491–494.
146. Jørgensen M, Norrman K, Krebs FC (2008) Stability/degradation of polymer solar cells, *Sol Energy Mater Sol Cells,* **92**, 686–714.

Chapter 10

Modeling OPV Performance—Morphology, Transport, and Recombination

Chris Groves
School of Engineering and Computing Sciences, Durham University, Durham DH1 3LE, United Kingdom
chris.groves@durham.ac.uk

This chapter gives an overview of the common simulation methods used to model the operation of organic photovoltaics (OPVs). How these techniques vary in the description of the internal OPV structure will be discussed as well as how they solve for the behavior of charges and excitons. Examples will be given from the literature where models have been used to provide insight into how morphology and energetic structure effect charge transport, charge recombination, charge injection, and OPV performance. The chapter concludes with a summary and outlook.

10.1 Introduction to Modeling and Charge Transport in Organic Materials

In the context of semiconductor devices, modeling is usually taken to mean the computer simulation of physical processes which

Organic Solar Cells: Fundamentals, Devices, and Upscaling
Edited by Barry P. Rand and Henning Richter

ISBN 978-981-4463-65-2 (Hardcover), 978-981-4463-66-9 (eBook)
www.panstanford.com

occur during device operation [1]. Models are an invaluable tool to better understand how OPV devices operate since they allow links to be made between the behavior of charges or excitons and device characteristics. As will be shown throughout this chapter, there are a number of models available that vary in the complexity of their descriptions of an OPV. The first task in OPV modeling is therefore to select a model that is appropriate to the intended use. This decision will involve considering the level of detail necessary to resolve the particular process of interest. In particular, how the model will represent the morphology of the OPV (Chapter 7), optical absorption (Chapter 8), charge generation and recombination (Chapter 9), and charge transport.

A clearly important part of OPV modeling is describing charge transport. Hence, before describing the various OPV modeling techniques, a brief overview of charge transport in organic materials will be given. Interested readers wishing to learn more about charge transport in organic materials are directed to reviews available in the literature [2–4].

10.1.1 Charge Transport—Hopping and Disorder

Interactions between organic molecules are relatively weak and so charge motion occurs by a series of non-adiabatic tunneling events, or "hops," between distinct localized states on different molecules. The rate at which a charge hops between an origin site i to a destination site j is given by the Marcus expression [5]:

$$v_{i-j} = v_0 \exp\left\{-\frac{(E_j - E_i + E_r)^2}{4E_r kT}\right\} \tag{10.1}$$

Here v_0 is the attempt to hop frequency, which is related to the electronic coupling between the origin and destination sites, k is Boltzmann's constant, and T is the temperature. E_i and E_j are the energies of the origin and destination sites including any electrostatic contributions, while E_r is the re-organization energy. The Marcus rate expression describes two contributions to the energetics of charge transfer. The first is the difference in the energies between the host and destination sites (i.e., $E_j - E_i$), which in turn is due to the difference in local HOMO and LUMO levels as well as electrostatic forces. The second arises because charges in organic materials polarize the surrounding molecule. The charge and surrounding polarization "cloud" is collectively termed a polaron, hence moving the charge also requires changing the electronic configuration of the origin and destination molecules. For the case of an organic material the energy required to reorganize the molecules, E_r, is twice the polaron binding energy [6]. In addition to hopping transport, organic materials are characterized by disorder in their energetic and positional structure.

Energetic disorder is the local variation of HOMO and LUMO levels, which is usually a consequence of the conformational freedom of the molecules or polymer chains. Sometimes, energetic disorder is referred to as diagonal disorder. Calculations predict a Gaussian distribution of the density of states (DoS) in conjugated polymers [3], although more complex DoS functions have been inferred from experiment [7]. Typically the width of the Gaussian DoS distribution is found to have a standard deviation σ around 100 meV [8]. Energetic disorder results in an undulating energy landscape which a charge must traverse to contribute toward current. The spatial arrangement of energetic disorder can therefore lead to charge trapping, the likelihood of which will depend upon the degree of disorder σ and whether the local HOMO and LUMO levels are spatially correlated, with more spatially correlated energetic disorder (e.g., as may occur in a crystal) leading to less trapping [9].

On the other hand, positional disorder describes the variation in the strength of electronic coupling between adjacent molecules, which in turn is a consequence of variations in their relative orientations and spacing. Positional disorder is sometimes referred to as off-diagonal disorder. Since it is often the case that materials used in OPVs are characterized by their molecular *dis*order rather than their order, positional disorder can lead to a wide range of charge transfer rates between molecules [10]. Both energetic and positional disorder can give rise to preferred routes through the film, or "filaments" [11], and dispersion [12] in the temporal behavior of charges. Spatial heterogeneity in charge transport can be seen quite vividly in conducting and photoconductive AFM measurements [13], while dispersive charge transport can be seen in time-of-flight experiments [14].

Electronic coupling can also exhibit a directional dependence that results in mobility anisotropy. This can be observed experimentally in highly ordered crystalline field-effect transistors (FETs) of small molecules [15] and conjugated polymers [16]. Even though OPVs are usually amorphous or contain only a small proportion of crystalline regions, the prospect of large differences in the efficiency of intra- and inter-molecular charge transport is a possibility. A consequence of the sensitivity of charge transport to molecular ordering is that ab initio calculations of mobility are challenging, although not impossible [10], since detailed information regarding chain packing and torsion angles is required. Local molecular ordering is also an important concern at the interfaces between donor and acceptor molecules in a blend since this can alter the local bandgap [17] as well as the nature of excited states at the heterojunction [18].

10.1.2 Charge Transport—Functional Dependencies

Hopping transport in organic materials displays different functional dependencies on temperature, charge density, and electric field when compared to band transport in inorganic materials. Broadly speaking, these functional dependencies arise due to hopping transport and the broadened DoS which is a consequence of energetic disorder. As hops are thermally activated (see Eq. (10.1)) charges naturally relax within the DoS. Charges at low energies within the DoS can be considered as trapped until they acquire sufficient thermal energy to move to the next molecule. The rate at which charges de-trap from low-lying parts of the DoS limits mobility, μ, since many hops are necessary to transit the device. Increasing the temperature therefore increases the energy available to the charge, and with it, the mobility. Mobility in inorganic materials, by contrast, reduces with increasing temperature due to increased phonon scattering. When a field is applied to an energetically disordered material with a Gaussian DoS, a Poole–Frenkel type dependence of mobility, i.e., $\mu \propto \sqrt{F}$, results [6]. Since both the field and temperature dependence of mobility are related to trapping and de-trapping of charges within the DoS, increasing the degree of energetic disorder σ typically increases the field and temperature dependence of mobility in organic materials [19].

Mobility also increases with charge-carrier density due to trapping within the DoS. This is because traps are single occupancy, and so as successive charges are trapped, the probability the next charge will be trapped is reduced. Indeed, experiments suggest that for some materials the apparent increase in mobility observed in single-carrier devices with increasing bias is actually due to the charge-density dependence of mobility rather than field dependence of mobility [20]. As an aside, increasing the density of trapped charges without limit could lead to scattering or blocking of mobile charges [21]; however, this appears to occur at much higher charge densities than those seen in OPVs [22].

10.2 Methods and Applications

Below the most common modeling methods used to simulate OPV device operation will be discussed, namely Kinetic Monte Carlo, drift–diffusion, and master equation techniques.

10.2.1 Kinetic Monte Carlo

Monte Carlo models seek to describe the behavior of individual charges and excitons that collectively give rise to OPV performance. This is unlike drift–diffusion and master equation techniques discussed later, which instead describe charges in terms of local populations. At each stage of the transit of the particle through the model, the Monte Carlo model must decide which of the competing processes a particle may participate in (e.g., hopping or recombination) will actually occur. To achieve this, the transition rates of all the competing processes are calculated and random numbers are used to pick the resulting behavior. It is the use of random numbers to determine behavior that gives Monte Carlo models their name. Monte Carlo models therefore resolve the underlying transport mechanisms that result in drift and diffusion of populations of charges. Additionally, since hops involve moving a particle from one place to another, Monte Carlo models also require an internal description of the spatial and energetic structure of the OPV.

Applying Monte Carlo techniques to hopping transport in energetically disordered systems was popularized by Bässler and coworkers [23]. Their work investigated many processes and parameters that are important to the operation of an OPV, such as geminate pair separation [24], charge injection [25], and mobility. Somewhat later, the first simulations of entire OPV devices were reported [26], and since then, an increasing number of groups have adopted the technique. This is motivated by their ability to examine in detail the effect of three-dimensional nanoscale structure (Chapter 7) and excited state dynamics (Chapter 9) upon OPV performance. Kinetic Monte Carlo OPV device models will be the focus of this section. However, it should be noted that Monte Carlo methods have been used in a variety of other applications, such as calculating ensemble parameters (e.g., mobility) very accurately [4,11,27] which can then be used in other models such as drift–diffusion.

10.2.1.1 Monte Carlo method and rate equations

First the process used to determine particle behavior and the relevant rate equations will be discussed. If the behavior of a particle in a given state, which we here denote as i, is to be determined, then the rates of transition from state i to all other allowed states j, R_{i-j}, must be calculated. In the context of an OPV, these rates may represent hopping from one location to another or recombination, for example. A "waiting time" is then generated for each process using the following equation:

$$\tau_{i-j} = \frac{-\ln(X)}{R_{i-j}} \tag{10.2}$$

Here X is a uniformly distributed random number between 0 and 1. The process with the shortest waiting is picked as the behavior for that particle.

In OPV simulations it is common that excitons are injected as the possible precursor to charges [26,28–31]. The following rate of exciton hopping is typically used to approximate Förster energy transfer [32]:

$$w_{i-j} = w_e \left(\frac{R_0}{r_{i-j}} \right)^6 f(E_j, E_i) \tag{10.3}$$

where

$$f(E_j,E_i)=\begin{cases}\exp\left(-\dfrac{E_j-E_i}{kT}\right) & :E_j>E_i\\ 1 & :E_j<E_i\end{cases} \tag{10.4}$$

Here w_e is the attempt to hop frequency, R_0 is the exciton localization radius, $r_{i\text{-}j}$ is the distance between sites i and j. $R_0 w_e^6$ is usually treated as a fitting parameter to give an exciton diffusion length in the region of ~10 nm, typical of conjugated polymers [33], for a measured exciton lifetime. Excitons may also decay, which is usually taken to occur at a constant rate taken from experiment [29,34]. Finally, excitons may dissociate into charges if they are incident upon a donor–acceptor heterojunction. The rate at which dissociation occurs is typically taken to be much faster than any competing process [29,26] due to the speed at which charge-transfer states are generated in donor–acceptor systems (Chapter 9).

Hopping of charges is typically described using either a Miller–Abrahams- or Marcus-type rate expression. The Miller–Abrahams-type expression is derived assuming weak electronic coupling and considers the emission or absorption of a single phonon [35]:

$$v_{i-j}=v_0\exp(-2\gamma r_{i-j})f(E_i,E_j) \tag{10.5}$$

Here γ characterizes the localization of the charge, and f is given by Eq. (10.4). The prefactor v_0 is either calculated from the mobility and other physical parameters of the material [26], or used as a fit parameter to match experimental data [36]. It should be noted that although Eq. (10.5) is almost ubiquitously referred to as the Miller–Abrahams expression, it is in fact a simplification and is most valid for $|E_i - E_j| \gg 2kT$ [19].

As discussed previously, charges in organic materials are expected to distort the electronic configuration of molecules surrounding the charge, collectively known as a polaron. Motion of the charge therefore implies changing the electronic configuration of both the origin and destination sites. This reconfiguration energy can be large, and so Miller–Abrahams, with its single phonon exchange assumption, may not be appropriate in all circumstances

[37]. To explicitly include the energy cost of molecular electronic reconfiguration, the Marcus expression of Eq. (10.1) should be used.

An important related concern is the maximum hopping range, R_{max}. The value of R_{max} used in the literature varies widely, spanning from nearest-neighbor hopping (i.e., $R_{i-j} = 1$ lattice unit) [6] to 9 nm [26]. The choice of R_{max} is determined by the trade-off between computational cost and degree of charge delocalization. Large R_{max} will be of greatest benefit to accuracy when the long-range hops are expected to compete efficiently with shorter-range hops, for example, as occurs in Mott-type variable range hopping at low temperatures [38].

10.2.1.2 Electrostatic interactions and energetic disorder

The site energies E_i and E_j in the above rate equations include contributions from electrostatic interactions with charges and image charges, the electric field, and energetic disorder. A charge situated at a distance r will contribute a potential:

$$\varphi(r) = \frac{q}{4\pi\varepsilon_0\varepsilon_r r} \tag{10.6}$$

Here, q is the charge of the electron or hole, ε_0 is the absolute permittivity, and ε_r is the relative permittivity (usually taken to be the average permittivity of the blend in question). Charges also induce an image charge in the electrode. The potential due to an image charge in a contact at a distance x is given by [25]

$$\varphi(x) = -\frac{q^2}{16\pi\varepsilon_0\varepsilon_r x} \tag{10.7}$$

The electric field modifies the difference in site energies by an amount $q\bar{F} \cdot \bar{r}_{i-j}$, where $\bar{r}_{i-j}$ is the hop vector. Energetic disorder is picked at random from a (typically Gaussian) distribution prior to a simulation beginning. Commonly, each simulation is repeated using different random choices from the same distribution to ensure that data are representative of the parameters as a whole and not a specific arrangement of energetic disorder [6,26].

10.2.1.3 Injection

Calculating the photocurrent of an OPV requires the rate of exciton creation to be defined. This can be obtained via optical modeling as discussed in Chapter 8, as in Ref. [36], or defined to be similar to absorption under AM1.5 conditions, as in Ref. [30]. Excitons created then diffuse following Eqs. (10.3) and (10.4) and may dissociate into geminate pairs if they reach a donor–acceptor heterojunction prior to relaxing. A simpler method of photoinjecting charges is to directly inject geminate charges on either side of a heterojunction [6]. However, this leads to charges being created at random energies within the DoS. Further work has shown that the energy at which geminate pairs are created can substantially affect their separation efficiency [39]. Hence one has to consider carefully whether geminate charges in the system of interest are created at random energies within the DoS, are relaxed [39], or possibly even hot [40]. Injection of charges from the contacts (i.e., the "dark" or diode current) must be included if the *J–V* curve and associated metrics are to be calculated. This can be achieved by calculating the rate of charge injection from the electrode Fermi level into the local HOMO or LUMO [6,25] via thermionic emission for each site adjacent to the electrode [6].

10.2.1.4 Queuing

The above discusses how waiting times for particle behavior and injection events are determined. Monte Carlo models form an event queue by arranging these planned events (hops, recombinations, injections, etc.) in order of increasing waiting time [26], the fastest of which is performed first. For example, if the shortest waiting time, τ_f, is a hop, then the simulation will perform the hop and reduce the waiting times of the remaining events by τ_f. The charge which has just been moved then has its behavior recalculated. Recall that charges interact via their Coulomb interactions. Hence for complete accuracy, whenever a charge is moved, introduced, or removed, the behavior of *all* other charges ought to be recalculated [41]. However, recalculating the behavior of all charges within the simulation can be computationally demanding. To overcome this problem, the so-called first reaction method (FRM) [42] is often used [26,29–31,34]. In the FRM the behavior of a charge is calculated once only when it is moved to its new position. If the local potential around a charge changes, for example, because another charge passes by, the behavior of the charge is *not* updated. This simplification leads to a substantial reduction in run time. Although the FRM is an approximation, it has been shown empirically to have minimal effect upon both the dynamics of geminate charge pair separation [6] and the absolute value of charge separation efficiency [41].

10.2.1.5 Morphology

A required part of a Monte Carlo model is a description of the OPV morphology in either two [43,44] or three dimensions [29,30], which is similar to that found in real devices. This is a significant challenge due to the complexity of the spin-coating/printing process, evaporation of the solvent, possible crystallization or aggregation of the components, as well as annealing.

The simplest technique to generate morphologies considers phase separation as being akin to an Ising spin-exchange process [45,46]. First, the simulation volume is divided into a Cartesian lattice, each site of which is randomly picked as either donor or acceptor in proportion to the donor–acceptor volume ratio, commonly $\alpha = 0.5$. Donor and acceptor sites are assigned "spins" of +1 and −1, respectively. This finely mixed initial morphology is then coarsened by considering spin exchange of these sites. The Hamiltonian for the energy contributed by site i is

$$W_i = -\frac{K}{2}\sum_j \left(\delta_{S_i,S_j} - 1\right) \tag{10.8}$$

where $\delta_{Si,Sj}$ is the Kronecker delta for spins occupying sites i and j, and the summation is over all nearest and next-nearest neighbor sites. K is the interaction energy, which is usually set to kT to ensure that phase separation is favored [34,26]. The energetic contribution is inversely proportional to site separation, so next-nearest neighbor sites are weighted by a factor $1/\sqrt{2}$. Successive pairs of neighboring sites are selected at random and, using Eq. (10.8), the difference in interaction energy were those sites to be swapped, ΔW_{ij}, is calculated. The probability a swap is accepted is given by

$$P(\Delta(W_{ij}) = \frac{\exp\left(-\Delta W_{ij}/kT\right)}{1+\exp\left(-\Delta W_{ij}/kT\right)} \tag{10.9}$$

A uniformly distributed random number between 0 and 1 is used to decide whether each swap is performed. Successive swaps simulate annealing of the blend and can be used to generate morphologies with a range of domain sizes. The description of Ising spin-exchange is used since it is a convenient method of simulating coarsening a blend morphology. However, while the broad features of blend evolution are recreated, the detailed thermodynamics and kinetics of phase separation are not considered.

The average domain size, d, can be estimated using the equation

$$d = \begin{cases} 6\alpha V/A & \alpha \le 0.5 \\ 6(1-\alpha)V/A & \alpha > 0.5 \end{cases} \tag{10.10}$$

where V is the simulation volume, and A is the total interfacial area between donor and acceptor sites [47]. A more accurate domain size can be obtained by instead calculating the pair distribution function [30].

Cahn-Hilliard modeling is another morphology modeling technique which instead considers the blend undergoing phase separation to be a polymer melt. Consequently, Cahn-Hilliard more completely describes the thermodynamics and kinetics of OPV blend formation [29,30]. The simulation volume, again divided into a Cartesian three-dimensional lattice, is initially assigned a homogeneous local composition equal to the donor–acceptor volume ratio plus small, randomly chosen perturbations to "seed" the morphology. The evolution of the blend morphology is described by the Cahn-Hilliard equation using a temperature quench to mimic the effects of blend coarsening during solvent evaporation. Relevant equations can be found in Ref. [29]. The simulation is allowed to proceed until the desired domain size is achieved. A single-temperature quench yields a unimodal domain size distribution [30], although two quenches can yield bimodal domain size distributions if required [48]. While aspects of blend formation are not explicitly included (e.g., solvent evaporation), Cahn-Hilliard techniques produce qualitatively similar morphologies to that seen in experiment [30], and enable the inclusion of surface wetting effects [29], and have the capability to be extended in the future.

Another approach to recreating a morphology is to describe the dynamic behavior of individual molecules. For example, by tuning the molecular interactions of polymer chains from homophyllic to heterophyllic [43], the connectivity of the chains can be affected. Furthermore, resolving individual polymer chains also provides the opportunity to differentiate between intra- and inter-molecular hopping rates [49].

10.2.1.6 Calculating Coulomb interactions

Techniques to calculate Coulombic interactions efficiently will now be discussed. Some Monte Carlo models [6,26] only calculate the contribution of a charge to the local potential up to a cut-off radius, r_C. While reducing the number of calculations performed, limiting the range of Coulomb interactions has been shown to underestimate photocurrent [28], particularly for small r_C. An alternative method is to only calculate the potential due to charges at positions where it is needed (i.e., the sites where charges are hopping from and to) [41,50]. Since the charge density within an OPV is rather low, only calculating the potential due to charges around the charges being moved means that longer range Coulomb interactions can be included whilst still allowing simulations of entire devices [39]. Since interactions between charges also require evaluation of the image charges in both contacts, variations of the Ewald sum technique can also be helpful in shortening run time [28].

Further reductions in run time can be achieved by coupling techniques to solve for the Poisson equation (equation 10.20) with Monte Carlo models [51,21]. Like the cut-off technique, contributions to potential due to point charges are included at short ranges, but in this case, long-range interactions are preserved by considering distant charges as one-dimensional sheets of charge [21]. Although this requires integration of a Poisson solver into the Monte Carlo algorithm, this technique realizes the benefits of reduced run time without the disadvantage of being unable to simulate space–charge limited devices [51]. However, care must be exercised to ensure that charges separated by small distances are not double-counted [21]. Further simplification is possible by use of coarse-graining techniques [52].

10.2.2 Drift–Diffusion

Drift–diffusion models, as the name suggests, describe the transport of populations of charges in terms of drift and diffusion components. Drift is the average motion of a population of charges in response to an electric field, while diffusion is thermal motion of charges which leads to net charge motion of charges when concentration gradients are present. These models themselves are a simplification of the Boltzmann transport equation, and have found extensive use in the simulation of inorganic semiconductor devices for many years [1]. In the context of organic devices, drift–diffusion models were first used to examine bilayer OPVs [53] and LEDs [54], before being modified to examine bulk heterojunction (BHJ) OPVs by Koster and coworkers [55]. Since then, drift–diffusion models have been used extensively to simulate OPVs. Commonly, drift–diffusion models consider the OPV in one dimension, with the donor–acceptor BHJ represented as a homogenous effective medium which transports both electrons and holes sandwiched between the electrodes. Hence drift–diffusion models are sometimes referred to as metal–insulator–metal models. Many variants of drift–diffusion OPV models have been developed, most notably including time dependence to allow analysis of transient phenomena [7,56,57], and allowing variation of parameters in one and two dimensions [56,58,59] to allow some analysis of the effect of morphology.

However, despite the diversity of drift–diffusion models available, their method of solution is very similar to that found described in the inorganic semiconductor literature (e.g., [1]). For simplicity, the time-independent (i.e., steady-state), one-dimensional version of the equations will briefly be discussed here. Particular attention will be given to how OPV drift–diffusion modeling differs from that for inorganic devices. Readers wishing to learn about drift–diffusion models in more detail are referred to the seminal paper of Koster et al. [55].

10.2.2.1 Current continuity and recombination

Recall that there are two modes of recombination in OPVs. Geminate recombination [60] involves a (geminate) charge pair from the same exciton. Non-geminate recombination, which is synonymous with bimolecular recombination [61], involves charges from different excitons or from the contacts. The current-continuity equations for electron and hole current densities, J_n and J_p, for OPVs must capture these recombination mechanisms:

$$\frac{\partial}{\partial x}J_n(x)=-\frac{\partial}{\partial x}J_p(x)=q[G(x)\eta_{ED}\bar{P}(x)-R(x)] \qquad (10.11)$$

The right-hand side of equation (10.11) describes the net generation rate of free charges. The first term in brackets describes the rate at which free photo-generated charges are created, where $G(x)$ is the local rate of photon absorption (Chapter 8), η_{ED} is the efficiency with which excitons dissociate, and $\bar{P}(x)$ is the local efficiency with which geminate pairs separate into free charges. The second term in brackets, $R(x)$, is the local rate at which free charges recombine bimolecularly. In most drift–diffusion simulations of BHJ OPVs it is assumed that the exciton dissociation efficiency, $\eta_{\mathrm{ED}} = 1$, while the local absorption rate of photons, $G(x)$, can be fitted to give an appropriate photocurrent or calculated via optical models as in Chapter 8. However, the functions used to describe P and R vary significantly depending on the recombination mechanisms considered. For example, some studies consider that geminate recombination does not occur, or that it has no field dependence [7,62], while others consider that geminate recombination does occur [55]. Further, some studies consider bimolecular recombination to be diffusion limited, i.e., Langevin type [55,61], while others consider it to be trap-assisted, i.e., Shockley–Read–Hall-type [7], or a combination [63]. The decision about what recombination mechanisms are appropriate will be informed by the OPV materials system in question.

First, the efficiency of geminate charge pair separation, $\bar{P}(x)$, will be discussed. Clearly if geminate recombination does not occur in a particular OPV then $\bar{P}(x) = 1$ and the field dependence of photocurrent in an OPV is due to bimolecular recombination, $R(x)$ only [7,62]. However, when geminate recombination does occur, the most common way of representing $\bar{P}(x)$ is using the so-called Onsager–Braun model. The original Onsager model [64] solved for the probability that a charge pair initially separated by a distance r would dissociate to infinity under the influence of an electric field, diffusion motion, and their mutual Coulombic attraction. This was later modified by Braun [60] to include a bound state with a finite lifetime, $1/k_r$, that was a precursor to recombination. Onsager–Braun is sometimes modified further [55] to include a distribution of initial separations after Goliber and Perlstein [65] to take some account of positional disorder within the film. However, it should be noted that Onsager–Braun does not include the effects of energetic disorder [39] or the effects of morphology [66]. As such, the parameters used in such models should be thought of as empirical fitting factors rather than lifetimes or molecular spacings that have direct correspondence with experiment. The probability of charge pair separation in the Onsager–Braun model is given by

$$P(r,F,T,x)=\frac{k_{\mathrm{d}}(r,F,T,x)}{k_{\mathrm{d}}(r,F,T,x)+k_r} \quad (10.12)$$

Here k_{d} is the dissociation rate, defined by Koster et al. [55] as

$$k_{\mathrm{d}}(r,T,F,x)=\frac{\partial R(x)}{4\pi r^3}\exp\left(-\frac{E_{\mathrm{bind}}}{kT}\right)J_1\left(2\sqrt{-2b}\right)/\sqrt{-2b} \quad (10.13)$$

where E_{bind} is the initial Coulombic binding energy of the charge pair, $b = q^3F/(8\pi\varepsilon_0\varepsilon_r k^2T^2)$, J_1 is the first-order Bessel function, and $R(x)$ is the Langevin bimolecular recombination rate discussed below. When a distribution of starting radii is considered, after Goliber and Perlstein [65], $P(x)$ must be averaged over this distribution before being used in equation (10.11). Note that the recombination probability in equation (10.12) is field dependent, hence the field dependence of photocurrent in OPVs when geminate recombination is present is due to both geminate *and* bimolecular recombination.

It should be noted that another formulation of $P(x)$ which more accurately describes diffusion-controlled recombination with a finite recombination rate than Onsager–Braun has been developed [67]. This alternative theory has latterly been developed to include energetic disorder with an exponential DoS [68]. However, no analytical descriptions of geminate recombination yet include realistic OPV morphology. Onsager–Braun therefore continues to find use in drift–diffusion models since it uses a convenient, simple function that can be made to fit experimental data with sufficient effort [39].

Returning to equation (10.11), the bimolecular recombination rate, $R(x)$, will now be discussed. Diffusion limited, or Langevin-type, bimolecular recombination is defined as having [55]

$$R(x)=\beta[n(x)p(x)-n_i^2] \quad (10.14)$$

Here n_i is the intrinsic charge carrier density due to thermal excitation across the diagonal bandgap. The recombination prefactor is typically given by the Langevin expression [69]:

$$\beta=\frac{q}{\varepsilon_0\varepsilon_{\mathrm{r}}}\left(\mu_{\mathrm{n}}+\mu_{\mathrm{p}}\right) \quad (10.15)$$

Here ε_r is a spatial average of the components' permittivity. Sometimes a variant of equation (10.15) is used in which the rate has been lowered to reflect experimental observations [61,70]. Furthermore, equation (10.15) may vary with position via the charge density or field dependence of mobility (see Section 10.3).

Shockley–Read–Hall recombination, by contrast, is the process of bimolecular recombination occurring via a trap state somewhere in the bandgap. The rate for this process in equilibrium when the Fermi level coincides with the electron-trap energy is given by [63]

$$R(x) = \frac{c_n c_p N_t [n(x)p(x) - n_e(x)p_e(x)]}{c_n [n(x) + n_e(x)] + c_p [p(x) + p_e(x)]} \tag{10.16}$$

Here C_n and C_p are the capture cross-sections for electrons and holes, N_t is the density of electron traps, and n_e and p_e are the density of free electrons and holes in equilibrium ($n_e p_e = n_i^2$). More complex trap distributions can be considered, as well as non-equilibrium transport (which is necessary to describe transient phenomena), but this requires a more detailed treatment which is discussed in Ref. [7].

10.2.2.2 Drift, diffusion, and the Einstein relation

Current density and charge density are related via the drift–diffusion equations for electron and hole current density:

$$J_n(x) = -q\left[\mu_n n(x)\frac{\partial \varphi(x)}{\partial x} + D_n \frac{\partial n(x)}{\partial x}\right] \tag{10.17}$$

$$J_p(x) = -q\left[\mu_p p(x)\frac{\partial \varphi(x)}{\partial x} + D_p \frac{\partial p(x)}{\partial x}\right] \tag{10.18}$$

The first term in brackets describes drift while the second describes diffusion. Here q is the electronic charge, μ is the mobility, φ is the local potential, and x is the perpendicular distance from the substrate. The subscripts n and p denote electrons and holes, respectively, and D is the diffusion coefficient, which can be taken from the Einstein relation:

$$D = \frac{\mu k T}{q} \tag{10.19}$$

However, due to the distributed nature of the DoS in organic materials, it is expected that diffusion will be enhanced with respect to the Einstein relation under conditions of large charge density or large energetic disorder [71]. This enhanced diffusion of charges has been included in some drift–diffusion models [72]. Other studies have rederived the Einstein relation for the case of correlated (dipolar) energetic disorder, and found the diffusion coefficient to be enhanced at high fields [73].

10.2.2.3 Electrostatics and boundary conditions

The remaining equations involve the electrostatic interactions between charges as codified by the Poisson equation:

$$\frac{\partial^2}{\partial x^2}\varphi(x) = \frac{q}{\varepsilon_0 \varepsilon_r}[n(x) - p(x)] \tag{10.20}$$

Here x is the perpendicular distance from the substrate, φ is the local potential, ε_r is a spatial average of the components' permittivity, while $n(x)$ and $p(x)$ are the local electron and hole concentrations, respectively. The boundary condition for potential is

$$\varphi(L) - \varphi(0) = \frac{E_g}{q} - V \tag{10.21}$$

Here V is the applied voltage assuming that the cathode is at $x = 0$ and the anode at $x = L$, and E_g is the diagonal bandgap between the acceptor LUMO and donor HOMO. Boundary conditions are also necessary to define the charge-carrier densities at the metal–organic interfaces. For an Ohmic contact, it is expected that there will be good alignment between the metal electrode and the center of the HOMO or LUMO as appropriate. In this case, if the density of available states is N, then [74]

$$n(0) = p(L) = N/2 \tag{10.22}$$

The charge-carrier density of the opposite carrier type can be estimated using Boltzmann statistics:

$$p(0) = n(L) = N\exp\left(-\frac{qE_g}{kT}\right) \tag{10.23}$$

It should be noted, however, that assuming a good alignment between the metal and HOMO and LUMO level leads to significant band bending within the device [72].

10.2.2.4 Solution of equations

Once the equations describing the various physical processes are defined, they must be solved self-consistently to determine the potential and charge densities throughout the device. This process is described in detail by Koster et al. [55] and others [72]. Briefly this involves an initial guess at charge densities and potentials at each grid point in x. Improvements to the potentials are made using the iterative Gummel [75] technique. Charge densities are then recalculated and the process repeated until a self-consistent solution for charge density and potential is obtained, whereupon current can be calculated. Since this technique is based on finite differences, the value of quantities in-between grid points must be evaluated. Since values of charge-carrier density can vary rapidly between grid points, it is necessary to ensure that numerical instabilities do not arise [55]. To overcome this, the technique of Sharfetter and Gummel [76] is often used.

The difficulty in performing a drift–diffusion simulation is therefore in robustly solving the coupled equations using numerical techniques. However, once implemented, the iterative techniques commonly applied are computationally efficient. It should be noted that drift-diffusion methods are underpinned by analytical expressions that can omit important detail in the underlying physics, particularly charge generation and recombination. Drift–diffusion methods can, however, be used with greater confidence in single-carrier diodes in which charge generation and recombination do not occur [72].

10.2.3 Master Equation

Though master equation approaches have not at the time of writing been used to simulate the performance of OPVs, the technique will be discussed briefly here since it represents a different approach to describing charge transport in organic devices. Like Monte Carlo models, transport is described by rate equations for hopping processes and can consider three-dimensional transport and morphology [47]. These hopping rates are used to describe the flux of occupation probability between sites, N, which in steady state is equal to zero:

$$\Sigma_j \left[v_{i-j} N_i (1-N_j) - v_{j-i} N_j (1-N_i) \right] = 0 \qquad (10.24)$$

Here v are the hopping rates in equations (10.1) or (10.5), while i and j again represent different sites. The first term in brackets describes the flux of probability (i.e., charges) from site i to the unoccupied portion of sites j, while the second term describes the flux of probability from sites j to the unoccupied portion of site i. Similar equations for other sites are written and solved simultaneously using an iterative technique [77]. Unlike a Monte Carlo model, the master equation describes charges by local occupation probabilities. Equation (10.24) shows that the flux of charges from one site to another is weighted by the probability of charges being available to make the hop, and the probability that an empty space is available to hop into. This approximates a prohibition of double site occupancy. However, beyond this, Coulomb interactions between individual charges are not included. Master equation techniques can therefore give rise to inaccuracies when predicting charge mobility at charge densities >0.01 carriers per site [78]. However, charge densities in OPV applications are typically much smaller than this. Parameterizations of charge-carrier mobility derived from master equation techniques [8] can usefully be employed in drift–diffusion models [72].

10.3 Charge Transport

Much effort has been expended to recreate functional dependencies of μ within simulations. Energetic disorder has been shown [9,23,79] to naturally give rise to a Poole–Frenkel dependence of mobility (i.e., $\propto \sqrt{F}$). However, the field range over which this occurs depends upon whether energetic disorder is spatially random [23] or spatially correlated [9,79], with the latter showing the larger field range over which Poole–Frenkel dependence is exhibited. The effect of positional disorder, by contrast, has also been shown to give mobility a *negative* field dependence at high fields [23]. This is because increasing electric field reduces the speed at which charges can back-track around obstacles presented by positional disorder. The field at which mobility begins to show a negative field dependence depends upon the degree of energetic and positional disorder [80]. Simulations have been successful in recreating the experimentally observed temperature dependence of mobility, in particular showing an increased temperature dependence of mobility with energetic disorder [19]. The Monte Carlo studies of Bässler [23] led to parameterizations of mobility as a function of temperature, positional disorder, and energetic disorder. The so-called Gaussian disorder model (GDM) is valid in the low-charge-density limit and has been shown to fit well the temperature dependence of mobility for some conjugated polymers [81].

Monte Carlo models offer the chance to examine charge transport in more detail than ensemble quantities like mobility. For example, it has been shown how energetic disorder naturally leads to dispersive charge transport as observed in time-of-flight experiments [82], in which the transit time of photo-generated charges traversing a (typically) thick film are measured. Modeling has also allowed inspection of where current flows within an organic device on the nanoscale, something that is challenging to achieve in experiment. Van der Holst et al. [83] used a three-dimensional master equation model to show how Gaussian disorder leads to filamentary transport through single-carrier devices, while Yu et al. used a similar technique to show that the paths taken by charges between the electrodes in single-carrier devices become straighter with increasing field [84]. Lyons et al. [29] used a Monte Carlo model to map the charge extraction from various regions of the top surface of an OPV, analogous to a photoconductive-AFM measurement [13,85]. It was shown that energetic disorder and BHJ morphology can lead to extremely heterogeneous current flow, with some areas of the blend film having 10^2–10^3 times more of the current-carrying burden than the film average.

The effect of BHJ morphology upon mobility was further examined by Groves et al. [58] and Koster [47]. The domain size and relative composition of the blend were varied, and it was found that reducing the composition of the transport fraction below 25% of the blend volume was shown to significantly reduce mobility since charges could not efficiently percolate through the network. Furthermore, smaller domains led to smaller mobilities, which was due to the more torturous routes charges had to take through the morphology. The effect of morphology on a molecular level upon mobility was investigated by Frost et al. [43] who used a Monte Carlo model and a morphology that considered the interactions between polymer chains. Depending upon the type of polymer chain interaction, aggregated or interpenetrated polymer networks could be generated. Aggregated morphologies had a lower percolation threshold and led to more dispersive transport than the interpenetrated counterpart.

Energetic disorder in materials also leads to charge-density dependence of mobility. Pasveer et al. [8] used a one-dimensional master equation model to parameterize the charge density, electric field, and temperature dependence of mobility, which is referred to as the extended Gaussian disorder model (EGDM). The EGDM has been shown to describe more successfully the *J*–*V* characteristics of single-carrier devices at a range of temperatures than the GDM which omits the charge-density dependence of mobility [72,86]. Furthermore, the EGDM has been used to show the dangers in using the Mott-Gurney expression to determine mobility in space–charge limited diodes [87]. The Mott-Gurney expression does not consider carrier diffusion or the charge-density dependence of mobility, which in turn leads to an overestimation of the apparent mobility. The EGDM assumes a spatially random distribution of energetic disorder; this was later extended to consider spatially correlated disorder to result in the extended correlated disorder model (ECDM) [88]. The modified electric-field dependence of mobility in the ECDM when compared to the EGDM may be important in the fitting of experimental data for some materials [72]. A broader overview of models to describe charge transport in energetically disordered materials is reviewed in Ref. [89].

Another consequence of energetic disorder is the possibility of relaxation within the DoS. Bässler [23] used Monte Carlo simulations to show how charges injected at random within the Gaussian DoS with a standard deviation σ relax over time. Eventually, an energy distribution that is Gaussian distributed with a standard deviation σ and a mean—σ^2/kT results. Monte Carlo models have already proven to be capable of modeling transient device characteristics, for example, Germs et al. [90] successfully modeled capacitance–voltage curves at a range of frequencies.

It should be noted, however, that the studies mentioned (with some exceptions) do not consider the effect of chain orientation, packing, or the differences between inter- and intra-chain hops upon mobility. Instead, hopping rates represent a spatial average. Monte Carlo studies, for example, Athanasopoulos et al. [10], have shown that small changes in local torsion angle and separation can significantly affect charge transport.

10.4 Recombination

10.4.1 Geminate Recombination

Geminate recombination is the process whereby charges originating from the same exciton recombine. This process is a particular concern in OPVs since the small dielectric constant of materials used in OPVs ($\varepsilon_R \sim 3$) leads to strong mutual Coulombic attraction between geminate charges (see Eq. (10.6)). As all photo-generated charges begin life as a geminate pair, they must avoid geminate recombination if they are to result in useful photocurrent. Geminate recombination is therefore usually discussed in terms of the geminate separation efficiency, η_{GS}, since larger values of η_{GS} lead to better OPV performance.

Onsager–Braun methods [60], as discussed in Section 10.2.2.1, are typically only useful in extracting qualitative changes in fitting parameters related to geminate recombination (e.g., recombination rate or initial separation) since the effects of morphology and energetic disorder are not explicitly included. The approximate nature of Onsager–Braun was highlighted by Albrecht and Bässler [24], who used a Monte Carlo model to predict η_{GS} in an energetically disordered material and compared the result to Onsager–Braun using the same input parameters. These data showed that Onsager–Braun underestimated η_{GS} since it does not account for relaxation of charges in the DoS during charge separation. Relaxation in the DoS also makes η_{GS} less sensitive to electric field, temperature, and separation radius than that predicted by Onsager–Braun. Although good fits could eventually be achieved between Onsager–Braun and Monte Carlo data, this was *after* the input parameters were altered from the "real" values used in the Monte Carlo model.

Albrecht and Bässler's [24] qualitative Monte Carlo study showed the principle of how relaxation in the DoS can assist η_{GS}, which was later shown to agree with experiment [91]. Later, van Eersel et al. [92] showed that η_{GS} in the particular case of MDMO-PPV:PCBM can be explained by relaxation into the DoS. These studies show the effect of disorder if charges are created at random positions within the DoS; however, this may not always be the case. For example, excitons relax in the DoS as they diffuse to the donor–acceptor interface [93], perhaps favoring the creation of relaxed charges [26], while in some donor–acceptor systems, the nature of energetic interactions can favor the creation of trapped charges [18]. Groves et al. [39] found that increasing the degree of trapping at the interface substantially reduced η_{GS}. This effect was found to be largely independent of the ensemble mobility of charges, highlighting the difficulty in describing η_{GS} in terms of macroscopic parameters such as mobility.

Another critical factor in determining η_{GS} is the morphology. This was first examined using a Monte Carlo model by Peumans and Forrest [94], who showed that restricting the transport of charges to donor or acceptor phases increases η_{GS}. This initial work assumed a bilayer geometry. This was later extended by Groves et al. [66] who additionally considered the effect of orientation of the bilayer normal to the electric field (θ), which is important since a variety of interface orientations will be present in a BHJ. Values of η_{GS} for orientations in the range $-90° \leq \theta \leq 90°$ were similar and optimal, but outside this range ($90° \leq |\theta| \leq 180°$) the values of η_{GS} were significantly reduced. In a BHJ, not only is there the possibility of poor alignment of the interface to the electric field, there is also the possibility of impeded charge separation due to fine phase separation. Increasing the domain size has been shown to significantly improve η_{GS} in BHJ [6,66,26] and gyroid [50] morphologies.

Monte Carlo simulations also allow detailed inspection of how charge separation occurs. For example, Groves et al. [66] tracked the behavior of charges as a function of their relative positions and found, surprisingly, that charges which had attained a separation of only 4–5 nm have substantially higher η_{GS} than the ensemble average. This is perhaps unexpected considering that the thermal capture radius (i.e., the separation between charges required for the attractive mutual Coulomb potential to drop to kT) for these simulations was 16 nm. This in turn suggests that the first few nanometers of the separation process are most important. Related to this, Lyons [30] showed that morphology in the region of the heterojunction is more critical to η_{GS} than morphology far from the interface.

The reports discussed so far do not resolve individual molecules, which prohibits examination of the effects of differences between intra- and inter-chain hops, or charge delocalization. The limiting case of efficient intra-chain transport is transport along a single chain. Rubel et al. [95] calculated η_{GS} for transport along a single chain and it was found that one-dimensional transport increased separation efficiency at low fields and inhibited it at high fields when compared to three-dimensional transport. Deibel et al. [49] also considered the effect of polymer chains, but instead examined the effect of delocalization of the charge along the chain by "sharing" the hole along a chain of finite length. Charge separation was therefore limited by inter-chain hops and the polymer chain length. It was found that increased charge delocalization increased the efficiency of charge separation.

Separately confining the transport of electrons and holes to their respective phases raises the issue of differing transport properties of the donor and acceptor phases. Offermans et al. [44] used a two-dimensional Monte Carlo model to show that increased disorder in one phase can assist charge separation, in a similar fashion to that shown by Bässler [24]. Peumans and Forrest [94], meanwhile, showed that increasing the mobility of one charge carrier made the kinetics of charge separation more competitive with those of recombination, and in turn increased η_{GS}. To underscore the importance of treating the behavior of charges individually, Groves et al. [66] performed Monte Carlo simulations of η_{GS} in which the sum of the electron and hole mobilities was constant but their ratios were different. It was found that the separation efficiency was higher for the simulations which had the largest mobility.

Now moving on to consider particular materials systems, it is noteworthy that the success of the standard Monte Carlo model (as described above) has had mixed success in obtaining quantitative agreement with experiment. For instance, the standard Monte Carlo model has been shown to give excellent quantitative agreement with all-polymer devices [36] (see Section 10.6). However, in other systems the agreement with the standard model described here is less good. For example, in P3HT:PCBM OPVs, the efficiency of charge separation is good despite a fast recombination rate of the order 10^9 s^{-1} [49]. In order to achieve the same high charge separation efficiencies seen in such devices, a standard Monte Carlo model requires a recombination rate of ~10^6 s^{-1}. This shows that Monte Carlo cannot be considered to be a "one-size-fits-all" solution, and that practitioners should be wary as to whether additional physics is required to describe a particular system. In the case of P3HT:PCBM, modifications to the standard Monte Carlo model have shown that delocalization of charges along polymer chains [49], the energetic structure of the interface [17], and long-range exciton dissociation [96] can all contribute toward the "anomalously" high charge separation efficiency in P3HT:PCBM.

Before finishing this section, it should be noted that geminate recombination occurs in the context of a device in which other factors may affect η_{GS}. In particular, varying the mobility of the charges can lead to poor charge extraction and space charge buildup [31]. Also, high-intensity irradiation may lead to competition between geminate and bimolecular recombination [6,50].

10.4.2 Non-geminate Recombination

Non-geminate, or bimolecular, recombination is the process whereby charges from different sources (e.g., different excitons, or from the contacts) meet one another and recombine. This process is therefore limited by the mobility of the charges involved. Bimolecular recombination is commonly described in terms of the Langevin rate coefficient, β, of equation (10.15) [69]. However, due to the complex morphology and energetic structure of an OPV, which is not considered in the derivation of the Langevin rate, the measured value of β is unsurprisingly different from the value calculated [97,98].

Koster et al. [61] using a one-dimensional drift–diffusion model to analyze data for P3HT:PCBM solar cells found that the value of β which gave a good fit to experiment was much smaller than the value calculated using equation (10.15). This was explained in terms of morphological trapping in which the slowest carrier traveling to a donor–acceptor interface limited the rate of recombination. The same OPV blend was investigated by MacKenzie et al. [7] who considered the possibility of energetic trapping using a one-dimensional drift–diffusion model that included trapping and detrapping from a broadened DoS. Both *J–V* and transient photocurrent and photovoltage measurements could only be fitted using a specific DoS function and Shockley–Read–Hall recombination.

Non-geminate recombination can be modeled directly using Monte Carlo methods. Marsh et al. [6] showed that Monte Carlo predicts the experimentally observed increase in non-geminate recombination with increasing charge carrier density, and also that the degree of non-geminate recombination depends sensitively upon the type and characteristic length scale of blended morphology. However, it is difficult to relate ensemble average efficiencies of non-geminate recombination to the underlying recombination process. Groves and Greenham [99] used Monte Carlo techniques to predict the effective recombination rate, β_{eff}, and compare to the expected value of β from equation (10.15) for a range of conditions that occur within an OPV. Blend morphologies were found to reduce recombination ($\beta_{eff} < \beta$) by factors between 2 and 20 depending upon the charge mobilities. The mechanism for this reduction is similar to that proposed by Koster, i.e., that the slower charge retarded the recombination process since both charges need to travel to an interface to recombine. The quantitative details were however different, with β_{eff} exhibiting a weaker dependence on the smallest mobility than implied by drift–diffusion modeling. Energetic disorder in the context of a blend morphology was found to reduce β_{eff} still further as this increased the probability that a charge would be trapped *en route* to recombination at the interface. The effect of two-dimensional transport, as may occur in crystallized lamellae (e.g., P3HT), was also examined. In addition to reducing β_{eff} from the three-dimensional value, two-dimensional transport was found to introduce a charge-carrier density dependence to β_{eff}. The reduction in recombination rate constant, γ, was found to be fitted well by the expression

$$\gamma = 1.5 n^{1/3} l \tag{10.25}$$

where l is the thickness of a slab within which the transport of charges is constrained, and n is the charge density. Note that equation (10.25) is valid in the range $\gamma < 1$. This result also has relevance for recombination of charges across a bilayer. Greenham and Bobbert [100] examined recombination across a bilayer more explicitly using both Monte Carlo and master equation methods, and showed that the recombination rate for a flat heterojunction is limited by the transport of charges to positions where they are opposite one another.

Groves and Greenham [99] also examined the effect of mobility anisotropy, as may occur in some molecular crystals [16] or along polymer chains [101]. It was found that β_{eff} for this circumstance can be fitted well by the empirical equation

$$\beta_{\text{eff}} = \frac{2q}{\varepsilon_0 \varepsilon_r} \sqrt{\mu_{\text{para}} \mu_{\text{perp}}} \tag{10.26}$$

Here μ_{para} is the mobility parallel to the high-mobility axis, while μ_{perp} is the mobility along the low-mobility axis. This strong dependence of recombination strength on mobility anisotropy is in agreement with other studies that considered the capture cross-section for bimolecular recombination rather than the rate constant [102,103].

The *J–V* characteristics of OPVs have also been analyzed considering non-geminate recombination as the sole, or only field-dependent, loss mechanism. Nelson [104,105] used a Monte Carlo model to show that recombination kinetics of MDMO-PPV:PCBM OPVs can be described by trap-limited diffusion of holes. Using the same model, it was shown that a linear dependence of short circuit photocurrent upon light intensity as seen in experiment can be obtained using bimolecular recombination only, which may have been unexpected considering the charge-density dependence of bimolecular recombination. A similar finding was also reported for the same OPV blend by Maturova et al. [62] using a two-dimensional drift–diffusion model. However, in this case, the model described transport in terms of a columnar morphology in which one phase could transport both carriers and one can only transport electrons.

10.5 Charge Injection and Extraction

Although modeling has not yet been used extensively to examine charge injection and extraction in OPVs specifically, many applicable reports are available from the organic LED literature [106]. Charge injection is traditionally described in terms of Fowler–Nordheim field-assisted tunneling or Richardson–Schottky thermionic emission models. Fowler–Nordheim considers injection as a single tunneling jump through a triangular barrier, whilst Richardson–Schottky considers injection via thermally assisted jumps over a barrier. However, in organic materials, hopping is the dominant transport mechanism and the organic material is energetically disordered. Alternative expressions to describe charge injection and extraction (recombination) for metal–organic interfaces have been developed, for example, the models of Arkhipov et al. [107,108], Burin and Ratner [106], and Scott and Malliaras [109]. The extraction process is shown to be somewhat analogous to that described by Onsager and Langevin, i.e., transport-limited recombination. These techniques and others have been used to successfully describe the field and temperature dependence of experimental data as well as being used in other device models. For example, Blakesley and Greenham [74] used a drift–diffusion model to examine how injected charge led to band bending and was able to show excellent agreement with ultraviolet photoelectron spectroscopy (UPS) data for a range of polymers.

Monte Carlo models can explicitly probe the effect of spatially varying energetic disorder on charge injection and extraction. Wolf et al. [25] predicted injection efficiency from a metal into a disordered organic material for a wide range of energy barriers, fields, and temperatures. Analogous to the case of geminate recombination, it was found that relaxation within the DoS resulted in a lower than expected temperature dependence of injection efficiency when plotted on an Arrhenius graph. These data were later shown to be in good agreement with experiments on a single-carrier tris(8-hydroxyquinolinato) aluminum (Alq_3) devices [110]. Charge extraction in bilayer OPVs with heterogeneous electrodes was examined by Zacher and Armstrong [111]. It was found that surprisingly large fractions of inactive electrode surface could be tolerated without significant impact on OPV performance provided the individual inactive regions were small and well dispersed on the electrode surface. However, large individual inactive regions of an electrode clustered together (as might occur for a conductive grid, for example) were shown to lead to space-charge buildup and concomitantly higher recombination.

One area that has yet to receive much attention in the OPV literature, but has been considered extensively in the LED literature, is the process of charge transport across organic heterojunctions. Although the HOMO and LUMO offsets in an OPV may be large, thermal excitation of the charges across a heterojunction is nonetheless possible. Houlli et al. [112] used a master equation technique to quantify the current flow over an organic heterojunction as a function of barrier height in an LED. Forward currents of 1 mA/cm^2 were found at fields of 0.5 MV/m and a barrier height of 0.3 eV. Modeling charge transport over an organic heterojunction using a Monte Carlo model is however computationally expensive. More efficient drift–diffusion modeling of charge transport over organic heterojunctions has been enabled by Cottaar et al. [113], who used Monte Carlo models to define accurate boundary conditions for simpler models.

10.6 Devices

Drift–diffusion models have been demonstrated to be effective at fitting experimental OPV *J–V* data, and in doing so, show qualitative changes in underlying physical processes. For example, Koster et al. [61] showed how fits to experimental data for P3HT:PCBM devices could only be achieved with a much lower bimolecular recombination rate than that predicted by the Langevin equation (largely due to the equation not being appropriate to OPVs as discussed in Section 10.4.2). This also showed the detrimental effect that bimolecular recombination can have upon the fill factor of OPVs. Similarly, fits to MDMO-PPV:PCNEPV-blended OPVs showed how inefficient geminate separation can lead to poor fill factors [114]. Changes in structure can also be modeled using drift–diffusion methods. For example, Lenes et al. [115] examined the thickness dependence of MDMO-PPV:PCBM OPVs and showed that the expected increase in photocurrent with thickness was not realized experimentally because of increased bimolecular recombination.

Quantitative agreement between Kinetic Monte Carlo and experiment was reported by Yan et al. [36]. In this paper, a series of bilayer *J–V* curves were simulated using experimental input data for all but one parameter, ν_0, which was fitted. This in turn showed the significant effect of interface roughness on the geminate separation process. Lyons et al. [29] also examined the effect of interfaces, but this time at the air surface of a BHJ blend, using a Monte Carlo model. Surface wetting layers are commonly reported in OPV blend systems due to differences in surface energy of the blend components, but it is difficult to ascertain their effect due to the conjoined nature of bulk and surface morphology. Lyons et al. [29] showed that the effect of surface wetting layers was smaller than expected considering the percolation threshold of cubic hopping systems. The reduction in photocurrent with abundance of the "wrong" material *en route* to the electrode was characterized and shown to agree quantitatively with a range of experiments on all-polymer OPVs.

10.7 Summary and Outlook

The wide range of OPV modeling techniques and findings which have been discussed in this chapter have demonstrated the valuable role that modeling can play in the ongoing development in the technology. In many cases it simply is not possible to measure experimentally a quantity or process in sufficiently isolated detail to understand how it impacts device performance. Herein also lays a problem, since quantitative modeling analysis requires sufficiently accurate descriptions of the physical processes or properties to feed into the model. This is comparatively straightforward in drift–diffusion models, for example, since one only has to describe average parameters. Master equation and Monte Carlo techniques, however, require great detail at miniscule length scales to achieve the same quantitative correspondence. This said, demonstrations of quantitative agreement between Monte Carlo and master equation models with experiment show that quantitative, even predictive models are possible. Further improvements in the descriptions of the morphology will only improve matters. Continued effort to improve the computational efficiency of models will be required to manage the increased level of detail required to make models still more quantitative.

References

1. Selberherr S (1984) *Analysis and simulation of semiconductor devices*, Springer-Verlag, Wien, New York.
2. Bredas JL, Norton JE, Cornil J, Coropceanu V (2009) Molecular understanding of organic solar cells: The challenges, *Acc Chem Res*, **42**, 1691–1699.
3. Coropceanu V, Cornil J, da Silva DA, Olivier Y, Silbey R, Bredas JL (2007) Charge transport in organic semiconductors, *Chem Rev*, **107**, 926–952.

4. Nelson J, Kwiatkowski JJ, Kirkpatrick J, Frost JM (2009) Modeling charge transport in organic photovoltaic materials, *Acc Chem Res,* **42**, 1768–1778.

5. Marcus RA (1993) Electron-transfer reactions in chemistry—theory and experiment (Nobel lecture), *Angew Chem-Int Ed Engl,* **32**, 1111–1121.

6. Marsh RA, Groves C, Greenham NC (2007) A microscopic model for the behavior of nanostructured organic photovoltaic devices, *J Appl Phys,* **101**, 083509.

7. MacKenzie RCI, Shuttle CC, Chablinyc ML, Nelson J (2012) Extracting microscopic device parameters from transient photocurrent measurements of P3HT:PCBM solar cells, *Adv Energy Mater,* **2**, 662–669.

8. Pasveer WF, Cottaar J, Tanase C, et al. (2005) Unified description of charge-carrier mobilities in disordered semiconducting polymers, *Phys Rev Lett,* **94**, 206601.

9. Gartstein YN, Conwell EM (1995) High-field hopping mobility in molecular-systems with spatially correlated energetic disorder, *Chem Phys Lett,* **245**, 351–358.

10. Athanasopoulos S, Kirkpatrick J, Martinez D, et al. (2007) Predictive study of charge transport in disordered semiconducting polymers, *Nano Lett,* **7**, 1785–1788.

11. Kwiatkowski JJ, Nelson J, Li H, Bredas JL, Wenzel W, Lennartz C (2008) Simulating charge transport in tris(8-hydroxyquinoline) aluminium (Alq3), *Phys Chem Chem Phys,* **10**, 1852–1858.

12. H Scher EM (1975) Anomalous transit time dispersion in amorphous solids, *Phys Rev B,* **12**, 2455–2477.

13. Groves C, Reid OG, Ginger DS (2010) Heterogeneity in polymer solar cells: Local morphology performance in organic photovoltaics studied with scanning probe microscopy, *Acc Chem Res,* **43**, 612–620.

14. Khan RUA, Poplavskyy D, Kreouzis T, Bradley DDC (2007) Hole mobility within arylamine-containing polyfluorene copolymers: A time-of-flight transient-photocurrent study, *Phys Rev B,* **75**, 035215.

15. Lee JY, Roth S, Park YM (2006) *App Phys Lett,* **88**, 252106.

16. Crossland EJW, Tremel K, Fischer F, et al. (2012) Anisotropic charge transport in spherulitic poly(3-hexylthiophene) films, *Adv Mater,* **24**, 839–844.

17. McMahon DP, Cheung DL, Troisi A (2011) Why holes and electrons separate so well in polymer/fullerene photovoltaic cells, *J Phys Chem Lett,* **2**, 2737–2741.

18. Huang YS, Westenhoff S, Avilov I, et al. (2008) Electronic structures of interfacial states formed at polymeric semiconductor heterojunctions, *Nat Mater,* **7**, 483–489.

19. Fishchuk II, Hertel D, Bässler H, Kadashchuk AK (2002) Effective-medium theory of hopping charge-carrier transport in weakly disordered organic solids, *Phys Rev B,* **65**, 125201.

20. Tanase C, Blom PWM, de Leeuw DM (2004) Origin of the enhanced space-charge-limited current in poly(p-phenylene vinylene), *Phys Rev B,* **70**, 1933202.

21. van der Holst JJM, van Oost FWA, Coehoorn R, Bobbert PA (2011) Monte Carlo study of charge transport in organic sandwich-type single-carrier devices: Effects of Coulomb interactions, *Phys Rev B,* **83**, 085206.

22. Tanase C, Meijer EJ, Blom PWM, de Leeuw DM (2003) Unification of the hole transport in polymeric field-effect transistors and light-emitting diodes, *Phys Rev Lett,* **91**, 216601.

23. Bassler H (1993) Charge transport in disordered organic photoconductors—a Monte-Carlo simulation study, *Phys Stat Sol B-Basic Res,* **175**, 15–56.

24. Albrecht U, Bassler H (1995) Yield of geminate pair dissociation in an energetically random hopping system, *Chem Phys Lett,* **235**, 389–393.

25. Wolf U, Arkhipov VI, Bassler H (1999) Current injection from a metal to a disordered hopping system. I. Monte Carlo simulation, *Phys Rev B,* **59**, 7507–7513.

26. Watkins PK, Walker AB, Verschoor GLB (2005) Dynamical Monte Carlo modelling of organic solar cells: The dependence of internal quantum efficiency on morphology, *Nano Lett,* **5**, 1814–1818.

27. Kwiatkowski JJ, Frost JM, Nelson J (2009) The effect of morphology on electron field-effect mobility in disordered C_{60} thin films, *Nano Lett,* **9**, 1085–1090.

28. Casalegno M, Raos G, Po R (2010) Methodological assessment of kinetic Monte Carlo simulations of organic photovoltaic devices: The treatment of electrostatic interactions, *J Chem Phys,* **132**, 094705.

29. Lyons BP, Clarke N, Groves C (2011) The quantitative effect of surface wetting layers on the performance of organic bulk heterojunction photovoltaic devices, *J Phys Chem C,* **115**, 22572–22577.

30. Lyons BP, Clarke N, Groves C (2012) The relative importance of domain size, domain purity and domain interfaces to the performance

of bulk-heterojunction organic photovoltaics, *Energy Environ Sci,* **5**, 7657–7663.

31. Yang F, Forrest SR (2008) Photocurrent generation in nanostructured organic solar cells, *ACS Nano,* **2**, 1022–1032.
32. Scheidler M, Lemmer U, Kersting R, et al. (1996) Monte Carlo study of picosecond exciton relaxation and dissociation in poly(phenylenevinylene), *Phys Rev B,* **54**, 5536.
33. Scully SR, McGehee MD (2006) Effects of optical interference and energy transfer on exciton diffusion length measurements in organic semiconductors, *J Appl Phys,* **100**, 034907.
34. Meng LY, Shang Y, Li QK, et al. (2010) Dynamic Monte Carlo simulation for highly efficient polymer blend photovoltaics, *J Phys Chem B,* **114**, 36–41.
35. Miller A, Abrahams E (1960) Impurity conduction at low concentrations, *Phys Rev,* **120**, 745–755.
36. Yan H, Swaraj S, Wang C, et al. (2010) Influence of annealing and interfacial roughness on the performance of bilayer donor/acceptor polymer photovoltaic devices, *Adv Funct Mater,* **20**, 4329–4337.
37. Pope M, Swenburg CE (1982) *Electronic processes in organic crystals,* Oxford University Press, Oxford.
38. Mott NF (1969) Conduction in non-crystalline materials, *Phil Mag,* **19**, 835–852.
39. Groves C, Blakesley JC, Greenham NC (2010) Effect of charge trapping on geminate recombination and polymer solar cell performance, *Nano Lett,* **10**, 1063–1069.
40. Arkhipov VI, Emelianova EV, Bassler H (1999) Hot exciton dissociation in a conjugated polymer, *Phys Rev Lett,* **82**, 1321–1324.
41. Groves C, Kimber RGE, Walker AB (2010) Simulation of loss mechanisms in organic solar cells: A description of the mesoscopic Monte Carlo technique and an evaluation of the first reaction method, *J Chem Phys,* **133**, 144110.
42. Lukkien JJ, Segers JPL, Hilbers PAJ, Gelten RJ, Jansen APJ (1998) Efficient Monte Carlo methods for the simulation of catalytic surface reactions, *Phys Rev E,* **58**, 2598–2610.
43. Frost JM, Cheynis F, Tuladhar SM, Nelson J (2006) Influence of polymer-blend morphology on charge transport and photocurrent generation in donor-acceptor polymer blends, *Nano Lett,* **6**, 1674–1681.

44. Offermans T, Meskers SCJ, Janssen RAJ (2005) Monte-Carlo simulations of geminate electron-hole pair dissociation in a molecular heterojunction: A two-step dissociation mechanism, *Chem Phys,* **308**, 125–133.

45. Adams CD, Srolovitz DJ, Atzman M (1993) Monte Carlo simulation of phase separation during thin film codeposition, *J Appl Phys,* **74**, 1707–1715.

46. Peumans P, Uchida S, Forrest SR (2003) Efficient bulk heterojunction photovoltaic cells using small-molecular-weight organic thin films, *Nature,* **425**, 158–162.

47. Koster LJA (2010) Charge carrier mobility in disordered organic blends for photovoltaics, *Phys Rev B,* **81**, 205318.

48. Henderson IC, Clarke N (2004) Two-step phase separation in polymer blends, *Macromolecules,* **37**, 1952–1959.

49. Deibel C, Strobel T, Dyakonov V (2009) Origin of the efficient polaron-pair dissociation in polymer-fullerene blends, *Phys Rev Lett,* **103**, 036402.

50. Kimber RGE, Walker AB, Schroder-Turk GE, Cleaver DJ (2010) Bicontinuous minimal surface nanostructures for polymer blend solar cells, *Phys Chem Chem Phys,* **12**, 844–851.

51. Meng LY, Wang D, Li QK, Yi YP, Bredas JL, Shuai ZG (2011) An improved dynamic Monte Carlo model coupled with poisson equation to simulate the performance of organic photovoltaic devices, *J Chem Phys,* **134**, 124102.

52. Casalegno M, Carbonera C, Luzzati S, Raos G (2012) Coarse-grained kinetic modelling of bilayer heterojunction organic solar cells, *Org Electron,* **13**, 750–761.

53. Barker JA, Ramsdale CM, Greenham NC (2003) Modeling the current-voltage characteristics of bilayer polymer photovoltaic devices, *Phys Rev B,* **67**, 9.

54. Crone BK, Davids PS, Campbell IH, Smith DL (2000) Device model investigation of bilayer organic light emitting diodes, *J Appl Phys,* **87**, 1974–1982.

55. Koster LJA, Smits ECP, Mihailetchi VD, Blom PWM (2005) Device model for the operation of polymer/fullerene bulk heterojunction solar cells, *Phys Rev B,* **72**, 085205.

56. Buxton GA, Clarke N (2006) Predicting structure and property relations in polymeric photovoltaic devices, *Phys Rev B,* **74**, 085207.

57. Hwang I, Greenham NC (2008) Modeling photocurrent transients in organic solar cells, *Nanotechnology,* **19**, 424012.

58. Groves C, Koster LJA, Greenham NC (2009) The effect of morphology upon mobility: Implications for bulk heterojunction solar cells with nonuniform blend morphology, *J Appl Phys,* **105**, 094510.

59. Williams J, Walker AB (2008) Two-dimensional simulations of bulk heterojunction solar cell characteristics, *Nanotechnology,* **19**, 424011.

60. Braun CL (1984) Electric field assisted dissociation of charge transfer states as a mechanism of photocarrier production, *J Chem Phys,* **80**, 4157–4161.

61. Koster LJA, Mihailetchi VD, Blom PWM (2006) Bimolecular recombination in polymer/fullerene bulk heterojunction solar cells, *App Phys Lett,* **88**, 3.

62. Maturova K, van Bavel SS, Wienk MM, Janssen RAJ, Kemerink M (2009) Morphological device model for organic bulk heterojunction solar cells, *Nano Lett,* **9**, 3032–3037.

63. Wetzelaer GAH, Kuik M, Nicolai HT, Blom PWM (2011) Trap-assisted and Langevin-type recombination in organic light-emitting diodes, *Phys Rev B,* **83**, 165204.

64. Onsager L (1938) *Phys Rev,* **54**, 554.

65. Goliber TE, Perlstein JH (1984) Analysis of photogeneration in a doped polymer system in terms of a kinetic model for electric-field-assisted dissociation of charge-transfer states, *J Chem Phys,* **80**, 4162–4167.

66. Groves C, Marsh RA, Greenham NC (2008) Monte Carlo modeling of geminate recombination in polymer-polymer photovoltaic devices, *J Chem Phys,* **129**, 114903.

67. Sano H, Tachiya M (1979) Partially diffusion-controlled recombination, *J Chem Phys,* **71**, 1276–1282.

68. Tachiya M, Seki K (2010) Theory of bulk electron-hole recombination in a medium with energetic disorder, *Phys Rev B,* **82**, 085201.

69. Langevin P (1903) *Ann Chim Phys,* **28**, 433.

70. Juška G, Arlauskas K, Stuchlik J, Osterbaka R (2006) Non-Langevin bimolecular recombination in low-mobility materials, *J Non-Crystal Sol,* **352**, 1167–1171.

71. Roichman Y, Tessler N (2002) Generalised Einstein relation for disordered semiconductors—implications for device performance, *App Phys Lett,* **80**, 1948–1950.

72. Blakesley JC, Clubb HS, Greenham NC (2010) Temperature-dependent electron and hole transport in disordered semiconducting polymers: Analysis of energetic disorder, *Phys Rev B,* **81**, 045210.
73. Parris PE, Dunlap DH, Kenkre VM (1997) Dispersive aspects of the high-field mobility of molecularly doped solids with dipolar disorder, *J Polym Sci B,* **35**, 2803–2809.
74. Blakesley JC, Greenham NC (2009) Charge transfer at polymer-electrode interfaces: The effect of energetic disorder and thermal injection on band bending and open-circuit voltage, *J Appl Phys,* **106**, 034507.
75. Gummel HK (1964) A self-consistent iterative scheme for one-dimensional steady state transistor calculations, *IEEE Trans Electron Dev,* **11**, 455–465.
76. Scharfetter DL, Gummel HK (1969) Large-signal analysis of a silicon read diode oscillator, *IEEE Trans Electron Dev,* **16**, 64–77.
77. Yu ZG, Smith DL, Saxena A, Martin RL, Bishop AR (2001) Molecular geometry fluctuation model for the mobility of conjugated polymers, *Phys Rev B,* **63**, 085202.
78. Zhou J, Zhou YC, Zhao JM, Wu CQ, Ding XM, Hou XY (2007) Carrier density dependence of mobility in organic solids: A Monte Carlo simulation, *Phys Rev B,* **75**, 4.
79. Novikov SV, Dunlap DH, Kenkre VM, Parris PE, Vannikov AV (1998) Essential role of correlations in governing charge transport in disordered organic materials, *Phys Rev Lett,* **81**, 4472.
80. Goldie DM (2000) The effect of spatial disorder on the mobility of charge carriers hopping through energetically disordered dipolar lattices, *J Non-Cryst Sol,* **266–269**, 294–298.
81. Andersson LM, Osikkowic W, Jakobsson FLE, et al. (2008) Intrinsic and extrinsic influences on temperature dependence of mobility in conjugated polymers, *Org Electron,* **9**, 569–574.
82. Chatten AJ, Tuladhar SM, Choulis SA, Bradley DDC, Nelson J (2005) Monte Carlo modeling of hole transport in MDMO-PPV:PCBM blends, *J Mater Sci,* **40**, 1393–1398.
83. van der Holst JJM, Uijttewaal MA, Balasubramanian R, et al. (2009) Modeling and analysis of the three-dimensional current density in sandwich-type single-carrier devices of disordered organic semiconductors, *Phys Rev B,* **79**, 085203.
84. Yu ZG, Smith DL, Saxena A, Martin RL, Bishop AR (2000) Molecular geometry fluctuation model for the mobility of conjugated polymers, *Phys Rev Lett,* **84**, 721–724.

85. Pingree LSC, Reid OG, Ginger DS (2009) Electrical scanning probe microscopy on active organic electronic devices, *Adv Mater,* **21**, 19–28.

86. van Mensfoort SLM, Vulto SIE, Janssen RAJ, Coehoorn R (2008) Hole transport in polyfluorene-based sandwich-type devices: Quantitative analysis of the role of energetic disorder, *Phys Rev B,* **78**, 085208.

87. van Mensfoort SLM, Coehoorn R (2008) Effect of Gaussian disorder on the voltage dependence of the current density in sandwich-type devices based on organic semiconductors, *Phys Rev B,* **78**, 085207.

88. Bouhassoune M, Mensfoort SLMv, Bobbert PA, Coehoorn R (2009) Carrier-density and field-dependent charge-carrier mobility in organic semiconductors with correlated Gaussian disorder, *Org Electron,* **10**, 437–445.

89. Coehoorn R, Pasveer WF, Bobbert PA, Michels MAJ (2005) Charge-carrier concentration dependence of the hopping mobility in organic materials with Gaussian disorder, *Phys Rev B,* **72**, 155206.

90. Germs WC, Van der Holst JJM, Van Mensfoort SLM, Bobbert PA, Coehoorn R (2011) Modeling of transient mobility in disordered organic semiconductors with a Gaussian density of states, *Phys Rev B,* **84**, 165210.

91. Barth S, Hertel D, Tak Y-H, Horhold HH (1997) Geminate pair dissociation in random organic systems, *Chem Phys Lett,* **274**, 165–170.

92. van Eersel H, Janssen RAJ, Kemerink M (2012) Mechanism for efficient photoinduced charge separation at disordered organic heterointerfaces, *Adv Funct Mater,* **22**, 2700–2708.

93. Athanasopoulos S, Emelianova EV, Walker AB, Beljonne D (2009) Exciton diffusion in energetically disordered organic materials, *Phys Rev B,* **80**, 195209.

94. Peumans P, Forrest SR (2004) Separation of geminate charge-pairs at donor-acceptor interfaces in disordered solids, *Chem Phys Lett,* **398**, 27–31.

95. Rubel O, Baranovskii SD, Stolz W, Gebhard F (2008) Exact solution for hopping dissociation of geminate electron-hole pairs in a disordered chain, *Phys Rev Lett,* **100**, 196602.

96. Caruso D, Troisi A (2012) Long-range exciton dissociation in organic solar cells, *Proc Natl Acad Sci,* **109**, 13498–13502.

97. Pivrikas A, Sariciftci NS, Juška G, Österbacka R (2007) A review of charge transport and recombination in polymer/fullerene organic solar cells, *Prog Photovolt,* **15**, 677–696.

98. Shuttle CG, O'Regan B, Ballantyne AM, Nelson J, Bradley DDC, Durrant JR (2008) Bimolecular recombination losses in polythiophene:fullerene solar cells, *Phys Rev B,* **78**, 113201.

99. Groves C, Greenham NC (2008) Bimolecular recombination in polymer electronic devices, *Phys Rev B,* **78**, 155205.

100. Greenham NC, Bobbert PA (2003) Two-dimensional electron-hole capture in a disordered hopping system, *Phys Rev B,* **68**, 245301.

101. Zaumseil J, Groves C, Winfield JM, Greenham NC, Sirringhaus H (2008) Electron-hole recombination in uniaxially aligned semiconducting polymers, *Adv Funct Mater,* **18**, 3630–3637.

102. Gartstein YN, Conwell EM, Rice MJ (1996) Electron-hole collision cross-section in discrete hopping systems, *Chem Phys Lett,* **249**, 451–458.

103. Ries B, Bassler H (1984) Monte Carlo study of bimolecular charge-carrier recombination in anisotropic lattices, *Chem Phys Lett,* **108**, 71–75.

104. Nelson J (2003) Diffusion-limited recombination in polymer-fullerene blends and its influence on photocurrent collection, *Phys Rev B,* **67**, 155209.

105. Nelson J, Choulis SA, Durrant J (2004) Charge recombination in polymer/fullerene photovoltaic devices, *Thin Solid Films,* **451–452**, 508–514.

106. Burin AL, Ratner MA (2003) Charge injection into disordered molecular films, *J Polym Sci B,* **41**, 2601–2621.

107. Arkhipov VI, Emelianova EV, Bassler H (2001) Charge carrier transport and recombination at the interface between disordered organic dielectrics, *J Appl Phys,* **90**, 2352–2357.

108. Arkhipov VI, Emelianova EV, Tak Y-H, Bassler H (1998) Charge-injection into light-emitting diodes: Theory and experiment, *J Appl Phys,* **84**, 848–856.

109. Scott JC, Malliaras GG (1999) Charge injection and recombination at the metal-organic interface, *Chem Phys Lett,* **299**, 115–119.

110. Barth S, Wolf U, Bassler H (1999) Current injection from a metal to a disordered hopping system. III. Comparison between experiment and Monte Carlo simulation, *Phys Rev B,* **60**, 8791–8797.

111. Zacher B, Armstrong NR (2011) Modeling the effects of molecular length scale electrode heterogeneity in organic solar cells, *J Phys Chem C,* **115**, 25496–25507.

112. Houlli H, Tutis E, Batistic I, Zuppiroli L (2006) Investigation of the charge transport through disordered organic molecular heterojunctions, *Appl Phys Lett,* **100**, 033702.

113. Cottaar J, Coehoorn R, Bobbert PA (2012) Modeling of charge transport across disordered organic heterojunctions, *Org Electron,* **13**, 667–672.

114. Mandoc MM, Veurman V, Koster LJA, de Boer B, Blom PWM (2007) Origin of reduced fill factor and photocurrent in MDMO-PPV:PCNEPV all-polymer solar cells, *Adv Funct Mater,* **17**, 2167–2173.

115. Lenes M, Koster LJA, Mihailetchi VD, Blom PWM (2006) Thickness dependence of the efficiency of polymer:fullerene bulk heterojunction solar cells, *Appl Phys Lett,* **88**, 243502.

Chapter 11

Modeling the Electronic and Optical Processes in Organic Solar Cells: Density Functional Theory and Beyond

Jean-Luc Brédas,[a] Veaceslav Coropceanu, Curtis Doiron, Yao-Tsung Fu, Thomas Körzdörfer, Laxman Pandey, Chad Risko, John Sears, Bing Yang, Yuanping Yi, and Cairong Zhang

School of Chemistry and Biochemistry, Center for Organic Photonics and Electronics, and Center for Computational Molecular Science and Technology, Georgia Institute of Technology, Atlanta, Georgia 30332-0400, USA

[a]Also affiliated with Department of Chemistry, King Abdulaziz University, Jeddah 21589, Saudi Arabia

jean-luc.bredas@chemistry.gatech.edu

We describe some of the computational methodologies that are being used in our research group to develop a better understanding of the geometric and electronic structure at the organic-organic interfaces present in the active layer. We focus, in particular, on the nature of the excitons that are formed upon photoexcitation of small-gap conjugated polymers and on the exciton-dissociation and charge-transfer processes at donor–acceptor interfaces. The

Organic Solar Cells: Fundamentals, Devices, and Upscaling
Edited by Barry P. Rand and Henning Richter

ISBN 978-981-4463-65-2 (Hardcover), 978-981-4463-66-9 (eBook)
www.panstanford.com

local morphology at the pentacene–C_{60} interface is discussed on the basis of molecular dynamics simulations.

11.1 Introduction

The present book offers a broad overview of the current state of the art in organic photovoltaics (OPV). In this chapter, we focus on some of the computational methodologies that are being used in our research group to develop a better understanding of the geometric and electronic structure at the organic–organic interfaces present in the active layer. The impact of the morphological nature of these interfaces on device performance is further discussed in Chapter 10.

The active layer within an OPV cell is composed of a minimum of two distinct materials, an electron-donating/hole-transport material **D** and an electron-accepting/electron-transport material **A**; these have, respectively, a small IP and a large electron affinity. Within this active layer, a combination of optical/electronic processes needs to be optimized for efficient OPV operation:

i. The initial step involves photoexcitation and formation of bound electron–hole pairs, i.e., excitons. It is important to keep in mind that exciton binding energies in π-conjugated systems usually reach several hundreds of milli electronvolts. Such high values for the binding energies come from a combination of low dielectric constant, strong electron–electron interactions, and large electron-vibration couplings, with the latter leading to substantial geometry relaxations upon excitation or ionization [1]. This is in contrast to the situation in inorganic semiconductors, where the exciton binding energy generally reaches only a few milli electronvolts and, as a result, the electron and the hole can freely separate at room temperature.
ii. Then, prior to decay back to the ground state, the excitons must be able to reach the **D/A** interface; this migration is not influenced by the presence of an electric field as the exciton is charge neutral. Note that, in a number of instances, a strong intermixing of the **D** and **A** species can occur within the active layer, for example, as a result of intercalation of fullerene

molecules in between the side chains of the polymers. In such situations, steps (i) and (ii) do effectively merge and the excitons can readily dissociate within less than 100 fs [2].

iii. For exciton dissociation to lead to efficient separation of charge carriers into the **D** component (holes) and **A** component (electrons), a process that is more complex than a simple electron transfer between **D** and **A** has to follow, since the hole and the electron need to overcome their Coulomb attraction.

iv. Once they have separated, the hole and the electron have to migrate toward the electrodes within their respective **D** and **A** phases and avoid bimolecular recombination.

v. Eventually, the carriers need to be collected efficiently at the electrodes.

Even though the *combination* of all the steps above has to be optimized, which generally requires a fine balancing act, it can be seen that a most critical process is the dissociation of excitons into electrons and holes at the **D/A** interface and the prevention of (geminate and bimolecular) charge recombination (for more extensive discussions on charge recombination, see Chapters 9 and 10).

From a computational standpoint, much progress has been achieved over the past decade on the description of processes (iv) and (v). With regard to charge transport, while purely phenomenological models were prominent up to the early 2000s, it has now become possible to describe rather accurately the microscopic/molecular parameters that govern charge carrier mobilities in organic materials [3–5]. These parameters correspond to the electronic couplings between adjacent molecules or polymer chain segments and the electron-vibration couplings that translate the strong sensitivity of the geometries of π-conjugated systems to excitation or ionization. Also, a comprehensive description has developed of the electronic structure at the interface between a (metal or conducting oxide) electrode and an organic layer [6–8]; in particular, the impact that an organic layer has on the modification of the electrode work function is now well understood.

Thus, in this contribution, we have chosen to discuss mainly processes (i) and (iii). Intriguingly, at the present time, a clear picture has still to emerge to describe the nature of the excitons in the small-gap polymers often used as **D** materials, or to depict the exciton dissociation process at the molecular level. In Section 11.2, we focus on a density functional theory (DFT) investigation, based on modern long-range corrected functionals, of the lowest excited states in small-gap polymers and in **D/A** complexes. Section 11.3 is devoted to a further discussion of the **D/A** complexes in terms of the rates of exciton dissociation into charge-transfer states and of charge recombination to the ground state. Finally, in Section 11.4, we describe the results of a molecular dynamics approach to depict the interfacial packing configurations (interfacial "morphology") when pentacene molecules interact with C_{60} molecules.

11.2 Density Functional Theory Description of the Lowest Excited States in Small-Gap Polymers and D/A Complexes

From our introductory discussion, it is clear that there is a major need for quantum-mechanical methodologies that can describe reliably the excited electronic states of the π-conjugated materials present in the active layers of OPV cells. DFT methods have definitely become a method of choice for many applications in organic electronics; indeed, DFT-based approaches provide the most computationally efficient route to calculations of a scale needed to answer many of the questions we are facing in the field, which can require the treatment of up to thousands of atoms at the quantum level. However, while the optical excitations in π-conjugated molecules and polymers can be readily computed at the time-dependent DFT (TD-DFT) level [9–11], conventional semi-local and standard hybrid functionals (such as the very popular B3LYP functional [12–14]) are known to have difficulties in describing these properties accurately. Here, we will highlight these challenges, illustrate how recent advances in DFT methodologies are bringing about some answers to these problems, and provide examples, whereby these modern approaches give new insights into the excited states of systems of relevance to OPV, such as small-gap polymers and pentacene–C_{60} complexes.

11.2.1 Recent Advances in DFT Methodologies

DFT is a powerful tool for modeling the complex physics behind the electronic structure of atoms, molecules, or solids. It simplifies the problem for an N-electron system by replacing a search for the optimal wave function Ψ in a $3N$-dimensional space with a search for the optimal electron density ρ in just three dimensions:

$$E[\Psi(\vec{x}_1,\vec{x}_2,\ldots,\vec{x}_N)]=E[\rho(\vec{r})]. \tag{11.1}$$

Our goal here is not to provide a detailed account of DFT. For that purpose, we refer to an excellent recent review by Yang and coworkers [15]. However, it is important to set the proper context that will allow the reader to grasp the basic concepts behind recent DFT developments of interest to the field of π-conjugated systems.

In the Kohn–Sham (KS) approach to DFT, the energy functional is partitioned into four terms. The first three are purely classical and are handled easily; they involve the non-correlated part of the kinetic energy, the external (i.e., nuclear potential) energy, and the classical Coulomb energy. The remaining term, referred to as the exchange-correlation (XC) energy functional, involves the quantum-mechanical interactions among electrons and, at this point, cannot be expressed exactly. This is the reason why the accuracy of DFT calculations intimately depends on the nature of the approximate exchange-correlation functional being used.

The simplest exchange-correlation functionals generally used in the chemistry community are the semi-local functionals, in which the XC energy is expressed as a function of the local density and its gradient (and optionally the kinetic energy density):

$$E_{XC}^{\text{semi-local}}[\rho(\vec{r})]=\int_{\mathbb{R}^3}\rho(\vec{r})\in_{XC}(\rho(\vec{r}),\nabla\rho(\vec{r}),\tau_s(\vec{r}))\mathrm{d}^3r, \tag{11.2}$$

usually in the context of the generalized-gradient approximation (GGA). While semi-local functionals are prized for their computational simplicity and efficiency, they can exhibit unacceptably large errors, necessitating the inclusion of some nonlocal terms. Generalized Kohn–Sham (GKS) functionals include nonlocality via the orbital-dependent exchange energy term, while still allowing for a single Slater-determinant representation of the wave function. The simplest improvement is the global hybrid ansatz, in which some fraction of the exact nonlocal exchange energy is added. In this ansatz, the XC functional is written (with E_X denoting the exchange energy and E_C, the correlation energy):

$$E_{XC}^{\text{global hybrid}}[\rho(\vec{r})] = \alpha E_X^{\text{exact}} + (1-\alpha)E_X^{\text{semi-local}}[\rho(\vec{r})] + E_C^{\text{semi-local}}[\rho(\vec{r})]. \tag{11.3}$$

While this generally leads to improvements for many applications, the inclusion of only a small fraction (typically $\alpha \sim 0.25$) of the nonlocal exchange can result in systematic errors. The most important deficiencies arising from the global hybrids, at least in the present context, are related to (i) the inaccurate description of relative and absolute energy levels (assuming that occupied and unoccupied GKS eigenvalues can be interpreted as ionization potentials and electron affinities, respectively); and (ii) the improper asymptotic dependence with distance of the energies of charge-transfer states obtained from linear-response TD-DFT calculations.

According to the ionization potential (IP) theorem [16], the highest occupied eigenvalue from an *exact* KS or GKS approach must equal the first IP. With approximate density functionals, it is typically found that the HOMO eigenvalue is too high in energy (the IP is too small) and the KS HOMO-LUMO gap is too small as compared to the fundamental gap (which is defined as the difference between the vertical IP and electron affinity). When compared to semi-local approximations, the GKS HOMO eigenvalues and HOMO-LUMO gaps obtained from global hybrid functionals are generally improved (with larger values of the IP and HOMO-LUMO gap). However, these values are sensitive to the fraction of exchange included and are still usually too small when employing the limited fractions of exchange found in common global-hybrid approximations.

In the case of charge-transfer (CT) states, we recall that the energy of the lowest charge-transfer state between two molecules should vary, asymptotically, as the inverse of the distance between the molecules, $1/r$, and should approach at very long distances the difference between the IP of the donor and the electron affinity of the acceptor. In fact, the semi-local and global hybrid functional approximations fail to describe properly what essentially amounts to the long-range electrostatic interactions between the excited electron and hole. In pure functionals (i.e., functionals that do not include any nonlocal exchange), the $1/r$ asymptotic dependence is entirely missing, while it is only partially accounted for in the global-hybrid approaches. This leads to a spurious lowering of long-range charge-transfer state energies, a problem that can give rise to difficulties in computing the excited states at the TD-DFT level for large systems even when no low-lying CT states are expected [17,18].

One solution to achieve the correct $1/r$ dependence and an alternative approach to incorporating some amount of the exact exchange, originally proposed by Savin [19], is to partition the Coulomb operator $1/r_{12}$ describing the interaction between electrons 1 and 2. This gives rise to the range-separated hybrid approaches, wherein the Coulomb operator is divided into short-range (SR) and long-range (LR) components, usually via a standard error function:

$$\frac{1}{r_{12}} = \frac{\operatorname{erf}(\omega r_{12})}{r_{12}} + \frac{\operatorname{erfc}(\omega r_{12})}{r_{12}}, \tag{11.4}$$

Where ω denotes the so-called range-separation parameter (which has a dimension of inverse length). By treating the SR and LR electron–electron interactions on a different footing, the range separation of the Coulomb operator allows one to utilize the advantages of well-established semi-local approximations to exchange and correlation, while at the same time circumventing many of their shortcomings. We note that in the calculation of the electronic band structure of solids, the range separation of the Coulomb operator allows one to effectively incorporate the long-range correlation effects by screening the long-range exchange, leading to computational savings as well as alleviating numerical difficulties that arise due to the divergence of the Coulomb operator when including the full $1/r$ potential [20–23]. In molecular systems, on the other hand, the motivations for incorporating the range separation are different. We focus on this second type of range-separated functionals, that are referred to as long range-corrected (LRC) hybrids. These functionals employ the full nonlocal exchange plus local correlation in the LR part, while the nonclassical interactions in the SR part are treated by standard semi-local or hybrid functionals. The motivation for this separation is to preserve the favorable properties of the semi-local or global hybrid functionals in the bonding region, while at the same time allowing for a correct treatment of the asymptotic region. This ansatz produces the correct $1/r$ asymptotic decay of the exchange potential, and often produces markedly improved results for difficult problems.

The correct description of long-range CT excitations was a primary motivation in the initial developments of LRC functional approximations. A number of studies on small representative systems demonstrated the potential for such an approach [24,25]. However, while the standard LRC functionals provide the correct $1/r$ dependence in the long range, the CT energy still frequently does not approach the correct asymptotic limit. This is a consequence of the fact that the GKS gap obtained from this class of functionals critically depends on the value chosen for the range-separation parameter ω, which is typically treated as a functional-dependent constant instead of a system-dependent parameter.

Several recent works [26–29] have shown that, by optimizing (tuning) the range-separation parameter ω for each system, tuned LRC functionals provide very good estimates of the fundamental gap of π-conjugated systems and are thus able to depict accurately the lowest energy features in ultraviolet photoelectron spectroscopy (UPS) and inverse photoemission spectroscopy (IPES) measurements (we note, however, that extension to the full spectra remains a challenge [30]). Approaches for determining an optimal ω value can include minimization of target functions of the form:

$$J_{\mathrm{IP}}(\omega) = |\,\varepsilon_{\mathrm{H}}^{\omega}(N) + E_{\mathrm{gs}}^{\omega}(N-1) - E_{\mathrm{gs}}^{\omega}(N)\,| \quad (11.5)$$

$$J_{\mathrm{EA}}(\omega) = |\,\varepsilon_{\mathrm{H}}^{\omega}(N+1) + E_{\mathrm{gs}}^{\omega}(N) - E_{\mathrm{gs}}^{\omega}(N+1)\,| \quad (11.6)$$

$$J_{\mathrm{gap}}(\omega) = J_{\mathrm{IP}}(\omega) + J_{\mathrm{EA}}(\omega) \quad (11.7)$$

where $\varepsilon_{\mathrm{H}}^{\omega}(N)$ is the HOMO energy for the N-electron system and $E_{\mathrm{gs}}^{\omega}(N)$ is the corresponding SCF total energy. So, tuning ω to minimize J_{IP} corresponds to minimizing the difference between the Koopmans' Theorem IP and the ΔSCF IP for the given semi-local functional and for the specific system, while minimizing J_{gap} will improve the description of the fundamental gap (which can prove important for describing CT state energetics). In most applications relevant for OPV, the choice of the target function

(J_{IP}, J_{EA}, or J_{gap}) turns out to have little impact on the optimized value for ω. The tuning is said to be physically motivated in the sense that it represents a constraint that should hold for the exact KS or GKS functional [16] and involves no fitting to experimental or theoretical reference data.

By tuning the range-separation parameter ω to the specific system of interest, ω is treated as a system-dependent parameter rather than a functional-dependent constant. As has recently been demonstrated for the case of molecular chains, the system-dependent ω value depends on both the system size and degree of conjugation [31]. In order to illustrate this behavior, in Fig. 11.1, we plot $1/\omega$, i.e., the characteristic length of the range separation obtained by IP-tuning of the LRC-hybrid functional ωPBE, on the one hand for highly conjugated polyenes and on the other hand for saturated alkanes, as a function of the number of repeat units. For the case of saturated alkanes, $1/\omega$ increases with chain length only up to $n = 8$, then saturating at around 3.5 bohr (1 bohr corresponds to 0.54 Å), which corresponds to a range-separation parameter of approximately 0.29 bohr^{-1}. In contrast, for the strongly conjugated polyenes, the characteristic length $1/\omega$ is seen to increase without having reached convergence after 20 double bonds (for $n = 20$, the optimal value of ω becomes as low as 0.12 bohr^{-1}). Interestingly, we found a correlation between the optimal value of $1/\omega$ at long chain lengths and the extent of conjugation in the system [31].

As will become clear below, the system dependence of ω provides many advantages for tuned range-separated hybrids as compared to standard implementations of LRC functionals. For example, the tuning procedure typically leads to a significant improvement in the calculated values of IPs, fundamental gaps, and optical excitation energies. However, there is a price that has to be paid for this improvement. By modifying ω with system size, one is effectively changing the functional and, as a result, the GKS Hamiltonian. Indeed, the decrease in ω with system size, as seen in Fig. 11.1, means a more extended short-range region and thus, effectively, a lesser contribution to the functional from the nonlocal exchange. Then, an important property of standard LRC hybrids is lost, i.e., size consistency. This has to be kept in mind when working with this class of functionals.

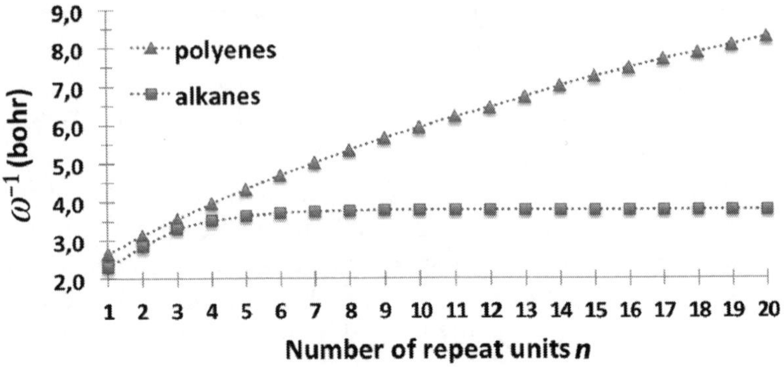

Figure 11.1 Characteristic length $1/\omega$ obtained from IP-tuning of the ωPBE functional for polyenes H-$(HC = CH)_n$-H (triangles) and alkanes H-$(H_2C$-$CH_2)_n$-H (squares) as a function of n. Adapted from Ref. [31].

11.2.2 Nature of the Lowest Excitations in Small-Gap Polymers

We now turn our attention to the description of the electronic structure of small optical-gap polymers comprised of alternating electron-rich and electron-poor moieties; such polymers have been used as the primary photon absorber and hole–transport material in several high-performing bulk-heterojunction solar cells [32–38]. The small optical gap in these materials is intended to increase the overlap between the polymer absorption profile and the incident solar radiation, especially in the near-infrared.

Using tuned LRC functionals (in particular the ωB97 [24] functional of Chai and Head-Gordon and the BNL [39] functional from Baer, Neuhauser, and Livshits), the absorption spectra and low-lying excited states in a series of small optical-gap polymers of relevance for OPV were examined to evaluate the performance of these functionals [40]. Importantly, the same polymers were previously studied with the popular B3LYP functional, [41] which allows an assessment of the differences brought by tuned LRC functionals. Here, we will focus on two representative polymers, see Fig. 11.2: poly[N-alkyl-2,7-carbazole-alt-5,5-(4',7'-di-2-thienyl-2',1',3'-benzothiadiazole)] (PCDTBT) [34] and (poly[4,8-bis-alkyloxy-benzo[1,2-*b* : 4,5-*b'*]dithiophene-2,6-diyl-

alt-4-(alkyl-1-one)-3-fluorothieno[3,4-*b*]thiophene-2,6-diyl]), PBDTTT-CF [36].

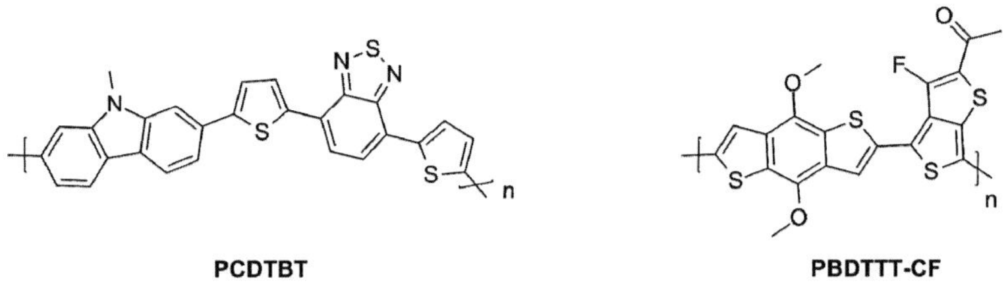

Figure 11.2 Chemical structures of PCDTBT and PBDTTT-CF. In the calculations, values of n from 1 to 6 were considered (each oligomer being capped by hydrogen atoms). Adapted from Ref. [40].

We note that, since the combination of electron-rich and electron-deficient moieties in these polymers is expected *a priori* to result in a CT character of the low-lying excitations along the polymer backbone, the tuning of the range-separation parameter was based on a gap-fitting procedure, similar to that used in the evaluation of a donor–acceptor molecular series [42].

A common means to study the electronic and optical properties of π-conjugated polymers is to employ an oligomer approach, where the length of the system under study is increased until a saturation point (e.g., conjugation length) is reached. Here, we studied systems ranging from the monomer (n = 1) through the hexamer (n = 6). The TD-DFT first transition energies ($S_0 \rightarrow S_1$) for the oligomers in question start to level off after about 5 repeat units for PBDTTT-CF and 3 repeat units for PCDTBT, an evolution similar to that observed for the saturation of $1/\omega$ with increasing oligomer length. From the analysis of the evolution of the range-separation parameter with chain length, it is found that the degree of conjugation in these systems is at least as large as in coplanar oligothiophene chains [40].

We note that the optimized ω values are very similar for both the ωB97 and BNL functionals, and that nearly identical absorption spectra are calculated using the tuned long-range correction parameters, see Fig. 11.3. These results confirm that, when tuning the range-separation parameter, the differences in the semi-local approximations to exchange and correlation do not significantly impact the description of the simulated optical properties [31].

The B3LYP $S_0 \rightarrow S_1$ energies are consistently 0.15–0.50 eV lower than the tuned ωB97 energies; these energy differences are largest for monomers and progressively reduce with the increase in oligomer size.

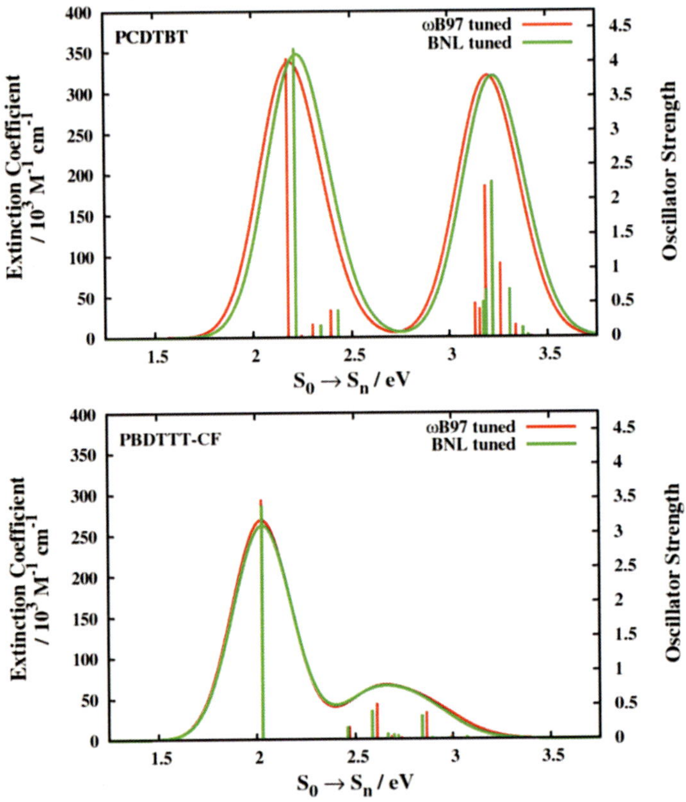

Figure 11.3 Calculated optical absorption spectra from tuned ωB97 and BNL for the tetramers of PCDTBT and PBDTTT-CF at the TD-DFT level. Adapted from Ref. [40].

The real comparison regarding performance, however, revolves around how well the computed absorption profiles reproduce the experimental optical absorption spectra. The experimental data were taken from the works of Blouin et al. [34] and Chen et al. [36]. Figure 11.4 shows the absorbance spectra of the systems compared with the computed spectra at the TD-DFT level with the ωB97 (tuned and default) and B3LYP functionals. In all systems, B3LYP underestimates the energy of the lowest absorption bands, as a result of too large a charge-transfer character of the excited states. Indeed, the mixing of charge-transfer configurations, which are too low in energy, with local excitations leads to an underestimation of the excited-state energies in the TD-DFT spectrum with B3LYP [17,18]. Using the default value for ω in the ωB97 functional provides for an overly localized description of the wave functions (related to too large a contribution from nonlocal exchange), resulting in an overestimation of the charge-transfer configuration energies and leading to too blue-shifted absorptions. Tuned ωB97 provides for the best comparison with the experimental absorption spectra, which can be related to an improved balance in the description of the local and CT excitations provided by the optimization of the range-separation parameter. We note that for the longest oligomers, we have also explored the impact of solvent polarity on the computed optical absorption spectra. Interestingly, the results are found to be very similar to the "gas-phase" results; only extremely small red shifts are observed.

In order to examine more closely the nature of the lowest electronic transitions, natural transition orbitals (NTOs) [43] for the $S_0 \rightarrow S_1$ excitations were evaluated. The NTOs depict the dominant hole and electron wave functions involved in a given electronic transition. Figure 11.5 illustrates the main hole–electron contributions and associated weights for the tetramer of PCDTBT. For B3LYP, the excitation is a contribution of two hole–electron pairs, where in each the hole is largely delocalized and the electron is localized over a benzothiadiazole unit. However, for tuned ωB97, the $S_0 \rightarrow S_1$ excitation is described by three hole–electron pairs, where each hole is localized within a specific thiophene–benzene (belonging to a benzothiadiazole segment)–thiophene segment and the electron localized within the benzothiadiazole unit. Importantly, the description of the $S_0 \rightarrow S_1$ transition as corresponding to excitations essentially localized within such dithienylbenzothiadiazole segments is consistent with the negligible solvatochromism recently reported for these systems [44]. In fact, the decreased absorption energy and increased oscillator strength in PCDTBT with increasing chain length can be seen to arise not from delocalized charge-transfer excitations involving an electron-rich carbazole but from *the coupling of the dithienylbenzothiadiazole-*

localized excitations along the chain. Although not always as strong, the $S_0 \rightarrow S_1$ NTO descriptions for the other oligomers exhibit similar differences between B3LYP and tuned ωB97. Importantly, the more balanced description from tuned ωB97 yields optical absorption spectra that best match experiment and *provides a more localized picture of the lowest singlet exciton* than anticipated from the electron-rich/electron-poor alternating nature of the copolymers.

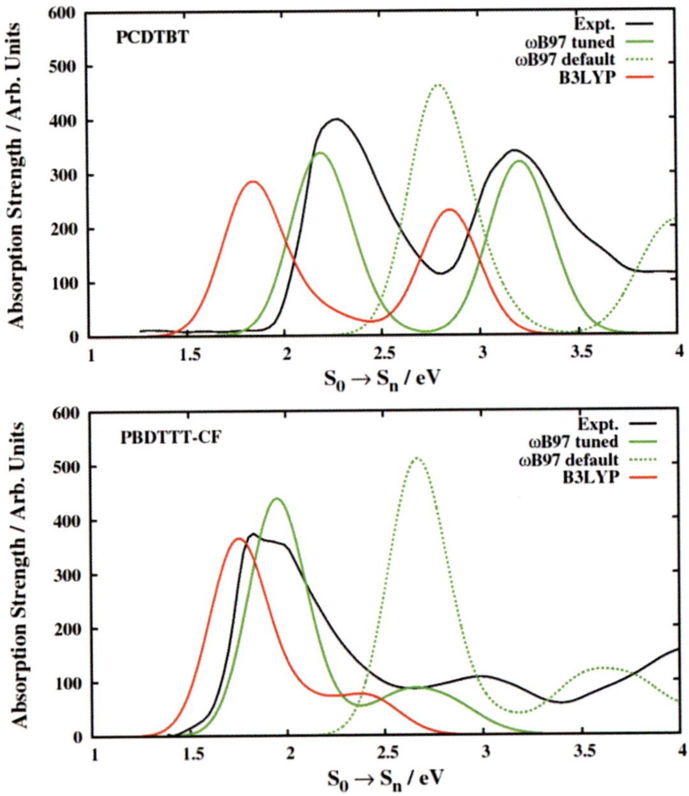

Figure 11.4 Calculated ("gas-phase") optical absorption spectra with the ωB97 (default and tuned ω) and B3LYP functionals for the tetramer of PCDTBT and hexamer of PBDTTT-CF compared to the digitized experimental data. Adapted from Ref. [40].

Thus, these calculations demonstrate the ability of tuned LRC functionals to provide a description of the lowest excited states in small-gap π-conjugated polymers and lead to better agreement with experimental optical absorption spectra vs. the B3LYP functional (which systematically underestimates the energy of charge-transfer configurations) or ("out-of-the-box") standard LRC functionals based on default values for the range-separation parameter. Without tuning, the standard LRC functionals give significantly blue-shifted absorption energies. Interestingly, upon careful optimization of the range-separation parameter, our results indicate that the simulated optical properties are nearly completely independent of the semi-local functional approximation. Overall, the tuned LRC functionals appear to offer an attractive alternative method to conventional DFT functionals for evaluating the excited-state properties of small-gap materials of interest for organic electronics applications.

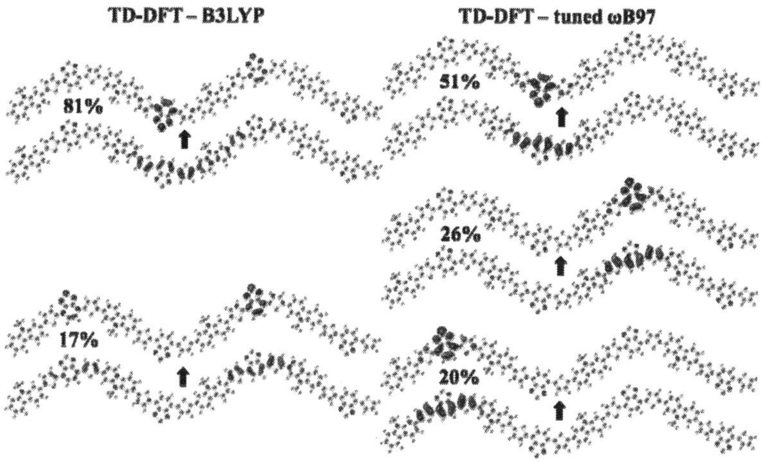

Figure 11.5 TD-DFT NTOs (isovalue surface 0.03 a.u.) for $S_0 \rightarrow S_1$ from B3LYP and tuned ωB97 for the tetramer of PCDTBT. For each pair, the hole wave function is at the bottom and the electron wave function at the top. The weight of each hole–electron pair in the description of the excitation is also included. Adapted from Ref. [40].

11.2.3 Donor/Acceptor Charge-Transfer Excitations

If the exciton is to result in the generation of photocurrent, it must dissociate upon reaching the **D/A** interface to ultimately form separated charge carriers. For reasons we discussed earlier, the description of the charge-transfer excited states has remained a challenge for DFT approaches. Being capable of accurately describing the charge-transfer states occurring at or near the **D/A** interface will provide valuable insight into the fundamental processes giving rise to photocurrent. In this section, we examine how the tuned LRC functionals can provide an improved description of the charge-transfer states in a prototypical **D/A** complex consisting of a pentacene molecule and a C_{60} molecule. We will then turn to larger models of the interface.

Using the gas-phase B3LYP/6–31G(d,p) geometries of pentacene and C_{60}, the range-separation parameter was tuned according to Eq. (11.7). The LRC functionals ωB97X and ωB97X-D from Chai and Head-Gordon were employed, along with the LRC variants of the PBE [45–48] and BLYP [49,50] functionals (LRC-ωPBE and LRC-BLYP). Again, we find that, while the results with the standard LRC functionals (without tuning) are sensitive to the default value of the range-separation parameter, the results and the values for ω upon tuning are largely insensitive to the choice of the semi-local functional approximation. Here, we focus the discussion on the performance of both the default and tuned LRC-BLYP functionals and compare the results to those for the pure BLYP and global-hybrid B3LYP functionals.

We first consider the individual molecular properties in the case of pentacene, see Table 11.1. The excitation energies are computed both at the linear-response TD-DFT level and within the Tamm-Dancoff approximation (TDA) [51–54], for reasons that will be discussed below. The optimized range-separation parameter for pentacene is significantly smaller than the default value proposed for LRC-BLYP; clearly, such a difference will have a pronounced impact on the computed properties. Both the long-range corrected functionals give a dramatic improvement in the agreement between the HOMO eigenvalue and the first IP of pentacene, which is measured experimentally to be around 6.59 eV [55]. In comparison, the B3LYP and BLYP HOMO eigenvalues differ by roughly 2–3 eV. This is a consequence of the many-electron self-interaction error (MSIE), which is typically much smaller in these systems with the LRC functionals even when using the default range-separation parameters. The tuning results in a significantly smaller MSIE and a further improvement in the performance, although we note that this is only strictly true for orbitals that are similar in nature to the orbitals (typically HOMO/LUMO) for which the tuning is performed, here π-type orbitals [30].

Table 11.1 Range-separation parameter ω (bohr^{-1}), orbital energies (eV), HOMO-LUMO gap (eV), and lowest TDDFT and TDA-TD-DFT excitation energies to the S_1 and T_1 states (eV) of pentacene; img refers to obtaining an imaginary value

					TDDFT		TDA-TD-DFT	
Functional	ω	**HOMO**	**LUMO**	**Gap**	S_1	T_1	S_1	T_1
LRC-BLYP default	0.47	−6.73	−0.58	6.15	2.60	img	2.99	0.94
LRC-BLYP tuned	0.21	−6.05	−0.89	5.16	2.28	0.71	2.65	1.09
B3LYP	–	−4.61	−2.40	2.21	1.95	0.59	–	–
BLYP	–	−3.88	−2.68	1.20	1.66	0.84	–	–

Considering the lowest excited states, the gas-phase absorption spectra for pentacene places the lowest singlet (S_1) state at ~2.31 eV, [56] while the first triplet (T_1) state is estimated to be at ~0.86 eV [57]. The TD-DFT computed triplet energy is found to be imaginary when using any of the default range-separation parameters. This effect has been explained by the presence of singlet–triplet instabilities in the ground-state solution with the LRC functionals, that result in the triplet energies being sensitive to the specific value of the range-separation parameter within the full linear-response TD-DFT approach. This sensitivity is removed with TDA, which provides for improved descriptions of the singlet and triplet excited states [58,59]. Within the full TD-DFT approach, the tuning of the range-separation parameter was observed to improve the description significantly for a series of linear acenes [58]. Indeed, whether the calculations are carried out at the TD-DFT or the TDA-TD-DFT level, the description of the lowest singlet and triplet excitation energies is improved by the tuning procedure.

Of course, the greatest value of the tuned LRC functionals is not simply with the description of the individual molecular properties but more so in their expected ability to describe the charge-transfer states and the localized (molecular) states on an equal footing in **D/A** complexes. The tuned LRC functionals provide for not only a quantitative but also a qualitative difference in the description of the excited states in these systems. As an example, in Fig. 11.6, we plot the NTOs describing the excitation to the S_2 state in a cofacial configuration of the pentacene–C_{60} complex, for B3LYP as well as for the LRC-BLYP functionals with default and tuned ω values [60]. The CT energy can be estimated by considering the difference in the experimental IP of pentacene (6.59 eV) and EA of C_{60} (2.68 eV) [61] and evaluating the Coulomb stabilization from the charges for the pentacene cation and the C_{60} anion (this assumes a full charge transfer and gives an upper estimate of the Coulomb stabilization for the complex in Fig. 11.6 of 1.8 eV; see Eqs. (11.13) and (11.14) in Section 11.3). Thus, the CT state for the complex should lie above 2.1 eV and approach asymptotically 3.9 = [6.59 – 2.68] eV at infinite separation.

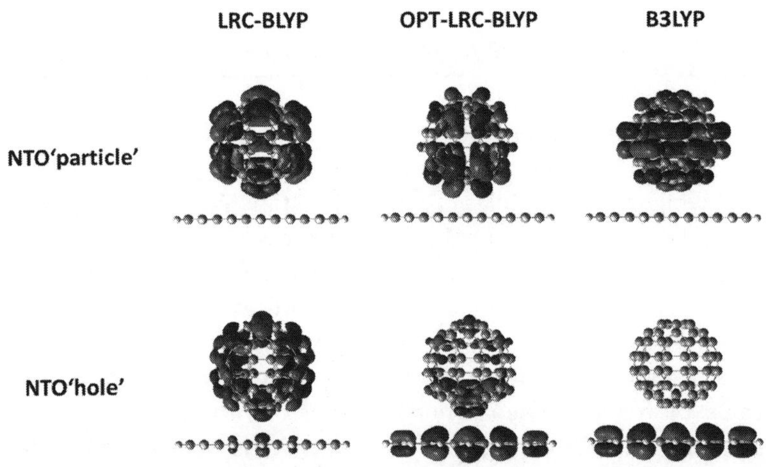

Figure 11.6 TD-DFT NTOs (isovalue surface 0.02 a.u.) for the $S_0 \rightarrow S_2$ transition in a cofacial model of pentacene–C_{60} from LRC-BLYP with default and optimized range-separation parameters as well as from B3LYP. In each case, the hole wave function is at the bottom and the particle (i.e., electron) wave function at the top.

B3LYP provides for a pure CT state with an energy that is too low by 1 eV at short distances and increases to only 1.4 eV at large distances, that is, about one-third of the expected value. The default LRC-BLYP describes an excitation that is almost purely localized on

C_{60}, with the CT state much higher in energy (3.2 eV) and approaching 5.3 eV at infinite separation; this is due to the improper description of the fundamental gap with the default ω. The tuned LRC-BLYP gives an excited state, that is dominated by CT character mixed with some contribution from the local C_{60} excitation, with a more reasonable description of the energetics: 2.3 eV at short distances and approaching 4 eV at infinite separation, which is very close to the experimental value of 3.9 eV. This description is independent of the nature of the semi-local functional approximation in the LRC functionals (energies for all excited states differ by less than 0.1 eV after tuning ω). Thus, the tuned LRC functionals appear to give an improved and balanced description of the local and CT excitations and, as we have recently demonstrated, of the triplet states in these materials at the TD-DFT level [58]. Such a balanced description of the local and CT excitations is difficult to reach with other modern DFT functional approaches.

Thus, in the case of a single pentacene–C_{60} complex, the description of the excited states and particularly the charge-transfer states is qualitatively different when using the tuned LRC approaches. While, as we have done above, the charge-transfer state is often depicted as involving just one **D** and one **A** molecule or polymer chain segment, this is not necessarily the case. Indeed, if the electron and hole are localized strictly on adjacent molecules at the interface, their strong electrostatic attractions are likely to create a large Coulombic barrier that has to be overcome to generate charge-separated carriers and avoid recombination losses at the interface. Recent work, both experimentally and theoretically, has examined the importance of delocalized charge-transfer excitations that may exist at or near the **D/A** interface [62,63]. A major consequence of such a delocalized nature of the CT state would then be to reduce the electrostatic attraction between hole and electron and thereby decrease the Coulombic barrier to charge separation.

Intriguingly, from a theoretical perspective, the differences observed in the description of the CT states of a single pentacene–C_{60} complex become even more pronounced as the system size increases to take account of more realistic interfacial structures. We have recently highlighted these differences with simple model systems by starting with the cofacial model system from Fig. 11.6 and successively adding more pentacenes in a cofacial fashion [64]. In Fig. 11.7, we display the HOMO orbital along with the dominant NTOs for the charge-transfer state in a system consisting of a single C_{60} acceptor with a cofacial stack of five pentacene donors. While B3LYP and the tuned ωB97X provide very similar distributions for the frontier energy levels, their descriptions of the charge-transfer state are completely different. The tuned ωB97X provides for a CT state, where the hole resides close to the interface and is delocalized over approximately three pentacenes; in contrast, B3LYP describes this state as delocalized at least to some extent over all five of the pentacenes. It is also important to note the small contribution of the hole at both ends of the pentacene stack with B3LYP. This can be seen as a consequence of the lack of correct asymptotic dependence of the charge-transfer energy discussed above. This gives rise to an increasing dipole moment in the charge-transfer state that grows with the size of the system. Such an effect can be observed even in B3LYP calculations on the ground state of interfacial structures between strong donors (such as TTF, tetrathiofulvalene) and strong acceptors (such as TCNQ, tetracyanoquinodimethane) [65].

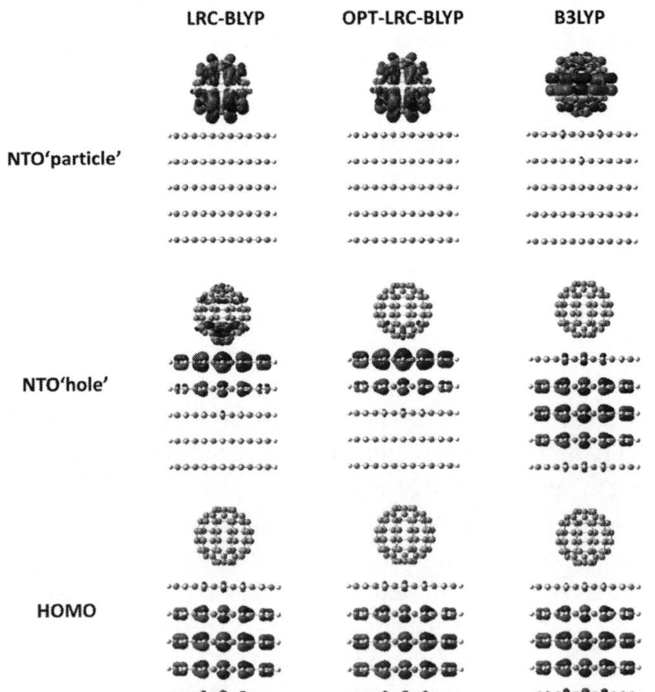

Figure 11.7 HOMO orbitals and TD-DFT NTOs (isovalue surface 0.02 a.u.) for the lowest CT state in a cofacial model consisting of 5 pentacenes interacting with C_{60}.

The tuned LRC functionals clearly offer numerous advantages in the qualitative and quantitative description of the relevant excited states at the **D/A** interface. They currently provide the most feasible approach to obtaining an accurate description of the energetics and the extent of delocalization of the charge-transfer states at these interfaces. These features are difficult to measure experimentally; yet these are parameters that will have significant impact upon the picture that emerges describing the pathway(s) from the exciton to the charge-separated carriers. We now address in more detail the exciton-dissociation and charge-recombination processes.

11.3 Exciton-Dissociation and Charge-Recombination Processes at Donor/Acceptor Interfaces

As reading the present book will make clear, the optimization of organic solar cells requires a fine balancing act. It involves finding the optimal compromise among a combination of material characteristics that sometimes happen to work in opposite directions. The local **D/A** interface geometry has an impact on the rates of exciton dissociation (which needs to be maximized) and of charge recombination (which decreases the photocurrent and needs to be minimized), as well as on the magnitude of the reverse saturation current in the dark (which needs to be minimized to increase the open-circuit voltage) [66,67]. Therefore, in order to understand and eventually to develop new strategies to increase the efficiency of organic solar cells, it is crucial to obtain a much better, fundamental understanding of the charge-separation (CS) and charge-recombination (CR) pathways.

In general, several pathways can be followed for both the CS and the CR processes. First, multiple donor and/or acceptor excited states could be populated by the incident light. Next, while often only the lowest charge-transfer (CT) state is considered in the literature, recent experimental [62,63] and theoretical [68–71] findings suggest that the CS process could actually involve more "efficient" excited charge-transfer states. The charge-recombination process could also take place via several pathways; for instance, CR via triplet states could in some cases be more of an issue than the direct transition back to the singlet ground state.

The electron-transfer rate of any CS or CR process depends on several factors such as electron-vibration coupling, electronic coupling between reactants and products, and relative free energies of reactants and products. However, to obtain reliable estimates of these parameters on the basis of quantum-chemical approaches represents in general a very difficult task. In particular, a very challenging proposition is the determination of the electronic coupling, since its value is determined by the overlap between the relevant donor and acceptor wave functions. Here, we first discuss a computational approach developed recently in our group [69,70] that allows the calculation of the electronic coupling between any two molecular (exciton) and charge-transfer states. In contrast to other approaches reported in the literature, the present method can be used to evaluate the electronic couplings between electronic states of any spin multiplicity. As a result, the electron-transfer rates involving singlet and triplet states can be computed. We then illustrate the application of this methodology to investigate the role that the interface geometry exerts on the efficiency of both exciton-dissociation and charge-recombination processes for model **D/A** complexes consisting of pentacene/C_{60}, oligothiophene/C_{60}, and oligothiophene/perylenediimide (PDI) [70,71].

11.3.1 Evaluation of the Electronic Couplings and Electron-Transfer Rates

The electronic coupling between the molecular (exciton) and charge-transfer states is a major parameter that controls the dynamics of the charge-separation and charge-recombination processes at **D/A** interfaces. The electronic coupling V_{IF} can be evaluated by a diabatic-state approach (involving the wave functions of the isolated components), that was recently developed in our group [69,70]:

$$V_{IF} = \langle \Psi^{I} | H | \Psi^{F} \rangle \tag{11.8}$$

Here, the initial and final states (Ψ^{I} and Ψ^{F}) are the local (molecular) excited state and charge-transfer state (Ψ^{LE} and Ψ^{CT}) for the exciton-dissociation process or the CT state and ground state for the charge-recombination process, respectively. The states Ψ^{LE} and Ψ^{CT} are spin-adapted diabatic states that are constructed as antisymmeterized products of the isolated donor and acceptor wave functions [72]:

$$\Psi_{ij}^{LE}(S,M)=\Sigma_{M_i,M_j}C_{S_iM_iS_jM_j}^{SM}\,|\Psi_i^D(S_i,M_i)\Psi_j^A(S_j,M_j)| \quad (11.9)$$

$$\Psi_{kl}^{CT}(S,M)=\Sigma_{M_k,M_l}C_{S_kM_kS_lM_l}^{SM}\,|\Psi_k^{D^+}(S_k,M_k)\Psi_l^{A^-}(S_l,M_l)| \quad (11.10)$$

where, S and M denote the total spin and spin projection of the diabatic states for the donor–acceptor complex; $\Psi_m^{D/A}$ is the m-th excited state of the isolated neutral donor or acceptor with spin S_m and spin projection M_m; and $\Psi_m^{D^+/A^-}$ corresponds to the oxidized donor or reduced acceptor. The $C_{S_kM_kS_lM_l}^{SM}$ terms represent the Clebsch–Gordan coefficients that result in the linear combination of the isolated wavefunction products to be an eigenfunction of the total spin. For the sake of simplicity, we computed the isolated excited and charged states and then the electronic couplings in the framework of the semiempirical intermediate neglect of differential overlap (INDO) Hamiltonian [73] coupled to a single configuration interaction (SCI) scheme.

When the CT states and/or the local excited states are orbitally degenerate (as in the case of C_{60} and its derivatives), an effective electronic coupling (V_{eff}) can be calculated by the following equation:

$$V_{eff}^2=\frac{1}{g_I}\Sigma_{i,j,k,l}\left\langle\Psi_{ij}^I\,|\,H\,|\,\Psi_{kl}^F\right\rangle^2, \quad (11.11)$$

where g_I is the orbital multiplicity of the initial diabatic state, and can be calculated as the product of the multiplicities of the initial monomer states ($g_i\times g_j$).

According to the semi-classical model of Marcus electron-transfer theory, the rates of exciton dissociation and charge recombination can be estimated as [74,75]:

$$k_{IF}=V_{IF}^2\sqrt{\frac{\pi}{\lambda k_BT\hbar^2}}\exp\left[-\frac{(\Delta G^0+\lambda)^2}{4\lambda k_BT}\right] \quad (11.12)$$

Here, k_B and $\hbar$ denote the Boltzmann and Planck constants, respectively; T is the temperature (set to 300 K in our calculations); λ represents the reorganization energy, and ΔG^0 the Gibbs free energy of the electron-transfer reaction.

If we ignore the entropy contribution, ΔG^0 can be estimated as the energy difference between the diabatic initial and final states.

The energy of the diabatic states is composed of the total energies of the isolated states of the donor and acceptor and the Coulomb energy between the donor and acceptor. The Coulomb energy can be computed as the sum of the Coulombic interactions between the atomic charges of the donor and acceptor; for instance, the Coulomb energy between the donor cation and acceptor anion can be written as

$$E_{\text{coul}}^{\text{D}^+/\text{A}^-} = \sum_{d\in \text{D}^+, a\in A^-} \frac{q_d q_a}{4\pi\varepsilon_0 \varepsilon r_{da}} \tag{11.13}$$

where, q_d and q_a are the partial charges (they can be obtained via an INDO Mulliken population analysis) on atom d of the donor cation and atom a of the acceptor anion, respectively; r_{da} is the distance between atoms d and a; ε_0 is the vacuum permittivity and ε corresponds to the medium. The results show that the Coulomb energies for the diabatic local (i.e., intramolecular) states of the **D/A** complexes are vanishingly small, while those for the diabatic CT states become very significant. Thus, using the total energy of the diabatic ground state as the reference, the energies of the local-excited states for the **D/A** complexes are approximated as the excitation energies of the isolated molecules; on the other hand, the energy of the lowest CT state must include the contribution from the Coulomb interactions:

$$E_{\text{CT}} = IP^{\text{D}} + EA^{\text{A}} + E_{\text{coul}}^{\text{D}^+/\text{A}^-} \tag{11.14}$$

where IP^{D} is the IP of the donor and EA^{A} the electron affinity of the acceptor.

11.3.2 The Pentacene–C_{60} Complex

Using this methodology, we have investigated the role of the pentacene/C_{60} interface geometry on the efficiency of both exciton-dissociation and charge-recombination processes, namely to rationalize some of the differences in performance between the bilayer and bulk-heterojunction architectures [70]. Figure 11.8a illustrates the model complex configurations we have considered (a more detailed discussion of the energetics of these configurations will be given in Section 11.4). We have calculated the electronic couplings and rates for exciton dissociation and charge recombination as a function of intermolecular distance and the rotation angle or lateral displacement of pentacene.

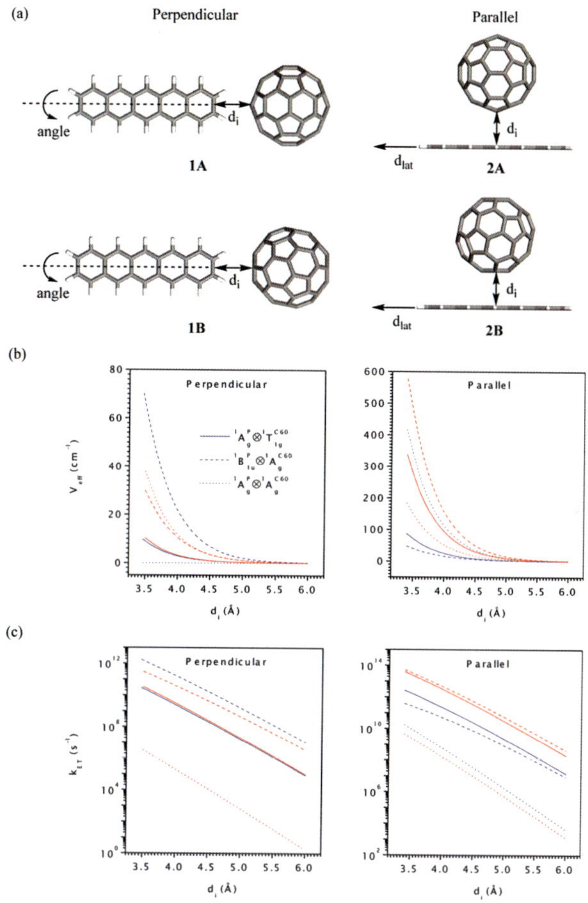

Figure 11.8 Illustration of the pentacene/C_{60} configurations (a) and dependence of the electronic couplings (b) and rates (c) for exciton dissociation [from molecular excited states ${}^1A_g{}^P\otimes{}^1T_{1g}{}^{C60}$ (solid lines) and ${}^1B_{1u}{}^P\otimes{}^1A_g{}^{C60}$ (dashed lines) to the lowest CT state, ${}^2B_{2g}{}^{P+}\otimes{}^2T_{1u}{}^{C60-}$] and charge recombination [from the lowest CT state to the ground state: ${}^2B_{2g}{}^{P+}\otimes{}^2T_{1u}{}^{C60-} \rightarrow {}^1A_g{}^P\otimes{}^1A_g{}^{C60}$ (dotted lines)], as a function of intermolecular distance, d_i; blue lines refer to orientations 1A and 2A and red lines to orientations 1B and 2B. In the calculations of rates, we make use of the experimental values for energies estimated from the condensed phase: the S_1 energy is 1.7 eV for C_{60} [76] and 1.85 eV for pentacene [77,78], IP(P) = 5.1 eV [79] and EA(C_{60}) = −3.5 eV [80]; the reorganization energy is set to 0.5 eV. In the case of orientation 1A, the CR rate is equal to zero due to the vanishing coupling. Adapted from Ref. [70].

The calculated electronic couplings involving the lowest CT state (CT_0) for the perpendicular and parallel configurations of the pentacene/C_{60} complex as a function of the intermolecular distance are shown in Fig. 11.8b. For the same intermolecular distance, the electronic couplings obtained in the case of parallel configurations are significantly larger than in the case of perpendicular orientations; for the intermolecular distance corresponding to the sum of the van der Waals radii of the closest atoms of pentacene and C_{60} molecules (which represents the smallest intermolecular distance considered here), the electronic coupling can be as large as 75 meV. Except for the case of orientation 1A (pentacene perpendicular to a C=C edge, see Fig. 11.8a) where due to symmetry reasons there is no interaction between the ground state and the CT_0 state, the respective coupling is significant in all other cases. The electronic couplings for all states show the expected exponential dependence on intermolecular distance.

The lowest excited CT states (CT_1,CT_2) arise from the first excited state of the radical anion of C_{60} and the cation of pentacene; from electronic absorption spectra, these two states are located 1.15 [81,82] and 1.31 eV [83], respectively, above the corresponding ground states of the charged species. The electronic couplings between the lowest molecular excited states and the lowest excited CT states (CT_1,CT_2) show trends very similar to those derived for CT_0, see Fig. 11.9. However, since in this simple complex, the CT_1 and CT_2 states are much higher in energy than the CT_0 state, the latter is expected to play the most significant role in the exciton dissociation process, as detailed below.

As shown in Fig. 11.8c, the calculated CT and CR rates underline that, for typical intermolecular distances of ~3.5–4.5 Å between adjacent molecules, the electron-transfer rates from local (intramolecular) excited states to the lowest CT state can be very large. These rates can reach ~10^{10}–10^{12} s^{-1} even in the case of perpendicular configurations of the complex that do not lead to large electronic couplings; the results indicate that for such orientations pentacene-based excitons dissociate at least ten times faster than fullerene-based excitons. When going from perpendicular to parallel configurations, there occurs a significant increase in the dissociation rate for fullerene-based excitons that comes from an increase in the related electronic coupling and driving force. In the case of electron transfer from the lowest excited state of pentacene, the CT rate significantly increases for the "hexagon"-type parallel geometry 2B (where a fullerene hexagon directly faces pentacene); for the "edge"-type configuration 2A (where a fullerene C-C bond directly faces pentacene), the rate is similar to that in the perpendicular orientation 1A. Thus, our calculations suggest that, irrespective of the geometrical configurations of the interface, both types of excitons, whether formed on the pentacene molecule or on the C_{60} molecule, are able to dissociate efficiently.

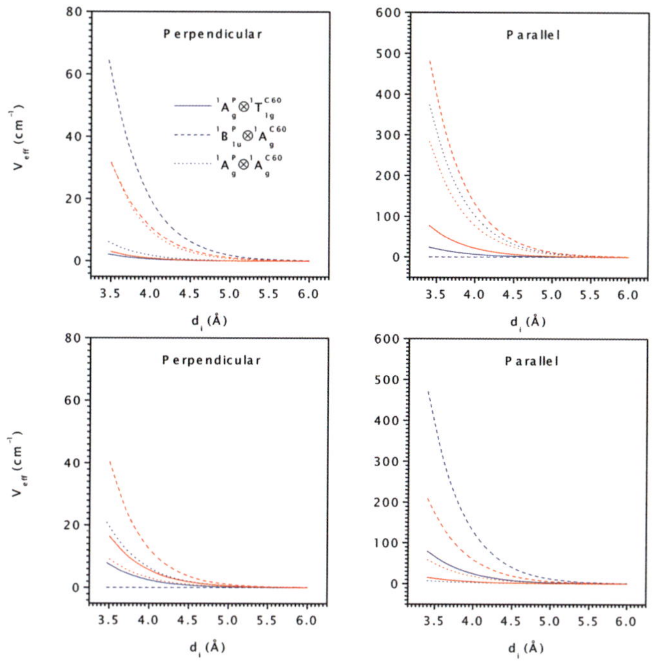

Figure 11.9 Electronic couplings of the ground state [$^1A_g{}^P \otimes {}^1A_g{}^{C60}$ (dotted lines)] and local-excited singlet states [$^1A_g{}^P \otimes {}^1T_{1g}{}^{C60}$ (solid lines) and $^1B_{1u}{}^P \otimes {}^1A_g{}^{C60}$ (dashed lines)] with the CT_1 (upper panels) and CT_2 (lower panels) states, as a function of intermolecular distance for the perpendicular orientations 1A and 1B and parallel orientations 2A and 2B illustrated in Fig. 11.8; blue lines refer to orientations 1A and 2A and red lines to orientations 1B and 2B. Adapted from Ref. [70].

It is also important to discuss the charge-recombination rates between the lowest CT state and the ground state. As seen from Fig. 11.8, the CR rates can be very large for parallel configurations of the complex, reaching values over 10^{10} s^{-1}. In the case of "hexagon"-type perpendicular configurations, the CR rates are below 10^7 s^{-1} even for small intermolecular distances. Moreover, in the case of "edge"-

type configurations, as a result of the weak electronic interactions, the CR rates become vanishingly small. Very small CR rates are also found for parallel but significantly displaced configurations of the complex.

Thus, our results suggest that in the case of pentacene/fullerene bilayer architectures, where the **D/A** molecular pairs are expected to be found mainly in perpendicular or displaced-parallel orientations, the CR processes from the CT state should play less of a significant role than in the case of bulk heterojunctions, where a substantial number of pentacene/fullerene complexes should be found in parallel configurations. In the latter case, the CR process could therefore compete efficiently with the dissociation of the CT states into mobile charge carriers and constitute an adverse factor limiting device performance.

A clear evidence of the dependence of power conversion efficiency on the heterojunction morphology was recently described in bilayer zinc phthalocyanine (ZnPc)/C_{60} solar cells [84]. Devices based on bilayers where C_{60} is deposited on top of a ZnPc layer at the surface of which the ZnPc molecules lie flat (parallel or face-on configurations) show a larger power conversion efficiency than those where the ZnPc molecules stand up (perpendicular configurations). Based on the experimental data and on calculations of the electronic couplings similar to those detailed above for pentacene/C_{60}, it was found that the main contributor to this difference is the more efficient exciton dissociation for face-on configurations.

11.3.3 Oligothiophene/Fullerene and Oligothiophene/Perylenediimide Complexes

Experiments have consistently shown that the performance of oligo- or polythiophene/PDI solar cells [85–91] is significantly lower (by nearly one order of magnitude) with respect to devices based on oligo- or polythiophene and fullerene derivatives [88,92–97]. In order to provide some possible understanding of the reasons for this difference, we carried out a comparative study of **D/A** complexes based on oligothiophene and PDI or fullerene [71]. We chose α-sexithienyl (6T) to represent oligothiophenes and polythiophene and C_{60} and *N,N′*-dimethyl-PDI to represent the fullerene and PDI derivatives, respectively. The electronic couplings and electron transfer rates were evaluated for the model intermolecular configurations illustrated in Fig. 11.10a.

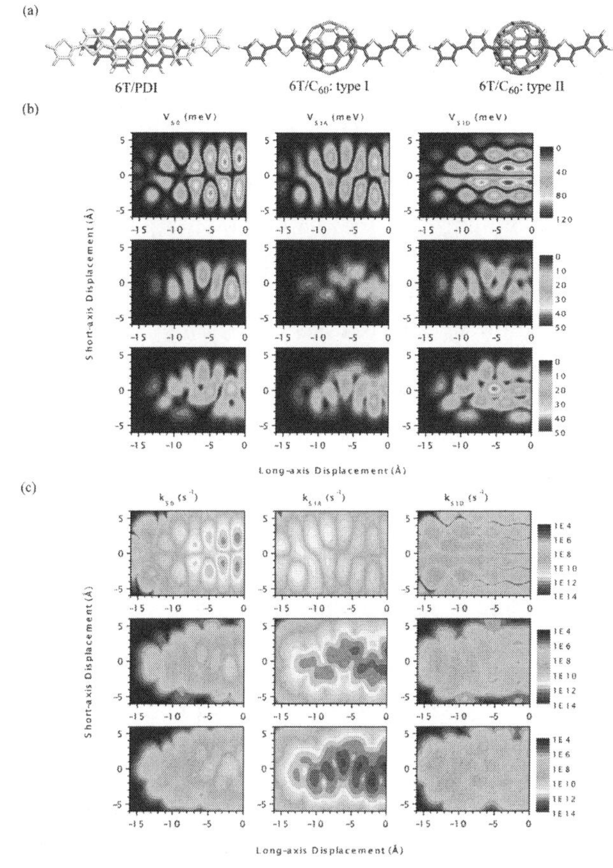

Figure 11.10 Illustration of the model intermolecular configurations for the 6T/PDI and 6T/C_{60} complexes (a), and evolution of the electronic couplings (b) and rates (c) for charge recombination from the lowest CT state to the ground state (V_{S0} and k_{S0}) and exciton dissociation from the lowest singlet state on PDI or C_{60} (V_{S1A} and k_{S1A}) and on 6T (V_{S1D} and k_{S1D}) to the lowest CT state, as a function of the displacements of 6T along the long and short axes; in (b) and (c), the upper, middle, and lower panels correspond to the 6T/PDI, 6T/C_{60} type I and II configurations, respectively. In the calculations of rates, according to experimental measurements in solid state, we consider S_1 energies of 1.7, 2.18, and 2.3 eV for C_{60} [76], PDI [98], and 6T [99–101], respectively, and the IP of 5.3 eV for 6T [102]; the EA values for C_{60} [79,80] and PDI [103,104] are taken in the range of –3.5 to –3.95 eV; here, EA = –3.8 eV. The reorganization energy is set to 0.5 eV. Adapted from Ref. [71].

The calculated electronic couplings involving the lowest CT state for the 6T/PDI and 6T/C_{60} complexes are depicted in Fig. 11.10b. The results confirm once again that the electronic couplings are very sensitive to both the relative molecular orientations and intermolecular distances. As the acceptor (PDI and C_{60}) slides on top of the molecular plane of 6T, the electronic coupling shows an oscillatory dependence on the displacements along the long and short axes of 6T, due to the alternation in the bonding/antibonding pattern of the involved molecular orbitals [3]. A larger extent of bonding/antibonding alternations for long-axis displacements with respect to short-axis displacements leads to a larger number of peaks and valleys in the electronic couplings. In most instances, the electronic couplings between PDI and 6T are significantly larger than those with C_{60} for both CT and CR processes due to the closer molecular packing between PDI and 6T. For 6T/PDI, the electronic couplings can be as large as 100 meV, while they only reach 40 meV in a very limited region for 6T/C_{60}. Similarly, slightly stronger electronic couplings appear in type-II configurations than in type-I configurations for 6T/C_{60} (see Fig. 11.10a for an illustration of these configurations).

For 6T/PDI, in general, the electronic couplings to the lowest CT state are slightly stronger for the ground state than for the lowest excited state of the acceptor (PDI), but weaker than for the lowest excited state of the donor (6T). For 6T/C_{60}, similar situations are found for these couplings in the case of type-II configurations, while in the case of type-I configurations the electronic couplings between the ground state and the CT state are in most instances the strongest.

The electron-transfer rates are shown in Fig. 11.10c. In a way similar to the electronic couplings, they also display alternating modulation as a function of the long- and short-axis displacements, although the trends in peak values are different. In most cases, the acceptor (PDI and C_{60})-based excitons dissociate more quickly than the donor (6T)-based excitons, even though the acceptor-based excitons have relatively smaller couplings with the CT state. This can be attributed to the value of ($\Delta G^0+\lambda$) term in Eq. (11.12) becoming closer to zero for the acceptor-based excitons. The exciton dissociation rates for 6T/C_{60} are calculated to be larger than those for 6T/PDI when the excitons are on the acceptor, while these rates are similar when the excitons are on the donor. For both 6T/PDI and 6T/C_{60} complexes, the dissociation rates can be very large and generally reach ~10^9–10^{13} s^{-1} for all excitons (whether they are generated on the donor or the acceptor). Thus, our calculations suggest that excitons are able to dissociate efficiently in both 6T/PDI and 6T/C_{60} solar cells.

Interestingly, for most system configurations, the decay of the lowest CT state to the ground state for complex 6T/C_{60} is substantially slower than the exciton dissociation. In contrast, for a large number of intermolecular configurations of 6T/PDI, the rate of decay of the lowest CT state back to the ground state is *similar or even faster* than the rate of formation of the CT state. When comparing the two systems, our calculations reveal that, in general, the decay of the CT state back to the ground state in 6T/PDI is more efficient with respect to 6T/C_{60}; this conclusion also holds if we vary the electron-affinity values in the range –3.95 to –3.5 eV. Thus, it appears that for 6T/PDI, the CR process could efficiently compete with the dissociation of the CT state into mobile charge carriers. To evaluate more generally the impact of the interfacial configurations that can be present in BHJ solar cells, we have also performed calculations where other intermolecular orientations (generated by rotating the PDI or 6T molecule) of the 6T/PDI and 6T/C_{60} complexes are considered. The results again demonstrate that the CR rates are much faster in 6T/PDI.

It is tempting to attribute the lower performance of the OPV devices based on oligothiophene/PDI to the large CR rates. However, it is useful to note that large CR rates should not necessarily limit device performance, provided bimolecular CR can be avoided. Indeed, as long as the rate at which the CT states dissociate into separated carriers is much larger than the CR rate, the latter process has little effect on the overall charge photogeneration efficiency. For instance, high charge photogeneration efficiencies have been recently observed in polythiophene/PDI blends [105]. The poor photovoltaic performances for devices based on these blends have then been attributed to unfavorable morphological structures that prevent efficient charge collection and lead to bimolecular CR [106].

Finally, we underline that while the present model provides useful insight for the understanding of geminate-recombination processes, its application to the description of exciton-dissociation processes remains limited. Indeed, as discussed in Section 11.2, electronic interactions between adjacent donor or acceptor sites can result in the formation of CT states that are effectively delocalized over several molecules or polymer repeat units. It was suggested recently that complete charge separation could take place from such more delocalized states, before they thermally relax to the more localized CT states that have been discussed above [62,63]. It is clear that the development of suitable models for the description of such delocalized states and of the related dynamical processes would represent a major step toward a detailed understanding of charge separation at **D/A** interfaces in organic solar cells. This is also the reason why it is critical to obtain a reliable description of the interfacial packing configurations ("local" morphology) between and among **D** and **A** molecules or chain segments, which is the focus of the next section.

11.4 Molecular Dynamics Description of the Pentacene–C_{60} Interface

We finally turn our attention to the nature of the morphology at the **D/A** interface, and in particular that of a relatively "simple" bilayer interface. The importance of the **D/A** interface has long been recognized, and recent experimental evidence has shone new light on the morphological complexity. This is particularly the case in polymer/fullerene bulk-heterojunction solar cells [107–118] and more recently in molecule/molecule and polymer/molecule bilayers [119–122]. The relative miscibility of the **D** and **A** components, the deposition protocol (e.g., vacuum deposition, solution casting, or spin coating to name but a few [123–125]), and a variety of post-processing procedures (e.g., solvent and thermal annealing or solvent additives) not only impact the properties of the "bulk" materials but can lead to potentially dramatic changes to the interface morphology [126].

The morphological complexity and variability then lead to difficulties in trying to grasp the nature of the underlying physical processes that take place at the **D/A** interface. As detailed in Section 11.2, the relevant electronic states, as well as the magnitude and direction of electrostatic interactions, are highly dependent on the molecular packing configurations between the **D** and **A** molecules or chain segments [68,70,71,84,127–133]. Also, Section 11.3 has shown that the local configurations at the interface between the **D** and **A** components have substantial impact on the strengths of the electronic couplings and the rates of charge separation and charge recombination. Hence, to achieve optimal OPV performance, one must be able to understand and ultimately control the morphology of these heterojunctions [126,134]. The question then becomes whether we can determine reliably the nature of the interactions and types of local molecular configurations at the **D/A** interface via computational approaches. To try and answer that question, we focus again on a bilayer structure with pentacene as the electron donor and C_{60} as the electron acceptor [66,135–142].

The relatively small size and rigid structural characteristics of these molecules has also allowed the pentacene–C_{60} interface to be considered in a number of theoretical investigations, ranging from electronic-structure calculations on molecular complexes, as described in Sections 11.2 and 11.3, to atomistic molecular dynamics (MD) simulations [70,127,128,143–148]. Using MD simulations (employing the MM3 force field of Allinger and coworkers [149]), we have probed on a larger scale the nature of the pentacene–C_{60} interface, with a particular emphasis on an architecture in which the C_{60} layer resides on top of the pentacene layer (the common motif used in conventional pentacene–C_{60} bilayer devices) [150]. We were particularly interested in delineating how the nature of the pentacene surface—either the (001) surface where the molecules stand upright (nearly perpendicular) to the surface or the (010) surface for which the molecules lie parallel to the surface—influences the characteristics of the interface with C_{60}. We also considered what consequence terraced (or stepped) pentacene surfaces may have; the formation of terraced surfaces is determined by the surface energy, roughness, and temperature of the substrate on which the pentacene film is deposited along with deposition temperature and rates [151–154].

As a preliminary step, it is thus useful to understand the strength of the interactions between pentacene and C_{60} over different scales. Energy-minimization calculations (MM3) were employed to analyze the binding energy of a single pentacene–C_{60} complex as a function of distance and pentacene orientation. The calculations indicate, as could be intuitively expected, that pentacene and C_{60} are more strongly bound when the pentacene face is parallel with the C_{60} cage (–9.44 and –8.92 kcal/mol when the pentacene center of mass is aligned with the five- and six-membered ring on the C_{60}, respectively) vs. perpendicular (–2.27 and –2.40 kcal/mol for interactions with the five- and six-membered ring on the C_{60}, respectively). Analysis of the cohesive (adsorption) energy of a single C_{60} on the pentacene surfaces of interest is consistent with this picture. The MM3 results show that C_{60} preferentially binds to the pentacene (010) face (33.5±2 kcal/mol) vs. the (001) face (18.9±2 kcal/mol); these trends also agree with previous estimates (that also take into account various polymorphs) [143,145,146], even though different pentacene models were used.

In addition, we evaluated the surface energies for pentacene and C_{60}, as stronger interfacial interactions between surfaces can be expected when the surface energies are most similar. The MM3 surface energies show trends similar to these for the complex binding energies and cohesive energy for a single C_{60} on a pentacene surface. For pentacene, the surface energies for the (001) and (010) are 53.0 and 91.8 mJ/m^2, respectively; periodic slab calculations employing DFT have determined the energies of these faces to be 49.7 and 76.9 mJ/m^2, respectively [155]. On the other hand, contact angle measurements have derived a surface energy of approximately 34–38 mJ/m^2 for the thin-film pentacene structure that is expected to present mainly an (001) surface with the pentacene molecules standing up [151,156]; the notable differences with respect to the calculations, however, can be due to the difficulty in identifying the exact geometrical parameters of the thin-film pentacene used in the experiment (including the phases, planes, and any surface reconstruction). For C_{60}, the calculated surface energies of the (001) and hexagonal-close packed (h.c.p.) faces are 98.5 and 88.0 mJ/m^2, respectively (again, there is deviation with the empirically determined C_{60} surface energy of approximately 42 mJ/m^2) [157–163]. As a first approximation based on the results from all these evaluations, it should be anticipated that C_{60} will interact more strongly with the pentacene (010) surface than the (001) surface, given their more similar surface energies.

To describe the morphology of the pentacene–C_{60} interface, we considered MD simulations with up to 105 fullerenes randomly distributed on the "pristine" pentacene (001) and pentacene (010) surfaces (Fig. 11.11). For pentacene (001), the presence of C_{60} leads to a structural change of the top layer. The fullerenes position themselves in the divots between the pentacene molecules. As the fullerene content increases, the first fullerene layer takes on the hexagonal closed pack configuration reported for fullerenes [146,160,163], with the additional layers taking on the same packing configuration.

When C_{60} is deposited on the pentacene (010) surface, however, there is a considerable rearrangement of the upper pentacene layers within a very short period of time due to the increased intermolecular interactions noted above. The fullerenes tend to sit in the middle portion of the pentacene molecules and eventually "burrow" or displace the pentacene molecules in the upper layer(s), consistent with previous results of C_{60} on the pentacene (010) surface [146]. Importantly, this leads to *interfacial disorder and intermixing* within the top pentacene layer, see Fig. 11.11. These results suggest that fullerenes deposited on the (010) surface should lead to a mixed structure right at the interface between the two materials, in stark contrast with what is seen for the pentacene (001) surface.

As mentioned above, the surface energy, roughness, and temperature of the substrate on which the pentacene film is deposited can have a considerable impact on the interface as terraced surfaces could be prevalent. To consider such situations, a series of terraced surfaces were constructed. Again, significant disorder and intermixing of the pentacene and C_{60} molecules is observed (Fig. 11.12). For the predominantly pentacene (001) structures, which previously showed minimal interactions between the pentacene and C_{60} layers, the pentacene molecules along the now open (0–10) and (100) faces readily disorder and intermix with the C_{60} molecules. There is also disruption of the molecular packing of the pentacene layers underneath the exposure. For terraces created in pristine pentacene (010) surfaces, the degree of disorder and interdiffusion is even larger, as one might expect from the results on the pristine surfaces.

The results of these simulations, in agreement with a growing number of experimental [84,120,121,134] and theoretical [122,126,130] results for polymer–fullerene blends and molecule-molecule and polymer–molecule bilayer architectures, reveal *large degrees of disorder and intermixing* among the pentacene and C_{60} molecules depending on the crystalline face on which the fullerenes are deposited and the presence of terraces. In particular, the details of the simulations are consistent with the results of time-of-flight secondary ion mass spectroscopy (TOF-SIMS) experiments on copper phthalocyanine (CuPc)–C_{60} interfaces that reveal an intermixed layer of 6.5 nm when C_{60} is deposited on CuPc (and 2.7 nm when the deposition order is reversed) [119], as well as a MD simulation study on pentacene interfaces with perylene-3,4,9,10-tetracarboxylic-3,4,9,10-dianhydride (PTCDA) [122].

The growing evidence of these complex "local" interface morphologies, as indicated above, leads to ambiguity pertaining to

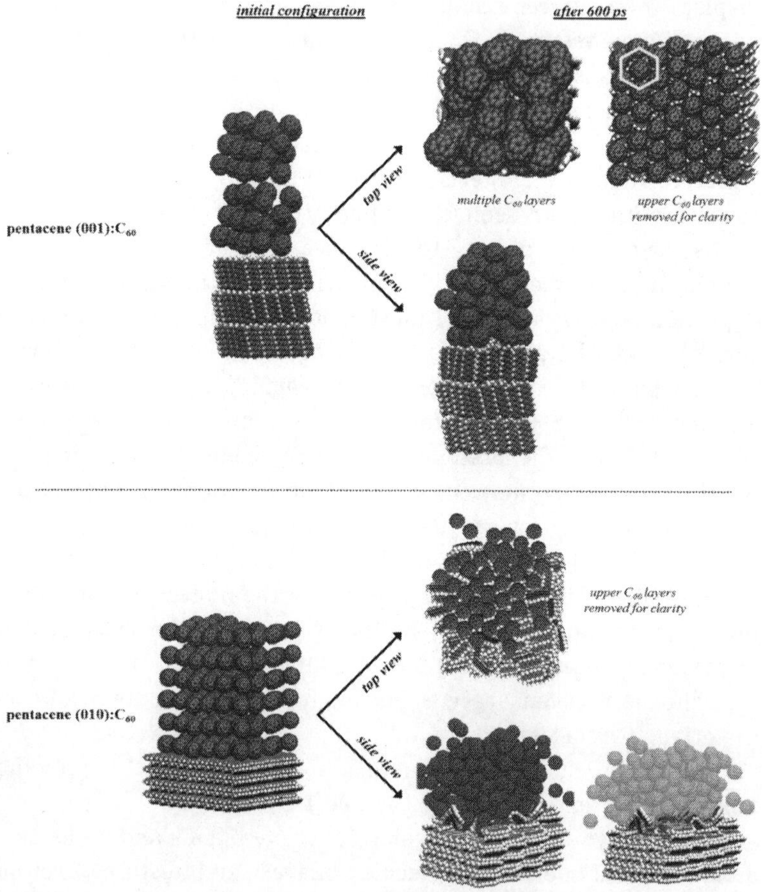

Figure 11.11 Initial configuration and snapshots of the side and top views after 600 ps simulations for the (top) pentacene (001):C_{60} interface and (bottom) pentacene (010):C_{60} interface. Note that some of the fullerenes have been removed or made transparent to aid visualization. Adapted from Ref. [150].

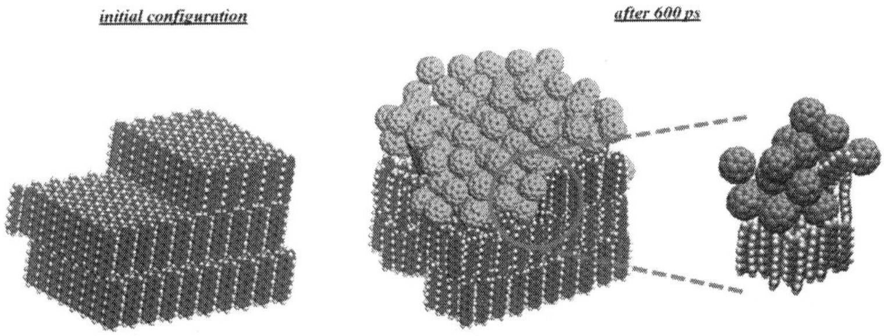

Figure 11.12 Representative example of the disorder and intermixing of pentacene and C_{60} on a terraced pentacene (001) surface.

the underlying physical mechanisms behind OPV operation and how one might design new molecular structures to develop appropriate active-layer architectures. It is currently not clear what role disorder and intermixing play, as such effects seemingly show different behavior as a function of the active-layer construct. Groves and coworkers [126] have elegantly shown that for BHJ architectures smooth (sharp) interfaces should lead to more efficient OPVs vs. those with rough (intermixed) interfaces. On the other hand, a number of researchers are intrigued by the electronic and optical properties of the amorphous, highly intermixed regions in BHJ cells and their ultimate function. Regarding their superb performance in BHJ cells, a recently hypothesized key attribute of fullerene derivatives is precisely their general ability to, at the same time, form well-ordered domains and intermix with polymers [116]. Also, in the case of bilayer architectures, Zimmerman et al. have shown that controlled disorder at the interface [134] can minimize charge recombination, while improved bulk crystallinity increases both exciton- and charge-transport efficiency.

11.5 Synopsis

The discussion presented in Sections 11.2 through 11.4 underlines that, when developing design rules regarding the synthesis of new molecular and polymeric materials for OPV, it is not enough to consider the intrinsic energetics of the charged and excited states of the individual components. Ultimately, there is a critical need to consider as well (i) the active-layer architecture, in which the material will be employed, (ii) the material miscibility not only in solution (if the active layer is to be solution processed) but also with the other active component, and (iii) the post-processing procedure, since each of these can present a tremendous impact on the OPV power conversion efficiency.

To comprehend all these aspects will require multiscale simulations—some of which have been outlined here—and controlled experiments to fully understand and appreciate the complexity of the bulk and interfacial structures/morphologies and how these impact the intrinsic molecular/polymer characteristics. It is only when armed with such a detailed knowledge that one could begin to rationalize design principles leading to more efficient OPV materials.

For multiscale simulations to reach an integrated description of the exciton-dissociation, charge-separation, and (geminate and bimolecular) charge-recombination processes, multiple challenges must be met. These include

i. The development of quantum-mechanical methodologies able to describe reliably the CT and CS states on a large (meso) scale involving many **D** and **A** chain segments/molecules. While popular DFT methodologies tend to overdelocalize the wave functions, and therefore underestimate the CT energies, Hartree–Fock-based methods suffer from too much localization. As we have seen in Section 11.2, modern long-range-corrected DFT functionals could provide a way forward.
ii. The development of hybrid quantum mechanics/molecular mechanics (QM/MM) methodologies where the MM part needs to provide an adequate description of the polarization phenomena occurring upon charge separation, which requires appropriate polarizable force fields.
iii. The development of molecular dynamics (MD) methodologies that, given the absence of experimental data with the necessary nanoscale (molecular) resolution, can provide adequate, reliable descriptions of the local morphologies at **D/A** interfaces. This is an important part of the problem indeed as the work we described in Section 11.3 has demonstrated that the rates of exciton dissociation

dramatically depend on the way **D** and **A** chain segments/ molecules pack at their interfaces.

iv. The development of Monte Carlo methodologies to describe the rates of the charge-transport and bimolecular charge-recombination processes as a function of the electrochemical potential within the active layer.

Acknowledgements

We gratefully acknowledge the support of various parts of this work by the Office of Naval Research (Award No. N00014-11-1-0211), the Center for Advanced Molecular Photovoltaics (Award No. KUS-C1-015-21, made by King Abdullah University of Science and Technology—KAUST), the National Science Foundation under the STC Program (Award No. DMR-0120967), and the Air Force Office of Scientific Research under the COMAS MURI Program (Award No. FA9550-10-1-0558). T.K. would like to thank the AvH Foundation for financial support through the Feodor-Lynen Program. B.Y. and C.Z. would like to thank the Chinese Visiting Scholar Program sponsored by the China Scholarship Council. The Center for Computational Molecular Science and Technology is funded through a National Science Foundation CRIF Award (Award No. CHE-0946869).

References

1. Bredas JL, Street GB (1985) Polarons, bipolarons, and solitons in conducting polymers, *Acc Chem Res,* **18**, 309–315.
2. Banerji N, Cowan S, Leclerc M, Vauthey E, Heeger AJ (2010) Exciton formation, relaxation, and decay in PCDTBT, *J Am Chem Soc,* **132**, 17459–17470.
3. Bredas JL, Calbert JP, da Silva Filho DA, Cornil J (2002) Organic semiconductors: A theoretical characterization of the basic parameters governing charge transport, *Proc Natl Acad Sci USA,* **99**, 5804–5809.
4. Bredas JL, Beljonne D, Coropceanu V, Cornil J (2004) Charge-transfer and energy-transfer processes in pi-conjugated oligomers and polymers: A molecular picture, *Chem Rev,* **104**, 4971–5004.
5. Coropceanu V, Cornil J, da Silva Filho DA, Olivier Y, Silbey R, Bredas JL (2007) Charge transport in organic semiconductors, *Chem Rev,* **107**, 926–952.

6. Heimel G, Romaner L, Zojer E, Bredas JL (2008) The interface energetics of self-assembled monolayers on metals, *Acc Chem Res,* **41**, 721–729.

7. Heimel G, Salzmann I, Duhm S, Koch N (2011) Design of organic semiconductors from molecular electrostatics, *Chem Mater,* **23**, 359–377.

8. Paramonov PB, Paniagua SA, Hotchkiss PJ, et al. (2008) Theoretical characterization of the indium tin oxide surface and of its binding sites for adsorption of phosphonic acid monolayers, *Chem Mater,* **20**, 5131–5133.

9. Runge E, Gross EKU (1984) Density-functional theory for time-dependent systems, *Phys Rev Lett,* **52**, 997.

10. Casida ME (1995) *Recent advances in density functional methods, Part I* (ed DPChong), 155–192 (World Scientific).

11. Casida ME, Gutierrez F, Guan JG, Gadea FX, Salahub D, Daudey JP (2000) Charge-transfer correction for improved time-dependent local density approximation excited-state potential energy curves: Analysis within the two-level model with illustration for H-2 and LiH, *J Chem Phys,* **113**, 7062–7071.

12. Becke AD (1993) Density-functional thermochemistry. 3. The role of exact exchange, *J Chem Phys,* **98**, 5648–5652.

13. Lee CT, Yang WT, Parr RG (1988) Development of the Colle-Salvetti correlation-energy formula into a functional of the electron-density, *Phys Rev B,* **37**, 785–789.

14. Stephens PJ, Devlin FJ, Chabalowski CF, Frisch MJ (1994) Ab initio calculation of vibrational absorption and circular dichroism spectra using density functional force fields, *J Phys Chem,* **98**, 11623–11627.

15. Cohen AJ, Mori-Sanchez P, Yang WT (2012) Challenges for density functional theory, *Chem Rev,* **112**, 289–320.

16. Janak JF (1978) Proof that $\partial E/\partial n_i = \varepsilon_i$ in density-functional theory, *Phys Rev B,* **18**, 7165.

17. Magyar RJ, Tretiak S (2007) Dependence of spurious charge-transfer excited states on orbital exchange in TDDFT: Large molecules and clusters, *J Chem Theory Comput,* **3**, 976–987.

18. Lange A, Herbert JM (2007) Simple methods to reduce charge-transfer contamination in time-dependent density-functional calculations of clusters and liquids, *J Chem Theory Comput,* **3**, 1680–1690.

19. Savin A, Flad H-J (1995) Density functionals for the Yukawa electron-electron interaction, *Int J Quantum Chem,* **56**, 327–332.

20. Heyd J, Scuseria GE, Ernzerhof M (2003) Hybrid functionals based on a screened Coulomb potential, *J Chem Phys,* **118**, 8207–8215.

21. Heyd J, Scuseria GE, Ernzerhof M (2006) Erratum: "Hybrid functionals based on a screened Coulomb potential", *J Chem Phys,* **124**, 219906.

22. Krukau AV, Vydrov OA, Izmaylov AF, Scuseria GE (2006) Influence of the exchange screening parameter on the performance of screened hybrid functionals, *J Chem Phys,* **125**, 224106.

23. Schimka L, Harl J, Kresse G (2011) Improved hybrid functional for solids: The HSEsol functional, *J Chem Phys,* **134**, 024116.

24. Chai J-D, Head-Gordon M (2008) Systematic optimization of long-range corrected hybrid density functionals, *J Chem Phys,* **128**, 084106.

25. Rohrdanz MA, Martins KM, Herbert JM (2009) A long-range-corrected density functional that performs well for both ground-state properties and time-dependent density functional theory excitation energies, including charge-transfer excited states, *J Chem Phys,* **130**, 054112.

26. Foster ME, Wong BM (2012) Nonempirically tuned range-separated DFT accurately predicts both fundamental and excitation gaps in DNA and RNA nucleobases, *J Chem Theory Comput,* **8**, 2682–2687.

27. Refaely-Abramson S, Baer R, Kronik L (2011) Fundamental and excitation gaps in molecules of relevance for organic photovoltaics from an optimally tuned range-separated hybrid functional, *Phys Rev B,* **84**, 075144.

28. Stein T, Kronik L, Baer R (2009) Prediction of charge-transfer excitations in coumarin-based dyes using a range-separated functional tuned from first principles, *J Chem Phys,* **131**, 244119.

29. Stein T, Kronik L, Baer R (2009) Reliable prediction of charge-transfer excitations in molecular complexes using time-dependent density functional theory, *J Am Chem Soc,* **131**, 2818–2820.

30. Körzdörfer T, Parrish RM, Marom N, Sears JS, Sherrill CD, Bredas JL (2012) Assessment of the performance of tuned range-separated hybrid density functionals in predicting accurate quasiparticle spectra, *Phys Rev B,* **86**, 205110.

31. Körzdörfer T, Sears JS, Sutton C, Bredas JL (2011) Long-range corrected hybrid functionals for pi-conjugated systems:

Dependence of the range-separation parameter on conjugation length, *J Chem Phys*, **135**, 204107.

32. Roncali J (1997) Synthetic principles for bandgap control in linear pi-conjugated systems, *Chem Rev*, **97**, 173–205.
33. Muhlbacher D, Scharber M, Morana M, et al. (2006) High photovoltaic performance of a low-bandgap polymer, *Adv Mater*, **18**, 2884–2889.
34. Blouin N, Michaud A, Leclerc M (2007) A low-bandgap poly(2,7-carbazole) derivative for use in high-performance solar cells, *Adv Mater*, **19**, 2295–2300.
35. Liang Y, Xiao S, Feng D, Yu L (2008) Control in energy levels of conjugated polymers for photovoltaic application, *J Phys Chem C*, **112**, 7866–7871.
36. Chen H-Y, Hou J, Zhang S, et al. (2009) Polymer solar cells with enhanced open-circuit voltage and efficiency, *Nat Photonics*, **3**, 649–653.
37. Park SH, Roy A, Beaupre S, et al. (2009) Bulk heterojunction solar cells with internal quantum efficiency approaching 100%, *Nat Photonics*, **3**, 297–303.
38. Zhang Y, Hau SK, Yip H-L, Sun Y, Acton O, Jen AK-Y (2010) Efficient polymer solar cells based on the copolymers of benzodithiophene and thienopyrroledione, *Chem Mater*, **22**, 2696–2698.
39. Livshits E, Baer R (2007) A well-tempered density functional theory of electrons in molecules, *Phys Chem Chem Phys*, **9**, 2932–2941.
40. Pandey L, Doiron C, Sears JS, Bredas JL (2012) Lowest excited states and optical absorption spectra of donor-acceptor copolymers for organic photovoltaics: A new picture emerging from tuned long-range corrected density functionals, *Phys Chem Chem Phys*, **14**, 14243–14248.
41. Risko C, McGehee MD, Bredas JL (2011) A quantum-chemical perspective into low optical-gap polymers for highly-efficient organic solar cells, *Chem Sci*, **2**, 1200–1218.
42. Karolewski A, Stein T, Baer R, Kummel S (2011) Tailoring the optical gap in light-harvesting molecules, *J Chem Phys*, **134**, 151101.
43. Martin RL (2003) Natural transition orbitals, *J Chem Phys*, **118**, 4775–4777.
44. Banerji N, Gagnon E, Morgantini P-Y, et al. (2012) Breaking down the problem: Optical transitions, electronic structure, and photoconductivity in conjugated polymer PCDTBT and in its separate building blocks, *J Phys Chem C*, **116**, 11456–11469.

45. Tawada Y, Tsuneda T, Yanagisawa S, Yanai T, Hirao K (2004) A long-range-corrected time-dependent density functional theory, *J Chem Phys,* **120**, 8425–8433.

46. Vydrov OA, Heyd J, Krukau AV, Scuseria GE (2006) Importance of short-range versus long-range Hartree-Fock exchange for the performance of hybrid density functionals, *J Chem Phys,* **125**, 074106.

47. Vydrov OA, Scuseria GE (2006) Assessment of a long-range corrected hybrid functional, *J Chem Phys,* **125**, 234109.

48. Vydrov OA, Scuseria GE, Perdew JP (2007) Tests of functionals for systems with fractional electron number, *J Chem Phys,* **126**, 154109.

49. Becke AD (1988) Density-functional exchange-energy approximation with correct asymptotic-behavior, *Phys Rev A,* **38**, 3098–3100.

50. Iikura H, Tsuneda T, Yanai T, Hirao K (2001) A long-range correction scheme for generalized-gradient-approximation exchange functionals, *J Chem Phys,* **115**, 3540–3544.

51. Cui GL, Yang WT (2010) Challenges with range-separated XC functionals in time-dependent density functional theory calculations, *Mol Phys,* **108**, 2745–2750.

52. Wang YL, Wu GS (2008) Improving the TDDFT calculation of low-lying excited states for polycyclic aromatic hydrocarbons using the Tamm-Dancoff approximation, *Int J Quantum Chem,* **108**, 430–439.

53. Hsu CP, Hirata S, Head-Gordon M (2001) Excitation energies from time-dependent density functional theory for linear polyene oligomers: Butadiene to decapentaene, *J Phys Chem A,* **105**, 451–458.

54. Hirata S, Head-Gordon M (1999) Time-dependent density functional theory within the Tamm-Dancoff approximation, *Chem Phys Lett,* **314**, 291–299.

55. Gruhn NE, da Silva DA, Bill TG, et al. (2002) The vibrational reorganization energy in pentacene: Molecular influences on charge transport, *J Am Chem Soc,* **124**, 7918–7919.

56. Heinecke E, Hartmann D, Muller R, Hese A (1998) Laser spectroscopy of free pentacene molecules (I): The rotational structure of the vibrationless S1 ← S0 transition, *J Chem Phys,* **109**, 906–911.

57. Burgos J, Pope M, Swenberg CE, Alfano RR (1977) Hetero-fission in pentacene-doped tetracene single-crystals, *Phys Stat Sol B,* **83**, 249–256.

58. Sears JS, Körzdörfer T, Zhang CR, Bredas JL (2011) Orbital instabilities and triplet states from time-dependent density

functional theory and long-range corrected functionals, *J Chem Phys*, **135**, 151103.

59. Peach MJG, Williamson MJ, Tozer DJ (2011) Influence of triplet instabilities in TDDFT, *J Chem Theory Comput*, **7**, 3578–3585.
60. Zhang CR, Sears JS, Coropceanu V, Bredas JL. Charge-transfer excitations in pentacene–C_{60} complexes from tuned long-range corrected hybrid functionals, *submitted.*
61. Wang XB, Woo HK, Wang LS (2005) Vibrational cooling in a cold ion trap: Vibrationally resolved photoelectron spectroscopy of cold C–60 anions, *J Chem Phys*, **123**, 051106.
62. Bakulin AA, Rao A, Pavelyev VG, et al. (2012) The role of driving energy and delocalized states for charge separation in organic semiconductors, *Science*, **335**, 1340–1344.
63. Cowan SR, Banerji N, Leong WL, Heeger AJ (2012) Charge formation, recombination, and sweep-out dynamics in organic solar cells, *Adv Funct Mater*, **22**, 1116–1128.
64. Yang B, Yi Y, Zhang CR, Sears JS, Coropceanu V, Bredas, JL. Charge delocalization of charge-transfer states at pentacene/C_{60} heterojunctions from tuned long-range corrected density functionals, *submitted.*
65. Van Regemorter T, Guillaume M, Sini G, et al. (2012) Density functional theory for the description of charge-transfer processes at TTF/TCNQ interfaces, *Theor Chem Acc*, **131**, 1273.
66. Potscavage WJ, Yoo S, Kippelen B (2008) Origin of the open-circuit voltage in multilayer heterojunction organic solar cells, *Appl Phys Lett*, **93**, 193308.
67. Perez MD, Borek C, Forrest SR, Thompson ME (2009) Molecular and morphological influences on the open circuit voltages of organic photovoltaic devices, *J Am Chem Soc*, **131**, 9281–9286.
68. Bredas JL, Norton JE, Cornil J, Coropceanu V (2009) Molecular understanding of organic solar cells: The challenges, *Acc Chem Res*, **42**, 1691–1699.
69. Kawatsu T, Coropceanu V, Ye A, Bredas JL (2008) Quantum-chemical approach to electronic coupling: Application to charge separation and charge recombination pathways in a model molecular donor–acceptor system for organic solar cells, *J Phys Chem C*, **112**, 3429–3433.
70. Yi Y, Coropceanu V, Bredas JL (2009) Exciton-dissociation and charge-recombination processes in pentacene/C_{60} solar cells: Theoretical

insight into the impact of interface geometry, *J Am Chem Soc*, **131**, 15777–15783.

71. Yi Y, Coropceanu V, Bredas JL (2011) A comparative theoretical study of exciton-dissociation and charge-recombination processes in oligothiophene/fullerene and oligothiophene/perylenediimide complexes for organic solar cells, *J Mater Chem*, **21**, 1479–1486.
72. Edmonds AR (1985) *Angular momentum in quantum mechanics*, 3rd Ed (Princeton University Press).
73. Ridley J, Zerner M (1973) Intermediate neglect of differential overlap technique for spectroscopy—pyrrole and azines, *Theor Chim Acta*, **32**, 111–134.
74. Marcus RA (1993) Electron transfer reactions in chemistry. Theory and experiment, *Rev Mod Phys*, **65**, 599–610.
75. Barbara PF, Meyer TJ, Ratner MA (1996) Contemporary issues in electron transfer research, *J Phys Chem*, **100**, 13148–13168.
76. Akimoto I, Ashida M, Kan'no K (1998) Luminescence from C_{60} single crystals in glassy phase under site-selective excitation, *Chem Phys Lett*, **292**, 561–566.
77. Jundt C, Klein G, Sipp B, Lemoigne J, Joucla M, Villaeys AA (1995) Exciton dynamics in pentacene thin-films studied by pump-probe spectroscopy, *Chem Phys Lett*, **241**, 84–88.
78. Faltermeier D, Gompf B, Dressel M, Tripathi AK, Pflaum J (2006) Optical properties of pentacene thin films and single crystals, *Phys Rev B*, **74**, 125416.
79. Hwang J, Wan A, Kahn A (2009) Energetics of metal-organic interfaces: New experiments and assessment of the field, *Mater Sci Eng R*, **64**, 1–31.
80. Schwedhelm R, Kipp L, Dallmeyer A, Skibowski M (1998) Experimental bandgap and core-hole electron interaction in epitaxial C-60 films, *Phys Rev B*, **58**, 13176–13180.
81. Kato T, Kodama T, Shida T, et al. (1991) Electronic absorption spectra of the radical anions and cations of fullerenes: C_{60} and C_{70}, *Chem Phys Lett*, **180**, 446–450.
82. Kato T, Kodama T, Shida T (1993) Spectroscopic studies of the radical anion of C_{60}. Detection of the fluorescence and reinvestigation of the ESR spectrum, *Chem Phys Lett*, **205**, 405–409.
83. Szczepanski J, Wehlburg C, Vala M (1995) Vibrational and electronic-spectra of matrix-isolated pentacene cations and anions, *Chem Phys Lett*, **232**, 221–228.

84. Rand BP, Cheyns D, Vasseur K, et al. (2012) The impact of molecular orientation on the photovoltaic properties of a phthalocyanine/fullerene heterojunction, *Adv Funct Mater,* **22**, 2987–2995.
85. Kuwabara K, Miyawaki K, Nawa K, Noma N, Shiroto Y (1992) Photovoltaic properties of oligothiophene perylene pigment layered cells, *Nippon Kagaku Kaishi*, 1168–1173.
86. Noma N, Tsuzuki T, Shirota Y (1995) Alpha-thiophene octamer as a new class of photoactive material for photoelectrical conversion, *Adv Mater,* **7**, 647–648.
87. Dittmer JJ, Marseglia EA, Friend RH (2000) Electron trapping in dye/polymer blend photovoltaic cells, *Adv Mater,* **12**, 1270–1274.
88. Shin WS, Jeong HH, Kim MK, et al. (2006) Effects of functional groups at perylene diimide derivatives on organic photovoltaic device application, *J Mater Chem,* **16**, 384–390.
89. Guo X, Bu L, Zhao Y, Xie Z, Geng Y, Wang L (2009) Controlled phase separation for efficient energy conversion in dye/polymer blend bulk heterojunction photovoltaic cells, *Thin Solid Films,* **517**, 4654–4657.
90. Ie Y, Uto T, Saeki A, Seki S, Tagawa S, Aso Y (2009) Photovoltaic performance and charge-carrier mobility of dendritic oligothiophene bearing perylene bis(dicarboximide) groups, *Synth Met,* **159**, 797–801.
91. Ie Y, Uto T, Yamamoto N, Aso Y (2009) Dendritic oligothiophene bearing perylene bis(dicarboximide) groups as an active material for photovoltaic device, *Chem Commun*, 1213–1215.
92. Padinger F, Rittberger RS, Sariciftci NS (2003) Effects of postproduction treatment on plastic solar cells, *Adv Funct Mater,* **13**, 85–88.
93. Li G, Shrotriya V, Huang J, Yao Y, Moriarty T, Emery K, Yang Y (2005) High-efficiency solution processable polymer photovoltaic cells by self-organization of polymer blends, *Nat Mater,* **4**, 864–868.
94. Nishizawa T, Tajima K, Hashimoto K (2008) The effect of crystallinity in donor groups on the performance of photovoltaic devices based on an oligothiophene-fullerene dyad, *Nanotechnology,* **19**, 424017.
95. Sakai J, Taima T, Saito K (2008) Efficient oligothiophene:fullerene bulk heterojunction organic photovoltaic cells, *Org Electron,* **9**, 582–590.
96. Xin H, Kim FS, Jenekhe SA (2008) Highly efficient solar cells based on poly(3-butylthiophene) nanowires, *J Am Chem Soc,* **130**, 5424–5425.

97. Sakai J, Taima T, Yamanari T, Saito K (2009) Annealing effect in the sexithiophene: C_{70} small molecule bulk heterojunction organic photovoltaic cells, *Sol Energy Mater Sol Cells,* **93**, 1149–1153.

98. Zhao H-M, Pfister J, Settels V, et al. (2009) Understanding ground- and excited-state properties of perylene tetracarboxylic acid bisimide crystals by means of quantum chemical computations, *J Am Chem Soc,* **131**, 15660–15668.

99. Janssen RAJ, Smilowitz L, Sariciftci NS, Moses D (1994) Triplet-state photoexcitations of oligothiophene films and solutions, *J Chem Phys,* **101**, 1787–1798.

100. Marks RN, Michel RH, Gebauer W, et al. (1998) The origin of photoluminescence from alpha-sexithienyl thin films, *J Phys Chem B,* **102**, 7563–7567.

101. Muccini M, Schneider M, Taliani C, et al. (2000) Effect of wave-function delocalization on the exciton splitting in organic conjugated materials, *Phys Rev B,* **62**, 6296–6300.

102. Hill IG, Kahn A, Soos ZG, Pascal JRA (2000) Charge-separation energy in films of [pi]-conjugated organic molecules, *Chem Phys Lett,* **327**, 181–188.

103. Sato N, Yoshida H, Tsutsumi K (1998) Photoemission and inverse photoemission studies on a thin film of N,N′-dimethylperylene-3,4,9,10-bis(dicarboximide), *J Electron Spectrosc Relat Phenom,* **88**, 861–865.

104. Zahn DRT, Gavrila GN, Gorgoi M (2006) The transport gap of organic semiconductors studied using the combination of direct and inverse photoemission, *Chem Phys,* **325**, 99–112.

105. Shoaee S, Clarke TM, Huang C, et al. (2010) Acceptor energy level control of charge photogeneration in organic donor/acceptor blends, *J Am Chem Soc,* **132**, 12919–12926.

106. Keivanidis PE, Howard IA, Friend RH (2008) Intermolecular interactions of perylene diimides in photovoltaic blends of fluorene copolymers: Disorder effects on photophysical properties, film morphology and device efficiency, *Adv Funct Mater,* **18**, 3189–3202.

107. Campoy-Quiles M, Ferenczi T, Agostinelli T, et al. (2008) Morphology evolution via self-organization and lateral and vertical diffusion in polymer:fullerene solar cell blends, *Nat Mater,* **7**, 158–164.

108. Mayer AC, Toney MF, Scully SR, et al. (2009) Bimolecular crystals of fullerenes in conjugated polymers and the implications of molecular mixing for solar cells, *Adv Funct Mater,* **19**, 1173–1179.

109. Collins BA, Gann E, Guignard L, He X, McNeill CR, Ade H (2010) Molecular miscibility of polymer–fullerene blends, *J Phys Chem Lett,* **1**, 3160–3166.

110. Chen D, Nakahara A, Wei D, Nordlund D, Russell TP (2010) P3HT/PCBM bulk heterojunction organic photovoltaics: Correlating efficiency and morphology, *Nano Lett,* **11**, 561–567.

111. Szarko JM, Guo J, Liang Y, et al. (2010) When function follows form: Effects of donor copolymer side chains on film morphology and BHJ solar cell performance, *Adv Mater,* **22**, 5468–5472.

112. Yin W, Dadmun M (2011) A new model for the morphology of P3HT/PCBM organic photovoltaics from small-angle neutron scattering: Rivers and streams, *ACS Nano,* **5**, 4756–4768.

113. Lu H, Akgun B, Russell TP (2011) Morphological characterization of a low-bandgap crystalline polymer:PCBM bulk heterojunction solar cells, *Adv Energy Mater,* **1**, 870–878.

114. Parnell AJ, Cadby AJ, Mykhaylyk OO, et al. (2011) Nanoscale phase separation of P3HT PCBM thick films as measured by small-angle x-ray scattering, *Macromolecules,* **44**, 6503–6508.

115. Collins BA, Tumbleston JR, Ade H (2011) Miscibility, crystallinity, and phase development in P3HT/PCBM solar cells: Toward an enlightened understanding of device morphology and stability, *J Phys Chem Lett,* **2**, 3135–3145.

116. Jamieson FC, Domingo EB, McCarthy-Ward T, Heeney M, Stingelin N, Durrant JR (2012) Fullerene crystallisation as a key driver of charge separation in polymer/fullerene bulk heterojunction solar cells, *Chem Sci,* **3**, 485–492.

117. Miller NC, Sweetnam S, Hoke ET, et al. (2012) Molecular packing and solar cell performance in blends of polymers with a bisadduct fullerene, *Nano Lett,* **12**, 1566–1570.

118. Treat ND, Brady MA, Smith G, et al. (2011) Interdiffusion of PCBM and P3HT reveals miscibility in a photovoltaically active blend, *Adv Energy Mater,* **1**, 82–89.

119. Sai N, Gearba R, Dolocan A, et al. (2012) Understanding the interface dipole of copper phthalocyanine (CuPc)/C_{60}: Theory and experiment, *J Phys Chem Lett,* **3**, 2173–2177.

120. Loiudice A, Rizzo A, Biasiucci M, Gigli G (2012) Bulk heterojunction versus diffused bilayer: The role of device geometry in solution p-doped polymer-based solar cells, *J Phys Chem Lett,* **3**, 1908–1915.

121. Loiudice A, Rizzo A, Latini G, Nobile C, de Giorgi M, Gigli G (2012) Graded vertical phase separation of donor/acceptor species for polymer solar cells, *Sol Energy Mater Sol Cells,* **100**, 147–152.

122. Poschlad A, Meded V, Maul R, Wenzel W (2012) Different interface orientations of pentacene and PTCDA induce different degrees of disorder, *Nanoscale Res Lett,* **7**, 248.

123. Brabec CJ, Durrant JR (2008) Solution-processed organic solar cells, *MRS Bull,* **33**, 670–675.

124. Helgesen M, Sondergaard R, Krebs FC (2010) Advanced materials and processes for polymer solar cell devices, *J Mater Chem,* **20**, 36–60.

125. Krebs FC (2009) Fabrication and processing of polymer solar cells: A review of printing and coating techniques, *Sol Energy Mater Sol Cells,* **93**, 394–412.

126. Lyons BP, Clarke N, Groves C (2012) The relative importance of domain size, domain purity, and domain interfaces to the performance of bulk-heterojunction organic photovoltaics, *Energy Environ Sci,* **5**, 7657–7663.

127. Verlaak S, Beljonne D, Cheyns D, et al. (2009) Electronic structure and geminate pair energetics at organic-organic interfaces: The case of pentacene/C_{60} heterojunctions, *Adv Funct Mater,* **19**, 3809–3814.

128. Linares M, Beljonne D, Cornil J, et al. (2010) On the interface dipole at the pentacene-fullerene heterojunction: A theoretical study, *J Phys Chem C,* **114**, 3215–3224.

129. Clarke TM, Durrant JR (2010) Charge photogeneration in organic solar cells, *Chem Rev,* **110**, 6736–6767.

130. McMahon DP, Cheung DL, Troisi A (2011) Why holes and electrons separate so well in polymer/fullerene photovoltaic cells, *J Phys Chem Lett,* **2**, 2737–2741.

131. Liu T, Cheung DL, Troisi A (2011) Structural variability and dynamics of the P3HT/PCBM interface and its effects on the electronic structure and the charge-transfer rates in solar cells, *Phys Chem Chem Phys,* **13**, 21461–21470.

132. Beljonne D, Cornil J, Muccioli L, Zannoni C, Bredas JL, Castet F (2011) Electronic processes at organic-organic interfaces: Insight from modeling and implications for opto-electronic devices, *Chem Mater,* **23**, 591–609.

133. Ojala A, Petersen A, Fuchs A, et al. (2012) Merocyanine/C_{60} planar heterojunction solar cells: Effect of dye orientation on exciton dissociation and solar cell performance, *Adv Funct Mater,* **22**, 86–96.

134. Zimmerman JD, Xiao X, Renshaw CK, et al. (2012) Independent control of bulk and interfacial morphologies of small molecular weight organic heterojunction solar cells, *Nano Lett,* **12**, 4366–4371.

135. Yoo S, Domercq B, Kippelen B (2004) Efficient thin-film organic solar cells based on pentacene/C_{60} heterojunctions, *Appl Phys Lett,* **85**, 5427–5429.

136. Pandey AK, Nunzi JM (2006) Efficient flexible and thermally stable pentacene/C_{60} small molecule-based organic solar cells, *Appl Phys Lett,* **89**, 213506.

137. Yang YC, Chang CH, Lee YL (2007) Complexation of fullerenes on a pentacene-modified Au(111) surface, *Chem Mater,* **19**, 6126–6130.

138. Al-Mahboob A, Sadowski JT, Fujikawa Y, Nakajima K, Sakurai T (2008) Kinetics-driven anisotropic growth of pentacene thin films, *Phys Rev B,* **77**, 035426.

139. Dougherty DB, Jin W, Cullen WG, Dutton G, Reutt-Robey JE, Robey SW (2008) Local transport gap in C_{60} nanochains on a pentacene template, *Phys Rev B,* **77**, 073414.

140. Conrad BR, Tosado J, Dutton G, et al. (2009) C_{60} cluster formation at interfaces with pentacene thin-film phases, *Appl Phys Lett,* **95**, 213302.

141. Dougherty DB, Jin W, Cullen WG, Reutt-Robey JE, Robey SW (2009) Striped domains at the pentacene:C_{60} interface, *Appl Phys Lett,* **94**, 023103.

142. Jin W, Dougherty DB, Cullen WG, Robey S, Reutt-Robey JE (2009) C_{60}-pentacene network formation by 2-D co-crystallization, *Langmuir,* **25**, 9857–9862.

143. Cantrell R, Clancy P (2008) A computational study of surface diffusion of C_{60} on pentacene, *Surf Sci,* **602**, 3499–3505.

144. Zheng Y, Pregler SK, Myers JD, Ouyang JM, Sinnott SB, Xue JG (2009) Computational and experimental studies of phase separation in pentacene:C_{60} mixtures, *J Vac Sci Technol B,* **27**, 169–179.

145. Cantrell R, Clancy P (2010) A molecular dynamics study of the effect of pentacene polymorphs on C_{60} surface adsorption and diffusional properties and the tendency to form nanowires, *Mol Simulat,* **36**, 590–603.

146. Cantrell RA, James C, Clancy P (2011) Computationally derived rules for persistence of C_{60} nanowires on recumbent pentacene bilayers, *Langmuir,* **27**, 9944–9954.

147. Minami T, Nakano M, Castet Fdr (2011) Nonempirically tuned long-range corrected density functional theory study on local and charge-transfer excitation energies in a pentacene/C_{60} model complex, *J Phys Chem Lett,* **2**, 1725–1730.

148. Muccioli L, D'Avino G, Zannoni C (2011) Simulation of vapor-phase deposition and growth of a pentacene thin film on C_{60} (001), *Adv Mater,* **23**, 4532–4536.

149. Lii JH, Allinger NL (1989) Molecular mechanics. The MM3 force field for hydrocarbons. 3. The van der Waals' potentials and crystal data for aliphatic and aromatic hydrocarbons, *J Am Chem Soc,* **111**, 8576–8582.

150. Fu Y-T, Risko C, Brédas JL (2013) Intermixing at the pentacene-fullerene bilayer interface: A molecular dynamics study, *Adv Mater,* **25**, 878–882.

151. Drummy LF, Miska PK, Alberts D, Lee N, Martin DC (2006) Imaging of crystal morphology and molecular simulations of surface energies in pentacene thin films, *J Phys Chem B,* **110**, 6066–6071.

152. Cheng H-L, Lin J-W (2010) Controlling polymorphic transformations of pentacene crystal through solvent treatments: An experimental and theoretical study, *Cryst Growth Des,* **10**, 4501–4508.

153. Desai TV, Woll AR, Engstrom JR (2012) Thin film growth of pentacene on polymeric dielectrics: Unexpected changes in the evolution of surface morphology with substrate, *J Phys Chem C,* **116**, 12541–12552.

154. Ruiz R, Choudhary D, Nickel B, et al. (2004) Pentacene thin film growth, *Chem Mater,* **16**, 4497–4508.

155. Northrup JE, Tiago ML, Louie SG (2002) Surface energetics and growth of pentacene, *Phys Rev B,* **66**, 121404.

156. Drummy LF, Martin DC (2005) Thickness-driven orthorhombic to triclinic phase transformation in pentacene thin films, *Adv Mater,* **17**, 903–907.

157. Feng M, Lee J, Zhao J, Yates JT, Petek H (2007) Nanoscale templating of close-packed C_{60} nanowires, *J Am Chem Soc,* **129**, 12394–12395.

158. Guo YJ, Karasawa N, Goddard WA (1991) Prediction of fullerene packing in C_{60} and C_{70} crystals, *Nature,* **351**, 464–467.

159. Ma X, Wigington B, Bouchard D (2010) Fullerene C_{60}: Surface energy and interfacial interactions in aqueous systems, *Langmuir,* **26**, 11886–11893.

160. Neel N, Kroger J, Berndt R (2006) Highly periodic fullerene nanomesh, *Adv Mater,* **18**, 174–177.

161. Pawlak R, Kawai S, Fremy S, Glatzel T, Meyer E (2011) Atomic-scale mechanical properties of orientated C_{60} molecules revealed by noncontact atomic force microscopy, *ACS Nano,* **5**, 6349–6354.

162. Xiao WD, Ruffieux P, Ait-Mansour K, et al. (2006) Formation of a regular fullerene nanochain lattice, *J Phys Chem B,* **110**, 21394–21398.

163. Zhang HL, Chen W, Huang H, Chen L, Wee ATS (2008) Preferential trapping of C_{60} in nanomesh voids, *J Am Chem Soc,* **130**, 2720–2721.

Part III

Technology, Lifetime, and Production

Chapter 12

Flexible Substrates and Barriers

Yulia Galagan
Holst Centre, 5656 AE Eindhoven, The Netherlands
yulia.galagan@tno.nl

The manufacturing of organic photovoltaic (OPV) devices starts from the selection of the substrate. The substrates can be flexible or rigid, be transparent or opaque, provide electrical properties, or be non-conducting. The choice of the substrate depends on the solar cell architecture, type of the electrode, solar cell manufacturing process, and, in general, the final requirements for the device. Furthermore, the required OPV manufacturing processes have a strong link to the needed substrate properties. Most plastic substrates cannot support very high temperature, limiting the maximum processing temperature for OPV layers. The ideal substrates for OPV applications present a unique balance of properties, which can provide simple low-cost manufacturing and long lifetime of the devices. A balance in the properties of substrates can guarantee the first step toward successful manufacturing of OPV devices. In addition to substrates, successful long-term functioning requires protection from environmental factors such as water and oxygen,

Organic Solar Cells: Fundamentals, Devices, and Upscaling
Edited by Barry P. Rand and Henning Richter

ISBN 978-981-4463-65-2 (Hardcover), 978-981-4463-66-9 (eBook)
www.panstanford.com

which can be provided by barrier layers. Together, substrates and barriers create "envelope" layers protecting OPV devices. The right choice of substrate and barrier will aid in the manufacturing and is key for successful commercialization of OPV modules. This chapter will make an overview of the main barriers and substrates suitable for OPV manufacturing, and will provide their main characteristics.

12.1 Substrates

The main substrates used for the manufacturing of organic photovoltaic devices are glass, plastic, and metal foil. The properties of typical substrate materials are given in Table 12.1.

Table 12.1 Comparison of PEN, stainless steel, and glass substrates

Property	Glass	PEN	Stainless steel
Weight, mg/m^2 (for 100-μm-thick film)	220	120	800
Transmission in the visible range (%)	92	90	0
Maximum process temperature (°C)	600	180	>600
TCE (ppm/°C)	5	16	10
Elastic modulus (GPa)	70	5	200
Permeability for oxygen and moisture	No	Yes	No
Coefficient of hydrolytic expansion (ppm/%RH)	0	11	0
Planarization necessary	No	No	Yes
Electrical conductivity	None	None	High
Thermal conductivity (W m^{-1} K^{-1})	1	0.1	16

Source: Ref. [1]

12.1.1 Metal Foil Substrates

Metal foil substrates [2–5] offer the advantages of higher process temperature capability, dimensional stability, and excellent barrier properties for oxygen and water. The disadvantages and limitations of the metal foil substrates include non-transparency and surface roughness. However, non-transparent metal foils can be successfully used in some device architectures, where illumination of the OPV devices occurs from the top side. The smoothness of the metal substrates can be increased by polishing [3] or by applying additional planarization layers. The planarization layer can be organic, inorganic, or a combination of the two. With high content of organic components, thicker layers of planarization material can be applied without forming cracks and a very smooth surface can be achieved. It is very important for OPV devices, because the electro-active layers are very thin (100–500 nm) and the chance to have shorts in the devices is increased with rougher surfaces.

The conductivity of metal substrates can be considered as both an advantage and disadvantage. Sometimes, the metal substrate can serve as a back contact, but very often the conductive property of the metal substrate is not used. In such cases, conducting substrates are insulated from devices by an additional planarization layer.

Stainless steel has been most commonly used in research because of its high resistance to corrosion and process chemicals, and its long record of application in amorphous silicon solar cells. Stainless steel substrates can tolerate process temperature as high as 1000 °C. These substrates are dimensionally stable and have excellent barrier properties against water and oxygen. Generally, stainless steel substrates are more durable than plastic substrates.

12.1.2 Flexible Plastic Substrates

Polymer foil substrates are highly flexible, can be inexpensive, and are roll-to-roll (R2R) compatible. Transparent plastic substrates have the advantage of being compatible with many organic solar cell architectures. The most common plastic substrates for fabrication of OPV devices are polyethylene terephthalate (PET) [6–11], polyethylene naphthalate (PEN) [6,7,9–12], polycarbonate (PC) [7,11,13,14], polyethersulfone (PES) [7,11,15], and polyimide (PI) [11,16–18]. The difference in chemical structure between these polymer materials is shown in Fig. 12.1. All of these polymer substrates have good light transmission in the visible range, except PI. PI (e.g., Kapton® from DuPont) absorbs in the visible region (has a yellow color), which makes it unsuitable for application in conventional device architectures. However, it is suitable for manufacturing top illuminated solar cells. The advantage of PI is its high process temperature capability (>350 °C), while other plastic substrates have limited process temperature capabilities, as shown in Table 12.2.

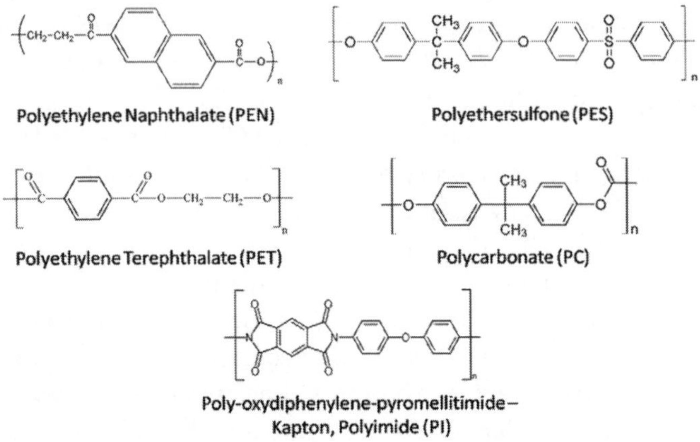

Figure 12.1. Chemical structures of polymers used for substrate manufacturing.

Table 12.2 Properties of plastic substrates [1]

Property	PET Melinex	PEN Teonex	PC Lexan	PES Sumilite	PI Kapton
T_g, °C	78	121	150	223	410
CTE* (−55 – 85 °C)	15	13	60–70	54	30–60
Transmission (400–700 nm), %	89	87	90	90	yellow
Moisture absorption, %	0.14	0.14	0.4	1.4	1.8
Young's modulus, GPa	5.3	6.1	1.7	2.2	2.5
Tensile strength, MPa	225	275	–	83	231
Density, g cm^{-3}	1.4	1.36	1.2	1.37	1.43
Refractive index	1.66	1.5–1.75	1.58	1.66	–
Birefringence, nm	46	–	14	13	–

*coefficient of thermal expansion

Another disadvantage of plastic substrates is the lack of dimensional stability during processing at elevated temperatures. There are no polymeric substrates which meet the extremely demanding requirements for low moisture and water permeability for OPV applications. The typical water and oxygen permeation rates of flexible plastic substrates are 1–10 g/m^2/day and 1–10 cm^3/m^2/day, respectively, relatively high with respect to those typically required for organic electronic devices, 10^{-3}–10^{-6} g/m^2/day and 10^{-3}–10^{-5} cm^3/m^2/day, respectively. Plastic substrates therefore require additional barrier layer coatings. Barrier coatings can reduce the absorption and permeability of atmospheric gasses, can raise the resistance to process chemicals, can strengthen the adhesion of the device layers, and can reduce the surface roughness. Despite some of these disadvantages, plastic substrates have a lower cost potential compared to metal foil substrates. Table 12.2 summarizes the properties of the most common plastic substrates which can be used in organic solar cell manufacturing.

The polymers can be further categorized into films that are semicrystalline (PET and PEN), amorphous and thermoplastic, and amorphous but solvent cast. Polymers with T_g of >140 °C that are semicrystalline tend generally to have melting points that are too high to allow the polymers to be melt-processed without significant degradation. PEN is the highest performance material available as a biaxially oriented semicrystalline film. The next category is polymers that are thermoplastic but non-crystalline. The polymers in this range are from PC, with a T_g of approximately 150 °C, to polyethersulphone (PES), with a T_g of ~220 °C. Although thermoplastic, these polymers may also be solvent cast to give high optical clarity. The third category is high T_g materials, that cannot be melt-processed. This group includes aromatic fluorene containing polyarylates (PAR) [19], and PI.

As many different types of plastic substrate have been proposed, the final step concerns the choice of the most suitable one. For example, it has been shown [20] that PET has a comparatively low price and can be delivered with a low average surface roughness (RA).

PEN provides a better thermal stability. Currently, the polymer film manufacturers work on improving the surface smoothness and thermal stability of PEN films as well as of thermally stabilized PET films [7]. However, both PEN and PET show an insufficient stability against moisture. PET has an increased brittleness after a damp-heat test, while PEN changes its optical properties. Also, the UV stability of both materials is insufficient for solar cell encapsulation. After irradiation with UV light, PEN changes its vapor transmission rate while PET loses its mechanical stability. Because of that, also ethylene-tetrafluoroethylene (ETFE) films are discussed as substrates for encapsulation. This material provides much better moisture and UV stability (no UV absorption measured at 300 nm wavelength) but is much more expensive than PET and a pretreatment is needed to provide a good barrier layer adhesion [7].

12.1.3 Glass Substrates

Plastic substrates offer the property of flexibility that is very important for R2R manufacturing. But at the same time, there are a lot of applications for solar cells where rigid substrates can be used—first of all, in solar windows. In this case, a glass substrate makes a logical choice. Glass has also many advantages over plastic substrates, being highly transparent over the visible spectrum, possessing a high homogeneity of the refractive index, and exhibiting a high UV resistance. Compared with plastic substrates, glass has more advanced thermal properties such as high temperature stability, high dimensional stability, and a low thermal expansion coefficient. Additionally, glass has a high chemical resistance and excellent barrier properties against water and oxygen. Good barrier properties of the glass substrate can probably significantly decrease the cost of the solar cells, as in such cases an additional barrier layer will not be required. Moreover, the good barrier properties of the glass can improve the lifetime of the solar cells. Due to the high mechanical stability, glass substrates have also advantage in high scratch resistance.

Plate glass becomes flexible when its thickness is reduced to several hundred micrometers. Ultra-thin glass sheets are superior to plastic films because of the inherent properties of glass, including gas barrier properties (which protect against gas and water), heat resistance, and transparency.

12.1.4 Substrate Properties

The substrates for OPV devices must satisfy numerous requirements, such as optical quality and transparency to allow light to reach the photoactive layer; substrate smoothness in the nanometer range to provide a surface that will promote high-quality deposition of subsequent layers and prevent the penetration of potential substrate spikes or irregularities into the device layers; the ability to support processing at high enough temperatures; good dimensional stability; good resistance to any chemicals used during processing; low water absorption.

i. *Optical properties:* Transparency is one of the main characteristics of substrates. Both transparent and non-transparent substrates are suitable for the manufacturing of organic solar cells. However, non-transparent substrates constrain the usage of some device architectures. The usage of non-transparent substrates requires transparent top electrodes with a transparent top-barrier layer. Devices on transparent substrates can have their transparent electrode either on top or on the bottom side of the devices or even on both sides, in the case of semitransparent solar cells. However, in most cases, the transparent substrate is usually combined with a transparent electrode. Such a device architecture requires high optical transparency of the electrode as well as the substrate. Also, the barrier or protection layer against moisture and oxygen (directly attached to the substrate or laminated afterward) should have high optical transparency. In addition, substrates for OPV devices should have low birefringence.
ii. *Surface roughness:* Thin-film devices are very sensitive to surface roughness. High roughness structures over short distances (i.e., short scale) must be avoided, as it can create shunts in the devices. However, an intermediate roughness over long distance (i.e., long scale) is acceptable.
iii. *Thermal and thermo-mechanical properties:* Some OPV fabrication processes, such as drying, annealing, and thermal sintering, require applying high temperatures. For fabrication, it is important that the maximum process temperature be lower than the glass transition temperature of the substrate polymer.
iv. *Dimensional stability:* Glass substrates have very high dimensional stability, but conventional processing on glass substrates cannot be directly transferred to plastic. The thermal contraction mismatch between the substrates and

the deposited device film and built-in stresses in these films leads to curving and changing in the in-plane dimension of the substrates. This change causes misalignment between the device layers. Plastic substrates will also change size on exposure to moisture. The first defense against these changes is the selection of substrates with low water absorption and low coefficients of hydroscopic expansion. Appropriate substrate selection and heat stabilization techniques can significantly reduce the degree of the changes, but they cannot eliminate them completely.

v. *Mechanical properties:* A high elastic modulus makes the substrate stiff, and a hard surface supports the device layer under impact.

vi. *Chemical properties:* The substrates should not release contaminants and should be inert against process chemicals.

vii. *Barrier properties:* Barrier properties of the substrates against permeation of atmospheric gases such as water and oxygen can be the biggest advantage of the substrate, as it can allow to avoid using two-barrier layers.

viii. *Electrical properties:* Some substrates, such as metal foil or ITO-coated plastic foils, are conductive. These conductive substrates may then serve as one electrode in the deposited device.

12.1.4.1 Electrical properties

Stainless steel and ITO-coated plastic substrates provide conductive platforms from which to fabricate devices, with ITO the most commonly used transparent electrode. For metal substrates, the intrinsic conductivity is not always used, with the metal substrate requiring a planarization layer. However, there are a few cases where the steel substrate has been used as an electrode in the solar cell architecture [21].

12.1.4.1.1 ITO-coated plastic substrates

Thin films of ITO [10,22–25] are widely used as transparent electrodes in organic photovoltaic devices [26–28], since it has relatively low electrical resistivity and good transparency in the visible range. Numerous deposition techniques have been utilized to prepare ITO films [29–32]. Now large-scale coatings of high-quality ITO films on glass are available in industry. Preparation of ITO films with high quality is highly requested in OPV applications. However, ITO films deposited on polymer substrates show high electrical resistivity because the substrate heating is limited due to their weak thermal resistance. Typically, the sheet resistance of ITO/glass substrate is 10–15 Ω/□, while the sheet resistance of ITO on PET substrates is around 60 Ω/□. High sheet resistance of ITO becomes apparent in a rapid decay of the efficiency with increasing the width of the solar cell (Fig. 12.2). The additional series resistance induced by the sheet resistance of the electrode can dramatically reduce the charge collection in the device.

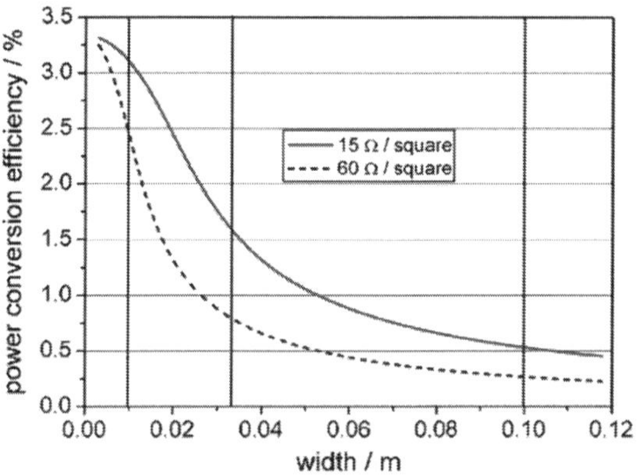

Figure 12.2 Calculated theoretical power conversion efficiency of a P3HT:PCBM-based single rectangular organic solar cell as a function of the width of the electrode. Two cases are shown, an ITO sheet resistance of 15 Ω/□ (solid line) and 60 Ω/□ (dashed line). Reproduced from [27] with permission, Copyright © 2006, Elsevier B.V.

Within realistic ranges of substrate sheet resistances (15–60 Ω/□), the overall device efficiency heavily depends on the width of the electrode. To keep the resistivity losses in the electrode as small as possible, the width has to be narrow with the contacts taken on the long side (for additional information, see Chapter 14).

The processing temperature during the deposition of ITO on plastic foils such as PEN and PET is typically limited to about 100 °C. Thus, the quality of ITO layer is not optimal. First of all, this results in a higher sheet resistance of ITO on plastic substrates, as discussed above. Additionally, low-temperature processing leads to an increased surface roughness, compared to ITO annealed at higher temperatures as for ITO on glass. The ITO-coated PET foil is typically characterized by the presence of a large number of spikes, which create unwanted shunts in the resulting OPV devices. Figure 12.3 depicts the AFM topology images of ITO on glass and ITO on PET foil.

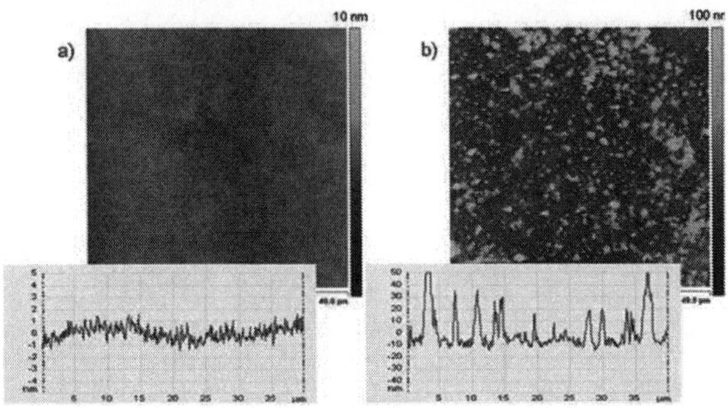

Figure 12.3 AFM topology image (10×10 μm) of (a) ITO-coated glass and (b) ITO-coated PET foil. Reproduced from [33] with permission, Copyright © 2010, Elsevier B.V.

Another disadvantage of TCO layers and ITO in particular is their brittleness. The ITO anode has a critical bending radius of about 8 mm, up to which the film can withstand the load without increasing in resistance [34]. However, the number of bending cycles, even with high bending radius, will affect the resistance of ITO. ITO breaks down rather quickly and shows an increase in resistance, which will negatively affect the performance of the organic solar cells. The optical microscopy images of Fig. 12.4 [10] (relating to 150 cycle tests) show that the ITO coating is damaged considerably after undergoing tensile and compressive stresses, degrading conductance. The number of visible cracks increases for smaller radii of curvature (thus, leading to an increase in resistance). For tensile stresses (Fig. 12.4), the cracks are less visible, although they are more numerous per square centimeter, compared to those arising from compressive stress.

Furthermore, the limited availability and supply of materials such as indium results in a high cost of ITO coatings. In Ref. [35], it was estimated that the cost of typical commercially available PET/ITO substrates, used for R2R manufacturing of organic solar cells

[36,37], takes more than half of the total material cost required for the manufacturing of OPV modules.

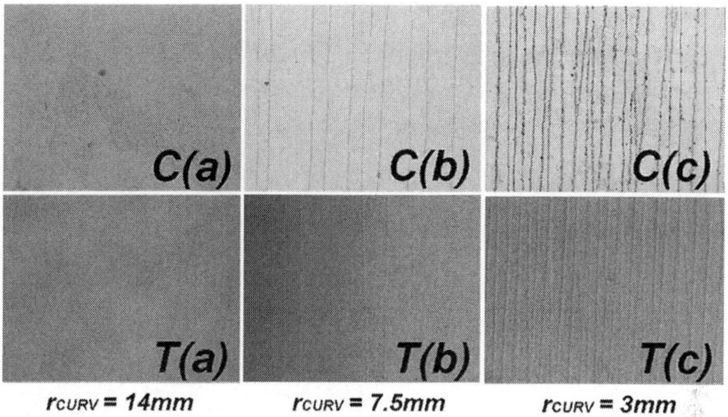

Figure 12.4 Optical microscopy images (width 500 μm) of the ITO surface (on PET) after 150 bending cycles under compressive (*C*) and tensile (*T*) stresses for different radii of curvature: (a) $r = 14$ mm; (b) $r = 7.5$ mm; (c) $r = 3$ mm. Reproduced from Ref. [10] with permission, Copyright © 2011, Wiley Periodicals Inc., A Wiley Company.

12.1.4.1.2 Alternative for ITO

The high price of indium in combination with the relatively low conductivity of ITO on flexible substrates (60 Ω/□) and low mechanical resistance create a need for finding alternative transparent electrodes [38–42]. Such alternative electrodes should combine good conductivity, high transparency, and relatively low price [43–45]. A number of alternative materials for transparent electrodes have been proposed and demonstrated. The most promising candidates are carbon nanotubes [42,46–49], graphene [42,50–52], Ag nanowire [16,53–55], or conducting polymers, for example, poly-(3,4-ethylenedioxythiophene)/poly-(styrenesulfonate) (PEDOT:PSS) [56–61]. However, the sheet resistance of high-conductive PEDOT:PSS is typically in the range of 100–200 Ω/□, which is even higher with respect to the sheet resistance of ITO on flexible substrates. This high sheet resistance strongly decreases the maximum size of solar cells with only PEDOT:PSS as a transparent electrode [62,63].

Aernouts et al. reported on the usage of PEDOT:PSS in combination with a metallic silver grid produced by a standard photographic technique [64]. The silver grid strongly increases the conductivity of the composite Ag-PEDOT:PSS electrode. Alternative approaches for grid deposition are lithographic methods [38,65,66], evaporation through a mask [67], or printing [36–38,62,63,68–71]. Recently, screen-printed silver grids, with a sheet resistance of 1 Ω/□, in combination with high-conducting PEDOT:PSS, have been reported as a good substitute for ITO electrodes in organic solar cells [63]. For further information about electrodes, the reader is referred to Chapter 5.

12.1.4.1.3 Stainless steel substrates

It is important to keep in mind that, despite their inherent conductivity, the electrical properties of metal substrates are not always used. Very often the conducting substrates are completely insulated from the OPV device by applying an additional insulating planarization layer. However, sometimes the metal substrate can serve as a back contact [21,72].

Holst Centre has demonstrated large-area ITO-free OPV devices on steel substrates (Fig. 12.5) [21]. The stainless steel was used as a substrate with excellent barrier properties and as a bottom electrode with very low sheet resistance. The low sheet resistance of the steel electrode allows manufacturing of large-area devices (5×10 cm^2) with high fill factor [21].

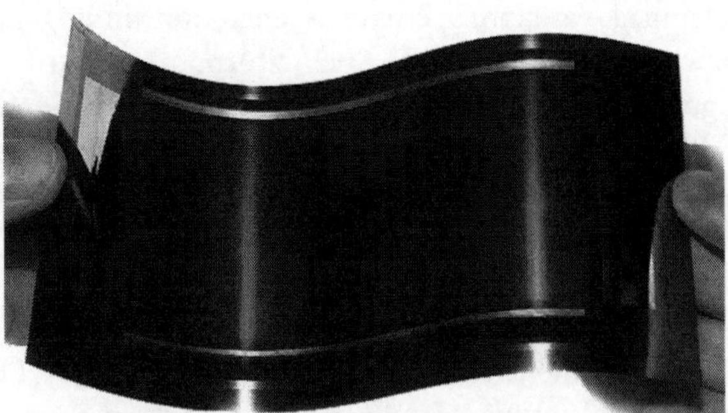

Figure 12.5 Large-area (5×10 cm^2) OPV devices on stainless steel substrate produced in Holst Centre. Reproduced from Ref. [21] with permission, Copyright © 2012, Elsevier B.V.

Relatively low roughness (R_a = 60 nm, R_q = 76 nm on the scanned area of 123 × 94 μm, Fig. 12.6) of polished stainless steel allows direct processing on this substrate. However, for industrial manufacturing and up-scaling of this process, several issues, such as electrical insulation of the cells and manufacturing of the modules, need to be addressed.

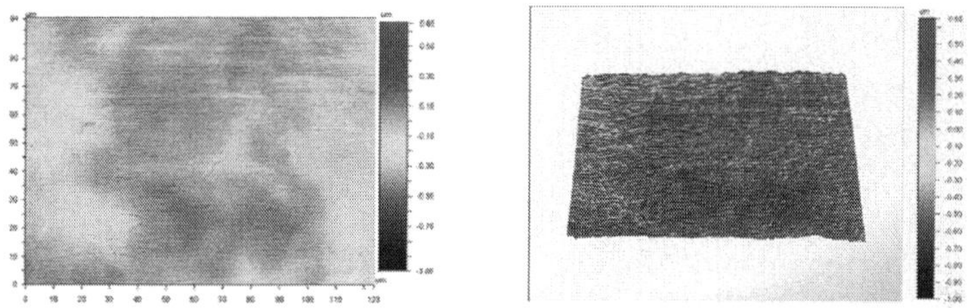

Figure 12.6 2D and 3D height view of the steel foil surface (type 1.4301 from Hasberg Schneider GmbH) obtained by white light interferometry at a magnification of 50×. The arithmetic mean roughness value (R_a) is 60 nm, the root-mean-square roughness (R_q) amounts to 76 nm on the scanned area of 123 × 94 μm. Reproduced from Ref. [21] with permission, Copyright © 2012, Elsevier B.V.

12.1.4.2 Surface quality

The surface quality of the substrates for OPV applications is determined by surface roughness (smoothness) and surface cleanliness [6]. Surface smoothness is largely defined by the internal cleanliness of the film substrate and typified by the presence of peaks with size from tens to hundreds of nanometers. Surface cleanliness is determined by the presence of dust and surface scratches. These defects can range in size up to tens of micrometers both laterally and vertically, and are unavoidable in a film that is handled and slit in a non-clean room environment. Some amount of dust particles can be removed by surface cleaning, but in order to eliminate all these types of defects, it is necessary to apply a planarizing coating which ideally should be done in a clean room environment. These coatings have the dual function of providing both a surface smoothness and also a certain surface hardness to prevent scratching of the surface during film processing.

12.1.4.3 Surface energy of the substrates

Surface energy of the substrate [73] is a critically important parameter for deposition of subsequent layers. The surface energy is a direct manifestation of intermolecular tension [74]. Molecules at the surface of the polymer are not surrounded by other molecules, so the molecular forces are unbalanced and the molecules have additional energy compared with the molecules inside the liquid or solid.

The surface or interfacial tension is expressed in $J \cdot m^{-2}$ ($N \cdot m^{-1}$) or in $erg \cdot cm^{-2}$ ($dyn \cdot cm^{-1}$). The surface energy of polymer substrates can be modified by flame treatment, corona discharge treatment, and ozone treatment, all of which impart some oxidation to the surface of the polymer and promote adhesion for printing of subsequent layers.

12.1.4.4 Dimensional stability

One of the main motivations for working on plastic substrates is that it opens up the possibility of R2R processing and the economic advantages that this process brings. Under these conditions, a winding tension will clearly be present and polymer film substrates with low moduli will be susceptible to internal deformation, particularly at elevated process temperatures [75–77].

However, in the region 120–160 °C, PEN is significantly stiffer and stronger with a modulus almost twice that of PET. Any small tension or load applied to the film in this temperature range will therefore impose strain in the material, which will be "frozen in" upon cooling and reappear as shrinkage upon reheating. In the implication of this, if the film is constrained through a thermal processing cycle, e.g., in an R2R process or by lamination to a rigid carrier, the strain introduced into the film may result in subsequent shrinkage on reheating. Figure 12.7 illustrates foil deformation during the cooling down cycle of free-laying PEN foil within the temperature range of 21–160 °C.

Zardetto et al. [10] have reported on the deformation of plastic substrates when they are heated. Figure 12.8a illustrates the radius of curvature (and its reciprocal) of the deformed substrate as a function of treatment temperature (30 min). Obviously, the smaller the radius, the larger the temperature-induced deformation undergone by the substrate. Results are presented for PET, HS-PET (heat-stabilized), HS-PEN, PET/ITO, and PEN/ITO. The higher temperature stability of HS-PET, in comparison to PET, is due to the heat-stabilization process that increases the dimensional stability of polymer films, because the internal strain in the film itself is relaxed by exposure to high temperature while under minimum line tension [78]. In addition, HS-PEN is more stable than HS-PET due to the presence of naphthalene rings instead of benzene rings, which increase the T_g of the material [7]. The behavior in temperature changes when one considers the same substrates coated with an ITO layer. First, the substrates show some (albeit not significant) curvature even at room temperature. This may be due to residual stresses contained in the foils arising from the deposition process of the ITO on the polymer [79]. The films start to bend more rapidly (i.e., at lower temperatures) than their neat PET and PEN counterparts. The presence of ITO leads to more pronounced deformation as the radii of curvature were about an order of magnitude greater than the neat polymer film counterparts. Above 150 °C, PET/ITO shows a more pronounced deformation compared to PEN/ITO, which shows a dramatic increase in curvature only above 180 °C. PET/ITO films actually tend to completely roll up at 220 °C (and to melt at 250 °C) as shown in Fig. 12.8b.

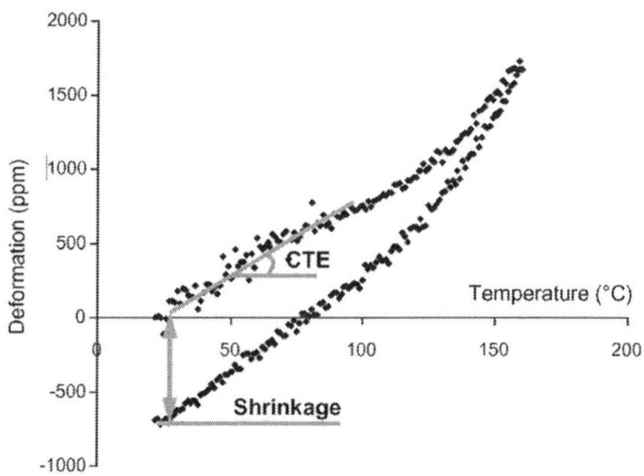

Figure 12.7 Heating up and cooling down cycle of free-laying PEN foil within the temperature range 21–160 °C. Reproduced from Ref. [76] with permission, Copyright © 2009, Elsevier B.V.

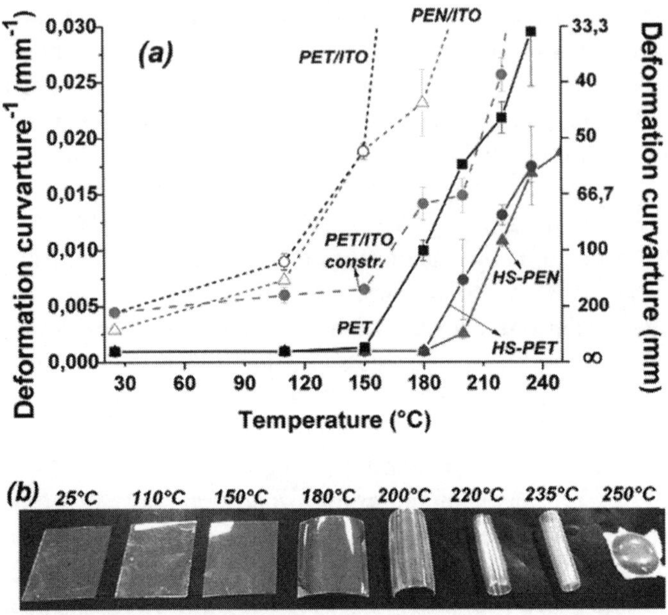

Figure 12.8 (a) Deformation of plastic substrates (PET, HS-PET, HS-PEN, PET/ITO, PEN/ITO) in terms of radius of curvature (and its reciprocal) as a function of treatment temperature (30 min). The "PET/ITO constr." Samples consist of PET/ITO mechanically constrained on two sides of the film onto rigid substrates. (b) Image of PET/ITO substrates after the thermal treatment. Reproduced from [10] with permission, Copyright © 2011, Wiley Periodicals, Inc.

12.1.4.5 Optical properties

Transparency of the substrate is a critical requirement for organic solar cells, where light has to penetrate to the device. Transmittance spectra for HS-PET and HS-PEN and their ITO-covered counterparts are reported in Fig. 12.9a in the 330–800 nm range [10]. Neat HS-PET films show high and constant transmittance over the visible range and around 90% (τ_{tot} = 90.5%), only decreasing appreciably in the UV (λ < 380 nm). HS-PEN films show a lower transmittance (τ_{tot} = 88.7%) especially in the blue region of the visible spectrum (transmittance drops markedly below 400 nm). The difference between the two materials is ascribed to the naphthalene rings in the PEN main chain, which start absorbing at longer wavelengths compared to the benzene rings in the PET. Note that non-heat-stabilized films are more transparent than the HS-PET.

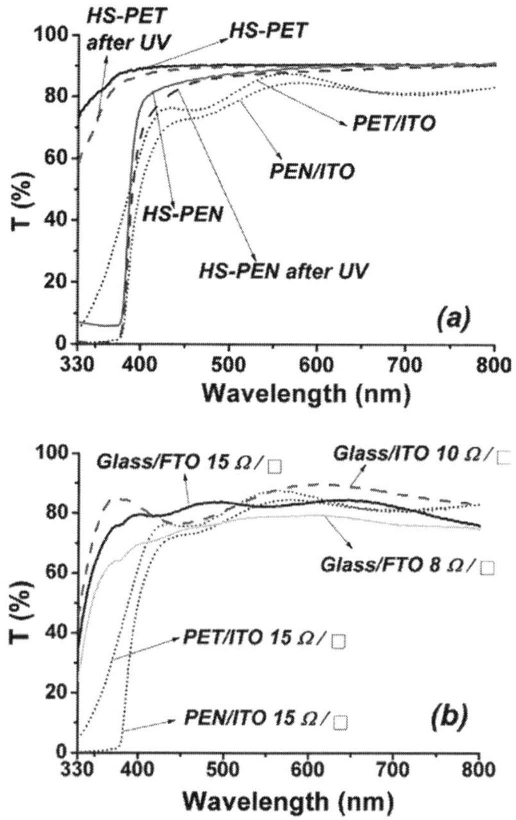

Figure 12.9 (a) UV–vis transmittance spectra for PET/ITO and PEN/ITO and also HS-PET and HS-PEN (before and after UV exposure). (b) UV–vis transmittance spectra for PET/ITO (15 Ω/□), PEN/ITO (15 Ω/□), glass/ITO (10 Ω/□), glass/FTO (8 Ω/□ and 15 Ω/□) conductive substrates. Reproduced from Ref. [10] with permission, Copyright © 2011, Wiley Periodicals, Inc.

The deposition of ITO on PET and PEN foils bring about an overall reduction of τ_{tot} (85.2% and 82.3%, respectively) particularly evident in the blue and especially in the UV part ($\lambda < 420$ nm) of the visible spectrum. The spectra show a maximum at 560 nm (87.3%) and 576 nm (84.6%) [10].

Figure 12.9b shows a comparison of the transmittance spectra of conductive plastic (PET/ITO, PEN/ITO), with those of conductive glass (glass/ITO—10 Ω/□ and glass/FTO—8 Ω/□ and 15 Ω/□). The glass substrate covered with ITO shows a total light transmittance of 86%, while glass covered with FTO 78.6% and 83% in the visible range, for the more and less conductive versions, respectively (note: the glass substrates covered with different conductive layers are also different in thickness). The spectra also show that glass substrates are more transparent in the UV range compared to the plastic films and also that FTO shows a flatter transmittance in the visible range.

12.1.4.6 Stability under UV exposure

Figure 12.9a shows a comparison between the transmission spectra of HS-PET, and HS-PEN before and after UV-light exposure (150 min). When HS-PEN is exposed to a significant flux of UV light, transparency decreases (τ_{tot} from 88.7% to 87.7%) especially in the blue range of wavelengths, 380 nm < λ < 500 nm. PET is more resistant to UV exposure (τ_{tot} from 90.6% to 90%). Indeed, Zardetto et al. [10] noticed a discernible yellowing of PEN after UV irradiation, not appreciable with PET (even if the transmission spectra in the violet range is slightly lower posttreatment even for PET). Yellowing is due to photo-oxidation, triggered by energetic UV photons leading to scission of bonds in the backbone of polymer, formation of new complexes that can lead to yellowing, loss of elasticity, and even cracks and material loss [80,81]. The difference is mainly due to the presence in PEN of the naphthalene structures (instead of the benzenes in PET), which are more easily affected by UV radiation [80]. Processes for UV stabilization of polymers and UV-stabilized films can be considered where increased UV stability is sought.

12.1.4.7 Solvent resistance

A wide range of solvents and chemicals can potentially be used when processing the various OPV layers. Amorphous polymers, in general, have poor solvent resistance compared to semicrystalline polymers (Table 12.3) [7]. This deficiency is overcome by the application of a hard coat to the amorphous resins, which significantly improves the solvent and chemical resistance to solvents such as IPA, acetone, methanol, THF, ethyl acetate, 98% sulfuric acid, glacial acetic acid, 30% hydrogen peroxide, and saturated bases such as sodium hydroxide. With polyethylene terephthalate and polyethylene naphthalate films, a hard coat is not required to provide solvent resistance.

Table 12.3 Table of solvent resistance

	Unit	PEN(Q65)	PET	PC	PES
Ketone	Acetone	Good	Good	(Fair)	(Unknown)
	MEK	Good	Good	(Fair)	(-)
Alcohol	Methanol	Good	Good	(Good)	(-)
	Ethanol	Good	Good	(Good)	(Good)
	Isopropanol	Good	Good	(Good)	(-)
	Butanol	Good	Good	(Good)	(-)
	Tetrachloroethane	Good	Good	(-)	(Good)
Acid	10% HCl	Good	Good	(-)	(Good)
	10% HNO_3	Good	Good	(-)	(Good)
	10% H_2SO_4	Good	Good	(-)	(Good)
	Acetic acid	Good	Good	(-)	(-)
Alkali	10% NaOH	Good	Fair	(-)	(Good)

Source: Ref. [7].

12.1.4.8 Impact of environmental conditions

With the knowledge of the solubility level of moisture in PEN films and its rate of diffusion as a function of temperature, it is possible to model the impact of various environmental conditions on moisture content changes and hence volumetric changes in the different thicknesses of film [6,7]. PET and PEN will typically pick up ~ 1000 ppm at an RH of 40% at 20 °C and it can take up to 12 h to reach equilibrium. Although PET and PEN will reach the same equilibrium level, at the initial stage PET picks up moisture about three times faster than PEN [6]. Preannealing of the substrate can greatly reduce shrinkage and careful process design can reduce the effect of swelling due to solvents and water.

12.2 Encapsulation and Barriers

The biggest advantage of organic photovoltaics over conventional silicon PVs is the potential of R2R manufacturing, which allows manufacturing of large-area and low-cost devices (see Chapter 13 for more information). However, many of the materials used in OPV devices are sensitive to the environment and limiting lifetime. These materials can be protected using substrates and barriers such as glass and metal, but it results in rigid devices. Typical plastic substrates offer little protection to oxygen and water, resulting in rapid degradation of the devices. In order to achieve long lifetime, good barrier against water and oxygen is required. Table 12.4 shows commercially available barrier technologies [82].

Table 12.4 Comparison of barrier types.

Barrier type	WVTR g/m^2/day	O_2TR cm^3/m^2/ day	Strengths	Weaknesses
Polymer	10^{+1}– 10^{+2}	10^{-1} – 10^{0}	- Excellent clarity - Flexible, may be creased without damage - Excellent toughness - High fatigue resistance	- High performance H_2O and O_2 barriers not usually found in a single polymer - Expensive - WVTR can change abruptly
Ceramic-coated polymer	10^{0} – 10^{-2}	3×10^{-1}	- Good clarity - Somewhat flexible because of thin deposition	- Brittle in tension - Fair fatigue resistance - Must not be creased
Multilayer ceramic-coated polymer	10^{-3}–10^{-6} (depends on layer count)	10^{-1}–10^{-4} (depends on layer count)	- Good clarity - Somewhat flexible because of thin deposition	- Brittle in tension - Lower fatigue resistance - Must not be creased or wrinkled - High cost for more than two barrier layers

(Continued)

Table 12.4 (*Continued*)

Metal foil < 25 μm thick	$< 10^{-1}$	$< 3 \times 10^{-1}$	- Economical - Somewhat flexible	- Opaque - Must not be creased or wrinkled - Pinholes may cause spot defects in non-reversible reaction products
Metal foil 25 μm thick	zero	zero	- Zero gas transmission at 25 μm thickness - Somewhat flexible	- Opaque - Must not be creased or wrinkled - Prone to wrinkles or tears unless supported by polymer laminate
Glass (for reference purposes)	zero	zero	- Zero gas transmission - Transparent - Scratch resistant	- Not (yet) processable roll-to-roll

Source: Ref. [82].

The most well-known barrier producers are Vitex Systems (Barix TM) [83], GE [84], 3M [85], Amcor [86], FujiFilm [87], Mitsubishi Plastic (X-barrier) [88], and Tera-Barrier [89]. Several novel encapsulation technologies, such as multilayer barrier with high performance, can serve important sectors of the PV market. These new encapsulation technologies have high importance for flexible PV sector (OPV in particular), especially for building integrated PV (BIPV) and other applications. Encapsulation is a key enabling technology for BIPV, which is expected to be the fastest growing sector of the PV industry over the next decade.

Barrier producers can create a larger market for their technologies by leveraging existing efforts for the PV sector. For example, encapsulation for organic light-emitting diodes (OLEDs) is already in the focus of commercialization efforts. Much of the work done for the OLED sector can be transferred to the OPV sector with relatively minor adjustments. For example, the optical and barrier performance requirements in OPV are significantly less severe than in the OLED industry, yet encapsulation in OPV requires different environmental stability characteristics and, likely, much longer lifetimes. Differentiation of encapsulation products to meet the specific needs of the OPV industry offers a way for firms to expand their market.

12.2.1 Requirements for the Barrier

It is not very well defined what the exact requirements for the barrier properties should be. It is generally assumed to be comparable to the gas permeation rates required for OLED, i.e., in the range of 10^{-6} g/m^2/day for moisture ingress [90,91]. The barrier specification for OLED can be partly explained by the presence of highly reactive low work function metals as a cathode. Additionally, the specification of a water vapor transmission rate (WVTR) of 10^{-6} g/m^2/day is often described as the amount of water that is capable to oxidize a 10-nm-thick layer of metal at the cathode–organic interface during the specified lifetime of the devices (10 years). However, this suggests homogeneous degradation of the cathode, where the actual penetration of water is limited to the area of the pinholes. Outside these pinholes the combination of the thin-film barrier and metal cathode prevents any water penetration into device.

Some studies point out that OPV barrier requirements might be less demanding to achieve manufacturing of OPV with significant lifetimes, i.e., in the order of several years [92–94]. Cros et al. [93], Hauch et al. [92,94], and Krebs [95] have demonstrated that OPV cells can be protected with medium barrier material with a WVTR of around 10^{-3} g/m^2/day, in case of OPV cells free from low work function metals.

Hermenau et al. have investigated the effect of barrier performance on lifetime of small-molecule organic solar cells [96]. It has been shown that small-molecule organic solar cells can be very sensitive to oxygen and water [97], which creates a need for an ultra-high barrier. In the case of doped small-molecule materials, it is possible to increase device lifetimes by moving the electron transport layers, which are more sensitive to water and oxygen, away from the permeable top contact. Another option is a reduction of the WVTR of the top electrode, especially by avoiding or closing pinholes.

The top electrode in conventional OPV devices most often consists of a thin layer of Ca or LiF (1–5 nm) covered with a relatively thick Al layer (100 nm). Aluminum is an excellent barrier against water; however, pinholes in Al are the main reason for degradation of OPV devices [96] (for additional information, see Chapter 15). Thus, for conventional device architecture, the degradation mechanism is similar to degradation of OLED. In this case, the barrier specification developed for OLEDs can also be applied to OPV, where the WVTR should be 10^{-6} g/m^2/day. Inverted structures typically utilize a top silver electrode, which is less reactive than Ca/Al or LiF/Al. That is why, it is assumed that inverted structure is more stable than conventional one [98–99]. However, silver electrode used in inverted architecture is also less dense than aluminum. It provides a large number of pinholes, which allow penetration of water and oxygen inside the device stack. Increasing of the electrode thickness decreases the amount of pinholes and improves stability of the devices [99]. Decoupling of pinholes in metal layers is also possible by deposition of a barrier layer.

Depending on the solar cell materials, solar cell architecture, manufacturing method, and method of applying the barrier, the requirements for WVTR for OPV barriers can be in the range of 10^{-3}–10^{-6} g/m^2/day. These transmission rates are several orders of magnitude smaller than is possible using any plastic substrates. For this reason, development of transparent barrier materials with low water transmission rate is required. The barrier materials should have the same driving aspects as organic devices like transparency, flexibility, and low-cost manufacturing. The requirements to WVTR also depend on the application of solar cells and desired lifetime of the devices.

12.2.2 Single-Layer Barrier

Various single-layer materials have been investigated as barriers. The high-density barrier films include silicon nitride, aluminum oxide, titanium oxide, and silicon oxide mainly deposited via evaporation, sputtering, or plasma vapor deposition [100–107]. A single-barrier layer is able to cover 90–99% of the pinholes (Fig. 12.10). Some pinholes, typically caused by particles, are not covered, and these provide an opening for water and oxygen ingress. Small holes in a barrier layer lead to a much higher permeation than a few large holes of the same total area [108]. It was moreover established that the barrier performance on polymer substrates depends on coating thickness [20,107], substrate roughness [109,110], presence of particles, and coating density [111]. Particularly, rougher surfaces require thicker coatings for maintaining barrier properties, which tend to promote cracking [112,113]. Therefore, it is essential to have a detailed understanding of the initial surface morphology of the polymer substrate, of its evolution during the deposition process, and the resulting final structure of the coating.

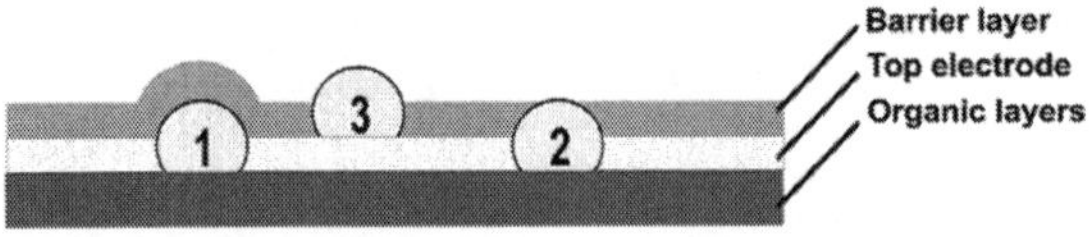

Figure 12.10 Schematic illustration of thin-film encapsulation (single barrier layer). 1 and 2 are the pinholes (caused by particles) in the top electrode (1 – covered by single barrier layer and 2 – not covered), 3 – pinhole in a barrier layer.

Extensive studies [114] of single-barrier layers show typical thin-film defect diameter of ≈1000 nm (1 μm), with spacing on the order of 0.1 mm (100 μm) or less for measured thin-film barrier structure. The data for different barrier materials are summarized in Table 12.5.

Table 12.5 Defect sizes and defect densities of single-layer inorganic barrier film deposited on polymeric substrates

Defect radius (μm)	Defect density (mm^{-2})	Coating material	Deposition method	Substrate	Reference
0.6	11–1100	SiO_2	PECVD	PET	[115]
0.6	5–1000	Si_3N_4	PECVD	PET	[115]
1.0	25–400	Al	Evaporated	PET	[116]
1.0	100–300	Al	Evaporated	PET	[117]
2–3	200	Al	Sputtering	PET	[118]
0.5–1.4	600	AlO_xN_y	Sputtering	PET	[100]
0.4	10–1000	Al	Evaporated	BOPP	[119]
0.5	700	AlO_xN_y	Sputtering	PET	[101]

Reproduced from Ref. [114] with permission, Copyright © 2004, American Institute of Physics.

Generally speaking, two kinds of defects are found in thin oxide films, namely microdefects and nanodefects [108]. On the one hand, microdefects are comprised of so-called pinholes and microcracks [117], with sizes of the order of 1 mm. Microdefects in the form of uncoated areas were detected in the case of PE substrates, which were related to the high roughness of the initial polymer surface [113]. A detailed characterization of microdefects was performed by Wertheimer and coworkers, using reactive ion etching in oxygen plasma followed by microscopy analyses [115,120,121].

The degradation mechanism, caused by water and oxygen diffusion through the pinholes, is easily observed in OLEDs by forming so-called "black spots." Formation of black spots in OLEDs makes these devices unsuitable for further application, even if the black spot area is not very large. In organic solar cells, black spots are not observable by the naked eye, but rather require applying a specialized imaging technique. The presence of black spots in OPV devices will decrease the active area, which will result in a decrease of device performance. However, for example, if black spots cover 5–10% of the active area, such OPV devices are still suitable for energy generation, whereas 5–10% of black spot area in OLEDs is not acceptable. This difference in OLED and OPV explains the potentially differing requirements in barrier properties for these two types of devices.

The imaging techniques that have been applied to visualizing black spots in OPV devices are laser-beam induced current (LBIC) scanning [71,122] and luminescence imaging [122,123], specifically photoluminescence (PLI) and electroluminescence (ELI). ELI imaging directly probes the reverse functionality of solar cells by passing a forward current through the device, driving the OPV like an OLED. As OPV devices operate as very poor emitters, highly sensitive and linearly responding detector arrays are required to detect quantitative signals. Degradation of non-encapsulated cells, monitored by ELI characterization caused by water and oxygen ingress through the pinholes in the aluminum electrode, is shown in Fig. 12.11 [122].

12.2.3 Multilayer Barrier Coatings

In order to avoid the continuous and propagating growth of pinholes during deposition of the barrier layer, a widely used approach is stacking [104,114,124–127]. This implies that the growth of the barrier layer [64] is interrupted by growing an intermediate layer. Each individual barrier layer has its own pinholes, but the pinholes are decoupled, i.e., they do not coincide in the lateral plane. In this way the penetration of water through the thin-film encapsulation is obstructed. There are two different ways of stacking: inorganic and organic stacking, the adjective is referring to the nature of the intermediate layer.

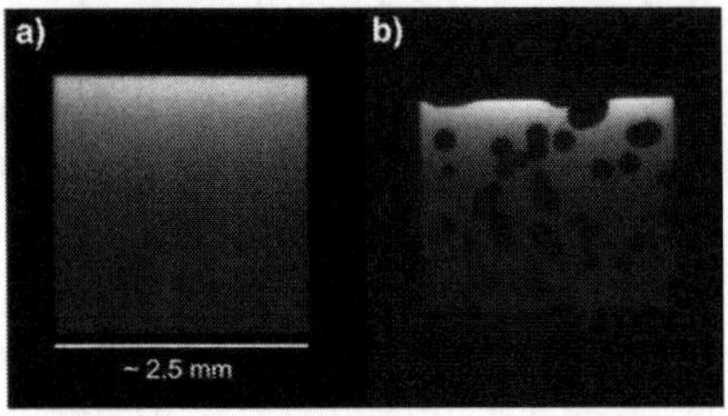

Figure 12.11 ELI characterization of the solar cells without encapsulation: initially at T100 (a) and after 42 hours under full sun simulation at T72 (b). All images were recorded under 1.5 V forward bias. Reproduced from Ref. [122] by permission of The Royal Society of Chemistry, Copyright © 2012.

A commonly used inorganic interlayer material is SiO_2 [125]. The thin SiO_2 is intended to cover the particles that remain uncovered by the first barrier inorganic layer. In this way particles that are not successfully covered by the first inorganic layer might be covered by the next one (Fig. 12.12). When the number of alternating layers is increased, the time of water/oxygen diffusion through pinholes might be sufficient to delay the formation of black spots outside the time frame of the specified lifetime of the device. Though good adhesion is the primary functionality of the intermediate layer, the water barrier ability of the intermediate layer is also helpful in the extension of the time delay of black spot formation. Thus, a stack of $(SiN–SiO_2)_n$–SiN, where $n > 5$, can provide excellent barrier properties with WVTR of 10^{-6} g/m^2/day [125].

Typical organic interlayer materials are polyacrylates. Vitex System [83] reported flexible encapsulation layers (trade name of Barix®) made of alternating Al_2O_3 and polyacrylate layers as shown in Fig. 12.13 [102,128,129]. The Al_2O_3 layers are reactively sputtered via an energetic plasma between polyacrylate layers, which are deposited via flash evaporation of the monomer followed by ultraviolet (UV) curing [129]. Multilayer barrier-coated polymers from Vitex have demonstrated a WVTR estimated to be equivalent to 2×10^{-6} g/m^2/day at ambient condition using the Ca test [130].

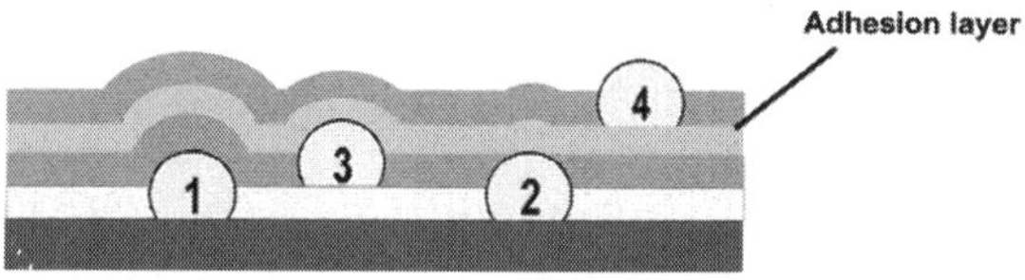

Figure 12.12 Schematic illustration of multilayer barrier coatings. 1 and 2 are the pinholes (caused by particles) in the top electrode (1 – covered by first barrier layer and 2 – not covered), 3 is a pinhole in a first barrier layer, 2 and 3 are the pinholes which are covered by second barrier layer, 4 is a pinhole in a top barrier layer.

Another common choice is silicon nitride (variably referred to as a-SiN, SiN_x, or SiN:H), applied by plasma-enhanced chemical vapor deposition (PECVD), as a water barrier [125,131]. The organic interlayer is intended to planarize all particles in or on top of the SiN-covered top electrode. In this way any pinholes in the subsequent SiN layer will be decoupled from pinholes in the first SiN layer (Fig. 12.14). In the case of organic stacking, any pinholes in the second SiN layer will be harmful, as penetration of water into the intermediate layer, followed by lateral water diffusion, is considered to be relevant for an organic layer. This means that the number of black spots is not reduced by organic stacking. All particles of type 2 (Fig. 12.14) will result in a black spot, but the time of black spots formation, determined by diffusion path, will be significantly delayed with respect to a single-layer barrier.

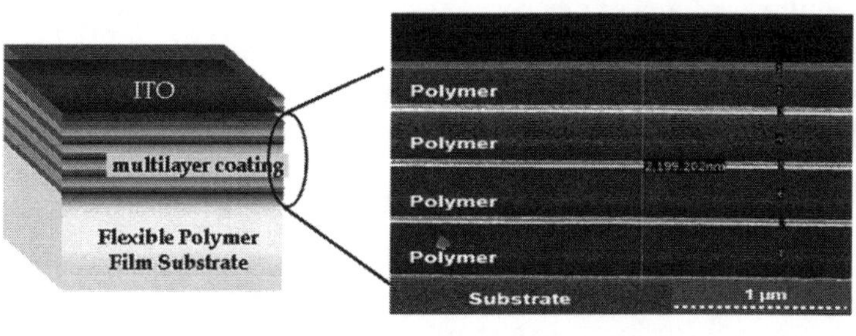

Figure 12.13 A cross-sectional scanning electron microscopy (SEM) of the multilayer barrier layer (inorganic/organic layers): four pairs of AlO_x/polymer multilayer deposited by the Vitex barrier deposition system. Reproduced from Ref. [102] with permission, Copyright © 2011, IOP Publishing Ltd.

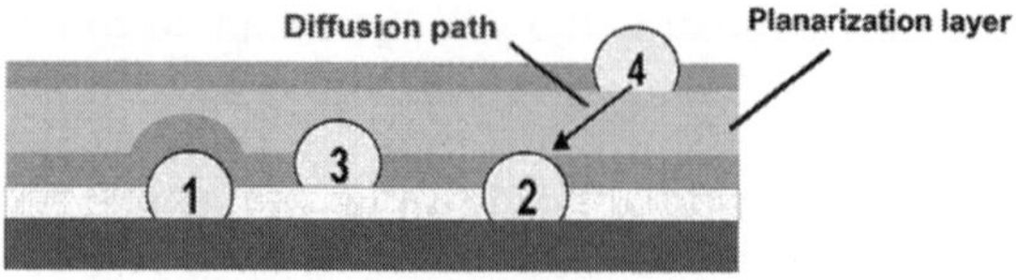

Figure 12.14 Schematic illustration of multilayer barrier coatings with organic planarization layer. 1 and 2 are the pinholes (caused by particles) in the top electrode (1 – covered by first barrier layer and 2 – not covered), 3 is a pinhole in a first barrier layer. 2 and 3 are the pinholes in a first barrier layer which are covered by second barrier layer. 4 is a pinhole in a top barrier layer. Diffusion path is the distance between the pinholes in first and second barrier layers.

In conclusion, the critical process parameters which can improve barrier performance are surface roughness, particle density, and the chosen thin-film deposition method [114,132]. Understanding and controlling these parameters can help to produce a high-performing barrier, suitable for OPV applications.

12.2.4 Mechanical Properties of Organic/Inorganic Barrier Films

Many demonstrations of electronic devices on flexible substrates involve the fabrication of traditional devices substituting the glass substrate for plastic. Often, smoothing layers are employed and, if necessary, the processing temperatures are reduced. However, there has been relatively little attention paid to the limits of flexibility of these devices. Mechanical failure in a device is determined by the weakest component. For those devices that use inorganic thin films, these brittle layers are likely to be the source of failure [130]. Barrier layers, which typically contain inorganic components, can be the weakest part in the devices under bending conditions [36].

When a sheet is bent, the outer surface experiences tensile stress and the inner surface experiences compressive stress. At the same time a plane inside the sheet (in the middle if the sheet is homogeneous), defined as the neutral plane, experiences no uniaxial stress at all. Films, deposited on one or both sides, shift the stress and strain distribution. However, for many flexible electronics devices, where the films are very thin relative to the substrate, the following simple approximation for the relationship between film strain and radius of curvature can be made:

$$\varepsilon f = d/2r,$$

where d is the thickness of the substrate and r is the radius of curvature. To make a rough estimate of the radius of curvature at failure, we can see that for a 100-μm substrate and for a typical failure strain of 0.5–1.0% for brittle films in tension, the minimum radius of curvature is 5–10 mm.

In Fig. 12.15 [131] the ratio of the measured WVTR of several samples before and after cyclic bending is shown as a function of the applied bending radius, as well as for varying amounts of bending cycles. The trend in Fig. 12.15 indicates that the WVTR is not affected for bending radii down to ~20 mm. There is no clear dependence on the number of bending cycles during the test. It indicates that if damage results from the bending, this already happens during the initial bending cycles, while no further damage results from further bending.

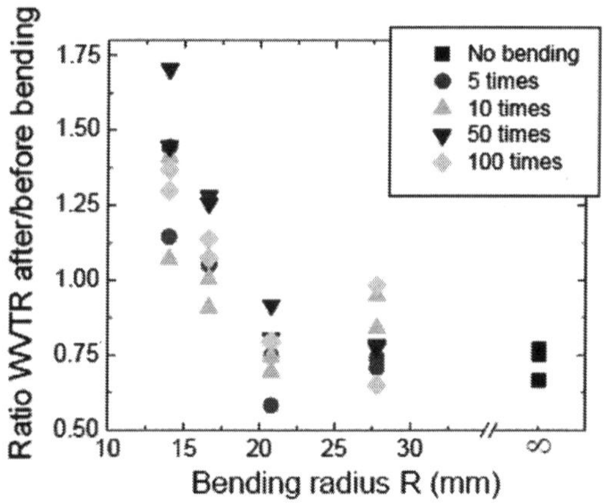

Figure 12.15 Increase in WVTR (ratio after/before bending) as a function of the bending radius (adapted from Ref. [131]).

Failure modes of layered materials involve the growth of microcracks under stress [130,133] and the details of the fracture mode depend on the substrate modulus, film adhesion, and film cohesion among other things. The most common failure modes for brittle films on flexible substrates are film cracking and channeling [134], as depicted in Fig. 12.16. The first mechanism is more common for films in tension with good adhesion between layers, while the second mechanism is more common for films in compression or when adhesion is poor.

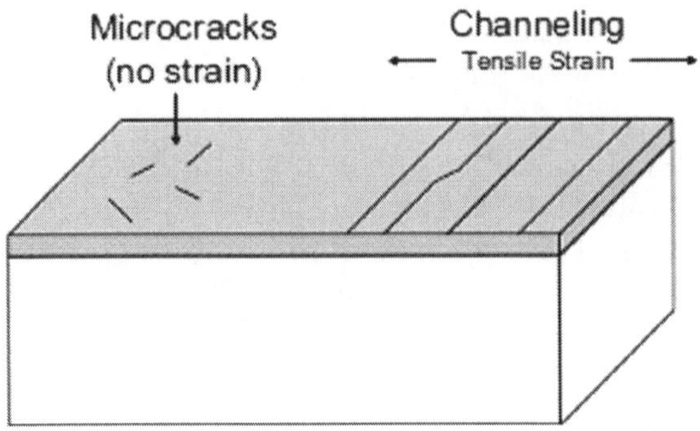

Figure 12.16 Primary failure model for brittle films on polymer substrates. Reproduced from Ref. [130] with permission, Copyright © 2006, Elsevier Ltd.

Early failure in brittle films is caused by preexisting defects. For oxide films on polymer substrates, it often happens because of defects in the substrate morphology. Levels of degradation of the multilayered barrier system can be lowered by improving the processing conditions and/or the material choice. In order to minimize residual strain and to enhance the level of adhesion between layers of the stack, a thin organic interlayer is the best choice.

12.2.5 Solution-Processable Barriers

As has been discussed above, multilayer barrier coatings can demonstrate excellent barrier properties. However, to fit the criteria of organic electronics, organic solar cells in particular, the barrier layer should demonstrate low manufacturing cost. That is why R2R solution processing of barrier layers is intensively investigated. Organic/inorganic hybrid nanocomposites, composed of organic polymer matrices (e.g., acrylics, epoxy, and silicone) and inorganic nano-fillers (e.g., silica (SiO_2), titanium oxide (TiO_2), zinc oxide (ZnO), and alumina (Al_2O_3) have recently attracted attention as a solution-processable barrier for organic optoelectronic devices

[135,136]. With a new kind of inorganic–organic polymer coating, it is possible to obtain high barrier properties with respect to the permeation rates of oxygen and water vapor [137–141]. New paths have been explored for the use of coatings combining different barrier functions. By systematically varying the inorganic and the organic network densities and the polarity of the hybrid polymer material matrices, it was possible to develop coating materials with barrier properties for water vapor and oxygen [142].

The hybrid polymers are accessible via the sol–gel technique [143]. The inorganic network is formed as a result of controlled hydrolysis and condensation of alkoxysilanes, organoalkoxysilanes, acryloxysilanes, or metal alcoholates. The organic network results from subsequent thermal or UV-induced polymerization of organofunctional groups. Due to the control of the inorganic and organic network densities and the insertion of specific functional groups to control the polarity of the resulting material, it was possible to develop high barrier coatings with excellent adhesion properties on a wide variety of polymer films. These high barrier coatings are also suitable as adhesives which can be used in laminates. The properties of the processable multilayer structures are preserved to a much higher extent than with other comparable, commercially available materials even under high mechanical and thermal stress and storage in humidity.

12.2.6 Methods of Applying Barriers

Depending on the organic solar cell architecture and substrate used for manufacturing, different encapsulation approaches can be selected for applying a barrier. In general, OPV devices can be produced on substrates with barrier properties, such as glass, metal foil, or plastic substrates with precoated barrier layers [63,144], or on substrates with weak barrier properties. In the second case applying two barrier layers, one on top and the other on the bottom of the device, is required. Encapsulation of the devices can go by different routes (Fig. 12.17).

One of the most developed and often used methods is encapsulation between two glass plates. The usage of glass/glass encapsulation can be done using a glass plate with a cavity, kind of glass cap, or with the usage of a plane glass. In the case of glass cap, which also can be substituted by a metal cap, the sealing material, usually epoxy glue, is applied around the perimeter of the cap (Fig. 12.17a). The cavity in the cap provides the additional possibility to use an extra oxygen or water getter inside of the encapsulation. Both glass and metal are excellent barriers against water and oxygen. The weakest point of such encapsulation is epoxy sealing, which can create paths for water/oxygen ingress. Minimizing diffusion of gases, it is possible to use an epoxy glue with low permeation rate and good adhesion to the substrate and cap materials, and by removing organic layers on the contact areas of the cap and substrate.

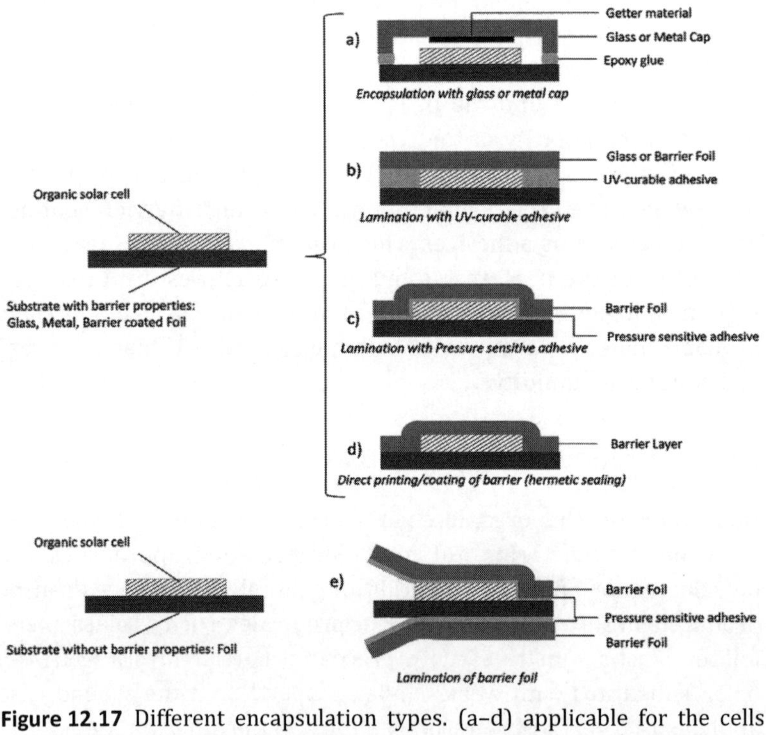

Figure 12.17 Different encapsulation types. (a–d) applicable for the cells on barrier substrates (e.g., glass of barrier foil), (e) applicable for OPV cells on plastic substrates without barrier properties. (a) encapsulation with glass or metal cap; (b) lamination with UV-curable adhesive; (c) lamination with pressure-sensitive adhesive; (d) direct printing/coating of barrier (thin-film encapsulation); (e) lamination of two barrier layers with pressure-sensitive adhesive.

It is also possible to encapsulate solar cells between two plane glass plates. In this case, adhesive or bonding material is usually deposited over the full area of the devices (Fig. 12.17b). The adhesive material, which is usually UV-curable epoxy resin, can be dispensed or printed on top of devices, covered with a second glass, and then UV cured. The main requirements for adhesive materials are good adhesion to the glass (or barrier foils) and compatibility with the device stack, especially the top electrode.

Lamination of flexible barriers with a pressure-sensitive adhesive material can be done for both glass and flexible substrates (Fig. 12.17c). Moreover, when flexible substrates with low barrier properties are used, the barrier should be added from both sides of the devices (Fig. 12.17 e). Lamination of the barrier is usually done by R2R processing, which always associates with bending, and high-bending radii are known to cause a decrease in barrier properties [36,131] (see Sec. 12.2.4). In addition to mechanical properties of the laminated barrier, the weak point of such encapsulation is direct contact of OPV components with adhesive material, and therefore the adhesive needs to be made from non-reactive components. It also should provide good mechanical properties to avoid delamination, contain very low (or not at all) amount of water and oxygen, and be as thin as possible while avoiding water and oxygen migration through this layer.

The most reliable encapsulation method is direct printing or coating of a barrier layer on top of the device stack. This encapsulation method is often called "thin-film encapsulation" (TFE) (Fig. 12.17d). The advantages of the TFE method are a thinner/lighter form factor and a higher flexibility of devices. With this encapsulation method a top barrier can mate with the bottom-barrier layer at the edge of devices, creating a hermetic seal which prevents side leakage into the device. However, there are still several issues to adopt the TFE methods, i.e., the process must be performed at relatively low temperatures and be compatible with OPV stacks and substrates.

12.2.7 Side Leakage

The one disadvantage that lamination or gluing of a barrier presents over other encapsulation strategies is side leakage. The barrier performance for such encapsulation types is partly limited by epoxy glue or adhesive attached to the barrier. The gases diffusing through this epoxy layer can reach the encapsulated device without even passing through the high-performance barrier material. In many potential applications for plastic electronic devices, inexpensive epoxies are not optimized for barrier performance and thus may have substantially high permeation of the atmospheric gases. As a result, diffusion through the epoxy may be large enough to result in significant flux of gases through the edge. For high-performance barrier materials this may constitute the limiting factor for encapsulation [145]. Therefore, it is desirable to use an encapsulation architecture that minimizes transport through epoxy layers and thereby enables the use of low-cost, relatively high-permeability epoxy, both for potential industrial applications and for laboratory prototypes.

The degradation of the solar cells which starts from the edges can be easily visualized using, for example, PLI imaging [122]. Figure 12.18 [122] displays PLI image of three solar cells produced on the same substrate and glass-to-glass sealed. The image demonstrates clear differences between intact active layer regions and those that suffer from oxygen and water ingress through the edges of the sealing.

Figure 12.18 Photoluminescence imaging of ISE-Fraunhofer-processed [146], glass-to-glass-sealed, polymer solar cell. The active layer degradation upon oxygen and water vapor ingress from the edges of the sealing is clearly visible. Reproduced from Ref. [122] by permission of The Royal Society of Chemistry, Copyright © 2012.

The ingress of the side leakage depends on the quality of the seal, the length of the path from the edge to the active parts of the device, thickness of the adhesive, and its properties. There are many methods that have been proposed to slow down or completely prevent side leakage. Thus, in Ref. [147] the side leakage was decreased by another lamination of the same barrier and adhesive around the cut edges of the OPV devices. Granstrom et al. [145] have proposed to use a "Russian Doll" encapsulation architecture utilizing pairs of free-standing barrier films and epoxy seals separated by nitrogen spacers, enabling the use of low-cost epoxy to attach two or more free-standing barrier films to a substrate with improved barrier performance. In addition to these, the seal of the edges can be done by printing additional adhesives, or joining the barriers by laser-melting or hot-melting processes [148].

12.2.8 Conclusions

Developments in the field of organic electronic devices demonstrate a need for high-quality flexible substrates and highly effective barriers. As the requirements for the performance of the barrier increase, the difficulties in the production of barrier materials also increase. There are three main research directions in the field of transparent barriers. The first is focused on the deposition of a single layer of inorganic materials such as silicon oxide, silicon nitride, silicon oxynitride, aluminum oxide, or their mixtures. The second direction is a multilayer structure which includes several barrier layers with organic or inorganic interlayers. Such types of barriers are currently the benchmark for performance and provide the best protection for OPV devices. The third direction is a coating of organic/inorganic nanocomposites. This is a very promising approach as it allows R2R printing or coating of the barrier film, which can guarantee low manufacturing cost. The main targets for barrier producers are high performance of the barrier (low water vapor and oxygen transmission rates) and low manufacturing cost.

The barrier, of course is not a stand-alone item. It must be integrated into the final device. That device may have other needs [82] such as anti-glare, UV resistance, scratch resistance (hard coat). It is possible to add such capabilities one by one. Doing so adds additional layers into the structure. Each additional layer requires another processing step and adds one or two material interfaces. Each interface affects the total light transmission as well as clarity and haze except in the unlikely event that the layers have matching refractive indices or a controlled sequence of refractive indices to maximize light transmission. It would be better to combine as many of the requirements as possible into a simpler overall structure. This would drive down costs and, by minimizing material interfaces and total thickness, should improve light transmission properties. Achieving these goals will enable long lifetime and ideally low-cost organic solar cells.

References

1. Lawrence DKJ, Brollier B, Claypole T, Burgin T (2004) Manufacturing platforms for printing organic circuit, In *Printed organic and molecular electronics* (Daniel R, Gamota PB, Krishna Kalyanasundaram, Jie Zhang, eds.), Kluwer Academic Publishers, pp. 161–346.
2. Chuang TK, Troccoli M, Kuo P-C, et al. (2007) Top-emitting 230 dots/in, active-matrix polymer light-emitting diode displays on flexible metal foil substrates, *Appl Phys Lett,* **90**, 151114.
3. Hu X, Song Z, Liu W, Qin F, Zhang Z, Wang H (2012) Chemical mechanical polishing of stainless steel foil as flexible substrate, *Appl Surf Sci,* **258**, 5798–5802.
4. Jeong JK, Jin DU, Shin HS, et al. (2007) Flexible full-color AMOLED on ultrathin metal foil, *IEEE Electron Dev Lett,* **28**, 389–391.
5. Xie Z, Hung LS, Zhu F (2003) A flexible top-emitting organic light-emitting diode on steel foil, *Chem Phys Lett,* **381**, 691–696.
6. MacDonald WA, Looney MK, MacKerron D, Eveson R, Rakos K (2008) Designing and manufacturing substrates for flexible electronics, *Plast Rub Compos,* **37**, 41–45.
7. MacDonald WA, Looney MK, Mackerron D, et al. (2007) Latest advances in substrates for flexible electronics, *J Soc Inform Disp,* **15**, 1075–1083.
8. MacDonald WA (2002) New advances in poly(ethylene terephthalate) polymerization and degradation, *Polym Int,* **51**, 923–930.
9. MacDonald WA (2004) Engineered films for display technologies, *J Mater Chem,* **14**, 4–10.
10. Zardetto V, Brown TM, Reale A, Di Carlo A (2011) Substrates for flexible electronics: A practical investigation on the electrical, film flexibility, optical, temperature, and solvent resistance properties, *J Polym Sci Part B: Polym Phys,* **49**, 638–648.

11. Choi MC, Kim Y, Ha CS (2008) Polymers for flexible displays: From material selection to device applications, *Prog Polym Sci (Oxford)*, **33**, 581–630.

12. Fonrodona M, Escarré J, Villar F, et al. (2005) PEN as substrate for new solar cell technologies, *Sol Energy Mater Sol Cells*, **89**, 37–47.

13. Vergnat C, Landais V, Combet J, et al. (2012) Comparing the growth of a molecular semiconductor on amorphous and semi-crystalline polycarbonate substrates, *Org Electron*, **13**, 1594–1601.

14. Yan M, Kim TW, Erlat AG, et al. (2005) A transparent, high barrier, and high heat substrate for organic electronics, *Proc IEEE*, **93**, 1468–1477.

15. Woo K, Kim D, Kim JS, Lim S, Moon J (2009) Ink-jet printing of Cu-Ag-based highly conductive tracks on a transparent substrate, *Langmuir*, **25**, 429–433.

16. Lim JW, Cho DY, Eun K, et al. (2012) Mechanical integrity of flexible Ag nanowire network electrodes coated on colorless PI substrates for flexible organic solar cells, *Sol Energy Mater Sol Cells*, **105**, 69–76.

17. Sun JY, Lu N, Yoon J, Oh KH, Suo Z, Vlassak JJ (2009) Inorganic islands on a highly stretchable polyimide substrate, *J Mater Res*, **24**, 3338–3342.

18. Lu YH, Hu ZZ, Bian JM, Wang YF (2011) Transparent polyimides and their ITO flexible conductive film, **239–242**, 1211–1214.

19. Andoh H, Teramoto T (1990) Polyarylates containing fluorene units, preparation and properties, *Am Chem Soc*, **31**(2), 677–678.

20. Fahlteich J, Fahland M, Schönberger W, Schiller N (2009) Permeation barrier properties of thin oxide films on flexible polymer substrates, *Thin Solid Films*, **517**, 3075–3080.

21. Galagan Y, Moet DJD, Hermes DC, Blom PWM, Andriessen R (2012) Large area ITO-free organic solar cells on steel substrate, *Org Electron*, **13**, 3310–3314.

22. Alzoubi K, Hamasha MM, Lu S, Sammakia B (2011) Bending fatigue study of sputtered ITO on flexible substrate, *IEEE/OSA J Disp Tech*, **7**, 593–600.

23. Chen HC, Huang KT, Lo YM (2012) Measurement of residual stress for ITO/PET substrates by the double beam shadow Moiré interferometer, *Appl Opt*, **51**, 1566–1571.

24. Hamasha MM, Dhakal T, Alzoubi K, et al. (2012) Stability of ITO thin film on flexible substrate under thermal aging and thermal cycling conditions, *IEEE/OSA J Disp Tech,* **8**, 383–388.

25. Socol G, Socol M, Stefan N, et al. (2012) Pulsed laser deposition of transparent conductive oxide thin films on flexible substrates, *Appl Surf Sci*, **260**, 42–46.

26. Al-Ibrahim M, Ambacher O, Sensfuss S, Gobsch G (2005) Effects of solvent and annealing on the improved performance of solar cells based on poly(3-hexylthiophene):fullerene, *Appl Phys Lett,* **86**, 201120.

27. Lungenschmied C, Dennler G, Neugebauer H, et al. (2007) Flexible, long-lived, large-area, organic solar cells, *Sol Energy Mater Sol Cells,* **91**, 379–384.

28. Yano H, Kouro D, Sasaki N, Muramatsu S (2009) Improvement of polymer/fullerene solar cells by controlling geometry of the ITO substrate surface, *Sol Energy Mater Sol Cells,* **93**, 976–979.

29. Kim D-H, Park M-R, Lee H-J, Lee G-H (2006) Thickness dependence of electrical properties of ITO film deposited on a plastic substrate by RF magnetron sputtering, *Appl Surf Sci,* **253**, 409–411.

30. Korösi L, Papp S, Dékány I (2011) Preparation of transparent conductive indium tin oxide thin films from nanocrystalline indium tin hydroxide by dip-coating method, *Thin Solid Films,* **519**, 3113–3118.

31. Li TL, Hsu SLC (2010) Preparation and properties of transparent conductive indium-tin-oxide/polyimide substrates via a sol-gel process, *Thin Solid Films,* **518**, 6761–6766.

32. Li ZH, Ren DY (2006) Preparation of ITO transparent conductive film by sol-gel method, *Trans Nonferr Met Soc China (Engl Ed),* **16**, 1358–1361.

33. Galagan Y, de Vries IG, Langen AP, et al. (2011) Technology development for roll-to-roll production of organic photovoltaics, *Chem Eng Proc: Process Intens,* **50**, 454–461.

34. Paetzold R, Heuser K, Henseler D, Roeger S, Wittmann G, Winnacker A (2003) Performance of flexible polymeric light-emitting diodes under bending conditions, *Appl Phys Lett,* **82**, 3342–3344.

35. Krebs FC, Tromholt T, Jørgensen M (2010) Upscaling of polymer solar cell fabrication using full roll-to-roll processing, *Nanoscale,* **2**, 873–886.

36. Krebs FC, Fyenbo J, Tanenbaum DM, et al. (2011) The OE-A OPV demonstrator anno domini 2011, *Energy Environ Sci,* **4**, 4116–4123.

37. Krebs FC, Fyenbo J, Jørgensen M (2010) Product integration of compact roll-to-roll processed polymer solar cell modules: Methods and manufacture using flexographic printing, slot-die coating and rotary screen printing, *J Mater Chem,* **20**, 8994–9001.

38. Galagan Y, Zimmermann B, Coenen EWC, et al. (2012) Current collecting grids for ITO-free solar cells, *Adv Energy Mater,* **2**, 103–110.

39. Medford AJ, Lilliedal MR, Jørgensen M, et al. (2010) Grid-connected polymer solar panels: Initial considerations of cost, lifetime, and practicality, *Opt Express,* **18**, A272–A285.

40. Krebs FC (2009) All solution roll-to-roll processed polymer solar cells free from indium-tin-oxide and vacuum coating steps, *Org Electron,* **10**, 761–768.

41. Krebs FC (2009) Roll-to-roll fabrication of monolithic large-area polymer solar cells free from indium-tin-oxide, *Sol Energy Mater Sol Cells,* **93**, 1636–1641.

42. Po R, Carbonera C, Bernardi A, Tinti F, Camaioni N (2012) Polymer- and carbon-based electrodes for polymer solar cells: Toward low-cost, continuous fabrication over large area, *Sol Energy Mater Sol Cells,* **100**, 97–114.

43. Espinosa N, García-Valverde R, Krebs FC (2011) Life-cycle analysis of product integrated polymer solar cells, *Energy Environ Sci,* **4**, 1547–1557.

44. Espinosa N, García-Valverde R, Urbina A, Krebs FC (2011) A life cycle analysis of polymer solar cell modules prepared using roll-to-roll methods under ambient conditions, *Sol Energy Mater Sol Cells,* **95**, 1293–1302.

45. Espinosa N, García-Valverde R, Urbina A, et al. (2012) Life cycle assessment of ITO-free flexible polymer solar cells prepared by roll-to-roll coating and printing, *Sol Energy Mater Sol Cells,* **97**, 3–13.

46. Capasso A, Salamandra L, Di Carlo A, Bell JM, Motta N (2012) Low-temperature synthesis of carbon nanotubes on indium tin oxide electrodes for organic solar cells, *Beilstein J Nanotechnol,* **3**, 524–532.

47. Dubacheva GV, Liang CK, Bassani DM (2012) Functional monolayers from carbon nanostructures—fullerenes, carbon nanotubes, and

graphene—as novel materials for solar energy conversion, *Coord Chem Rev*, **256**, 2628–2639.

48. Müller-Meskamp L, Sachse C, Kim YH, Furno M, May C, Leo K (2012) Transparent conductive oxides and alternative transparent electrodes for organic photovoltaics and OLEDs, **24**, 24–31.
49. Ratier B, Nunzi JM, Aldissi M, Kraft TM, Buncel E (2012) Organic solar cell materials and active layer designs—improvements with carbon nanotubes: A review, *Polym Int*, **61**, 342–354.
50. Battumur T, Mujawar SH, Truong QT, et al. (2012) Graphene/carbon nanotubes composites as a counter electrode for dye-sensitized solar cells, *Cur Appl Phys*, **12**, e49–e53.
51. Lee S, Yeo JS, Ji Y, et al. (2012) Flexible organic solar cells composed of P3HT:PCBM using chemically doped graphene electrodes, *Nanotechnology*, **23**, 344013.
52. Yeh MH, Sun CL, Su JS, et al. (2012) A low-cost counter electrode of ITO glass coated with a graphene/Nafion® composite film for use in dye-sensitized solar cells, *Carbon*, **50**, 4192–4202.
53. Ajuria J, Ugarte I, Cambarau W, Etxebarria I, Tena-Zaera R, Pacios R (2012) Insights on the working principles of flexible and efficient ITO-free organic solar cells based on solution processed Ag nanowire electrodes, *Sol Energy Mater Sol Cells*, **102**, 148–152.
54. Lin J, Gu W, Cui Z (2012) *Santa Clara, CA*.
55. Van De Groep J, Spinelli P, Polman A (2012) Transparent conducting silver nanowire networks, *Nano Lett*, **12**, 3138–3144.
56. Zhou Y, Li F, Barrau S, Tian W, Inganäs O, Zhang F (2009) Inverted and transparent polymer solar cells prepared with vacuum-free processing, *Sol Energy Mater Sol Cells*, **93**, 497–500.
57. Zhou Y, Cheun H, Choi S, Potscavage JWJ, Fuentes-Hernandez C, Kippelen B (2010) Indium tin oxide-free and metal-free semi-transparent organic solar cells, *Appl Phys Lett*, **97**, 153304.
58. Hau SK, Yip HL, Zou J, Jen AKY (2009) Indium tin oxide-free semi-transparent inverted polymer solar cells using conducting polymer as both bottom and top electrodes, *Org Electron*, **10**, 1401–1407.
59. Ahlswede E, Mühleisen W, Bin Moh Wahi MW, Hanisch J, Powalla M (2008) Highly efficient organic solar cells with printable low-cost transparent contacts, *Appl Phys Lett*, **92**, 143307.
60. Chang Y-M, Wang L, Su W-F (2008) Polymer solar cells with poly(3,4-ethylenedioxythiophene) as transparent anode, *Org Electron*, **9**, 968–973.

61. Rahman ZA, Sulaiman K, Shuhaimi A, Rusop M (2012) PEDOT:PSS *thin film as transparent electrode In ITO-free organic solar cell*, Vol. **501**, Seremban, Negeri Sembilan, pp. 252–256.

62. Manor A, Katz EA, Andriessen R, Galagan Y (2011) Study of organic photovoltaics by localized concentrated sunlight: Towards optimization of charge collection in large-area solar cells, *Appl Phys Lett*, **99**, 173305.

63. Galagan YJm, Rubingh JE, Andriessen R, et al. (2011) ITO-free flexible organic solar cells with printed current collecting grids, *Sol Energy Mater Sol Cells*, **95**, 1339–1343.

64. Aernouts T, Vanlaeke P, Geens W, et al. (2004) Printable anodes for flexible organic solar cell modules, *Thin Solid Films*, **451–452**, 22–25.

65. Zou J, Yip H-L, Hau SK, Jen AKY (2010) Metal grid/conducting polymer hybrid transparent electrode for inverted polymer solar cells, *Appl Phys Lett*, **96**, 203301.

66. Tvingstedt K, Inganäs O (2007) Electrode grids for ITO free organic photovoltaic devices, *Adv Mater*, **19**, 2893–2897.

67. Glatthaar M, Niggemann M, Zimmermann B, et al. (2005) Organic solar cells using inverted layer sequence, *Thin Solid Films*, **491**, 298–300.

68. Krebs FC (2009) Fabrication and processing of polymer solar cells: A review of printing and coating techniques, *Sol Energy Mater Sol Cells*, **93**, 394–412.

69. Galagan Y, Coenen EWC, Sabik S, et al. (2012) Evaluation of ink-jet printed current collecting grids and busbars for ITO-free organic solar cells, *Sol Energy Mater Sol Cells*, **104**, 32–38.

70. Søndergaard R, Hösel M, Angmo D, Larsen-Olsen TT, Krebs FC (2012) Roll-to-roll fabrication of polymer solar cells, *Mater Today*, **15**, 36–49.

71. Krebs FC, Søndergaard R, Jørgensen M (2011) Printed metal back electrodes for R2R fabricated polymer solar cells studied using the LBIC technique, *Sol Energy Mater Sol Cells*, **95**, 1348–1353.

72. Chen FC, Wu JL, Lee CL, Huang WC, Chen HMP, Chen WC (2009) Flexible polymer photovoltaic devices prepared with inverted structures on metal foils, *IEEE Electron Dev Lett*, **30**, 727–729.

73. Petrie EM (2006) Chapter 2: Theories of adhesion, In *Handbook of adhesives and sealants*, 2nd ed, McGraw-Hill, New York.

74. Auras R, Harte B, Selke S (2004) An overview of polylactides as packaging materials, *Macromol Biosci*, **4**, 835–864.

75. Van Den Berg D, Barink M, Giesen P, Meinders E, Yakimets I (2011) Hygroscopic and thermal micro deformations of plastic substrates for flexible electronics using digital image correlation, *Polym Testing,* **30**, 188–194.

76. Yakimets I, Barink M, Goorhuis M, Giesen P, Furthner F, Meinders E (2010) Micro-deformation of flexible substrate for electronic devices during handling prior to lithography patterning, *Microelectron Eng,* **87**, 641–647.

77. Yakimets I, MacKerron D, Giesen P, et al. (2010) *Polymer substrates for flexible electronics: Achievements and challenges*, Vol. **93–94**, Bangkok, pp. 5–8.

78. Sarma KR, Chanley C, Dodd S, et al. (2003).

79. Leterrier Y, Médico L, Demarco F, et al. (2004) Mechanical integrity of transparent conductive oxide films for flexible polymer-based displays, *Thin Solid Films,* **460**, 156–166.

80. Scheirs J, Gardette JL (1997) Photo-oxidation and photolysis of poly(ethylene naphthalate), *Polym Degrad Stab,* **56**, 339–350.

81. Fechine GJM, Souto-Maior RM, Rabello MS (2007) Photodegradation of multilayer films based on PET copolymers, *J Appl Polym Sci,* **104**, 51–57.

82. Barker E. Combining barrier technology with other important properties in flexible electronics, *White paper, FLEXcon.*

83. http://www.vitexsys.com

84. http://www.ge-energy.com/index.jsp

85. http://news.3m.com/press-release/company/3m-announces-commercial-availability-3m-ftb3-barrier-film

86. http://www.amcor.com/

87. http://www.fujifilm.com/news/n090224.html

88. http://www.x-barrier.com/en/index.html

89. http://www.tera-barrier.com/technology.html

90. Moro RJV (2008) Barrier films for photovoltaics. In *Organic photovoltaics—materials, device physics, and manufacturing technologies* (CJ Brabec, VDU Scherf, eds.), Wiley-VCH, Weinheim.

91. Dennler G, Lungenschmied C, Neugebauer H, Sariciftci NS, Labouret A (2005) Flexible, conjugated polymer-fullerene-based bulk-heterojunction solar cells: Basics, encapsulation, and integration. *J Mater Res,* **20**, 3224–3233.

92. Hauch JA, Schilinsky P, Choulis SA, Childers R, Biele M, Brabec CJ (2008) Flexible organic P3HT:PCBM bulk-heterojunction modules with more than 1 year outdoor lifetime, *Sol Energy Mater Sol Cells,* **92**, 727–731.

93. Cros S, De Bettignies R, Berson S, et al. (2011) Definition of encapsulation barrier requirements: A method applied to organic solar cells, *Sol Energy Mater Sol Cells,* **95**, S65–S69.

94. Hauch JA, Schilinsky P, Choulis SA, Rajoelson S, Brabec CJ (2008) The impact of water vapor transmission rate on the lifetime of flexible polymer solar cells, *Appl Phys Lett,* **93**, 103306.

95. Krebs FC (2008) Air stable polymer photovoltaics based on a process free from vacuum steps and fullerenes, *Sol Energy Mater Sol Cells,* **92**, 715–726.

96. Hermenau M, Schubert S, Klumbies H, et al. (2012) The effect of barrier performance on the lifetime of small-molecule organic solar cells, *Sol Energy Mater Sol Cells,* **97**, 102–108.

97. Hermenau M, Riede M, Leo K, Gevorgyan SA, Krebs FC, Norrman K (2011) Water and oxygen induced degradation of small molecule organic solar cells, *Sol Energy Mater Sol Cells,* **95**, 1268–1277.

98. Lloyd MT, Peters CH, Garcia A, et al. (2011) Influence of the hole-transport layer on the initial behavior and lifetime of inverted organic photovoltaics, *Sol Energy Mater Sol Cells,* **95**, 1382–1388.

99. Lloyd MT, Olson DC, Berry JJ, et al. (2010) Enhanced lifetime in unencapsulated organic photovoltaics with air stable electrodes, *Conf Rec IEEE Photovol Spec Conf,* art. no. 5614676, 1060–1063.

100. Erlat AG, Henry BM, Grovenor CRM, Briggs AGD, Chater RJ, Tsukahara Y (2004) Mechanism of water vapor transport through PET/AlO_xN_y gas barrier films, *J Phys Chem B,* **108**, 883–890.

101. Erlat AG, Henry BM, Ingram JJ, et al. (2001) Characterisation of aluminium oxynitride gas barrier films, *Thin Solid Films,* **388**, 78–86.

102. Park JS, Chae H, Chung HK, Lee SI (2011) Thin film encapsulation for flexible AM-OLED: A review, *Semiconductor Sci Technol,* **26**, 034001.

103. Ummartyotin S, Juntaro J, Sain M, Manuspiya H (2011) Si-O barrier technology for bacterial cellulose nanocomposite flexible displays, *Carbohydr Polym,* **86**, 337–342.

104. Chen TN, Wuu DS, Wu CC, Chiang CC, Chen YP, Horng RH (2006) High-performance transparent barrier films of SiO_x/SiN_x stacks on flexible polymer substrates, *J Electrochem Soc,* **153**, F244–F248.

105. Fahlteich J, Schönberger W, Fahland M, Schiller N (2011) Characterization of reactively sputtered permeation barrier materials on polymer substrates, *Surf Coat Technol,* **205 (Supplement 2)**, S141–S144.

106. Low H, Xu Y (2005) Moisture barrier of Al_xO_y coating on poly(ethylene terephthalate), poly(ethylene naphthalate) and poly(carbonate) substrates, *Appl Surf Sci,* **250**, 135–145.

107. Bieder A, Gruniger A, von Rohr R (2005) Deposition of SiO_x diffusion barriers on flexible packaging materials by PECVD, *Surf Coat Technol,* **200**, 928–931.

108. Leterrier Y (2003) Durability of nanosized oxygen-barrier coatings on polymers, *Prog Mater Sci,* **48**, 1–55.

109. Bichler C, Kerbstadt T, Langowski HC, Moosheimer U (1999) Plasma-modified interfaces between polypropylene films and vacuum roll-to-roll coated thin barrier layers, *Surf Coat Technol,* **112**, 373–378.

110. Wang BC, Tropsha Y, Montgomery DB, Vogler EA, Spontak RJ (1999) Enhanced barrier performance of SiO_x-modified polymer substrates: Some morphological considerations, *J Mater Sci Lett,* **18**, 311–315.

111. Agres L, Ségui Y, Delsol R, Raynaud P (1996) Oxygen barrier efficiency of hexamethyldisiloxane/oxygen plasma-deposited coating, *J Appl Polym Sci,* **61**, 2015–2022.

112. Bordet H, Ignat M, Dupeux M (1998) Analysis of the mechanical response of film on substrate systems presenting rough interfaces, *Thin Solid Films,* **315**, 207–213.

113. Benmalek M, Dunlop HM (1995) Inorganic coatings on polymers, *Surf Coat Technol,* **76–77**, 821–826.

114. Graff GL, Williford RE, Burrows PE (2004) Mechanisms of vapor permeation through multilayer barrier films: Lag time versus equilibrium permeation, *J Appl Phys,* **96**, 1840–1849.

115. Sobrinho ASdS, Czeremuszkin G, Latreche M, Wertheimer MR (2000) Defect-permeation correlation for ultrathin transparent barrier coatings on polymers, *J Vac Sci Technol A: Vac Surf Films,* **18**, 149–157.

116. Jamieson EHH, Windle AH (1983) Structure and oxygen-barrier properties of metallized polymer film, *J Mater Sci,* **18**, 64–80.

117. Chatham H (1996) Oxygen diffusion barrier properties of transparent oxide coatings on polymeric substrates, *Surf Coat Technol,* **78**, 1–9.

118. Hanika M, Langowski HC, Moosheimer U, Peukert W (2003) Inorganic layers on polymer films—influence of defects and morphology on barrier properties, *Chem Eng Technol*, **26**, 605–614.

119. Hanika M, Langowski HC, Peukert W (2003) Simulation and verification of defect-dominated permeation mechanisms in multiple structures of inorganic and polymeric barrier layers, *Proceedings, Annual Technical Conference—Society of Vacuum Coaters*, 592–599.

120. Da Silva Sobrinho AS, Czeremuszkin G, Latrèche M, Dennler G, Wertheimer MR (1999) A study of defects in ultra-thin transparent coatings on polymers, *Surf Coat Technol*, **116–119**, 1204–1210.

121. Da Silva Sobrinho AS, Czeremuszkin G, Latrèche M, Wertheimer MR (1999) A study of defects in ultra-thin transparent coatings on polymers, *Appl Phys A: Mater Sci Process*, **68**, 103–105.

122. Rösch R, Tanenbaum DM, Jørgensen M, et al. (2012) Investigation of the degradation mechanisms of a variety of organic photovoltaic devices by combination of imaging techniques—the ISOS-3 inter-laboratory collaboration, *Energy Environ Sci*, **5**, 6521–6540.

123. Rösch R, Krebs FC, Tanenbaum DM, Hoppe H (2012) Quality control of roll-to-roll processed polymer solar modules by complementary imaging methods, *Sol Energy Mater Sol Cells*, **97**, 176–180.

124. Schaepkens M, Kim TW, Erlat AG, et al. (2004) Ultrahigh barrier coating deposition on polycarbonate substrates, *J Vac Sci Technol A: Vac Surf Films*, **22**, 1716–1722.

125. Tanase C, Van De Weijer P, Lifka H, Rietjens G (2008) Thin film encapsulated transparent organic light emitting diodes, *Materials Research Society Symposium Proceedings*, **1091**, 50–55.

126. Affinito JD, Gross ME, Coronado CA, Graff GL, Greenwell IN, Martin PM (1996) A new method for fabricating transparent barrier layers, *Thin Solid Films*, **290–291**, 63–67.

127. Charton C, Schiller N, Fahland M, Holländer A, Wedel A, Noller K (2006) Development of high barrier films on flexible polymer substrates, *Thin Solid Films*, **502**, 99–103.

128. Chwang AB, Rothman MA, Mao SY, et al. (2003) Thin film encapsulated flexible organic electroluminescent displays, *Appl Phys Lett*, **83**, 413–415.

129. Suen CS, Chu X (2008) Multilayer thin film barrier for protection of flex-electronics, *Solid State Technol*, **51**, 36–39.

130. Lewis J (2006) Material challenge for flexible organic devices, *Mater Today*, **9**, 38–45.

131. van Assche F, Rooms H, Young E, et al. (2008) Thin-film barrier on foil for organic LED lamps, *AIMCAL Fall Technical Conference and 22nd International Vacuum Web Coating Conference.*

132. Affinito JD, Eufinger S, Gross ME, Graff GL, Martin PM (1997) PML/oxide/PML barrier layer performance differences arising from use of UV or electron beam polymerization of the PML layers, *Thin Solid Films,* **308–309**, 19–25.

133. Servati P, Nathan A (2005) Orientation-dependent strain tolerance of amorphous silicon transistors and pixel circuits for elastic organic light-emitting diode displays, *Appl Phys Lett,* **86**, 033504.

134. Hutchinson JW, Suo Z (1991) *Mixed mode cracking in layered materials*, Vol. **29**, pp. 63–191.

135. Chung MH, Lin JS, Hsieh TE, et al. (2011) Preparation of organic/inorganic hybrid nanocomposites by ultraviolet irradiation and their packaging applications for organic optoelectronic devices, *Appl Surf Sci,* **257**, 9142–9151.

136. Gaume J, Taviot-Gueho C, Cros S, Rivaton A, Thérias S, Gardette JL (2012) Optimization of PVA clay nanocomposite for ultra-barrier multilayer encapsulation of organic solar cells, *Sol Energy Mater Sol Cells,* **99**, 240–249.

137. Amberg-Schwab S, Hoffmann M, Bader H, Gessler M (1998) Inorganic-organic polymers with barrier properties for water vapor, oxygen and flavors, *J Sol-Gel Sci Technol,* **13**, 141–146.

138. Amberg-Schwab S, Katschorek H, Weber U, et al. (2003) Inorganic-organic polymers as migration barriers against liquid and volatile compounds, *J Sol-Gel Sci Technol,* **26**, 699–703.

139. Amberg-Schwab S, Katschorek H, Weber U, Hoffmann M, Burger A (2000) Barrier properties of inorganic-organic polymers: Influence of starting compounds, curing conditions and storage-scaling-up to industrial application, *J Sol-Gel Sci Technol,* **19**, 125–129.

140. Marini M, Pilati F, Saccani A, Toselli M (2008) Improvement of thermo-oxidative stability of electrically insulating polymeric materials by organic-inorganic hybrid coating, *Polym Degrad Stab,* **93**, 1170–1175.

141. Toselli M, Pilati F, Marini M, Doghieri F, De Angelis MG, Minelli M (2008) Oxygen permeability of novel organic-inorganic coatings: II. Modification of the organic component with a hydrogen-bond forming polymer, *Eur Polym J,* **44**, 3256–3263.

142. Minelli M, De Angelis MG, Doghieri F, Marini M, Toselli M, Pilati F (2008) Oxygen permeability of novel organic-inorganic coatings:

I. Effects of organic-inorganic ratio and molecular weight of the organic component, *Eur Polym J,* **44**, 2581–2588.

143. Jitianu A, Klein LC (2009) *Sol–gel hybrids for electronic applications: Hermetic coatings for microelectronics and energy storage hybrid nanocomposites for nanotechnology,* (Merhari L, ed.), Springer US, pp. 429–453.

144. Angmo D, Hösel M, Krebs FC (2012) All solution processing of ITO-free organic solar cell modules directly on barrier foil, *Sol Energy Mater Sol Cells,* **107**, 329–336.

145. Granstrom J, Gällstedt M, Villet M, Moon JS, Chatterjee T (2010) High performance encapsulation structures utilizing Russian Doll architectures, *Thin Solid Films,* **518**, 5282–5287.

146. Zimmermann B, Würfel U, Niggemann M (2009) Longterm stability of efficient inverted P3HT:PCBM solar cells, *Sol Energy Mater Sol Cells,* **93**, 491–496.

147. Tanenbaum DM, Dam HF, Rösch R, Jørgensen M, Hoppe H, Krebs FC (2012) Edge sealing for low cost stability enhancement of roll-to-roll processed flexible polymer solar cell modules, *Sol Energy Mater Sol Cells,* **97**, 157–163.

148. Muller-Meskamp L, Fahlteich J, Krebs FC (2012) Barrier technology and application, In *Stability and degradation of organic and polymer solar cells,* (Krebs FC, ed.), John Wiley & Sons, p. 368.

Chapter 13

Large-Area Processing of Organic Photovoltaics

Roar R. Søndergaard, Markus Hösel, and Frederik C. Krebs
Department of Energy Conversion and Storage, Technical University of Denmark, 4000 Roskilde, Denmark
frkr@dtu.dk

13.1 Introduction

Organic photovoltaics (OPVs) have for small-scale devices (<1 cm^2) now reached levels with respect to both efficiency and stability where they will soon be able to compete with amorphous silicon solar cells pending continued development at the current rate. With reported efficiencies up to 9–10% [25,38] and with numerous reports on improved outdoor stabilities [21,27,54,63], the field has reached the point where it is essential that more effort is allocated toward the original idea behind OPVs, which is high-throughput production of inexpensive organic solar cells on flexible substrates—preferably using roll-to-roll (R2R) processing. Today, most research is performed through preparation of small-scale devices on rigid substrates, and although the transfer to larger areas and flexible substrates might seem straightforward, this is definitely not the case. Many of the

Organic Solar Cells: Fundamentals, Devices, and Upscaling
Edited by Barry P. Rand and Henning Richter

ISBN 978-981-4463-65-2 (Hardcover), 978-981-4463-66-9 (eBook)
www.panstanford.com

processing techniques used in the laboratory do not comply with high-throughput production. Take as an example spin coating. This is by far the most common way of applying the different thin layers in organic solar cells, but the technique does not scale well to larger areas, is not R2R compatible, and far too much material is wasted in the process. The use of vacuum-deposited electrodes is another problem, as high vacuum requires extra energy and compromises throughput, which consequently brings up the cost of the final device. Alternative ways such as printable electrodes must be sought. As silver is presently the only real candidate for printing, the geometry of the device has to have an "inverted" structure, thus changing the buildup of the more commonly used "normal" geometry.

Another issue is the knowledge and experience of how to process and optimize larger areas, which is currently only a fraction of the same for small-scale devices—and there is a huge difference between processing on a small rigid substrate compared to a flexible substrate which moves continuously with speeds up to 20 m/min.

When wanting to move toward large-area manufacture, we virtually need to change all of the parameters that scientists have spent years to optimize on a smaller scale (processing methods, solvents, drying procedures, the materials, additives, the substrate, etc.), and it would be naïve to believe that such transfers can be performed without consequences. A good argument supporting this is the fact that new high-performing materials for years have dominated the small-scale device aim of high efficiencies, but none of these materials have so far proven efficient in R2R-processed large area solar cells—and it is not for lack of trying.

A vast variety of film-forming techniques are available today and among these several could be suitable for the processing of one or several layers in OPVs. So far, the main developments in OPV processing have not evolved around the technology itself, but more around the need for solution processing, which opens the possibility for inexpensive, high-throughput processing. As a consequence, the OPV technology is very much defined by the available processing methods, and great care must therefore be taken when choosing our future ways of preparing OPVs, as research in processing methods that do not have the potential to be up-scaled might very well end up as useless efforts when looking at the real goal—large-scale production of low-cost alternative energy.

In this chapter, we aim to give insight into some of the techniques which have proven successful in R2R processing of organic solar cells along with R2R techniques that could potentially be good candidates for future use.

Because of the multilayered structure of the OPV device and the various properties of the inks to be processed, the choice of several different techniques might be available for one layer, whereas only very specific techniques can be applied for others. Figure 13.1 shows schematic illustrations of a variety of R2R compatible techniques that have been employed in the fabrication of organic solar cells and which will be covered in this chapter.

In addition to the actual fabrication of the solar cells, issues such as encapsulation and device testing will also be touched upon.

13.2 Established Noncontact Coating Techniques

A huge amount of different and very sophisticated coating techniques exist, but for simplicity we are only going to deal with the two, which have already been used in the preparation of OPVs, knife coating and slot-die coating, both of which are noncontact techniques with continuous wet-film formation via a meniscus. The film is the result of constant feeding of ink to a standing meniscus between the "coating head" and the moving substrate. Because they are noncontact techniques, minimal physical damage/stress to what is being coated upon is obtained, and generally very even layers can be prepared.

13.2.1 Knife Coating

Knife coating is very similar to laboratory-scale doctor blading. In knife coating, the wet film is formed on the web as this passes a stationary knife (often also referred to as a blade) in front of which a reservoir of ink is constantly fed. The wet layer thickness is defined by the distance between the web and the knife and to some extent also the web speed (the thickness can roughly be estimated as half the gap height). The method is very suitable for continuous coating of large unpatterned areas, and can be performed at high speed

(>10 m/min). Reports can be found on the processing of ZnO as a hole-blocking layer [41], PEDOT:PSS as an electron-blocking layer [32], and the active layer [64].

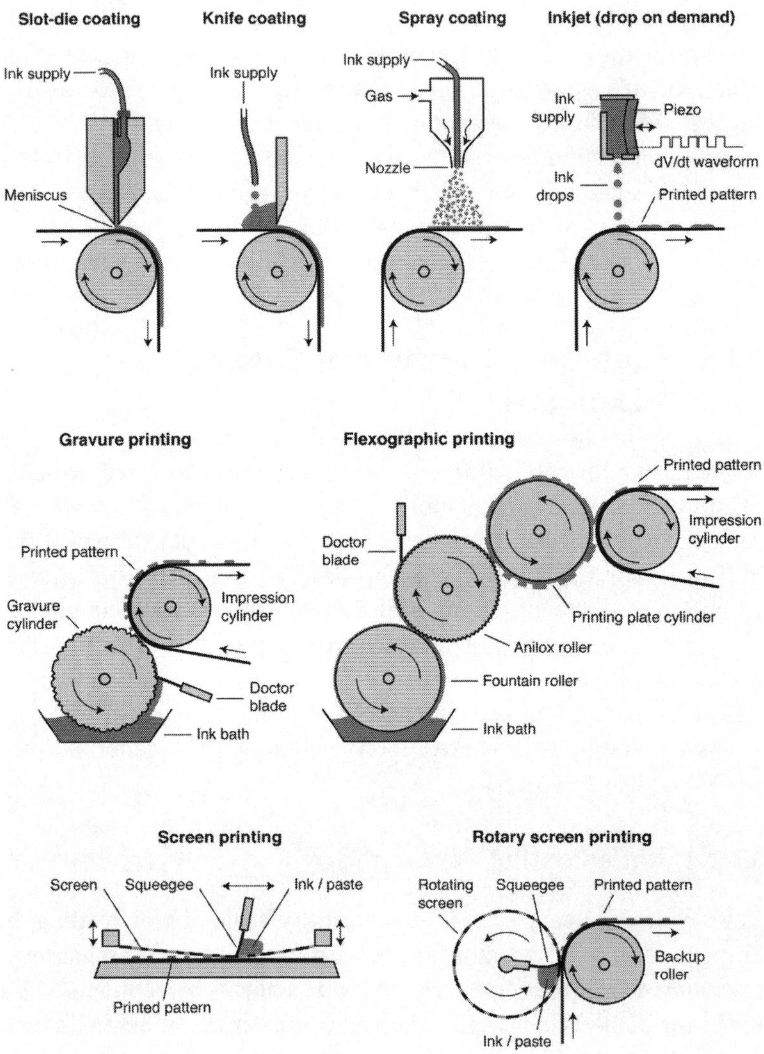

Figure 13.1 Schematic illustrations of a variety of coating and printing techniques.

13.2.2 Slot-Die Coating

In slot-die coating the ink is supplied through a slot in the coating head which is typically fed with a pump to ensure control over the ink flow. The wet layer thickness can be controlled with high precision and is defined by the web speed, the width of the coated area, and the pumping rate. Because of the high control it can be specified as a pre-metered coating technique. If the coating head is furthermore fitted with a meniscus guide, 1D patterning in the form of stripes is possible. An example of slot-die coating is shown in Fig. 13.2, illustrating the strength of R2R slot-die coating. Very tightly spaced and highly homogeneous lines can be prepared simultaneously with high precision at high speeds. The technology allows for processing speeds up to 200 m/min, but so far the best results in OPV have been obtained at processing speeds of 1–6 m/min.

Figure 13.2 Picture of slot-die coating with a meniscus guide. The meniscuses (indicated with an arrow) are clearly visible and several identical lines can be processed simultaneously.

Slot-die coating is currently by far the most widely used technique for processing of layers in organic solar cells and has mainly been used for processing of hole-blocking layers, active layers, and electron-blocking layers [3,4,7,8,10,18,28,41–43,47,48,51,53,60,62,64,65,67], although not all reported cases were performed on R2R systems.

Techniques like double-slot-die coating and differentially pumped slot-die coating are among the more sophisticated for slot-die coating. In double-slot-die coating, two inks are processed simultaneously from the same coating head, thus saving a processing step. Larsen-Olsen et al. used this technique for processing an aqueous suspension of P3HT:PCBM and PEDOT:PSS simultaneously at a web speed of 1 m/min [50]. Although the resulting cells showed relatively poor performance (PCE: 0.03%) it approaches key elements for future large-scale production—reducing energy payback times of the OPV as well as the time/cost of manufacture.

Differentially pumped slot-die coating, a technique developed by Alstrup et al. [2], has proven a powerful and fast way of screening donor/acceptor ratios when wanting to optimize the active layer composition and layer thickness [3,8,62]. Two components of the final ink (e.g., a polymer and an acceptor or a mixture of donor/acceptor with a predefined ratio and a solvent) are pumped simultaneously, but with varying speeds, into the coating head, where the two components are mixed before leaving the slot-die head. Figure 13.3 illustrates the principles behind double-slot-die coating and differentially pumped slot-die coating.

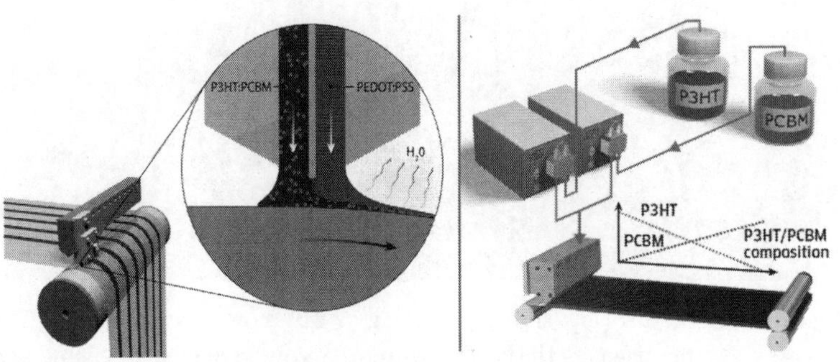

Figure 13.3 Illustrations of the principles of double-slot-die coating (left) and differentially pumped slot-die coating (right).

Analyzing the performance of the final solar cells, prepared along the film presenting a gradual change in a parameter along the length of it, then allows for determining the optimal conditions/compositions for this type of solar cell. Because of a very small dead volume in the coating head it is possible to perform this with very small amounts of material, making the technique extremely useful for fast screening of new materials—something that generally takes weeks to analyze can be performed in a matter of hours.

13.3 Printing Techniques

As opposed to the described coating techniques most printing techniques are contact techniques with the only exception being inkjet printing, which has a somewhat misleading name. "Printing" implies by definition the transfer of a motif through physical contact of the carrier of the motif and a substrate. Printing techniques are as a consequence 2-dimentional patterning methods. Several printing techniques are illustrated in Fig. 13.1.

13.3.1 Screen Printing

In screen printing, the motif to be printed is defined by the vacant 2D pattern in an otherwise filled mesh in a screen. As a squeegee moves relative to the screen in the presence of ink, the latter is pushed through the open area of the screen and onto the substrate. The method inherently allows for the formations of very thick wet layers (10–500 μm) and thus also very thick dry films. The technique is generally only useful for rather viscous inks with thixotropic (shear thinning) properties, as low-viscosity inks will simply pass through the mesh by gravity.

Screen printing can be performed R2R using two types of techniques—flat-bed screen printing and rotary screen printing. Flat-bed screen printing is a step-wise process where the screen is lowered upon the substrate followed by passing of a squeegee which by pushing the ink through the unfilled parts of the mesh ensures transfer of the motif. After raising the screen, the web is moved forward and the process starts again. Very large areas can be processed in this way (up to 10 m^2), but caution must be taken, as factors like solvent evaporation from the ink, which is left open to the ambient on the screen, can change the printing conditions in an undesirable way.

In rotary screen printing the screen is folded into a tube that rotates with the same speed as the web. The squeegee is in this case stationary and is placed inside the screen, which allows for continuous printing of the screen motif upon each rotation. As the ink is situated inside the screen it is much less exposed to the surroundings compared to flat-bed. Rotary screen printing is by far superior to flat-bed printing with respect to speed, edge definition/resolution as well as wet thickness and is a true R2R printing technique. That said, the rotary screen is considerably more expensive than its flat-bed counterpart, it is more difficult to operate/adjust, and from a more practical point of view also much more difficult to clean because of the restricted access inside the screen. Although the procedure is very reliable when adjusted correctly, much longer run-in is required in order to get to this point, and the technique is thus not as suitable for small-scale laboratory work as the cheaper flat-bed alternative.

Screen printing has proven especially useful for printing of front [19] and back electrodes [3,4,8,10,28,42,43,47,48,53,62,65], but has also been used for the processing of the active layer [44,66] and PEDOT:PSS [41,44]. Figure 13.4 shows pictures of the use of screen printing for silver back electrodes.

Figure 13.4 Pictures of flat-bed screen printing (left) and rotary screen printing (right).

13.3.2 Flexoprinting

In flexographic (flexo) printing the ink is transferred from a relief on the printing plate, which is typically made out of rubber or a photopolymer as illustrated in Fig. 13.1. The printing plate is continuously supplied with ink from a ceramic anilox roller which has a surface full of small holes/microcavities. The size/depth of these cavities controls the amount of ink that can be transferred to the printing plate. The anilox itself is supplied with ink from a fountain roller which is immersed in an ink bath, and excess ink is removed from the anilox by use of a doctor blade, ensuring that ink is only present inside the cavities. When using inks with more volatile solvents it can be useful to employ a chambered doctor blade system as it minimizes problems of solvent exchange with the surroundings. Figure 13.5 shows pictures of the processing of a silver grid by flexo printing.

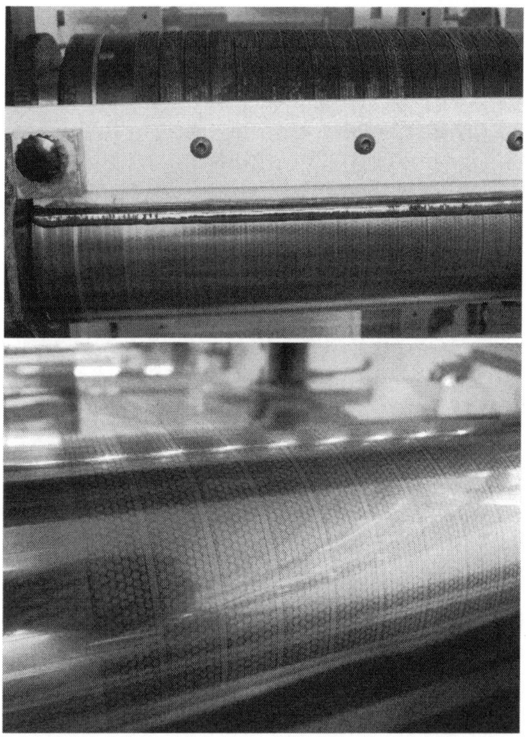

Figure 13.5 Pictures of flexographic printing of silver grid. **Top**: Flexo roll setup (here without the fountain roll) showing how the excess silver ink is removed from the anilox. The negative imprint, caused by ink pickup from the printing plate cylinder seen in the background, can be observed on the anilox. **Bottom**: The final flexo-printed silver grid on PET-foil.

Flexo printing has only been used on a few occasions for the preparations of organic solar cells [33,42,65], but the potential for very high processing speeds (>25 m/min) makes this technique an interesting candidate for future processing of OPVs. A rule of thumb in flexographic printing is "the faster the better." Yu et al. recently showed that silver grid structures as front electrode can be processed with speeds of 25 m/min [65], and also Deganello et al. have used flexo printing to create a microscale conductive network although no solar cells were prepared in this case [11]. Among other uses of flexo printing for OPVs are deposition of a wetting agent on top of the active layer before further processing [42] and for the printing of PEDOT:PSS [33].

13.3.3 Gravure Printing

Gravure printing is like flexo printing, a technique where very high processing speeds can be obtained (up to around 15 m/s). The technique is suitable for low-viscosity inks and is widely used for the printing of high volumes of catalogues and magazines. As opposed to flexo printing the ink transfer does not happen from a relief but through tiny engraved cavities forming a pattern in the gravure cylinder. As the web is brought into contact with the gravure cylinder by pressure from a softer impression cylinder, the ink is transferred through surface tension transfer from the cavities to the moving web. The shape of the pattern and the volume of the engraved cavities in the cylinder define the final motif. The gravure cylinder is continuously filled by immersion into an ink bath, and excess ink is removed by a doctor blade. Similar to flexo printing a chambered blade system can at times be used with advantage. Very high processing speeds can, as mentioned, be obtained with gravure printing, but processing is very dependent on ink rheology, the pressure from the impression cylinder, and the web speed. Great care must therefore be taken to optimize processing conditions and the surface tension of the used ink.

The use of gravure for the processing of OPVs has only been reported in a few cases. Among the more exotic is a recent publication by Hübler et al. who prepared OPV on paper [33]. Gravure was first used for the application of glue to a paper substrate, and the glue-patterned substrate was then brought into contact with a Zn-covered transfer foil, ensuring Zn coverage only on the parts that contained glue. Oxidation of the Zn surface forms a thin layer of ZnO which acts as a hole-blocking layer. Subsequently, the active layer was printed on top of this also using gravure printing and was finalized by flexo printing of PEDOT:PSS as a transparent back electrode. The entire process was performed R2R and the architecture avoids the use of the expensive indium-tin-oxide (ITO), but the final cells (max. PCE: 1.3% at 600 W/m^2) are very small (0.09 cm^2).

Kopola et al. have reported two cases of processing of normal device structure OPV using a desktop gravure printability tester (single sheets—not R2R) to process PEDOT:PSS and P3HT:PCBM of single cells (PCE: 2.8%, 0.19 cm^2) [39] and small modules (5 cells in series, PCE: 1.9%, 9.6 cm^2) [40] on PET. In both reports, the back electrode was processed by evaporating first calcium and then silver. Figure 13.6 illustrates how important tuning of the ink properties and processing conditions are to obtain a smooth homogeneous film.

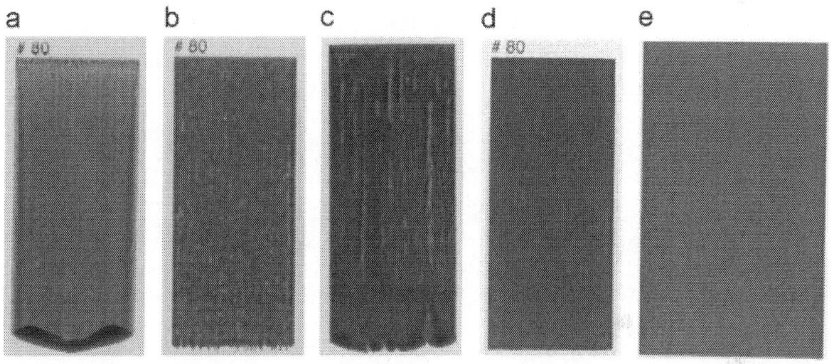

Figure 13.6 Pictures of final active layer films formed with different ink formulations and under different processing speeds. The different smoothness of the films illustrates well the importance of tuning both ink formulation and processing conditions when using gravure printing. Reprinted with permission from Ref. [39] © 2010 Elsevier B.V.

13.4 "Printing" and Coating through Droplets and Brushes

Finding ways to categorize the various film forming techniques is not always simple. The naming of inkjet printing is an example of this, which, although "printing" implies otherwise, is a completely noncontact technique. In the following we will present a few techniques which are all noncontact, and which are slightly difficult to categorize along with the others.

13.4.1 Inkjet Printing

Inkjet printing is a fully-digitalized two-dimensional technique, where the substrate is divided into pixels with a certain resolution. A pixel will, during printing, either be left blank or receive ink droplets with different size or quantity, thus defining the desired pattern and layer thickness. The most widely used system for inkjet printing employs piezoelectric materials which, upon applying a voltage, will expand and cause a droplet of ink to be ejected from a nozzle. "Drop-on-demand" (DOD), as the technique is called, requires many nozzles, which previously has limited its industrial use, but systems with high resolution (>600 DPI) which are capable of processing at high speeds (>50 m/min) are now commercially available.

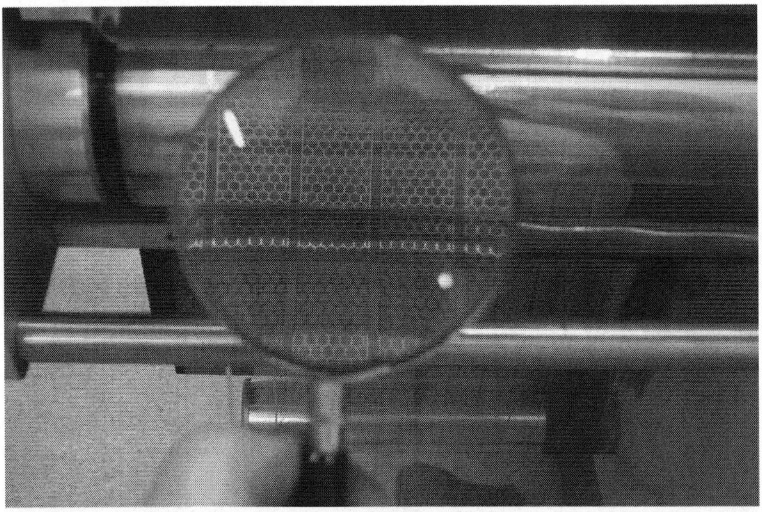

Figure 13.7 Example of R2R inkjet-printed silver front grid. The printer used can print at 600 DPI in the web width and 1200 DPI in the web length at processing speed >50 m/min.

The real strength of the technique lies within the flexibility of having a digital master. Unlike the other available 2D techniques, there is no need for preparing a physical printing master—pattern modifications and completely new patterns can be made with a computer and can be used immediately. The technique has some limitations to what ink formulations can be used and this is probably a main reason why its use in the preparation of OPVs has so far been scarce. Besides the preparation of R2R-patterned front grid lines [65], the technique has mainly been restricted to the preparation of small-scale devices (0.03–1 cm^2) on glass and has been used to process PEDOT:PSS [13], P3HT:PCBM [1,30,31], or both [12,49]. A picture of a R2R inkjet-printed silver grid can be seen in Fig. 13.7.

13.4.2 Spray Coating

Spray coating is like inkjet printing based on the formation of small droplets, but unlike inkjet printing no precise patterning can be obtained with this technique as it is inherently zero-dimensional. The most typically used pneumatic systems are based on a spray nozzle where pressurized air or gas (e.g., nitrogen) breaks up the ink into droplets. Several factors such as surface tension, viscosity, and density of the ink as well as the nozzle design and the gas flow properties will influence the result of the final atomizing process. When shot from the nozzle the kinetic energy of the droplets will help them to spread upon impact with the substrate. The quality of the final film is highly dependent on the match in polarity between the substrate and the ink used, but it also relies on processing conditions such as spray distance, coating speed, as well as the number of "spray passings" and layers.

Other types of droplet generation in spray coating are ultrasonication with a directed carrier gas and electro-spraying. The method is highly R2R compatible with processing of large areas, but as patterning can only be obtained by use of a shadow mask it is questionable whether the technique will be too impractical outside the lab.

Spray coating has been used on several occasions in the preparation of OPVs on a smaller scale for the processing of the active layer [9,23,57], silver back electrodes [24,26], hole transport layer and active layer [22,55] and electron transport, active and hole transport layers [52]. Some of these have demonstrated cells with PCE comparable to those processed from spin coating, giving promise to this technique.

However, no real R2R processing has been carried out yet, but small lab-scale roll coaters have been developed [10], and these have been used to explore spray coating of the polymeric layers in electrochromic devices [34].

13.4.3 Brush Painting

The use of brushes to create a film is probably among the oldest techniques that exist and at the same time it is one of the most difficult to describe in detail, as it is a mixture of a multitude of techniques. In an ink-soaked brush the ink is retained within the fibers via the surface tension of the ink. When the brush is brought into "contact" with the substrate a meniscus is formed between the fibers and the substrate, and the ink is pulled out of the brush as it moves over the substrate. Depending on the brush size, the technique allows for some patterning in which case it is a combination of contact, meniscus, and two-dimensional patterning. The technique has yet to be used in R2R systems but it is plausible that R2R brush methods can be developed. Brush painting has been used on several occasions in the preparation of small-scale OPV devices [29,36,37].

13.5 Lamination

Once the entire solar cell stack has been processed, it needs to be encapsulated (for further information about encapsulation, see Chapter 12), foremost to ensure operational stability by protection from oxygen and water, but also to protect the delicate layers from mechanical damage when handling the device. Several different lamination types exist, each of which has its own pros and cons, and all of which rely on relatively simple systems to bring two foils together in a finalizing step by applying pressure as they pass between two rolls.

13.5.1 Cold Lamination

Cold lamination is typically carried out using a pressure-sensitive adhesive that is first lined on the laminate in a lamination process, followed by continuous lamination of the substrate carrying the solar cell stack with the now adhesive-covered laminate by pressing the two foils together, while removing the liner from the laminate in the process [47]. The technique is very simple to approach and can be carried out with ease at relatively high speeds (>20 m/min). The main challenge of the technique is the handling of the sticky laminate.

Generally, relatively thick adhesive layers (typically around 50 μm) are used in order to ensure good adhesion between the two foils. It is possible to go as thin as 20 μm adhesive but such thin adhesive layers can be challenging to handle. The major weakness of cold lamination is the rather thick adhesive layer needed compared to other techniques.

13.5.2 Hot Lamination

Hot melt lamination uses a technique similar to cold lamination, but here the barrier/laminate already carries the adhesive material, which becomes adhesive when heated. The solar cell substrate and the laminate are then unified by passing the two webs between a set of hot rollers, which melts the adhesive forming a tight seal.

Hot lamination is probably the easiest of the available techniques to handle and compared to cold lamination, much thinner adhesive thickness can be used, although care must be taken not to use a too thin adhesive layer when processing solar cells with a rough surface topography.

13.5.3 UV-Lamination

In UV-lamination, a UV-curable adhesive or glue is added to the laminate just prior to joining the webs between two rollers. This is immediately followed by exposure to UV-light, which causes the adhesive to cure. As a printing or coating step is needed in the actual lamination process, UV lamination must be categorized as the most complicated of the here-mentioned encapsulation types. That being said, it is also the most versatile with respect to the thickness of the adhesive (1–100 μm) as it is only limited by the employed coating/printing process and the curability of the adhesive.

13.6 Device Testing

During processing, inline monitoring techniques such as cameras provide 2D information using transmission, reflection, and dark field imaging, and can reveal details of the printed or coated film that can be used to observe and guide the process. Unfortunately, such monitoring cannot provide information on the actual performance of the final device. Once laminated the final roll will in most cases need to be tested, and as a 20–100 m long foil generally means testing hundreds or thousands of solar cells or solar cell modules it is practical to be able to do this in an automated process while the devices are still on the roll. Besides traditional current-voltage and spectral response measurements, which can tell you if a device is working but not what is potentially wrong, a number of 2D imaging techniques are available, such as light beam-induced current (LBIC) mapping [46], dark lock-in thermographic imaging [6], and photoluminescence and electroluminescence imaging [59,61]. Figure 13.8 shows a R2R characterizer and an example of an LBIC image.

Figure 13.8 **Left**: A photograph of a R2R tester allowing for measuring of IV characteristics and for 2D imaging. **Middle** and **right**: A damaged solar cell module and an example of an LBIC image of the same device. Reprinted with permission from Ref. [4] © 2012 Elsevier B.V.

13.7 Demonstrations of Device Integration

The final goal of large-scale manufacturing is of course the full integration of the manufactured devices into a product. Generally, the experience is still very limited and only a few examples of application and demonstrations of organic solar cells on a larger scale have so far been reported, and commercially the only available product has been Konarka's solar bag with a small solar module on the front. One of the earliest demonstrations was that of the "solar hat" where more than 2000 solar modules with a size of 130 cm^2

were produced using only screen printing [44]. The modules were then inserted into a pocket on top of a hat and used to charge a small radio. In the project "Lighting Africa," a reading light was provided by using organic solar cell modules (20×25 cm) to charge a small lamp with a rechargeable battery [45], an idea which was later reused as a miniaturization in the production of more than 10000 credit card-sized OE-A flashlight demonstrators [48]. Figure 13.9 shows all three examples.

Figure 13.9 Picture showing from the left: the "solar hat," the "lighting Africa" reading light, and the OE-A flashlight demonstrator.

An interesting development from the solar hat to the OE-A demonstrator is the progressive use of R2R techniques to integrate the final solar cell into a product. Where all assembly of the solar hat was carried out by hand, the OE-A demonstrator was a fully automated process. Although the final assembly of the OE-A demonstrator was carried out sheet-to-sheet, handling of 15 units simultaneously, they had all undergone more than 30 separate processing steps using either R2R or sheet handling. Full exploitation of the potential of R2R-processed solar cells will first be seen when product integration can be carried out at speeds that are similar to those of the solar cell manufacture.

13.8 Summary and Outlook

Organic solar cells are at the verge where the technology has to prove itself capable of transfer from the protective surrounding of the laboratory and into large-scale production systems. To do so involves a shift of the general focus of the field from the preparation of small high-performing solar cells to adapting processing procedures to what is feasible on a larger scale. Any effort that does not have a realistic chance of such a transfer is prone to be wasted. Further approaches toward bringing the final cost down are still necessary, and finding a substitute for ITO is one of the main goals in this respect. Refining the processing procedures to be more environmentally friendly by substituting the present chlorinated and toxic solvents, preferably with water, is another important issue. If the imagined production, which lies in the scale of GW_{peak}/day, is to be realized, the environmental impact could be catastrophic if such precautions are not taken into account at an early stage.

Tools such as life-cycle analysis (LCA) [14–16,20] and financial analysis of R2R-processed solar cells [5,35,56,58] must furthermore be used to direct science if the final goal of high-throughput production is to be realized. Recent LCA studies show that by avoiding scarce elements, avoiding vacuum steps, and using only solar heat and solar energy during processing, energy payback times can be brought down to a few days [17]. This demonstrates the need to move toward processing of higher volumes in order to gain the necessary knowledge that will allow the field to confront these new challenges.

References

1. Aernouts TT, Aleksandrov C, Girotto J, Genoe J, Poortmans (2008) Polymer based organic solar cells using ink-jet printed active layers, *Appl Phys Lett,* **92**, 033306.
2. Alstrup JM, Jørgensen AJ, Medford FC, Krebs (2010) Ultra fast and parsimonious materials screening for polymer solar cells using differentially pumped slot-die coating, *ACS Appl Mater Interf,* **2**, 2819–2827.
3. Amb CM, Craig MR, Koldemir U, et al. (2012) Aesthetically pleasing conjugated polymer:fullerene blends for blue-green solar cells via roll-to-roll processing, *ACS Appl Mater Interf,* **4**, 1847–1853.

4. Angmo D, Hösel M, Krebs FC (2012) All solution processing of ITO-free organic solar cell modules directly on barrier foil, *Sol Energy Mater Sol Cells*, **107**, 329–336.
5. Azzopardi B, Emmott CJM, Urbina A, Krebs FC, Mutale J, Nelson J (2011) Economic assessment of solar electricity production from organic-based photovoltaic modules in a domestic environment, *Energy Environ Sci*, **4**, 3741–3753.
6. Bachmann J, Buerhop-Lutz C, Deibel C, et al. (2010) Organic solar cells characterized by dark lock-in thermography, *Sol Energy Mater Sol Cells*, **94**, 642–647.
7. Blankenburg L, Schultheis K, Schache H, Sensfuss S, Schrödner M, (2009) Reel-to-reel wet coating as an efficient up-scaling technique for the production of bulk-heterojunction polymer solar cells, *Sol Energy Mater Sol Cells*, **93**, 476–483.
8. Bundgaard E, Hagemann O, Manceau M, Jørgensen M, Krebs FC (2010) Low bandgap polymers for roll-to-roll coated polymer solar cells, *Macromolecules*, **43**, 8115–8120.
9. Chen LM, Hong Z, Kwan WL, et al. (2010) Multi-source/component spray coating for polymer solar cells, *ACS Nano*, **4**, 4744–4752.
10. Dam HF, Krebs FC (2012) Simple roll coater with variable coating and temperature control for printed polymer solar cells, *Sol Energy Mater Sol Cells*, **97**, 191–196.
11. Deganello D, Cherry JA, Gethin DT, Claypole TC (2010) Patterning of micro-scale conductive networks using reel-to-reel flexographic printing, *Thin Solid Films*, **518**, 6113–6116.
12. Eom SH, Park H, Mujawar SH, et al. (2010) High efficiency polymer solar cells via sequential inkjet-printing of PEDOT:PSS and P3HT:PCBM inks with additives, *Org Electron*, **11**, 1516–1522.
13. Eom SH, Senthilarasu S, Uthirakumar P, et al. (2009) Polymer solar cells based on inkjet-printed PEDOT:PSS layer, *Org Electron*, **10**, 536–542.
14. Espinosa N, García-Valverde R, Urbina A, Krebs FC (2011) A life cycle analysis of polymer solar cell modules prepared using roll-to-roll methods under ambient conditions, *Sol Energy Mater Sol Cells*, **95**, 1293–1302.
15. Espinosa N, García-Valverde R, Krebs FC (2011) Life-cycle analysis of product integrated polymer solar cells, *Energy Environ Sci*, **4**, 1547–1557.
16. Espinosa N, García-Valverde R, Urbina A, et al. (2012) Life cycle assessment of ITO-free flexible polymer solar cells prepared by roll-to-roll coating and printing, *Sol Energy Mater Sol Cells*, **97**, 3–13.

17. Espinosa N, Hösel M, Angmo D, Krebs FC (2012) Solar cells with one-day energy payback for the factories of the future, *Energy Environ Sci,* **5**, 5117–5132.
18. Galagan Y, de Vries IG, Langen AP, et al. (2011) Technology development for roll-to-roll production of organic photovoltaics, *Chem Eng Process,* **50**, 454–461.
19. Galagan Y, Rubingh JM, Andriessen R, et al. (2011) ITO-free flexible organic solar cells with printed current collecting grids, *Sol Energy Mater Sol Cells,* **95**, 1339–1343.
20. García-Valverde R, Cherni JA, Urbina A (2010) Life cycle analysis of organic photovoltaic technologies, *Prog Photovolt Res Appl,* **18**, 535–558.
21. Gevorgyan SA, Medford AJ, Bundgaard E, et al. (2011) An inter-laboratory stability study of roll-to-roll coated flexible polymer solar modules, *Sol Energy Mater Sol Cells,* **95**, 1398–1416.
22. Girotto C, Moia D, Rand BP, Heremans P (2011) High-performance organic solar cells with spray-coated hole-transport and active layers, *Adv Funct Mater,* **21**, 64–72.
23. Girotto C, Rand BP, Genoe J, Heremans P (2009) Exploring spray coating as a deposition technique for the fabrication of solution-processed solar cells, *Sol Energy Mater Sol Cells,* **93**, 454–458.
24. Girotto C, Rand BP, Steudel S, Genoe J, Heremans P (2009) Nanoparticle-based, spray-coated silver top contacts for efficient polymer solar cells, *Org Electron,* **10**, 735–740.
25. Green MA, Emery K, Hishikawa Y, Warta W, Dunlop ED (2012) Solar cell efficiency tables (version 40), *Prog Photovolt Res Appl,* **20**, 606–614.
26. Hau SK, Yip HL, Leong K, Jen AK, (2009) Spraycoating of silver nanoparticle electrodes for inverted polymer solar cells, *Org Electron,* **10**, 719–723.
27. Hauch JA, Schilinsky P, Choulis SA, Childers R, Biele M, Brabec CJ (2008) Flexible organic P3HT:PCBM bulk-heterojunction modules with more than 1 year outdoor lifetime, *Sol Energy Mater Sol Cells,* **92**, 727–731.
28. Helgesen M, Carlé JE, Andreasen B, et al. (2012) Rapid flash annealing of thermally reactive copolymers in a roll-to-roll process for polymer solar cells, *Polym Chem,* **3**, 2649–2655.
29. Heo SW, Lee JY, Song HJ, Ku JR, Moon DK (2011) Patternable brush painting process for fabrication of flexible polymer solar cells, *Sol Energy Mater Sol Cells,* **95**, 3041–3046.

30. Hoth CN, Choulis SA, Schilinsky P, Brabec CJ (2007) High photovoltaic performance of inkjet printed polymer:fullerene blends, *Adv Mater,* **19**, 3973–3978.

31. Hoth CN, Schilinsky P, Choulis SA, Brabec CJ (2008) Printing highly efficient organic solar cells, *Nano Lett,* **8**, 2806–2813.

32. Hoth CN, Steim R, Schilinsky P, et al. (2009) Topographical and morphological aspects of spray coated organic photovoltaics, *Org Electron,* **10**, 587–593.

33. Hübler A, Trnovec B, Zillger T, et al. (2011) Printed paper photovoltaic cells, *Adv Energy Mater,* **1**, 1018–1022.

34. Jensen J, Dam HF, Reynolds JR, Dyer AL, Krebs FC (2012) Manufacture and demonstration of organic photovoltaic-powered electrochromic displays using roll coating methods and printable electrolytes, *J Polym Sci Part B,* **50**, 536–545.

35. Kalowekamo J, Baker E (2009) Estimating the manufacturing cost of purely organic solar cells, *Sol Energy,* **83**, 1224–1231.

36. Kim SS, Na SI, Jo J, Tae G, Kim DY (2007) Efficient polymer solar cells fabricated by simple brush painting, *Adv Mater,* **19**, 4410–4415.

37. Kim SS, Na SI, Kang SJ, Kim DY (2010) Annealing-free fabrication of P3HT:PCBM solar cells via simple brush painting, *Sol Energy Mater Sol Cells,* **94**, 171–175.

38. Konarka Technologies Inc., www.konarka.com

39. Kopola P, Aernouts T, Guillerez S, et al. (2010) High efficient plastic solar cells fabricated with a high-throughput gravure printing method, *Sol Energy Mater Sol Cells,* **94**, 1673–1680.

40. Kopola P, Aernouts T, Sliz R, et al. (2011) Gravure printed flexible organic photovoltaic modules, *Sol Energy Mater Sol Cells,* **95**, 1344–1347.

41. Krebs FC (2009) Polymer solar cell modules prepared using roll-to-roll methods: Knife-over-edge coating, slot-die coating and screen printing, *Sol Energy Mater Sol Cells,* **93**, 465–475.

42. Krebs FC, Fyenbo J, Jørgensen M (2010) Product integration of compact roll-to-roll processed polymer solar cell modules: Methods and manufacture using flexographic printing, slot-die coating and rotary screen printing, *J Mater Chem,* **20**, 8994–9001.

43. Krebs FC, Gevorgyan SA, Alstrup J (2009) A roll-to-roll process to flexible polymer solar cells: Model studies, manufacture and operational stability studies, *J Mater Chem,* **19**, 5442–5451.

44. Krebs FC, Jørgensen M, Norrman K, et al. (2009) A complete process for production of flexible large area polymer solar cells entirely using screen printing—first public demonstration, *Sol Energy Mater Sol Cells,* **93**, 422–441.
45. Krebs FC, Nielsen TD, Fyenbo J, Wadstrøm M, Pedersen MS (2010) Manufacture, integration and demonstration of polymer solar cells in a lamp for the "Lighting Africa" initiative, *Energy Environ Sci,* **3**, 512–525.
46. Krebs FC, Søndergaard R, Jørgensen M (2011) Printed metal back electrodes for R2R fabricated polymer solar cells studied using the LBIC technique, *Sol Energy Mater Sol Cells,* **95**, 1348–1353.
47. Krebs FC, Tromholt T, Jørgensen M (2010) Upscaling of polymer solar cell fabrication using full roll-to-roll processing, *Nanoscale,* **2**, 873–886.
48. Krebs FC, Fyenbo J, Tanenbaum DM, et al. (2011) The OE-A OPV demonstrator anno domini 2011, *Energy Environ Sci,* **4**, 4116–4123.
49. Lange A, Wegener M, Boeffel C, Fischer B, Wedel A, Neher D (2010) A new approach to the solvent system for inkjet-printed P3HT:PCBM solar cells and its use in devices with printed passive and active layers, *Sol Energy Mater Sol Cells,* **94**, 1816–1821.
50. Larsen-Olsen TT, Andreasen B, Andersen TR, et al. (2012) Simultaneous multilayer formation of the polymer solar cell stack using roll-to-roll double slot-die coating from water, *Sol Energy Mater Sol Cells,* **97**, 22–27.
51. Larsen-Olsen TT, Machui F, Lechene B, et al. (2012) Round-robin studies as a method for testing and validating high-efficiency ITO-free polymer solar cells based on roll-to-roll-coated highly conductive and transparent flexible substrates, *Adv Energy Mater,* DOI: 10.1002/aenm.201200079.
52. Lewis JE, Lafalce E, Toglia P, Jiang X (2011) Over 30% transparency large area inverted organic solar array by spray, *Sol Energy Mater Sol Cells,* **95**, 2816–2822.
53. Manceau M, Angmo D, Jørgensen M, Krebs FC (2011) ITO-free flexible polymer solar cells: From small model devices to roll-to-roll processed large modules, *Org Electron,* **12**, 566–574.
54. Medford AJ, Lilliedal MR, Jørgensen M, et al. (2010) Grid-connected polymer solar panels: Initial considerations of cost, lifetime, and practicality, *Opt Express,* **18**, A272–A285.
55. Na SI, Yu BK, Kim SS, et al. (2010) Fully spray-coated ITO-free organic solar cells for low-cost power generation, *Sol Energy Mater Sol Cells,* **94**, 1333–1337.

56. Nielsen TD, Cruickshank C, Foged S, Thorsen J, Krebs FC (2010) Business, market and intellectual property analysis of polymer solar cells, *Sol Energy Mater Sol Cells,* **94**, 1553–1571.

57. Park SY, Kang YJ, Lee S, et al. (2011) Spray-coated organic solar cells with large-area of 12.25 cm^2, *Sol Energy Mater Sol Cells,* **95**, 852–855.

58. Powell C, Bender T, Lawryshyn Y (2009) A model to determine financial indicators for organic solar cells, *Sol Energy,* **83**, 1977–1984.

59. Rösch R, Krebs FC, Tanenbaum DM, Hoppe H (2012) Quality control of roll-to-roll processed polymer solar modules by complementary imaging methods, *Sol Energy Mater Sol Cells,* **97**, 176–180.

60. Schrödner M, Sensfuss S, Schache H, et al. (2012) Reel-to-reel wet coating by variation of solvents and compounds of photoactive inks for polymer solar cell production, *Sol Energy Mater Sol Cells,* **107**, 283–291.

61. Seeland M, Roesch R, Hoppe H (2011) Luminescence imaging of polymer solar cells: Visualization of progressing degradation, *J Appl Phys,* **109**, 064513.

62. Søndergaard R, Manceau M, Jørgensen M, Krebs FC (2012) New low-bandgap materials with good stabilities and efficiencies comparable to P3HT in R2R-coated solar cells, *Adv Energy Mater,* **2**, 415–418.

63. Søndergaard R, Makris T, Lianos P, et al. (2012) The use of polyurethane as encapsulating method for polymer solar cells—an inter laboratory study on outdoor stability in 8 countries, *Sol Energy Mater Sol Cells,* **99**, 292–300.

64. Wengeler L, Schmidt-Hansberg B, Peters K, Scharfer P, Schabel W (2011) Investigations on knife and slot die coating and processing of polymer nanoparticle films for hybrid polymer solar cells, *Chem Eng Process,* **50**, 478–482.

65. Yu J-S, Kim I, Kim J-S, et al. (2012) Silver front electrode grids for ITO-free all printed polymer solar cells with embedded and raised topographies, prepared by thermal imprint, flexographic and inkjet roll-to-roll processes, *Nanoscale,* **4**, 6032–6040.

66. Zhang B, Chae H, Cho SM (2009) Screen-printed polymer:fullerene bulk-heterojunction solar cells, *Jpn J Appl Phys,* **48**, 020208.

67. Zimmermann B, Schleiermacher H, Niggemann M, Wuerfel U (2011) ITO-free flexible inverted organic solar cell modules with high fill factor prepared by slot die coating, *Sol Energy Mater Sol Cells,* **95**, 1587–1589.

Chapter 14

Module Design, Fabrication, and Characterization

Roland Rösch and Harald Hoppe
Institute of Physics, Ilmenau University of Technology, Ilmenau 98693, Germany
roland.roesch@tu-ilmenau.de

Cost-effective production of polymer solar modules is ensured by fast, large-area printing techniques in roll-to-roll processes. With these processes each necessary layer of a polymer solar cell module can be applied on a flexible substrate. Nevertheless, further steps are required for efficient mass production of polymer solar modules. At first, an efficient module design that maximizes power generation and minimizes power losses on a given area is required. To translate the theoretical specification into real applications, structuring processes have to be applied that enable structuring widths within the range of ~100 μm. Finally, an efficient characterization of the produced modules has to be applied in order to evaluate their quality and stability. With regard to fast in-line inspection during production, optical imaging methods are currently of interest. This chapter will give a comprehensive overview, from basics to advanced concepts, on polymer solar module design, fabrication challenges, and postproduction characterization techniques.

Organic Solar Cells: Fundamentals, Devices, and Upscaling
Edited by Barry P. Rand and Henning Richter

ISBN 978-981-4463-65-2 (Hardcover), 978-981-4463-66-9 (eBook)
www.panstanford.com

14.1 Motivation

Conjugated polymer–based organic photovoltaics has matured considerably in terms of power conversion efficiency (PCE) and lifetime during the past decade. Efficiencies for cells processed in laboratories are currently exceeding 10% [1–3], while values for the lifetime of such cells of several thousand hours under artificial illumination were reported multiple times [4–9]. These achievements together with the basic advantages of thin film photovoltaics, that is, lightweight, flexibility, potential semitransparency, tunable color, and high production speed, allow polymer photovoltaics to enter first markets. Possible applications for this third-generation photovoltaic technology are the same as for all thin film photovoltaics: from small gadget applications, where flexibility, low cost, and light weight are required for economic self-sustaining power generation, to building-integrated photovoltaics (BIPVs), where the potential semitransparency and tunable color provide additional value to building façades, besides providing the ability of power generation and thus an ecological and economic amortization [10]. In Fig. 14.1 several long-term existing gadget applications from the first wave of thin film photovoltaics are depicted.

Figure 14.1 Several examples of gadget applications of thin film photovoltaic technologies. In this field of application, lightweight and high flexibility are essential for economic self-sustaining power generation. Reprinted with permission from [11].

These constitute exemplarily some early economically viable applications. The first commercialization of polymer-based photovoltaics took place in bags, in which a buffer battery was charged and used as an energy bank for mobile electronic applications such as cell phones.

Figure 14.2 shows such an early realization of market outreach.

Figure 14.2 Early applications of polymer-based solar modules for powering small-scale fixed installations and mobile electronic applications.

According to the widely accepted roadmap developed by the Organic Electronics Association, the next step in commercialization concerns the integration of organic-based photovoltaics (OPV) into buildings, either into the façade or into windows. Figure 14.3 shows a façade of a building fully covered with opaque and semitransparent thin film photovoltaics. In this case the semitransparent modules were used to cover the windows.

One important issue, that every thin film photovoltaic technology requires to overcome before it goes into market, is the small power that a single solar cell can provide. Although a single polymer-based solar cell with very large area could be produced

[12,13], its potentially high current output is strongly limited by inner serial resistances, which scale proportional to the device's area [14–17]. On the other hand, a single photovoltaic device delivers always a very small photovoltage [18,19], which is always smaller than the absorber's bandgap and thus in general limited to ~1 V for visible light absorption.

Figure 14.3 Thin film solar cell technology power generating system on the whole face of a building. Semitransparent modules are used for and integrated into the windows. Reprinted with permission from [11].

Since all applications require larger voltages, a serial interconnection of many cells to a so-called "module" is essential for efficient solar power application on commercially realizable scales. Likewise to stacking of batteries, resulting in the addition of the individual cell's voltage, modules of serially connected solar cells provide useful voltage ranges depending on the application. A further benefit of the higher voltage is that for the same power output the currents remain smaller, resulting in less Joule losses within the electric circuit.

The major disadvantage of a serially interconnected solar module is its lower "geometric fill factor (GFF)," defined by the area of photoactive regions compared with the total module area that includes the space required for interconnection of the individual solar cells. Furthermore, serial interconnections realized by Ohmic contacts between different electrode materials may induce additional inner serial resistances of the power converting device.

In this chapter, we begin with a discussion of the design rules for efficient polymer solar modules based on electrical simulations. Next, several roll-to-roll compatible and thus high-throughput processing techniques will be compared and discussed in terms of their applicability for solar module production. Thereafter, standard characterization and advanced imaging methods, specifically useful for the analysis of large-area solar modules, will be presented and the chapter closes by presenting state-of-the-art solar module lifetimes.

14.2 Architecture of Polymer Solar Cells

Polymer solar cells with "conventional" architecture receive their light through the substrate and therefore feature a transparent conductive electrode on the substrate that is collecting and transporting the holes [20]. As the charge carrier mobility of the photoactive layer in polymer solar cells is comparatively small with respect to inorganic photoactive materials [21], the transparent electrode needs to cover the whole substrate in order to collect all photogenerated holes. Typical values for charge carrier mobilities of organic semiconducting materials are in the order of ~10^{-6} to 10^{-2} cm^2/Vs. In case of insufficient selectivity for holes, a hole transport layer (HTL) is applied on top of this electrode. As a result, charge carrier recombination at this contact gets reduced considerably and the rectification of the current–voltage characteristics gets improved. The next layer is the photoactive layer, in which light absorption and charge photogeneration take place. For OPV, it usually consists of an electron donor and an electron acceptor that are intimately intermixed within a so-called bulk heterojunction [22]. Upon light absorption, at first Coloumbically bound excitons are created, which will be separated due to the energy offset for the electrons at the heterojunction, yielding free electrons and holes within the acceptor and donor, respectively. Atop the photoactive layer an optional electron transport layer (ETL) follows, into which the electrons are collected selectively. The final layer that completes a fully operational solar cell is the back contact, which in this architecture has to extract and transport the electrons. Of course this contact needs to be transparent as well if semitransparency for the whole device is desired.

A common alternative to the conventional architecture is called the "inverse" architecture [23]. The main difference from the conventional architecture is that HTL and ETL exchange their place and thus the transparent contact on top of the substrate has to transport the electrons, while the back contact has to collect the holes. The major advantage of this inverse architecture is that the back contact—as it has to extract the holes—is usually a metal with high work function, for example, silver or gold, that is much more resistant against oxidation than low-work-function metals commonly used for electron extraction [4,24].

In Fig. 14.4 cross sections of the complete layer stacks for both architectures are displayed.

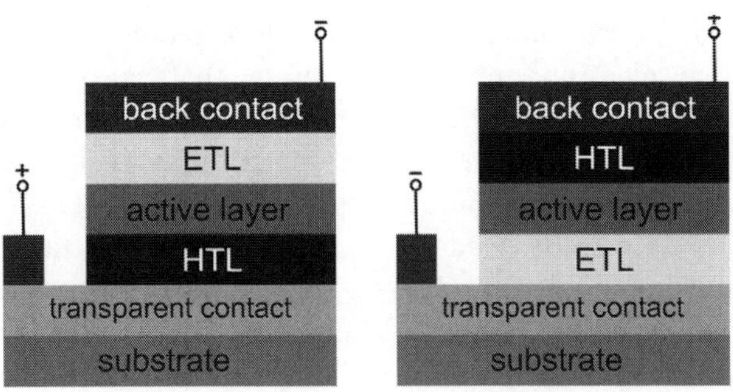

Figure 14.4 Cross-sectional schematics through the layer stack of polymer solar cells with conventional architecture (a) and inverse architecture (b) are shown. The images are not true to scale.

14.3 Material Properties

14.3.1 Photoactive Materials

In polymer-based photovoltaic devices, commonly a blend of two materials is used for photon to charge carrier conversion, due to the fact that the light absorbing material, for example, a conjugated polymer, possesses a very large exciton binding energy [25]. Therefore, a combination of two organic materials is applied, since at the heterojunction a sufficient energy level offset enables exciton dissociation via photo-induced electron transfer [26]. Subsequently, both electrons and holes are transported to their respective electrode within spatially separated materials. Since exciton lifetimes and consequently exciton diffusion lengths within common organic semiconductors are very limited (~1 ns and 10 nm, respectively), donor and acceptor require to be intimately intermixed on a scale comparable to the exciton diffusion length, in order to warrant large charge generation efficiency. This intimate mixture between donor and acceptor is called a "bulk heterojunction" [27]. The final requirement concerns the efficient extraction of charge carriers through percolating transport paths within each of the hole and electron transporting materials. The major advantage of such a bulk heterojunction lies in the fact that light absorption throughout the whole depth of the active layer yields charge generation [28], whereas the largest challenge is the formation of well-percolating transport phases for both signs of charge carriers.

In the following, two conjugated polymers commonly applied as electron donors in polymer solar cells will be exemplarily introduced. The most thoroughly investigated organic semiconducting polymer is poly(3-hexylthiophene) (P3HT). Polythiophenes have been investigated for about two decades as organic semiconductors [29], and P3HT has already been applied for one decade within polymer solar cells and is today a well-established reference material [30,31]. However, due to its large bandgap of about 2 eV, it is strongly limited in its maximum PCE [32]. To overcome the strict limitations in PCE, many new low-bandgap materials have been developed during the last years. One of them, poly[[9-(1-octylnonyl)-9H-carbazole-2,7-diyl]-2,5-thiophenediyl-2,1,3-benzothiadiazole-4,7-diyl-2,5-thiophenediyl] (PCDTBT) [33] became another widespread material providing PCEs of up to ~7% [34]. Besides these two materials, today a good number of polymers can yield above 8% PCE.

As electron acceptor materials mostly C_{60} and C_{70} derivatives are used, since they are soluble in the same solvents as these conjugated polymers. In Fig. 14.5 the chemical structures of PCDTBT, P3HT, and the two most common fullerene derivatives, namely phenyl-CXX-butyric acid methyl ester (PCBM), where XX equals 61 or 71, are shown.

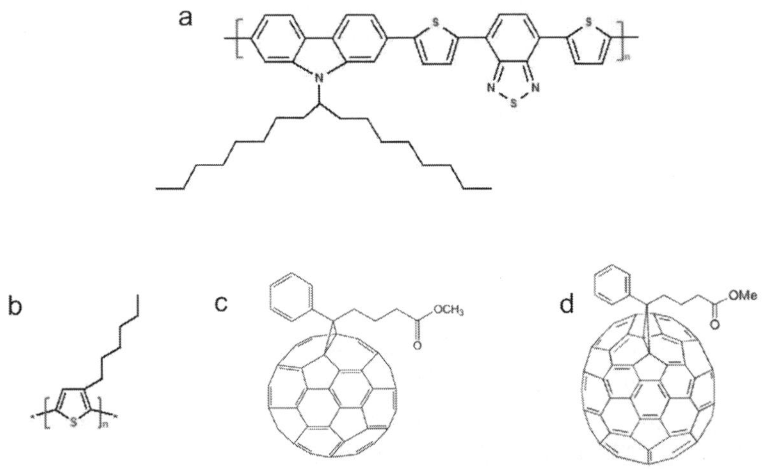

Figure 14.5 Chemical structures of PCDTBT (a), P3HT (b), $PC_{60}BM$ (1-(3-methoxycarbonyl)-propyl-1-phenyl-(6,6)C61) (c), and $PC_{70}BM$ (1-(3-methoxycarbonyl)-propyl-1-phenyl-(6,6)C71) (d).

14.3.2 Charge Carrier Transport/Blocking Layers

As shown in Fig. 14.4 the so-called charge carrier transport layers connect the photoactive layer with the electrodes of a polymer solar cell. Their task is to provide barrier-free, for example, Ohmic, electrical contact between the specific charge transport energy levels of the photoactive layer and the respective charge collecting electrodes. In general, the lowest unoccupied molecular orbital (LUMO) of the acceptor and the highest occupied molecular orbital (HOMO) of the donor in the bulk heterojunction correspond to the conduction band and the valance band of inorganic semiconductors and are thus responsible for the charge carrier transport of electrons and holes. These energy levels then have to be matched by the electron and hole transport layers. Furthermore, due to their selectivity, the charge transport layers prevent the wrong sign of charge carrier from reaching the charge extracting electrode and hence minimize shunting and recombination losses within the photovoltaic device.

ETLs are not always used, since some low-work-function metals like aluminum already allow a barrier-free electrical contact to the LUMO of PCBM. Zinc oxide [35], titanium oxide [36], and 1,10-bathophenanthroline (BPhen) [37] (refer to Fig. 14.6) are some examples of ETLs.

HTLs can also be made from organic or inorganic materials, for example, metal oxides. The most commonly used material is poly(3,4-ethylenedioxythiophene) : poly(styrenesulfonate) (PEDOT : PSS) (refer Fig. 14.6). It is provided as an aqueous dispersion and thus processable without vacuum. Nevertheless, metal oxides such as molybdenum oxide [38], nickel oxide [39], or vanadium oxide [40] also perform well or even superior compared with PEDOT : PSS. For more information about buffer layers, see Chapter 4.

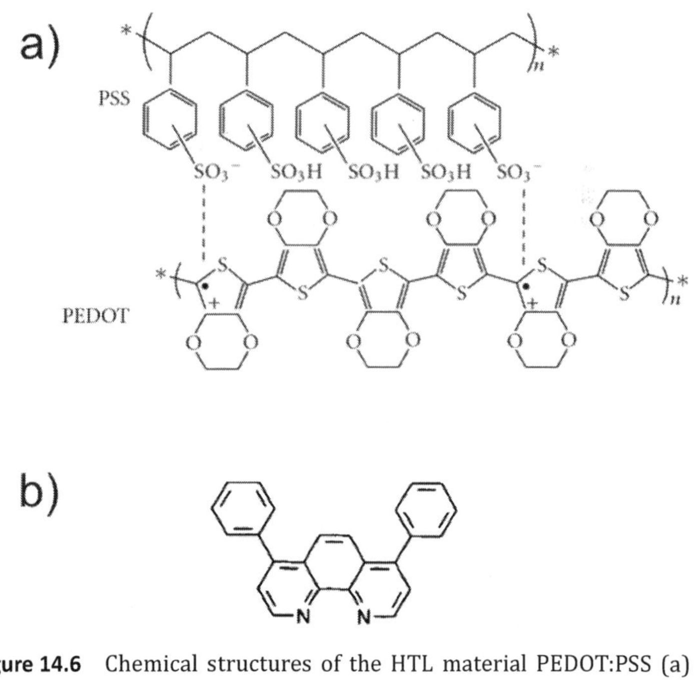

Figure 14.6 Chemical structures of the HTL material PEDOT:PSS (a) and the ETL material BPhen (b).

14.3.3 Electrodes

Electrodes in polymer solar cells consist of opaque metals like aluminum, silver, or gold at the reflective back contact, and out of semitransparent, that is, good compromise between transparency and conductivity, materials or material combinations for the front contact, facing the light incidence. The best compromise between semitransparency and conductivity comes from indium tin oxide (ITO) among all metal oxides. The major technical disadvantage of this material is that it has to be deposited in vacuum under high temperature in order to yield reasonably small sheet resistances. Alternative, solution-based and thus vacuum-free semitransparent electrodes can be made out of highly conductive PEDOT:PSS derivates, graphene, carbon nanotubes, printed silver meshes and nanowires; see, for example, [28,41–43].

Besides a potential transparency, one main feature to be considered for electrode materials is their sheet resistance. While sheet resistances for metals are in the range of 0.1 Ω/□ and lower, the values for semitransparent materials are often two to three orders of magnitude larger [44]. For more information about electrodes, see Chapter 5.

14.4 Monolithic Polymer Solar Modules

In the same manner as accumulators or batteries, solar modules can be built up by either parallel or serial interconnection of single solar cells. While in a parallel interconnection all the collected currents from each cell add up, a serially interconnected module adds up all the voltages. Therefore, serially interconnected modules are in general more efficient, since internal dissipative Joule losses P_{loss} are proportional to the resistance R and the square of the current I^2:

$$P = R \cdot I^2 \tag{14.1}$$

Typically, silicon solar modules are built up by connecting each separately produced single cell through wires. This is justified because the width of Si ingots, and thus each product out of it, is limited strictly in size.

On the contrary, polymer solar cells and modules are mostly made of soluble components. This allows building up these modules as a whole on one and the same substrate, which is called "monolithic" design (based on the Greek word *monolithos*, which means "single stone") [45–47]. This concept does not require any time-consuming patching of the individual cells to a module. A schematic cross section of a simple monolithically and serially interconnected polymer solar module is shown in Fig. 14.7. Between every two cells, there is an area for the serial interconnection that cannot contribute to the power conversion. The GFF is now defined as the ratio between the area that can be utilized for power conversion (the active area) and the total area of the module (i.e., the total substrate). Assuming a stripe design for the subcells with length l in current transport direction and a distance Δl between them, the GFF becomes

$$\mathrm{GFF} = \frac{\text{Active area}}{\text{Total area}} = \frac{l}{l + \Delta l} \tag{14.2}$$

A conceptual disadvantage of this design on the other hand is the need for deposition and structuring of each layer subsequently. In Fig. 14.8 a sequence of deposition and structuring steps for processing a monolithic polymer solar module with the layer stack of substrate/transparent front contact/HTL/active layer/back contact is shown. A big challenge is not to damage the underlying layers upon the structuring process. Common methods for structuring in the laboratory are wet chemical etching (e.g., for ITO), mechanical scribing of the HTL/active layer/ETL stack, and evaporation through a shadow mask for creating the metal back electrode. Alternatively, deposition methods that inherently provide a structured layer, for example, screen or flexographic printing, can be applied for avoiding additional efforts in structuring.

Figure 14.7 Schematic cross section of a monolithically built-up polymer solar module with conventional architecture. Proportions are not true to scale.

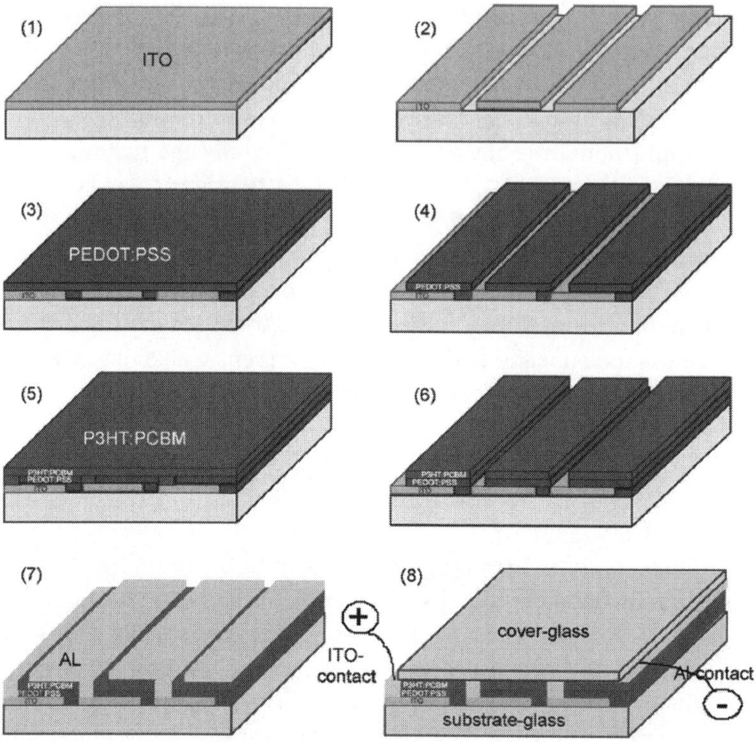

Figure 14.8 Sequential processing of a monolithic polymer solar module. After deposition of each layer, it has to be structured without damaging the underlying layers. In this example, glass was used as substrate, ITO as semitransparent front contact, PEDOT:PSS as HTL, P3HT:PCBM as active layer and aluminum as back contact. Sealing the device under a cover glass is done for protection of the device. Reprinted with permission from [45].

14.5 Efficient Module Design

If besides the PCE of the photoactive layer only the GFF would have to be considered, the performance of a monolithic solar module could be optimized by maximizing the ratio between the solar cell length and the sum of cell length and cell distance (compare Eq. 14.2). In general, the module efficiency η_{module} can be deduced from the cell efficiency η_{cell} as follows:

$$\eta_{\text{module}} = \eta_{\text{cell}} \cdot \frac{l}{l+\Delta l} = \eta_{\text{cell}} \cdot \text{GFF} \qquad (14.3)$$

However, due to the limited lateral conductivity of the current collecting electrodes, the cell efficiency becomes a function of the solar cell length l ($\eta_{\text{cell}} = \eta_{\text{cell}}(l)$), and dissipative power losses within the electrodes are to be taken into account as well. Hence the optimum solar cell length is limited by the current generation and the sheet resistance of the lower conductive charge collecting electrode [14,44,48–51]. Furthermore, for infinitesimal solar cell distances Δl, also the contact resistance between both electrodes critically increases and becomes performance limiting [44]. In conclusion, the optimum balance between GFF, solar cell length, and solar cell distances has to be determined.

Hence the maximum PCE of a monolithic polymer solar module depends not only on the active materials used, but also on the geometric design of the module. In Fig. 14.9 a cross section of such a module is depicted. Between each subcell with length l there is a cell distance Δl for interconnecting the cells, which does not contribute to power conversion.

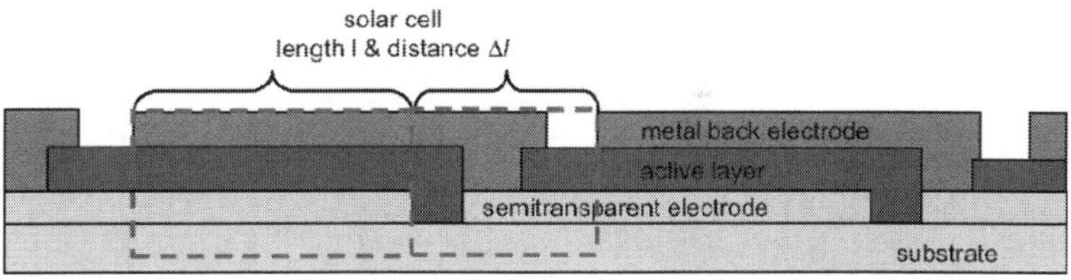

Figure 14.9 Cut-out of a cross section of a monolithic polymer solar module exposing the serial interconnection. Reprinted with permission [44].

Considering the current transport direction within the monolithic solar module, the optimization problem can be limited to one subcell and one interconnection region due to periodic boundary conditions (compare with Fig. 14.10). For simplicity, the interconnection region between the individual solar cells consists of three parts of equal length: the contact pad length and the two electrode bridge lengths.

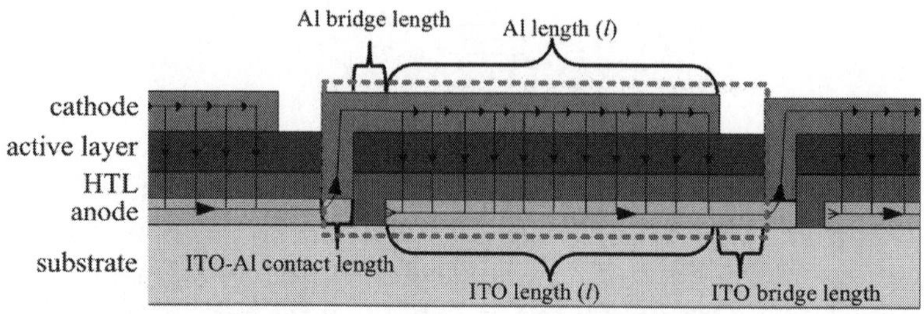

Figure 14.10 Confining the optimization problem for the module performance to one subcell and one interconnection region using periodic boundary conditions. For simplicity, the interconnection region consists of three parts of equal length (here, ITO-Al contact pad length, Al bridge length, and ITO bridge length). Reprinted with permission from [44].

In order to calculate all resistive power losses, the problem can now be divided into the following contributions:

- the sheet resistance of the back electrode,
- the sheet resistance of the front electrode,
- the contact resistance between both electrodes within the interconnection region,
- the serial resistance of the photoactive layer.

Since the current collected by and transported within the electrodes depends on the position x of the subcell in current transport direction, the power dissipation based on sheet resistances within the electrodes has to be divided into infinitesimal contributions:

$$dP(x,V) = I(x,V)^2 \cdot dR, \text{ with } dR = R_\square \frac{dx}{w} \tag{14.4}$$

with V the applied voltage, dx an infinitesimal length within current transport direction, and w the solar cell width orthogonal to current transport direction. Integrating over the whole solar cell length (between $x = 0$ and $x = l$) leads to

$$P(V) = I(V)^2 \cdot l \cdot \frac{R_\square}{3w} \tag{14.5}$$

From this equation an effective sheet resistance R_{eff} can be defined for each of the electrodes as

$$R_{\text{eff}} = R_{\square}\frac{l+\Delta l}{3w} \tag{14.6}$$

The contacting length between the front and back contact was set as one-third of the total cell distance, leading to the contact resistance as

$$R_{\text{contact}} = \rho_{\text{c}}\frac{3}{\Delta l \cdot w} = \rho_{\text{c}}\frac{3}{A_{\text{interconnect}}} \tag{14.7}$$

with ρ_c the specific contact resistance between the two electrode materials. The parallel resistance of the active layer simply scales with the cell area:

$$R_{\text{p}} = \rho_{\text{P}}\frac{1}{l \cdot w} = \rho_{\text{P}}\frac{1}{A_{\text{cell}}} \tag{14.8}$$

with ρ_P as the specific parallel resistance of the active material.

To calculate the photovoltaic parameters of a subcell under the influence of all above-mentioned serial and parallel resistances, the implicit Shockley equation for one diode can be used:

$$I = I_0\left\{\exp\left(\frac{e}{nkT}(V - IR_{\text{S}})\right) - 1\right\} + \frac{V - IR_{\text{S}}}{R_{\text{P}}} - I_{\text{SC}} \tag{14.9}$$

where I, I_0, and I_{SC} are the total current, saturation current and short-circuit current, respectively, V the applied voltage, R_S and R_P the serial and parallel resistances of each periodic unit, T the temperature, k the Boltzmann constant, e the elementary charge, and n the diode ideality factor. To solve this equation all input parameters, except the solar cell length l and the distance Δl, were determined experimentally by characterization of single solar cells and various resistance measurements. The most important experimental values are the sheet and contact resistances, which have to be determined individually for each investigated material system [44]. The solar cell width is a parameter within this calculation and can be set constant for the optimization (here 5 cm), whereas the solar cell length and distance are varied to investigate the geometric influence on the solar cell PCE.

In Figs 14.11 and 14.12, results of this simulation for a single solar cell with ITO front contact, aluminum back contact, and

P3HT:PCBM as photoactive material and a virtual interconnection region are shown. It is evident that the PCE is mainly controlled by the cell length and almost independent from the contact distance or interconnection length except for very small or very large values. Therefore, the PCE and the fill factor follow roughly inversely the shape of the serial resistance.

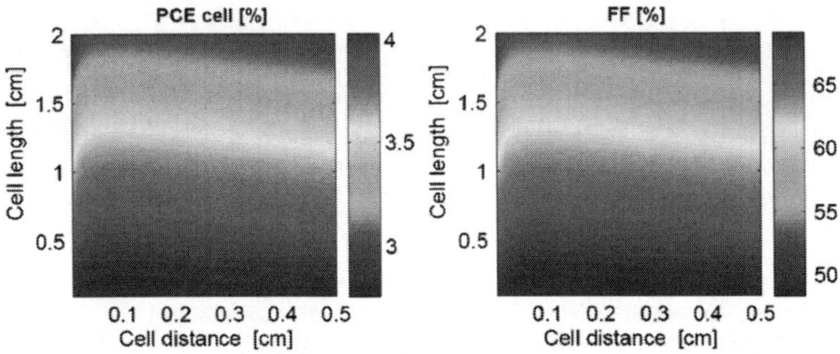

Figure 14.11 Calculated PCE and FF of a single solar cell for a set of cell lengths and cell distances. Since only solar cells are considered, no area losses are taken into account. Reprinted with permission from [44].

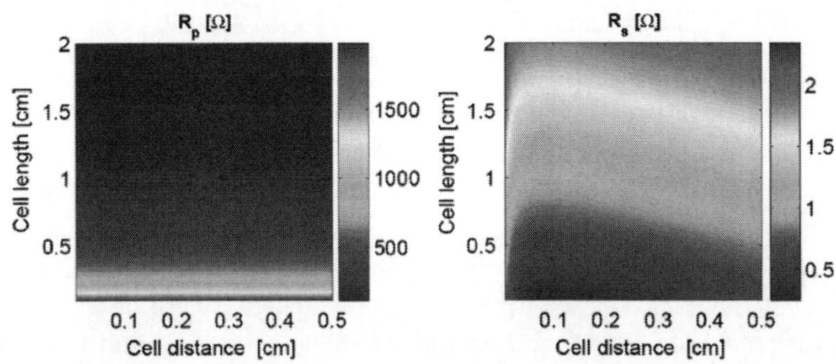

Figure 14.12 Calculated parallel (R_P) and serial resistances (R_S) of a single solar cell for a set of cell lengths and cell distances. Whereas R_P only depends on the cell length, R_S depends on both, the cell length and cell distance. Reprinted with permission from [44].

To elucidate the individual contribution of all parts that were considered in the serial resistance, each of them were calculated separately. The result is drawn in Fig. 14.13. While the aluminum sheet resistance is negligibly small, the ITO sheet resistance, both for the active area and within the interconnection region, results in major contributions to the total series resistance. The contact resistance between the front and back contact on the other hand only plays a role for very small cell distances below 100 μm for this material system.

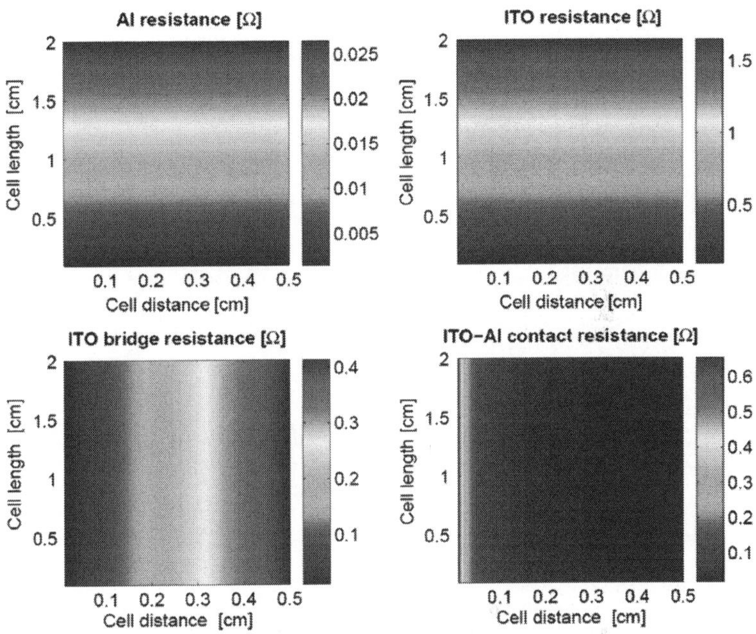

Figure 14.13 Geometric dependencies for all contributions to the serial resistance of a single solar cell. While the aluminum sheet resistance can be neglected, the ITO sheet resistance has a major impact for long cell lengths and interconnections ("bridge"). The contact resistance between both electrodes has a remarkable influence only for very short cell distances. Reprinted with permission from [44].

Once all resistive contributions are accounted for, the PCEs can be calculated for a set of solar cell lengths and distances for the solar submodule. For the module efficiency, also areal losses have to be taken into account. In the following, a set of three different substrate front contact material combinations are discussed: ITO on glass, ITO on plastic foil, and high-conductivity PEDOT : PSS on plastic foil. In Tables 14.1 and 14.2, the experimentally determined sheet and specific contact resistances for these semitransparent electrodes in combination with aluminum as a back electrode are shown.

Table 14.1 Experimentally determined sheet resistances for three different semitransparent electrode materials and aluminum [44]

Material	Layer thickness (nm)	Sheet resistance (Ω/□)
Aluminum	135	0.16 ± 0.01
ITO on glass	150	12.5 ± 0.6
ITO on PET	150	50 ± 3
High-conductivity PEDOT:PSS	150	100 ± 20

Table 14.2 Experimentally determined specific contact resistances between three different semitransparent electrode materials and aluminum [44]

Aluminum vs.	ρ_c (mΩ cm^2)
ITO on glass	11 ± 1
ITO on PET	17 ± 1
High-conductivity PEDOT:PSS	200 ± 20

Due to the different sheet and contact resistances, these electrode combinations resulted in different module PCEs, as shown in Fig. 14.14. The influence of the dissipative Joule losses for all material combinations is reflected by the dramatic loss of PCE for long cell lengths and distances, in contrast to the single solar cell. This becomes more pronounced for the semitransparent electrodes exhibiting larger sheet resistances.

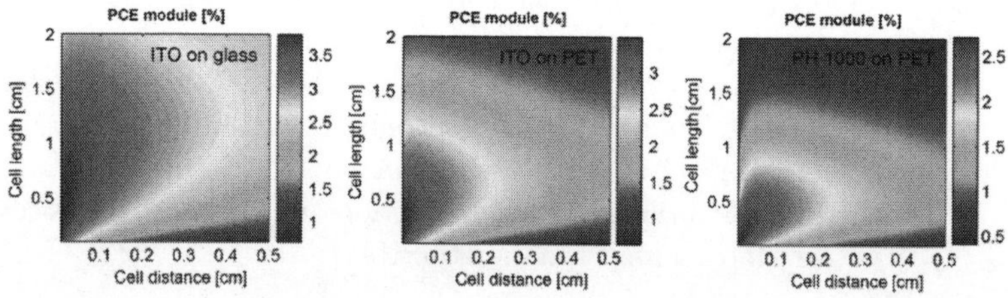

Figure 14.14 Comparison of the geometric dependency of the PCE of a polymer solar module for three different substrate front electrodes in combination with aluminum as back electrode: ITO on glass (left), ITO on plastic foil (center), and highly conductive PEDOT:PSS on plastic foil (right). The sheet resistance and contact resistance increase from left to right, making the area and resistive power losses more pronounced. Reprinted with permission from [44].

For example, for a cell length of 0.5 cm and a cell distance of 0.05 cm, which is feasible by laser structuring processes (refer to Section 14.7), maximum module PCEs of 3.5%, 3.1%, and 2.4% can be obtained for ITO on glass, ITO on plastic foil, and highly conductive PEDOT:PSS on plastic foil, respectively.

In conclusion, these simulations lead to two important design rules for monolithic thin film solar modules:

- Sheet resistances of and contact resistances between electrodes should be as small as possible.
- Individual solar cell lengths and solar cell interconnection distances have to be adapted to the electrode resistances.

In general, the findings presented in this section are transferable to any thin film photovoltaic material system by explicitly considering the material properties of electrodes and those of the photoactive layer expressed by the Shockley equation.

14.6 Characterization Methods

14.6.1 Current–Voltage Characterization

The most essential but also simplest characterization method for solar cells and modules is the measurement of their current–voltage (*I–V*) characteristics. The *I–V* characteristics of a solar cell under illumination can in general be described by the implicit Shockley equation (compare with Eq. 14.9). Even though it has been developed on the basis of an illuminated p–n junction, it is also well applicable for the description of OPV devices.

A typical *I–V* characteristic in the dark and under illumination for a single solar cell is shown in Fig. 14.15, where I_{SC} denotes the short-circuit current, V_{OC} the open-circuit voltage, and FF the fill factor. It is the ratio between $V_{MPP} \times I_{MPP}$ and $V_{OC} \times I_{SC}$, where V_{MPP} and I_{MPP} are the corresponding voltage and current at the "maximum power point" located in the fourth quadrant. The PCE of a solar cell is calculated as follows:

$$\mathrm{PCE} = \frac{P_{out}}{P_{in}} = \frac{\mathrm{FF} \cdot V_{OC} \cdot I_{SC}}{P_{in}} \tag{14.10}$$

The incoming power P_{in} coincides with the incoming light intensity, whose spectrum should be as close to the standard AM1.5 solar spectrum as possible. Here AM1.5 stands for "air mass 1.5," that is, a path through 1.5 times the thickness of the atmosphere, usually achieved for solar incidence at an angle of 48.2°.

The photovoltaic parameters I_{SC}, V_{OC}, and FF underlie different limitations:

- The short-circuit current is mainly limited by charge generation and recombination within the photoactive layer and at the interfaces to charge transport layers and the electrodes.
- The open-circuit voltage is mainly limited by the energetic separation of the two charge transporting energy levels of the donor and the acceptor. Further losses may arise from the contacts and charge recombination.
- The fill factor is mainly influenced by the serial and parallel resistances, the mobility–lifetime product of the charge carriers and an imbalance in charge carrier mobilities and thus space charge regions.

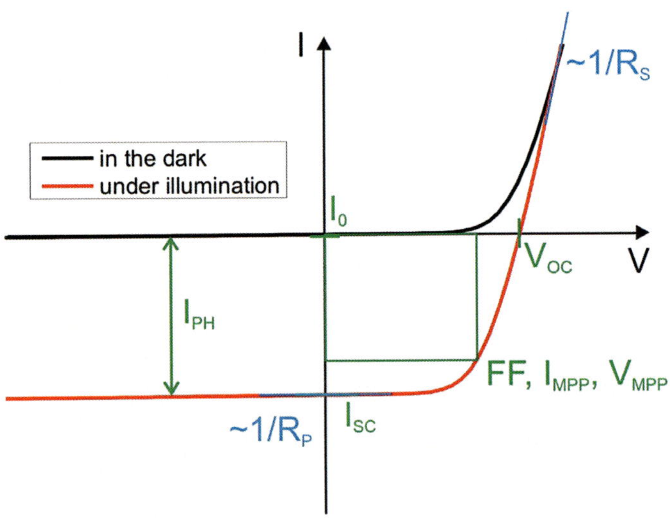

Figure 14.15 Qualitative representation of an I–V characteristic in the dark and under illumination of a solar cell, shown together with all important photovoltaic parameters.

What can now be understood from an I–V characterization of a solar module? Just like in the case of serially stacked batteries, in monolithic solar modules the series connection results in the addition of voltages supplied by each of the solar cells. In the ideal case, each identical subcell provides the same open-circuit voltage to the module, which results in a total value of

$$V_{\text{OC,module}} = N \cdot V_{\text{OC,cell}} \tag{14.11}$$

where N is the number of subcells in the module. As a consequence of the serial interconnection, the current running through the whole module has to be identical to the one running through each subcell. Therefore, the overall current is limited by the worst performing subcell in the module.

As detailed in Section 14.5, the PCE of a solar module is dependent on the whole module area. Since additional area is required for subcell interconnections and spacing at the edges, the module power conversion efficiency (MPCE) is strictly smaller than the corresponding PCE of its constituent solar cells. In order to separate influences arising from area losses and the quality of the individual solar cells, for example, for scientific evaluations, often the PCE of a solar module with respect to the total active area of all subcells is considered besides the ultimate MPCE.

For illustration, an image of a serially interconnected polymer solar module with four subcells is shown together with the corresponding module area and the active area of one subcell in Fig. 14.16. The corresponding I–V characteristic of this solar module, measured under solar simulation with 1000 W/m^2, is depicted in Fig. 14.17. The calculated PCE corresponding to the total module area as well as to one subcell area is also depicted. As the ratio between total active area (4×4 cm^2) and the total module area (25 cm^2) is 16/25 or ~2/3, the resulting MPCE (2%) is about 2/3 of the cell PCE (3%).

Measurement of the I–V characteristics is done by a source measurement unit, where the voltage is swept and remeasured simultaneously with the resulting current. This process must not be done too fast in order to avoid charging effects arising from the area-dependent capacitance of the device under test. For example, the noise seen on the I–V characteristics of the solar module in this case (Fig. 14.17) stems from capacitive couplings during the measurement itself.

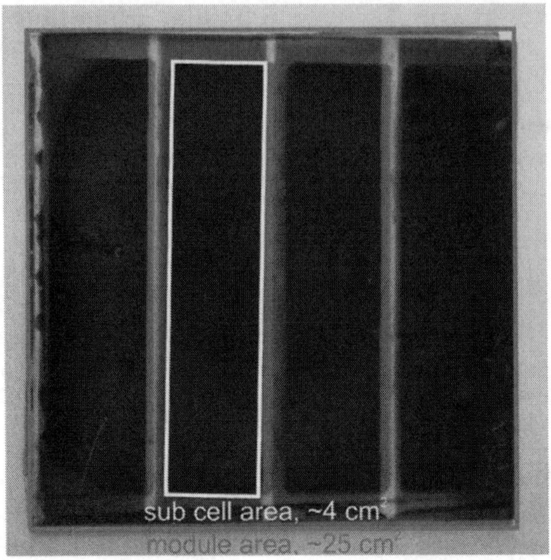

Figure 14.16 Photographic image of a polymer solar module on glass substrate consisting of four subcells. The total module area and the area of a single subcell are highlighted.

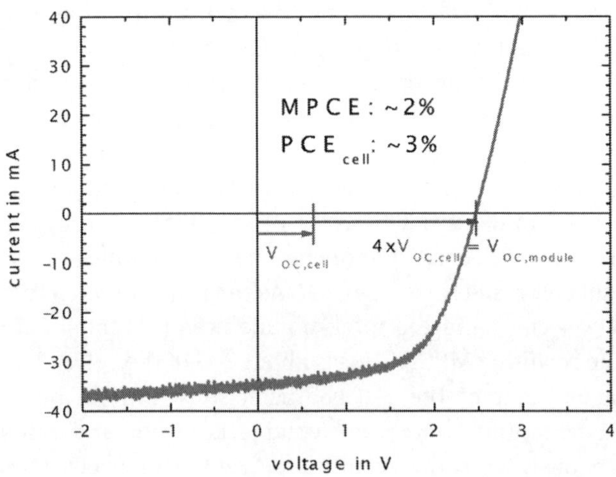

Figure 14.17 Measured *I*–*V* characteristics under 1 sun illumination intensity and the PCEs corresponding to the total module area and to the total subcell area of the polymer solar module as shown in Fig. 14.16. From the open-circuit voltage of the module (~2.4 V), the average value of one subcell (~0.6 V) can be derived.

14.6.2 Imaging Methods

Whereas laboratory solar cell devices often have an area between 1 mm^2 and 1 cm^2, solar modules are designed for large-scale applications. In fact, the laboratory scale devices are designed such that the probability of finding a defect within the solar cell active area is minimized; however, this cannot be prevented from the beginning for large-scale solar modules. In addition, the higher complexity of a module device due to the incorporation of, for example, serial interconnections may result in a number of novel defects, unseen in single solar cells. Unfortunately such defects can neither be detected nor distinguished nor characterized exactly and unambiguously by simple *I–V* characterization. Hence the need for characterization methods providing locally resolved analysis is rising with upscaling of the device area and its complexity. Various imaging methods are already well established in the two-dimensional quality control of inorganic solar cells or other items of high-throughput mass production.

Among these methods, luminescence imaging [52–59], lock-in thermography [60–62], and light beam induced current (LBIC) [63–66] are of high relevance. Furthermore, all of these methods have also proven their applicability to polymer solar cells and modules [15,16,24,67–76].

In the following, these three main imaging characterization methods will be introduced in combination with single examples.

Luminescence imaging makes use of the capability of the organic semiconductor or the respective semiconductor heterojunction to yield radiative de-excitation processes. Excitation can be done either by light or by injection of a current. The experimental setup for performing luminescence imaging is displayed in Fig. 14.18.

In photoluminescence imaging (PLI) the whole device is preferably homogeneously illuminated by a light source with spectral distribution matching the active layer absorption, for example, by light emitting diodes. Electrons within pristine phases of the donor or acceptor semiconductor are lifted into an excited state (HOMO–LUMO transition), followed by de-excitation upon radiative (and nonradiative) recombination within the same material. This process is generally called photoluminescence. In order to detect only the light emitted from the semiconductor, a cut-off filter is placed in front of a sensitive camera, for example, a charge-coupled device silicon detector. As solar cells are designed in such a way that photoinduced charge transfer is maximized and photoluminescence is minimized, even at careful choices of the exciting light source and the blocking filter, reflected excitation light is in competition with the true photoluminescence.

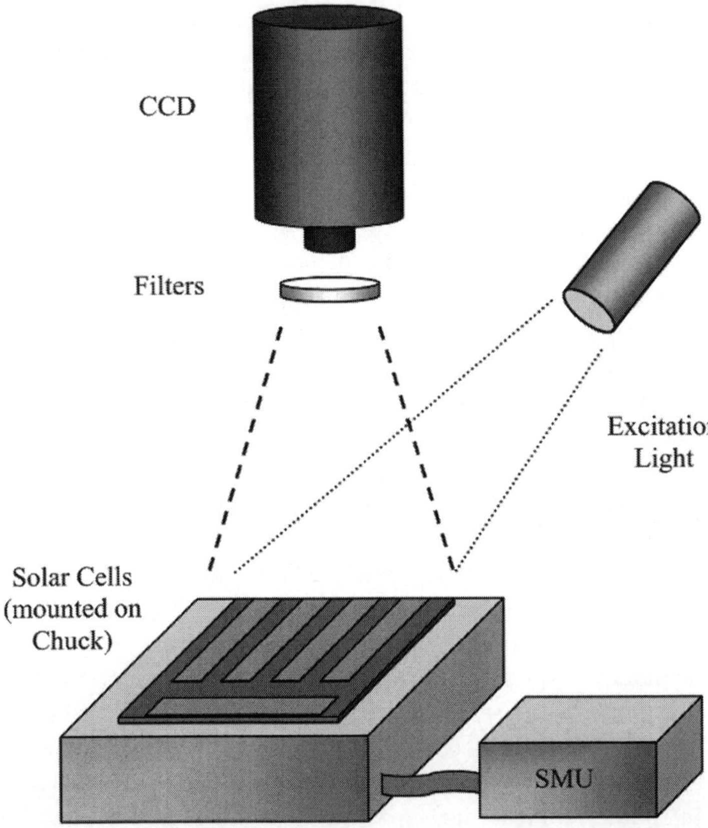

Figure 14.18 Scheme of measurement setup for luminescence imaging. Either light or current can be used as excitation source. Reprinted with permission from [72].

For electroluminescence imaging (ELI), the photovoltaic device is driven by a DC forward bias voltage above the V_{OC} of the subcell, yielding electrons injected from the cathode and holes injected from the anode. A minor fraction of these charge carriers can recombine radiatively within percolating pristine phases or across the semiconductor heterojunction. The latter process involves so-called charge transfer complexes between holes in the donor and electrons in the acceptor [19,76,77]. As the current is driven through the whole photovoltaic device, series resistance effects control the lateral electroluminescence intensity pattern depending on the voltage of device operation [72]. Since electroluminescence requires locally functioning electrode–active layer interfaces, it constitutes a good method to test this sensitive part of the device.

LIT is a highly sensitive method for detection of local device heating upon excitation. Excitation can be done either with illumination (illuminated lock-in thermography, ILIT) [67] or by current injection (dark lock-in thermography, DLIT) [15]. Upon the resulting current I, heat is generated through power dissipation (P) over a resistor R (Joule losses) according to Eq. 14.1.

The lock-in principle allows detection of temperature differences well below the milli-Kelvin (mK) range. However, it requires a periodically applied excitation, hence a function generator in combination with an amplifier is used to generate either AC light emission or pulsed electric currents to drive the device at proper frequencies. For sensing of the local heating, a high-frequency computer controlled infrared camera is employed, which images the whole device simultaneously. A schematic of the LIT measurement setup is shown in Fig. 14.19. In general, the two modes of LIT can be compared with those of luminescence imaging, with the difference that no visible light emission, but rather heat radiation is imaged. This especially enables the detection of shunts in the device that generally cannot be distinguished by luminescence imaging.

The measurement principle of LBIC is based on point-by-point scanning of a focused monochromatic light beam with typical beam diameters of 100 μm across the active area of the device, which limits also the spatial resolution of this method to ~100 μm. To be able to detect small locally generated photocurrents, the photovoltaic device is usually held under short-circuit conditions and the excitation light is usually modulated by a frequency, allowing precise photocurrent detection by the lock-in technique [76]. Especially for large-area modules sensitive signal detection has to be achieved by the use of lock-in technique, since their potential noise is high due to the area. By simultaneous registration of the spatial position of the light spot on the device, the LBIC image is collected and constructed by a computer. Darker regions in the LBIC image thus correspond to smaller photocurrents, which have to be correlated with local inhomogeneities due to, for example, processing failures or device degradation. As the generation of LBIC images requires light absorption and current generation and extraction, this technique monitors the complete device functionality at locally high but globally low excitation levels. The experimental setup for performing LBIC measurements is depicted in Fig. 14.20.

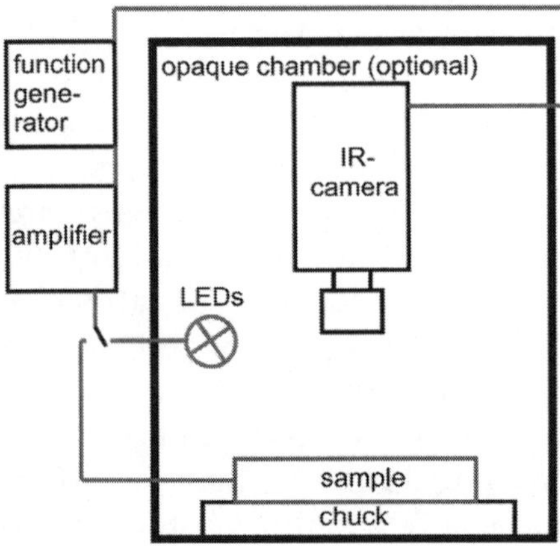

Figure 14.19 Experimental setup for lock-in thermography. Excitation can be done either by light or by electrical current. Reprinted with permission from [78].

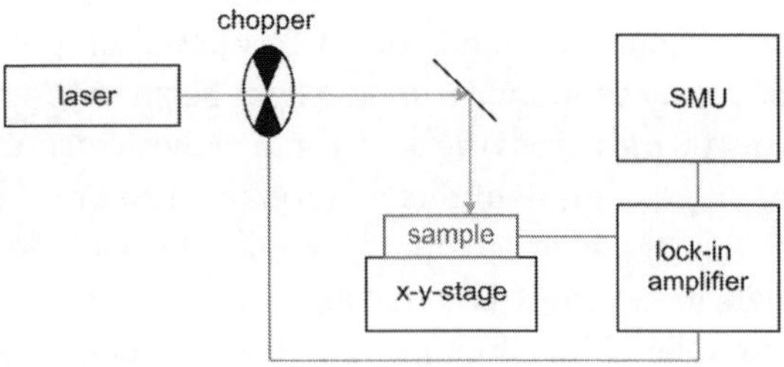

Figure 14.20 Experimental setup for LBIC measurement. Instead of a laser a white light source together with a monochromator can be used as well.

In addition to the brief introduction of luminescence imaging, LIT, and LBIC, two examples of application of these methods are shown.

In Fig. 14.21 both modes of luminescence imaging for quality control of the same polymer solar module as shown in Fig. 14.16 are presented [79]. Several defects and inhomogeneities of this handmade polymer solar module with the size of 2 inch × 2 inch on a glass substrate [45] are observable in both luminescence imaging modes, but with different contrast:

- holes in the cathode accidentally introduced by scratching (red ellipses) are similarly well recognized in both images
- handling-induced defects (blue rectangles) are slightly better visible by ELI
- defects in the active layer due to dust and nondissolved material (orange pentagons) are better visible by ELI
- inhomogeneities in active layer film thickness in the corners, arising from the spin coating process, are better visible by PLI.

For a thorough investigation of the degradation behavior of photovoltaic devices, a combination of ELI and PLI is necessary for a comprehensive analysis of the photoactive layer and the electrical contact influence on device performance; see, for example, [72].

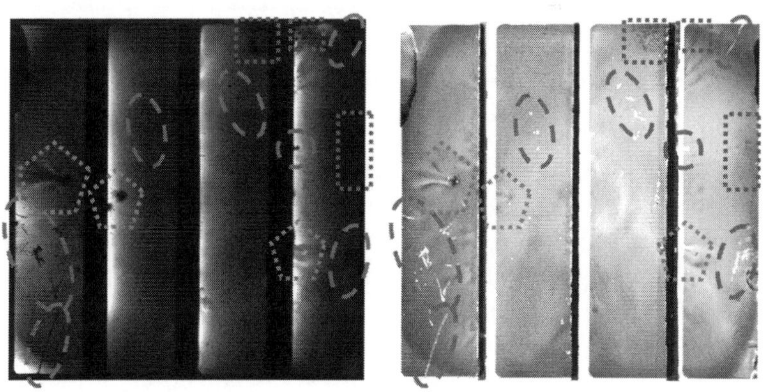

Figure 14.21 Electroluminescence image (left) and photoluminescence image (right) of a 2 inch × 2 inch polymer solar module on glass, already shown in Fig. 14.16. Various defects are marked and can be observed in both images [79].

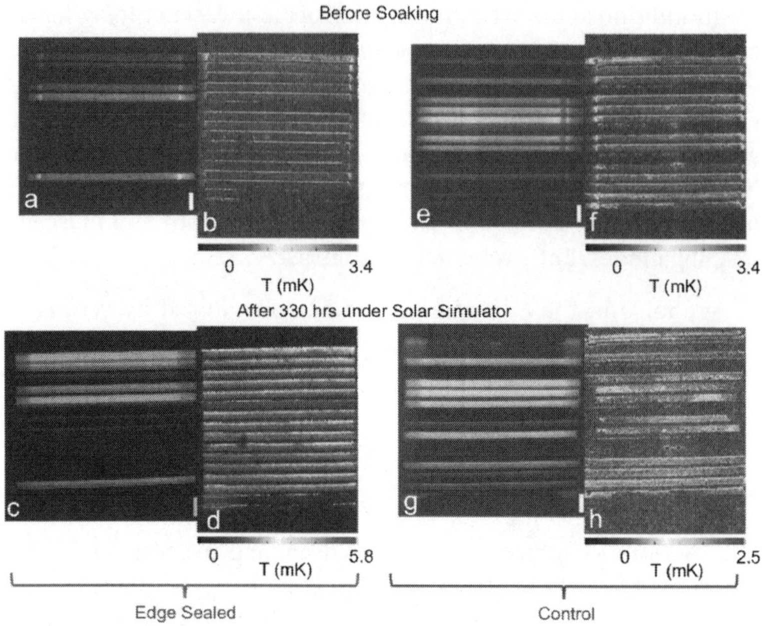

Figure 14.22 LBIC (a, e, c, g) and DLIT (b, f, d, h) images of the same edge sealed (left, a–d) and control (right, e–h) samples without additional edge sealing, both before (upper, a, b, e, f) and after (lower, c, d, g, h) 330 h of light soaking. Localized degradation features are visible for the control device without edge sealing upon DLIT characterization after 330 h of light soaking (h) [75].

Figure 14.22 shows an example of the application of DLIT and LBIC for studying the degradation behavior of inverted-type flexible polymer solar modules [75]. For comparison of the results between DLIT and LBIC it needs to be stated that the characterizations generally take place under two different excitation regimes: in the case of LBIC, a strong local but weak global excitation due to overall small local currents is governing the characterization, whereas for DLIT a strong global excitation by current injection involves larger currents and therefore higher resistive losses. Two batches of printed polymer solar modules on PET substrates, consisting of 16 subcells, were treated under harsh light soaking degradation conditions [80]. It should be noted that the stressing under the solar simulator induces two principal processes for these kinds of inverted devices: (i) photoinduced degradation and (ii) solarization upon light soaking, that is, the improvement of the injection and extraction properties of the semitransparent ZnO–ITO electrode upon filling of trap states [81]. Both solar modules, one with and the other without additional edge sealing, were investigated by DLIT and LBIC before and after 330 h of stressing, respective light soaking. Electrical characterization already revealed that the edge-sealed samples were more stable than the control samples without additional edge sealing.

DLIT and LBIC images were able to confirm this conclusion by local information about the differences in degradation between both batches. While the additionally edge sealed samples revealed almost no change in device functionality, DLIT images revealed that the control samples without additional edge sealing suffered from local failure near the edges. Since LBIC images of the control samples did not exhibit any features at the device's edges, the photovoltaic charge generation capability of the active layer remained intact at this comparatively low-level excitation. Hence it can be concluded that the degradation of the devices in this particular study affected mostly the electrical contacts or the interface between them and the photoactive layer. Presumably the electrode degradation took place due to edge-in diffusion of reactive species like oxygen and water, confirming the need for additional edge sealing.

14.7 Roll-to-Roll Processing—Technology and Challenges

Cost-effective production of polymer photovoltaic devices on a large scale requires precise thin film deposition techniques that allow deposition of thin films with a thickness of a few hundred nanometers, but are much more efficient than classical laboratory techniques like spin coating, drop casting, or doctor blading. These techniques indeed allow very precise deposition but are very wasteful concerning either time or material or both, when larger device areas need to be addressed. Furthermore, large-area devices with a serially interconnected subcell structure are desired for economic applications, for example, like BIPVs. Therefore, precise large-area coating and structuring processes are necessary for successful commercialization of polymer photovoltaics. Fortunately, all functional layers in a polymer solar module are in principle coatable from a liquid phase, for example, solution, dispersion, or emulsion [12]. Hence, classical printing techniques that are already used for several hundred years especially for reproduction of books and newspapers are well suited. They enable high-throughput large-area deposition in combination with simultaneous structuring of a lacquer onto a substrate, well suitable for the design of monolithic solar modules. Examples of printing techniques that were already used in the processing of polymer photovoltaics are gravure printing, flexographic offset printing, inkjet printing, screen printing, and rotary screen printing [12,82,83].

Another approach to high-throughput production of homogeneous thin films is the deposition of the liquid layers by coating techniques that cover the whole substrate area. Knife coating, curtain coating, slide coating, and slot die coating are famous examples of coating techniques that were already applied to polymer photovoltaics processing [12,82,83]. The advantage of these coating techniques may be found in the comparatively high uniformity of the layers at a large scale, due to the fact that the edges of coating—resulting in rims and other artifacts—are exclusively at the sides of the roll on which the film is cast. Unfortunately, for the realization of monolithic solar modules these deposition techniques require a subsequent structuring process for each of the dried layers by a separate processing step. For more detailed information on coating techniques, the interested readers are directed to Chapter 13.

Since most of the printing and coating techniques mentioned earlier are capable of continuous roll-to-roll production, the advantages and disadvantages of each technique have to be weighed up against each other to choose the best printing or coating technique for the desired application. In the following, three of the above-mentioned deposition techniques are compared considering their applicability of producing polymer photovoltaic devices on the module scale. Selected are the techniques with the highest potential throughput of at least 10 m/s web feed rate: gravure printing, flexographic offset printing, and slot die coating. In Table 14.3, each of them is rated with reference to six parameters that are to be considered for the production of polymer photovoltaics. In Fig. 14.23 the schematic working principles of these techniques are shown.

Furthermore, the main difference between structured (printing) and unstructured (coating) deposition of large areas of thin films is depicted in Fig. 14.24. While printing allows a structured but disrupted deposition, coating results in continuous but unpatterned layer formation.

Table 14.3 Three different large-area lacquer deposition techniques compared with each other with reference to six essential parameters for polymer photovoltaics processing

	Gravure printing	Flexographic offset printing	Slot die coating
Thickness control	–	+/–	+
Viscosity range	–	–	+
Speed	+/–	+/–	+
Structuring	+	+	–
Maintenance	–	–	+
Cleanliness	–	–	+

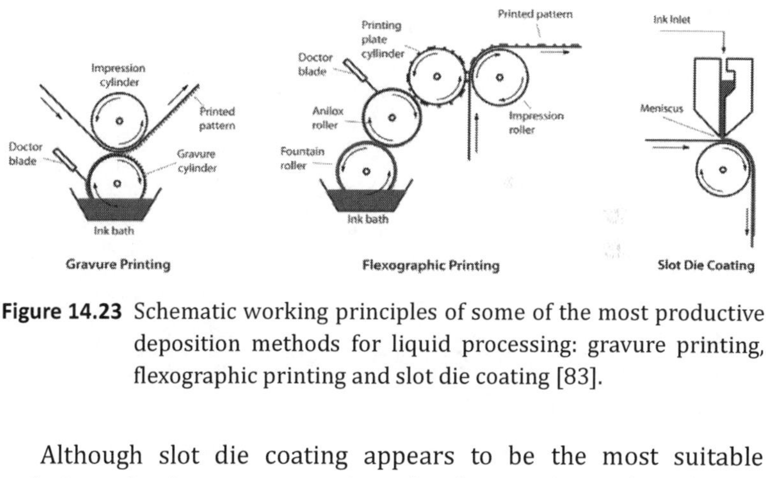

Figure 14.23 Schematic working principles of some of the most productive deposition methods for liquid processing: gravure printing, flexographic printing and slot die coating [83].

Although slot die coating appears to be the most suitable technique for large-area coating of polymer photovoltaic layers, its main disadvantage, the lack of structuring capability, has to be overcome in order to make use of this technique. Whereas it is already possible to combine several lines of paralleled slots fed from a single liquid container, this approach is limited in the lateral spacing between the deposited ribbons of material, and on each of the ribbons edge effects diminish the overall film homogeneity and quality.

Figure 14.24 Photographic images of flexographic offset printing of solution processed metal electrodes (top) and slot die coating a HTL (bottom) during processing [46, 83].

In Fig. 14.25, the principle layout for structuring of a polymer-based thin film solar module is displayed. The structuring of the transparent front contact is called "P1," of the photoactive layer "P2," and of the back electrode "P3," in accordance with all inorganic thin film structuring processes.

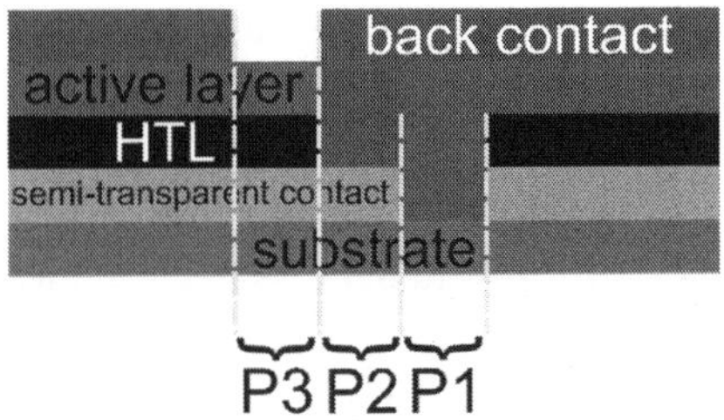

Figure 14.25 Structuring steps P1, P2, and P3 for a polymer-based thin film solar module.

In the following, we highlight the combined application of slot die coating for unstructured deposition with laser ablation as a structuring process, as it enables high-throughput (fast web speed), homogeneous film formation and very fine structuring as demanded for a high GFF and thus high module performances (MPCE) (compared with Section 14.5) [45,84–88].

Since modern state-of-the-art solid-state pulsed lasers enable repetition frequencies up to 100 kHz in combination with laser foci spot sizes of 10–100 μm, processing speeds between 1 and 10 m/s are within reach. Thus the web speeds that can be obtained by laser ablation are compatible and in the same range as those known from slot die coating, and the combination of both techniques does not harm productivity. On the other hand, laser focus spot sizes between 10 and 100 μm indeed enable very fine structuring and thus minimal area losses required for the serial interconnection of the solar cells within the monolithic solar module. In fact, solar cell distances between 100 and 500 μm are easily accessed; however, this puts a high demand on the registry of adjacent structuring lines and thus the process control in the roll-to-roll production. Further challenges to roll-to-roll processing are found in the limited focal depth of the laser spot, which puts a high demand on controlling the distance between the focusing optics and the film on the substrate.

Even though it is principally feasible to do the laser structuring process orthogonal to the web direction by application of fast *x*–*y* scanners, this approach will limit the web speed and will be more challenging in terms of reproducibility. Hence, parallel operation of the laser scribing process requires either splitting up one laser beam into multiple beams (increasing the need for pulse energies into the range of 100 μJ or higher) or a number of lasers in the same range as the number of serially connected solar cell are required in the production process. Both approaches place high demands concerning the investment. Another factor influencing the cost of ownership is based on the fact that shorter laser pulses generally lead to higher quality of the structuring lines, as the shorter light–matter interaction leads to smaller energy affected zones at the scribing line. Shorter pulse durations result in a smaller penetration depth of the laser pulse and thus better controlled ablation with less residue and debris. However, short pulse duration may involve overlapping of several laser pulses due to their relative weak ablation capability limiting the web speed, and state-of-the-art femtosecond lasers are significantly more expensive than pico- or even nanosecond lasers, and thus a trade-off between costs and quality has to be made. On the other hand, longer laser pulses increase the need for different laser wavelengths, as the ablation process becomes more absorption specific in that case. Once again this produces additional parameters to be considered when equipping module production with such a laser ablation structuring process.

The most detrimental disadvantage of this structuring technique is the fact that it can easily damage underlying layers by too harsh laser pulses and last but not least it tends to easily produce large amounts of debris and "walls" of residual material directly next to the structured line.

As an example of an insufficient P1 structuring by laser ablation, Fig. 14.26 shows a microscopic top view image together with a tactile profilometer scan of a laser scribe of ITO on top of a PET substrate [45]. This particular laser scribing process was based on a microsecond laser pulse and resulted in huge expulsion of residual material that condensed at the edges of the laser scribe. Such a laser scribe profile is unacceptable since it inevitably results in shunting of adjacent solar cells within the module.

In contrast to the example mentioned earlier, in Fig. 14.27 three images of high quality P1, P2, and P3 laser structured layers are shown. Structuring widths are approximately 20 and 50 μm, respectively. Compared with the example mentioned earlier, much shorter laser pulses of less than 1 ns pulse duration were used for the scribing process. It is interesting to note that the removal of the ITO (P1) and the organic PEDOT:PSS and photoactive layer (P2) required a large overlap between adjacent laser ablation spots, whereas the aluminum back electrode (P3) could be removed by a minimal overlapping of subsequent laser pulses. In general, a higher need for overlapping of ablations zones can be correlated with stronger adhesion of the layer to be structured.

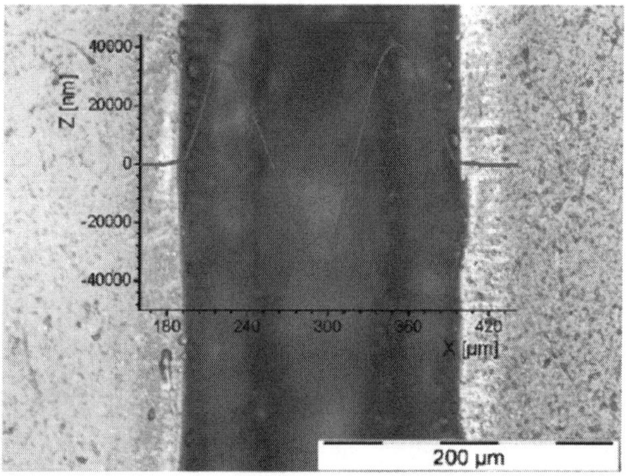

Figure 14.26 Microscopic top view image together with a tactile profilometer scan of a laser scribe with huge "walls" due to the throw-off of residual material at the edges of the scribe [45].

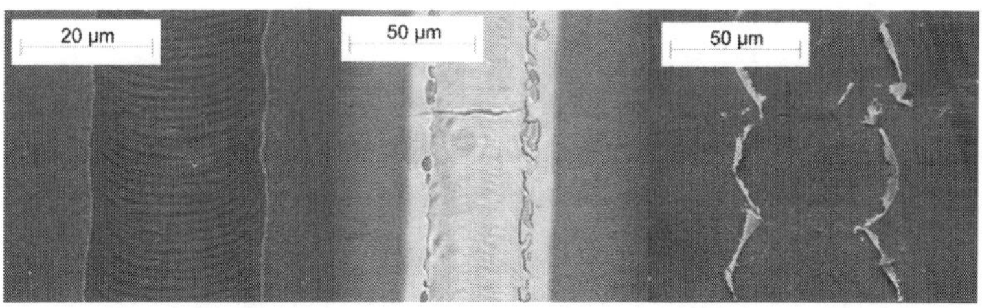

Figure 14.27 Scanning electron microscopy top view images (left and right) and microscopy image (center) of high-quality ITO (P1, left), PEDOT:PSS and photoactive layer (P2, center), and aluminum (P3, right) laser structuring processes [89].

As an example of the challenge of realizing all three structuring steps (P1–P3) subsequently, Fig. 14.28 depicts a DLIT image of a fully laser structured polymer solar module on glass substrate exhibiting ~1.5% MPCE. On the image several processing deficiencies resulting in inhomogeneous laser ablation and thus electrical interconnections can be observed, including shunting over the serial connection. However, MPCE between 0.5% and 3% could be obtained so far in fully or partly laser structured solar modules, providing a positive perspective for future developments [89].

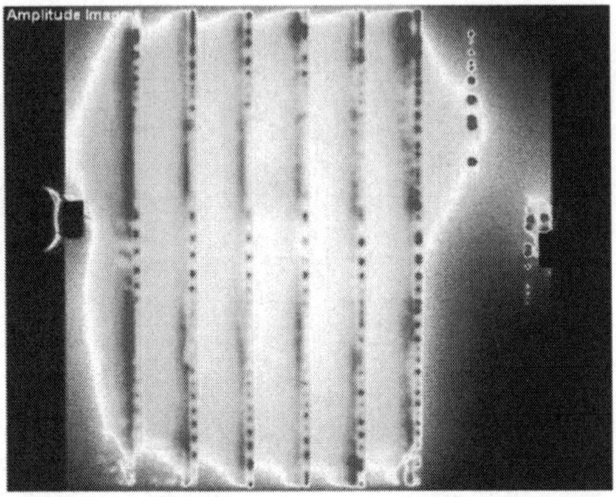

Figure 14.28 Forward bias DLIT image of a fully laser structured polymer solar module on glass substrate exhibiting ~1.5% MPCE. Two out of eight submodules on the right-hand side are completely shunted by insufficient laser scribing, while the other interconnections also show inhomogeneous patterning, expressed by the red dots [89].

14.8 Lifetime Measurements of Polymer Solar Modules

Stability is a core requirement for the commercial application of polymer solar modules. Even though it is experimentally not always easy to access, the knowledge about the product lifetime of polymer solar modules under real stress conditions is essential for its marketing, since they have to compete with inorganic photovoltaics that provide lifetime warranties of 20 years. Intriguingly, to date there is no international standard to determine the lifetime of photovoltaic modules or cells under operating conditions. The existing standards demand an operating test for ~60 h and they apply only for inorganic thin film photovoltaic technologies [90], concentrator photovoltaics [91], or silicon photovoltaics [92].

To redress these unsatisfying circumstances, a large-scale collaborative effort within the International Summit on OPV Stability

(ISOS) community resulted recently in the publication of a catalog of experimental conditions specifically suited for determining the lifetime of organic solar cells and modules [93]. In this document, an agreement about five different degradation scenarios is summarized: dark storage, outdoor, laboratory weathering (under illumination), thermal cycling and solar–thermal humidity cycling with three levels of complexity each (basic, intermediate, and advanced). Although this catalog of standard does not constitute the scale of an International Electrotechnical Commission (IEC) norm, the research community accepts these testing protocols in order to provide a better comparability of photovoltaic stability [75,80,94–96].

The first comprehensive study of the stability of polymer solar modules under realistic stress conditions was performed by Hauch et al. in 2006 and 2007 [7]. The investigated devices were printed monolithic polymer solar modules in inverse architecture with P3HT : PCBM as active layer material. The modules were sealed with a transparent barrier film with a water vapor transmission rate of 0.03 g/m^2 day. For determining the device's lifetime, defined as the time that passes by until the device loses 20% of its initially specified performance, the solar modules were placed on the rooftop (in Lowell, MA), facing south and the voltage was monitored over a constant load resistor during the entire experiment. In Fig. 14.29 the used measurement setup and in Fig. 14.30 the obtained results, that is, the normalized power output over degradation time, are shown. The authors observed several interesting trends: first the module power output raised by roughly 40% with respect to the starting value, next there was a period where the power output followed the temperature and finally the final ~8 months of the experiment the power output decreased linearly with time to 80% of the starting value.

The in situ recorded data were restricted to dissipative power loss over a constant load resistor that was unlikely at the maximum power point for the whole experiment and no photovoltaic parameters were measured. The conclusion the authors drew from that experiment was that P3HT : PCBM constitutes a reasonable light stability for "first outdoor applications" [7]. It may be speculated that this result motivated and promoted the first commercial product based on solar modules produced by Konarka: the solar bag (as shown in Fig. 14.2).

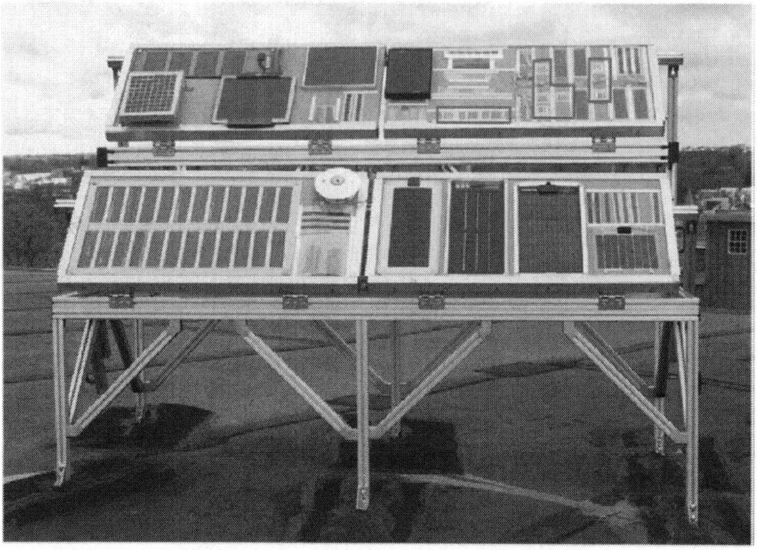

Figure 14.29 Outdoor testing setup in Lowell, MA, at ~42.6° N altitude. The solar cells and modules were facing south and placed at ~42° against the horizontal. The red framings highlight the samples investigated in this study [7].

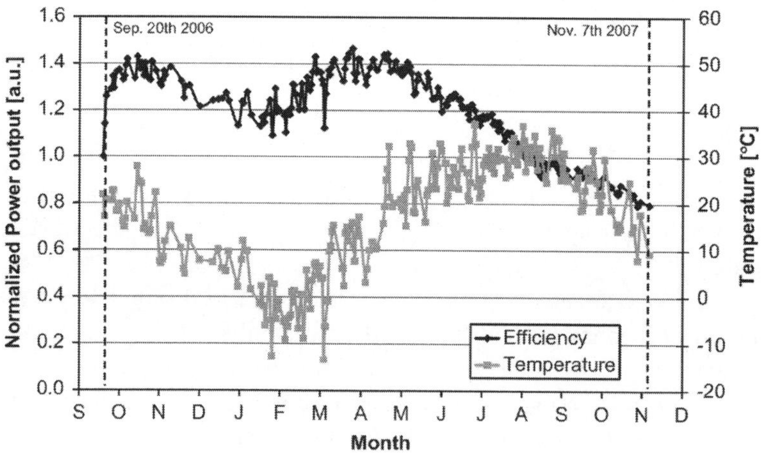

Figure 14.30 Normalized power output as determined from the current measured over a constant resistor and temperature over time for a polymer solar module placed on a rooftop for ~14 months [7].

Recently, an international consortium of more than 20 research institutes and university work groups has conducted a round robin about the stability evaluation of intentionally simply sealed polymer solar modules of inverse architecture [94]. Among a number of different tests according to the ISOS protocols, the indoor light stability test (ISOS-L) revealed interesting results concerning the stressing conditions imposed by different light sources: the application of halogen, metal halide, xenon arc, and sulfur plasma lamps resulted in largely varying lifetimes. In Fig. 14.31, the normalized photovoltaic parameters over the stressing time for the various lamps are shown. While the results for Xe arc and metal halide lamps were principally comparable, halogen and sulfur plasma lamps overestimated the lifetime compared with that found in outdoor measurements. This raises important conclusions for the applicability and comparability of specific stability tests under artificial illumination.

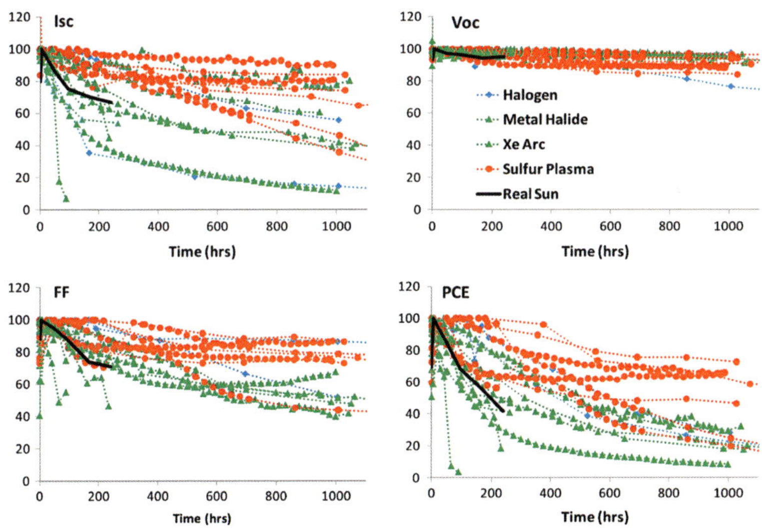

Figure 14.31 Normalized PV parameters for 14 identically constructed polymer solar modules aged in 14 different laboratories and 5 different illumination conditions. Measurements with sulfur plasma lamps and halogen lamps tend to overestimate the device's stability [94].

In 2012 some remarkable progress in reporting stability was achieved by Konarka, as they demonstrated passing of the IEC 61646 testing conditions for their latest generation of polymer solar modules (http://www.konarka.com/index.php/site/pressreleasedetail/konarkas_next_generation_organic_photovoltaic_cells_first_opv_technology_to).

At present, an international work group within the IEC is developing a standardization of stability testing procedures beyond the before-mentioned scientific ISOS testing levels, in order to meet the stronger requirements for an industrialization of polymer solar modules.

14.9 Summary and Outlook

In this chapter we have shown that the design of efficient polymer-based solar modules has to be closely linked to the constituent material properties. As the conductivity in organic materials is principally low, solar cells based on them require fully covering charge collecting electrodes. This in turn dictates the use of semitransparent electrodes that in general exhibit comparatively high sheet resistances. As a consequence, properly designed serially interconnected solar cells minimize dissipative power losses. As these serial connections can be realized on a single substrate, the design is easily scalable. For practical upscaling, indeed high-throughput processing methods such as printing or coating, which originally have been developed for other applications and for the deposition of other materials, can be readily employed but need further optimization. In combination with an advanced materials development, these production technologies provide a pathway for progressively boosting the commercialization of solution processed OPVs in the near future.

Acknowledgments

We are grateful to Hans Melissant for consultations about large-area processing. We thank Alexander Horn, Uwe Stute and Javier Gonzales for fruitful discussions about laser ablation. Furthermore, we appreciate financial support by the German Federal Ministry of Education and Research in the framework of PPP (contract number 13N9843).

References

1. Green MA, Emery K, Hishikawa Y, Warta W, Dunlop ED (2012) Solar cell efficiency tables (version 39), *Prog Photovolt*, **20**, 12–20.
2. Li G, Zhu R, Yang Y (2012) Polymer solar cells, *Nat Photonics*, **6**, 153–161.
3. Service RF (2011) Outlook brightens for plastic solar cells, *Science*, **332**, 293–293.
4. Zimmermann B, Würfel U, Niggemann M (2009) Longterm stability of efficient inverted P3HT:PCBM solar cells, *Sol Energy Mater Sol Cells*, **93**, 491–496.
5. Peters CH, Sachs-Quintana IT, Kastrop JP, Beaupre S, Leclerc M, McGehee MD (2011) High efficiency polymer solar cells with long operating lifetimes, *Adv Energy Mater*, **1**, 491–494.
6. Rösch R, Eberhardt K-R, Engmann S, Gobsch G, Hoppe H (2013) Polymer solar cells with enhanced lifetime by improved electrode stability and sealing, *Sol Energy Mater Sol Cells*, **117**, 59–66.
7. Hauch JA, Schilinsky P, Choulis SA, Childers R, Biele M, Brabec CJ (2008) Flexible organic P3HT:PCBM bulk-heterojunction modules with more than 1 year outdoor lifetime, *Sol Energy Mater Sol Cells*, **92**, 727–731.
8. Krebs FC, Norrman K (2007) Analysis of the failure mechanism for a stable organic photovoltaic during 10000 h of testing, *Prog Photovoltaics*, **15**, 697–712.
9. Voroshazi E, Verreet B, Aernouts T, Heremans P (2011) Long-term operational lifetime and degradation analysis of P3HT:PCBM photovoltaic cells, *Sol Energy Mater Sol Cells*, **95**, 1303–1307.
10. Nielsen TD, Cruickshank C, Foged S, Thorsen J, Krebs FC (2010) Business, market and intellectual property analysis of polymer solar cells, *Sol Energy Mater Sol Cells*, **94**, 1553–1571.
11. Wakisaka K, Kuwano Y (1998) Terrestrial applications of amorphous silicon solar cells, *Prog Photovoltaics*, **6**, 207–217.
12. Krebs FC, Tromholt T, Jørgensen M (2010) Upscaling of polymer solar cell fabrication using full roll-to-roll processing, *Nanoscale*, **2**, 873–886.

13. Jin H, Tao C, Velusamy M, Aljada M, Zhang YL, Hambsch M, Burn PL, Meredith P (2012) Efficient, large area ITO-and-PEDOT-free organic solar cell sub-modules, *Adv Mater*, **24**, 2572–2577.
14. Muhsin B, Renz J, Drue KH, Gobsch G, Hoppe H (2009) Influence of polymer solar cell geometry on series resistance and device efficiency, *Phys Stat Sol A-Appl Mater Sci*, **206**, 2771–2774.
15. Bachmann J, Buerhop-Lutz C, Deibel C, Riedel I, Hoppe H, Brabec CJ, Dyakonov V (2010) Organic solar cells characterized by dark lock-in thermography, *Sol Energy Mater Sol Cells*, **94**, 642–647.
16. Seeland M, Rösch R, Hoppe H (2012) Quantitative analysis of electroluminescence images from polymer solar cells, *J Appl Phys*, **111**, 024505.
17. Manor A, Katz EA, Tromholt T, Hirsch B, Krebs FC (2011) Origin of size effect on efficiency of organic photovoltaics, *J Appl Phys*, **109**, 074508.
18. Würfel P (2009) *Physics of solar cells: From basic principles to advanced concepts*, Weinheim: Wiley-VCH.
19. Vandewal K, Tvingstedt K, Gadisa A, Inganas O, Manca JV (2009) On the origin of the open-circuit voltage of polymer-fullerene solar cells, *Nat Mater*, **8**, 904–909.
20. Tang CW (1986) 2-Layer organic photovoltaic cell, *Appl Phys Lett*, **48**, 183–185.
21. Dimitrakopoulos CD, Mascaro DJ (2001) Organic thin-film transistors: A review of recent advances, *IBM J Res Dev*, **45**, 11–27.
22. Sariciftci NS, Braun D, Zhang C, Srdanov VI, Heeger AJ, Stucky G, Wudl F (1993) Semiconducting polymer-buckminsterfullerene heterojunctions—diodes, photodiodes, and photovoltaic cells, *Appl Phys Lett*, **62**, 585–587.
23. Glatthaar M, Niggemann M, Zimmermann B, Lewer P, Riede M, Hinsch A, Luther J (2005) Organic solar cells using inverted layer sequence, *Thin Solid Films*, **491**, 298–300.
24. Rösch R, Tanenbaum DM, Jørgensen M, Seeland M, Bärenklau M, Hermenau M, Voroshazi E, Lloyd MT, Galagan Y, Zimmermann B, Würfel U, Hosel M, Dam HF, Gevorgyan SA, Kudret S, Maes W, Lutsen L, Vanderzande D, Andriessen R, Teran-Escobar G, Lira-Cantu M, Rivaton A, Uzunoglu GY, Germack D, Andreasen B, Madsen MV, Norrman K, Hoppe H, Krebs FC (2012) Investigation of the degradation mechanisms of a variety of organic photovoltaic devices by combination of imaging techniques—the ISOS-3 inter-laboratory collaboration, *Energy Environ Sci*, **5**, 6521–6540.

25. Gregg BA, Hanna MC (2003) Comparing organic to inorganic photovoltaic cells: Theory, experiment, and simulation, *J Appl Phys*, **93**, 3605–3614.

26. Sariciftci NS, Smilowitz L, Heeger AJ, Wudl F (1992) Photoinduced electron-transfer from a conducting polymers to buckminsterfullerene, *Science*, **258**, 1474–1476.

27. Hoppe H, Sariciftci NS (2006) Morphology of polymer/fullerene bulk heterojunction solar cells, *J Mater Chem*, **16**, 45–61.

28. Hoppe H, Sariciftci NS (2004) Organic solar cells: An overview, *J Mater Res*, **19**, 1924–1945.

29. Assadi A, Svensson C, Willander M, Inganas O (1998) Field-effect mobility of poly(3-hexylthiophene), *Appl Phys Lett*, **53**, 195–197.

30. Padinger F, Rittberger RS, Sariciftci NS (2003) Effects of postproduction treatment on plastic solar cells, *Adv Funct Mater*, **13**, 85–88.

31. Dang MT, Hirsch L, Wantz G (2011) P3HT:PCBM, best seller in polymer photovoltaic research, *Adv Mater*, **23**, 3597–3602.

32. Scharber MC, Wuhlbacher D, Koppe M, Denk P, Waldauf C, Heeger AJ, Brabec CL (2006) Design rules for donors in bulk-heterojunction solar cells—towards 10% energy-conversion efficiency, *Adv Mater*, **18**, 789–794.

33. Blouin N, Michaud A, Leclerc M (2007) A low-bandgap poly(2,7-carbazole) derivative for use in high-performance solar cells, *Adv Mater*, **19**, 2295–2300.

34. Chu TY, Alem S, Tsang SW, Tse SC, Wakim S, Lu JP, Dennler G, Waller D, Gaudiana R, Tao Y (2011) Morphology control in polycarbazole based bulk heterojunction solar cells and its impact on device performance, *Appl Phys Lett*, **98**, 253301.

35. Umeda T, Shirakawa T, Fujii A, Yoshino K (2003) Improvement of characteristics of organic photovoltaic devices composed of conducting polymer-fullerene systems by introduction of ZnO layer, *Jpn J Appl Phys Part 2: Lett*, **42**, L1475–L1477.

36. Hayakawa A, Yoshikawa O, Fujieda T, Uehara K, Yoshikawaa S (2007) High performance polythiophene/fullerene bulk-heterojunction solar cell with a TiO_x hole blocking layer, *Appl Phys Lett*, **90**, 163517.

37. Hermenau M, Riede M, Leo K, Gevorgyan SA, Krebs FC, Norrman K (2011) Water and oxygen induced degradation of small molecule organic solar cells, *Sol Energy Mater Sol Cells*, **95**, 1268–1277.

38. Hancox I, Sullivan P, Chauhan KV, Beaumont N, Rochford LA, Hatton RA, Jones TS (2010) The effect of a MoO_x hole-extracting layer on the performance of organic photovoltaic cells based on small molecule planar heterojunctions, *Org Electron*, **11**, 2019–2025.
39. Irwin MD, Buchholz B, Hains AW, Chang RPH, Marks TJ (2008) p-Type semiconducting nickel oxide as an efficiency-enhancing anode interfacial layer in polymer bulk-heterojunction solar cells, *Proc Natl Acad Sci USA*, **105**, 2783–2787.
40. Shrotriya V, Li G, Yao Y, Chu CW, Yang Y (2006) Transition metal oxides as the buffer layer for polymer photovoltaic cells, *Appl Phys Lett*, **88**, 073508.
41. Wu JB, Becerril HA, Bao ZN, Liu ZF, Chen YS, Peumans P (2008) Organic solar cells with solution-processed graphene transparent electrodes, *Appl Phys Lett*, **92**, 263302.
42. Pasquier AD, Unalan HE, Kanwal A, Miller S, Chhowalla M (2005) Conducting and transparent single-wall carbon nanotube electrodes for polymer-fullerene solar cells, *Appl Phys Lett*, **87**, 203511.
43. Lee JY, Connor ST, Cui Y, Peumans P (2008) Solution-processed metal nanowire mesh transparent electrodes, *Nano Lett*, **8**, 689–692.
44. Hoppe H, Seeland M, Muhsin B (2012) Optimal geometric design of monolithic thin-film solar modules: Architecture of polymer solar cells, *Sol Energy Mater Sol Cells*, **97**, 119–126.
45. Muhsin B, Renz J, Drue KH, Gobsch G, Hoppe H (2009) Efficient polymer solar cell modules, *Synth Met*, **159**, 2358–2361.
46. Krebs FC (2009) Roll-to-roll fabrication of monolithic large-area polymer solar cells free from indium-tin-oxide, *Sol Energy Mater Sol Cells*, **93**, 1636–1641.
47. Zimmermann B, Glatthaar M, Niggemann M, Riede MK, Hinsch A, Gombert A (2007) ITO-free wrap through organic solar cells—a module concept for cost-efficient reel-to-reel production, *Sol Energy Mater Sol Cells*, **91**, 374–378.
48. Choi S, Potscavage WJ, Kippelen B (2009) Area-scaling of organic solar cells, *J Appl Phys*, **106**, 054507.
49. Lyu HK, Jeong S, Sim JH, Woo S, Kim KP, Shin JK, Han YS (2011) Electrical power loss model for large-area monolithic organic photovoltaic module, *Curr Appl Phys*, **11**, S166–S170.
50. Lyu HK, Sim JH, Woo SH, Kim KP, Shin JK, Han YS (2011) Efficiency enhancement in large-area organic photovoltaic module using theoretical power loss model, *Sol Energy Mater Sol Cells*, **95**, 2380–2383.

51. Lyu HK, Woo SH, Jeong S, Kim KP, Shin JK (2012) Investigation of power loss and device efficiency using a three-dimensional finite element simulation on organic solar cell geometry, *Sol Energy Mater Sol Cells*, **96**, 66–70.

52. Breitenstein O, Khanna A, Augarten Y, Bauer J, Wagner JM, Iwig K (2010) Quantitative evaluation of electroluminescence images of solar cells, *Phys Stat Sol-Rapid Res Lett*, **4**, 7–9.

53. Fuyuki T, Kondo H, Yamazaki T, Takahashi Y, Uraoka Y (2005) Photographic surveying of minority carrier diffusion length in polycrystalline silicon solar cells by electroluminescence, *Appl Phys Lett*, **86**, 262108.

54. Giesecke JA, Michl B, Schindler F, Schubert MC, Warta W (2011) Minority carrier lifetime of silicon solar cells from quasi-steady-state photoluminescence, *Sol Energy Mater Sol Cells*, **95**, 1979–1982.

55. Haunschild J, Glatthaar M, Kasemann M, Rein S, Weber ER (2009) Fast series resistance imaging for silicon solar cells using electroluminescence, *Phys Stat Sol-Rapid Res Lett*, **3**, 227–229.

56. Kampwerth H, Trupke T, Weber JW, Augarten Y (2008) Advanced luminescence based effective series resistance imaging of silicon solar cells, *Appl Phys Lett*, **93**, 202102.

57. Kasemann M, Grote D, Walter B, Kwapil W, Trupke T, Augarten Y, Bardos RA, Pink E, Abbott MD, Warta W (2008) Luminescence imaging for the detection of shunts on silicon solar cells, *Prog Photovoltaics*, **16**, 297–305.

58. Trupke T, Pink E, Bardos RA, Abbott MD (2007) Spatially resolved series resistance of silicon solar cells obtained from luminescence imaging, *Appl Phys Lett*, **90**, 093506.

59. Würfel P, Trupke T, Puzzer T, Schaffer E, Warta W, Glunz SW (2007) Diffusion lengths of silicon solar cells from luminescence images, *J Appl Phys*, **101**, 123110.

60. Busse G, Wu D, Karpen W (1992) Thermal wave imaging with phase sensitive modulated thermography, *J Appl Phys*, **71**, 3962–3965.

61. Isenberg J, Riepe S, Glunz SW, Warta W (2003) Imaging method for laterally resolved measurement of minority carrier densities and lifetimes: Measurement principle and first applications, *J Appl Phys*, **93**, 4268–4275.

62. Breitenstein O, Langenkamp M (2003) *Lock-in thermography—basics and use for functional diagnostics of electronic components*, Heidelberg: Springer.

63. Hari Rao CV, Bates HE, Ravi KV (1976) Electrical effects of SiC inclusions in EFG silicon ribbon solar cells, *J Appl Phys*, **47**, 2614–2619.

64. Zook JD (1980) Evaluation of silicon-on-ceramic using light-beam-induced currents, *Solar Cells*, **1**, 273–274.

65. Donolato C (1983) Theory of beam induced current characterization of grain-boundaries in polycrystalline solar-cells, *J Appl Phys*, **54**, 1314–1322.

66. Damaskinos S (1989) An improved laser scanning technique for evaluation of solar-cells—application to $CuInSe_2$ (CdZn)s devices, *Sol Cells*, **26**, 151–158.

67. Bachmann J, Buerhop-Lutz C, Steim R, Schilinsky P, Hauch JA, Zeira E, Christoh JB (2012) Highly sensitive non-contact shunt detection of organic photovoltaic modules, *Sol Energy Mater Sol Cells*, **101**, 176–179.

68. Hoppe H, Bachmann J, Muhsin B, Drue KH, Riedel I, Gobsch G, Buerhop-Lutz C, Brabec CJ, Dyakonov V (2010) Quality control of polymer solar modules by lock-in thermography, *J Appl Phys*, **107**, 014505.

69. Hoyer U, Pinna L, Swonke T, Auer R, Brabec CJ, Stubhan T, Li N (2011) Comparison of electroluminescence intensity and photocurrent of polymer based solar cells, *Adv Energy Mater*, **1**, 1097–1100.

70. Hoyer U, Wagner M, Swonke T, Bachmann J, Auer R, Osvet A, Brabec CJ (2010) Electroluminescence imaging of organic photovoltaic modules, *Appl Phys Lett*, **97**, 233303.

71. Lloyd MT, Peters CH, Garcia A, Kauvar IV, Berry JJ, Reese MO, McGehee MD, Ginley DS, Olson DC (2011) Influence of the hole-transport layer on the initial behavior and lifetime of inverted organic photovoltaics, *Sol Energy Mater Sol Cells*, **95**, 1382–1388.

72. Seeland M, Rösch R, Hoppe H (2011) Luminescence imaging of polymer solar cells: Visualization of progressing degradation, *J Appl Phys*, **109**, 064513.

73. Seeland M, Rösch R, Hoppe H (2012) Imaging techniques for studying OPV stability and degradation, in: Krebs FC (Ed.) *Stability and degradation of organic and polymer solar cells* FC, Hoboken: Wiley.

74. Seeland M, Rösch R, Muhsin B, Gobsch G, Hoppe H (2012) Electroluminescence as characterization tool for polymer solar cells and modules, *Energy Procedia*, **31**, 167–172.

75. Tanenbaum DM, Dam HF, Rösch R, Jørgensen M, Hoppe H, Krebs FC (2012) Edge sealing for low cost stability enhancement of roll-to-roll processed flexible polymer solar cell modules, *Sol Energy Mater Sol Cells*, **97**, 157–163.

76. Krebs FC, Søndergaard R, Jørgensen M (2011) Printed metal back electrodes for R2R fabricated polymer solar cells studied using the LBIC technique, *Sol Energy Mater Sol Cells*, **95**, 1348–1353.

77. Tvingstedt K, Vandewal K, Gadisa A, Zhang FL, Manca J, Inganas O (2009) Electroluminescence from charge transfer states in polymer solar cells, *J Am Chem Soc*, **131**, 11819–11824.

78. Rösch R, Krebs FC, Tanenbaum DM, Hoppe H (2012) Quality control of roll-to-roll processed polymer solar modules by complementary imaging methods, *Sol Energy Mater Sol Cells*, **97**, 176–180.

79. Seeland M, Rösch R, Muhsin B, Gobsch G, Hoppe H. Electroluminescence as characterization tool for polymer solar cells and modules, *Energy Procedia*, DOI: 10.1016/j.egypro.2012.11.179.

80. Tanenbaum DM, Hermenau M, Voroshazi E, Lloyd MT, Galagan Y, Zimmermann B, Hösel M, Dam HF, Jørgensen M, Gevorgyan SA, Kudret S, Maes W, Lutsen L, Vanderzande D, Wurfel U, Andriessen R, Rösch R, Hoppe H, Teran-Escobar G, Lira-Cantu M, Rivaton A, Uzunoglu GY, Germack D, Andreasen B, Madsen MV, Norrman K, Krebs FC (2012) The ISOS-3 inter-laboratory collaboration focused on the stability of a variety of organic photovoltaic devices, *RSC Adv*, **2**, 882–893.

81. Lilliedal MR, Medford AJ, Madsen MV, Norrman K, Krebs FC (2010) The effect of post-processing treatments on inflection points in current-voltage curves of roll-to-roll processed polymer photovoltaics, *Sol Energy Mater Sol Cells*, **94**, 2018–2031.

82. Krebs FC (2009) Fabrication and processing of polymer solar cells: A review of printing and coating techniques, *Sol Energy Mater Sol Cells*, **93**, 394–412.

83. Sondergaard R, Hosel M, Angmo D, Larsen-Olsen TT, Krebs FC (2012) Roll-to-roll fabrication of polymer solar cells, *Mater Today*, **15**, 36–49.

84. Dennler G, Lungenschmied C, Neugebauer H, Sariciftci NS, Labouret A (2005) Flexible, conjugated polymer-fullerene-based bulk-heterojunction solar cells: Basics, encapsulation, and integration, *J Mater Res*, **20**, 3224–3233.

85. Schoonderbeek A, Schutz V, Haupt O, Stute U (2010) Laser processing of thin films for photovoltaic applications, *J Laser Micro Nanoeng*, **5**, 248–255.

86. Rösch R, Muhsin B, Bärenklau M, Schoonderbeek A, Gobsch G, Richter L, Kling R, Teckhaus D, Hoppe H (2009) Towards roll-to-roll processing of flexible polymer solar cell modules, EUPVSEC Proceeding, 10.4229/24thEUPVSEC2009-1DV.4.59.

87. Schoonderbeek A, Bärenklau M, Rösch R, Teckhaus D, Muhsin B, Haupt O, Hoppe H, Stute U (2010) Laser structuring for monolithic series connection of organic thin-film solar cells, EUPVSEC Proceeding, 10.4229/25thEUPVSEC2010-1DO.9.4.

88. González Moreno J, Bärenklau M, Schoonderbeek A, Muhsin B, Haupt O, Rösch R, Gobsch G, Teckhaus D, Hoppe H, Stute U (2011) Thin-film organic solar modules—Processing and laser ablation, EUPVSEC Proceeding, 10.4229/26thEUPVSEC2011-1DV.3.13.

89. Bärenklau M, Muhsin B, Gonzalez J, Rösch R, Horn A, Gobsch G, Stute U, Hoppe H (2011) Polymer solar modules: Laser structuring and quality control by lock-in thermography, MRS online proceedings library, 1390.

90. Thin-film terrestrial photovoltaic (PV) modules—design qualification and type approval, IEC 61646 (2008).

91. Concentrator photovoltaic (CPV) modules and assemblies—design qualification and type approval, IEC 62108 (2007).

92. Crystalline silicon terrestrial photovoltaic (PV) modules—design qualification and type approval, IEC 61215 (2005).

93. Reese MO, Gevorgyan SA, Jørgensen M, Bundgaard E, Kurtz SR, Ginley DS, Olson DC, Lloyd MT, Moryillo P, Katz EA, Elschner A, Haillant O, Currier TR, Shrotriya V, Hermenau M, Riede M, Kirov KR, Trimmel G, Rath T, Inganäs O, Zhang FL, Andersson M, Tvingstedt K, Lira-Cantu M, Laird D, McGuiness C, Gowrisanker S, Pannone M, Xiao M, Hauch J, Steim R, DeLongchamp DM, Rösch R, Hoppe H, Espinosa N, Urbina A, Yaman-Uzunoglu G, Bonekamp JB, van Breemen A, Girotto C, Voroshazi E, Krebs FC (2011) Consensus stability testing protocols for organic photovoltaic materials and devices, *Sol Energy Mater Sol Cells*, **95**, 1253–1267.

94. Gevorgyan SA, Medford AJ, Bundgaard E, Sapkota SB, Schleiermacher HF, Zimmermann B, Wurfel U, Chafiq A, Lira-Cantu M, Swonke T, Wagner M, Brabec CJ, Haillant OO, Voroshazi E, Aernouts T, Steim R, Hauch JA, Elschner A, Pannone M, Xiao M, Langzettel A, Laird D, Lloyd MT, Rath T, Maier E, Trimmel G, Hermenau M, Menke T,

Leo K, Rosch R, Seeland M, Hoppe H, Nagle TJ, Burke KB, Fell CJ, Vak D, Singh TB, Watkins SE, Galagan Y, Manor A, Katz EA, Kim T, Kim K, Sommeling PM, W.Verhees JH, Veenstra SC, Riede M, Christoforo MG, Currier T, Shrotriya V, Schwartz G, Krebs FC (2011) An inter-laboratory stability study of roll-to-roll coated flexible polymer solar modules, *Sol Energy Mater Sol Cells*, **95**, 1398–1416.

95. Hermenau M, Schubert S, Klumbies H, Fahlteich J, Muller-Meskamp L, Leo K, Riede M (2012) The effect of barrier performance on the lifetime of small-molecule organic solar cells, *Sol Energy Mater Sol Cells*, **97**, 102–108.

96. Jørgensen M, Norrman K, Gevorgyan SA, Tromholt T, Andreasen B, Krebs FC (2012) Stability of polymer solar cells, *Adv Mater*, **24**, 580–612.

Chapter 15

Stability of Organic Photovoltaic Cells: Failure Mechanisms and Operational Stability

Eszter Voroshazi
imec, Kapeldreef 75, Heverlee, 3001, Belgium
eszter.voroshazi@imec.be

One of the remaining challenges for the commercialization of organic photovoltaics (OPVs) is their unknown and/or insufficient device reliability. Fundamental insights into the failure mechanisms gathered in the last years paved the way toward the current state-of-the-art result of over 10 years of operational lifetime. Despite the remarkable progress in the last years, our understanding of device failure and methodology to establish device lifetime remains limited and sometimes even incoherent. To overcome these challenges, the topic of device stability is poised to become one of the most critical aspects of OPV research and development. This chapter first presents to the reader a comprehensive review of our current understanding of failure mechanisms and their mitigation routes. This is followed by a summary of operational lifetime studies to highlight the progress and remaining challenges in the field.

Organic Solar Cells: Fundamentals, Devices, and Upscaling
Edited by Barry P. Rand and Henning Richter

ISBN 978-981-4463-65-2 (Hardcover), 978-981-4463-66-9 (eBook)
www.panstanford.com

15.1 Introduction

Swift progress in the performance of organic photovoltaics (OPVs) has led to over 10% efficient devices with both solution- and vacuum-processed cells [44]. By reaching this threshold in performance, the industrial relevance of OPVs is clearly demonstrated. However, further prerequisites such as device reliability and durability are indispensable for practical applications. The market requirement of 5–15 years of operational lifetime appeared elusive a few years ago, but recent state-of-the-art devices are about to reach this target [27,208–210]. Significant efforts of both the academic and industrial research communities in the last years enabled this progress.

Unlike inorganic semiconductor technologies, where failure mechanisms are well known, in OPVs, the exploration of degradation pathways is one of the biggest challenges of device reliability. In contrast to inorganic PV and other electronic devices, the large diversity of materials and device architectures in this research field further complicate the quest to enumerate all failure routes. Therefore, most OPV device reliability studies focus on the origin of electrical device performance loss upon exposure to stress factors, i.e., stress-based testing [204]. The fundamental understanding derived from these studies paves the way toward stable OPVs. Currently, fewer investigations are aimed at the evaluation of operational lifetime. This latter aspect is also explained by the fact that operational lifetime studies require sophisticated tools dedicated to long-term experiments as well as a reliable device production and packaging technology.

In this chapter, we devote a large section to review the current understanding of failure mechanisms. We aim to provide the reader with an insight into various failure routes of both polymer and small-molecule-based devices. In the second part of the chapter, we summarize the current practices to evaluate the stability of devices and we present the state of the art in operational stability of different OPV technologies. Finally, the chapter is concluded with a discussion of the potential for OPVs to meet market requirements.

15.2 Failure Mechanisms in Photovoltaic Cells

Intrinsic failure mechanisms determine the inherent device stability as these processes are triggered even in a perfectly sealed device where oxygen and humidity ingress is eliminated. In contrast, extrinsic degradation mechanisms are induced by exposure to ambient air. Although investigation of intrinsic failure mechanisms is critical to assess the ultimate lifetime of OPVs, extrinsic failure processes should also be well understood. On the one hand, the mitigation of extrinsic mechanisms via proper packaging is an obvious step to improve device stability. On the other hand, one should consider that extrinsic failure routes cannot be entirely suppressed in most devices or modules due to a nonzero permeability of plastic barriers or edge sealants. Thus, the understanding of extrinsic failure mechanisms remains relevant to describe packaging requirements. In this chapter, we enumerate both types of degradation mechanisms layer by layer as illustrated in Fig. 15.1. Firstly, we focus on the photoactive layer, and afterward discuss the failure of the buffer and electrode layers. We limit the discussion to the most common materials to propose a close to comprehensive description of the current understanding, as well as to comment on topics still under debate. The outline of the chapter is motivated by the common methodology applied in OPV device engineering. Both device optimization and reliability studies are based on systematic variations of the layers. Numerous mitigation strategies are directly included in this section, as most stability studies compare devices with differing stability to elucidate the origin of the failure.

Under operating conditions, devices are subjected to weathering stress factors, namely solar light, heat, and ambient atmosphere. Additionally, electrical stress as well as other environmental stresses can be listed among potential initiators of failure. Beyond electrical failure, the mechanical integrity and fatigue upon bending of flexible devices has received little attention until now. However, we will also provide the reader an insight into these emerging reliability topics. The interested reader is also referred to numerous recent reviews on this topic [47,70,71,89,203].

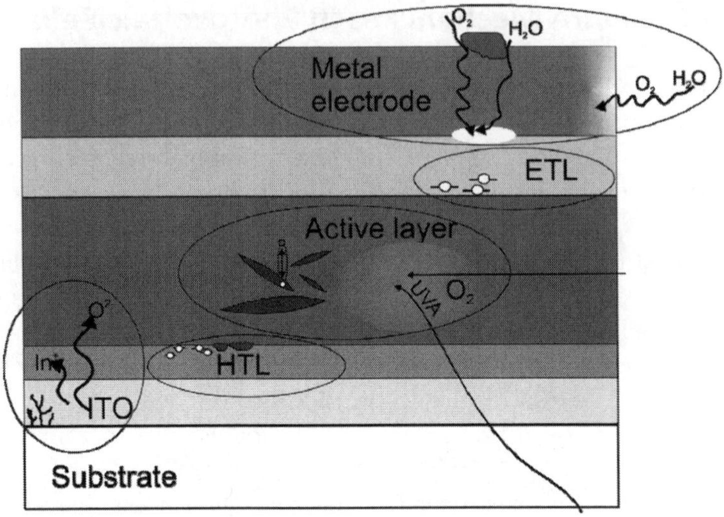

Figure 15.1 Schematic representation of failure mechanisms in conventional architecture OPV cells utilizing an ITO anode.

15.2.1 Degradation Mechanisms of the Photoactive Layer

Considerably different properties and deposition methods of polymer and small-molecule donor molecules motivate us to discuss in distinct sections the stability of their heterojunction with fullerenes.

15.2.1.1 Degradation mechanisms in polymer:fullerene blends

Both physical and chemical degradation of the polymer:fullerene blend can be triggered by solar light. The first failure route is induced by IR radiation, i.e., elevated heat, while UV and visible radiation mostly cause chemical degradation. Electrical properties of the molecules can be, sometimes reversibly, altered upon exposure to air in the absence of light.

15.2.1.2 Physical degradation of the polymer:fullerene blend morphology

Solar radiation raises device temperature close to 70 °C in outdoor applications [39]. This motivated the recent implementation of 65 °C and 85 °C as standard testing temperatures in OPV reliability protocols [131]. The amorphous and semicrystalline materials in OPVs become diffusive above their glass transition temperature (T_g) leading to the reorganization of the blend morphology. Upon optimized deposition and post-deposition treatments, the donor and fullerene domains are nanometer sized and form an intermixed and percolating network. Continued annealing of the blend above its T_g induces demixing and phase segregation of the donor polymer and fullerene phases. Crystallization and reordering of the separate phases can also be observed in some cases [6]. These processes result in increased domain sizes that reduce the donor–acceptor interface area and can also interrupt the percolating pathways for charge carriers. Consequently, both the exciton collection and charge transport rapidly degrade [6]. The extent of the phase segregation is determined by both the material properties and temperature. The solid-state miscibility of the two phases and their diffusion coefficient in the matrix confined by electrodes are the most critical parameters [15]. For example, the immiscibility of MDMO-PPV and PCBM drives this blend to complete demixing. However, steric hindrance limits its progression until equilibrium [16,191,193,196]. Indeed, it has been shown that the temperature-dependent behavior of this blend follows the classical spinodal decomposition model [16,131]. Spinodal decomposition refers to the demixing of one phase into two distinct phases. In contrast, the partial miscibility of amorphous P3HT and $PC_{60}BM$ has proven to change the kinetics of its phase segregation [83]. In P3HT:$PC_{60}BM$ blends, the degradation is triggered already in operating conditions as P3HT has a low T_g of 15–30 °C, while the T_g of $PC_{60}BM$ is 139 °C. Upon blending the two materials, the T_g remains only 45–50 °C [197]. Moreover, the blend undergoes cold crystallization at 110–130 °C. Measurement of these thermal characteristics of the materials and their blend is presented in Fig. 15.2. Glass transition is a second-order phase transition between the hard and malleable state of the material; hence, it is detected by a change in the slope of heat flow. Cold crystallization corresponds to crystal formation from the glassy but still, solid-state material contrary to melt crystallization. Cold crystallization is an exothermic process and therefore, detected as deviation below the heating curve in endothermic heat flow differential scanning calorimetry (DSC). The melting of the crystals is an endothermic process, and hence, the heat flow deviates above the linear slope.

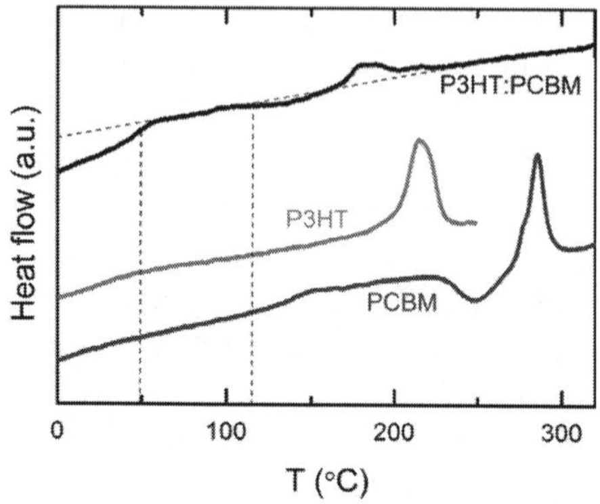

Figure 15.2 Differential calorimetry heating scans of P3HT, PCBM, and their mixtures. For all materials the second cycle is measured at a heating rate of 500 °C/min.

The softening of the polymer matrix permits the diffusion of the commonly used $PC_{60}BM$ and $PC_{70}BM$ molecules and formation of aggregates, which can even crystallize above 100 °C [6,102,163,191]. The kinetics of $PC_{60}BM$ crystallization can be divided into two stages: in the first stage, fast nucleation takes place, while in the second stage, the $PC_{60}BM$ crystals grow in height and then expand laterally [25,191]. The density of $PC_{60}BM$ nuclei is controlled by the initial phase segregation, which depends on the surface energy of the underlying transport layer [55]. Reduced absorption of $PC_{60}BM$, due to increased light scattering by the aggregates, is a characteristic signature of the demixing process [171]. The large-scale aggregation of $PC_{60}BM$ in a device aged at elevated temperature is illustrated in Fig. 15.3. The crystallization process can be reduced in certain blends by substituting bulkier fullerene molecules such as bis-fullerenes or larger mono-adduct C_{60} derivatives for $PC_{60}BM$ [13,115,195]. Minor alteration of the molecular weight and regioregularity can also alter the thermal behavior of the blend. Furthermore, morphology engineering by additives also appears to impact the degradation rate [87].

Current understanding of the phase segregation is derived from studies conducted on P3HT:$PC_{60}BM$ blends aged at temperatures between 90 °C and 150 °C [6,25,141]. Annealing devices over 100 °C for 50–100 hr typically result in a more than 30–50% performance drop [6,25,141]. These experiments can be considered as accelerated lifetime tests, as they exceed the temperature under operating conditions. However, the relevance of acceleration factors extracted following the methodology developed for inorganic semiconductors is often questioned [16,141].

Indeed, in accelerated tests above 100 °C, the degradation mechanism is altered compared to operational conditions as the crystallization of P3HT and $PC_{60}BM$ is triggered [6,102].

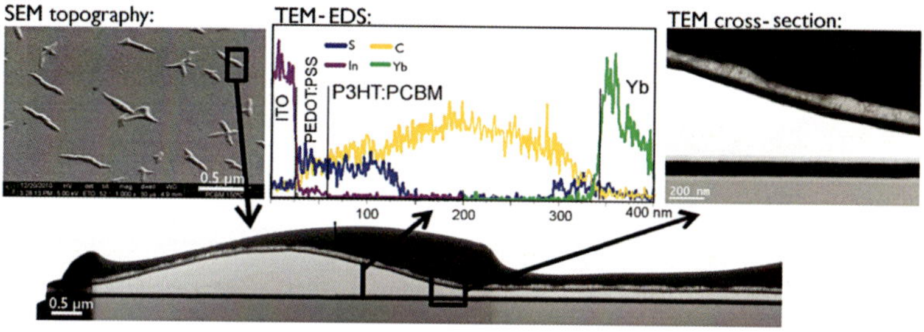

Figure 15.3 Electron microcopy images of a conventional P3HT:PCBM device aged at 150 °C for 100 hr. Top view of $PC_{60}BM$ aggregates in an SEM image (upper left) and bright-field TEM cross-section images over an aggregate (bottom and upper right). Elemental analysis recorded with TEM-EDS across the device stack (upper middle).

This failure path has raised considerable concerns about the stability of bulk heterojunction OPV cells. However, combination of increased T_g and suppressed fullerene crystallization can significantly improve the thermal stability of polymer:fullerene blends.

Considerable efforts are dedicated to the synthesis of thermally stable photoactive materials. One of the most common approaches is the cross-linking of organic compounds by different routes. Recently a cross-linked donor–acceptor-type polymer yielded over 4% efficient devices, which retained their performance at high-temperature annealing for 72 hr [46]. Alternatively, the use of novel polymers specifically designed to have high T_g has failed to combine increase in thermal stability and high performance [6]. The modification of the ubiquitous $PC_{60}BM$ is also a relevant approach. Two concepts have been briefly reported: the epoxy-based cross-linking via the side chains and the change of the solubilizing groups to render the molecule bulkier and hence less prone to crystallization [28,195].

Finally, cross-linking of donor and acceptor together has also been proven to stabilize the morphology of P3HT:$PC_{60}BM$ blends, although at the expense of initial device performance [41,79,80,198]. Some of the novel generation high-efficiency polymers also possess an elevated T_g. For example, the T_g of PCDTBT:$PC_{70}BM$ blends is 130 °C making this photoactive layer intrinsically more thermally stable [11]. Therefore, temperature-induced morphology reorganization is not intrinsically limiting OPV device lifetime, but rather requires a careful selection of thermally stable materials.

Unlike electrical device performance, the cohesion of the photoactive layer can be improved by thermal annealing. Indeed, the cohesion of the P3HT:$PC_{60}BM$ blend limits the mechanical stability of devices in a conventional architecture. The amount of $PC_{60}BM$ in the bulk heterojuntion also determines the cohesive strength of the layer. An initial study indicated that increased fullerene content of the bulk heterojunction lowers the cohesive strength of the layer [31]. This failure mechanism indicates a potential weakness of novel generation polymer:fullerene blends employing elevated fullerene loading. The implication of intercalation between polymer and fullerene molecules should be further incorporated in this current understanding of polymer:fullerene cohesive strength [7].

15.2.1.3 Chemical degradation of the polymer:fullerene blend

Chemical degradation of organic molecules under UV illumination in the presence of air is considered as one of the most critical failure mechanisms. Chemical degradation results in the modification of backbone of the organic semiconductors, which irreversibly alters their optical and electrical properties. The most visual consequence of this failure mechanism is the fading of the original color, hence degradation mechanism is often referred to as photobleaching.

This process can be suppressed in inert atmosphere, thus device lifetime of packaged samples is rarely limited by this failure pathway. The encapsulation of the organic layers by an evaporated metal cathode and an additional hermetic packaging has indeed proven to limit O_2 ingress. Notably, good quality electrodes have a water vapor and oxygen transmission rate (WVTR and OTR) of 10^{-3} g/cm^2/day [59]. Reduced photobleaching of the organic layer protected by the electrodes can be clearly observed on a severely degraded substrate with P3HT:$PC_{60}BM$. In between the metal electrodes, i.e., without protection, the polymer blend became transparent, while under the electrode the absorption remained unchanged (Fig. 15.4). Nevertheless, insight into the photo-oxidation mechanism is required both to develop intrinsically photostable materials and to define the encapsulation barrier limit.

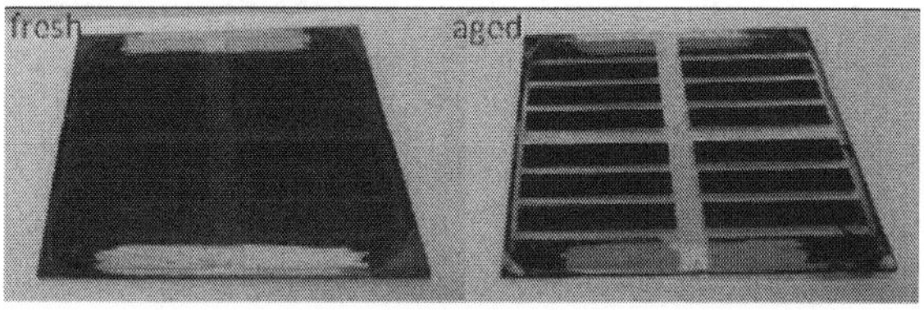

Figure 15.4 Photograph of a fresh (left) and an aged (right) substrate with several P3HT:$PC_{60}BM$ solar cells. Horizontal patches correspond to metal electrodes defining the active area. The aged sample was exposed to AM1.5G solar radiation in air for 150 hr.

Studies of the photo-oxidation process conducted on uncapped organic layers are summarized in the following paragraphs. One should keep in mind that *direct* air exposure of the organic layers considerably accelerates the process; hence quantitative correlation to device degradation studies cannot be directly extracted.

The pathways of photodegradation can be divided into two categories. The first one is triggered by superoxide radicals (O_2^-) that initiate radical reaction of the conjugated polymer. These superoxides are formed by electron transfer from the polymer to the oxygen molecule. The second mechanism, mostly observed in solution and not in thin films, involves the attack of the conjugated polymer by singlet oxygen ($^1O_2$). This excited molecule is formed by electron transfer from the triplet state of the excited polymer to the ground state of the O_2 molecule [3].

Initially, the photo-oxidation of PPV derivatives has been investigated in-depth. In these early studies, it was demonstrated that the photo-oxidation of PPV derivatives occurs via radical reaction and the influence of singlet O_2 can be excluded [104,135,143]. The photobleaching retarding effect of $PC_{60}BM$ blended in the polymer films has already been noted in these early studies. The photo-oxidation process reduces light absorption in the active layer, but this cannot entirely explain the lowering of photocurrent. The reduced conjugation length and resulting oxidation products also hamper charge transport through the development of deep traps [124]. The absence of oxygen only slows the photodegradation but does not restrain it. This indicates the intrinsic instability of PPV derivatives [9,10].

Photo-oxidation in polythiophenes in thin films also follows a two-step process as presented in Fig. 15.5a [2,108]. Comparable to PPV derivatives, radical cations form by oxidation of the n-hexyl side chain in the first step. Subsequently, the sulfur atom of the thiophene ring is oxidized leading to reduced conjugation at the backbone and bleaching of the sample as shown in Fig. 15.5b. Activation of the same two-step process under thermo-oxidation conditions, i.e., elevated temperature in darkness and air atmosphere, further proves that singlet oxygen is not required to initiate the process. Contrary to PPV, in the absence of light only reversible oxygen doping of the P3HT is observed and no photobleaching is noted. Samples illuminated in the absence of oxygen for over 1 year exhibited nearly no change and hence demonstrated the long-term inherent stability of this polymer [135]. This confirms that in P3HT only the *simultaneous* presence of light and oxygen triggers photo-oxidation. Quantitative influences of several environmental factors on the photo-oxidation have been characterized in a recent comprehensive study [60]. One of the critical points to retain is that humidity alone does not contribute to the oxidation of P3HT but it does act as a catalyst in the presence of oxygen [60,119,148].

Further insight in the photo-oxidation process has been provided by detecting the formation of a reversible charge transfer complex between P3HT and O_2, prior to the chemical degradation of the side chains. This complex is created in a matter of minutes upon simultaneous exposure of P3HT or PCDTBT to air and UV or visible light [3,17]. The long lifetime of the complex suggests that it is stabilized by a chemical interaction; however, it can be reversed by elimination of either light or O_2 [3]. Furthermore, the complex formation has also been detected in polymer and fullerene blends confirming previous conclusions that photo-oxidation is not suppressed by the presence of $PC_{60}BM$ [3].

Investigation of the photostability of P3HT blends with bis-fullerenes indicated the increase of the degradation rate as the energy difference between the LUMO of the donor and acceptor decreased [62,168]. The authors attributed this to the energetically favorable superoxide formation in blends of P3HT with bis- and tris-fullerene adducts. The charge transfer state lies above the energy level of singlet oxygen in these compounds, which favors the electron transfer to the latter molecule. This observation hints at a possible fundamental trade-off between device efficiency and stability.

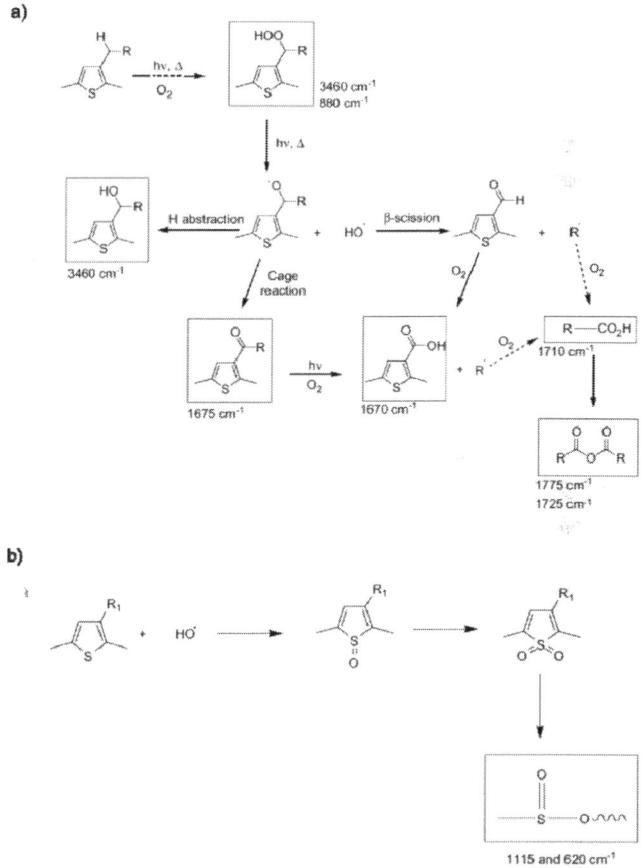

Figure 15.5a Oxidation mechanism of the P3HT monomer unit in two steps: (a) reaction of the akyl side chain, (b) reaction of the thiophene ring. Reproduced with permission from [108].

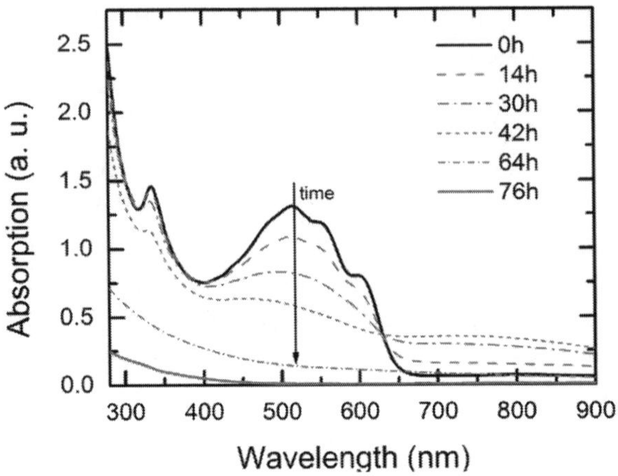

Figure 15.5b Evolution of the absorption spectrum of the P3HT:PCBM blend exposed to AM1.5G solar radiation in air.

As presented in the previous chapters, important efforts are dedicated to synthesis and understanding of the structure–property relationship in novel donor-type polymers to increase device efficiency. However, currently only one reported study explored the photostability of these novel push–pull polymers [106]. This study established a stability classification of the most frequently used monomer units as shown in Fig. 15.6. Furthermore, the conclusion that photo-oxidation of the polymers is mostly initiated through the side chains appears to be a general observation. Therefore, the photostability of the polymer could be enhanced by the design of thermo-cleavable side chains [107]. In such polymers, the solubilizing side chains are detached from the backbone by a post-deposition thermal annealing step. The layer becomes insoluble following this step and thus the detached side chains can be removed by solvent rinsing. Although these polymers have a limited efficiency, their photostability is considerably increased compared to their uncleaved counterparts [107].

The acceleration of photo-ageing in various polymers indicated that the chemical process remains identical and scales linearly with light intensity up to an extreme 100 sun illumination. This accelerated study confirmed that the photodegradation is reduced in P3HT compared to PPVs, while PCDTBT proves to be even more stable [105,147,168]. These results confirmed the superior inherent stability of commonly employed P3HT compared to PPV derivatives, thus confirming its industrial relevance. Another important implication of the intrinsic stability of P3HT in air is its processability in ambient conditions. From the first investigations on high-efficiency polymers, PCDTBT emerges as a molecule combining both light and heat stability.

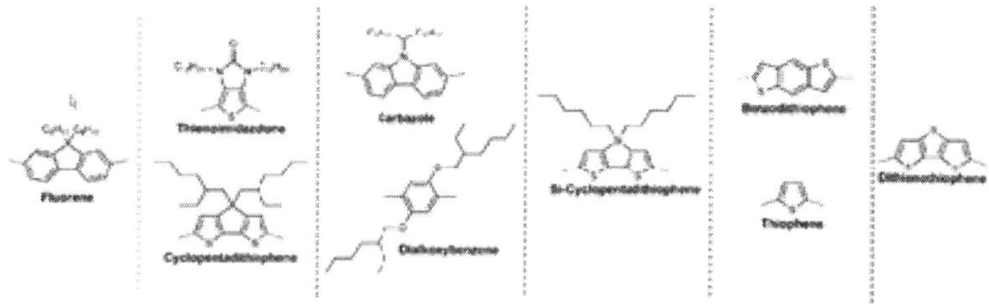

Figure 15.6 Common monomer units of p-type semiconducting polymers in increasing order of photostability. Reproduced with permission from [106].

Photoconductance measurements of P3HT:$PC_{60}BM$ blends exposed to illumination in ambient atmosphere yielded, surprisingly, that conductivity of the blend decreased faster than that of a pure polymer layer [133]. This was the first report indicating that $PC_{60}BM$ is also prone to photo-oxidation. Detection of the oxidized moieties further confirmed this degradation pathway. Since the LUMO level of oxidized $PC_{60}BM$ is significantly lowered compared to unaltered $PC_{60}BM$, these degraded molecules act as deep electron traps [133], and indeed the formation of C_{60} epoxide has been reported earlier [18]. Since this reaction occurs on the C_{60} cage independent of the solubilizing side chain, this photodegradation most probably occurs in all fullerenes.

Recently, material purity and structural defects have emerged as possibly critical, and long neglected, factors influencing photodegradation of polymers. Chemical impurities such as remaining metal catalysts or other reagents as well as structural defects might determine the degradation rate of polymers. Recently, Liang et al. reported that passivation of structural defects in P3HT reduces the rate of photo-oxidation [45,93]. Improved electrical bias stability has also been achieved in OTFTs by attaching an end group on the polymer chains [14]. Furthermore, intrinsic defects in PCDTBT are surmised to trigger device degradation under continuous illumination [125].

Alteration of the electrical properties of polymer:fullerene blends can also arise from oxygen adsorption in the absence of light. Contrary to photochemical and physical degradation mechanisms, these failure routes are partially reversible. Light and increased oxygen content accelerate the adsorption of oxygen in the polymer matrix, while vacuum background and heating above the glass transition temperature of the polymer promote desorption [94].

The reversible hole doping of P3HT by oxygen has been first characterized in OFETs [94,114]. Simultaneous reduction of carrier mobility has also been reported in OPVs, contrary to OFETs, where only carrier density increases upon air exposure [138,139]. This mobility drop originates from the formation of deep traps; however, blending with $PC_{60}BM$ seems to mitigate this process [117,138,139]. The increased carrier density of P3HT has also shown to hinder geminate pair separation, as it induces the formation of space charge regions screening the electric field [145,146]. Recently the role of ozone, also present in most ageing studies conducted in air, has proved to be non-negligible. Similarly to oxygen, ozone can alter the carrier density in P3HT [1,8].

Besides air exposure, oxygen can also originate from desorptive processes in metal oxide layers [97,169].

15.2.1.4 Degradation mechanisms of small-molecule : fullerene blends

Although phthalocyanines and C_{60} are employed in various applications, their stability in photovoltaic devices has been less investigated than solution-processed organic semiconductors. We outline in this part the major differences and similarities in their degradation compared to polymer:fullerene blends. Phthalocyanines and porphyrins present a good chemical and thermal stability, unlike common p-type polymers [147]. Similar photostability is indirectly observed by Hancox et al. in a study of interface layers [51]. Device stability under solar illumination in air or in N_2 differed little, suggesting that photobleaching is not a dominating failure mechanism in devices. This observation was confirmed for four commonly used donor materials: ZnPc, SubPc, CuPc, and pentacene [51]. Results of this study are presented in Fig. 15.7. Similarly, no electrical degradation has been noted in devices with ZnPc illuminated with laser light in air [82].

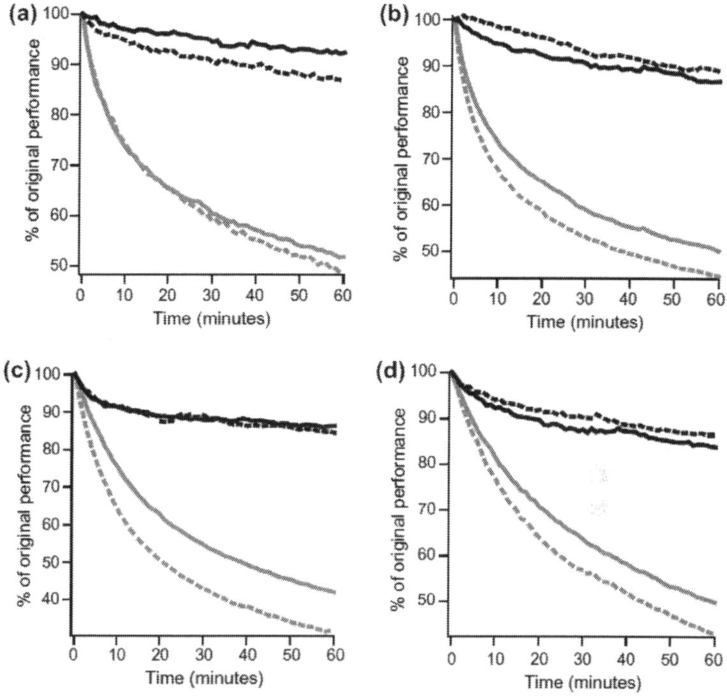

Figure 15.7 Device degradation data for percentage decrease of η_P under constant illumination at 1 sun for 60 min for (a) ITO/d nm MoO_x/20 nm ClAlPc/40 nm C_{60}/8 nm BCP/Al, (b) ITO/d nm MoO_x/14 nm SubPc/40 nm C_{60}/8 nm BCP/Al, (c) ITO/d nm MoO_x/30 nm CuPc/40 nm C_{60}/8 nm BCP/Al, (d) ITO/d nm MoO_x/43 nm pentacene/40 nm C_{60}/8 nm BCP/Al. For each grey and black line d = 0 and 5 nm, with solid lines and dashed lines representing N_2 and air atmosphere, respectively. Reproduced with permission from [51].

Only one report suggested that the photo-oxidation of organic layers is a dominant failure mechanism upon concomitant exposure to light and air [161]. This observation was reported by Sullivan et al., who assigned the fast reduction in photocurrent to chemical degradation of the pentacene molecule [161]. Since similar reduction in photocurrent is observed in pentacene/C_{60} bilayer cells exposed to air in dark, the photodegradation might not be the sole cause of device failure [130].

As with p-type polymers, exposure of phthalocyanines to oxygen increases charge-carrier mobility but introduces trap states [172]. Similar process has also been confirmed in pentacene layers investigated in OTFT. Like in P3HT, dry oxygen induces a reversible degradation via formation of deep charge traps [73,176,177]. Ambient humidity further increases the energetic disorder in pentacene and hence contributes to reduction of the mobility [72]. Similarly to polymers, synergetic effect of oxygen and light initiates a photo-oxidation process attacking the most reactive central ring of the pentacene molecule [4]. However, photostable pentacene derivatives have been developed by passivating the central benzene ring through substitution of the H atom by a bulkier side chain [43,75]. Synergistic effect of light and air in ZnPc triggers the formation of superoxide radicals or charge transfer complexes [64]. Firstly, this process leads to reversible p-doping of phthalocyanines. Additionally, the superoxide migration to the donor/acceptor interface under the influence of an external electric field has been suggested by Huisman et al. [64]. This process reduces the internal quantum efficiency of the solar cell via exciton quenching [64].

Electrical stability of C_{60} in air has been extensively investigated in OTFTs [110,165]. Exposure to both O_2 and humidity induces shallow trap states that severely reduce the electron mobility. Upon annealing, device performance can be partially reversed in OTFTs, indicating that only physical adsorption occurs and no chemical bonds are formed between the molecules [49,110]. Synergistic interaction of light exposure and air triggers the irreversible chemical degradation of C_{60} as observed in polymer:fullerene blends [18,49,118]. In the absence of air, C_{60} is mostly considered intrinsically stable [49]. However, Eklund et al. reported the random dimerization of both C_{60} and C_{70} molecules under illumination in inert atmosphere [33]. The covalent bonds formed between C_{60} molecules could be reversed by thermal annealing. Modification of the electrical properties of the molecules has not been characterized during this photochemical process.

Photodegradation of C_{60} in devices has been observed under low-wavelength light illumination by Lessmann et al. in a planar heterojunction ZnPc/C_{60} device [92]. In contrast, no chemical degradation of C_{60} has been observed by TOF-SIMS in similar ZnPc:C_{60} devices exposed to solar illumination in air [58]. Contradictory conclusions of these two studies might originate from different degradation pathways in planar and bulk heterojunctions. A short stability study of an alternative acceptor molecule, Cl_6SubPc, suggested that non-fullerene acceptors might allow for superior stability in comparison to C_{60} [160].

Small-molecule donors and C_{60} are sublimed at elevated temperature, which intuitively suggests that their thermal stability is increased, compared to polymer:fullerene cells. Solubilizing side chains attached to polymer monomers lower their glass transition temperature, explaining the superior thermal stability of thin films composed of small molecules. Consequently, no report suggests major morphology reorganization in the bulk of sublimated thin films.

Thermal stability of active layers has only been investigated indirectly in a few studies. One of the early ageing studies conducted at 85–95 °C reported a significant performance loss over 100 h. However, the authors later elucidated that the performance loss originated from encapsulation failure. At elevated temperature, the increased water vapor transmission rate (WVTR) of the packaging accelerates ambient air-induced failure mechanism [38,205]. After improved packaging, Heliatek reported a lifetime of over 10000 hr at 85 °C on single-junction and tandem cells with ZnPc:C_{60} active layers [208].

Besides reorganization in the bulk of the layer, alteration in molecular orientation at the interfaces can also greatly alter electrical and optical properties of photoactive molecules. Modification of molecular ordering at the interface has been linked to lower device performance in pentacene/C_{60} cells. Sullivan et al. suggested that local heating at the ITO/pentacene interface was triggered by UV absorption in the ITO [161]. Introduction of a buffer layer between ITO and the donor material proved to stabilize device performance. Similar observations were reported for other common donor materials, CuPc, ZnPc, and SubPc [51]. The buffer layer in these cases surmised to mitigate the harmful influence of indium tin oxide (ITO). Further details of this failure process are described in Section 15.2.3.1 on the influence of ITO on device failure.

Performance loss under varied light intensities suggested that the total number of extracted carriers determines the failure rate in small-molecule-based OPVs [57,205]. Although failure rate is correlated with the number of carriers as it is shown in Fig. 15.8, excitons appear to be the origin of the degradation. This relationship is *in*valid if the variation in number of extracted charges is due to alteration in light spectrum, instead of light intensity [57], as illumination with higher energy radiation UV light appears to accelerate the ageing process. More systematic investigation on the influence of UV light on the ageing of small-molecule devices indicated that lifetime of the device can be doubled by using a cut-off filter of 360 nm instead of 320 nm and further doubled with a 420 nm cut-off filter [202]. Undeniably, the presence of photo-excited species generated from high-energy radiation triggers the failure of small-molecule devices in inert atmosphere, although the exact mechanism of the process remains unclear.

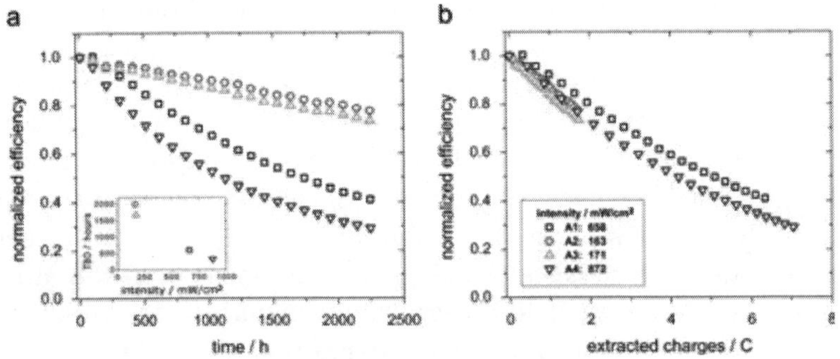

Figure 15.8 Normalized efficiency is shown versus time (left side) and versus the integrated number of extracted charges (right) for devices type A and white illumination. The inset in figure (a) shows the extracted lifetimes (T_{80}) for the four different intensities. Reproduced with permission from [57].

15.2.2 Degradation Mechanisms of the Buffer Layers

15.2.2.1 Degradation mechanisms of hole transport layers

In most conventional device architectures, a hole transport layer is incorporated between the ITO and the OPV layer. This layer significantly enhances device performance besides impeding any detrimental interaction between ITO and the photoactive layer (for further details see Chapter 4).

Overwhelmingly in polymer-based devices, but also for small-molecule-based devices, PEDOT:PSS is a commonly applied interface material between ITO and the photoactive layer. Deposition of PEDOT:PSS from a water-based solvent has a critical impact as demonstrated by the comparison of PANI:PSS processed from water and isopropanol. The latter version of PANI:PSS significantly improves air stability by suppressing interface traps at the HTL and active layer interface [32]. Furthermore, the removal of residual humidity from the PEDOT:PSS by thermal annealing has proven critical to improve the stability of small-molecule OLEDs and OPVs [36,68]. Exposure of PEDOT:PSS layer to ambient air should be minimized upon annealing. Due to its hygroscopic nature, this polymer layer rapidly reabsorbs humidity, which reduces its conductivity [32,116]. In addition, this process also accelerates the oxidation of the low work function metal cathode, as indicated by comparison with samples prepared with MoO_3 as HTL [162,178,187]. Considerable influence of the HTL on the device stability of conventional architecture P3HT:$PC_{60}BM$ cells is shown in Fig. 15.9. Shelf lifetime of cells with MoO_3 in comparison to reference cells with PEDOT:PSS is increased ten-fold at elevated humidity levels [178]. Besides ambient air, both light and heat affect the electrical properties of this polymeric transport layer. The synergistic effect of UV radiation and oxygen exposure leads to rapid photo-oxidation of the sulfur in the thiophene unit of the polymer, which interrupts conjugation within the conducting PEDOT segment [34,109]. Elevated temperature of over 100 °C triggers the structural reorganization of PEDOT and PSS domains, which also alters its electrical conductivity [174]. Although most of the above-described failure routes were studied in a separate PEDOT:PSS layer, they have also been detected in devices [78].

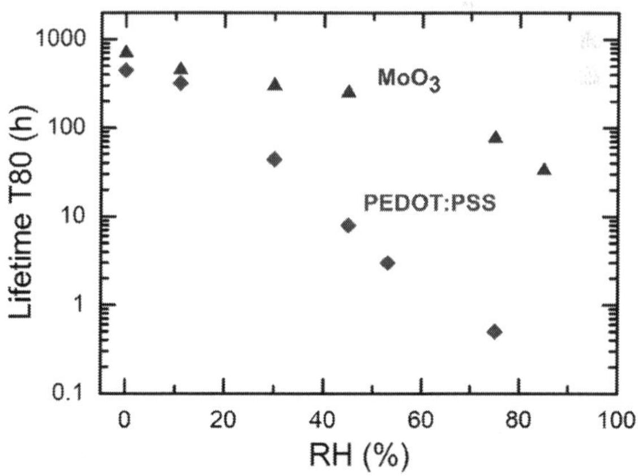

Figure 15.9 Lifetime of conventional P3HT : PCBM cells with MoO_3 and PEDOT:PSS as HTL as a function of ambient humidity. Nonencapsulated cells are stored at room temperature in ambient atmosphere with varied humidity level. Adapted from [178].

In addition to the intrinsic failure of the PEDOT:PSS layer, its detrimental interaction with the photoactive layer has also been reported. In conventional devices, Norrman et al. detected the reaction of PSS with the donor polymer under illumination [121].

In inverted cells, PEDOT:PSS is deposited on the photoactive layer, and consequently its interface failure mechanisms are altered. Unlike in conventional cells, the device failure of inverted cells under illumination in air has been linked to degradation of the PEDOT:PSS layer. One of the studies revealed that phase segregation inside the PEDOT:PSS layer is the first critical step in the device failure. Subsequently, the PEDOT phase is selectively oxidized by O_2, which not only reduces the conductivity of the transport layer, but also triggers a chemical degradation of the photoactive layer [119]. Another study compared the lifetime of inverted P3HT:PCBM devices with and without PEDOT:PSS under continuous illumination. Although the initial performance of the device without PEDOT:PSS was poor, it remained operational over 800 hr. In the device with PEDOT:PSS, the growth of inactive areas around pinholes has been detected leading to rapid failure [98]. Lastly, the impact of dispersion-stabilizing cations and other unknown additives is also often cited as a potential source of instability [119].

Several cross-linkable polymer alternatives to PEDOT:PSS have been reported to efficiently prolong the device lifetime of OLEDs [149,188]. A cross-linked polymer transport layer integrated in conventional architecture MDMO-PPV:PCBM cells exhibited improved chemical and morphological stability [48]. Unfortunately, the limited availability of these polymers currently restrains their use.

Besides the electrical stability of PEDOT:PSS, its mechanical stability has also been investigated. Most adhesion studies on polymer:fullerene cells focused on this HTL as its interface with P3HT:PCBM is the first to delaminate in the inverted device stack [31]. In conventional cells, the interface of the two layers is strengthened by an advantageous interaction between the organic layers, which is not present in inverted cells [29,63]. The initially low surface adhesion of inverted polymer:fullerene cells is further decreased in the presence of humidity. Cohesive failure of the PEDOT:PSS layer is triggered due to weakened secondary bonds in the presence of humidity [30].

Most stability investigations of evaporated hole transport layers are focused on their heat stability, and little is known about their air and light sensitivity. This is explained by the fact that most insights were gained by OLED durability studies. Device stability is strongly determined in OLEDs by morphological instability of the hole transport layer. Although the source of heating differs compared to OPVs, its impact is identical. High operational current induces Joule heating in OLEDs, while in OPVs solar radiation heats the devices to 60–70 °C.

To enhance the thermal stability of the layer, the T_g of the molecules should be above the operating temperature [167]. Hence several commonly used materials, such as (*N*,*N*′-diphenyl-*N*,*N*′-bis(3-methylphenyl)-[1,1′-biphenyl]-4,4′-diamine) MeO-TPD, with a T_g of 55–60 °C do not allow for intrinsically stable devices. It is critical to note that only the thermal stability of OLEDs is improved with these HTL layer, not their overall operational stability. Thermal stability of MeO-TPD also limits the lifetime of solar cells at ageing temperatures above its T_g [38]. Besides structural reorganization of the HTL, it is surmised that the diffusion of the dopants is enabled above the T_g of the layer. Continuously illuminated ZnPc/C_{60} cells, where the substrate temperature is kept below 50 °C, exhibited stable performance over 3000 hr even with MeO-TPD [58]. Despite the stable device performance in this study, the diffusion and reaction of the $C_{60}F_{36}$ dopant used in MeO-TPD was observed with TOF-SIMS analysis [58]. In contrast, a similar device aged 65 °C, above the T_g of MeO-TPD, exhibited a lifetime of 150 hr. As a stable alternative, BF-DPB (T_g = 160 °C) was suggested by Hermenau et al. since ZnPc/C_{60} cells with this HTL showed no performance loss at 65 °C [202]. Finally, as all the above-mentioned studies were carried out on encapsulated cells, we have little insight into the chemical and physical degradations of these layers in air.

Development of various high work function transition metal oxides, such as MoO_3, WO_3, V_2O_5, and NiO, proposes an interesting alternative to organic hole transport layers [158,200]. Among these materials, MoO_3 has been widely used to enhance the performance and stability of OLEDs and OPVs under operating conditions [50,68]. These oxides have been integrated in the fabrication of both solution-processed and evaporated cells, owing to their flexibility in synthesis and deposition [194].

Initial stability tests of these layers indicated their improved air stability in devices compared to PEDOT:PSS, which encourages their further development [42,162,200]. However, the release of oxygen under illumination from the oxides might lead to detrimental interaction in the organic layers in direct contact [97,168]. Additionally, the direct exposure of MoO_3 to air, and probably other oxide layers, induces alteration of its energy level and conductivity [66]. Although these oxide layers might improve device stability, their sensitivity toward oxygen and humidity should nevertheless be considered in order to establish OPV packaging requirements [35].

15.2.2.2 Degradation mechanisms of the electron transport layers

An interface layer between the organic photoactive layer and electron extracting contact is introduced to improve power conversion efficiency. In conventional devices, this layer is inserted between the organic layer and evaporated metal contact. Numerous reasons motivate the insertion of this layer as outlined in Chapter 4. In small-molecule-based devices, the ETL is a thin layer of a wide bandgap organic layer such as bathocuproine (BCP), bathophenanthroline (BPhen), or tris(8-hydroxyquinolinato) aluminum (Alq3). Alternatively, a thicker doped organic layer can be employed. Incorporation of the ETL is required in both conventional and inverted architecture devices. In contrast, in solution-processed devices the ETLs are often oxide layers, mostly incorporated in inverted devices and not in conventional cells.

Firstly, we discuss the stability of evaporated organic ETLs, which were the topic of the first reliability investigations on small-molecule devices. Afterward, we describe the stability of more recently developed metal oxide ETLs.

Instability of the BCP ETL in ambient atmosphere was observed in early OPV studies and was deemed one of the most critical degradation mechanisms of small-molecule cells [127]. Comparison of several ETL layers in CuPc/C_{60} bilayer cells suggested that the use of 1,3,5-tris(*N*-phenylbenzimiazole-2-yl) benzene (TPBi) and Alq3 can improve device stability [155,186]. Crystallization of Alq3 has also been observed in OLEDs upon air exposure [136,157,186]. Similarly, Wang et al. reported that substitution of BCP by TPBi and BPhen prolongs device lifetime in air [183]. Both authors ascribed the improvement to reduced air permeation through the alternative ETLs. Peumans et al.

suggested doping of BCP by 3,4,9,10-perylenetetracarboxylic bisbenzimidazole (PTCBI) to suppress its crystallization [128]. Recently, bilayers of BPhen and $SnCl_2Pc$ have also proven to enhance stability [199]. Similarly, the use of Alq3 doped with Mg has proven to improve device stability by suppressing the crystallization of Alq3. Despite various alternatives, device lifetime in these conventional cells remained only maximum 50 hr [155,183,186]. Crystallization of BCP triggered in humid environment is surmised to facilitate the ingress of oxygen and humidity [127], allowing for the subsequent oxidation of the BCP/C_{60} interface and the bulk of C_{60} leads and a decrease of photocurrent. Both exposure to oxygen and humidity seems to alter the band bending at the C_{60}/BCP interface [190]. Local lowering of the energy level creates electron traps leading to lower carrier mobility as well as increased leakage current [190]. This description of the interface energetics, presented in Fig. 15.10, correlates well with previously observed device degradation studies reporting degradation of nearly all device parameters upon air exposure [155]. In agreement with these results, TOF-SIMS profiling revealed the incorporation of molecular oxygen and to less extent water at BPhen/C_{60} interface [58]. Since dominant failure mechanisms such as the crystallization of BCP or Alq3 and oxidation Al electrode are accelerated in a humid environment, device degradation is accordingly faster [136,189].

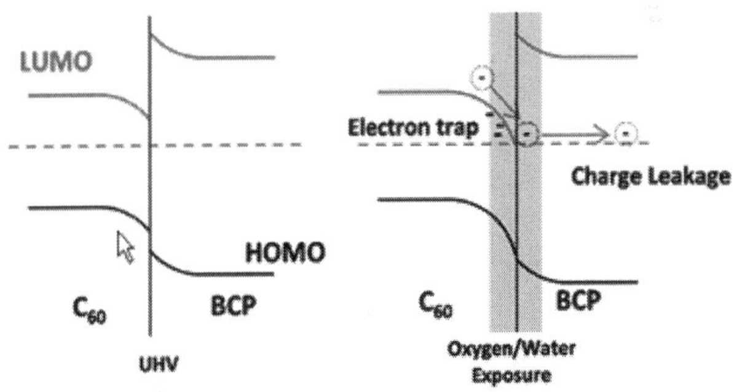

Figure 15.10 Schematic of energy level at the interface of C_{60}/BCP in inert conditions and upon ambient air exposure. Reproduced with permission from [190].

Since in inverted device architectures, the ETL/C_{60} interface is shielded from air exposure by multiple layers, device stability is slightly improved [156,175,182].

Transition metal oxide transport layers, employed in polymer:fullerene cells to eliminate buffer layer failure in air, have not yet been explored in small-molecule OPVs. Therefore, air-induced failure in small-molecule-based cells can only be mitigated by encapsulation of the devices. The air sensitivity of the ETL requires barriers with WVTR of less than 10^{-5} g/m^2/day [59].

Thermal stability of these transport layers in OPVs is little known, but it has been explored in OLEDs. The operational stability of OLEDs could be enhanced upon replacement of the ETLs with low T_g such as BCP and BPhen. Both materials have $T_g = 85$ °C leading to morphological changes and dopant migration under standard OLED operating conditions [202]. Replacement or mixing of these ETLs with higher T_g alternatives such as bis(2-methyl-8-quinolinolato-N1,O8)-(1,1'-Biphenyl-4-olato)aluminum (BAlq) ($T_g = 92$ °C), Alq3 ($T_g = 172$ °C), and TPBi ($T_g = 120$ °C) directly improved the heat resistance of the devices [20].

In exclusion of air, small-molecule devices exhibit stable device operation under solar illumination, suggesting that the above-discussed ETLs are light stable. Heliatek reported an extrapolated lifetime of over 20 years on tandem cells employing BPhen as ETL [208]. Only one report by Shiokawa et al. mentions decomposition of naphthalene tetracarboxylic dianhydride (NTCDA) under illumination even in exclusion of air [152,164]. Replacement of the NTCDA by a thick In/ZnO:Al_2O_3 bilayer ETL enables 1000 hr stable device operation under 1 sun illumination [61]. A laser-induced desorption/ionization TOF-SIMS study allowed them to evaluate the intrinsic photostability of the layers. Comparison of the commonly applied ETL layers, namely BCP, BPhen, Alq3, and BAlq, suggested that the latter compound presents outstanding stability [140].

Recently, doped C_{60} or BCP layers have also been reported as efficient electron transport layers [58,111]. Development of these layers has been motivated by the need for thick transport layers in tandem cells. To our knowledge, no studies explored their stability.

In conventional polymer:fullerene devices, the metal electrode is often directly deposited on the organic layer. It is only recently that the incorporation of a solution-processed metal oxide layer as ETL has been explored in inverted cells. Substituting transition metal oxides for the reactive low work function metal significantly improves the air stability of devices.

The use of either TiO_x or ZnO as ETL is explored in numerous studies [37,90,99]. Although these metal oxide layers are stable in air, they are not inert to air. Adsorption of O_2, H_2O, and other atmospheric gases in metal oxides has been the subject of intense study [12]. Molecular oxygen acts as an electron scavenger to form O_2^-, which alters the electron density of the oxide [12]. In the meantime, water generates surface hydroxyl groups on the surface of TiO_2, which open up reaction pathways [12]. The influence of these effects on the stability of OPVs has not been explored. Indirectly, a recent study suggested that reduced porosity of ZnO, i.e., reduced air ingress, enhanced device stability [37]. Several reports describe the reversible oxygen adsorption in ZnO, which reduces its conductivity. Illumination with UV light or heat triggers the desorption of O_2 and reverses the electrical properties of ZnO. This treatment is often called photodoping [95,151]. The oxygen scavenging nature of metal oxides is surmised to slow the photo-oxidation of organic layers and hence stabilize device efficiency [90]. Synergistic effect of ambient humidity and heat has been detrimental to TiO_x in inverted cells, while devices with SnO_x layer exhibited stable performance [170]. Current stability studies of OPVs suggest that metal oxides match the stability requirements. However, investigation of these layers and devices is still in the early stage. Moreover, reliability limitation of ZnO detected in damp heat condition ageing of CIGS modules raises concerns about this material in general [65].

15.2.3 Degradation Mechanisms Linked to the Electrodes

15.2.3.1 Degradation mechanisms of indium tin oxide

The ubiquitously applied transparent electrode, ITO, has been linked to several failure routes in both polymer and small-molecule OPVs and in OLEDs. These intrinsic degradation mechanisms originate from the migration of both In^+ and O^{2-} species into the organic layers [173]. Diffusion of In as the origin of OLED failure of both polymer and small-molecule-based devices has been proposed in the early investigations and since then remains a debated topic. In solution-processed devices, the etching of the ITO layer by the acidic PEDOT:PSS hole transport layer has been proposed [24,69,150].

Controlled In doping into the active matrix, indeed, proved that In is a luminescence quencher in OLEDs, but only at high concentration [91]. However, Kok et al. reported no improvement in the stability of OLEDs with the neutralized PEDOT:PSS [24]. The pH increase was achieved by addition of NaOH, which modifies the PEDOT:PSS properties in a similar way as electrochemical dedoping. Even though significant improvements in device stability have been achieved by altering the ITO treatment and by incorporating interface layers, these layers do not necessarily only block In^+ diffusion, but they also change the interface energetics. Improvement of charge injection has proven to enhance OLED device stability mostly by equilibrating the electron and hole carrier concentration [153].

Similarly to OLEDs, in OPVs the impact of In cations remains debated. In one case, In cation diffusion through all the layers of an OPV has been reported [86]. In agreement with these observations, a recent study reported stabilized device performance upon replacement of ITO by a combination of PEDOT:PSS and a Ag grid [206]. Numerous counterarguments have challenged these previous studies. For example, stable device performance is recorded in small-molecule cells, despite In diffusion in the photoactive layer detected by TOF-SIMS [58].

Simultaneously, oxygen out-diffusion from ITO and consequent oxidation of the contacting organic layers were reported both in polymer and small-molecule-based devices [51,142]. This process is accelerated by the UV light present in solar radiation [185]. Moreover, a change in work function of the ozone-treated ITO over time has also been surmised to hinder charge extraction [137]. The incorporation of solution-processed or sublimed transport layer(s) preventing direct contact between the ITO and the active layer seems to successfully alleviate this failure route. Alternatively, surface treatments by self-assembled monolayers or plasma treatment have been applied to alleviate the degradation of the ITO/organic interface [101].

Besides the constantly increasing price of ITO, its brittle nature is an equally critical issue as the final OPV products aim to be flexible [180]. Several conducting polymers, carbon or metal thin film, and nanowire-based alternative electrodes have been proposed to circumvent this failure (for more information on electrodes, see Chapter 5) [50,56].

Despite these suspected electrical and mechanical failures of ITO, it has not been replaced due to the limited commercial availability of alternative transparent electrodes. As importance of device reliability increases, the stability of suitable ITO alternatives should be evaluated in the early stage of the development.

15.2.3.2 Degradation mechanisms of metal electrodes

Low work function metal electrodes have been the standard for OPV until recently. Owing to the reactivity of these metals, the dominant failure mechanisms in these devices is the rapid oxidation of the metal upon air exposure.

These oxidized metal layers are insulators, significantly hindering the extraction of photogenerated carriers and hence device operation. The considerable impact of this degradation route has earlier been described in OLEDs. Upon oxidation of the electrode both dark, nonemissive spots and shrinkage at edges of the light-emitting area are observed [112,136]. Formation of these dark spots is attributed to pinholes in the metal layer formed by dust or other fabrication defects, leaving holes in the metal allowing rapid ingress of oxygen and humidity [81,96,103]. All studies confirm that the dark spot formation is only limited to the oxidation of the metal electrode and not of the underlying organic layer. The oxidation induced by humidity via an electrochemical process proceeds three times faster than that triggered by oxygen [23,136]. Several specialized reviews discussing this topic can provide further insights [5,153].

Similar oxidation phenomena of the metal electrode have been also reported in OPVs, where the inactive area can be detected by local photocurrent mapping [67,98,144]. Microscopic scale diffusion pathways of O_2 and H_2O have been investigated by isotope labeling. This investigation showed that H_2O penetrates between the grain boundaries of the metal electrode causing a homogeneous incorporation. On the contrary, O_2 only diffuses through large-scale defects [58,120].

Like in OLEDs, composite electrodes employing a thin layer of LiF, Ca, Ba, or Mg in contact with the organic layer have been introduced to enhance device efficiency of polymer:fullerene devices [134]. However, the alkaline earth metals have a lower reduction potential than Al, meaning that they oxidize rapidly in the presence of both oxygen and water. Owing to the porous and nonprotective nature of their oxide layers, the oxidation progresses until the entire metal is consumed. In contrast, Al rapidly forms a thin protective oxide layer. The oxidation simultaneously alters the thickness of the layers, which implies that it can induce void formation in stacked layers. Confirmation of the void formation in devices employing Ca/Al electrodes is presented in Fig. 15.11, using cross-sectional TEM analysis coupled with energy-dispersive X-ray spectroscopy (EDS) [99].

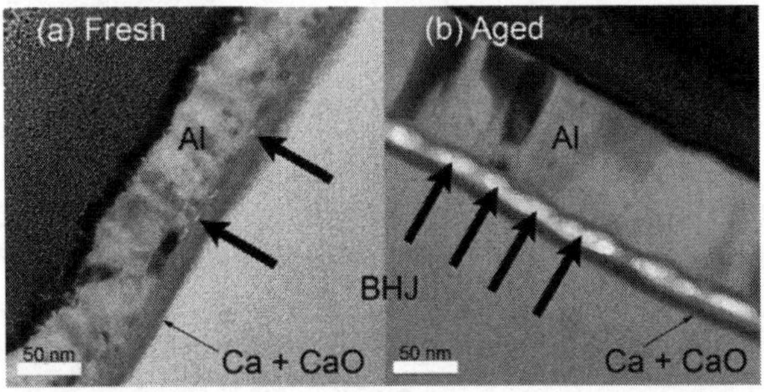

Figure 15.11 TEM image for (a) freshly prepared and (b) aged ITO/BHJ/Ca/Al devices. The active organic is labeled BHJ, the aluminum layer of the electrode is identified as Al, and the calcium layer, which likely contains a significant fraction of calcium oxide, is labeled Ca+CaO. Bold arrows indicate regions of void formation at the Ca/Al interface that enlarge as the device ages. Reproduced with permission from [99].

Despite the intrinsic chemical reactivity of alkaline earth metals, they have been shown to improve device stability. In the case of PPV derivatives, this can be explained by the chemical reaction and formation of covalent bonds between the PPV derivative and Al [100]. This reaction can be induced during fabrication by the higher deposition temperature of Al compared to Ca or Mg. Similarly, Ag in contact with P3HT forms AgS directly upon evaporation leading to a change in the work function, hence the introduction of a buffer layer is beneficial [99]. Studies conducted under continuous illumination further indicated the superior stability of Ca-based composite electrodes compared to pure Al or Ag cathodes [22,122,134].

Several studies reported the photodegradation of the buried interface between the organic and low work function metal layer under 1 sun illumination in inert atmosphere. Various mechanisms have been cited as the origin of the failure under this condition: formation of interface states, morphology reorganization, and chemical reaction of the low work function metal [122,134,179]. One of the implications of interface states can be the reversible accumulation of charges at the electrode/active layer interface [77]. Several independent studies reported the use of an evaporated C_{60} layer to eliminate or delay this mechanism; however, the origin of the improvement remains unclear [76,85,123,179]. One possible explanation might be the passivation of this interface via a reaction between C_{60} and the low work function metal. Indeed, in a previous study, reaction between Al and C_{60} has been detected [118].

In contrast, in small-molecule-based devices, light exposure destroys the initially formed organic-metal bonds at the interface. Beyond reduced interface adhesion, charge extraction and injection between the two layers are also compromised. Moreover, it has also been suggested that increased interface bonding also improves air stability of Alq3-based OLEDs [129]. The authors suspected that organic/metal adhesion might be a critical and long neglected factor for also determining electrical stability [184,129].

Important progress in air stability of polymer cells has been achieved by the replacement of the reactive Ca by metal oxide transport layers such as TiO_x, CrO_x, or ZnO serving as an ETL [37,90,181]. Furthermore, the concept of inverted devices has been introduced to further reduce the sensitivity of OPVs toward ambient atmosphere. In this architecture, the ITO coated with an ETL extracts the electrons, while holes are collected at the top. In these devices, the work function of the top electrode should align to the HOMO of the donor, and hence have a high work function. This motivated the replacement of the commonly used Ca/Al electrode by Ag or PEDOT:PSS/Ag electrodes. The intrinsic air stability of this metal layer allows for prolonged air stability in inverted polymer cells. Several independent reports demonstrated the long-term air stability of these cells in shelf-life conditions [52,84,88,99]. Similar measurements carried out in our group also confirmed the considerably improved stability of inverted cells in comparison to conventional cells as shown in Fig. 15.12.

First studies combining air exposure and illumination suggest that such organic/HTL or HTL/electrode interfaces are still unstable. These first studies suggested that the commonly applied PEDOT:PSS might be the origin of this failure as detailed in Section 15.2.2.1 [159].

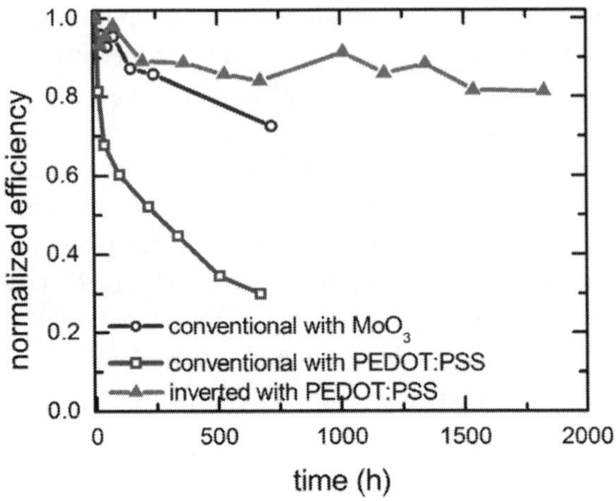

Figure 15.12 Device lifetime measured in shelf-life conditions in ambient air with 40% RH. Conventional devices had the following structure ITO/PEDOT:PSS or MoO_3/P3HT:PCBM/Yb/Al, while inverted cells were composed of ITO/ZnO/P3HT:PCBM/PEDOT:PSS/Ag/Al. Values are averages of 20 cells on two substrates.

Besides its electrical function, the electrode also serves as an integrated encapsulation of the organic layers. Considerably reduced photo-oxidation rate of the organic layer under the metal electrode clearly demonstrated the importance of this effect (see Section 15.2.1.3). Thermally evaporated 100-nm-thick Ag electrodes have a WVTR of 8×10^{-4} g/m^2/day [19,59]. As in the barrier layer, both pinholes and the density of the electrode layer determine the pace of oxygen and humidity ingress. Increasing the electrode thickness and the use of a composite electrode with a Ag/Al stack can considerably prolong device lifetime by suppressing both ingress pathways [207].

15.2.4 Summary

This section enumerating the multiple failure mechanisms of various materials, interfaces, and devices might appear at the first glance dauntingly long, but the reader should keep in mind that not all processes necessarily limit device lifetime. Operational device stability is generally determined by a few dominating failure mechanisms, while other alterations in the layers have a negligible influence. Moreover, degradation pathways observed in separate layers or in a certain device stack might be completely suppressed in another configuration. This insight into the degradation mechanisms brings great progress in both device stability and development of accelerated tests. Firstly, the development of stable interface layers and photoactive layers undeniably benefit from the fundamental insights. Secondly, knowing the failure mechanisms, the methodology of future OPV reliability studies could shift focus from stress-based ageing to more efficient failure mode testing [205]. This latter approach has been successfully implemented in reliability testing of other PV technologies and semiconductor devices.

15.3 Current Device Stability of OPVs

As a result of an in-depth understanding of the failure mechanisms, outlined in the first part of this chapter, the adapted mitigation strategies brought great progress in device reliability. First OPV cells barely lasted a few days under shelf-life conditions in the laboratory, while current devices withstand harsh outdoor testing for several years.

Development of impermeable packaging has played a critical role and will continue to determine the progress in device reliability. Since rigid and flexible encapsulation technologies have been described in detail in Chapter 12, we will only focus here on their impact on operational device stability. One of the most important steps was the determination of the device-specific packaging requirement in WVTR [19,59].

Firstly, we discuss the challenges to define adequate reliability standards and also describe the currently accepted protocols. Afterward, we propose an overview of the current status of operational device stability. We will devote special attention to hero results as they give an insight into the potential of OPVs to deliver an over 5 years lifetime required for consumer electronic products and to reach over 15 years of lifetime demanded for building integrated applications.

15.3.1 Current Standards and Protocols for Lifetime Evaluation

Unlike crystalline Si cells that have a clearly defined final application as a rooftop PV generation system, OPVs are considered both for indoor and outdoor exploitation. As the final application determines the lifetime measurement conditions (and required lifetime), the definition of specific reliability standard(s) for OPVs is probably a bit premature. Although certain types of OPVs might be deployed in similar applications as their inorganic thin film or Si counterparts, still the IEC standards developed for inorganic modules are not necessarily relevant and applicable to OPVs. This is explained by the fact that reliability standards are established based on in-depth understanding of the failure mechanisms. Thus, the qualification tests in standards are tailored to detect the appearance of major known failure mechanisms. For example, in Si PV modules the interconnections between cells is one of the most critical failure processes as the cells are intrinsically light, heat, and air stable. Therefore, the tests included in the IEC 61215 standard—warranting the long-term outdoor operation stability of Si PV cells—evaluate among other things the mitigation of this particular failure mechanism. Differences in thin-film PV module failure imposed the definition of an adapted standard, IEC 61646. Development of these standards for inorganic PV modules relied not only on existing knowledge of the degradation processes, but also on reliability standards of other semiconducting devices. The significant differences in the properties of organic semiconductors as well as the device architecture and fabrication processes of OPVs justify the development of novel reliability standards.

In an effort to work toward OPV standards, relevant definitions have been initiated at the International Summit on Organic Photovoltaic Stability (ISOS) in 2008. Our currently incomplete knowledge of failure mechanisms inhibits the definition of standard with lifetime predictive value. In the meantime, the academic and industrial OPV research communities established recommended practices for reliability testing [132]. The early development stage of the technology motivated the definition of numerous testing conditions, comprising outdoor as well as indoor tests. An overview of the different tests is proposed in Table 15.1. Recently, an initiative has been launched to describe these recommended practices also in an IEC standard.

Contrary to reliability standards applying qualification tests that only report pass or fail, these practices encourage the detailed description of the performance loss. The lifetime of a PV cell or module, OPV included, is defined as the time that the device efficiency decreases to 80% of its starting power conversion efficiency. Both the initial performance (PCE_0) and the "stabilized" power conversion efficiency (PCE_{S0}) can be considered as the starting value as represented on the schematic of Fig. 15.13. For clarity, lifetime reports should clearly state which starting efficiency is considered: T_{80} if the PCE_0 was considered as starting value and T_{S80} if the device was stabilized to PCE_{S0} prior to lifetime measurement.

Besides precise measurement of PCE_0 and PCE_{S0} values, the ageing conditions should also be included in the report according to the recommendations. The importance of this latter aspect has been demonstrated, by comparing the lifetime of devices processed by one research group and aged in comparable ageing conditions in different

Table 15.1 Overview of the different lifetime measurement protocols suggested by ISOS. Five distinct tests are outlined and for each of them different levels of equipment sophistication are proposed.

Test type		**Dark**	
Test ID	**ISOS-D-1 Shelf**	**ISOS-D-2 High temp.**	**ISOS-D-3 Damp heat**
Light source	None	None	None
Temperature	Ambient	65 °C/85 °C	65 °C/85 °C
Relative humidity	Ambient	Ambient (low)	85%
Environment	Ambient	Oven	Environmental chamber
Characterization light source	Solar simulator or sunlight	Solar simulator or sunlight	Solar simulator or sunlight
Load	Open circuit	Open circuit	Open circuit

(*Continued*)

Table 15.1 (*Continued*)

Test type	Outdoor		
Test ID	**ISOS-O-1**	**ISOS-O-2**	**ISOS-O-3**
Light source	Sunlight	Sunlight	Sunlight
Temperature	Ambient	Ambient	Ambient
Relative humidity	Ambient	Ambient	Ambient
Environment	Ambient	Ambient	Ambient
Characterization light source	Solar simulator	Sunlight	Solar simulator and sunlight
Load	MPP or open circuit	MPP or open circuit	MPP

Test type	Laboratory weathering		
Test ID	**ISOS-L-1**	**ISOS-L-2**	**ISOS-L-3**
Light source	Solar simulator	Solar simulator	Solar simulator
Temperature	Ambient	65 °C/85 °C	65 °C/85 °C
Relative humidity	Ambient	Ambient	Near 50%
Environment/ setup	Light only	Light and temperature	Light and temperature and R.H.
Characterization light source	Solar simulator	Solar simulator	Solar simulator
Load	MPP or open circuit	MPP or open circuit	MPP

Test type	Thermal cycling		
Test ID	**ISOS-T-1**	**ISOS-T-2**	**ISOS-T-3**
Light source	None	None	None
Temperature	Between room temp. and 65 °C/85 °C	Between room temp. and 65 °C/85 °C	−40 °C to +85 °C
Relative humidity	Ambient	Ambient	Near 55%
Environment/ setup	Hot plate/oven	Oven/environmental chamber	Environmental chamber
Load	Open circuit	Open circuit	Open circuit

(*Continued*)

Table 15.1 (*Continued*)

Test type	Solar-thermal-humidity cycling		
Test ID	**ISOS-LT-1**	**ISOS-LT-2**	**ISOS-LT-3**
Light source	Solar simulator	Solar simulator	Solar simulator
Temperature	Linear or step ramping between room temperature and 65 °C	Linear or step ramping between 5 °C and 65 °C	Linear or step ramping between –25 °C and 65 °C
Relative humidity	Monitored, uncontrolled	Monitored, controlled at 50% beyond 40 °C	Monitored, controlled at 50% beyond 40 °C
Environment/ setup	Weathering chamber	Environmental chamber with sun simulator	Environmental chamber with sun simulator and freezing
Characteri-zation light source	Solar simulator	Solar simulator	Solar simulator
Load	MPP or open circuit	MPP or open circuit	MPP or open circuit

Source: The table is adapted from [132].

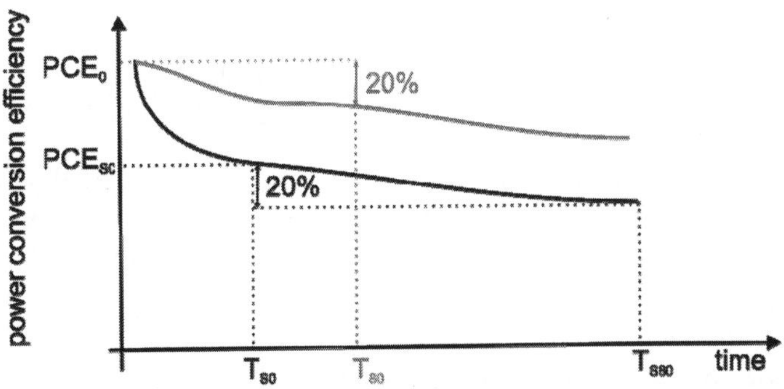

Figure 15.13 Schematic definition of device lifetime calculated from either initial or stabilized power conversion efficiency.

locations [40]. One of the biggest divergences in lifetime was recorded on devices aged under continuous illumination. Especially the spectrum of the illumination source determined not only the pace but also the failure mechanisms. The lifetime of modules aged under light sources with little or no UV illumination was not only longer, but also the origin of the performance loss was modified. Light sources with UV light (including outdoor exposure) triggered the decrease of FF in the modules, while this parameter remained mostly stable under UV-free illumination. This deviation between results reported by different research groups also draws the attention to the need of testing OPVs under real-world conditions. Currently only a few published reports described the durability of OPVs under outdoor conditions [53,154,166]. However, under long-term real-world conditions ageing of OPVs is indispensable to validate currently defined lifetime testing practices. This will allow on the longer term that these research guidelines evolve into industrial reliability standards.

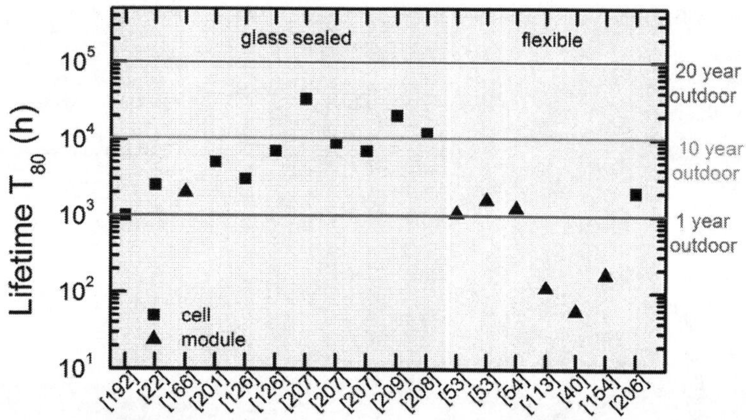

Figure 15.14 Operational device lifetime results measured on glass (left) or flexible (right) substrate sealed cell (squares) and modules (triangles). Lifetime under continuous illumination with 1 sun intensity is considered.

15.3.2 State-of-the-Art Operational Stability of OPVs

Operational lifetime studies follow the evolution of device performance under continuous illumination at constant temperature and/or humidity levels. Despite considerable deviations from outdoor weather conditions, these studies yield the most relevant insight into the durability of OPVs.

Correlation of these studies with outdoor measurements has confirmed that similar performance loss is recorded as a function of cumulated light dose at comparable temperature and humidity levels [40]. This agrees well with the fact that in different types of devices the light-induced failure rate scales linearly with the illumination intensity [57,169]. Therefore, a direct correlation between laboratory and outdoor performance can be established based on average annual irradiation data. For example, in Western Europe, the annual radiation dose is ±1000 kWh/m^2, which equals to 1000 hr of illumination with 1000 W/m^2 intensity solar simulator [40]. This equivalence became the reference in the OPV community to derive lifetime estimations. Since little is known about the spectral dependence of failure mechanisms, currently this aspect is neglected in the equivalence. Later, similar to the coating reliability standards, the dose of UV radiation probably will also be considered for equivalence.

Currently, several discrepancies can be noted between lifetime studies reported in the literature. Most studies employ 1000 W/m^2 intensity lamps with varying spectrum [4]. Since the temperature and humidity levels during ageing have only been recently set by ISOS protocols, most investigations have been conducted under different conditions. In addition, the quality of device packaging significantly differs between studies, and most importantly the permeability of the sealing is nearly never quantified. One study carefully measured WVTR rate of the same foil at elevated humidity level and combined with 40 °C or 65 °C and found that the WVTR is 0.06 and 0.2 g/m^2/day, respectively [54]. Therefore, in the following comparison of ageing studies the reader should keep in mind that concurrent to illumination and temperature, the ambient O_2 and H_2O ingress rates are also considerably different between studies.

Despite these complications of direct comparisons, we present here an overview of operational and outdoor lifetime studies in chronological order. This allows us to illustrate the progress in packaging and device technology between first studies reported in 2005 until today's promising results [192,208–210]. Owing to the determinant role of the permeability of the packaging, we discuss separately devices sealed between two glass or flexible substrates in

Tables 15.2 and 15.3, respectively. Moreover, lifetime results included in the tables are also summarized in Fig. 15.15, with the aim to provide a reader a global overview of the current status. Lifetime in this graph corresponds to the number of hours under continuous illumination with an intensity equivalent to 1 sun. Therefore, outdoor lifetime test results or results expressed in degradation rates are converted in equivalent number of continuous operation under 1 sun.

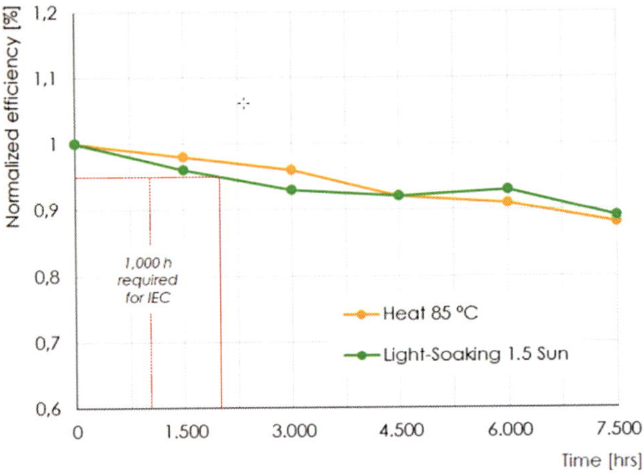

Figure 15.15 Normalized power conversion efficiency of small-molecule tandem cells as a function of time under continuous illumination (green) and at elevated temperature in dark (orange). These encapsulated cells were prepared by Heliatek with proprietary materials [210].

Initial lifetime studies employed conventional architecture P3HT:PCBM cells sealed between glass plates. Lifetime of these devices is estimated to be 2000–3000 hr under continuous illumination. Interestingly, the first study of Yang et al. in 2005 on this material system already reached a comparable performance to the results of Peters et al. reported in 2012 [126,192]. Comparing the different studies one can note that difference in UV content of the light source had less impact on device lifetime than sample/chamber temperature. Yang et al. observed a stabilized performance for 1000 hr after an initial 20% drop in performance [192]. Combining continuous illumination with increased temperature resulted in a performance drop of 50% over 600 hr in similar P3HT:PCBM cells [22]. Based on the acceleration factors, the authors estimated that device lifetime at room temperature can reach over 2500 hr [21,22]. Glass-sealed Plextronics modules with a P3HT : fullerene active layer showed a 10% decrease over 1000 hr continuous 1 sun illumination. The authors estimated a lifetime to be over 3500 hr considering day–night cycles [166]. This result was validated by outdoor measurements showing only 10–15% decrease over 5 months (3400 hr) [166]. In these studies, the initial efficiency is considered to be the stabilized performance. The same modules also exhibited over 3000 hr lifetime in shelf-life conditions. Outdoor testing of glass-encapsulated OPV modules in another climate, that of a hot and dry desert, revealed a rapid irreversible failure in a month, which is equivalent to 160 hr of 1 sun illumination [74]. In a recent study, conventional architecture and glass-sealed P3HT:PCBM devices are estimated to reach over 3 years of lifetime under a UV-filtered light source. This estimation is obtained by linear extrapolation from an over 4000 hr long measurement after burn-in of the cells [126].

Zimmerman et al. was the first to establish the lifetime of inverted and ITO-free devices [201]. Devices exhibited a 50% performance decrease over 5000 hr under continuous UV-free illumination [201]. An alternative technological solution to replace ITO with an Ag grid integrated in PEDOT:PSS has also proven stable in devices [206]. Samples with Ag grid showed a stable performance over 1000 hr under continuous illumination [206].

However, all lifetime studies conducted on flexible and intrinsically permeable substrates employed inverted architecture devices (Table 15.3). The enhanced air stability of this device architecture facilitates a prolonged lifetime. Additionally, nearly all studies focused on stability of cells with P3HT:PCBM bulk heterojunctions deposited by a printing technique. Konarka reported a comparison of devices with food packaging quality encapsulation (WVTR of 0.03 $g/m^2/day$) aged both outdoor and indoor [54]. The cells in outdoor testing maintained 80% of their initial value after 14 months. In parallel, devices were aged in laboratory, under continuous illumination at 65 °C, to demonstrate that 600–800 hr of continuous ageing is equivalent to 1 year outdoor lifetime in western European climate [55]. Similar modules exhibited 1250 hr lifetime at 65 °C and 1 sun illumination

Table 15.2 Summary of operational lifetime studies on rigid samples with glass substrate and glass packaging.

Substrate type	Device stack	Ageing conditions			Lifetime	Ref. / Year
		Illumination	Temperature	Atmosphere		
Glass	ITO/MDMO-PPV:PCBM/Ca/Ag	1 sun (halogen lamp)	4 °C, 60 °C, 85 °C, 105 °C	Air		[141] 2004
Glass	ITO/PEDOT:PSS/P3HT:PCBM/LiF/Ag	1 sun (tungsten halogen)	70 °C	Inert	20% performance drop over 1000 hr	[192] 2005
Glass	ITO/PEDOT:PSS/P3HT:PCBM/Ca/Ag	1 sun (xenon lamp)	70 °C	Inert	50% performance drop over 600 hr (equivalence of 2500 h at 25 °C)	[22] 2006
Glass	ITO/PEDOT:PSS/P3CT/PCBM/Al	1 sun (outdoor)	–	Air	50% performance drop over 160 hr cumulated light exposure	[74] 2006
Glass	ITO/PEDOT:PSS/ P3CT/C_{60}/Al	1 sun (metal halide lamp)	7 °C	Vacuum	80% performance loss over 10,000 hr	[86] 2007
Glass	ITO/MeOTPD:NDP2/ZnPc/C_{60}/C_{60}:NPN1/Al	1.85 sun (halogen lamp without UV)	50 °C	Air	3% performance loss over 1400 hr	[38] 2008
Glass	ITO/HTL/P3HT:PCBM/Cathode modules	1 sun (xenon lamp)	–	Air	10% performance loss after 1000 hr (after stabilization of performance)	[166] 2008

Glass	ITO/HTL/ZnPc:C_{60}/ DCV6:C_{60}/ETL/Cathode	1.5 sun (tunsgten halogen lamp)	50 °C	Air (< 30 %RH)	T_{80} = 8551 hr extrapolated from initial efficiency	[208] 2009
Glass	Undisclosed materials in a mini-module	1.5 sun (tunsgten halogen lamp)	50 °C	Air (< 30 %RH)	T_{80} = 7 years extrapolated from initial efficiency	[208] 2009
Glass	ITO or PEDOT:PSS/ Ag grid/P3HT:PCBM/ LiF/Al	1 sun	45 °C		Stable device performance over 1000 hr	[206] 2010
Glass	ITO/HTL/P3HT:PCBM/ Ca/Ag	1 sun (sulfur plasma lamp without UV)	37 °C	Air	3000 hr lifetime (extrapolated from stabilized efficiency)	[126] 2012
Glass	ITO/HTL/ PCDTBT:PCBM/Ca/Ag	1 sun (sulfur plasma lamp without UV)	37 °C	Air	7000 hr lifetime (extrapolated from stabilized efficiency)	[126] 2012
Glass	ITO/BF-DPB:$C_{60}F_{36}$/ ZnPc/ZnPc:C_{60}/C_{60}/ Bphen/Al	1sun (LED lamp without UV)	45 °C	Air 5.5%RH	T_{80} = ~ 1800 hr from initial efficiency	[59] 2012
Glass	Cells reported by Polyera	1 sun (metal halide lamp)	–	Air	5% performance drop after 3000 hr	[209] 2012 See Fig. 15.16
Glass	Cells reported by Heliatek	1.5 sun (tunsgten halogen lamp)	–	Air	T_{80} = 13,600 hr extrapolated from initial efficiency	[210] 2012 See Fig. 15.15

Source: Most results are reported in the literature and only the latest results are presented at scientific conferences

Table 15.3 Summary of operational lifetime studies on flexible samples with plastic barrier foil or on rigid substrate but not encapsulated

Substrate e type	Device stack	Ageing conditions: Illumination	Ageing conditions: Temperature	Ageing conditions: Atmosphere	Lifetime	Ref./Year
Flexible (with WVTR of 1.2 10^{-3} g/m2/day)	Inverted P3HT:PCBM modules	1 sun (outdoor)	–	Air	20% performance loss over 14 months	[53] 2008
Flexible (with WVTR of 1.2 10^{-3} g/m^2/day)	Inverted P3HT:PCBM modules	1 sun (xenon lamp)	65 °C	Air	15% performance loss after 1200 hr	[53] 2008
Flexible (with WVTR of 0.06g/m^2/day)	Inverted P3HT:PCBM modules	1 sun (xenon lamp)	65 °C	Air	20% performance loss after 1250 hr	[54] 2008
Flexible (semipermeable barrier)	Inverted P3HT:PCBM modules	1 sun (outdoor)		Air	40% performance loss over 1000 hr outdoor exposure	[40] 2011
Flexible (laminated between glass and Al sheets)	Inverted P3HT:PCBM modules	1 sun (outdoor)	–	Air	54% performance loss over 6 months	[113] 2010
Flexible (semipermeable barrier)	Inverted P3HT:PCBM modules	1 sun (outdoor)		Air	40% performance loss over 4.5 months (3000 hr) outdoor exposure	[154] 2012
Nonencapsulated	ITO/ZnO/P3HT:PCBM/PEDOT:PSS/Ag/Al	1 sun (sulfur plasma lamp without UV)	45 °C	Air	After initial drop of 25%, stable performance over 2000 hr	[207] 2010

with xenon light [55]. Earlier studies reported much shorter half-lifetimes as devices were prepared with the more air-sensitive MDMO-PPV polymer [26]. These results indicate that flexible substrates require both air-stable electrodes and photoactive materials.

For comparison, flexible cells processed at Risø and tested outdoors exhibited 20% performance decrease over 37 days [113]. The barrier foil used in this study also contained an integrated UV filter. Similar flexible cells produced at Risø with semipermeable barriers and outdoor tested at various locations degraded more than 50% after 300 hr exposure [40]. Identical cells under continuous illumination in laboratory conditions (ISOS-L-2) exhibited comparable stability, proving that ageing under xenon or metal halide lamps correlates well with outdoor testing.

One of the most promising reports on inverted P3HT:PCBM cells measured a stable electrical performance on nonencapsulated cells exposed to UV-filtered illumination over a duration of 2000 hr [98].

Recent lifetime studies launched the study of novel and high-efficiency polymer:fullerene blends. Conventional architecture PCDTBT:$PC_{70}BM$ cells sealed between glass plates exhibited a stable performance after an initial burn-in [125]. This stable electrical performance under UV light-free illumination indicated that cells could reach a 7-year-long operation outside. Further optimization of the active material by Polyera resulted in record performance polymer:fullerene cells [208]. Device performance loss was less than 5% after 3000 hr of continuous simulated illumination containing also UV light (Fig. 15.15). With this degradation rate, one can expect over 10 years of lifetime in outdoor conditions.

Small-molecule-based cells prepared by sublimation have been reported to reach long operational lifetimes. Single-junction devices with a photoactive layer composed of ZnPc and C_{60} were tested with a glass sealing by Franke et al. as well as by Hermenau et al. [38,59]. In both cases, the devices degraded 20% after 1800 hr of illumination in the absence of UV radiation. Tandem small-molecule cells composed of similar organic compounds exhibited an extraordinarily stable device performance over 4000 hr at 1.5 sun illumination intensity [208]. These tandem cells were projected to reach T_{80} after over 33,000 h, warranting over 20 years of operational lifetime in western Europe. Latest lifetime tests conducted by Heliatek on a tandem cell with 8.3% initial efficiency suggested that the T_{80} of the cell is reached after 13,600 hr (Fig. 15.16) [210]. In outdoor conditions, the authors estimate the lifetime of this cell to be over 20 years. First module tests indicated similarly encouraging results with T_{80} of over 7000 hr [208].

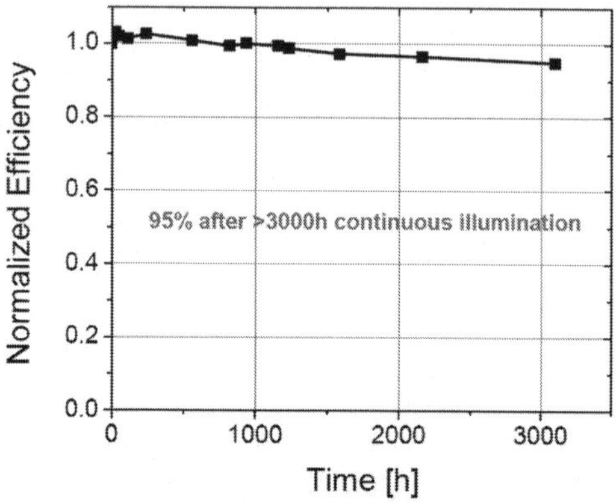

Figure 15.16 Normalized power conversion efficiency of polymer:fullerene cells as a function of time under continuous illumination. These encapsulated cells were prepared by Polyera with proprietary materials [209].

15.4 Conclusions and Outlook

Although progress in device stability might be more complex to track than device performance, it should be noted that tremendous advances have been reported in recent years. Early measurements suggested that device lifetime was limited to a few hours while the latest results prove that a lifetime of five years or more can be achieved with the current device technology. This progress has been driven by the insights gained from fundamental studies investigating the origin of cell failure. To our current understanding, the critical failure mechanisms in OPVs are the following: electrical and mechanical degradation of ITO, photo-oxidation of fullerenes and certain conjugated polymers, structural reorganization in polymer and fullerene blends, and the oxidation of the low work function organic and/or metal buffer layers and electrodes. In most cases, the detailed description of the failure process(es) allowed to define a mitigation route through integration of novel transport and photoactive layers.

Owing to the excellent light and thermal stability of evaporated small-molecule-based cells, their lifetime is predicted to reach over 20 years. Recently, solution-processed polymer:fullerene-based cells demonstrated a comparable stability of over 10 years. In both cases, the development of the rigid and flexible packaging technology has also significantly contributed to these achievements.

Among all scientific results, the definition of comparable testing conditions also played a key role in the advance of the OPV stability research. The recommended testing practices defined by the specialists have been published and they will soon be captured in an IEC standard. As OPVs are foreseen to integrate in a wide variety of applications, numerous operating conditions could and should be identified. This is why both the description of recommended practices and the IEC standard propose several testing conditions to cover all failure mechanisms and operating conditions. It is important to note that these practices are intended "to guide the community through technology improvement" and quantitative predictions are out of their scope.

Despite the great progress, one rapidly notes the numerous unexplored aspects of OPV reliability and durability. One of the most critical topics to be studied is the correlation between (accelerated) laboratory ageing studies and real-life operational lifetime. This implies that influence of cyclic variations in the stress levels should be understood as well as neglected stress factors should be explored. Furthermore, the material and device structure dependence of current studies and models is a considerable limitation. Relevant generalization of failure models will constitute a key step toward quantitative predictions.

Stability investigation of OPVs is at its start and will certainly become one of the most critical research topics of the field inspiring the development of novel materials and deepening our understanding in device physics. Despite the early stage of this research field, the necessity to forecast device lifetime is eminent for the rapid commercialization of OPVs. Both the rapid progress in the last years and stability of small-molecule OLEDs suggest that OPV technology can already satisfy the stability requirements of mobile power generation applications. More importantly, in the future this technology has also the potential to match the demand of long-term stability in building integrated applications.

References and Citations

a) Journal references

1. Abad J, Espinosa N, Garcia-Valverdec R, Colcheroa J, Urbina A (2011) The influence of UV radiation and ozone exposure on the electronic properties of poly-3-octyl-thiophene thin films, *Sol Energy Mater Sol Cells*, **95**(5), 1326–1332.
2. Abdou MSA, Holdcroft S (1993) Mechanisms of photodegradation of poly(3-alkylthiophenes) in solution, *Macromolecules*, **26**(11), 2954–2962.
3. Aguirre A, Meskers SCJ, Janssen RAJ, Eglehaaf HJ (2011) Formation of metastable charges as a first step in photo-induced degradation in π-conjugated polymer:fullerene blends for photovoltaic applications, *Org Electron*, **12**(10), 1657–1662.
4. Angelis FD, Gaspari M, Procopio A, Cuda G, Fabrizio ED (2009) Direct mass spectrometry investigation on pentacene thin film oxidation upon exposure to air, *Chem Phys Lett*, **468**(4–6), 193–196.
5. Aziz H, Popovic ZD (2004) Degradation phenomena in small-molecule light-emtting devices, *Chem Mater*, **16**, 4522–4532.
6. Bertho S, Janssen G, Cleij TJ, et al. (2008) Effect of temperature on the morphology and photovoltaic stability of bulk heterojunction polymer:fullerene solar cells, *Sol Energy Mater Sol Cells*, **92**(7), 753–760.
7. Cates Miller NC, Cho E, Gysel R, et al. (2012) Factors governing intercalation of fullerenes and other small molecules between the side chains of semiconducting polymers used in solar cells, *Adv Energy Mater*, **2**, 1208–1217.
8. Chabinyc ML, Street RA, Northrup JE (2007) Effects of molecular oxygen and ozone on polythiophene-based thin-film transistors, *Appl Phys Lett*, **90**(12), 123508.
9. Chambon S, Rivaton A, Gardette JL, Firon M (2007) Photo- and thermal degradation of MDMO-PPV: PCBM blends, *Sol Energy Mater Sol Cells*, **91**(5), 394–398.
10. Chambon S, Rivaton A, Gardette JL, Firon M (2008) Durability of MDMO-PPV and MDMO-PPV:PCBM blends under illumination in absence of oxygen, *Sol Energy Mater Sol Cells*, **92**(7), 785–792.
11. Cho S, Seo JH, Park SH, Beaupre S, Leclerc M, Heeger AJ (2009) A thermally stable semiconducting polymer, *Adv Mater*, **22**(11), 1253–1257.
12. Chen H, Nanayakkara CE, Grassian VH (2012) Titanium dioxide photocatalysis in atmospheric chemistry, *Chem Rev*, **112**(11), 5919–5948.

13. Cheng YJ, Liao HM, Chang CY, Kao WS, Wu CE, Hsu HS (2011) Di (4-methylphenyl)methano-C_{60} bis-adduct for efficient and stable organic photovoltaics with enhanced open-circuit voltage, *Chem Mater,* **23**(17), 4056–4062.

14. Choi HH, Lee WH, Cho K (2012) Bias-stress-induced charge trapping at polymer chain ends of polymer gate-dielectrics in organic transistors, *Adv Funct Mater,* **22**(22), 4833–4839.

15. Collins B, Tumbleston JR, Ade H (2011) Miscibility, crystallinity and phase development in P3HT/PCBM solar cells: Toward an enlightened understanding of device morhology and stability, *J Phys Chem Lett,* **2**(24), 3135–3145.

16. Conings B, Bertho S, Vandewal K, Senes A, D'Haen J, Manca J, Janssen RAJ (2010) Modeling the temperature induced degradation kinetics of short circuit current in organic bulk heterojunction solar cells, *Appl Phys Lett,* **96**(4), 163301.

17. Cook S, Fufube A, Katoh R (2012) Matter of minutes degradation of poly(3-hexylthiophene) under illumination in air, *J Mater Chem,* **22**, 4282–4289.

18. Creegan KM, Robbins JL, Robbins WK, et al. (1992) Synthesis and characterization of $C_{60}O$, the first fullerene epoxide, *J Am Chem Soc,* **114**(3), 1103–1105.

19. Cros S, de Bettignies R, Berson S, et al. (2011) Definition of encapsulation barrier requirements: A method applied to organic solar cells, *Sol Energy Mater Sol Cells,* **95**, 65–69.

20. D'Andrade BW, Forrest SR, Chwang AB (2003) Operational stability of electrophosphorescent devices containing p and n doped transport layers, *Appl Phys Lett,* **83**, 3858.

21. de Bettignies R, Guillerez S (2006) Cathode composition greatly influences the lifetime of organic photovoltaic cells, *SPIE Newsroom* DOI: 10.1117/2.1200606.0307.

22. de Bettignies R, Leroy J, Firon M, Sentein C (2006) Accelerated lifetime measurements of P3HT:PCBM solar cells, *Synth Met,* **156**, 510–513.

23. de Brito BC, Smits ECP, van Hal PA, et al. (2008) Ultralow power microfuses for write-once read-many organic memory elements, *Adv Mater,* **20**(19), 3750–3753.

24. de Kok MM, Buechel M, Vulto SIE, et al. (2004) Modification of PEDOT:PSS as hole injection layer in polymer LEDs, *Phys Stat Sol (a),* **201**, 1342–1359.

25. Demir F, der Brande NV, Mele BV, et al. (2011) Isothermal crystallization of P3HT:PCBM blends studied by RHC, *J Therm Anal Calorim,* **105**(3), 845–849.

26. Dennler G, Lungenschmied C, Neugebauer H, et al. (2006) A new encapsulation solution for flexible organic solar cells, *Thin Solid Films*, **511–512**, 349–353.

27. Dennler G, Scharber, MC, Brabec CJ, (2009) Polymer:fullerene bulk heterojunction solar cells, *Adv Mater*, **21**(13), 1323–1338.

28. Drees M, Hoppe H, Winder C, et al. (2005) Stabilization of the nanomorphology of polymer-fullerene "bulk heterojunction" blends using a novel polymerizable fullerene derivative, *J Mater Chem*, **15**, 5158–5163.

29. Dupont SR, Voroshazi E, Heremans P, Dauskardt RH (2012) The role of processing, layer thickness and material type on the adhesion properties in inverted P3HT:PCBM-based polymer solar cells, *Org Electron*, submitted.

30. Dupont SR, Voroshazi E, Heremans P, Dauskardt RH (2012) The effect of temperature, solar irradiation and humidity on the adhesion/cohesion properties of P3HT:PCBM based inverted polymer solar cells, *IEEE J Photovolt*, accepted.

31. Dupont SR, Oliver M, Krebs FC, Dauskardt RH (2012) Interlayer adhesion in roll-to-roll processed flexible inverted polymer solar cells, *Sol Energy Mater Sol Cells* **97**, 171–175.

32. Ecker B, Nolasco JC, Pallares J, et al. (2011) Degradation effects related to the hole transport layer in organic solar cells, *Adv Funct Mater*, **21**(14), 2705–2711.

33. Eklund PC, Rao AM, Zhou P, Wang Y, Holden JM (1992) Photochemical transformation of C_{60} and C_{70} films, *Thin Solid Films*, **257**(2,1), 185–203.

34. Elschner A (2011) The spectral sensitivity of PEDOT:PSS films, *Sol Energy Mater Sol Cells*, **95**(5), 1333–1338.

35. Erol A, Okurb S, Comba B, Mermerc M, Arikan M (2010) Humidity sensing properties of ZnO nanoparticles synthesized by sol-gel process, *Sensors Actuators B*, **145**(1), 174–180.

36. Fehse K, Meerheim R, Walzer K, Leo K, Lovenisch W, Elschner A (2008) Lifetime of organic light emitting diodes on polymer anodes, *Appl Phys Lett*, **93**(8), 083303.

37. Ferreira SR, Lu P, Lee YJ, Davis RJ, Hsu JWP (2011) Effect of zinc oxide electron transport layers on performance and shelf life of organic bulk heterojunction devices, *J Phys Chem C*, **115**(2), 13471–13475.

38. Franke R, Maenning B, Petrich A, Pfeiffer M (2008) Long-term stability of tandem solar cells containing small organic molecules, *Sol Energy Mater Sol Cells*, **92**, 732–735.

39. Garner BL, Papillo PJ (1962) Accelerated outdoor exposure testing in evaluation of ultraviolet light stabilizers for plastics, *Ind Eng Chem Prod Res Dev,* **1**, 249–253.

40. Gevorgyan SA, Medford AJ, Bundgaard E, et al. (2011) An inter-laboratory stability study of roll-to-roll coated flexible polymer solar modules, *Sol Energy Mater Sol Cells,* **95**, 1398–1416.

41. Gholamkhass B, Holdcroft S (2010) Toward stabilization of domains in polymer bulk heterojunction films, *Chem Mater,* **22**, 5371–5376.

42. Girotto C, Voroshazi E, Cheyns D, Heremans P, Rand BP (2011) Solution-processed MoO_3 thin films as a hole-injection layer for organic solar cells, *ACS Appl Mater Interfaces,* **3**(9), 3244–3247.

43. Gorodetsky AA, Cox M, Tremblay NJ, Kymissis I, Nuckolls C (2009) Solar cells from a solution-processable pentacene with improved air stability, *Chem Mater,* **21**, 4090–4092.

44. Green MA, Emery K, Hishikawa Y, Warta W, Dunlop ED (2011) Solar cell efficiency tables (version 39), *Prog Photovolt Res Appl,* **20**(1), 12–20.

45. Gregg BA (2009) Charged defects in soft semiconductors and their influence on organic photovoltaics, *Soft Matter,* **5**, 2985–2989.

46. Griffini G, Douglas JD, Piliego C, et al. (2011) Long-term thermal stability of high-efficiency polymer solar cells based on photocross linkable donor-acceptor conjugated polymers, *Adv Mater,* **23**(14), 1660–1664.

47. Grossiord N, Kroon JM, Andriessen R, Blom PWM (2012) Degradation mechanisms in organic photovoltaic devices, *Org Electron,* **13**(3), 432–456.

48. Hains AW, Liu J, Martinson ABF, Irwin MD, Marks TJ (2010) Anode interfacial tuning via electron-blocking/hole-transport layers and indium tin oxide surface treatment in bulk-heterojunction organic photovoltaic cells, *Adv Funct Mater,* **20**(4), 595–606.

49. Hamed A, Sun YY, Tao YK, Meng RL, Hor PH (1993) Effects of oxygen and illumination on the in situ conductivity of C_{60} thin films, *Phys Rev B,* **47**, 10873–10880.

50. Han S, Lim S, Kim H, Cho H, Yoo S (2010) Versatile multilayer transparent electrodes for ITO-free and flexible organic solar cells, *IEEE J Sel Top Quant,* **16**(6), 1656–1664.

51. Hancox I, Sullivan P, Chauhan KV, et al. (2010) The effect of a MoO_x hole-extracting layer on the performance of organic photovoltaic cells based on small molecule planar heterojunctions, *Org Electron,* **11**(12), 2019–2025.

52. Hau SK, Yip HL, Baek NS, Zou J, O'Malley K, Jen AKY (2008) Air-stable inverted flexible polymer solar cells using zinc oxide nanoparticles as an electron selective layer, *Appl Phys Lett*, **92**(25), 253301.

53. Hauch JA, Schilinsky P, Choulis SA, Childers R, Biele M, Brabec CJ (2008) Flexible organic P3HT:PCBM bulk heterojunction modules with more than 1 year outdoor lifetime, *Sol Energy Mater Sol Cells*, **92**(7), 727–731.

54. Hauch JA, Schilinsky P, Choulis SA, Rajoelson S, Brabec CJ (2008) The impact of water vapor transmission rate on the lifetime of flexible polymer solar cells, *Appl Phys Lett*, **93**, 103306.

55. He C, Germack DS, Kline RJ, et al. (2011) Influence of substrate on crystallization in polythiophene/fullerene blends, *Sol Energy Mater Sol Cells*, **95**(5), 1375–1381.

56. Hecht DS, Kaner RB (2011) Solution-processed transparent electrodes, *MRS Bull*, **36**, 749–754.

57. Hermenau M, Scholz S, Leo K, Riede M (2011) Total charge amount as indicator for the degradation of small molecule organic solar cells, *Sol Energy Mater Sol Cells*, **95**(5), 1278–1283.

58. Hermenau M, Riede M, Leo K, Gevorgyan SA, Krebs FC, Norrman K (2011) Water and oxygen induced degradation of small molecule organic solar cells, *Sol Energy Mater Sol Cells*, **95**(5), 1268–1277.

59. Hermenau M, Schubert S, Klumbies H, et al. (2012) The effect of barrier performance on the lifetime of small-molecule organic solar cells, *Sol Energy Mater Sol Cells*, **97**, 102–108.

60. Hintz H, Egelhaaf HJ, Luer L, Hauch J, Peisert H, Chasse T (2011) Photodegradation of P3HT: A systematic study of environmental factors, *Chem Mater*, **23**(2), 145–154.

61. Hiramoto M, Shiokawa H (2010) One month continuous operation of organic p-i-n solar cells covered with a zinc oxide protection layer, *Appl Phys Express*, 082301.

62. Hoke ET, Sachs-Quintana IT, Lloyd MT, et al. (2012) The role of electron affinity in determining whether fullerenes catalyze or inhibit photooxidation of polymers for solar cells, *Adv Energy Mater*, **2**(11), 1351–1357.

63. Huang DM, Mauger SA, Friedrich S, et al. (2011) The consequence of interface mixing on organic photovoltaic device characteristics, *Adv Funct Mater*, **21**(9), 1657–1665.

64. Huisman CL, Goossens A, Schoonman J (2002) Photodoping of zinc phthalocyanine: Formation, mobility, and influence of oxygen, *J Phys Chem B*, **106**, 10578–10584.

65. Igalson M, Wimbor M, Wennerberg J (2002) The change of the electronic properties of CIGS devices induced by the "damp heat" treatment, *Thin Solid Films*, **403–404**, 320–324.

66. Irfan Ding H, Gao Y, et al. (2010) Energy level evolution of air and oxygen exposed molybdenum trioxide films, *Appl Phys Lett*, **96**(24), 243307.

67. Jeranko T, Tributsch H, Sariciftci NS, Hummelen JC (2004) Patterns of efficiency and degradation of composite polymer solar cells, *Sol Energy Mater Sol Cells*, **83**(2–3), 247–262.

68. Jiang XY, Zhang ZL, Cao J, Khan MA, Haq K, Zhu WQ (2007) White OLED with high stability and low driving voltage based on a novel buffer layer MoO_x, *J Phys D: Appl Phys*, **40**, 5553–5557.

69. Jong MP, van Ijzendoorn LJ, de Voigt MJA (2000) Stability of the interface between indium-tin-oxide and poly(3,4-ethylenedioxythiophene)/poly(styrenesulfonate) in polymer light-emitting diodes, *Appl Phys Lett*, **77**(14), 2255.

70. Jørgensen M, Norrman K, Gevorgyan SA, Tromholt T, Andreasen B, Krebs FC (2012) Stability of polymer solar cells, *Adv Mater*, **24**(5), 580–612.

71. Jørgensen M, Norrman K, Krebs FC (2008) Stability/degradation of polymer solar cells, *Sol Energy Mater Sol Cells*, **92**(7), 686–714.

72. Jurchescu OD, Baas J, Palstra TTM (2005) Electronic transport properties of pentacene single crystals upon exposure to air, *Appl Phys Lett*, **87**(5), 052102.

73. Kalb W, Mattenberger K, Batlogg B (2008) Oxygen-related traps in pentacene thin films: Energetic position and implications for transistor performance, *Phys Rev B*, **78**, 035334.

74. Katz EA, Gevorgyan SA, Orynbayev MS, Krebs FC (2006) Out-door testing and long-term stability of plastic solar cells, *Eur Phys J Appl Phys*, **36**, 307–311.

75. Kaur IW, Jia W, Kopreski RP, et al. (2008) Substituent effects in pentacenes: Gaining control over HOMO-LUMO gaps and photo-oxidative resistances, *J Am Chem Soc*, **130**(48), 16274–16286.

76. Kawano K, Adachi C (2010) Reduced initial degradation of bulk heterojunction organic solar cells by incorporation of stacked fullerene and lithium fluoride interlayers, *Appl Phys Lett*, **96**, 053307.

77. Kawano K, Adachi C (2009) Evaluating carrier accumulation in degraded bulk heterojunction organic solar cells by a thermally stimulated current technique, *Adv Funct Mater*, **19**(24), 3934–3940.

78. Kawano K, Pacios R, Poplavzkyy D, Nelson J, Bradley DC, Durrant JR (2006) Degradation of organic solar cells due to air exposure, *Sol Energy Mater Sol Cells*, **90**(20), 3520–3530.

79. Kim BJ, Miyamoto Y, Ma B, Frechet JMJ (2009) Photocrosslinkable polythiophenes for efficient, thermally stable, organic photovoltaics, *Adv Funct Mater*, **19**, 2273–2281.

80. Kim HJ, Han AR, Cho CH (2012) Solvent-resistant organic transistors and thermally stable organic photovoltaics based on cross-linkable conjugated polymers, *Chem Mater*, **24**, 215–221.

81. Kim JS, Ho PKH, Murphy CE, Baynes N, Friend RH (2002) Nature of non-emissive black spots in polymer light-emitting diodes by in-situ micro-Raman spectroscopy, *Adv Mater*, **14**(3), 206–209.

82. Kim TM, Kim JW, Shim H, Kim J (2012) High efficiency and high photo-stability zinc-phthalocyanine based planar heterojunction solar cells with a double interfacial layer, *Appl Phys Lett*, **101**(11), 113301.

83. Kozub DR, Vakshouri K, Orme LM, Wang C, Hexemer A, Gomez ED (2011) Polymer crystallization of partially miscible polythiophene/fullerene mixtures controls morphology, *Macromolecules*, **44**(14), 5722–5726.

84. Krebs FC (2008) Air stable polymer photovoltaics based on a process free from vacuum steps and fullerenes, *Sol Energy Mater Sol Cells*, **92**(7), 715–726.

85. Krebs FC, Carl JE, Cruys-Bagger N, et al. (2005) Lifetimes of organic photovoltaics: Photochemistry, atmosphere effects and barrier layers in ITO-MEH-PPV:PCBM-aluminium, *Sol Energy Mater Sol Cells*, **86**(4), 499–516.

86. Krebs FC, Norrman K (2007) Analysis of the failure mechanisms for a stable organic photovoltaic during 10000h testing, *Prog Photovolt Res Appl*, **15**, 697–712.

87. Kumar A, Hong Z, Sista S, Yang Y (2011) The critical role of processing and morphology in determining degradation rates in polymer solar cells, *Adv Energy Mater*, **1**(1), 124–131.

88. Kuwabara T, Nakayama T, Uozumi K, Yamaguchi T, Takahashi K (2008) Highly durable inverted-type organic solar cell using amorphous titanium oxide as electron collection electrode inserted between ITO and organic layer, *Sol Energy Mater Sol Cells*, **92**(11), 1476–1482.

89. Lee JU, Jung JW, Jo JW, Jo WH (2012) Degradation and stability of polymer-based solar cells, *J Mater Chem*, **22**, 24265–24283.

90. Lee K, Kim J, Park S, Kim S, Cho SS, Heeger AJ (2007) Air-stable polymer electronic devices, *Adv Mater*, **19**(18), 2445–2449.

91. Lee ST, Zhao ZQ, Hung LS (1999) Metal diffusion from electrodes in organic light-emitting diodes, *Appl Phys Lett*, **75**(10).

92. Lessmann R, Hong Z, Scholz S, Maennig B, Riede MK, Leo K (2010) Aging of flat heterojunction zinc phthalocyanine/fullerene C_{60} organic solar cells, *Org Electron*, **11**(4), 539–543.

93. Liang Z, Reese MO, Gregg BA (2011) Chemically treating poly(3-hexylthiophene) defects to improve bulk heterojunction photovoltaics, *ACS Appl Mater Interfaces*, **3**, 2042–2050.

94. Liao HH, Yang CM, Liu CC, Horng SF, Meng HF, Shy JT (2008) Dynamics and reversibility of oxygen doping and de-doping for conjugated polymer, *J Appl Phys*, **103**(10), 104506.

95. Lilliedal MR, Medford AJ, Madsen MV, Norrman K, Krebs FC (2010) The effect of post-processing treatments on inflection point in current-voltage curves of roll-to-roll processed polymer photovolatics, *Sol Energy Mater Sol Cells*, **94**(12), 2018–2031.

96. Lim SF, Wang W, Chua SJ (2002) Understanding dark spot formation and growth in organic light-emitting devices by controlling pin-hole size and shape, *Adv Funct Mater*, **12**(8), 513–518.

97. Lira-Cantu M, Krebs FC (2006) Hybrid solar cells based on MEH-PPV and thin film semiconductor oxides (TiO_2, Nb_2O_5, ZnO, CeO_2 and CeO_2-TiO_2): Perfomance improvement during long-time irradation, *Sol Energy Mater Sol Cells*, **90**(14), 2076–2086.

98. Lloyd MT, Peters CH, Garcia A, et al. (2011) Influence of the hole transport layer on the initial behaviour and lifetime of inverted organic photovoltaics, *Sol Energy Mater Sol Cells*, **95**(5), 1382–1388.

99. Lloyd MT, Olson DC, Lu P, et al. (2009) Impact of contact evolution on the shelf life of organic solar cells, *J Mater Chem*, **19**, 7638–7642.

100. Logdlund M, Bredas JL (1994) Theoretical studies of the interaction between aluminum and poly(p-phenylenevinylene) and derivatives, *J Chem Phys*, **101**(5).

101. Lo MF, Ng TW, Mo HW, Lee CS (2013) Direct Threat of UV-ozone treated indium tin oxide substrate to the stabilities of common organic semiconductors, *Adv Funct Mater*, **23**(13), 1718–1723.

102. Lungenschmied C, Dennler G, Neugebauer H, et al. (2007) Flexible, long-lived, large-area, organic solar cells, *Sol Energy Mater Sol Cells*, **91**(5), 379–384.

103. Luo SC, Chung HH, Pashuck ET, Douglas EP, Holloway PH (2005) Formation of bubbles on electrical contacts to polymer light-emitting diode devices, *Thin Solid Films*, **478**, 326–331.

104. Ma L, Wang X, Wang B, et al. (2002) Photooxidative degradation mechanism of model compounds of poly(p-phenylenevinylenes) (PPVs), *Chem Phys*, **285**, 85–94.

105. Madsen MV, Thomas Tromholt T, Norrman K, Krebs FC (2013) Concentrated light for accelerated photo degradation of polymer materials, *Adv Energy Mater*, **3**(4), 424–427.

106. Manceau M, Bundgaard E, Carle JE, et al. (2011) Photochemical stability of p-conjugated polymers for polymer solar cells: A rule of thumb, *J Mater Chem*, **21**, 4132–4141.

107. Manceau M, Helgesen M, Krebs FC (2010) Thermo-cleavable polymers: Materials with enhanced photochemical stability, *Polym Degrad Stabil*, **95**(12), 2666–2669.

108. Manceau M, Rivaton A, Gardette JL, Guillerez S, Lemaitre N (2009) The mechanism of photo- and thermal oxidation of poly(3-hexylthiophene) (P3HT) reconsidered, *Polym Degrad Stabil*, **94**(6), 898–907.

109. Marciniak S, Crispin X, Uvdala K, et al. (2004) Light induced damage in poly(3,4-ethylenedioxythiophene) and its derivatives studied by photo-electron spectroscopy, *Synth Met*, **141**(1–2), 67–73.

110. Matsushima T, Yahiro M, Adachi C (2007) Estimation of electron traps in carbon-60 field-effect transistors by a thermally stimulated current technique, *Appl Phys Lett*, **91**(10), 103505.

111. Mityashin A, Cheyns D, Rand BP, Heremans P (2012) Understanding metal doping for organic electron transport layer, *Appl Phys Lett*, **100**, 053305.

112. McElvain J, Antoniadis H, Hueschen MR, et al. (1996) Formation and growth of black spots in organic light-emitting diodes, *J Appl Phys*, **80**(10), 6002–6007.

113. Medford AJ, Lilliedal MR, Jorgensen M, et al. (2010) Grid-connected polymer solar panels: Initial considerations of cost, lifetime, and practicality, *Opt Express*, **18**(23), 272–285.

114. Meijer EJ, Detcheverry C, Baesjou PJ, et al. (2003) Dopant density determination in disordered organic field-effect transistors, *J Appl Phys*, **93**(8), 4831–4835.

115. Meng X, Zhang W, Tan Z, et al. (2012) Dihydronaphthyl-based [60] fullerene bisadducts for efficient and stable polymer solar cells, *Chem Commun*, **48**(3), 425–427.

116. Nardes AM, Kemerinka M, de Kok MM, Vinken E, Maturova K, Janssen RAJ (2008) Conductivity, work function, and environmental

stability of PEDOT:PSS thin films treated with sorbitol, *Org Electron*, **9**(5), 727–734.

117. Neugebauer H, Brabec CJ, Hummelen JC, Sariciftci NS (2000) Stability and photodegradation mechanisms of conjugated polymer/fullerene plastic solar cells, *Sol Energy Mater Sol Cells*, **61**, 35–42.

118. Nishinaga J, Aihara T, Yamagata H, Horikoshia Y (2005) Mechanical and optical characteristics of Al-doped C_{60} films, *J Cryst Growth*, **28**(1–4), 633–637.

119. Norrman K, Madsen MV, Gevorgyan SA, Krebs FC (2010) Degradation patterns in water and oxygen of an inverted polymer solar cell, *J Am Chem Soc*, **132**(47), 16883–16892.

120. Norrman K, Gevorgyan SA, Krebs FC (2009) Water-induced degradation of polymer solar cells studied by $H_2{}^{18}O$ labeling, *ACS Appl Mater Interfaces*, **1**(1), 102–112.

121. Norrman K, Larsen NB, Krebs FC (2006) Lifetimes of organic photovoltaics: Combining chemical and physical characterisation techniques to study degradation mechanisms, *Sol Energy Mater Sol Cells*, **90**(17), 2793–2814.

122. Paci B, Generosi A, Albertini VR, et al. (2006) Controlling photo induced degradation in plastic photovoltaic cells: A time-resolved energy dispersive x-ray reflectometry study, *Appl Phys Lett*, **89**, 043507.

123. Paci B, Generosi A, Albertini VR, Perfetti P, de Bettignies R, Sentein C (2009) The role of C_{60} barrier layer in improving the performances of efficient polymer-based photovoltaic devices: An AFM/EDXR time-resolved study, *J Phys Chem C*, **113**(45), 19740–19747.

124. Pacios R, Chatten AJ, Kawano K, Durrant JR, Bradley DDC, Nelson J (2006) Effects of photo-oxidation on the performance of poly[2-methoxy-5-(3,7-dimethyloctyloxy)-1,4-phenylenevinylene]:[6,6]-phenyl C_{61}-butyric acid methyl ester solar cells, *Adv Funct Mater*, **16**(16), 2117–2126.

125. Peters CH, Sachs-Quintana IT, Mateker WR, et al. (2012) The mechanism of burn-in loss in a high efficiency polymer solar cell, *Adv Mater*, **24**(5), 663–668.

126. Peters CH, Sachs-Quintana IT, Kastrop JP, et al. (2011) High efficiency polymer solar cells with long operating lifetimes, *Adv Energy Mater*, **1**(4), 491–494.

127. Peumans P, Bulovic V, Forrest SR (2000) Efficient photon harvesting at high optical intensities in ultrathin organic double-heterostructure photovoltaic diodes, *Appl Phys Lett*, **76**(19), 2650.

128. Peumans P, Yakimov A, Forrest SR (2003) Small molecular weight organic thin-film photodetectors and solar cells, *Appl Phys Lett*, **93**(7), 3693.

129. Phatak R, Tsui TY, Aziz H (2012) Dependence of dark spot growth on cathode/organic interfacial adhesion in organic light emitting devices, *J Appl Phys*, **111**, 054512.

130. Postcavage WJ, Yoo S, Domercq B, Kippelen B (2007) Encapsulation of pentacene/C_{60} organic solar cells with Al_2O_3 deposited by atomic layer deposition, *Appl Phys Lett*, **90**, 253511.

131. Ray B, Alam MA (2011) A compact physical model for morphology induced intrinsic degradation of organic bulk heterojunction solar cell, *Appl Phys Lett*, **99**(3), 033303.

132. Reese MO, Gevorgyan SA, Jorgensen M, et al. (2011) Consensus stability testing protocols for organic photovoltaic materials and devices, *Sol Energy Mater Sol Cells*, **95**(5), 1253–1267.

133. Reese MO, Nardes AM, Rupert BL, et al. (2010) Photoinduced degradation of polymer and polymer:fullerene active layers: Experiment and theory, *Adv Funct Mater*, **20**(20), 3476–3483.

134. Reese MO, White MS, Rumbles G, Ginley DS, Shaheen, SE (2008) Optimal negative electrodes for poly(3-hexylthiophene):[6,6]-phenyl C_{61}-butyric acid methyl ester bulk heterojunction photovoltaic devices, *Appl Phys Lett*, **92**(5), 053307.

135. Rivaton A, Chambon S, Manceau M, Gardette JL, Lamaitre N, Guillerez S (2010) Light-induced degradation of the active layer of polymer-based solar cells, *Polym Degrad Stabil*, **95**(3), 278–284.

136. Schaer M, Nuesch F, Berner D, Leo W, Zuppiroli L (2001) Water vapor and oxygen degradation mechanisms in organic light emitting diodes, *Adv Funct Mater*, **11**(2), 116–121.

137. Schafer S, Petersen A, Wagner TA, et al. (2011) Influence of the indium tin oxide/organic interface on open-circuit voltage, recombination, and cell degradation in organic small-molecule solar cells, *Phys Rev B*, **83**(16), 165311.

138. Schafferhans J, Baumann A, Deibel C, Dyakonov V (2008) Trap distribution and the impact of oxygen-induced traps on the charge transport in poly(3-hexylthiophene), *Appl Phys Lett*, **93**(9), 093303.

139. Schafferhans J, Baumann A, Wagenpfahl A, Deibel C, Dyakonov V (2010) Oxygen doping of P3HT:PCBM blends: Influence on trap states, charge-carrier mobility and solar cell performance, *Org Electron*, **11**(10), 1693–1700.

140. Scholz S, Corten C, Walzer K, Kuckling D, Leo K (2007) Photochemical reactions in organic semiconductor thin films, *Org Electron*, **8**(6), 709–717.

141. Schuller S, Schilinsky P, Hauch J, Brabec CJ (2004) Determination of the degradation constant of bulk heterojunction solar cells by accelerated lifetime measurements, *Appl Phys A*, **79**(1), 37–40.

142. Scott JC, Kaufman JH, Brock PJ, DiPietro R, Salem J, Goitia JA (1995) Degradation and failure of MEH-PPV light-emitting diodes, *J Appl Phys*, **79**(5), 2745.

143. Scurlock RD, Wang B, Ogilby PR, Sheats JR, Clough RL (1995) Singlet oxygen as reactive intermediate in the photodegradation of an electroluminescent polymer, *J Am Chem Soc*, **117**, 10194–10202.

144. Seeland M, Rosch R, Hoppe H (2011) Luminescence imaging of polymer solar cells: Visualization of progressing degradation, *J Appl Phys*, **109**(6), 064513.

145. Seemann A, Egelhaaf JH, Brabec CJ, Hauch JA (2009) Influence of oxygen on semi-transparent organic solar cells with gas permeable electrodes, *Org Electron*, **10**(8), 1424–1428.

146. Seemann A, Sauermann T, Lungenschmied C, Armbruster O, Egelhaaf SBJH, Hauch JA (2011) Reversible and irreversible degradation of organic solar cell performance by oxygen, *Sol Energy*, **85**(6), 1238–1249.

147. Senge MO, Fazekas M, Notaras EGA, et al. (2007) Nonlinear optical properties of porphyrins, *Adv Mater*, **19**(19), 2737–2777.

148. Sengupta E, Domanski AL, Weber SAL, et al. (2011) Photoinduced degradation studies of organic solar cell materials using Kelvin probe force and conductive scanning force microscopy, *J Phys Chem C*, **115**(40), 19994–20001.

149. Shao Y, Gong X, Heeger AJ, Liu M, Jen AKY (2009) Long-lifetime polymer light-emitting electrochemical cells fabricated with crosslinked hole-transport layers, *Adv Mater*, **21**(19), 1972–1975.

150. Sharma A, Andersson G, Lewis DA (2011) Role of humidity on indium and tin migration in organic photovoltaic devices, *Phys Chem Chem Phys*, **13**, 4381–4387.

151. Sharma P, Sreenivas K, Rao KV (1993) Analysis of ultraviolet photoconductivity in ZnO films prepared by unbalanced magnetron sputtering, *J Appl Phys*, **93**(7), 3963.

152. Shiokawa H, Yokoyama M, Hiramoto M (2007) Long-term operation tests of organic solid-state solar cells, *Surf Rev Lett*, **14**(4), 539–542.

153. So F, Kondakov D (2010) Degradation mechanisms in small-molecule and polymer organic light-emitting diodes, *Adv Mater*, **22**(34), 3755–3898.

154. Søndergaard RR, Makris T, Lianos P, et al. (2012) The use of polyurethane as encapsulating method for polymer solar cells—An inter laboratory study on outdoor stability in 8 countries, *Sol Energy Mater Sol Cells*, **99**, 292–300.

155. Song QL, Li FY, Yang H, et al. (2005) Small-molecule organic solar cells with improved stability, *Chem Phys Lett*, **416**, 42–46.

156. Song QL, Wang ML, Obbard EG, Sun XY, Ding XM, Hou XY (2006) Degradation of small-molecule organic solar cells, *Appl Phys Lett*, **89**, 251118.

157. Song ZL, So SK, Qiu Y, Zhu Y, Cao L (2001) Dynamic SIMS characterization of interface structure of Ag/Alq_3/NPB/ITO model devices, *Surf Interface Anal*, **32**(1), 102–105.

158. Steim R, Kogler FR, Brabec CJ (2010) Interface materials for organic solar cells, *J Mater Chem*, **20**, 2499–2512.

159. Suh YJ, Park SY, Lee TH, Chung WS, Kim KK, Kim MJ (2010) TEM study of stability of inverted photovoltaic cells, *Microsc Microanal*, **16**, 1378–1379.

160. Sullivan P, Duraud A, Hancox l, Beaumont N, et al. (2011) Halogenated boron sub-phthalocyanines as light harvesting electron acceptors in organic photovoltaics, *Adv Energy Mater*, **1**(3), 352–355.

161. Sullivan P, Jones TS (2008) Pentacene/fullerene (C_{60}) heterojunction solar cells: Device performance and degradation mechanisms, *Org Electron*, **9**, 656–660.

162. Sun Y, Takacs CJ, Cowan SR, et al. (2011) Efficient, air-stable bulk heterojunction polymer solar cells using MoO_x as the anode interfacial layer, *Adv Mater*, **23**(19), 2226–2230.

163. Swinnen A, Haeldermans I, vande Ven M, et al. (2007) Tuning the dimensions of C_{60}-based needlelike crystals in blended thin films, *Adv Funct Mater*, **16**(6), 760–765.

164. Tachikawa H, Kawabata H (2008) A density functional theory study on the degradation mechanim of thin film of organic semiconductor by water molecules, *Thin Solid Films*, **516**, 3287–3293.

165. Tapponnier A, Biaggio I, Gunter P (2005) Ultrapure C_{60} field-effect transistors and the effects of oxygen exposure, *Appl Phys Lett*, **86**(11), 112114.

166. Tipnis R, Bernkopf J, Jia S, et al. (2009) Large-area organic photovoltaic module—fabrication and performance, *Sol Energy Mater Sol Cells*, **93**(4), 442–446.

167. Tokito S, Tanaka H, Noda K, Okada A, Taga Y (1997) Thermal stability in oligomeric triphenylamine/tris(8-quinolinolato) aluminium, *Appl Phys Lett*, **70**, 1929.

168. Tromholt T, Madsen MV, Carle JE, Helgesen M, Krebs FC (2012) Photochemical stability of conjugated polymers, electron acceptors and blends for polymer solar cells resolved in terms of film thickness and absorbance, *J Mater Chem*, **22**, 7592–7601.

169. Tromholt T, Manceau M, Helgesen M, Carle JE, Krebs FC (2011) Degradation of semiconducting polymers by concentrated sunlight, *Sol Energy Mater Sol Cells*, **95**(5), 1308–1314.

170. Trost S, Zilberberg K, Behrendt A, Riedl T (2012) Room-temperature solution-processed SnO_x as an electron extraction layer for inverted organic solar cells with superior thermal stability, *J Mater Chem*, **22**, 16224–16229.

171. Turkovic V, Engmann S, Gobsch G, Hoppe H (2012) Methods in determination of morphological degradation of polymer:fullerene solar cells, *Synth Met*, **161**(23–24), 2534–2539.

172. Twarowski A (1982) Oxygen doping of zinc phthalocyanine thin films, *J Chem Phys*, **77**, 5840–5846.

173. Vestweber H, Riess W (1997) Highly efficient and stable organic light-emitting diodes, *Synth Met*, **91**(3), 181–185.

174. Vitoratos E, Sakkopoulos S, Dalas E, et al. (2009) Thermal degradation mechanisms of PEDOT:PSS, *Org Electron*, **10**(1), 61–66.

175. Vivo P, Jukola J, Ola ja M, Chukharev V, Lemmetyinen H (2008) Influence of Alq_3/Au cathode on stability and efficiency of a layered organic solar cell in air, *Sol Energy Mater Sol Cells*, **92**, 1416–1420.

176. Vollmer A, Jurchescu O, Arfaoui I, et al. (2005) The effect of oxygen exposure on pentacene electronic structure, *Eur Phys J E*, **17**, 339–343.

177. Vollmer A, Weiss H, Rentenberger S, Salzmann I, Rabe J, Koch N (2006) The interaction of oxygen and ozone with pentacene, *Surf Sci*, **600**(18), 4004–4007.

178. Voroshazi E, Verreet B, Andrea B, Muller R, Nuzzo DD, Heremans P (2011) Influence of cathode oxidation via the hole extraction layer in polymer:fullerene solar cells, *Org Electron*, **12**(5), 736–744.

179. Voroshazi E, Verreet B, Aernouts T, Heremans P (2010) Long-term operational lifetime and degradation analysis of P3HT:PCBM photovoltaic cells solar energy materials and solar cells, *Sol Energy Mater Sol Cells*, **95**(5), 1303–1307.

180. Wang GF, Tao XM, Wang RX (2008) Flexible organic light-emitting diodes with a polymeric nanocomposite anode, *Nanotechnology*, **19**(14), 145201.

181. Wang M, Tang Q, An J, et al. (2010) Performance and stability improvement of P3HT:PCBM-based solar cells by thermally evaporated chromium oxide (CrO_x) interfacial layer, *ACS Appl Mater Interfaces*, **2**(10), 2699–2702.

182. Wang ML, Song QL, Wu HR, et al. (2007) Small-molecular organic solar cells with C_{60}/Al composite anode, *Org Electron*, **8**, 445–449.

183. Wang N, Yu J, Zang Y, Huang J, Jiang Y (2010) Effect of buffer layers on the performance of organic photovoltaic cells based on copper phthalocyanine and C_{60}, *Sol Energy Mater Sol Cells*, **94**(2), 263–266.

184. Wang Q, Luo Y, Aziz H (2012) Photodegradation of the organic/metal cathode interface in organic light-emitting devices, *Appl Phys Lett*, **96**(6), 063309.

185. Wang Q, Williams G, Aziz H (2012) Photo-degradation of the indium tin oxide (ITO)/organic interface in organic optoelectronic devices and a new outlook on the role of ITO surface treatments and interfacial layers in improving device stability, *Org Electron*, **13**(10), 2075–2082.

186. Wu HR, Song QL, Wang ML, et al. (2007) Stable small-molecule organic solar cells with 1,3,5-tris(2-n-phenylbenzimidazolyl) benzene as an organic buffer, *Thin Solid Films*, **515**, 8050–8053.

187. Yamanari T, Taima T, Sakai J, Tsukamoto J, Yoshida Y (2010) Effect of buffer layers on stability of polymer-based organic solar cells, *Jpn J Appl Phys*, **49**, 01AC02.

188. Yan H, Scott BJ, Huang Q, Marks TJ (2004) Enhanced polymer light-emitting diode performance using a crosslinked-network electron-blocking interlayer, *Adv Mater*, **16**(2), 1948–1953.

189. Yang HB, Song QL, Gong C, Li, CM (2010) The degradation of indium tin oxide/pentacene/fullerene/tris-8-hydroxy-quinolinato aluminum/aluminum heterojunction organic solar cells: By oxygen and moisture? *Sol Energy Mater Sol Cells*, **94**, 846–849.

190. Yang QD, Ng TW, Lo MF, Wang FY, Wong NB, Lee CS (2012) Effect of water and oxygen on the electronic structure of the organic photovoltaic, *J Phys Chem C*, **116**, 10982–10985.

191. Yang X, Alexeev A, Michels MAJ, Loos J (2005) Effect of spatial confinement on the morphology evolution of thin poly(p-phenylenevinylene)/methanofullerene composite films, *Macromolecules*, **38**(10), 4289–4295.

192. Yang X, Loos J, Veenstra SC, et al. (2005) Morphology of high-performance polymer solar cells, *Nano Lett*, **5**(4), 579–583.

193. Yang X, van Duren RAJ, Janssen JKJ, Michels MAJ, Loos J (2004) Morphology and thermal stability of active layer in poly(p-phenylenevinylene)/methanofullerene plastic photovoltaic devices, *Macromolecules*, **37**(6), 2151–2158.

194. Yip HL, Jen AKY (2012) Recent advances in solution-processed interfacial materials for efficient and stable polymer solar cells, *Energy Environ Sci*, **5**, 5994–6011.

195. Zhang Y, Yip HL, Acton O, Hau SK, Huang F, Jen AKY (2009) A simple and effective way of achieving highly efficient and thermally stable bulk-heterojunction polymer solar cells using amorphous fullerene derivatives as electron acceptor, *Chem Mater*, **21**(13), 2598–2600.

196. Zhao J, Bertho S, Vandenbergh J, et al. (2011) Phase behaviour of PCBM blends with different conjugated polymers, *Phys Chem Chem Phys*, **13**(26), 12285–12292.

197. Zhao J, Swinnen A, van Assche G, Manca J, Vanderzande D, van Mele B (2009) Phase diagram of P3HT/PCBM blends and its implication for the stability of morphology, *J Phys Chem B*, **113**(6), 1587–1591.

198. Zhu Z, Hadjikyriacou S, Waller D, Gaudiana R (2004) Stabilization of film morphology in polymer:fullerene heterojunction solar cells, *J Macromol Sci A*, **41**(12), 1467–1487.

199. Zhuang T, Su Z, Liu Y, et al. (2012) Improvement of both efficiency and working lifetime in organic photovoltaic devices by using bathophenanthroline/tin(iv)phthalocyanine dichloride as bilayer exciton blocking layer, *Appl Phys Lett*, **100**, 243902.

200. Zilberberg K, Trost S, Schmidt H, Riedl T (2011) Solution-processed vanadium pentoxide as charge extraction layer for organic solar cells, *Adv Energy Mater*, **1**(3), 377–381.

201. Zimmermann B, Wurfel U, Niggemann M (2009) Longterm stability of efficient inverted P3HT:PCBM solar cells, *Sol Energy Mater Sol Cells*, **93**(4), 491–496.

b) Book references

202. Hermenau M, Riede M, Leo K (2012) *Degradation and stability of organic and polymer solar cells*, Krebs FC (Ed.) Chapter 5: Degradation of small-molecule-based OPV (John Wiley Sons Ltd.), 109–143.

203. Krebs FC (Ed.) (2012) *Degradation and stability of organic and polymer solar cells*, John Wiley Sons Ltd.

204. Rausand M, Hoylan A (2004) *System reliability theorie, models, statistical methods, and applications*, Wiley series in probability and statistics, John Wiley Sons Ltd.

c) Proceedings references

205. Hermenau M, Karl L, Riede M (2010) Comparison of different conditions for accelerated ageing of small molecule organic solar cells, Proceedings of SPIE, Organic Photovoltaics, Vol. **7722**.

206. Galagan Y, Andriessen R, Rubingh E, et al. (2010) Towards fully printed organic photovoltaics: processing and stability, *LOPE-C*, Frankfurt, Germany.

207. Lloyd MT, Olson DC, Berry JJ, et al. (2010) Enhanced lifetime in unencapsulated organic photovoltaics with air stable electrodes, 35th IEEE Photovoltaic Specialists Conference (PVSC), 001060–001063.

208. Schwartz G, Maennig B, Sonntag S, et al. (2009) Efficient, long-term stable organic vacuum deposited tandem solar cells, Proceedings of SPIE, Organic Photovoltaics, Vol. **7146**.

209. Facchetti A (2012) Materials development for printed organic photovoltaics. New photoactive layers and role of the interfaces, *Bayern Innovativ*, Wurzburg, Germany.

210. Pffeifer M (2012) Production of multi-junction organic photovoltaic cells and modules, *Bayern Innovativ*, Wurzburg, Germany.

Index